T0191536

Lecture Notes in Computer Science 12822

More information about this subseries at http://www.springer.com/series/7412

Josep Lladós · Daniel Lopresti ·
Seiichi Uchida (Eds.)

Document Analysis and Recognition – ICDAR 2021

16th International Conference
Lausanne, Switzerland, September 5–10, 2021
Proceedings, Part II

 Springer

Editors
Josep Lladós 🆔
Universitat Autònoma de Barcelona
Barcelona, Spain

Daniel Lopresti 🆔
Lehigh University
Bethlehem, PA, USA

Seiichi Uchida 🆔
Kyushu University
Fukuoka-shi, Japan

ISSN 0302-9743   ISSN 1611-3349   (electronic)
Lecture Notes in Computer Science
ISBN 978-3-030-86330-2   ISBN 978-3-030-86331-9   (eBook)
https://doi.org/10.1007/978-3-030-86331-9

LNCS Sublibrary: SL6 – Image Processing, Computer Vision, Pattern Recognition, and Graphics

This Springer imprint is published by the registered company Springer Nature Switzerland AG
The registered company address is: Gewerbestrasse 11, 6330 Cham, Switzerland

Foreword

Our warmest welcome to the proceedings of ICDAR 2021, the 16th IAPR International Conference on Document Analysis and Recognition, which was held in Switzerland for the first time. Organizing an international conference of significant size during the COVID-19 pandemic, with the goal of welcoming at least some of the participants physically, is similar to navigating a rowboat across the ocean during a storm. Fortunately, we were able to work together with partners who have shown a tremendous amount of flexibility and patience including, in particular, our local partners, namely the Beaulieu convention center in Lausanne, EPFL, and Lausanne Tourisme, and also the international ICDAR advisory board and IAPR-TC 10/11 leadership teams who have supported us not only with excellent advice but also financially, encouraging us to setup a hybrid format for the conference.

We were not a hundred percent sure if we would see each other in Lausanne but we remained confident, together with almost half of the attendees who registered for on-site participation. We relied on the hybridization support of a motivated team from the Lule University of Technology during the pre-conference, and professional support from Imavox during the main conference, to ensure a smooth connection between the physical and the virtual world. Indeed, our welcome is extended especially to all our colleagues who were not able to travel to Switzerland this year. We hope you had an exciting virtual conference week, and look forward to seeing you in person again at another event of the active DAR community.

With ICDAR 2021, we stepped into the shoes of a longstanding conference series, which is the premier international event for scientists and practitioners involved in document analysis and recognition, a field of growing importance in the current age of digital transitions. The conference is endorsed by IAPR-TC 10/11 and celebrates its 30th anniversary this year with the 16th edition. The very first ICDAR conference was held in St. Malo, France in 1991, followed by Tsukuba, Japan (1993), Montreal, Canada (1995), Ulm, Germany (1997), Bangalore, India (1999), Seattle, USA (2001), Edinburgh, UK (2003), Seoul, South Korea (2005), Curitiba, Brazil (2007), Barcelona, Spain (2009), Beijing, China (2011), Washington DC, USA (2013), Nancy, France (2015), Kyoto, Japan (2017), and Syndey, Australia (2019).

The attentive reader may have remarked that this list of cities includes several venues for the Olympic Games. This year the conference was be hosted in Lausanne, which is the headquarters of the International Olympic Committee. Not unlike the athletes who were recently competing in Tokyo, Japan, the researchers profited from a healthy spirit of competition, aimed at advancing our knowledge on how a machine can understand written communication. Indeed, following the tradition from previous years, 13 scientific competitions were held in conjunction with ICDAR 2021 including, for the first time, three so-called "long-term" competitions addressing wider challenges that may continue over the next few years.

Other highlights of the conference included the keynote talks given by Masaki Nakagawa, recipient of the IAPR/ICDAR Outstanding Achievements Award, and Mickaël Coustaty, recipient of the IAPR/ICDAR Young Investigator Award, as well as our distinguished keynote speakers Prem Natarajan, vice president at Amazon, who gave a talk on "OCR: A Journey through Advances in the Science, Engineering, and Productization of AI/ML", and Beta Megyesi, professor of computational linguistics at Uppsala University, who elaborated on "Cracking Ciphers with 'AI-in-the-loop': Transcription and Decryption in a Cross-Disciplinary Field".

A total of 340 publications were submitted to the main conference, which was held at the Beaulieu convention center during September 8–10, 2021. Based on the reviews, our Program Committee chairs accepted 40 papers for oral presentation and 142 papers for poster presentation. In addition, nine articles accepted for the ICDAR-IJDAR journal track were presented orally at the conference and a workshop was integrated in a poster session. Furthermore, 12 workshops, 2 tutorials, and the doctoral consortium were held during the pre-conference at EPFL during September 5–7, 2021, focusing on specific aspects of document analysis and recognition, such as graphics recognition, camera-based document analysis, and historical documents.

The conference would not have been possible without hundreds of hours of work done by volunteers in the organizing committee. First of all we would like to express our deepest gratitude to our Program Committee chairs, Joseph Lladós, Dan Lopresti, and Seiichi Uchida, who oversaw a comprehensive reviewing process and designed the intriguing technical program of the main conference. We are also very grateful for all the hours invested by the members of the Program Committee to deliver high-quality peer reviews. Furthermore, we would like to highlight the excellent contribution by our publication chairs, Liangrui Peng, Fouad Slimane, and Oussama Zayene, who nego-tiated a great online visibility of the conference proceedings with Springer and ensured flawless camera-ready versions of all publications. Many thanks also to our chairs and organizers of the workshops, competitions, tutorials, and the doctoral consortium for setting up such an inspiring environment around the main conference. Finally, we are thankful for the support we have received from the sponsorship chairs, from our valued sponsors, and from our local organization chairs, which enabled us to put in the extra effort required for a hybrid conference setup.

Our main motivation for organizing ICDAR 2021 was to give practitioners in the DAR community a chance to showcase their research, both at this conference and its satellite events. Thank you to all the authors for submitting and presenting your out-standing work. We sincerely hope that you enjoyed the conference and the exchange with your colleagues, be it on-site or online.

September 2021

Andreas Fischer
Rolf Ingold
Marcus Liwicki

Preface

It gives us great pleasure to welcome you to the proceedings of the 16th International Conference on Document Analysis and Recognition (ICDAR 2021). ICDAR brings together practitioners and theoreticians, industry researchers and academics, representing a range of disciplines with interests in the latest developments in the field of document analysis and recognition. The last ICDAR conference was held in Sydney, Australia, in September 2019. A few months later the COVID-19 pandemic locked down the world, and the Document Analysis and Recognition (DAR) events under the umbrella of IAPR had to be held in virtual format (DAS 2020 in Wuhan, China, and ICFHR 2020 in Dortmund, Germany). ICDAR 2021 was held in Lausanne, Switzerland, in a hybrid mode. Thus, it offered the opportunity to resume normality, and show that the scientific community in DAR has kept active during this long period.

Despite the difficulties of COVID-19, ICDAR 2021 managed to achieve an impressive number of submissions. The conference received 340 paper submissions, of which 182 were accepted for publication (54%) and, of those, 40 were selected as oral presentations (12%) and 142 as posters (42%). Among the accepted papers, 112 had a student as the first author (62%), and 41 were identified as coming from industry (23%). In addition, a special track was organized in connection with a Special Issue of the International Journal on Document Analysis and Recognition (IJDAR). The Special Issue received 32 submissions that underwent the full journal review and revision process. The nine accepted papers were published in IJDAR and the authors were invited to present their work in the special track at ICDAR.

The review model was double blind, i.e. the authors did not know the name of the reviewers and vice versa. A plagiarism filter was applied to each paper as an added measure of scientific integrity. Each paper received at least three reviews, totaling more than 1,000 reviews. We recruited 30 Senior Program Committee (SPC) members and 200 reviewers. The SPC members were selected based on their expertise in the area, considering that they had served in similar roles in past DAR events. We also included some younger researchers who are rising leaders in the field.

In the final program, authors from 47 different countries were represented, with China, India, France, the USA, Japan, Germany, and Spain at the top of the list. The most popular topics for accepted papers, in order, included text and symbol recognition, document image processing, document analysis systems, handwriting recognition, historical document analysis, extracting document semantics, and scene text detection and recognition. With the aim of establishing ties with other communities within the concept of reading systems at large, we broadened the scope, accepting papers on topics like natural language processing, multimedia documents, and sketch understanding.

The final program consisted of ten oral sessions, two poster sessions, three keynotes, one of them given by the recipient of the ICDAR Outstanding Achievements Award, and two panel sessions. We offer our deepest thanks to all who contributed their time and effort to make ICDAR 2021 a first-rate event for the community. This year's ICDAR had a large number of interesting satellite events as well: workshops, tutorials, competitions, and the doctoral consortium. We would also like to express our sincere thanks to the keynote speakers, Prem Natarajan and Beta Megyesi.

Finally, we would like to thank all the people who spent time and effort to make this impressive program: the authors of the papers, the SPC members, the reviewers, and the ICDAR organizing committee as well as the local arrangements team.

September 2021

Josep Lladós
Daniel Lopresti
Seiichi Uchida

Organization

Organizing Committee

General Chairs

Andreas Fischer	University of Applied Sciences and Arts Western Switzerland, Switzerland
Rolf Ingold	University of Fribourg, Switzerland
Marcus Liwicki	Luleå University of Technology, Sweden

Program Committee Chairs

Josep Lladós	Computer Vision Center, Spain
Daniel Lopresti	Lehigh University, USA
Seiichi Uchida	Kyushu University, Japan

Workshop Chairs

Elisa H. Barney Smith	Boise State University, USA
Umapada Pal	Indian Statistical Institute, India

Competition Chairs

Harold Mouchère	University of Nantes, France
Foteini Simistira	Luleå University of Technology, Sweden

Tutorial Chairs

Véronique Eglin	Institut National des Sciences Appliquées, France
Alicia Fornés	Computer Vision Center, Spain

Doctoral Consortium Chairs

Jean-Christophe Burie	La Rochelle University, France
Nibal Nayef	MyScript, France

Publication Chairs

Liangrui Peng	Tsinghua University, China
Fouad Slimane	University of Fribourg, Switzerland
Oussama Zayene	University of Applied Sciences and Arts Western Switzerland, Switzerland

Sponsorship Chairs

David Doermann	University at Buffalo, USA
Koichi Kise	Osaka Prefecture University, Japan
Jean-Marc Ogier	University of La Rochelle, France

Local Organization Chairs

Jean Hennebert   University of Applied Sciences and Arts Western
                                  Switzerland, Switzerland
Anna Scius-Bertrand   University of Applied Sciences and Arts Western
                                  Switzerland, Switzerland
Sabine Süsstrunk   École Polytechnique Fédérale de Lausanne,
                                  Switzerland

Industrial Liaison

Aurélie Lemaitre   University of Rennes, France

Social Media Manager

Linda Studer   University of Fribourg, Switzerland

Program Committee

Senior Program Committee Members

Apostolos Antonacopoulos   University of Salford, UK
Xiang Bai   Huazhong University of Science and Technology,
                                  China
Michael Blumenstein   University of Technology Sydney, Australia
Jean-Christophe Burie   University of La Rochelle, France
Mickaël Coustaty   University of La Rochelle, France
Bertrand Coüasnon   University of Rennes, France
Andreas Dengel   DFKI, Germany
Gernot Fink   TU Dortmund University, Germany
Basilis Gatos   Demokritos, Greece
Nicholas Howe   Smith College, USA
Masakazu Iwamura   Osaka Prefecture University, Japan
C. V. Javahar   IIIT Hyderabad, India
Lianwen Jin   South China University of Technology, China
Dimosthenis Karatzas   Computer Vision Center, Spain
Laurence Likforman-Sulem   Télécom ParisTech, France
Cheng-Lin Liu   Chinese Academy of Sciences, China
Angelo Marcelli   University of Salerno, Italy
Simone Marinai   University of Florence, Italy
Wataru Ohyama   Saitama Institute of Technology, Japan
Luiz Oliveira   Federal University of Parana, Brazil
Liangrui Peng   Tsinghua University, China
Ashok Popat   Google Research, USA
Partha Pratim Roy   Indian Institute of Technology Roorkee, India
Marçal Rusiñol   Computer Vision Center, Spain
Robert Sablatnig   Vienna University of Technology, Austria
Marc-Peter Schambach   Siemens, Germany

Srirangaraj Setlur   University at Buffalo, USA
Faisal Shafait   National University of Sciences and Technology, India
Nicole Vincent   Paris Descartes University, France
Jerod Weinman   Grinnell College, USA
Richard Zanibbi   Rochester Institute of Technology, USA

Program Committee Members

Sébastien Adam
Irfan Ahmad
Sheraz Ahmed
Younes Akbari
Musab Al-Ghadi
Alireza Alaei
Eric Anquetil
Srikar Appalaraju
Elisa H. Barney Smith
Abdel Belaid
Mohammed Faouzi Benzeghiba
Anurag Bhardwaj
Ujjwal Bhattacharya
Alceu Britto
Jorge Calvo-Zaragoza
Chee Kheng Ch'Ng
Sukalpa Chanda
Bidyut B. Chaudhuri
Jin Chen
Youssouf Chherawala
Hojin Cho
Nam Ik Cho
Vincent Christlein
Christian Clausner
Florence Cloppet
Donatello Conte
Kenny Davila
Claudio De Stefano
Sounak Dey
Moises Diaz
David Doermann
Antoine Doucet
Fadoua Drira
Jun Du
Véronique Eglin
Jihad El-Sana
Jonathan Fabrizio

Nadir Farah
Rafael Ferreira Mello
Miguel Ferrer
Julian Fierrez
Francesco Fontanella
Alicia Fornés
Volkmar Frinken
Yasuhisa Fujii
Akio Fujiyoshi
Liangcai Gao
Utpal Garain
C. Lee Giles
Romain Giot
Lluis Gomez
Petra Gomez-Krämer
Emilio Granell
Mehdi Hamdani
Gaurav Harit
Ehtesham Hassan
Anders Hast
Sheng He
Jean Hennebert
Pierre Héroux
Laurent Heutte
Nina S. T. Hirata
Tin Kam Ho
Kaizhu Huang
Qiang Huo
Donato Impedovo
Reeve Ingle
Brian Kenji Iwana
Motoi Iwata
Antonio Jimeno
Slim Kanoun
Vassilis Katsouros
Ergina Kavallieratou
Klara Kedem

Contents – Part II

Mobile Text Recognition

Document Analysis for Social Good

Indexing and Retrieval of Documents

Physical and Logical Layout Analysis

Recognition of Tables and Formulas

NLP for Document Understanding

Document Analysis for Literature Search

Towards Document Panoptic Segmentation with Pinpoint Accuracy: Method and Evaluation

Rongyu Cao[1,2(✉)], Hongwei Li[3], Ganbin Zhou[4], and Ping Luo[1,2,5]

[1] Key Lab of Intelligent Information Processing of Chinese Academy of Sciences, Institute of Computing Technology, CAS, Beijing 100190, China

[2] University of Chinese Academy of Sciences, Beijing 100049, China
{caorongyu19b,luop}@ict.ac.cn

[3] Research Department, P.A.I. Ltd., Beijing 100025, China
lihw@paodingai.com

[4] WeChat Search Application Department, Tencent Ltd., Beijing 100080, China
ganbinzhou@tencent.com

[5] Peng Cheng Laboratory, Shenzhen 518066, China

Abstract. In this paper we study the task of document layout recognition for digital documents, requiring that the model should detect the *exact physical object region* without missing any text or containing any redundant text outside objects. It is the vital step to support high-quality information extraction, table understanding and knowledge base construction over the documents from various vertical domains (e.g. financial, legal, and government fields). Here, we consider *digital documents*, where characters and graphic elements are given with their exact texts, positions inside document pages, compared with image documents. Towards document layout recognition with pinpoint accuracy, we consider this problem as a document panoptic segmentation task, that each *token* in the document page must be assigned a class label and an instance id. Considering that two predicted objects may intersect under traditional visual panoptic segmentation method, like Mask R-CNN, however, document objects never intersect because most document pages follow manhattan layout. Therefore, we propose a novel framework, named document panoptic segmentation (DPS) model. It first splits the document page into *column regions* and groups tokens into *line regions*, then extracts the textual and visual features, and finally assigns class label and instance id to each line region. Additionally, we propose a novel metric based on the intersection over union (IoU) between the tokens contained in predicted and the ground-truth object, which is more suitable than metric based on the area IoU between predicted and the ground-truth bounding box. Finally, the empirical experiments based on PubLayNet, ArXiv and Financial datasets show that the proposed DPS model obtains 0.8833, 0.9205 and 0.8530 mAP scores on three datasets. The proposed model obtains great improvement on mAP score compared with Faster R-CNN and Mask R-CNN models.

© Springer Nature Switzerland AG 2021
J. Lladós et al. (Eds.): ICDAR 2021, LNCS 12822, pp. 3–18, 2021.
https://doi.org/10.1007/978-3-030-86331-9_1

1   Introduction

Although a large quantity of digital documents, in the form of HTML, WORD and PDF files, can be electronically accessed in the Web, these documents are mostly for human reading, but not machine-readable to support downstream applications with quantitative analysis. In other words, electronic access is a necessary but insufficient condition of machine readability [2]. Document layout recognition is the cornerstone task for this electronic access, which aims to automatically transform semi-structured information (e.g. tokens, vectors and rasters that scattered on the page) to structured physical objects (e.g. tables, graphs, paragraphs, etc.). Document layout recognition is critical for a variety of downstream applications such as information extraction, table understanding and knowledge base building over digital documents [4,15,35].

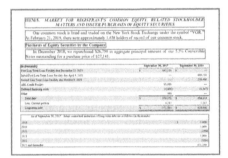 

(a) ground-truth physical objects   (b) predicted physical objects

Fig. 1. Examples on a part of a document page to illustrate how imprecisely predicted document objects will harm various downstream tasks. Red rectangles and green rectangles in (a) and (b) represent the tokens contained in paragraph and table objects. (Color figure online)

Since various downstream tasks, such as information extraction and table understanding, depend on the result of the predicted physical objects, it requires high accuracy for document layout recognition task. In other words, the document layout recognition model should detect the *exact physical object region* without missing any text or containing any redundant text outside objects. Otherwise, it will cause harmful impact on subsequent tasks. Figure 1 depicts several examples about how imprecisely-predicted physical objects will harm various downstream tasks. For example, ground-truth tables are two green solid boxes in Fig. 1(a) and the true relevant time of quantity "471,250" is "September 30, 2017" in downstream table understanding task. We assume that the model predicts two tables as one table by mistake as shown in Fig. 1(b). Thus, the relevant time of "471,250" might be recognized as "September 30, 2016" since "September 30, 2016" will be probably recognized as the column header of "471,250". So, the reason for the error is predicting two ground-truth tables as one table by mistake. Another example is that "Purchases of Equity Securities by the Company" has been merged into other paragraphs by mistake. Thus, in downstream table-of-content recognition task, the true heading "Purchases of Equity Securities by the Company" will be lost. Therefore, pinpoint accuracy of physical objects is indispensable for document layout recognition task. Next, we introduce three key aspects towards pinpoint accuracy: problem formation, solution and evaluation metric.

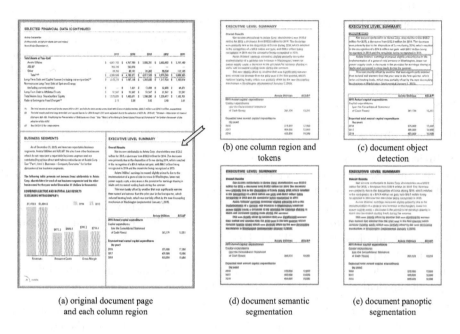

(a) original document page and each column region

(b) one column region and tokens

(c) document object detection

(d) document semantic segmentation

(e) document panoptic segmentation

Fig. 2. For a given document page (a), we select the right bottom column region (b). Then, we show the ground-truth of object detection (c), semantic segmentation (d) and panoptic segmentation (e) for document layout recognition. Transparent yellow boxes in (b) represent the tokens. Solid red boxes and green boxes in (c) represent paragraph and table objects. Transparent red boxes and green boxes in (d) and (e) represent the tokens contained in paragraph and table objects. Note that, we use different shades of the red color in (e) to distinguish different instances of paragraph objects. (Color figure online)

Problem Formation. Previous studies mostly regard the document layout recognition problem as an object detection or a semantic segmentation task [9,16–18,28,38,39]. Document object detection is defined as recognizing each physical object with the class label and its position by delineating it with a bounding box, as shown in Fig. 2(c). However, in the inference phase, two predicted objects may overlap since there are no restrictions among predicted bounding boxes. Note that, different from that visual objects in the natural image may overlap sometimes, physical objects on document page never overlap. This limitation may cause that some tokens belong to multiple physical objects, which will harm downstream tasks. Here, tokens refer to English words or Chinese characters on the digital document page.

On the other hand, document semantic segmentation aims to simply assign a class label to each pixel in a document image, or each token in a digital document. This way fixes the above limitation since each token can only be assigned to one class label. However, it causes another limitation - it cannot specify instances. In other words, it cannot distinguish adjacent objects of the same class as shown in Fig. 2(d).

To combine the merits and overcome the limitations of these two definitions, we propose a new formation called document panoptic segmentation, inspired by the previous study [13] in the computer vision field. We formulate document panoptic segmentation as that each token in the document page must be assigned a class label and an instance id, as shown in Fig. 2(e). Thus, tokens with the same class label and instance id constitute a physical object. This definition guarantees that each token belongs to only one physical object, meanwhile, adjacent document objects can be separated via distinct instance id.

Solution. In this paper, we assume that document page follows manhattan layout: that is, physical object can be segmented by a set of horizontal and vertical line [24]. Thus, for a given document page, we propose a novel framework, named document panoptic segmentation (DPS) model, which first segments the page into *column regions* to avoid interference from different columns, as shown in Fig. 2(a), inspired by [19]. Then, DPS further splits each column region into an ordered sequence of *line regions*. Finally, a network predicts the class label and instance id for each token and can group these line regions into different types of physical objects. The framework of solution is detailed in Sect. 3.

Evaluation Metric. Previous studies on document layout recognition mostly use the intersection over union area (area-IoU) between the predicted object and the matched ground-truth object as the evaluation metric [6,9,16–19,38,39]. However, area-IoU based metric only reflects the overlap of the pixels between the predicted object and the ground-truth object, which cannot faithfully reflect whether the predicted object matches the ground-truth with pinpoint accuracy. To this end, we propose token-IoU, which calculates the intersection over union between the tokens contained in predicted object and the ground-truth object. Then, the mAP value is calculated based on token-IoU as the final metric. The example and experiment (detailed in Sect. 4) show that the proposed token-IoU is more suitable than area-IoU for document layout recognition task.

In this paper, we use the PubLayNet Dataset and additionally collect two private datasets - ArXiv Dataset and Financial Dataset. ArXiv Dataset contains scientific publication document pages from ArXiv and Financial Dataset contains financial document pages from annual reports, prospectuses, research reports, etc. The empirical experiments based on these three datasets show that the proposed DPS model obtains 0.8833, 0.9205 and 0.8530 mAP scores on three datasets. However, Faster R-CNN model obtains 0.8415, 0.8451 and 0.8133 mAP scores and Mask R-CNN model obtains 0.8587, 0.8500 and 0.8209 mAP scores on three datasets. Therefore, the proposed model obtains great improvement on mAP score compared with baseline Faster R-CNN and Mask R-CNN models.

Our contributions in this paper are as follows:

- To the best of our knowledge, we first consider the document layout recognition problem as a document panoptic segmentation task.
- Towards document panoptic segmentation with pinpoint accuracy, we propose a novel and more rigorous evaluation metric.
- We propose a neural-based model and the empirical experiments based on three datasets show that the proposed model outperforms other baselines.

2   Related Work

Generally, all previous studies focused on document layout can be grouped into four classes, object detection, semantic segmentation, sequence labeling and graph learning method.

Object detection method deals with detecting instances of semantic objects of a certain class in images or videos [1]. Considering physical objects in the document as the target semantic objects, this type of method mostly applies the existing object detection framework to regress the bounding box of table regions. Along this line, Gilani et al. [6] add distance transformations to the input image and apply Faster R-CNN [27] to detect table regions. Li et al. [16] use Feature Pyramid Networks to predict physical object regions and apply three layer alignment modules for cross-domain object detection in documents.

Semantic segmentation is a pixel-labeling technology that maps every pixel of an image or tokens of a document into a class [21,30]. Then, a method to frame the region of table pixels is applied. Yang et al. [38] apply FCN [21] for classifying pixels and generate bounding box of table region by applying rule-based methods that rely on heuristics or additional PDF metadata. He et al. [9] first predict the contour of each table and apply connected components analysis (CCA) to make the contour more clear. Li et al. [18] use LayoutLM [37] framework to predict each token in the document a class label. Thus, adjacent tokens with the same class label can be grouped into a physical object.

Sequence labeling attempts to assign labels to a sequence of observations, where each observation in the sequence refers to the text line on the document page. Luong et al. [22] split text lines on the document page and apply CRFs to classify each text line into figure, table and other physical objects. Li et al. [19] split text lines and employ a verification network to classify each text line, then employ a pairwise network to merge adjacent text lines with the same type.

Graph network method considers texts inside page as nodes and the relation between texts as edges, then applies spatial-based [20] or spectral-based [12] graph network to classify each node and edge. Koci et al. [14] propose a remove and conquer algorithm and combine spectral-based graph network for detecting tables in spreadsheets. Riba et al. [29] apply a spectral-based graph network to classify nodes and edges, then group nodes that with the same label and with high edge confidence as a predicted table.

3   Document Panoptic Segmentation Model

3.1   System Overview

We aim to assign a class label and an instance id for each token on the document page. After obtaining the input document page, we first segment it into column cell regions. For each column cell region, we further split it into a sequence of line regions from top to bottom. Then, a FCN and LayoutLM is used to extract local and global features to obtain the representation of each line region. Another Unet is used to extract the relation information between line regions. Finally, each line region is assigned a class label with BIS tag via Softmax layer. After that, line regions can be merged to get different physical objects.

3.2   Column Region Segmentation

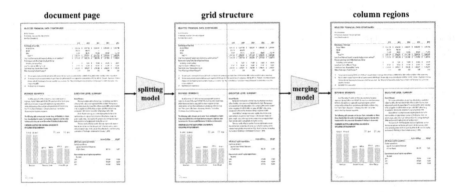

Fig. 3. An example about column region segmentation on a document page. On the right part, red solid boxes represent each column region on the document page. (Color figure online)

Previous column region segmentation methods [19, 22] design heuristic rules to segment column regions. However, for some complex document pages, these methods might obtain low accuracy. Here, column region is a part of the document page, in which the bounding box of two arbitrary physical objects must not intersect on the horizontal projection. In this paper, we consider that column region segmentation in the document page is analogous to parsing cell structure in table regions. Thus, we use state-of-art table structure recognition method in [34] to segment column regions. In detail, first, the splitting model predicts the basic grid of the page, ignoring regions that horizontally or vertically span multiple regions. Then, the merging model predicts which grid elements should be merged to recover spanning column regions. Figure 3 illustrates this process with a document page as an example.

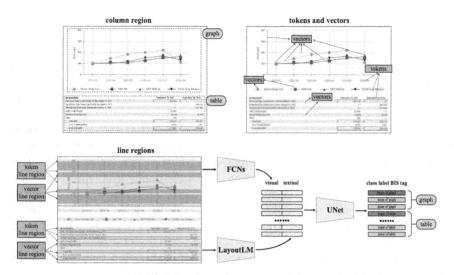

Fig. 4. Examples about line region splitting and sequence labeling. In the left top part, the purple and green dashed boxes represent graph and table object. In the right top part, the yellow box represent each token region. Some vectors such as polygons, curves and blocks are pointed with red arrows. In the left bottom part, the yellow and red boxes represent token line regions and vector line regions. (Color figure online)

3.3 Line Region Splitting

For each column region extracted in the previous step, we assume that the bounding box of two arbitrary physical objects in one column region must not intersect on the horizontal projection. Thus, if we further finely split the column region into a sequence of line regions from top to bottom, each physical object will contain continuous several line regions and each line region will be assigned to only one physical object. Considering that the basic components on the PDF document page are tokens and vectors. Here, tokens refer to English words or Chinese characters with the bounding box, vectors refer to some graphical element, such as polygons, curves and colorful blocks. Some physical objects could contain some tokens and vectors at the same time, such as graphs or tables, as shown in the example of Fig. 4.

Now we split the column region into line regions as follows. First, for each pixel line, if any token box exists in it, we name this pixel line as a token pixel line. Else, if no token box exists but any vector exists in this pixel line, we name it as a vector pixel line. Else, if no token box or vector exist in this pixel line, we name it as a blank pixel line. Then, we group continuous token pixel lines as token line region and group continuous vector pixel line as vector line region as shown in the left bottom part of Fig. 4. Thus, we obtain a sequence of token line regions and vector line regions from top to bottom in the column region. Each of these line regions contains some tokens or vectors.

3.4   Line Region Sequence Labeling

Now, the problem is to design a sequence labeling model over line regions. First, we use a fully convolutional network (FCN) [21] to obtain a feature map with the same size of the document page. After pooling the feature embeddings in each line region, we can obtain the fixed-length vector to represent the visual information of each line region.

Meanwhile, for each line region, we use a LayoutLM [37] layer to obtain the fixed-length vector to represent the textual information of each line region. In detail, for each line region, the input of LayoutLM is a collection of tokens and vectors contained in this line region. LayoutLM adds text embeddings and 2-d position embeddings of each token and vector and uses a Transformer to obtain the fixed-length vector of each token and vector. Here, we use the overall sequence vector (the vector of [CLS] token) as the output of the LayoutLM layer.

Then, we concatenate the visual vector and textual vector of each line region as the inner line feature. To further capture the relationship between each line region, we use a UNet [30] layer to obtain cross line features.

Finally, a Softmax layer is used to assign the class label with BIS tag to each line region. Each class corresponds to 3 BIS tags, and each one represents begin, inner and single, respectively. In other words, if a line region is the beginning line of a table object, its tag is "begin of table". Other line regions contained in this table object corresponds to the tag "inner of table". If a table object contains only one line region, the tag of this line region is "single of table". Thus, if there are 10 class labels, i.e. 10 types of physical objects, the number of labels is 30. According to these class labels with the BIS tag, line regions can be merged to get different physical objects. Note that, since we use 3 BIS tags, the adjacent objects of the same class can be split separately.

3.5   Page Representation

Inspired by Katti et al. [11], we make the page representation to integrate both visual and textual features in the document page. The page representation is a tensor with the size of $H \times W \times 3$, where H and W refer to the original height and width of the document page. First, we initialize the page representation with the image of the document page. To add semantic information of tokens, we pre-train the 3-dimension text embedding of each token by skip-gram model [23] and normalize the value in each dimension between 0 to 255. Thus, we can regard that different tokens distinguish their semantics according to their different "colors". Then, for each token, the area covered by the token on the page representation is filled with its text embedding. In other words, pixels the belong to the same token will share the same text embedding. All remaining pixels that do not belong to any tokens are filled with the color vector of the original document image. Therefore, we combine visual and textual features into one tensor as the page representation, which is the input of the DPS model.

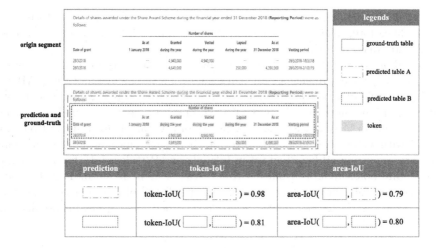

Fig. 5. An example to illustrate the difference between area-IoU and token-IoU. The blue dash-dot and purple solid bounding box represent two predicted tables. Intuitively, the blue dash-dot table is better than the purple solid table. (Color figure online)

4 Evaluation Metric

To adopt to pinpoint accuracy, we propose a new evaluation measure by considering tokens in document pages. For example, in a document page, we assume that there are n physical objects predicted by the model, namely $P = \{t_1^p, t_2^p, \cdots, t_n^p\}$, and m ground-truth objects, namely $G = \{t_1^g, t_2^g, \cdots, t_m^g\}$. Here, we assume these objects are of the same class for simplification. For each pair (t_i^p, t_j^g), we can calculate two values, namely token-IoU and area-IoU, as follow:

$$
\begin{aligned}
\text{token-IoU}(t_i^p, t_j^g) &= \frac{|\text{token}(t_i^p) \cap \text{token}(t_j^g)|}{|\text{token}(t_i^p) \cup \text{token}(t_j^g)|}, \\
\text{area-IoU}(t_i^p, t_j^g) &= \frac{\text{intersection-area}(t_i^p, t_j^g)}{\text{union-area}(t_i^p, t_j^g)},
\end{aligned}
\tag{1}
$$

where, $\text{token}(t)$ represents the set of tokens contained in the bounding box of object t, intersection-area(t_i, t_j) and union-area(t_i, t_j) represent intersection area and union area of two bounding boxes. Note that, token-IoU and area-IoU are both bound between 0 and 1. Area-IoU is widely used in traditional object detection tasks [7, 25–27].

Afterward, we will explain why token-IoU is more suitable than area-IoU to reflect the accuracy of detection for digital documents. As shown in Fig. 5, the predicted table A (the dot-dashed blue box) only misses little tokens and predicted table B (dashed purple box) misses the whole text line below the table, thus the predicted table B is apparently worse than predicted table A. However, they have similar area-IoU value (around 0.8), which means that area-IoU cannot distinguish the quality of these two different predictions. In contrast, predicted table B obtains much lower token-IoU (0.81) than predicted table A (0.98). Therefore, token-IoU is more suitable than area-IoU for document layout recognition task, intuitively.

Based on token-IoU or area-IoU, we can further calculate mean average precision (mAP) for a sequence of IoU thresholds. Then, the mean of the mAP on all objects categories is computed as the final score. ICDAR 2021 document layout recognition competition[1] and COCO competition[2] both use mAP based on area-IoU, aka mAP@area-IoU, for a sequence of IoU thresholds ranging from 0.5 to 0.95 with a step size of 0.05. However, in this paper, since our goal is pinpoint accuracy, we use mAP based on token-IoU, aka mAP@token-IoU, for a sequence of IoU thresholds ranging from 0.9 to 1.0 with a step size of 0.01. Note that, when setting IoU threshold as 1.0, only the predicted object that contains the same tokens as the ground-truth object is regarded as a successful prediction, which the most rigorous metric (Fig. 6).

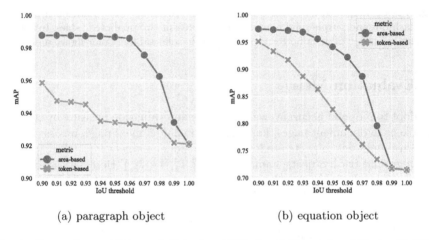

(a) paragraph object   (b) equation object

Fig. 6. Comparing mAP@area-IoU and mAP@token-IoU metric of Faster R-CNN model on paragraph and equation classes.

Finally, we design an experiment to further analyze the advantages of token-IoU compared with area-IoU. First, following [39], we use the Faster R-CNN model to detect physical objects on each document page in the ArXiv Dataset. After obtaining the predicted bounding box of each physical object, we employ post-processing to adjust the predicted bounding box. In detail, for a predicted bounding box b, we first find out all the tokens contained in b. Then, we shrink b to the bounding box b' that exactly bounds the contained token areas tightly. In other words, the left edge of b' is the leftmost edge of these tokens and the top, right and bottom edge are analogous. Then, we use the mAP@area-IoU

[1] https://icdar2021.org/competitions/competition-on-scientific-literature-parsing/.
[2] https://cocodataset.org/.

and mAP@token-IoU for IoU thresholds ranging from 0.9 to 1.0 with a step size of 0.01. Along with the increase of the IoU threshold, the mAP@area-IoU and mAP@token-IoU both decline. From the perspective of the rate of decline, area-IoU declines slowly when the threshold close to 0.9, but declines steeply when the threshold close to 1.0. In contrast, token-IoU declines uniformly from 0.9 to 1.0 threshold. That is to say, compared with area-IoU, token-IoU can objectively represent the different performance of the model under coarse or pinpoint accuracy. Therefore, in the experiments, we will use mAP@token-IoU as the final evaluation metric.

5   Experiments and Results

5.1   Dataset

There exist some public available document layout recognition datasets, however, many datasets are not suitable for pinpoint table detection in this paper. ICDAR 2013 Table Competition Dataset [8], UNLV Table Dataset [31] and Marmot Dataset [5] contain 128, 427 and 2000 document pages, respectively. However, these datasets only contain few thousands of document pages. TableBank Dataset [17] contains more than 700,000+ document images in all. Although document images can be converted to digital pages via executing optical character recognition (OCR). In this scenario, we employ Tesseract OCR Engine [33] to extract characters and words in these document images. However, due to the low resolution of these images, the produced OCR results are too poor to obtain pinpoint accuracy. DocBank Dataset [18] contains 5,000+ PDF documents, thus each token can be easily obtained according to PDF metadata. DocBank Dataset only provides the class label of each word but does not provide the bounding box of each physical object. In this scenario, it cannot distinguish two adjacent paragraphs or tables. Thus, DocBank Dataset can only handle document semantic segmentation problem, however, it cannot handle document instance segmentation or document panoptic segmentation. In this work, we focus on document panoptic segmentation problem on documents with complex layouts. Therefore, we do not use DocBank Dataset.

PubLayNet Dataset [39] contains 300,000+ document pages and releases PDF format of each document page recently. Meanwhile, the bounding box of each physical object is also provided, thus it is suitable for pinpoint document layout recognition task. Additionally, we also collect another two datasets - ArXiv Dataset and Financial Dataset. ArXiv Dataset contains 300,000+ scientific publication document pages from ArXiv and Financial Dataset contains 200,000+ financial document pages from annual reports, prospectuses, research reports, etc. For ArXiv Dataset, we use the similar method in [17,32] to automatically annotate the bounding box of each physical object. For Financial Dataset, each document page is assigned to at least two annotators for annotating the bounding box of each physical object. If the results on a document page are different, another senior annotator will address the conflicts and output the final answer. These datasets provide PDF format of each document and annotate bounding box of each physical object. The detailed information of each dataset are shown in Table 1. Thus, in this paper, we use PubLayNet Dataset, ArXiv Dataset and Financial Dataset for experiments towards document panoptic segmentation with pinpoint accuracy.

Table 1. The statistic information of different datasets.

Dataset	#class	#document	#page	#physical object
PubLayNet	4	125,621	335,640	3,208,989
ArXiv	9	23,770	313,424	4,583,502
Financial	5	2,998	228,036	2,352,714

5.2   Baseline Methods and Proposed Method

In this paper, we compare our proposed DPS model with an object detection based model [39] and an instance segmentation based model [18] as follows.

- **Faster R-CNN.** Zhong et al. [39] employ Faster R-CNN [27] to predict the bounding box of each physical object. Faster R-CNN is an object detection method. It is complemented by Detectron2 [36] and set as the baseline model in ICDAR 2021 document layout recognition competition. We use the same hyper-parameters and other configurations. We also use the page representation introduced in Sect. 3.5 as the input.
- **Mask R-CNN.** Zhong et al. [39] also employ Mask R-CNN [10] to predict the bounding box of each physical object on the document page. Mask R-CNN is an instance segmentation method. It is complemented by Detectron2 [36]. We use the same hyper-parameters and other configurations. We also use the page representation introduced in Sect. 3.5 as the input.
- **DPS.** We propose document panoptic segmentation (DPS) model towards document layout recognition with pinpoint accuracy. The framework is introduced in Sect. 3. The input page representation is the same as Faster R-CNN and Mask R-CNN.

5.3   Experimental Results

In this section, we compare the proposed DPS model with two baseline models on three datasets. Since our goal is pinpoint accuracy, we use mAP@token-IoU under 1.0 threshold as the evaluation metric. Note that, under this metric, only the predicted object that contains the same tokens as the ground-truth object is regarded as a success prediction. In other words, we require that the predicted object must not miss any tokens or contain any redundant tokens compared with the ground-truth object. Although this metric is very rigorous, it can guarantee

Table 2. Final results of different physical objects on three datasets.

(a) The PubLayNet Dataset

model	average	text	title	list	table
Mask R-CNN	0.8415	0.9051	0.9666	0.6807	0.8136
Faster R-CNN	0.8587	0.9057	**0.9707**	0.7290	0.8294
DPS	**0.8833**	**0.9062**	0.9273	**0.8194**	**0.8804**

(b) The ArXiv Dataset

model	average	table	graph	paragraph	header	footer	equation	reference	topic	caption
Faster R-CNN	0.8451	0.8031	0.6482	0.9209	0.9591	0.9993	0.7145	0.6755	0.9138	**0.9711**
Mask R-CNN	0.8500	0.8110	0.6513	0.9285	0.9534	**0.9941**	0.7155	0.7018	0.9245	0.9698
DPS	**0.9205**	**0.8261**	**0.7468**	**0.9689**	**0.9700**	0.9800	**0.9483**	**0.9383**	**0.9483**	0.9581

(c) The Financial Dataset

model	average	table	graph	paragraph	header	footer
Faster R-CNN	0.8133	0.8110	0.4505	0.9147	0.9295	0.9608
Mask R-CNN	0.8209	0.8293	0.4648	0.9206	0.9320	0.9580
DPS	**0.8530**	**0.8648**	**0.5358**	**0.9501**	**0.9345**	**0.9798**

the results of document layout recognition will not have bad effects for the downstream tasks.

The final results are shown in Table 2. For each dataset, we list the mAP@token-IoU under 1.0 threshold of each object class. Clearly, the DPS model obtains 0.8833, 0.9205 and 0.8530 mAP scores, which obtains 0.0418 and 0.0246 improvement compared with Faster R-CNN and Mask R-CNN in Pub-LayNet Dataset, obtains 0.0754 and 0.0705 improvement in ArXiv Dataset, and obtains 0.0397 and 0.0321 improvement in Financial Dataset. Note that, for list objects in PubLayNet Dataset, equation and inference objects in Arxiv Dataset the DPS model obtains great improvement compared with baselines. Since these objects have relatively small regions and appear more densely on the document page, Faster R-CNN and Mask R-CNN cannot predict them with pinpoint accuracy. However, the DPS model can predict these objects more accurately due to considering line regions sequentially. For some objects like footer, caption and title objects, the DPS model cannot obtain better performance, because there exists some errors in column region segmentation stage. Overall, the proposed DPS model obtains great improvement compared with Faster R-CNN and Mask R-CNN.

Additionally, we further explore the overlap ratio of physical objects in different models. For Faster R-CNN and Mask R-CNN, we retain the physical objects whose confidence score is more than 0.5. For the DPS model, we retain all the physical objects. Then, for each model, we judge each physical object whether it overlaps with other physical objects and calculate the overlap ratio of physical objects. The experimental results show that 3.7% and 3.4% physical objects are overlapped in Faster R-CNN and Mask R-CNN model. However, in our proposed DPS model, this ratio is 0. That is to say, the DPS model tackles the problem of overlapping between predicted physical objects.

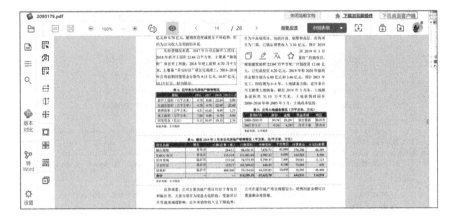

Fig. 7. PDFlux performs document panoptic segmentation for a PDF document.

5.4   Application

The core model in this study is also adopted in a PDF parser software, PDFlux[3], which is dedicated to accomplishing the task of document panoptic segmentation. For disclosure documents in the financial field, we carried out a large number of manual annotations and trained the proposed model. After years of polishing, it can now handle various types of documents, including financial annual reports, prospectuses, audit reports, scientistic publications, industry reports, etc.

Figure 7 shows the screenshot of PDFlux. A user only needs to upload a PDF document to PDFlux and click the "recognizing" button. Then, PDFlux will automatically execute all the modules in document panoptic segmentation. First, PDFlux recognizes the column region for each document page. Through the column lines, users can clearly see what contents are contained in each column region. Then, PDFlux extracts the physical objects in each column region, including paragraphs, tables, graphs, page headers, page footers, and so on. The user can locate each physical object by moving the cursor, and perform different operations such as copying texts, editing texts, translation, etc. on each physical object. Especially for the table object, PDFlux displays the in-depth analysis results on the table, such as row and column lines, merged cells, cell alignment and so on. Additionally, PDFlux also recognizes whether two paragraphs or two tables across two pages or two columns need to be merged. In the future, PDFlux aims to recover the logical hierarchy of the entire document [3]. Finally, the entire document can be converted to WORD/EPUB/MOBI format for users to download. In order to facilitate further analysis and understanding of PDF documents, PDFlux also provides external SaaS services in the form of REST API.

---

[3] http://pdflux.com/.

6   Conclusion

In this paper, we focus on document layout recognition and propose a document panoptic segmentation framework towards pinpoint accuracy. We also propose a novel evaluation metric, named token-IoU, for this task, which is more suitable for pinpoint accuracy. Furthermore, we use three datasets and compare the proposed DPS model with Faster R-CNN and Mask R-CNN. The results show that DPS model outperforms other two baselines on three datasets.

Acknowledgements. The research work supported by the National Key Research and Development Program of China under Grant No. 2017YFB1002104, the National Natural Science Foundation of China under Grant No. 62076231, U1811461. We thank Xu Wang and Jie Luo (from P.A.I Tech) for their kind help. We also thank anonymous reviewers for their valuable comments and suggestions.

References

1. Object detection. https://en.wikipedia.org/wiki/Object_detection
2. Bauguess, S.W.: The role of machine readability in an AI world (2018). https://www.sec.gov/news/speech/speech-bauguess-050318
3. Cao, R., Cao, Y., Zhou, G., Luo, P.: Extracting variable-depth logical document hierarchy from long documents: method, evaluation, and application. J. Comput. Sci. Technol. (2021)
4. Cao, Y., Li, H., Luo, P., Yao, J.: Towards automatic numerical cross-checking: extracting formulas from text. In: WWW (2018)
5. Fang, J., Tao, X., Tang, Z., Qiu, R., Liu, Y.: Dataset, ground-truth and performance metrics for table detection evaluation. In: DAS (2012)
6. Gilani, A., Qasim, S.R., Malik, I., Shafait, F.: Table detection using deep learning. In: ICDAR (2017)
7. Girshick, R.: Fast R-CNN. In: ICCV (2015)
8. Göbel, M., Hassan, T., Oro, E., Orsi, G.: ICDAR 2013 table competition. In: ICDAR (2013)
9. He, D., Cohen, S., Price, B., Kifer, D., Giles, C.L.: Multi-scale multi-task FCN for semantic page segmentation and table detection. In: ICDAR (2018)
10. He, K., Gkioxari, G., Dollár, P., Girshick, R.: Mask R-CNN. In: ICCV (2017)
11. Katti, A.R., et al.: Chargrid: towards understanding 2D documents. In: EMNLP (2018)
12. Kipf, T.N., Welling, M.: Semi-supervised classification with graph convolutional networks. In: ICLR (2017)
13. Kirillov, A., He, K., Girshick, R., Rother, C., Dollar, P.: Panoptic segmentation. In: CVPR (2019)
14. Koci, E., Thiele, M., Lehner, W., Romero, O.: Table recognition in spreadsheets via a graph representation. In: DAS (2018)
15. Li, H., Yang, Q., Cao, Y., Yao, J., Luo, P.: Cracking tabular presentation diversity for automatic cross-checking over numerical facts. In: KDD (2020)
16. Li, K., et al.: Cross-domain document object detection: benchmark suite and method. In: CVPR (2020)
17. Li, M., Cui, L., Huang, S., Wei, F., Zhou, M., Li, Z.: TableBank: table benchmark for image-based table detection and recognition (2019)

18. Li, M., et al.: Docbank: A benchmark dataset for document layout analysis. arXiv (2020)
19. Li, X.H., Yin, F., Liu, C.L.: Page object detection from pdf document images by deep structured prediction and supervised clustering. In: ICPR (2018)
20. Li, Y., Tarlow, D., Brockschmidt, M., Zemel, R.: Gated graph sequence neural networks. In: ICLR (2016)
21. Long, J., Shelhamer, E., Darrell, T.: Fully convolutional networks for semantic segmentation. In: CVPR (2015)
22. Luong, M.T., Nguyen, T.D., Kan, M.Y.: Logical structure recovery in scholarly articles with rich document features. Int. J. Digit. Libr. Syst. (2010)
23. Mikolov, T., Chen, K., Corrado, G., Dean, J.: Efficient estimation of word representations in vector space. In: ICLR (2013)
24. Nagy, G., Seth, S.C.: Hierarchical representation of optically scanned documents. In: Conference on Pattern Recognition (1984)
25. Redmon, J., Divvala, S., Girshick, R., Farhadi, A.: You only look once: unified, real-time object detection. In: CVPR (2016)
26. Redmon, J., Farhadi, A.: Yolo9000: better, faster, stronger. In: CVPR (2017)
27. Ren, S., He, K., Girshick, R., Sun, J.: Faster R-CNN: towards real-time object detection with region proposal networks. In: NeurIPS (2015)
28. Riba, P., Dutta, A., Goldmann, L., Fornés, A., Ramos, O., Lladós, J.: Table detection in invoice documents by graph neural networks. In: ICDAR (2019)
29. Riba, P., Dutta, A., Goldmann, L., Fornés, A., Ramos, O., Lladós, J.: Table detection in invoice documents by graph neural networks (2019)
30. Ronneberger, O., Fischer, P., Brox, T.: U-net: convolutional networks for biomedical image segmentation. In: MICCAI (2015)
31. Shahab, A., Shafait, F., Kieninger, T., Dengel, A.: An open approach towards the benchmarking of table structure recognition systems. In: DAS (2010)
32. Siegel, N., Lourie, N., Power, R., Ammar, W.: Extracting scientific figures with distantly supervised neural networks. In: JCDL (2018)
33. Smith, R.: An overview of the tesseract OCR engine. In: ICDAR (2007)
34. Tensmeyer, C., Morariu, V.I., Price, B., Cohen, S., Martinez, T.: Deep splitting and merging for table structure decomposition. In: ICDAR (2019)
35. Wu, S., et al.: Fonduer: Knowledge base construction from richly formatted data. In: SIGMOD (2018)
36. Wu, Y., Kirillov, A., Massa, F., Lo, W.Y., Girshick, R.: Detectron2 (2019). https://github.com/facebookresearch/detectron2
37. Xu, Y., Li, M., Cui, L., Huang, S., Wei, F., Zhou, M.: LayoutLM: pre-training of text and layout for document image understanding. In: KDD (2020)
38. Yang, X., Yumer, E., Asente, P., Kraley, M., Kifer, D., Giles, C.L.: Learning to extract semantic structure from documents using multimodal fully convolutional neural networks. In: CVPR (2017)
39. Zhong, X., Tang, J., Yepes, A.J.: PublayNet: largest dataset ever for document layout analysis. In: ICDAR (2019)

A Math Formula Extraction and Evaluation Framework for PDF Documents

Ayush Kumar Shah⬤, Abhisek Dey⬤, and Richard Zanibbi⁽✉⁾⬤

Rochester Institute of Technology, Rochester, NY 14623, USA
{as1211,ad4529,rxzvcs}@rit.edu

Abstract. We present a processing pipeline for math formula extraction in PDF documents that takes advantage of character information in born-digital PDFs (e.g., created using LaTeX or Word). Our pipeline is designed for indexing math in technical document collections to support math-aware search engines capable of processing queries containing keywords and formulas. The system includes user-friendly tools for visualizing recognition results in HTML pages. Our pipeline is comprised of a new state-of-the-art PDF character extractor that identifies precise bounding boxes for non-Latin symbols, a novel Single Shot Detector-based formula detector, and an existing graph-based formula parser (QD-GGA) for recognizing formula structure. To simplify analyzing structure recognition errors, we have extended the LgEval library (from the CROHME competitions) to allow viewing all instances of specific errors by clicking on HTML links. Our source code is publicly available.

Keywords: Math formula recognition · Document analysis systems · PDF character extraction · Single Shot Detector (SSD) · Evaluation

1 Introduction

There is growing interest in developing techniques to extract information from formulas, tables, figures, and other graphics in documents, since not all information can be retrieved using text [3]. Also, the poor accessibility of technical content presented graphically in PDF files is a common problem for researchers, and in various educational fields [19].

In this work, we focus upon improving automatic formula extraction [21]. Mathematical expressions are essential in technical communication, as we use them frequently to represent complex relationships and concepts compactly. Specifically, we consider PDF documents, and present a new pipeline that extracts symbols present in PDF files with exact bounding box locations where available, detects formula locations in document pages (see Fig. 1), and then recognizes the structure of detected formulas as Symbol Layout Trees (SLTs). SLTs represent a formula by the spatial arrangement of symbols on writing lines [23], and may be converted to LaTeX or Presentation MathML.

© Springer Nature Switzerland AG 2021
J. Lladós et al. (Eds.): ICDAR 2021, LNCS 12822, pp. 19–34, 2021.
https://doi.org/10.1007/978-3-030-86331-9_2

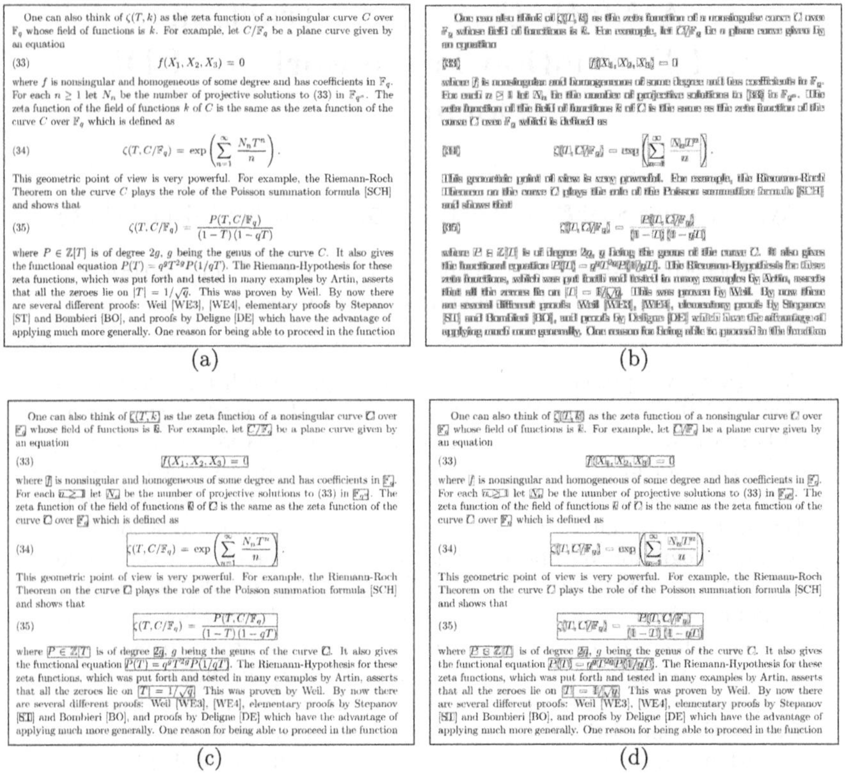

(a)

One can also think of $\zeta(T,k)$ as the zeta function of a nonsingular curve C over $\mathbb{F}_q$ whose field of functions is k. For example, let $C/\mathbb{F}_q$ be a plane curve given by an equation

$$(33) \qquad f(X_1, X_2, X_3) = 0$$

where f is nonsingular and homogeneous of some degree and has coefficients in $\mathbb{F}_q$. For each $n \geq 1$ let N_n be the number of projective solutions to (33) in $\mathbb{F}_{q^n}$. The zeta function of the field of functions k of C is the same as the zeta function of the curve C over $\mathbb{F}_q$ which is defined as

$$(34) \qquad \zeta(T, C/\mathbb{F}_q) = \exp\left(\sum_{n=1}^{\infty} \frac{N_n T^n}{n}\right).$$

This geometric point of view is very powerful. For example, the Riemann-Roch Theorem on the curve C plays the role of the Poisson summation formula [SCH] and shows that

$$(35) \qquad \zeta(T, C/\mathbb{F}_q) = \frac{P(T, C/\mathbb{F}_q)}{(1-T)(1-qT)}$$

where $P \in \mathbb{Z}[T]$ is of degree $2g$, g being the genus of the curve C. It also gives the functional equation $P(T) = q^g T^{2g} P(1/qT)$. The Riemann-Hypothesis for these zeta functions, which was put forth and tested in many examples by Artin, asserts that all the zeroes lie on $|T| = 1/\sqrt{q}$. This was proven by Weil. By now there are several different proofs: Weil [WE3], [WE4], elementary proofs by Stepanov [ST] and Bombieri [BO], and proofs by Deligne [DE] which have the advantage of applying much more generally. One reason for being able to proceed in the function

(b)

(encoded/distorted rendering of the text shown in panel (a))

(c)

One can also think of $\zeta(T,k)$ as the zeta function of a nonsingular curve C over $\mathbb{F}_q$ whose field of functions is k. For example, let $C/\mathbb{F}_q$ be a plane curve given by an equation

$$(33) \qquad f(X_1, X_2, X_3) = 0$$

where f is nonsingular and homogeneous of some degree and has coefficients in $\mathbb{F}_q$. For each $n \geq 1$ let N_n be the number of projective solutions to (33) in $\mathbb{F}_{q^n}$. The zeta function of the field of functions k of C is the same as the zeta function of the curve C over $\mathbb{F}_q$ which is defined as

$$(34) \qquad \zeta(T, C/\mathbb{F}_q) = \exp\left(\sum_{n=1}^{\infty} \frac{N_n T^n}{n}\right).$$

This geometric point of view is very powerful. For example, the Riemann-Roch Theorem on the curve C plays the role of the Poisson summation formula [SCH] and shows that

$$(35) \qquad \zeta(T, C/\mathbb{F}_q) = \frac{P(T, C/\mathbb{F}_q)}{(1-T)(1-qT)}$$

where $P \in \mathbb{Z}[T]$ is of degree $2g$, g being the genus of the curve C. It also gives the functional equation $P(T) = q^g T^{2g} P(1/qT)$. The Riemann-Hypothesis for these zeta functions, which was put forth and tested in many examples by Artin, asserts that all the zeroes lie on $|T| = 1/\sqrt{q}$. This was proven by Weil. By now there are several different proofs: Weil [WE3], [WE4], elementary proofs by Stepanov [ST] and Bombieri [BO], and proofs by Deligne [DE] which have the advantage of applying much more generally. One reason for being able to proceed in the function

(d)

One can also think of $\zeta(T,k)$ as the zeta function of a nonsingular curve C over $\mathbb{F}_q$ whose field of functions is k. For example, let $C/\mathbb{F}_q$ be a plane curve given by an equation

$$(33) \qquad f(X_1, X_2, X_3) = 0$$

where f is nonsingular and homogeneous of some degree and has coefficients in $\mathbb{F}_q$. For each $n \geq 1$ let N_n be the number of projective solutions to (33) in $\mathbb{F}_{q^n}$. The zeta function of the field of functions k of C is the same as the zeta function of the curve C over $\mathbb{F}_q$ which is defined as

$$(34) \qquad \zeta(T, C/\mathbb{F}_q) = \exp\left(\sum_{n=1}^{\infty} \frac{N_n T^n}{n}\right).$$

This geometric point of view is very powerful. For example, the Riemann-Roch Theorem on the curve C plays the role of the Poisson summation formula [SCH] and shows that

$$(35) \qquad \zeta(T, C/\mathbb{F}_q) = \frac{P(T, C/\mathbb{F}_q)}{(1-T)(1-qT)}$$

where $P \in \mathbb{Z}[T]$ is of degree $2g$, g being the genus of the curve C. It also gives the functional equation $P(T) = q^g T^{2g} P(1/qT)$. The Riemann-Hypothesis for these zeta functions, which was put forth and tested in many examples by Artin, asserts that all the zeroes lie on $|T| = 1/\sqrt{q}$. This was proven by Weil. By now there are several different proofs: Weil [WE3], [WE4], elementary proofs by Stepanov [ST] and Bombieri [BO], and proofs by Deligne [DE] which have the advantage of applying much more generally. One reason for being able to proceed in the function

Fig. 1. Detection of symbols and expressions. The PDF page shown in (a) contains encoded symbols shown in (b). (c) shows formula regions identified in the rendered page image, and (d) shows symbols located in each formula region.

There are several challenges in recognizing mathematical structure from PDF documents. Mathematical formulas have a much larger symbol vocabulary than ordinary text, formulas may be embedded in textlines, and the structure of formulas may be quite complex. Likewise, it is difficult to visualize and evaluate formula structure recognition errors - to address this, we present an improved evaluation framework that builds upon the LgEval library from the CROHME handwritten math recognition competitions [10, 12, 14]. It provides a convenient HTML-based visualization of detection and recognition results, including the ability to view *all* individual structure recognition errors organized by ground truth subgraphs. For example, the most frequent symbol segmentation, symbol classification, and relationship classification errors are automatically organized in a table, with links that take the user directly to a list of specific formulas with the selected error.

The contributions of this work include:

1. A new open-source formula extraction pipeline for PDF files,[1]
2. new tools for visualization and evaluation of results and parsing errors,
3. a PDF symbol extractor that identifies precise bounding box locations in born-digital PDF documents, and
4. a simple and effective algorithm which performs detection of math expressions using visual features alone.

We summarize related work for previous formula extraction pipelines in the next Section, present the pipeline in Sect. 3, our new visualization and evaluation tools in Sect. 4, preliminary results for the components in the pipeline in Sect. 5, and then conclude in Sect. 6.

2   Related Work

Existing mathematical recognition systems use different detection and recognition approaches. Lin et al. identify three categories of formula detection methods, based on features used [7]: character-based (OCR-based), image-based, and layout-based. Character-based methods use OCR engines, where characters not recognized by the engine are considered candidates for math expression elements. Image-based methods use image segmentation, while layout-based detection uses features such as line height, line spacing, and alignment from typesetting information available in PDF files [7], possibly along with visual features. Likewise, for parsing mathematical expressions, syntax (grammar-based) approaches, graph search approaches, and image-based RNNs producing LATEX strings as output are common.

This Section summarizes the contributions and limitations of some existing math formula extraction systems, and briefly describes our system's similarities and differences.

Utilizing **OCR and recognition as a graph search problem**, Suzuki et al.'s INFTY system [18] is perhaps the best-known commercial mathematical formula detection and recognition system. The system uses simultaneous character recognition and math-text separation based on two complementary recognition engines; a commercial OCR engine not specialized for mathematical documents and a character recognition engine developed for mathematical symbols, followed by structure analysis of math expressions performed by identifying the minimum cost spanning-trees in a weighted digraph representation of the expression. The system obtains accurate recognition results using a graph search-based formula structure recognition approach.

Using symbol information from PDFs directly rather than applying OCR to rendered images was first introduced by Baker et al. [2]. They used a syntactic pattern recognition approach to recognize formulas from PDF documents using an expression grammar. The grammar requirement and need for manually

---

[1] https://www.cs.rit.edu/~dprl/software.html.

segmenting mathematical expressions from text make it less robust than INFTY. However, the system was faster by avoiding rendering and analyzing document images, and improved accuracy using PDF character information. In later work, Sorge et al. [16] were able to reconstruct fonts not embedded in a PDF document, by mapping unicode values to standard character codes where possible. They then use CC analysis to identify characters with identical style and spacing from a grouping provided by pdf2htmlEX, allowing exact bounding boxes to be obtained. Following on Baker et al.'s approach to PDF character extraction [2], Zhang et al. [24] use a dual extraction method based on a PDF parser and an OCR engine to supplement PDF symbol extraction, recursively dividing and reconstructing the formula based on symbols found on the main baseline for formula structure analysis. Later, Suzuki et al. improved the recognition rate in INFTYReader [18], by also utilizing extracted PDF character information from PDFMiner [19].

Some PDF characters are composed of multiple glyphs, such as large braces or square roots (commonly drawn with a radical symbol connected to a horizontal line). These 'compound' characters were identified by Baker et al. [2] using overlapping bounding boxes in modern PDF documents containing Type 1 fonts.

Image-based detection using RNN-based recognition was first proposed by Deng et al. [4], inspired by RNN-based image captioning techniques. A more recent example of this approach is Phong et al.'s [15], which uses a YOLO v3 network based on a Darknet-53 network consisting of 53 convolutional layers for feature extraction and detection, and an advanced end to end neural network (Watch, Attend and Parse (WAP)) for recognition. The parser uses a GRU with attention-based mechanisms, which makes the system slow due to the pixel-wise computations. Also, error diagnosis in these recurrent image-based models is challenging, since there is not a direct mapping between the input image regions and the LATEX output strings. To address this issue for the CROHME 2019 handwriten math competition, evaluation was performed using the LgEval library, comparing formula trees over recognized symbols (e.g., as represented in LATEX strings), without requiring those symbols to have assigned strokes or input regions [10]. This alleviates, but does not completely resolve challenges with diagnosing errors for RNN-based systems.

This Paper. We introduce algorithms to create a unified system for detecting and recognizing mathematical expressions. We first introduce an improved PDF symbol extractor (SymbolScraper) to obtain symbol locations and identities. We also present a new Scanning Single Shot Detector (ScanSSD) for math formulas using visual features, by modifying the Single Shot Detector (SSD) [8] to work with large document images. For structure recognition, we use the existing Query-driven Global Graph Attention (QD-GGA) model [9], which uses CNN based features with attention. QD-GGA extracts formula structure as a maximum spanning tree over detected symbols, similar to [18]. However, unlike [18], the system trains the features and attention modules concurrently in a feed-forward pass for multiple tasks: symbol classification, edge classification, and segmentation resulting in fast training and execution.

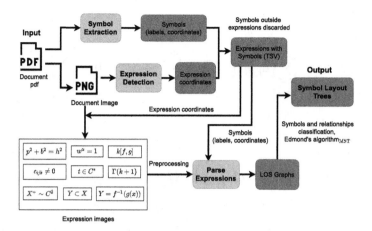

Fig. 2. Mathematical Expression Extraction Pipeline. *Symbol Extraction* outputs symbols and their bounding boxes directly from drawing commands in the PDF file, avoiding OCR. However, formula regions and formula structure are absent in PDF, and so are identified from rendered document pages (600 dpi).

Some important distinctions between previous work and the systems and tools presented in this paper include:

1. Our system focuses solely on formula extraction and its evaluation.
2. Grammars and manual segmentation are not required (e.g., per [2])
3. Recognition is graph-based, with outputs generated more quickly and with more easily diagnosed errors than RNN models (e.g., [15]).
4. Symbol information from born-digital PDFs is used directly; where absent, characters are recognized from connected components (CCs) in images.
5. Structure recognition errors can be directly observed in graphs grounded in input image regions (e.g., CCs), and the LgEval library [10,13,20,22] has been extended to visualize errors both in isolation, and in their page contexts.

3 Formula Detection Pipeline Components

We describe the components of our formula processing pipeline in this Section. As seen in Fig. 2, the extraction pipeline has three major components: 1) symbol extraction from PDF, 2) formula detection, and 3) formula structure recognition (parsing). The final outputs are Symbol Layout Trees (SLTs) corresponding to each detected mathematical expression in the input PDF documents, which we visualize as both graphs and using Presentation MathML (see Fig. 5).

3.1 SymbolScraper: Symbol Extraction from PDF

Unless formula images are embedded in a PDF file (e.g., as a .png), born-digital PDF documents provide encoded symbols directly, removing the need for character recognition [2]. In PDF documents, character locations are represented by

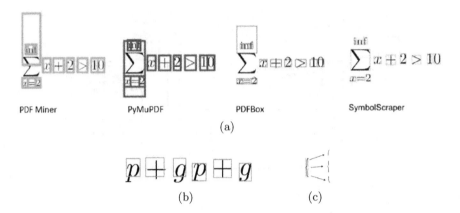

Fig. 3. Symbol extraction from PDF. (a) Symbol locations in a PDF formula from various tools. (b) correcting bounding box translations from glyph data. (b) Compound character: a large brace ('{') drawn with three characters.

their position on writing lines and in 'words,' along with their character codes and font parameters (e.g., size). Available systems such as PDFBox, PDF Miner, and PyMuPDF return the locations of characters using the drawing commands provided in PDF files directly. However, they return inconsistent locations (see Fig. 3(a))(a).

We resolve this problem by intercepting the rendering pipeline, and using the character outlines (*glyphs*) provided in embedded font profiles in PDF files. Glyphs are vector-based representations of how characters are drawn using a series of lines, arcs, and 'pen' up/down/move operations. These operations detail the outlines of the lines that make up the character. Glyphs are represented by a set of coordinates with *winding rules* (commands) used to draw the outline of a character. The glyph along with font scaling information is used to determine a character's bounding box and relative position in the PDF. Our symbol extractor extends the open-source PDFBox [1] library from Apache.

Character Bounding Boxes. For high-quality page rendering, individual squares used to represent the character outlines (i.e., glyphs) are known as 'em squares.' The em square is typically 1000 or 2048 pixels in length on each side, depending upon the font type used and how it is rendered.[2] Glyphs are represented by vector drawing commands in the em square 'glyph space,' which has a higher resolution (e.g., a 1000×1000) than that used to render the character in page images (in 'text space'). To obtain the bounding box for a character on a page, we need to: 1) convert glyph outline bounding boxes in glyph space to a bounding box in 'text space' (i.e. page coordinates), and 2) translate the box based on the *varying* origin location within each glyph's em square representation. The difference in origin locations by glyph reflects the different writing

---

[2] You can see the various ways we obtain the em square value by looking at the `getBoundingBox()` method in our `BoundingBox` class.

line positions for symbols, and allows the whole box to be used in representing character shapes (maximizing their level of detail/resolution).

Glyph outlines are translated to page coordinates using character font size and type information. A scaling matrix for the character, defined by the font size, text scaling matrix, and current transformation matrix is used for the conversion. PDF files provide a baseline starting point coordinate (b_x, b_y) for characters on writing lines, but this provided coordinate does not take character kerning, font-weight, ascent, descent, etc., into account, as seen in Fig. 3(a). To produce the appropriate translations for character bounding boxes, we use the textline position of the character provided within the em square for each glyph (in glyph space). We use displacement (δ_x, δ_y) of the bottom left coordinate within the glyph em square to obtain the corrective translation. Next, we scale the glyph space coordinates to text space coordinates and then use coordinate list G from the character's glyph, and find the width and height using the minimum and maximum coordinates of G.

The extracted character bounding boxes are very precise even for non-latin characters (e.g., $\sum$) as seen in Fig. 3(a) at far right. Likewise, the recognized symbol labels are accurate. Some documents contain characters embedded in images, in which case these characters must be identified in downstream processing. We currently use font tables provided by PDFBox for mapping characters to glyphs. Using these font tables, we have been able to retrieve glyphs for the following font families: TrueType, Type 1, Type 2, and CID. Font information is extracted using the `getFont()` method in the PDFBox `TextPosition` class. There are cases where the bounding boxes do not align perfectly with characters, due to the use of rarer font types not handled by our system (e.g., Type 3, which seems to be commonly used for raster fonts in OCR output). We hope to handle Type 3 in the future. A rare type of fonts, not handled by the current.

Compound Characters. Compound characters are formed by two or more characters as shown for a large brace in Fig. 3(c). For simplicity, we assume that bounding boxes for sub-character glyphs intersect each other, and merge intersecting glyphs into a single character. As another common example, square roots are represented in PDF by a radical symbol and a horizontal line. To identify characters with multiple glyphs, we test for bounding box intersection of adjacent characters, and intersecting glyphs are merged into a compound character. The label *unknown* is assigned to compound characters other than roots, where the radical symbol is easily identified.

Horizontal Lines. Horizontal lines are drawn as strokes and not characters. To extract them, we override the `strokepath()` method of the `PageDrawer` class in the PDFBox library [1]. By overriding this method, we obtain the starting and ending coordinates of horizontal lines directly. Some additional work is needed to improve the discrimination between different line types, and other vector-based objects could be extracted similarly (e.g., boxes, image BBs).

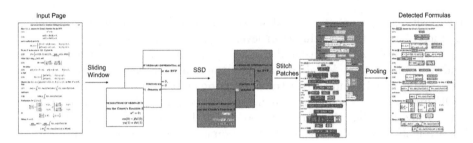

Fig. 4. ScanSSD architecture. Heatmaps illustrate detection confidences with $gray \approx 0$, $red \approx 0.5$, $white \approx 1.0$. Purple and green bounding boxes show formula regions after stitching window-level detections and pooling, respectively. (Color figure online)

3.2 Formula Detection in Page Images

We detect formulas in document page images using the Scanning Single Shot Detector (ScanSSD [11]). ScanSSD modifies the popular Single Shot Detector (SSD [8]) to work with sliding windows, in order to process large page images (600 dpi). After detection is performed in windows, candidate detections are pooled, and then finally regions are selected by pixel-level voting and thresholding.

Figure 4 illustrates the ScanSSD workflow. First, a sliding window samples overlapping page regions and applies a standard 512×512 Single-Shot Detector to locate formula regions. Non-Maximal Suppression (NMS) is used to select the highest confidence regions from amongst overlapping detections in each window. Formulas detected within each window have associated confidences, shown using colour in the 3rd stage of Fig. 4. As seen with the purple boxes in Fig. 4, many formulas are found repeatedly and/or split across windows.

To obtain page-level detections, we stitch window-level detections on the page, and then use a voting-based pooling method to produce output detections (green boxes in Fig. 4). Details are provided below.

Sliding Windows and SSD. Starting from a 600 dpi page image we slide a 1200×1200 window with a vertical and horizontal stride (shift) of 120 pixels (10% of window size). Windows are roughly 10 text lines in height. The SSD is trained using ground truth math regions cropped at the boundary of each window, after scaling and translating formula bounding boxes appropriately.

There are four main advantages to using sliding windows. The first is data augmentation: only 569 page images are available in our training set, which is *very* small for training a deep neural network. Our sliding windows produce 656,717 sub-images. Second, converting the original page image directly to 300×300 or 512×512 loses a great deal of visual information, and detecting formulas using sub-sampled page images yielded low recall. Third, windowing provides multiple chances to detect a formula. Finally, Liu et al. [8] mention that SSD is challenged when detecting small objects, and formulas with just one or two characters are common. Using high-resolution sub-images increases the relative size of small formulas, making them easier to detect.

The original SSD [8] architecture arranges initial hypotheses for rectangular dections unfiromly over grids at multiple resolutions. At each point in a grid, aspect ratios (width:height) of $\{1/3, 1/2, 1, 2, 3\}$ are used for initial hypotheses, which are translated and resized during execution of the SSD neural network. However, many formulas have aspect ratios greater than 3, and so we also include the wider default boxes sizes used in TextBoxes [6], with aspect ratios of 5, 7, and 10. In our early experiments, these wider default boxes increased recall for large formulas.

The filtered windowed detections are post-processed to tightly fit the connected components that they contain or intersect. The same procedure is applied after pooling detections at the page level, which we describe next.

Page-Level Detection Pooling. SSD detections within windows are stitched together on the page, and then each detection region votes at the pixel level (see Fig. 4). Detections are merged using a simple pixel-level voting procedure, with the number of detection boxes intersecting each pixel used as a coarse detection confidence value. Other confidence values, such as the sum of confidences, average confidence, and maximum confidence produced comparable or worse results while being more expensive to compute.

After voting, pixel region intersection counts are thresholded (using $t \geq 30$), producing a binary image. Connected components in the resulting image are converted to bounding boxes, producing the final formula detections. We then expand and/or shrink the detections so that they are cropped around the connected components they contain and touch at their border. The goal is to capture entire characters belonging to a detection region, without additional padding. Before producing final results, we discard large graphical objects like tables and charts, using a threshold for the ratio of height and area of the detected graphical objects to that of the document page.

Identifying Extracted Symbols in Formula Regions. We identify overlapping regions between detected formula regions and extracted symbol bounding boxes, discarding symbols that lie outside formula regions as shown in Fig. 1(d). We then combine formula regions and the symbols they contain, and write this to tab-separated variable (TSV) file with a hierarchical structure. The final results of formula detection and symbol extraction are seen in Fig. 1(d).

3.3 Recognizing Formula Structure (Parsing)

We use QD-GGA [9] for parsing the detected mathematical formulas, which involves recognizing a hierarchical graph structure from the expression images. In these graphs, symbols and unlabeled connected components act as nodes, and spatial relationships between symbols are edges. The set of graph edges under consideration are defined by a line-of-sight (LOS) graph computed over connected components. The symbol bounding boxes and labels produced by SymbolScraper are used directly in extraction results, where available (see Fig. 5), avoiding the need for character-level OCR. Otherwise, we use CCs extraction and allow QD-GGA [9] to perform symbol recognition (segmentation and classification).

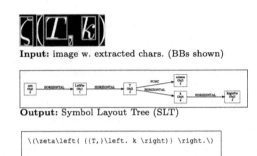

Input: image w. extracted chars. (BBs shown)

Output: Symbol Layout Tree (SLT)

\(\(\zeta\left({{T,}\left. k \right)} \right.\)

SLT in LaTeX

SLT in Presentation MathML

Fig. 5. Parsing a formula image. Formula regions are rendered and have characters extracted when they are provided in the PDF. QD-GGA produces a Symbol Layout Tree as output, which can be translated to LaTeX and Presentation MathML.

Symbol segmentation and classification are decided first, using provided locations and labels where available, and for connected components in the LOS graph defined by binary 'merge' relationships between image CCs, and then choosing the symbol class with the highest average label confidence across merged CCs. A Maximum Spanning Tree (MST) over detected symbols is then extracted from the weighted relationship class distributions using Edmond's arborescence algorithm [5] to produce the final formula interpretation.

QD-GGA trains CNN-based features and attention modules concurrently for multiple tasks: symbol classification, edge classification, and segmentation of connected components into symbols. A graph-based attention module allows multiple classification queries for nodes and edges to be computed in one feedforward pass, resulting in faster training and execution.

The output produced is an SLT graph, containing the structure as well as the symbols and spatial relationship classifications. The SLT may be converted into Presentation MathML or a LaTeX string (see Fig. 5).

4   Results for Pipeline Components

We provide preliminary results for each pipeline component in this section.

SymbolScraper. 100 document pages were randomly chosen from a collection of 10,000 documents analyzed by SymbolScraper. Documents were evaluated based on the percentage of detected characters that had correct bounding boxes. 62/100 documents had all characters detected without error. An additional 9 documents had over 80% of their characters correctly boxed, most with just one or two words with erroneous bounding boxes. 6 documents were not born-digital (producing no character detections); 10 were OCR'd documents, represented entirely in Type 3 fonts that SymbolScraper could not process, and 9 failed due to a bug in the image rendering portion of the symbol extractor. The remaining 4 documents had fewer than 80% of their character bounding boxes correctly detected.

Table 1. Formula detection results for TFD-ICDAR2019

	IOU ≥ 0.75			IOU ≥ 0.5		
	Precision	Recall	F-score	Precision	Recall	F-score
ScanSSD	**0.774**	**0.690**	**0.730**	**0.851**	**0.759**	**0.802**
RIT 2	0.753	0.625	0.683	0.831	0.670	0.754
RIT 1	0.632	0.582	0.606	0.744	0.685	0.713
Mitchiking	0.191	0.139	0.161	0.369	0.270	0.312
Samsung[‡]	0.941	0.927	0.934	0.944	0.929	0.936

[‡] Used character information

For documents with more than 80% of their text properly extracted, the number of pages with character extractions without error is 94.7% (71 of 75). We compare our SymbolScraper tool against PDFBox, PDFMiner and PyMuPDF. As shown in Fig. 3(a), only SymbolScraper recovers the precise bounding boxes directly from the PDF document. SymbolScraper is much faster than other techniques that obtain character bounding boxes using OCR or image processing. In our preliminary experiments, running SymbolScraper using a single process on a CPU takes around 1.7 s per page on a laptop computer with modest resources.

ScanSSD. The top of Table 1 shows formula detection results obtained by ScanSSD along with participating systems in the ICDAR 2019 Typeset Formula Detection competition [10]. Systems are evaluated using a coarse detection threshold with $IoU \geq 50\%$, and a fine detection threshold with $IoU \geq 75\%$. We use the implementation provided by the competition [10] to calculate intersections for box pairs, as well as precision, recall and f-scores.

ScanSSD outperforms all systems that did not use character locations and labels from ground truth in their detection system. Relative to the other image-based detectors, ScanSSD improves both precision and recall, but with a larger increase in recall scores for IOU thresholds of both 0.75 (fairly strict) and 0.5 (less strict). The degradation in ScanSSD performance between IOU thresholds is modest, losing roughly 7% in recall, and 7.5% precision, indicating that detected formulas are often close to their ideal locations.

Looking a bit more closely, over 70% of ground truth formulas are located at their ideal positions (i.e., with an IOU of 1.0). If one then considers the character level, i.e., the accuracy with which characters inside formulas are detected, we obtain an f-measure of 92.6%. The primary cause of the much lower formula detection rates are adjacent formulas merged across textlines, and splitting formulas at large whitespace gaps (e.g., for variable constraints). We believe these results are sufficiently accurate for use in prototyping formula search engines applied to collections of PDF documents. Examples of detection results are visible in Figs. 1 and 6.

Running the test set on a desktop system with a hard drive (HDD), 32 GB RAM, an Nvidia RTX 2080 Ti GPU, and an AMD Ryzen 7 2700 processor, our PyTorch-based implementation took a total of 4 h, 33 min, and 31 s to process 233

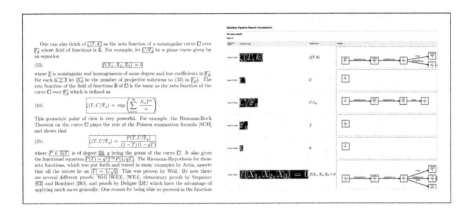

Fig. 6. HTML visualization for formulas extracted from a sample PDF page with detected formula locations (left), and a table (right) showing extracted formulas and recognition results as rendered MathML and SLT graphs.

pages including I/O operations (average: 70.4 secs/page). While not problematic for early development, this is too slow for a large corpus. We identify possible accelerations in the conclusion (see Sect. 6).

When using an SSD on whole document pages in a single detection window (e.g., 512×512), very low recall is obtained because of the low resolution, and generally bold characters were detected as math regions.

QD-GGA. We evaluated QD-GGA on InftyMCCDB-2[3], a modified version of InftyCDB-2 [17]. We obtain a structure recognition rate of 92.56%, and an expression recognition rate (Structure + Classification) of 85.94% on the test set (6830 images). Using a desktop system with a hard drive (HDD), 32 GB RAM, an Nvidia GTX 1080 GPU, and an Intel(R) Core(TM) i7-9700KF processor, QD-GGA took a total of 26 min, 25 s to process 6830 formula images (average: 232 ms/formula).

5 New Tools for Visualization and Evaluation

An important contribution of this paper is new tools for visualizing recognition results and structure recognition errors, which are essential for efficient analysis and error diagnosis during system development and tuning. We have created these in the hopes of helping both ourselves and others working in formula extraction and other graphics extraction domains.

Recognition Result Visualization (HTML + PDF). We have created a tool to produce a convenient HTML-based visualization of the detection and recognition results, with inputs and outputs of the pipeline for each PDF document page. For each PDF page, we summarise the results in the form of a table,

---

[3] https://zenodo.org/record/3483048#.XaCwmOdKjVo.

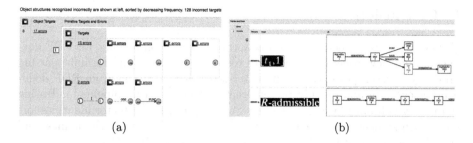

$$(a) \qquad\qquad\qquad\qquad (b)$$

Fig. 7. Error analysis (errors shown in red). (a) Main error table organized by decreasing frequency of errors. (b) Specific instances where 'l' is misclassified as 'one,' seen after clicking on the '10 errors' link in the main table. (Color figure online)

which contains the document page image with the detected expressions at the left and a scrollable (horizontally and vertically) sub-table with each row showing the expression image, the corresponding MathML rendered output, and the SLT graphs (see Fig. 6).

We created HTMLs for easy evaluation of the entire pipeline, to identify component(s) (symbol extraction, expression detection, or structure recognition) producing errors on each page. This makes it easier to diagnose the causes of errors, identify the strengths and weaknesses of different components, and improve system designs.

Improved Error Visualization. The LgEval library [13,14] has been used for evaluating formula structure recognition both for recognition over given input primitives (e.g., strokes) and for output representations without grounding in input primitives [10]. We extend LgEval's error summary visualization tool, `confHist`, to allow viewing all instances of specific errors through HTML links. These links take the user directly to a list of formulas containing a specific error. The tool generates a table showing all error types for ground truth SLT subtrees of a given size, arranged in rows and sorted by frequency of error, as shown in Fig. 7(a) (for single symbols). Each row contains a sub-table with all primitive level target and error graphs, with errors shown in red. Errors include missing relationships and nodes, segmentation errors, and symbol and relationship classification errors - in other words, all classification, segmentation, and relationship errors.

New links in the error entries of the HTML table open HTML pages, containing all formulas sharing a specific error along with additional details arranged in a table. This includes all expression images containing a specific error along with their SLT graph showing errors highlighted in red Fig. 7(b). The new tool helps to easily identify frequent recognition errors in the contexts where they occur. For example, as seen in Fig. 7(b), we can view all expression images in which 'l' is misclassified as 'one' by clicking the error entry link (10 errors) in Fig. 7(a) and locate the incorrect symbol(s) using the SLT graphs.

Both visualization tools load very quickly, as the SLT graphs are represented in small PDF files that may be scaled using HTML/javascript widgets.

6 Conclusion

We have presented a new open-source formula extraction pipeline for PDF documents, and new tools for visualizing recognition results and formula parsing errors. SymbolScraper extracts character labels and locations from PDF documents accurately and quickly by intercepting the rendering pipeline and using character glyph data directly. The new ScanSSD formula detector identifies formulas with sufficient accuracy for prototyping math-aware search engines, and very high accuracy at the character level. The existing QD-GGA system is used to parse formula structure in detected formula regions.

In the future, we would like to design an end-to-end trainable system for both formula detection and parsing, and also extend our system to handle more PDF character encodings (e.g., Type 3 fonts). Future work for SymbolScraper includes improved filtering of horizontal bars, integrating compound characters classification within the tool, and faster implementations (e.g., using parallelization). For detection using ScanSSD, we are looking at ways to accelerate the non-maximal suppression algorithm and fusion steps, and to improve the page-level merging of windowed detections to avoid under-segmenting formulas on adjacent text lines, and over-merging of formulas with large whitespace gaps (e.g., caused by variable constraints to the right of a formula).

We believe that our pipeline and tools will be useful for others working on extracting formulas and other graphics type from documents. Our system will be available as open source before the conference.

Acknowledgements. A sincere thanks to all the students who contributed to the pipeline's development: R. Joshi, P. Mali, P. Kukkadapu, A. Keller, M. Mahdavi and J. Diehl. Jian Wu provided the document collected used to evaluate SymbolScraper. This material is based upon work supported by the Alfred P. Sloan Foundation under Grant No. G-2017-9827 and the National Science Foundation (USA) under Grant Nos. IIS-1717997 (MathSeer project) and 2019897 (MMLI project).

References

1. Apache: PDFBOX - a Java PDF library. https://pdfbox.apache.org/
2. Baker, J.B., Sexton, A.P., Sorge, V.: A linear grammar approach to mathematical formula recognition from PDF. In: Carette, J., Dixon, L., Coen, C.S., Watt, S.M. (eds.) CICM 2009. LNCS (LNAI), vol. 5625, pp. 201–216. Springer, Heidelberg (2009). https://doi.org/10.1007/978-3-642-02614-0_19
3. Davila, K., Joshi, R., Setlur, S., Govindaraju, V., Zanibbi, R.: Tangent-V: math formula image search using line-of-sight graphs. In: Azzopardi, L., Stein, B., Fuhr, N., Mayr, P., Hauff, C., Hiemstra, D. (eds.) ECIR 2019. LNCS, vol. 11437, pp. 681–695. Springer, Cham (2019). https://doi.org/10.1007/978-3-030-15712-8_44
4. Deng, Y., Kanervisto, A., Ling, J., Rush, A.M.: Image-to-markup generation with coarse-to-fine attention. In: Precup, D., Teh, Y.W. (eds.) Proceedings of the 34th International Conference on Machine Learning, ICML 2017, Sydney, NSW, Australia, 6–11 August 2017. Proceedings of Machine Learning Research, vol. 70, pp. 980–989. PMLR (2017). http://proceedings.mlr.press/v70/deng17a.html

5. Edmonds, J.: Optimum branchings. J. Res. Nat. Bureau Stand. Sect. B Math. Math. Phys. **71B**(4), 233 (1967). https://doi.org/10.6028/jres.071B.032. https://nvlpubs.nist.gov/nistpubs/jres/71B/jresv71Bn4p233_A1b.pdf

6. Liao, M., Shi, B., Bai, X., Wang, X., Liu, W.: TextBoxes: A fast text detector with a single deep neural network. In: Thirty-First AAAI Conference on Artificial Intelligence (2017)

7. Lin, X., Gao, L., Tang, Z., Lin, X., Hu, X.: Mathematical formula identification in PDF documents. In: 2011 International Conference on Document Analysis and Recognition, pp. 1419–1423, September 2011. https://doi.org/10.1109/ICDAR.2011.285. iSSN: 2379-2140

8. Liu, W., et al.: SSD: single shot MultiBox detector. In: Leibe, B., Matas, J., Sebe, N., Welling, M. (eds.) ECCV 2016. LNCS, vol. 9905, pp. 21–37. Springer, Cham (2016). https://doi.org/10.1007/978-3-319-46448-0_2

9. Mahdavi, M., Sun, L., Zanibbi, R.: Visual parsing with query-driven global graph attention (QD-GGA): preliminary results for handwritten math formula recognition. In: 2020 IEEE/CVF Conference on Computer Vision and Pattern Recognition Workshops (CVPRW), pp. 2429–2438. IEEE, Seattle, June 2020. https://doi.org/10.1109/CVPRW50498.2020.00293. https://ieeexplore.ieee.org/document/9150860/

10. Mahdavi, M., Zanibbi, R., Mouchere, H., Viard-Gaudin, C., Garain, U.: ICDAR 2019 CROHME + TFD: Competition on recognition of handwritten mathematical expressions and typeset formula detection. in: 2019 International Conference on Document Analysis and Recognition (ICDAR), pp. 1533–1538. IEEE, Sydney, September 2019. https://doi.org/10.1109/ICDAR.2019.00247. https://ieeexplore.ieee.org/document/8978036/

11. Mali, P., Kukkadapu, P., Mahdavi, M., Zanibbi, R.: ScanSSD: Scanning single shot detector for mathematical formulas in PDF document images. arXiv:2003.08005 [cs], March 2020

12. Mouchère, H., Viard-Gaudin, C., Zanibbi, R., Garain, U.: ICFHR2016 CROHME: Competition on recognition of online handwritten mathematical expressions. In: 2016 15th International Conference on Frontiers in Handwriting Recognition (ICFHR), pp. 607–612, October 2016. https://doi.org/10.1109/ICFHR.2016.0116. iSSN: 2167-6445

13. Mouchère, H., Viard-Gaudin, C., Zanibbi, R., Garain, U., Kim, D.H., Kim, J.H.: ICDAR 2013 CROHME: Third international competition on recognition of online handwritten mathematical expressions. In: 2013 12th International Conference on Document Analysis and Recognition, pp. 1428–1432, August 2013. https://doi.org/10.1109/ICDAR.2013.288. iSSN: 2379-2140

14. Mouchère, H., Zanibbi, R., Garain, U., Viard-Gaudin, C.: Advancing the state of the art for handwritten math recognition: The CROHME competitions, 2011–2014. Int. J. Doc. Anal. Recogn. (IJDAR) **19**(2), 173–189 (2016). https://doi.org/10.1007/s10032-016-0263-5

15. Phong, B.H., Dat, L.T., Yen, N.T., Hoang, T.M., Le, T.L.: A deep learning based system for mathematical expression detection and recognition in document images. In: 2020 12th International Conference on Knowledge and Systems Engineering (KSE), pp. 85–90, November 2020. https://doi.org/10.1109/KSE50997.2020.9287693. iSSN: 2164-2508

16. Sorge, V., Bansal, A., Jadhav, N.M., Garg, H., Verma, A., Balakrishnan, M.: Towards generating web-accessible STEM documents from PDF. In: Proceedings of the 17th International Web for All Conference, W4A 2020, pp. 1–5. Association for Computing Machinery, New York, April 2020. https://doi.org/10.1145/3371300.3383351

17. Suzuki, M., Uchida, S., Nomura, A.: A ground-truthed mathematical character and symbol image database. In: Eighth International Conference on Document Analysis and Recognition (ICDAR 2005), vol. 2, pp. 675–679 (2005). https://doi.org/10.1109/ICDAR.2005.14

18. Suzuki, M., Tamari, F., Fukuda, R., Uchida, S., Kanahori, T.: INFTY: an integrated OCR system for mathematical documents. In: Proceedings of the 2003 ACM Symposium on Document Engineering, DocEng 2003, pp. 95–104. Association for Computing Machinery, New York, November 2003. https://doi.org/10.1145/958220.958239

19. Suzuki, M., Yamaguchi, K.: Recognition of E-Born PDF including mathematical formulas. In: Miesenberger, K., Bühler, C., Penaz, P. (eds.) ICCHP 2016. LNCS, vol. 9758, pp. 35–42. Springer, Cham (2016). https://doi.org/10.1007/978-3-319-41264-1_5

20. Zanibbi, R., Pillay, A., Mouchere, H., Viard-Gaudin, C., Blostein, D.: Stroke-based performance metrics for handwritten mathematical expressions. In: 2011 International Conference on Document Analysis and Recognition, pp. 334–338, September 2011. https://doi.org/10.1109/ICDAR.2011.75. iSSN: 2379-2140

21. Zanibbi, R., Blostein, D.: Recognition and retrieval of mathematical expressions. Int. J. Doc. Anal. Recogn. (IJDAR) **15**(4), 331–357 (2012). https://doi.org/10.1007/s10032-011-0174-4

22. Zanibbi, R., Mouchère, H., Viard-Gaudin, C.: Evaluating structural pattern recognition for handwritten math via primitive label graphs. In: Document Recognition and Retrieval XX, vol. 8658, p. 865817. International Society for Optics and Photonics, February 2013. https://doi.org/10.1117/12.2008409

23. Zanibbi, R., Orakwue, A.: Math search for the masses: Multimodal search interfaces and appearance-based retrieval. In: Kerber, M., Carette, J., Kaliszyk, C., Rabe, F., Sorge, V. (eds.) CICM 2015. LNCS (LNAI), vol. 9150, pp. 18–36. Springer, Cham (2015). https://doi.org/10.1007/978-3-319-20615-8_2

24. Zhang, X., Gao, L., Yuan, K., Liu, R., Jiang, Z., Tang, Z.: A symbol dominance based formulae recognition approach for PDF documents. In: 2017 14th IAPR International Conference on Document Analysis and Recognition (ICDAR), vol. 01, pp. 1144–1149, November 2017. https://doi.org/10.1109/ICDAR.2017.189. iSSN: 2379-2140

Toward Automatic Interpretation
of 3D Plots

Laura E. Brandt[1]([✉]) [iD] and William T. Freeman[1,2] [iD]

[1] MIT CSAIL, Cambridge, MA, USA
{lebrandt,billf}@mit.edu
[2] The NSF AI Institute for Artificial Intelligence and Fundamental Interactions,
Cambridge, MA, USA

Abstract. This paper explores the challenge of teaching a machine how
to reverse-engineer the grid-marked surfaces used to represent data in 3D
surface plots of two-variable functions. These are common in scientific and
economic publications; and humans can often interpret them with ease,
quickly gleaning general shape and curvature information from the simple
collection of curves. While machines have no such visual intuition, they
do have the potential to accurately extract the more detailed quantitative
data that guided the surface's construction. We approach this problem by
synthesizing a new dataset of 3D grid-marked surfaces (SurfaceGrid) and
training a deep neural net to estimate their shape. Our algorithm successfully recovers shape information from synthetic 3D surface plots that have
had axes and shading information removed, been rendered with a variety
of grid types, and viewed from a range of viewpoints.

Keywords: 3D plot interpretation · Figure analysis · Data
extraction · Shape-from-x · Graphics recognition · Neural network ·
Computer vision

1 Introduction

Suppose you encounter a chart in a published paper and want to automate the
interpretation of the data it represents so you can use it in your own research. If
the chart were a 2D one like those shown in Fig. 1 (*e.g.* a pie chart or a scatter
plot), you would be able to use any of an array of available tools like [14,29] to
recover the data yourself. However if the chart were instead a 3D surface plot like
that shown in Fig. 2, you would be out of luck. Humans have the ability to easily
perceive qualitative information such as shape and curvature from grid-marked
surfaces like these, but to recover quantitative data from a surface plot by hand
would be an arduous (and likely doomed) task. Machines on the other hand
lack our visual intuition about surface contours. But properly designed they
have the potential to do something humans cannot: reverse-engineer the surface
to recover the original data that created it. This work explores the challenge of
teaching a machine how to reverse-engineer 3D grid-marked surfaces. This visual
perception task is of inherent interest, and supports the goal of recovering data
from published 3D plots.

© Springer Nature Switzerland AG 2021
J. Lladós et al. (Eds.): ICDAR 2021, LNCS 12822, pp. 35–50, 2021.
https://doi.org/10.1007/978-3-030-86331-9_3

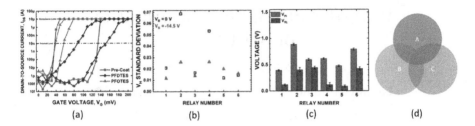

Fig. 1. 2D charts. Tools and algorithms exist to extract data from 2D charts like (a–c) experimental results from a micro-switch study [32], and (d) Venn diagrams. No such methods exist for 3D charts like the surface plot in Fig. 2.

Internet search engine providers like Google and Microsoft have interest in tools that can look up and recover data from published sources. Motivated by the amount of information stored in tables on the Internet and in documents, Google AI recently published the TAPAS approach to automated table parsing, which allows the user to query published tables with questions phrased in natural language [9,31]. An aim of such research is to improve user ability to comb the Internet or other large databases for published data based on their understanding of its content rather than a lucky guess of its associated keywords, and there is demand among scientists for tools to help them find published figures and papers based on their data content [22,49]. A model able to reverse-engineer published 3D plots would enable the vast digital library of technical publications containing such figures to be automatically processed and indexed based on the data they present, improving the quality and efficiency of scientific literature surveys.

There exist several kinds of 3D plots in published literature. This paper is concerned specifically with plots of two-variable functions like the potential energy surface shown in Fig. 2. To interpret such a figure, an interpreter must be able to do several things. It must be able to detect the figure in the first place. It has to segregate that figure into labels, axes, and the visual representation of the data (that is, the grid-marked deformed surface of the plot). Then it must reverse-engineer that visual representation to recover the data that was used to produce it. This involves perceiving the shape of these grid-marked surfaces, determining the perspective from which the surface plot was projected when the figure was rendered, undoing that perspective projection to recover the original data as a function $z(x, y)$, and calibrating that data using the axes and annotations earlier extracted from the figure.

This work is focused on the specific task of recovering shape information from grid-marked surfaces akin to 3D surface plots with axes removed. Figure detection, segregation, de-projection, and calibration are discussed further in the concluding section of this paper. Marr, Stevens, and others [28,41] have approached the more general problem of shape-from-surface contours by making a variety of assumptions about surface symmetries and constraints (*e.g.* that surfaces are locally cylindrical [28,41]). These are conditions that often do not

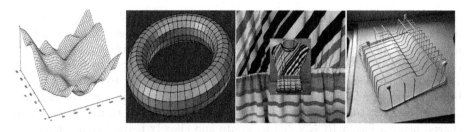

Fig. 2. Grid-marked surfaces are common both in technical publications and everyday life, and humans can easily interpret shapes ranging from simple toroids to ripples in fabric. This paper works toward the automated interpretation (by computer) of 3D plots like the potential energy surface at left [50], using neural models that might someday extend to the analysis of more general grid-marked surfaces such as this torus, shirt, and dish rack.

hold in 3D surface plots, and we instead approach our shape recovery problem with an artificial neural net.

The main contributions of this work are

1. A deep neural net trained to recover shape information from 3D surface plots without axes or shading information, synthesized with a variety of grid types and viewed from a range of viewpoints.
2. SurfaceGrid: a 98.6k-image dataset of such 3D plot surfaces, synthesized side-by-side with their corresponding depth maps and underlying data.
3. An experimental analysis of the neural model and training method, including an ablation study and performance evaluation.

2   Related Work

Automatic chart interpretation is an active area of research and there exist a large number of semi-automatic tools like [14,29,35,37], often called "plot digitizers", that can extract data from published 2D charts like those shown in Fig. 1 (bars, pies, *etc.*). However, these tools generally require that the human user manually perform key steps like axis calibration or feature identification. This requirement of human input for every image limits the application of such tools to the occasional reverse-engineering of single figures or to applications for which the number of man-hours required is deemed worth it.

Work has been done toward fully automating the processes of 2D chart interpretation. Yanping Zhou and Tan [55] used edge tracing and a Hough transform-based technique to extract bars from bar charts. Huang *et al.* and Liu *et al.* [12,23] used vectorized edge maps to extract features from bar, 2D line, and other 2D charts. Huang and Tan [11], Lu *et al.* [25], and Shao and Futrelle [40] used these kinds of features to classify chart types, and Lu *et al.* [26] used them to extract and re-plot data from smooth 2D line plots. Prasad *et al.* [33] measured similarity between the scale-invariant feature transform (SIFT) [24] and histograms of oriented gradients (HOG) [3] and used it to train a support vector machine (SVM) [1] to classify chart types. ReVision [39] used a combination of feature identification and patch clustering to not only classify figures but also reverse-engineer their data to enable re-visualization to other chart formats, and Jung et al. and Dai et al. [2,16] similarly trained classifiers to categorize published charts in addition to extracting features and text. Last year, Fangfang Zhou et al. [53] took a fully neural network-based approach to interpreting bar charts, using a Faster-RCNN [36] to locate and classify textual chart elements, and an attentional encoder-decoder to extract numerical information. To our knowledge the prior work focuses entirely on 2D charts, leaving the problem of interpreting 3D surface plots like that in Fig. 2 unaddressed.

Shape-from-X. Many kinds of cues have been used to teach machines to reconstruct shape from images including symmetry [6,30], shading [10,51], stereo [17,45] and multi-view [8], motion [5,44,54], silhouette [20], and texture [48]. Our work was particularly inspired by Marr's discussion of the human ability to perceive shape from from simple grid- and line- marked surfaces [28], which cited Stevens as an example of a computational attempt to reconstruct shape from surface contour cues by making the simplifying assumption that all instances of curvature are approximately cylindrical [41]. Weiss [47] proposed an alternative algorithm that modeled surface contours as springs with energy to be minimized in solving for surface shape, while Knill [19] proposed that image contours ought be interpreted with a human-inspired geodesic constraint, and Ulupinar and Nevatia [43] introduced a method for reconstructing the shape of generalized cylinders from median lines by assuming generic viewpoint [7] and certain symmetries. Mamassian and Landy [27] conducted a psychophysical study of the stability of human interpretations of simple grid-marked surfaces, and modeled their process as a simple Bayesian observer. More recently, Pumarola et al. [34] used a convolutional neural net to estimate the meshgrids of synthesized deformed surfaces "made" with various materials (fabric, metal, etc.), and Li et al. [21] introduced a tool (BendSketch) for sketch-based 3D model design capable of generating 3D models from sketches using an input collection of digital pen annotations from pre-defined categories of surface contours.

3   Method

3.1   Network Architecture

The task of reverse-engineering shape information from a picture of a 3D surface plot can be approached as an image translation task (Fig. 3), and we did so using the Pix2Pix neural image translation module [15]. Pix2Pix is a conditional generative adversarial network (cGAN) consisting of a UNet [38] generator that learns to map (in our case) from an input surface plot to an output "fake" depth map, and a PatchGAN [15] discriminator that takes a depth map as input and learns to identify whether it is real or merely a generated estimate by looking at small patches of local structure. The discriminator provides feedback to the

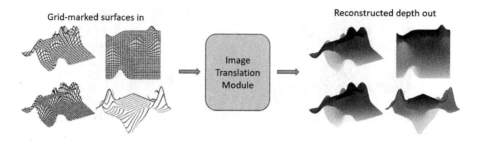

Fig. 3. Our Final model used an image translation module [15] to reverse-engineer grid-marked surfaces and recover depth maps representing their underlying data.

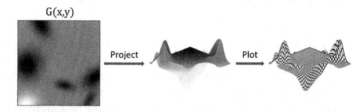

Fig. 4. Dataset generation was a three-step process. First multivariable functions $G(x, y)$ were created by summing synthesized Gaussians. These were then projected as depth maps and plotted from specified camera viewpoints. These projected plot-depth map pairs were used to train our model.

generator during training, so that the later learns simultaneously to "fool" the discriminator and to minimize the mean absolute error between estimated and true depth map. We refer readers to Isola *et al.*'s paper for further detail [15].

3.2   The SurfaceGrid Dataset

A sizable dataset is required to train a neural net, and the creation of a good one is a challenge. While images of grid-marked surfaces could in principle be gathered from "the wild", only rarely is truth data available. This necessitated the side-by-side synthesis of a new dataset: SurfaceGrid.

We produced a dataset of what are effectively mathematical surface plots without axes or shading, paired with depth maps. This allowed us to use the following 3-step procedure for synthesizing a large dataset (Fig. 4):

1. randomly generate a set of multivariable functions $G(x, y)$,
2. project these functions as depth maps from camera viewpoints of choice, and
3. plot these functions with a variety of grid parameters, projected from the same viewpoints as in Step 2.

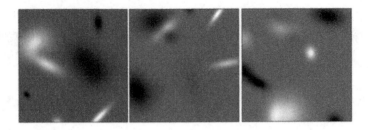

Fig. 5. Multivariable functions $G(x, y)$ were generated for our new dataset by summing Gaussians with randomly-synthesized parameters. Gaussians are radial basis functions, and sums of them can be used to model any two-variable function. These $G(x, y)$ were subsequently projected as depth maps and plotted.

In the first step we randomly generated $1 \leq N \leq 10$ Gaussians $g_i(x, y)$ with standard deviations $8 \leq \sigma_x, \sigma_y \leq 512$ pixels, rotated by a random angle in the xy plane, and shifted to a random mean μ anywhere on the domain $0 \leq x, y \leq 512$ pixels. The individual Gaussian amplitudes were chosen to normalize their areas, and could be positive or negative with equal probability. These N Gaussians $g_i(x, y)$ were then summed to produce the final multivariable function $G(x, y) = \sum_{i=1}^{N} g_i(x, y)$. Figure 5 shows examples.

In the second step we took these functions and computed their corresponding depth maps as projected from a set of camera viewpoints chosen to cover an octant of space. Specifically, in spherical coordinates (r, θ, ϕ) we chose a camera distance $r = 1024$ pixels from the origin and azimuth and elevation angles $\theta, \phi \in [0, 30, 60]°$. We generated an additional viewpoint of $(\theta, \phi) = (45, 22.5)°$.

In the third and final step we generated grid-marked surface plots of these functions using lines $\delta x = \delta y = 3$ pixels wide and spaced $\Delta x = \Delta y \in [17, 25, 34, 42]$ pixels apart for effective sampling resolutions of $\Delta X \equiv \Delta x + \delta x = \Delta Y \equiv \Delta y + \delta y \in [20, 28, 37, 45]$ pixels. These numbers were chosen somewhat arbitrarily based on a human judgement of what looked "reasonable but different", and it is worth noting here that the $\Delta x = \Delta y = 42$ pixel case was often guilty of undersampling the data. We also produced line- (not grid-) marked surfaces, one type of grid with different line spacings in the two directions ($\Delta X = 20 \neq \Delta Y = 28$ pixels), and grids rotated by azimuthal angles $\theta_g \in [30, 50, 60]°$.

Overall, 76.6k images of grid-marked surfaces, 20k corresponding projected depth maps, and 2000 underlying multivariable functions were generated for a total of 98.6k images in this dataset (Fig. 6). The grid-marked surfaces shown here and throughout the paper are inverted for printing purposes from the raw data in the dataset which is in inverse (white-on-black). All images are single-channel, 512×512, and available online at https://people.csail.mit.edu/lebrandt.

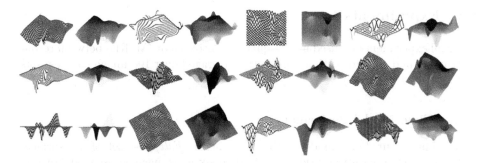

Fig. 6. The SurfaceGrid dataset is our new 98.6k-image dataset. It contains synthesized binary 3D surfaces marked with grids and lines of varying resolutions, viewed from a multitude of camera positions, and paired with their associated ground truth depth maps. The underlying multivariable functions are also included. The dataset is available online at https://people.csail.mit.edu/lebrandt.

3.3   Training Details

Published 3D surface plots come in many forms; the data being plotted, surface marking resolutions, and rendering viewpoints all vary greatly. In this work we used sums of radial basis functions (Gaussians) in an effort to efficiently cover the enormous domain of published data with a small number of parameters. Any two-variable function can be modeled by summed Gaussians, if a sufficient number are used. We needed to do similarly with the domains of surface markings and viewpoints.

We could have trained our model on examples rendered with randomly-generated markings and viewpoints, but were concerned that it would over-train in a way difficult to detect via experiment. When training examples are randomly-generated, randomly-generated test examples are very likely to have a "near neighbour" in the training set that is near enough to result in a low mean error, but far enough that the rules being used to reverse-engineer the surface's shape are not general. We therefore built up our training set by adding *specific* categories of grid-marked surfaces to a very narrow *Baseline* training set. By training models and testing as we went (using test parameters very different from training examples, *e.g.* viewpoint shifted from 30 to 60° azimuth), we were able to see how general the rules learned by the model were. We will discuss this further in Sect. 4.2.

We used the Pix2Pix architecture introduced in [15] and implemented for PyTorch in [56], using an Adam [18] optimizer with learning rate = 0.0002 and momentum parameters $\beta_1 = 0.5$ and $\beta_2 = 0.999$. The training set size was 1.8k image pairs for all models, created via a shuffled sampling of surfaces corresponding to functions 0–1799. Mean absolute error between reconstruction and ground truth was computed over a 100-image pair validation set (functions 1800–1899). Training was stopped (and the model rolled back) when the validation error had been increasing continuously for 50 epochs. Our *Final* model trained for 274 epochs.

4   Experiments

For all experiments we used mean-squared relative error (MSRE) between reconstruction and ground truth as our metric, averaged over 100-image pair test sets
(functions 1900–1999).

4.1   Final Model Performance

We applied our *Final* model both to a general (*Full*) test set of 3D surfaces
marked with a variety of grid types, and to six more narrowly-scoped test sets
that isolated specific surface types. The narrowest *Base* test set consisted solely
of surfaces marked with 20 × 20 pixel grids and projected from camera viewpoints
that had appeared in the training set. We excluded viewpoints with extreme
(very high/low) elevation angles.

Five test sets added a single class of surfaces to the *Base* set. The *Resolutions*
and *Angled Grids* test sets added new grid resolutions and angles, respectively.
The *Viewpoints* dataset added surfaces viewed from new generic viewpoints.
The *Lines* dataset added surfaces marked with lines instead of grids. Finally,
the *Accidental Viewpoints* dataset restored the previously-excluded viewpoints
with extreme elevation angles. The *Full* test set included images of all these
types, including images that combined multiple of these effects. (*E.g.* line-marked
surfaces viewed from new viewpoints, or surfaces marked with angled lines.)

Table 1 reports the MSRE for our *Final* model when applied to these test
sets, and Fig. 7 shows characteristic results. The *Final* model generalized well to
viewpoints and line/grid resolutions not seen during training, though it struggled
with surfaces viewed from extreme elevation angles (Fig. 8).

Table 1. Final model performance on narrowly-scoped test sets. The *Final* model
generalized well to new grid resolutions and angles, line-marked surfaces, and generic
viewpoints. It struggled with surface plots viewed from extreme elevation angles (*Acc.
Viewpoints*), but overall generalized well to the *Full* test set (last row). None of the
reconstruction errors exceeded 0.5% MSRE.

Test set	MSRE ($\times 10^{-2}$)
Base	0.106
Resolutions	0.107
Angled Grids	0.124
Lines	0.213
Viewpoints	0.281
Acc. Viewpoints	0.424
Full	0.327

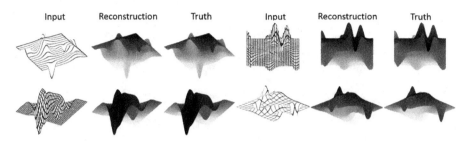

Fig. 7. **Final model performance** when applied to various input grid- and line-marked images. It performed well on a variety of grid resolutions, angles, and generic camera viewpoints. Top row inputs had similar parameters as during training. Bottom row inputs tested (L) new camera viewpoint and (R) new viewpoint and grid resolution.

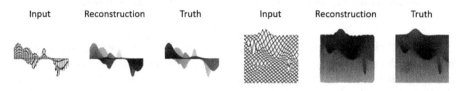

Fig. 8. **Limitations.** The *Final* model struggled to generalize to extreme-elevation viewpoints, especially when the grid was angled. There is an inherent depth ambiguity in edge-on projections at low elevations because occlusion cues are lacking. At high elevations, curvature cues are lost and, in the case of angled grids, plot boundaries interpreted as jagged. Note that data is rarely plotted from such extreme elevations.

4.2 Impact of Training Set Components on Performance

We compared the performance of our *Final* model with that of four variants trained on reduced datasets of increasing complexity (Fig. 9). The naïve *Baseline* model was trained using only 20×20 grid-marked surfaces viewed from a single generic viewpoint of $\theta = \phi = 30°$. The *Viewpoints* training set added general viewpoints $\theta, \phi \in [0, 30, 60]$ to the *Baseline* training set (chosen to cover an octant of space). Elevation $\phi = 90°$ was neglected because grid-marked surfaces lack curvature cues when viewed directly from above, and azimuths $\theta \geq 90°$ were excluded because they have equivalent viewpoints in $0 \leq \theta < 90°$. Adding these eight viewpoints to the dataset allowed our model to recover shape from arbitrary generic viewpoints, but it could not handle general grid resolutions.

We found that training on a single additional class of input images was sufficient: surfaces marked with *Lines* (not grids) of a new spacing (we somewhat arbitrarily chose 37 pixels). Even though this class of training data was *line*-marked, it enabled the model to generalize to completely new *grid* resolutions. As an experiment we added surfaces with *Angled Grids* that had been rotated by an azimuthal angle $\theta_g \in [30, 60]$, which enabled generalization to new grid angles but lowered overall performance. Angled grids are also quite rare in 3D surface plots. Figure 9 shows examples from these reduced training sets.

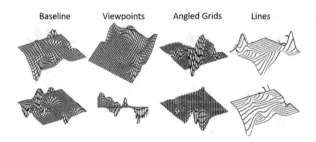

Fig. 9. Reduced training sets. Example images from the four reduced training sets discussed in Sect. 4.2. *Baseline* contained only 20×20 pixel grids viewed from a single viewpoint. *Viewpoints* added additional general viewpoints. *Angled Grids* and *Lines* added angled grids and line-marked surfaces, respectively, to the *Viewpoints* dataset. The *Final* model used a training set containing all of these image types.

We applied these models both to a general (*Full*) test set of 3D surfaces marked with a variety of grid types, and to three more narrowly-scoped test sets. The *Base* and *Full* test sets were the same as in Sect. 4.1. The *General Grids* test set added new grid resolutions and angles, along with line-marked surfaces, to the *Base* test set. Alternatively, the *General Views* test set added new viewpoints without diversifying the surface markings themselves.

Table 2 compares the MSRE of the reconstructions produced by these five models when applied to these four test sets. Table 3 reports the relative impact of each addition to the training set on model performance, expressed as a percent improvement in MSRE on these four test sets. Figure 10 gives a few examples of the impact of these training set additions on the model's ability to generalize to new grid resolutions and viewpoints.

Table 2. Impact of different additions to the training set (down) on model performance when applied to specific types of images (across), reported as MSRE x10^{-2}. Our *Final* model generalizes best to the *Full* and *General Grids* test sets, and performs within 2% and 4% of the best on the *General Views* and *Base* sets, respectively.

↓ Training \ Test →	Base	Gen. Grids	Gen. Views	Full
Baseline	0.237	0.804	0.723	0.951
Viewpoints	0.127	1.43	0.385	1.18
Angled Grids	0.109	1.49	**0.346**	1.20
Lines	**0.102**	0.183	0.355	0.329
Final	0.106	**0.148**	0.353	**0.327**

Table 3. Relative impact of different additions to the training set (down) on model performance when applied to specific types of images (across), reported as a % improvement in MSRE. Positive numbers are good and the largest improvements in each category are in bold. *Angled Grids* were not particularly beneficial inclusions, while *Lines* were particularly important for enabling generalization to new grid resolutions and improving overall performance on the *Full* test set.

↓ Training \ Test →	Base	Gen. Grids	Gen. Views	Full
Baseline	–	–	–	–
Viewpoints	**46.4**	−77.7	**46.8**	−24.3
Angled Grids	14.2	−4.15	10.0	−1.95
Lines	19.7	**87.2**	7.80	**72.2**
Final	−3.92	19.3	0.564	0.608

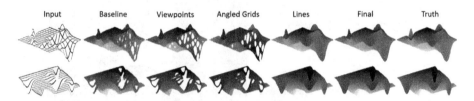

Fig. 10. Relative impact of different additions to the training set on reconstructions. The addition of line-marked surfaces to the training set was particularly impactful (third-to-last column). The bottom row input had similar parameters as during training. The top row input tested a new viewpoint and grid resolution.

4.3 Performance on Real Plots and Wireframe Models

We applied our *Final* model to real plots taken from publications [28,50]. In these cases we lacked truth data and so could not compute the MSRE for our reconstructions, but Fig. 11 shows that our model successfully extracted general shape and relative depth information from these plots.

We also applied our neural model to wireframe models from [13] and binarized images of physical wireframe objects from [4,46]. Figure 12 shows that even for the object models, which look qualitatively very different from the synthetic training data, our neural model still manages to extract general relative depth information. Notice also that unlike the training data, the real-world wireframe objects do not have opaque surfaces between the surface contours, meaning that contours on front and back surface are equally visible. Despite this, our *Final* model generates reasonable interpretations of entire sections of these images.

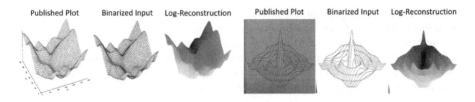

Fig. 11. Performance on real plots. Our *Final* model successfully extracted general shape and relative depth information from published plots taken from [28,50], even though these plots would clearly require more than the 10 Gaussians summed in our synthetic dataset to model. Note that pre-processing was required to binarize the images and remove plot axes, and that the results are un-calibrated.

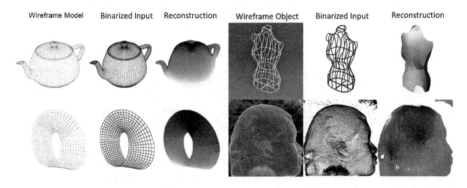

Fig. 12. Performance on wireframe models and objects. We also applied our neural model to a variety of wireframe models from [13] and real-world objects from [4,46] and found that despite clear differences between them and the training data, our model extracted meaningful shape information.

All of these surfaces would clearly require far more than ten Gaussian basis functions to fully model. They also featured different viewpoints and line spacings than the synthetic training data. The ability of our model to recover meaningful information from these images suggests that our carefully constructed training set successfully taught the neural net general rules for computing local shape from surface contours, suggesting high promise for use in automatic 3D plot and graphical model interpretation.

5   Conclusion

In order to reverse-engineer real published 3D plots, several problems remain to be solved. A front-end system to detect figures within documents and segregate them into axes, labels, and the plotted surface will be necessary. Additionally, the recovered depth maps must be calibrated and de-projected (to estimate the function $z(x,y)$) according to the axes. Since axes are orthogonal, the general procedure for de-projecting images can be simplified [42,52].

Our model successfully recovered depth maps from synthetic 3D surface plots without axis or shading information. It generalized to such surfaces rendered with line/grid resolutions and viewpoints that it had not encountered before (Fig. 7) with less than 0.5% mean-squared relative error (Table 1). When applied to real published plots, the quantitative data extracted by our model was un-calibrated; but as Fig. 11 clearly shows, it is possible to generate qualitative interpretations of the results. Our model was also able to extract meaningful information from pictures of wireframe models and objects (Fig. 12).

We measured how different types of training set data (categorized by surface marking type/resolution and camera viewpoint) affected model performance and ability to generalize, and identified which ones were most (and least) important. Some kinds of data boosted performance far more than we expected; for example we found that training on *line*-marked surfaces (in addition to a single-resolution grid-marked dataset) was responsible for a >70% improvement in overall model performance (Table 3) *and* the ability to generalize to new *grid* resolutions (Fig. 10). We hope our study will help inform future training set design.

Automatic recovery of numerical data from published 3D plots is an exciting and difficult problem, and our work successfully demonstrates that a very simple neural net architecture is capable of accurately reverse-engineering 3D surface plots if it is carefully trained, the axes are identified and removed, and a calibration obtained. We establish a baseline of performance on this task, present SurfaceGrid (a new 98.6k-image dataset of grid-marked surfaces and their associated depth map truths, Fig. 6), and invite other researchers to contribute to this exciting effort.

Acknowledgement. This work is supported by the National Science Foundation (NSF) under Cooperative Agreement PHY-2019786 (The NSF AI Institute for Artificial Intelligence and Fundamental Interactions, http://iaifi.org) and by the NSF Graduate Research Fellowship Program under Grant No. 1745302. Any opinions, findings, and conclusions or recommendations expressed in this material are those of the authors and do not necessarily reflect the views of the National Science Foundation.

References

1. Cortes, C., Vapnik, V.: Support-vector networks. Mach. Learn. **20**, 273–297 (1995)
2. Dai, W., Wang, M., Niu, Z., Zhang, J.: Chart decoder: generating textual and numeric information from chart images automatically. J. Vis. Lang. Comp. **48**, 101–109 (2018)
3. Dalal, N., Triggs, B.: Histograms of oriented gradients for human detection. In: IEEE Conference Computer Vision and Pattern Recognition, pp. 886–893 (2005)
4. Donaldson, L.: Jaume Plensa Exhibition at Yorkshire Sculpture Park. https://www.pinterest.co.uk/pin/166985098655171747/
5. Faugeras, O., Luong, Q.T., Papadopoulo, T.: The Geometry of Multiple Images: The Laws that Govern the Formation of Multiple Images of a Scene and Some of Their Applications. MIT Press (2001)

6. François, A.R., Medioni, G.G., Waupotitsch, R.: Mirror symmetry → 2-view stereo geometry. Image Vis. Comput. **21**(2), 137–143 (2003)
7. Freeman, W.T.: The generic viewpoint assumption in a framework for visual perception. Nature **368**, 542–545 (1994)
8. Godard, C., Aodha, O.M., Brostow, G.J.: Unsupervised monocular depth estimation with left-right consistency. In: IEEE Conference Computer Vision and Pattern Recognition, pp. 270–279 (2017)
9. Herzig, J., Nowak, P.K., Müller, T., Piccinno, F., Eisenschlos, J.M.: Tapas: weakly supervised table parsing via pre-training. Ann. Assoc. Comp. (2020)
10. Horn, B.K., Brooks, M.J.: Shape From Shading. MIT Press (1989)
11. Huang, W., Tan, C.L.: A system for understanding imaged infographics and its applications. In: ACM Symposium on Document Engineering, pp. 9–18 (2007)
12. Huang, W., Tan, C.L., Leow, W.K.: Model-based chart image recognition. In: Lladós, J., Kwon, Y.-B. (eds.) GREC 2003. LNCS, vol. 3088, pp. 87–99. Springer, Heidelberg (2004). https://doi.org/10.1007/978-3-540-25977-0_8
13. Hull, R.: Wireframes gallery. https://github.com/rm-hull/wireframes/blob/master/GALLERY.md
14. Huwaldt, J.A.: Plot digitizer (2015). http://plotdigitizer.sourceforge.net/
15. Isola, P., Zhu, J.Y., Zhou, T., Efros, A.A.: Image-to-image translation with conditional adversarial networks. In: IEEE Conference Computer Vision Pattern Recognition, pp. 55967–5976 (2017)
16. Jung, D., et al.: Chartsense: Interactive data extraction from chart images. CHI Conference Human Factors in Computing Systems, pp. 6706–6717 (2017)
17. Kim, B., Burger, P.: Depth and shape from shading using the photometric stereo method. CVGIP: Image Underst. **54**(3), 416–427 (1989)
18. Kingma, D.P., Ba, J.: Adam: a method for stochastic optimization. In: International Conference on Learning Representations (2015)
19. Knill, D.C.: Perception of surface contours and surface shape: from computation to psychophysics. JOSA A **9**(9), 1449–1464 (1992)
20. Koenderink, J.J.: What does the occluding contour tell us about solid shape? Perception **13**(3), 321–330 (1984)
21. Li, C., Pan, H., Liu, Y., Tong, X., Sheffer, A., Wang, W.: BendSketch: modeling freeform surfaces through 2D sketching. ACM Trans. Graph. **36**(4), 1–14 (2017)
22. Liechti, R., et al.: SourceData: a semantic platform for curating and searching figures. Nature Methods **14**, 1021–1022 (2017)
23. Liu, R., Huang, W., Tan, C.L.: Extraction of vectorized graphical information from scientific chart images. In: International Conference on Document Analysis and Recognition, pp. 521–525 (2007)
24. Lowe, D.G.: Distinctive image features from scale-invariant keypoints. Int. J. Comput. Vis. **60**(2), 91–110 (2004)
25. Lu, X., Mitra, P., Wang, J.Z., Giles, C.L.: Automatic categorization of figures in scientific documents. In: ACM/IEEE Conference on Digital Libraries, pp. 129–138 (2006)
26. Lu, X., Wang, J.Z., Mitra, P., Giles, C.L.: Automatic extraction of data from 2-D plots in documents. In: International Conference on Document Analysis and Recognition, pp. 188–192 (2007)
27. Mamassian, P., Landy, M.S.: Observer biases in the 3d interpretation of line drawings. Vis. Res. **38**(18), 2817–2832 (1998)
28. Marr, D.: Vision: A Computational Investigation into the Human Representation and Processing of Visual Information. MIT Press (1982)

29. Mitchell, M., et al.: Engauge digitizer software (2019). http://markummitchell. github.io/engauge-digitizer/
30. Mukherjee, D.P., Zisserman, A.P., Brady, M., Smith, F.: Shape from symmetry: detecting and exploiting symmetry in affine images. Philos. Trans. Roy. Soc. London Ser. A Phys. Eng. Sci. **351**(1695), 77–106 (1995)
31. Müller, T.: Using neural networks to find answers in tables (2020). https://ai. googleblog.com/2020/04/using-neural-networks-to-find-answers.html
32. Osoba, B., et al.: Variability study for low-voltage microelectromechanical relay operation. IEEE Trans. Elec. Dev. **65**(4), 1529–1534 (2018)
33. Prasad, V.S.N., Siddiquie, B., Goldbeck, J., Davis, L.S.: Classifying computer generated charts. In: Content-Based Multimedia Indexing Workshop, pp. 85–92 (2007)
34. Pumarola, A., Agudo, A., Porzi, L., Sanfeliu, A., Lepetit, V., Moreno-Noguer, F.: Geometry-aware network for non-rigid shape prediction from a single view. In: IEEE Conference Computer Vision and Pattern Recognition (2018)
35. Quintessa: Graph grabber (2020). https://www.quintessa.org/software/ downloads-and-demos/graph-grabber-2.0.2
36. Re, S., He, K., Girshick, R., Sun, J.: Faster R-CNN: towards real-time object detection with region proposal networks. In: Advances in Neural Information Processing Systems (2015)
37. Rohatgi, A.: Webplotdigitizer (2020). https://automeris.io/WebPlotDigitizer
38. Ronneberger, O., Fischer, P., Brox, T.: U-net: convolutional networks for biomedical image segmentation. In: Navab, N., Hornegger, J., Wells, W.M., Frangi, A.F. (eds.) MICCAI 2015. LNCS, vol. 9351, pp. 234–241. Springer, Cham (2015). https://doi.org/10.1007/978-3-319-24574-4_28
39. Savva, M., Kong, N., Chhajta, A., Fei-Fei, L., Agrawala, M., Heer, J.: Revision: automated classification, analysis and redesign of chart images. In: ACM User Interface Software and Technology, pp. 393–402 (2011)
40. Shao, M., Futrelle, R.P.: Recognition and classification of figures in PDF documents. In: Liu, W., Lladós, J. (eds.) GREC 2005. LNCS, vol. 3926, pp. 231–242. Springer, Heidelberg (2006). https://doi.org/10.1007/11767978_21
41. Stevens, K.A.: The visual interpretation of surface contours. Artif. Intell. **17**(1–3), 47–73 (1981)
42. Tsai, R.Y.: An efficient and accurate camera calibration technique for 3D machine vision. In: IEEE Conference Computer Vision and Pattern Recognition, pp. 364–374 (1986)
43. Ulupinar, F., Nevatia, R.: Perception of 3-D surfaces from 2-D contours. IEEE Trans. Pattern Anal. Mach. Intell. **15**(1), 3–18 (1993)
44. Ummenhofer, B., et al.: Demon: depth and motion network for learning monocular stereo. IEEE Conference Computer Vision and Pattern Recognition, pp. 5038–5047 (2017)
45. de Vries, S.C., Kappers, A.M., Joenderink, J.J.: Shape from stereo: a systematic approach using quadratic surfaces. Perc. Psychophys. **53**, 71–80 (1993)
46. Wang, P., Liu, L., Chen, N., Chu, H.K., Theobalt, C., Wang, W.: Vid2Curve: simultaneous camera motion estimation and thin structure reconstruction from an RGB video. ACM Trans. Graph. **39**(4) (2020)
47. Weiss, I.: 3D shape representation by contours. Comp. Vis. Graph. Image Proc. **41**(1), 80–100 (1988)
48. Witkin, A.P.: Recovering surface shape and orientation from texture. Artif. Intell. **17**(1–3), 17–45 (1981)
49. Wu, J., et al.: Citeseerx: AI in a digital library search engine. In: AAAI, pp. 2930–2937 (2014)

50. Zamarbide, G.N., et al.: An ab initio conformational study on captopril. J. Mol. Struc.: THEOCHEM, 666–667, 599–608 (2003)
51. Zhang, R., Tsai, P.S., Cryer, J.E., Shah, M.: Shape-from-shading: a survey. IEEE Trans. Pattern Anal. Mach. Intell. **21**(8), 690–706 (1999)
52. Zhang, Z.: A flexible new technique for camera calibration. IEEE Trans. Pattern Anal. Mach. Intell., 1330–1334 (2000)
53. Zhou, F., et al.: Reverse-engineering bar charts using neural networks. J. Vis. **24**(2), 419–435 (2020). https://doi.org/10.1007/s12650-020-00702-6
54. Zhou, T., Brown, M., Snavely, N., Lowe, D.G.: Unsupervised learning of depth and ego-motion from video. In: IEEE Conference Computer Vision and Pattern Recognition, pp. 1851–1858 (2017)
55. Zhou, Y.P., Tan, C.L.: Hough technique for bar charts detection and recognition in document images. In: International Conference on Image Processing, pp. 605–608 (2000)
56. Zhu, J.Y., Park, T., Isola, P., Efros, A.A.: Unpaired image-to-image translation using cycle-consistent adversarial networks. In: International Conference on Computer Vision (2017). https://github.com/junyanz/pytorch-CycleGAN-and-pix2pix

Document Summarization and Translation

Can Text Summarization Enhance the Headline Stance Detection Task? Benefits and Drawbacks

Marta Vicente[(✉)], Robiert Sepúlveda-Torrres, Cristina Barros, Estela Saquete, and Elena Lloret

Department of Software and Computing Systems, University of Alicante, Alicante, Spain
{mvicente,rsepulveda,cbarros,stela,elloret}@dlsi.ua.es

Abstract. This paper presents an exploratory study that analyzes the benefits and drawbacks of summarization techniques for the headline stance detection. Different types of summarization approaches are tested, as well as two stance detection methods (machine learning *vs* deep learning) on two state-of-the-art datasets (Emergent and FNC–1). Journalists' headlines sourced from the Emergent dataset have demonstrated with very competitive results that they can be considered a summary of the news article. Based on this finding, this work evaluates the effectiveness of using summaries as a substitute for the full body text to determine the stance of a headline. As for automatic summarization methods, although there is still some room for improvement, several of the techniques analyzed show greater results compared to using the full body text—Lead Summarizer and PLM Summarizer are among the best-performing ones. In particular, PLM summarizer, especially when five sentences are selected as the summary length and deep learning is used, obtains the highest results compared to the other automatic summarization methods analyzed.

Keywords: NLP document understanding · Misleading headlines · Stance detection · Summarization · Information retrieval · Document classification

1 Introduction

Nowadays, there is huge problem concerning information overload, and particularly concerning news [34], which could lead to infoxication [16,42] or infobesity [9]. People have access to more information than ever before, and this has a negative impact on managing information effectively, for society at large.

Within this context, it has emerged the disinformation problem [43], which is a threat to accurate journalism and democracy. The desire to get clicks or share immediate content has the drawback of transmitting information of lesser quality because the

This research work has been partially funded by Generalitat Valenciana through project "SIIA: Tecnologias del lenguaje humano para una sociedad inclusiva, igualitaria, y accesible" (PROMETEU/2018/089), by the Spanish Government through project "Modelang: Modeling the behavior of digital entities by Human Language Technologies" (RTI2018-094653-B-C22), and project "INTEGER - Intelligent Text Generation" (RTI2018-094649-B-I00). Also, this paper is also based upon work from COST Action CA18231 "Multi3Generation: Multi-task, Multilingual, Multi-modal Language Generation".

© Springer Nature Switzerland AG 2021
J. Lladós et al. (Eds.): ICDAR 2021, LNCS 12822, pp. 53–67, 2021.
https://doi.org/10.1007/978-3-030-86331-9_4

stories that are released first are often less accurate as there is less time to verify sources. Furthermore, an alarming claim made in a headline tempts people to share the news feed without having read the entire story. This often results in stories going viral due to an attractive headline, despite the lack of accurate information in the body text. For example, the following headline[1] : *"It is becoming pretty clear that the Hunan coronavirus is an engineered bio-weapon that was either purposely or accidentally released."* was promoting a conspiracy theory.

According to MIT [52], false information is 70% more likely to be shared than true information. This phenomenon is corroborated by the current pandemic caused by COVID-19. This disinformation spreads faster than the virus itself and plethora of contradictory and nonsensical recommendations related to cures that are claimed to be miraculous are published, yet they are ineffective [15]. For instance, the headline *"Will Lemons and Hot Water Cure or Prevent COVID-19?"*, which was demonstrated to be wrong shortly after its publication[2]. There is no doubt that these statements can have dangerous implications for public health, like those that might result from misleading headlines such this one: *"A vaccine meant for cattle can be used to fight COVID-19"*.[3] Given the very serious consequences that the spread of fake news, and specifically misleading claims made in headlines, can have for today's society, it is essential to find fast and effective methods for determining dissonances between the headline and body text of published news.

In this respect, Natural Language Processing (NLP) can be leveraged in order to provide strategies for effective text understanding, thus helping to address infoxication and disinformation. Among the NLP tasks, techniques aimed at condensing information able to retrieve and extract the gist of it—the case of Text Summarization—can be very usefully integrated into fact-checking systems to jointly fight against the spread of inaccurate, incorrect or false information.

Given the previous context, the main objective of this paper is to present an exploratory study by which we analyze the potentials and limitations of current text summarization techniques for tackling the disinformation problem, specifically, headline stance detection. Stance detection involves estimating the relative perspective (or stance) of two pieces of text relative to a topic, claim or issue[4] [2].

Therefore, our initial hypothesis is that the use of summaries for the stance detection task is beneficial and useful. This manner, if summaries are good enough, they will allow us to reduce the input information to the most important facts, and verify whether or not the claim made in a headline is appropriate. To check our hypothesis, different summarization strategies will be studied as a mean to facilitate the verification of a claim made in a headline, only using the resulting summary instead of having to check the full body text or introductory paragraphs of the body text. Our experiments and analysis will be guided by different research questions to allow us to validate our hypothesis, finally showing that some summarization techniques could be effective and useful for supporting the stance detection task. Nevertheless, some of the results suggest

---

[1] Published at Technocracy News https://www.technocracy.news/.

[2] https://www.snopes.com/fact-check/lemons-coronavirus/.

[3] https://www.snopes.com/fact-check/cattle-vaccine-covid-19/.

[4] http://www.fakenewschallenge.org/.

that our approach still holds certain limitations, thus indicating that there still remains some room for improvement.

2   Related Work

Related work is divided into two parts: i) the state of the art in text summarization to indicate its potential in other areas and ii) NLP works dealing with misleading headlines.

Automatic summaries have been also proven effective for a wide range of NLP tasks to be used as a component of more complex systems. This includes Text Classification [27,44,49], Question Answering [29] and Information Retrieval [37,40]. In these tasks, different techniques and approaches were used to create automatic summaries that were used as a substitute of the original documents. In this manner, the underlying tasks (e.g. Text Classification or Question Answering) directly employ those generated summaries—shorter and informative enough—to meet their objectives, while at the same time optimizing resources. Additionally, the use of summaries were also successful in reducing the length of the full piece of news and determining whether a news item is real or not [18,46].

Regarding misleading headlines, this type of headlines tend to be more focused on attracting the reader's attention —with little regard for accuracy— thus leading to mis- or disinformation through erroneous/false facts or headline/body text dissonance [13].

Misleading headlines significantly misrepresent the findings reported in the news article [14] by exaggerating or distorting the facts described in it. The reader only discovers the inconsistencies after reading the body text [53]. Some important nuances that are part of the body text are missing in the claim made in the headline, causing the reader to come to the wrong conclusion. An example of a misleading headline is shown below: *"China seeks for court's approval to kill the over 20,000 coronavirus patients to avoid further spread of the virus"*[5]. Detecting this type of headline is addressed as a stance detection problem (i.e. Emergent project [47] with the Emergent dataset [21] or the Fake News Challenge [2]). In the literature, the problem of misleading headlines has been tackled with both Machine Learning (ML) [8,21,53] and Deep Learning (DL) [1,10,23,39,41] approaches, with similar results.

However, to the best of our knowledge, there is no research in the specific context of analyzing the usefulness of text summarization for addressing the problem of detecting misleading headlines, this is, determining the stance of the headline with respect to its body text. The relationship identification between headlines and its source news items becomes more challenging, as headlines are an important element as they represent a snapshot of the news item. Therefore, if they are purposely written to deceive or mislead readers, they will contain distorted or inaccurate information, which needs to be detected and classified as quickly as possible.

This leads us to propose a novel study where state-of-the-art summarization methods—including extractive, abstractive and hybrid approaches—are analyzed in the context of stance headline detection. Our purpose is to verify whether the inclusion of

---

[5] Published on Feb. 5, 2020, the website AB-TC (aka City News) and fact-checked as false in https://www.snopes.com/fact-check/china-kill-coronavirus-patients/?collection-id=240413.

more concise text fragments containing the relevant information of the news in the corresponding classification task proves to be beneficial. In this way, while considering its advantages, we also analyze its disadvantages.

3 Text Summarization Approaches

Text summarization approaches can be divided into extractive and abstractive [30]. Extractive approaches focus on detecting the most relevant information in a text, which is then copy-pasted verbatim into the final summary. By contrast, abstractive approaches detect the relevant information, and then, a more elaborate process is performed whereby the information is fused, compressed, or even inferred. Besides these two, the development of hybrid approaches, which combine extractive and abstractive methods, is also possible [28].

In this section, the different summarization approaches used for this research are described. In particular, extractive, abstractive and hybrid state-of-the-art summarization approaches that are able to obtain a summary automatically were selected. Our selection was performed so as to ensure that the proposed methods employed different strategies, such as graph-based, discourse-aware, statistical, and DL techniques, providing a variety of methods to be later compared. Some summarization methods and strategies were selected based on the research conducted in [20], where it was determined which are the best summarisation techniques in terms of effectiveness and efficiency in the journalistic domain.

3.1 Extractive Summarization

The following systems were chosen for the extractive summarization methods:

Lead Summarizer. This method verbatim extracts the first sentences of the body of the news article up to a specific length as the resultant summaries. Although it is a very simple and basic method, this approach is very competitive for automatic summarization systems because according to the structure of a news document, its most important information is normally provided at the beginning [17,38], becoming a strong competitive baseline within the summarization community [54].

TextRank Summarizer. TextRank [32] is a graph-based ranking model for text processing which has been used in these experiments to summarise text by extracting the most relevant sentences of the source. This method builds a graph associated with a plain text, being the graph vertices representative of the sentences to be ranked. A weight is computed for each of the graph edges indicating the strength of the connection between the sentences pairs/vertices linked by them. These connections are determined by the similarity between the text sentences measured by their overlapping content.

Once the graph is built, a weighted graph-based ranking is performed in order to score each of the text sentences. The sentences are then sorted in reversed order of their score. Finally, the top ranked sentences are selected to be included in the final summary.[6]

---

[6] For this research, the implementation used was obtained from: https://github.com/miso-belica/sumy/blob/master/sumy/summarizers/text_rank.py.

PLM Summarizer. This summarization method relies on the potential of Positional Language Models (PLMs)—a language model that efficiently captures the dependency of non-adjacent terms—to locate the relevant information within a document [51]. The basic idea behind PLMs is that for every position within a document, it is possible to calculate a score for each word from the vocabulary. This value represents the relevance of such term in that precise position, taking into account the distance to other occurrences of the same word along the document so that the closer to that position the terms appear, the higher the score obtained. Therefore, the model expresses the significance of the elements considering the whole text as their context, rather than being limited to the scope of a single sentence.

As PLMs have been successfully applied in information retrieval [31,55] and language generation [50], we hypothesize that they can be beneficial also for selecting relevant information. Therefore, to obtain an extractive summary from the source text, first, a vocabulary needs to be defined, composed of named entities and synsets[7] of content words (nouns, verbs and adjectives). A language analysis tool, Freeling [33] in this case, needs to be employed to identify such elements. After building the PLMs, a representation of the text is obtained, involving both the vocabulary and the positions of its elements. Next, a seed is selected, i.e., a set of words that can be significant for the text and will help the system to discard irrelevant parts of the discourse. Finally, the processing of the PLM results against the seed helps us to compute scores for the text positions, conditioned by the information contained in the seed. These values are aggregated into a Score Counter (*SC*), and positions with highest scores are retrieved as the most relevant. The sentences to which these positions belong are then extracted as relevant candidates for the summary. The *SC* provides a value computed for each position in the sentence, allowing us to obtain a score for every candidate sentence adding up the values for its positions. The summary will comprise the best scored candidate sentences up to the amount of sentences required by the user.

3.2 Abstractive Summarization

Regarding the abstractive summarization methods, we selected a recent pure abstractive method, relying on natural language generation techniques, which has been proven to effectively work for abstractive summary generation [4]. Specifically, we used an adaptation of the approach HanaNLG [3], a hybrid natural language generation approach that is easily adaptable to different domains and tasks. HanaNLG has been adapted to the automatic summarization task for conducting our current experiments. The text generation in this approach is performed following an over-generation and ranking strategy that is based on the use of Factored Language Models—[6] trained over documents that were previously processed with Freeling to annotate factors—, linguistic resources—VerbNet [45] and WordNet [19]—and a set of seed features (e.g. phonemes, sentiments, polarities, etc.) that will guide the generation process in terms of vocabulary.

In order to adapt HanaNLG for generating abstractive summaries (from now on, **HanaNLG Summarizer**), keywords are used as seed feature. This seed feature allows the generation of sentences related to a specific keyword or topic. These topics must

---

[7] Synsets are identifiers that denote a set of synonyms.

be relevant, so it is necessary to establish a method to determine the relevant topics from which the sentence is going to be generated. This will allow the generation of summaries based on the essential information of the source documents. In this case, named entities are used as the keywords for the generation of summaries since they represent relevant information related to the names of persons, locations, organizations, etc., of a news article.

3.3 Hybrid Summarization

Finally, **Fast Abstractive Summarizer** [12] was selected as the hybrid summarization method, which exploits the benefits of extractive and abstractive paradigms. This method proposes a two-step strategy that does not need documents preprocessing. First, salient sentences are selected, and second, they are rewritten to generate the final summary. The rewriting process consists of compressing or paraphrasing the sentences initially chosen as relevant. For both stages, neural models are used. For the detection of salient information, a combination of reinforcement learning and pointer networks are used. Then, for the rewriting module, a simple encoder-aligner-decoder is employed to which a copy mechanism is added to help directly copy some out of-vocabulary words.[8]

4   Stance Detection Approaches

As explained in Sect. 2, determining whether a claim made in a headline corresponds to the body text is approached here as a stance detection task. We apply traditional ML and DL techniques to measure the validity of applying summarization to the problem.[9]

4.1   Deep Learning Stance Detection Approach

The design of the stance classifier that we evaluated was inspired by the Enhanced Sequential Inference Model (ESIM) [11]. This model, commonly used in Recognizing Textual Entailment task, was adapted to the stance detection task. For using this model, we relied on the implementation of [1]. This system, based on [23], obtained competitive results in the Fever Shared Task [48]. Our modifications are publicly accessible at https://github.com/rsepulveda911112/DL_Stance_Detection.

In this work the inputs of the model are: i) a claim/headline of the article to be classified, and ii) the summary obtained using the methods explained in the previous section. Each word from both inputs is represented as a vector. The model was experimented with different word embeddings independently, as well as the combination of them. The best results were obtained when concatenating two of them, in our case Fast-Text [7] and Glove [36]. Specifically, LSTMs are applied in the first layers of the ESIM model in order to obtain an inference from the processing of the claim and summary. Once the inference is done, this output vector is passed to a pooling layer and finally to the softmax classification layer.

---

[8] From the implementation available at: http://www.github.com/ChenRocks/fast_abs_rl.

[9] DL is a specific type of ML but we use this nomenclature to indicate a difference between non-DL approaches and DL ones.

4.2   Machine Learning Stance Detection Approach

The ML approach used in this work was proposed by [21], and it is publicly available at https://github.com/willferreira/mscproject. They treated stance classification as a 3-way classification task using a logistic regression classifier as the ML model with L1 regularization [35]. They extracted different features from the input, namely, presence of question marks, bag of words representation, cosine similarity, the minimum distance from the root of the sentence to common refuting and hedging/reporting words, negation and paraphrase alignment, and subject-verb-object triples matching.

5   Evaluation Environment

5.1   Datasets

To validate if summaries are effective enough in the context of headline stance detection, the experiments are conducted over two resources, publicly available: Emergent and Fake News Challenge dataset. Both consist of a headline/claim and a body text between which a stance relation can be detected. More detailed information about the datasets is presented next, as well as the alignment of the labels in both datasets.

Emergent Dataset.[10] Developed by [21], this dataset contains 300 rumoured claims and 2,595 news articles, along with a real headline for each article. They were collected and labelled by journalists with one of the following tags: *for*, if the article states that the claim is true; *against*, if it states that the claim is false and *observing*, when the claim is reported in the article, but without assessment of its veracity (see Table 1).

Table 1. Description of the Emergent dataset: documents and distribution of assigned labels.

	News bodies	Headlines	Claims	For	Against	Observing	Total examples
Train	2,048	2,023	240	992	304	775	2,071
Test	523	513	60	246	91	187	524
Complete dataset	2,571	2,536	300	1,238	395	962	2,595

Fake News Challenge Dataset (FNC-1). The FNC-1 dataset was developed in the context of the Fake News Challenge.[11] Originally, its instances were labeled as *agree*—the body agrees with the claim made in the headline, *disagree*—the body disagrees with the claim made in the headline, *discuss*—the body discusses the same topic as the claim made in the headline, but does not take a position—and *unrelated*—the body text is not related with the headline. However, in order to replicate our experimental environment with this dataset, we discarded the *unrelated* instances, thus extracting a subset from the original FNC-1 composed of all the examples labeled as *agree*, *disagree* and *discuss*. The equivalence between labels in both datasets regarding their meaning is *for* ≃ *agree*, *against* ≃ *disagree and observing* ≃ *discuss*.

---

[10] https://github.com/willferreira/mscproject.
[11] http://www.fakenewschallenge.org/.

Table 2. Description of the FNC-1 subset: documents and distribution of assigned labels.

	News bodies	Claim headlines	Agree	Disagree	Discuss	Total examples
Train	1,683	1,648	3,678	840	8,909	13,427
Test	904	894	1,903	697	4,464	7,064
Complete dataset	2,587	2,542	5,581	1537	13,373	20,491

The distribution of documents (bodies, headlines and assignments) is presented in Table 2. Unlike the Emergent dataset described above, real headlines, written by journalists, were not provided, only claims whose stance relation with the body text must be determined.

5.2 Experiments

To test the effectiveness of using news summaries to determine whether the headline (claim) fits the content of the news (stance detection), a battery of experiments is proposed in this paper. All the experiments are conducted using the two models for the stance detection task explained in Sect. 4 (ML and DL) and over the two different datasets described in Sect. 5.1 (Emergent and FNC-1). Since the experiments are designed to test which type of summarization approach and summary length would be more appropriate for this task, they were grouped into two types:

- *Validation of the type of summarization approach*: In this experiment, we test the extractive (*Lead Summarizer*, *TextRank Summarizer* and *PLM Summarizer*), abstractive (*HanaNLG Summarizer*) and hybrid (*Fast Abstractive Summarizer*) summarization approaches, in order to determine the most appropriate approach for the stance detection task.
- *Validation of the summary length*: Neural models can have a negative impact on efficiency when processing long texts, so previous studies either used the first sentence of the text [24] or a specific fragment [26] to address this problem. In view of this, three different summary lengths are proposed for the experiments: One, three and five sentences, in order to determine the summary length which provides most benefits to the stance detection task, and therefore is more appropriate. These summary lengths would mean an average compression ratio[12] of 12%, 37% and 62%, in the case of one, three and five sentences, respectively.

Finally, we generated 346,290 summaries for both datasets (23,086 documents for both datasets × 5 summarization approaches × 3 summary lengths), publicly accessible at http://www.github.com/rsepulveda911112/FNC_Emergent_summary_dataset.

---

[12] Compression ratio means how much of the text has been kept for the summary and it is calculated as the length of the summary divided by the length of the document [25].

In addition to the summarization approaches described in Sect. 3, we also include two additional configurations that will be used for comparative purposes:

- *Using the whole body text (Full body)*: This task is performed on both datasets and both stance detection approaches using the whole body text instead of using summaries as input for the models. This will allow us to check if summarization brings a benefit to the stance detection task and how effective it is.
- *Using a human-written headline (Upper bound)*: This task is performed on both stance detection approaches, using a real headline written by a journalist as input, considered as a perfect summary. This upper bound is only possible for the Emergent dataset since FNC-1 dataset does not provide a real headlines for the articles.

5.3 Evaluation Metrics and Results

To address the performance of the experiments in line with the category of the headline, following [22], this paper incorporates both a measure of F_1 class-wise, and a macro-averaged F_1 (F_1m) as the mean of those per-class F scores. Both datasets used in the experiments are significantly imbalanced, especially at the against-disagree classes. The advantage of this measure is that it is not affected by the size of the majority class.

After generating the corresponding summaries for all the summarization approaches and their different lengths, and testing them with the two stance detection models using the aforementioned evaluation metrics, Tables 3 and 4 show the results obtained, for the ML, and the DL approaches, respectively. The results which are better than those obtained using the full body text are highlighted in bold typeface. Moreover, the best result for each column is indicated with a star (*).

Table 3. *ML Model with summary results.* Emergent (left) and FNC-1 (right) experiments, showing class-wise F_1 performance and macro-averaged F_1 (F_1m).

Experiment	Emergent F_1 (%)			F_1m	FNC-1 F_1 (%)			F_1m
	For	Against	Observing		Agree	Disagree	Discuss	
Full body	68.03	42.10	54.49	54.88	44.88	**13.71***	75.35	**44.65**
Upper bound	81.53	74.53	68.23	74.76	–	–	–	–
Lead Summarizer-1	67.24	36.36	51.17	51.59	**46.73**	0.56	**77.66**	41.65
Lead Summarizer-3	66.79	**53.06**	54.09	**57.98**	**50.34**	1.13	**78.05**	43.17
Lead Summarizer-5	**68.92**	**54.16**	**57.68**	**60.25***	**50.91***	7.39	**75.94**	**44.75***
TextRank Summarizer-1	65.54	**43.24**	**55.64**	54.81	41.18	1.13	73.52	38.59
TextRank Summarizer-3	67.39	**43.05**	53.46	54.61	**48.52**	2.71	73.17	41.47
TextRank Summarizer-5	66.42	40.90	50.13	52.49	**49.15**	6.95	74.83	43.64
PLM Summarizer-1	64.20	21.42	44.24	43.29	28.98	0.0	**77.53**	35.50
PLM Summarizer-3	**68.27**	**43.20**	**55.49**	**55.65**	39.40	0.28	**76.66**	38.78
PLM Summarizer-5	**68.32**	**43.54**	**58.29***	**56.72**	36.43	2.72	**78.23***	39.13
HanaNLG Summarizer-1	59.66	2.17	41.24	43.43	25.50	0.0	**75.39**	33.63
HanaNLG Summarizer-3	61.98	12.96	44.63	39.86	41.86	11.69	72.99	42.18
HanaNLG Summarizer-5	62.93	17.24	47.48	42.55	41.09	11.90	73.46	42.15
Fast Abstractive Summ.-1	**69.14**	35.93	52.59	52.55	35.19	0.0	**75.82**	37.00
Fast Abstractive Summ.-3	**70.73***	**54.66***	55.13	**60.17**	**46.05**	1.13	71.75	39.26
Fast Abstractive Summ.-5	66.66	**44.92**	47.05	52.88	42.81	0.56	68.27	37.21

Table 4. *DL Model with summary results.* Emergent(left) and FNC-1(right) experiments, showing class-wise F_1 performance and macro-averaged F_1 ($F_1 m$).

Experiment	Emergent F_1 (%)			$F_1 m$	FNC-1 F_1 (%)			$F_1 m$
	For	Against	Observing		Agree	Disagree	Discuss	
Full body	67.08	26.44	58.03	50.52	56.98	23.02	80.96	53.65
Upper bound	77.47	67.09	71.31	71.96	–	–	–	–
Lead Summarizer-1	63.71	**33.33**	51.25	49.43	47.77	14.59	77.93	46.76
Lead Summarizer-3	**69.14***	**43.66**	54.64	**55.81**	55.70	21.85	**81.21***	52.92
Lead Summarizer-5	61.20	**27.80**	45.57	44.86	53.72	**24.65**	79.80	52.73
TextRank Summarizer-1	56.50	18.29	54.13	42.97	46.40	23.49	79.76	49.96
TextRank Summarizer-3	63.49	**29.94**	48.80	47.41	55.90	**27.01**	79.67	**54.19**
TextRank Summarizer-5	61.07	16.51	57.53	45.04	**57.32***	29.56	77.69	**54.85**
PLM Summarizer-1	52.43	**44.73***	**59.16***	**52.10**	53.66	9.36	77.86	46.96
PLM Summarizer-3	62.84	**30.33**	54.91	49.36	45.62	21.78	80.36	49.25
PLM Summarizer-5	**67.56**	**44.72**	55.70	**55.99***	54.84	**34.44***	79.16	**56.15***
HanaNLG Summarizer-1	61.45	22.38	47.42	43.75	49.29	13.86	75.08	46.07
HanaNLG Summarizer-3	59.63	14.03	55.58	43.08	45.20	11.12	76.12	44.15
HanaNLG Summarizer-5	59.91	19.17	50.73	43.27	51.03	7.79	76.65	45.15
Fast Abstractive Summ.-1	55.79	**31.81**	46.30	44.63	51.17	22.18	76.43	49.92
Fast Abstractive Summ.-3	65.18	**40.20**	41.40	48.93	49.92	22.30	77.47	49.89
Fast Abstractive Summ.-5	58.94	22.01	55.38	45.77	49.98	18.80	78.52	49.10

5.4   Discussion

Given our initial hypothesis that if summaries are good enough they will allow us to reduce the input information to the most important facts, enabling verification as to the appropriateness of a claim made in a headline, the discussion of the results obtained for the $F_1 m$ is guided by the following evaluation questions:

1. Can summaries be effectively used instead of the whole body text for the headline stance detection task? A summary can definitely provide an improvement over using the full text. The human-written headline —created by a professional journalist—, which is only available for the Emergent dataset [21], has been shown to be the best possible summary, with a positive impact on the headline stance detection task when compared to the full body text. These results are very promising, so the hypothesis of using summaries instead of the whole body text is consistent. The headline generated by professional journalist is considered the upper bound, and since the summaries automatically generated imply errors in their creation, this explains why the results of our experiments with the automatic summarization methods are lower.

Using automatic summarization, the results are similar or better than using the full text. This is a positive finding that means that summarization is appropriate for this task, as long as the summarization method is effective enough. Although it is true that the improvement of summaries is very slightly noted in the experiments with ML concerning FNC-1 dataset (only the Lead Summarizer-5 slightly improves the results over the baseline body), this is explained by the inherent nature of the ML method applied here, which was initially developed for the Emergent dataset. When the ML method is transferred to another dataset it does not work as well as a more general approach. As can be observed, the ML experiments for FNC-1 (Table 3) are leading to much worse results than the experiments for the DL approach (Table 4). This latter approach does not depend on the ad-hoc dataset features implemented for the ML approach, and it is therefore more generalizable, allowing some of the summarization methods to improve the results on using the full body text. In addition, the dataset of FNC-1 has a very imbalanced disagree class with respect to the other classes, preventing correct learning in the case of the ML approach, in addition to incorporating ad-hoc features to the Emergent dataset results with quite low values of performance.

One drawback of using summaries in ML when the approach is using so many ad-hoc features is that reducing the input implies reducing the capacity of the system to learn categories with a very small number of examples as the case of disagree in the experiments of FNC-1 dataset. The implication being using the whole body text would increase the number of features obtained, thereby improving performance. Nevertheless, as reported in the results, current summarization approaches can use the generated summaries to replace the full body text, without greatly worsening the performance of the stance detection models compared to when using the full body text. In particular, it is worth stressing that some of the summarization approaches (e.g. PLM Summarizer) results are better than with the full body text, even using a much smaller portion of text.

2. Which is the best summarization approach for the headline stance detection task? In general, the extractive systems obtain better results than the abstractive or hybrid ones, although the results of the hybrid system (Fast Abstractive Summarizer) are very competitive as well. In particular, the good results of the Fast Abstractive Summarizer are due to the fact that it combines an extractive module with an abstractive one and therefore, the extractive part that selects relevant phrases is working correctly. The good results obtained by the Lead Summarizer method, which simply extracts the first sentences, has its explanation in the typical structure of journalist produced news content, which [5,38] described as an inverted pyramid. This implies that the most important information of a news item is expressed at the beginning of the body text and the content's significance decreases towards the end of the news item. In the case of the experiments carried out with DL, the best summaries were obtained with the PLM approach on both datasets. As already indicated, the DL method is a more generalizable approach, since it does not use any dataset ad-hoc features, and in this respect, DL could be more appropriate for the stance detection task.

3. Which is the most appropriate length for a summary to be used for the headline stance detection task? There is no clear conclusion on what summary length is more appropriate, since it greatly depends on the summarization method and the stance detection approach used. Reducing the text in a compression ratio range between 12% and 60% seems to be a good compression rate for the tested summarization approaches, except for the abstractive methods. This is because abstractive methods face more challenges for this task, and sometimes the sentences generated may be shorter or even meaningless in themselves, or not perfect, in contrast to the extractive methods, which extract verbatim the most relevant sentences from the text.

Independently of the summary lengths involved, Lead Summarizer and PLM Summarizer indicate the most stable performance for both of the datasets compared to other methods. Additionally, when the length of the summary is increased, the results become more robust for the three possible configurations offered by both approaches. These methods are significant because they consider the structure of the discourse. Regardless of whether this structure is defined in terms of the inverted pyramid theory, as in the case of the Lead Summarizers approach, or is expressed by the representation of the PLM technique, the results indicate that taking into account the document's structure is beneficial for preserving the relevant information in the generated summary, thus having great impact on performance. Therefore, to support the stance detection task, PLM Summarizer could be considered a simple yet very effective summarization method. It can capture relevant information when the information follows a specific structure determined by the document's genre —as in the case of news stories—, but it is also equally applicable to other types of documents and genres.

6   Conclusion and Future Work

This paper propose an exploratory study through which different summarization techniques applied to the headline stance detection task were analyzed to determine the most suitable approach for the task, thus evaluating the effectiveness of using summaries instead of full article bodies. The use of summaries was tested considering two different methods (ML and DL), over two datasets, the Emergent and the FNC-1 dataset.

The results obtained indicate that the quality of the summarization method applied influences its effectiveness in helping to detect the headline stance. Apart from the headline created by journalists, considered an upper bound, it is difficult to identify one summarization technique that stands out in detecting all the classes, for both datasets and approaches, achieving the best scores in each case. Nonetheless, the analysis shows that several methods can generate summaries that become good substitutes of the full text, thus helping and benefiting the headline stance detection task. All considered, the results point to using PLM summarization techniques—which are genre-structure independent—rather than the Lead Summarizer because they produced very competitive and stable results for the stance detection task and provided the additional advantage of using a very small portion of the news article instead of the whole body text.

Our exploratory analysis clearly demonstrates that summaries are useful, and therefore, from this starting point, various approaches and techniques could be further tested to more precisely determine in which cases and how a summary, capturing the most important meaning, could be used as a substitute for the original document. Keeping this principle in mind, also as future steps, it would be possible to analyse how the summarisers behave regarding specific words or expressions, such as opinion-bearing words, or even to condition the summarisers in that sense.

Despite the promising results, there is still room for improvement and therefore, in the short term we plan to continue working in this line to analyze learning strategies to automatically decide which summarization technique would be more appropriate for the headline stance detection task, as well as for the wider context of fake news problem.

Furthermore, we are also interested in studying the reverse problem in the medium-long term, i.e., how to integrate fact-checking methods in summarization approaches to prevent the automatic generation of misleading or incongruent information.

References

1. Alonso-Reina, A., Sepúlveda-Torres, R., Saquete, E., Palomar, M.: Team GPLSI. Approach for automated fact checking. In: Proceedings of the Second Workshop on Fact Extraction and VERification, pp. 110–114. Association for Computational Linguistics (2019)
2. Babakar, M., et al.: Fake news challenge - I (2016). http://www.fakenewschallenge.org/. Accessed 21 Jan 2021
3. Barros, C., Lloret, E.: HanaNLG: a flexible hybrid approach for natural language generation. In: Proceedings of the International Conference on Computational Linguistics and Intelligent Text Processing (2019)
4. Barros, C., Lloret, E., Saquete, E., Navarro-Colorado, B.: NATSUM: narrative abstractive summarization through cross-document timeline generation. Inf. Process. Manag. **56**(5), 1775–1793 (2019)
5. Benson, R., Hallin, D.: How states, markets and globalization shape the news the French and US national press, 1965–97. Eur. J. Commun. **22**, 27–48 (2007)
6. Bilmes, J.A., Kirchhoff, K.: Factored language models and generalized parallel backoff. In: Proceedings of the Conference of the North American Chapter of the Association for Computational Linguistics, pp. 4–6. Association for Computational Linguistics (2003)
7. Bojanowski, P., Grave, E., Joulin, A., Mikolov, T.: Enriching word vectors with subword information. Trans. Assoc. Comput. Linguistics **5**, 135–146 (2017)
8. Bourgonje, P., Moreno Schneider, J., Rehm, G.: From clickbait to fake news detection: an approach based on detecting the stance of headlines to articles. In: Proceedings of the 2017 EMNLP Workshop: Natural Language Processing Meets Journalism, pp. 84–89. ACL (2017)
9. Bulicanu, V.: Over-information or infobesity phenomenon in media. Int. J. Commun. Res. **4**(2), 177–177 (2019)
10. Chaudhry, A.K., Baker, D., Thun-Hohenstein, P.: Stance detection for the fake news challenge: identifying textual relationships with deep neural nets. In: CS224n: Natural Language Processing with Deep Learning (2017)
11. Chen, Q., Zhu, X., Ling, Z.H., Wei, S., Jiang, H., Inkpen, D.: Enhanced LSTM for natural language inference. In: Proceedings of the 55th Annual Meeting of the Association for Computational Linguistics (Volume 1: Long Papers), pp. 1657–1668 (2017)
12. Chen, Y.C., Bansal, M.: Fast abstractive summarization with reinforce-selected sentence rewriting. In: Proceedings of the 56th Annual Meeting of the Association for Computational Linguistics, (Volume 1: Long Papers), pp. 675–686 (2018)
13. Chen, Y., Conroy, N.K., Rubin, V.L.: News in an online world: The need for an "automatic crap detector". In: Proceedings of the Association for Information Science and Technology, vol. 52, no. 1, pp. 1–4 (2015)
14. Chesney, S., Liakata, M., Poesio, M., Purver, M.: Incongruent headlines: yet another way to mislead your readers. Proc. Nat. Lang. Process. Meets J. **2017**, 56–61 (2017)
15. Colomina, C.: Coronavirus: infodemia y desinformación (2017). https://www.cidob.org/es/publicaciones/serie_de_publicacion/opinion_cidob/seguridad_y_politica_mundial/coronavirus_infodemia_y_desinformacion. Accessed 21 Jan 2021
16. Dias, P.: From "infoxication" to "infosaturation" : a theoretical overview of the congnitive and social effects of digital immersion. In: Primer Congreso Internacional Infoxicación : mercado de la información y psique, Libro de Actas, pp. 67–84 (2014)
17. van Dijk, T.: News As Discourse. Taylor & Francis. Routledge Communication Series (2013)

18. Esmaeilzadeh, S., Peh, G.X., Xu, A.: Neural abstractive text summarization and fake news detection. CoRR (2019). http://arxiv.org/abs/1904.00788
19. Fellbaum, C.: WordNet: An Electronic Lexical Database. MIT Press (1998)
20. Ferreira, R., et al.: Assessing sentence scoring techniques for extractive text summarization. Expert Syst. Appl. **40**(14), 5755–5764 (2013)
21. Ferreira, W., Vlachos, A.: Emergent: a novel data-set for stance classification. In: Proceedings of the Conference of the North American Chapter of the Association for Computational Linguistics, pp. 1163–1168. Association for Computational Linguistics (2016)
22. Hanselowski, A., et al.: A retrospective analysis of the fake news challenge stance-detection task. In: Proceedings of the 27th International Conference on Computational Linguistics, pp. 1859–1874. Association for Computational Linguistics, August 2018
23. Hanselowski, A., et al.: UKP-Athene: multi-sentence textual entailment for claim verification. In: Proceedings of the First Workshop on Fact Extraction and VERification, pp. 103–108 (2018)
24. Hayashi, Y., Yanagimoto, H.: Headline generation with recurrent neural network. In: Matsuo, T., Mine, T., Hirokawa, S. (eds.) New Trends in E-service and Smart Computing. SCI, vol. 742, pp. 81–96. Springer, Cham (2018). https://doi.org/10.1007/978-3-319-70636-8_6
25. Hovy, E.: Text summarization. In: Mitkov, R. (ed.) The Oxford Handbook of Computational Linguistics, pp. 583–598. Oxford University Press, Oxford (2004)
26. Huang, Z., Ye, Z., Li, S., Pan, R.: Length adaptive recurrent model for text classification. In: Proceedings of the ACM on Conference on Information and Knowledge Management, pp. 1019–1027. Association for Computing Machinery (2017)
27. Jeong, H., Ko, Y., Seo, J.: How to improve text summarization and classification by mutual cooperation on an integrated framework. Expert Syst. Appl. **60**, 222–233 (2016)
28. Kirmani, M., Manzoor Hakak, N., Mohd, M., Mohd, M.: Hybrid text summarization: a survey. In: Ray, K., Sharma, T.K., Rawat, S., Saini, R.K., Bandyopadhyay, A. (eds.) Soft Computing: Theories and Applications. AISC, vol. 742, pp. 63–73. Springer, Singapore (2019). https://doi.org/10.1007/978-981-13-0589-4_7
29. Lloret, E., Llorens, H., Moreda, P., Saquete, E., Palomar, M.: Text summarization contribution to semantic question answering: new approaches for finding answers on the web. Int. J. Intell. Syst. **26**(12), 1125–1152 (2011)
30. Lloret, E., Palomar, M.: Text summarisation in progress: a literature review. Artif. Intell. Rev. **37**(1), 1–41 (2012)
31. Lv, Y., Zhai, C.: Positional language models for information retrieval. In: Proceedings of the 32Nd International ACM SIGIR, pp. 299–306. ACM (2009)
32. Mihalcea, R., Tarau, P.: TextRank: bringing order into text. In: Proceedings of the 2004 Conference on Empirical Methods in Natural Language Processing, pp. 404–411. Association for Computational Linguistics (2004)
33. Padró, L., Stanilovsky, E.: Freeling 3.0: towards wider multilinguality. In: Proceedings of the Language Resources and Evaluation Conference. ELRA (2012)
34. Park, C.S.: Does too much news on social media discourage news seeking? Mediating role of news efficacy between perceived news overload and news avoidance on social media. Soc. Media Soc. **5**(3), 1–12 (2019)
35. Pedregosa, F., et al.: Scikit-learn: machine learning in Python. J. Mach. Learn. Res. **12**, 2825–2830 (2011)
36. Pennington, J., Socher, R., Manning, C.D.: Glove: global vectors for word representation. In: Conference on Empirical Methods on Natural Language Processing 2014, vol. 14, pp. 1532–1543 (2014)
37. Perea-Ortega, J.M., Lloret, E., Ureña-López, L.A., Palomar, M.: Application of text summarization techniques to the geographical information retrieval task. Expert Syst. Appl. **40**(8), 2966–2974 (2013)

38. Pöttker, H.: News and its communicative quality: the inverted pyramid—when and why did it appear? J. Stud. **4**(4), 501–511 (2003)
39. Rakholia, N., Bhargava, S.: Is it true?-Deep learning for stance detection in news. Technical report. Stanford University (2016)
40. Raposo, F., Ribeiro, R., Martins de Matos, D.: Using generic summarization to improve music information retrieval tasks. IEEE/ACM Trans. Audio Speech Lang. Process. **24**(6), 1119–1128 (2016)
41. Riedel, B., Augenstein, I., Spithourakis, G.P., Riedel, S.: A simple but tough-to-beat baseline for the Fake News Challenge stance detection task. CoRR abs/1707.03264 (2017). http://arxiv.org/abs/1707.03264
42. Rodríguez, R.F., Barrio, M.G.: Infoxication: implications of the phenomenon in journalism. Revista de Comunicación de la SEECI **38**, 141–181 (2015). https://doi.org/10.15198/seeci.2015.38.141-181
43. Rubin, V.L.: Disinformation and misinformation triangle. J. Doc. **75**(5), 1013–1034 (2019)
44. Saggion, H., Lloret, E., Palomar, M.: Can text summaries help predict ratings? A case study of movie reviews. In: Bouma, G., Ittoo, A., Métais, E., Wortmann, H. (eds.) NLDB 2012. LNCS, vol. 7337, pp. 271–276. Springer, Heidelberg (2012). https://doi.org/10.1007/978-3-642-31178-9_33
45. Schuler, K.K.: VerbNet: a broad-coverage, comprehensive verb lexicon. Ph.D. thesis, University of Pennsylvania (2005)
46. Shim, J.-S., Won, H.-R., Ahn, H.: A study on the effect of the document summarization technique on the fake news detection model **25**(3), 201–220 (2019)
47. Silverman, C.: Lies, Damn Lies and Viral Content (2019). https://academiccommons.columbia.edu/doi/10.7916/D8Q81RHH. Accessed 21 Jan 2021
48. Thorne, J., Vlachos, A., Christodoulopoulos, C., Mittal, A.: FEVER: a large-scale dataset for fact extraction and verification. In: Proceedings of the Conference of the North American Chapter of the Association for Computational Linguistics, pp. 809–819. Association for Computational Linguistics (2018)
49. Tsarev, D., Petrovskiy, M., Mashechkin, I.: Supervised and unsupervised text classification via generic summarization. Int. J. Comput. Inf. Syst. Ind. Manag. Appl. MIR Labs **5**, 509–515 (2013)
50. Vicente, M., Barros, C., Lloret, E.: Statistical language modelling for automatic story generation. J. Intell. Fuzzy Syst. **34**(5), 3069–3079 (2018)
51. Vicente, M., Lloret, E.: A discourse-informed approach for cost-effective extractive summarization. In: Espinosa-Anke, L., Martín-Vide, C., Spasić, I. (eds.) SLSP 2020. LNCS (LNAI), vol. 12379, pp. 109–121. Springer, Cham (2020). https://doi.org/10.1007/978-3-030-59430-5_9
52. Vosoughi, S., Roy, D., Aral, S.: The spread of true and false news online. Science **359**(6380), 1146–1151 (2018)
53. Wei, W., Wan, X.: Learning to identify ambiguous and misleading news headlines. In: Proceedings of the 26th International Joint Conference on Artificial Intelligence, pp. 4172–4178. AAAI Press (2017)
54. Widyassari, A.P., Affandy, A., Noersasongko, E., Fanani, A.Z., Syukur, A., Basuki, R.S.: Literature review of automatic text summarization: research trend, dataset and method. In: International Conference on Information and Communications Technology, pp. 491–496 (2019)
55. Yan, R., Jiang, H., Lapata, M., Lin, S.D., Lv, X., Li, X.: Semantic v.s. positions: utilizing balanced proximity in language model smoothing for information retrieval. In: Proceedings of the Sixth International Joint Conference on Natural Language Processing, pp. 507–515 (2013)

The Biased Coin Flip Process
for Nonparametric Topic Modeling

Justin Wood[(✉)], Wei Wang, and Corey Arnold

University of California, Los Angeles, CA 90095, USA
{juwood03,weiwang}@cs.ucla.edu, cwarnold@ucla.edu

Abstract. The Dirichlet and hierarchical Dirichlet processes are two
important techniques for nonparametric Bayesian learning. These learning techniques allow unsupervised learning without specifying traditionally used input parameters. In topic modeling, this can be applied to
discovering topics without specifying the number beforehand. Existing
methods, such as those applied to topic modeling, usually take on a
complex sampling calculation for inference. These techniques for inference of the Dirichlet and hierarchal Dirichlet processes are often based
on Markov processes that can deviate from parametric topic modeling.
This deviation may not be the best approach in the context of nonparametric topic modeling. Additionally, since they often rely on approximations they can negatively affect the predictive power of such models. In
this paper we introduce a new interpretation of nonparametric Bayesian
learning called the biased coin flip process—contrived for use in the context of Bayesian topic modeling. We prove mathematically the equivalence of the biased coin flip process to the Dirichlet process with an
additional parameter representing the number of trials. A major benefit
of the biased coin flip process is the similarity of the inference calculation
to that of previous established parametric topic models—which we hope
will lead to a more widespread adoption of hierarchical Dirichlet process
based topic modeling. Additionally, as we show empirically the biased
coin flip process leads to a nonparametric topic model with improved
predictive performance.

Keywords: Topic modeling · Bayesian nonparametrics · Latent
Dirichlet allocation

1 Introduction

Bayesian nonparametric learning is a form of Bayesian learning that involves
model inference without some traditionally used parameter(s). For example,
in topic modeling, it is assumed that the number of topics is given as input
beforehand. If we would like to discover topics for a corpus without knowing the
number of topics a priori, an applicable model for topic discovery would be a
model which utilizes Bayesian nonparametric learning. In nonparametric topic
modeling often some form of the Dirichlet process is used to infer topics. The

© Springer Nature Switzerland AG 2021
J. Lladós et al. (Eds.): ICDAR 2021, LNCS 12822, pp. 68–83, 2021.
https://doi.org/10.1007/978-3-030-86331-9_5

Dirichlet process is not only useful to topic modeling but also useful to many learning tasks where some set of input parameters are unknown [7,8,17,32].

Nonparametric learning is advantageous when some set inputs parameters are unknown. One solution to finding the correct set of inputs is to use a brute-force method to try every possible or probable input parameter. Then after the model is learned, some type of scoring metric would need to be used to compare different runs. This however is, if not intractable, then extremely inefficient. The Dirichlet process rectifies this by allowing for an infinite set of possibilities; all while performing inference in a relative efficient time.

The Dirichlet process can discover predictions using an infinite set of inputs by gradually decreasing the probability a new input parameter is expanded. In the context of topic modeling this takes on the form of the number of topics. During inference, a new topic is created with a decreasing probability. Though it is theoretically possible for an infinite number of topics, each new topic decreases the chance of creating another topic so that the number of topics in practice converge to a finite number.

The Dirichlet process is often expanded to include two Dirichlet processes, with one being an input to the other in what is known as a hierarchal Dirichlet process (which is a different concept altogether from hierarchal topic modeling [4]). The hierarchal Dirichlet process is the distribution used in the theoretical generative model for nonparametric topic modeling. This process complicates the original non hierarchal process in it's generative model and inference calculation. In nonparametric topic modeling inference is done using approximation techniques [13,38] or more esoteric techniques to sampling [15,34]. The approximation techniques are limited to how well the approximation fits the underlying calculation. Which can lead to the possibility of less-than-optimal results. Additionally, the more esoteric sampling techniques require a greater cost to understanding the material.

Are the existing methods serving to impede adoption of nonparametric Bayesian topic modeling? Indeed, it does appear that in topic modeling, parameter based Dirichlet methods are more popular than Dirichlet process methods[1]. In this paper we seek to improve the inference inference capability of existing nonparametric topic modeling leading to better predictive performance. By changing a slight yet fundamental detail in how the process generates data we can frame the inference as a sub routine of the already adopted, well documented, and less complex latent Dirichlet allocation inference [5]. To help understand the intuition behind the new inference calculation it helps to interpret the Dirichlet process in a new light, in what we introduce as the biased coin flip process.

2   Background

2.1   Dirichlet Process

The Dirichlet process [10] is a probability distribution that describes a method of generating a series of separate probability distributions. Each subsequent

---

[1] Based on Google Scholar index of research publications in 2019.

probability distribution is generated with a decreasing probability. The process is parameterized by some underlying distribution and a variable or variables that determine when to generate a new distribution.

Though each generated probability distribution is created with a decreasing probability, the process has no stopping criteria. One way to think of the likelihood of generation is to take the previous probability of generating a new distribution and multiplying it by some probability value. This continuous multiplication of probabilities asymptotically tends to zero however will never actually be zero. For this reason, the Dirichlet process is often thought of as the infinite length Dirichlet distribution.

To understand the generative process better, several analogies have been created that describe the process. In the Chinese restaurant process a customer sits at an existing table (assigned to an existing generated distribution) with probability equal to the number of customers already seated at the respective table, or is seated a completely new table (newly generated distribution) with probability equal to some scaling parameter γ. Another such analogy is that of the stick-breaking process. Each time a distribution is requested a stick that was initially at length 1 is broken with a percentage length drawn from the beta distribution parametrized by $(1, \gamma)$. Each stick break length then represents the probability that the corresponding probability distribution is return on subsequent requests. The third most common view of the Dirichlet process is the Pólya urn scheme, where a ball is drawn from a bag and fill the bag with the ball as well as a duplicate of the ball. The bag represents a distribution that resembles the Dirichlet process distribution.

We define a sample from the Dirichlet process as:

$$G_i \sim DP(\gamma \cdot G_0) \tag{1}$$

With G_0 being the underlying distribution. It follows then that the posterior takes the form of:

$$P(G|\boldsymbol{G}) \propto DP(\gamma \cdot G_0 + \sum \boldsymbol{G}) \tag{2}$$

2.2 Latent Dirichlet Allocation

Latent Dirichlet allocation [5] is an unsupervised technique to discovering topics in a corpus. The topics are composed of a discrete distribution over the vocabulary. Each document is also assigned a distribution over the set of topics.

The corpus (C) is assumed to be generated according to a generative model that samples the mixture of words (words per topic) from one Dirichlet distribution parameterized by β and each document (topics per document) is sampled from a Dirichlet distribution parameterized by α. To build a document, after the topic to document mixture is sampled θ, for each word we sample a topic assignment (z) and for that sampled topic we sample a word from the corresponding topic's mixture over words (ϕ). This is given as:

```
 1: for k ← 1 to K do
 2:     Choose φ_k ~ Dir(β)
 3: end for
 4: for c ← 1 to |C| do
 5:     Choose θ_c ~ Dir(α)
 6:     Choose N_c ~ Poisson(d*)
 7:     for n ← 1 to N_c do
 8:         Choose z_{n,c} ~ Multinomial(θ)
 9:         w_{n,c} ~ Multinomial(φ_{z_{n,c}})
10:     end for
11: end for
```

With d^* defined as the average length for each document.

To learn the parameters of the model a Gibbs sampler can be built based on the following update equation:

$$P(z_i = j | z_{-i}, w) \propto \frac{n^{w_i}_{-i,j} + \beta}{n^{(\cdot)}_{-i,j} + V\beta} \cdot \frac{n^{d_i}_{-i,j} + \alpha}{n^{(d_i)}_{-i} + K\alpha} \tag{3}$$

With n being count matrices for word counts in a topic or topic counts in a document, V the size of the vocabulary of the corpus, K the number of topics, w the vector of words in document i and d the vector of documents of the corpus.

2.3   Hierarchical Dirichlet Processes

The concept of nonparametric topic modeling was shown to be possible using a hierarchical Dirichlet process [35]. By doing so topics were able to be discovered without supplying the number of topics a priori. The solution utilizes the concept of the Chinese restaurant franchise, which is a hierarchical interpretation of the Chinese restaurant process. Inference was done using Markov chain Monte Carlo algorithms. Another interpretation of the Chinese restaurant franchise is the Chinese district process, whereby a customer is searching for a table inside a number of restaurants in a district [26].

The Pitman-Yor process can also form the basis for n-gram models. For example, a hierarchical model based off the Pitman-Yor process has also been shown useful in language models [33]. In this model inference is based on an approximation as gives results like established smoothing methods.

Like the Chinese restaurant franchise is the Indian buffet process [12]. The underlying distribution of the Indian buffet process has been shown to be a beta process [37]. This interpretation can be useful in document classification and other machine learning tasks.

Outside of traditionally used Markov chain Monte Carlo methods for inference of hierarchical Dirichlet processes [27], inference can be done using collapsed variational inference [36]. This technique has advantages over Gibbs sampling and is general to the hierarchal Dirichlet process. Furthermore, other techniques such as slice sampling have also been used in place of Gibbs sampling [34].

Additionally, adoptions of the Chinese restaurant process have been proposed [1] which are useful in evolutionary clustering or for tree construction [11]. Connecting data across data prevalence is applied using a compound Dirichlet process [40]. And similar works can be applied to a nonparametric Pachinko allocation [19]. Another field of use is in genomic applications where discrete nonparametric priors can be used in place of the previously used priors [20]. Lastly, the use of hierarchical Dirichlet processes is beneficial to labeling topic models which suffer from a problem of interpretability [30].

Since hierarchal Dirichlet processes are a well established and not so emerging technique, much of the work does not involve the hypothetical interpretation of the process to improve upon the sampling. Recent work focuses more on the application of the hierarchal Dirichlet processes [6,18,21,23,31] and less on estimation and sampling techniques [3].

3   Methods

We describe the biased coin flip process in the context of a bank deciding how to partition a set of coins. To begin with a teller at the bank sets aside an infinite number of bags labeled numerically starting from 1. Each bag is also associated with a coin flip bias h and a sample from a distribution G_0. Additionally, we assume there is a row of coins on a table and there exists a way to assign a uniform bias to each coin on the table. The process begins at the first bag, B_1. The bank teller takes the bias associated with B_1, h_1 as the bias to make each coin on the table flip heads with probability h_1. Next the teller takes the first coin on the table and flips it. If the coin lands on heads, the coin is placed in bag 1 (B_1). If the coin lands on tails the coin is placed back on the table, never to be flipped again in this initial step. After the flip of the first coin the teller moves to the next coin on the table and repeats the process until all coins have been flipped. At this point we say the teller is done with bag 1, then proceeds to take the bias out of bag 2 and sets all the coin's biases to h_2. The teller proceeds the same procedure with all coins left on the table. This process is repeated until all coins are off the table. An algorithmic description of this would be as follows:

```
 1: for i ← 1 to ∞ do
 2:     Choose φᵢ ~ G₀
 3:     Choose hᵢ ~ Beta(1, γ)
 4:     Bᵢ ← {}
 5: end for
 6: for i ← 1 to ∞ do
 7:     for all cⱼ ∈ C do
 8:         Choose f ~ Bernoulli(hᵢ)
 9:         if f = 1 then
10:             C ← C \ {cⱼ}
11:             Bᵢ ← Bᵢ ∪ {cⱼ}
12:         end if
13:     end for
14: end for
```

At first glance it may not appear as though the binary coin flip process is equivalent to n draws from a Dirichlet process. We prove equivalence below:

$$B = \bigcup_{i=1}^{\infty} \{\phi_i\} \times C_i \tag{4}$$

$$C_i^* = C_{i-1}^* \setminus C_i \tag{5}$$

$$C_i = \{c_j \in C_{i-1}^* \mid f_{i,j} = 1\} \tag{6}$$

$$f_{i,j} \sim Bernoulli(h_i) \tag{7}$$

$$h_i \sim Beta(1, \gamma) \tag{8}$$

$$\phi_i \sim G_0 \tag{9}$$

alternatively, we can write this as:

$$G_n = \sum_{i=1}^{\infty} \phi_i \cdot C_i \tag{10}$$

$$C_i = (1 - \sum_{j=1}^{i-1} C_j) \cdot f_i \tag{11}$$

$$f_{i,j} \sim Bernoulli(h_i) \tag{12}$$

$$h_i \sim Beta(1, \gamma) \tag{13}$$

$$\phi_i \sim G_0 \tag{14}$$

Where $\mathbf{1}$ is a vector of all 1's. Since $C_{i,j} = 1$ with probability $h_i \cdot \prod_{k=1}^{i-1} 1 - h_k$ we can rewrite the above as:

$$G_n = \sum_{i=1}^{\infty} \phi_i \cdot f_i \tag{15}$$

$$f_{i,j} \sim Bernoulli(h_i \cdot \prod_{k=1}^{i-1} 1 - h_k) \tag{16}$$

$$h_i \sim Beta(1, \gamma) \tag{17}$$

$$\phi_i \sim G_0 \tag{18}$$

And thus each $G_{n,j}$ will be ϕ_i with probability $h_i \cdot \prod_{k=1}^{i-1} 1 - h_k$; this becomes a discrete distribution over each ϕ_i, and can be reformulated as:

$$G_{n,j} = \sum_{i=1}^{\infty} h_i \cdot \prod_{k=1}^{i-1} (1 - h_k) \delta_{\phi_i} \tag{19}$$

which has previously been established as equivalent to the Dirichlet process [25].

The biased coin flip process is equivalent to the Dirichlet process in the same way that the Chinese restaurant process, the stick-breaking process or the Pólya urn scheme is equivalent to the Dirichlet process. It represents an alternative view. We maintain the benefit of this view is that it guides the thought process of the Dirichlet process away from a single draw to a series of draws (coins on a table). In this way it represents a departure from existing interpretations, such as the stick-breaking process or the Chinese district process [26]. Another important difference is this interpretation leads to a novel yet familiar inference calculation. To establish these points more succinctly—the BCP is equivalent to the Dirichlet process, yet represents an alternate view. This alternate view distinguishes itself from existing views and contributes in two important ways: (1) the BCP view frames the process as a series of draws, as opposed to a single draw, and (2) this view allows for inference to be done in a similar way to LDA.

So given an input of $C = c_1 c_2 c_3 \ldots c_n$ we are tasked to find the matrix z with $z_{ij} \in \{H, T\}$ and ϕ which will take the form of

$$p(z, \phi | C, \gamma) = \frac{p(z, \phi, C | \gamma)}{p(C | \gamma)} \qquad (20)$$

But this looks strikingly like existing Dirichlet process inference equations, and indeed in this form the biased coin flip process is not of much use.

The major advantage of the bias coin flip process occurs when we are asked to find z_t and ϕ_t given $C_t \subseteq C$. From the biased coin flip analogy this is equivalent to finding the assignments of H and T for the coins left on the table for the tth bag; as well as the tth bag's distribution ϕ_t. If the underlying distribution of G_0 is a Dirichlet distribution parameterized by β, the probability becomes:

$$p(z_i, \phi_i | C_i, \gamma, \beta) = \frac{p(z_i, \phi, C_i | \gamma, \beta)}{p(C_i | \gamma, \beta)} \qquad (21)$$

Which thought of in a different way is the exact same calculation as finding the topic assignments of a single document with $K = 2$, $\gamma = \alpha = 1$, $w = C_t$ and $V = 2$. In fact a Gibbs sampler would then be indistinguishable from Eq. 3.

We now have all the components necessary to build the Gibbs sampler for the entire bias coin flip process. If we assume one time step to be flipping all the coins on the table for a single bag, then for each bag B_t we calculate the probability of a coin belonging to bag B_t using Eq. 3 multiplied by the probability the previous $t - 1$ flips were all tails. This allows for a recursive multiplication of the previous probabilities that the current coin lands on tails. We can split the head and previous tail probabilities calculation for the ith coin at time t as

$$P(z_i = 1 | z_{-i}, w_t, p_{t-1}) \propto P(z_i = 1 | z_{-i}, w_t) \cdot p_{t-1} \qquad (22)$$

and cumulated tail probability as

$$p_t \propto P(z_i = 0 | z_{-i}, w_t) \cdot p_{t-1} \qquad (23)$$

To account for an infinite amount of time steps we consider the bag that contains the very last coin from the table (b'). At this point any remaining time step follows a monotone probability calculation of

$$P(z_i = 1|z_{-i}, w_t, p_{t-1}, t) \propto \frac{1}{V} \cdot \frac{1}{K\alpha} \cdot p_{b'} \cdot p_e^{t-1-b'} \tag{24}$$

and p_e as

$$p_e \propto \frac{1}{V} \cdot \frac{\gamma}{K\alpha} \tag{25}$$

To aggregate the mass for all bags $t > b'$ we can take the improper integral which equates to

$$\int P(z_i = 1|z_{-i}, w_t, p_{t-1}, t)dt \tag{26}$$

This can be further simplified by normalizing the posterior conditionals when $t > b'$

$$P(z_i = 1|z_{-i}, w_t) = \frac{P(z_i = 1|z_{-i}, w_t)}{\sum P(z_i|z_{-i}, w_t)} = \frac{1}{1+\gamma} \tag{27}$$

and

$$P(z_i = 0|z_{-i}, w_t) = p_e = \frac{\gamma}{1+\gamma} \tag{28}$$

with the total mass as

$$\int P(z_i = 1|z_{-i}, w_t, p_{t-1}, t)dt = \frac{-p_{b'} \cdot p_e}{\gamma \cdot \ln(p_e)} \tag{29}$$

3.1   Hierarchical Biased Coin Flip Process

In the hierarchical biased coin flip process, we choose a biased coin flip process as the base distribution. The "parent" process then will take a Dirichlet distribution as it's base distribution. We can extend our bank analogy by adding a central branch. The bank teller at the local branch continues with setting side an infinite number of bags. But instead of generating the distributions for each bag, the teller must call the central branch to get the distribution. For each call to the central branch the bankers place a coin on its table. And before any of the local branches were even created the central branch had already generated an infinite number of bags each with their associated biases and draws from the base distribution G_0.

In a topic modeling analogy each local branch represents a document, and each local branch coin is a word. To the central branch each bag is a topic (ϕ_i) and each coin is a bag of words corresponding to a particular topic assignment in a particular document (θ_{ij}). To describe it algorithmically:

```
1: C* ← {}
2: for i ← 1 to ∞ do
3:     Choose φ*_i ~ G_0
4:     Choose h*_i ~ Beta(1, ζ)
5:     B*_i ← {}
6: end for
7: for j ← 1 to D do
8:     for k ← 1 to ∞ do
```

```
 9:          C* ← C* ∪ {c_jk}
10:          Choose h_jk ~ Beta(1, γ)
11:          B_jk ← {}
12:      end for
13: end for
14: for i ← 1 to ∞ do
15:      for all c_jk ∈ C* do
16:          Choose f ~ Bernoulli(h_i*)
17:          if f = 1 then
18:              C* ← C* \ {c_jk}
19:              B_i* ← B_i* ∪ {c_jk}
20:              φ_jk ← φ_i*
21:          end if
22:      end for
23: end for
24: for j ← 1 to D do
25:      for k ← 1 to ∞ do
26:          for all c_l ∈ C_j do
27:              Choose f ~ Bernoulli(h_jk)
28:              if f = 1 then
29:                  C_j ← C_j \ {c_l}
30:                  B_jk ← B_jk ∪ {c_l}
31:              end if
32:          end for
33:      end for
34: end for
```

In the hierarchical biased coin flip process, the Gibbs sampler equations remain the same except for the conditional posterior at the central branch level. This must consider the word bag instead of a single token. This reduces to

$$P(z_i = j | z_{-i}, w) \propto \prod_{w_i \in B_i^*} \frac{n_{-B_i^*,j}^{w_i} + \beta}{n_{-B_i^*,j}^{(\cdot)} + V\beta} \cdot \frac{n_{-i,j}^{d_i} + \alpha_j}{n_{-i}^{(d_i)} + K\alpha} \tag{30}$$

To calculate the infinite mass sum the only changes to Eq. 29 are to p_e:

$$p_e = \frac{\zeta}{1 + \zeta^{b'}} \tag{31}$$

With ζ being the scaling parameter at the central branch level.

4   Results

Having already proved the theoretical equivalence to the Dirichlet process, we seek to do the same empirically. Since the biased coin flip process does not rely on

Table 1. Datasets used in the evaluation of the biased coin flip process and a description of the number of documents in the coporus (D) and topics (K) used in the corpus.

	Description	D	K
CiteULike-180	Manually tagged scholarly papers	182	1,660
SemEval-2010	Scientific articles with manually assigned key phrases	244	3,107
NLM500	A collection of PubMed documents and MeSH terms	203	1,740
RE3D	A set of labeled relationship and entity extraction documents	98	2,200
Reuters-21578	Manually labeled documents from the 1987 Reuters newswire	21,578	2,700
Wiki-20	20 Computer Science papers annotated from Wikipedia articles	20	564
FAO-30	Manually annotated documents from the Food and Agriculture Organization of the UN	30	650

approximations it is entirely possible that it can outperform existing methods. We show this is indeed the case in terms of predictive ability. Next, we show the ability to accurately discover the correct number of topics given a generated corpus. The datasets used in evaluation are described by Table 1.

4.1 Perplexity

We seek to compare the ability of the biased coin flip process to predict held out data with established methods, the Infinite-LDA (INF) [13] ,the Nonparametric Topic Model (NTM) [38] and the hierarchical Dirichlet process [35].

Experimental Setup. For all models we set the hyperparameters for γ as 1, and β as $200/V$, with V being the size of the vocabulary. We then run each topic model for 1000 iterations. Each data set was cleaned to strip out stop words, excess punctuation and frequent and infrequent terms. Additionally, since all baseline models can update their respective hyperparameters during inference [9, 35], we add these models to our baseline comparison. For the perplexity analysis we take roughly 80% of the corpus for training and test on the remaining 20%.

Experimental Results. After the 1000 iterations had completed, we compare the ability for each model to predict the held-out data. We calculate perplexity to be:

$$\sqrt[T]{\prod p(w_i|D_t)} \tag{32}$$

With T the sum of all tokens in the test corpus, w_i the word at token i in document D_t.

As we show in Table 2 and Table 3, out of all the models BCP performs the best by a substantial amount. We hypothesis this is due to a more direct inference calculation that considers two sets of concentration parameters: one for the local and central branch. This surprising result emphasizes the importance of the inference calculation when performing nonparametric topic modeling. Additionally, we find that optimization does have much of an effect. In some datasets predictive power is better, while in others it is worse.

4.2 K-Finding

We propose a way to test the topic discovery capability of each model is to generate a document using the hierarchical Dirichlet process's generative model and storing the number of topics generated (K). Then we compare the found number of topics for each model run on the generated dataset.

Experimental Setup. For each dataset we take a histogram of the words as the Dirichlet hyperparameter input for a new topic to be created. We set the corpus size to 1000 documents and take the average document size as a sample from the Poisson distribution having a Poisson centering parameter of 100. With the sampled number of words, we sample from the hierarchal Dirichlet process to get a topic distribution. We then sample a word from the returned topic distribution. This process is continued for all 1000 documents. We repeat this corpus generation process for different values of γ and ζ, each ranging from 0.1 to 4.0. Additionally, we consider a model to "timeout" if the number of discovered topics exceeds 1000. At this point a heat map score of 0 is assigned to the run. We do this because at extreme topic counts, the computation time becomes infeasible for the current implementation of the models.

Table 2. Perplexity of the biased coin flip process (BCP) compared against baseline methods.

	BCP	Inf-LDA	NTM	HDP
CiteULike-180	**2262**	36742	71244	13873
SemEval-2010	**6591**	64963	54397	92635
NLM500	**4306**	54333	66652	20846
Re3d	**134**	312	323	1492
Reuters-21578	1591	1168	**860**	3160
Wiki-20	**337**	1645	2152	3406
FAO-30	**314**	4353	4143	5878

Table 3. Perplexity of the biased coin flip process (BCP) compared against baseline methods with optimized parameters.

	BCP	Inf-LDA-Opt	NTM-Opt	HDP-Opt
CiteULike-180	**2262**	40695	68435	81111
SemEval-2010	**6591**	43192	49309	92635
NLM500	**4306**	46870	69815	20846
Re3d	**134**	312	366	7834
Reuters-21578	**1591**	1890	900	3093
Wiki-20	**337**	2087	1844	20582
FAO-30	**314**	6391	6766	16578

Experimental Results. After the bias coin flip process and the two baseline models, we also run the two baseline models with parameter updating. Each model is run with the scaling parameters equal to what generated the corpus. We present the results as a heat map, show by Fig. 1. To calculate the heat map values (M) we define a metric of similarity that must account for up to an infinite distance from the true value (K). We use the sigmoid function to map the negative K, positive infinity range into the interval $[0, 1]$. However, we want to want to reward answers that are close to the target value more so then answers that are extremely far. The sigmoid function is too sensitive at values close to 1 and quickly jumps to the outer bounds at higher values. For this we take the difference as a percentage of the target value. We formulate this as:

$$E_{\hat{k}} = \frac{|K - \hat{k}|}{K} \tag{33}$$

$$M = 2 \cdot \left| \frac{-1 + exp(-E_{\hat{k}})}{2 + 2 \cdot exp(-E_{\hat{k}})} \right| \tag{34}$$

This trivial example that underscores some of the difficulties in using previous hierarchical Dirichlet processes. We would expect each model to discover the K topics within a reasonable error. However, as we can see from the heat map, only BCP reliably does. Inf-LDA has the tendency to increasingly add topics, making the error from the target larger as the number of iterations increase. Likewise, with the Nonparametric Topic Model outside of the diagonal—though not to the same effect as Inf-LDA. Much like the perplexity results, it is the author's intuition that a more direct inference calculation is leading to superior results. It may also be to the act of including both scaling parameters—as the biased coin flip process uses the same amount as the stick-breaking process—which was how the corpora were generated. It does appear that for NTM that when scaling parameters were the same the results improve. However, for Inf-LDA this is not the case. Additionally, the parameter updates should rectify this deficiency but fail to do so.

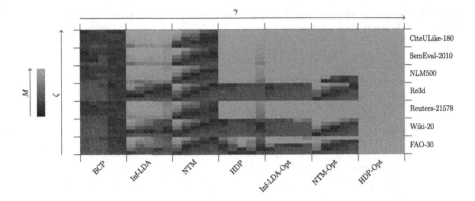

Fig. 1. Heat map showing the error in finding K topics.

4.3   Gaussian Mixture Models

To test the BCP on non topic modeling tasks we seek to find accurate predictive densities for real-world data sets. These datasets are assumed to be generated from a mixture of Gaussian distributions. The three datasets used in analysis are Faithful [2], Stamp [16], and Galaxies [29]. For each dataset we follow established techniques [14,22] of using our nonparametric model to estimate a Gaussian mixture density. We assume a fixed a fixed variance and take 30 samples of the mixture density at various iterations after an initial 1,000 iterations. The densities of the samples from the BCP are plotted against their respective kernel density estimates in Fig. 2. As we can see from Fig. 2, the densities from our model closely resembles that of the density from kernel density estimation. This similarity suggests our model to be useful in tasks outside of topic modeling such as estimating Gaussian mixture models.

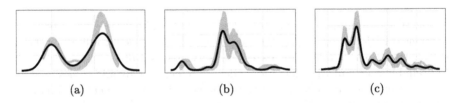

(a)   (b)   (c)

Fig. 2. 30 density estimates taken from the BCP shown in gray plotted against a kernel density estimation for the Faithful (a), Galaxies (b) and Stamp (c) datasets.

5   Conclusion

This paper introduces a novel way to think about Dirichlet and hierarchical Dirichlet processes. Thinking about the process as a series of coin flips, where each round we partition the coins into bags and what's left on a table, we can see the similarity to an established method for inference—latent Dirichlet allocation. Because this method is based on topic modeling inference, it may lead to better results in the context of discovering topics.

The downside of the technique presented in this paper is the increase in execution time. Since we are performing two sets of Gibbs sampling, one at the local branch level and one at the central branch level, we ultimately need more computations than baseline methods. It is left as an open research area to find improvements or improved efficiencies. Although not implemented in this paper, we do acknowledge the ability for concurrent processing of our approach on the different branch levels [24, 28, 39, 41]. Additionally, an interesting area to investigate would be optimization of the two scaling parameters. This would then allow for the model to be completely parameter free.

We maintain that the downsides of our approach are outweighed by the upsides. The reliance on previous established parametric topic modeling inference calculations leads to a theoretical advantage to existing techniques. And as we show empirically the biased coin flip process does seem to yield better results. In both prediction of held out data and finding the appropriate number of topics, our model improves upon existing methods.

References

1. Ahmed, A., Xing, E.P.: Dynamic non-parametric mixture models and the recurrent Chinese restaurant process: with applications to evolutionary clustering. In: Proceedings of the SIAM International Conference on Data Mining, SDM 2008, 24–26 April 2008, Atlanta, Georgia, USA, pp. 219–230 (2008)
2. Azzalini, A., Bowman, A.W.: A look at some data on the old faithful geyser. J. Roy. Stat. Soc. Ser. C (Appl. Stat.) **39**(3), 357–365 (1990)
3. Bacallado, S., Favaro, S., Power, S., Trippa, L.: Perfect sampling of the posterior in the hierarchical pitman-YOR process. Bayesian Anal. **1**(1), 1–25 (2021)
4. Blei, D.M., et al.: Hierarchical topic models and the nested Chinese restaurant process. In: Advances in Neural Information Processing Systems 16 [Neural Information Processing Systems, NIPS 2003, 8–13 December 2003, Vancouver and Whistler, British Columbia, Canada], pp. 17–24 (2003)
5. Blei, D.M., et al.: Latent Dirichlet allocation. J. Mach. Learn. Res. **3**, 993–1022 (2003)
6. Camerlenghi, F., Lijoi, A., Prünster, I.: Survival analysis via hierarchically dependent mixture hazards. Ann. Stat. **49**(2), 863–884 (2021)
7. Christensen, R., Johnson, W.: Modelling accelerated failure time with a Dirichlet process. Biometrika **75**(4), 693–704 (1988)
8. Diana, A., Matechou, E., Griffin, J., Johnston, A., et al.: A hierarchical dependent Dirichlet process prior for modelling bird migration patterns in the UK. Ann. Appl. Stat. **14**(1), 473–493 (2020)
9. Escobar, M.D., West, M.: Bayesian density estimation and inference using mixtures. J. Am. Stat. Assoc. **90**(430), 577–588 (1995)
10. Ferguson, T.S.: A Bayesian analysis of some nonparametric problems. Ann. Stat., 209–230 (1973)

11. Finkel, J.R., Grenager, T., Manning, C.D.: The infinite tree. In: ACL 2007, Proceedings of the 45th Annual Meeting of the Association for Computational Linguistics, 23–30 June 2007, Prague, Czech Republic (2007)
12. Griffiths, T.L., Ghahramani, Z.: The Indian buffet process: an introduction and review. J. Mach. Learn. Res. **12**, 1185–1224 (2011)
13. Heinrich, G.: Infinite LDA implementing the HDP with minimum code complexity (2011)
14. Ishwaran, H., James, L.F.: Approximate Dirichlet process computing in finite normal mixtures: smoothing and prior information. J. Comput. Graph. Stat. **11**(3), 508–532 (2002)
15. Ishwaran, H., James, L.F.: Generalized weighted Chinese restaurant processes for species sampling mixture models. Statistica Sinica, 1211–1235 (2003)
16. Izenman, A.J., Sommer, C.J.: Philatelic mixtures and multimodal densities. J. Am. Stat. Assoc. **83**(404), 941–953 (1988)
17. Krueger, R., Rashidi, T.H., Vij, A.: A Dirichlet process mixture model of discrete choice: comparisons and a case study on preferences for shared automated vehicles. J. Choice Modelling **36**, 100229 (2020)
18. Lehnert, L., Littman, M.L., Frank, M.J.: Reward-predictive representations generalize across tasks in reinforcement learning. PLoS Comput. Biol. **16**(10), e1008317 (2020)
19. Li, W., et al.: Nonparametric Bayes pachinko allocation. In: UAI 2007, Proceedings of the Twenty-Third Conference on Uncertainty in Artificial Intelligence, Vancouver, BC, Canada, 19–22 July 2007, pp. 243–250 (2007)
20. Lijoi, A., Prünster, I., Walker, S.G., et al.: Bayesian nonparametric estimators derived from conditional Gibbs structures. Ann. Appl. Probab. **18**(4), 1519–1547 (2008)
21. Masumura, R., Asami, T., Oba, T., Sakauchi, S.: Hierarchical latent words language models for automatic speech recognition. J. Inf. Process. **29**, 360–369 (2021)
22. McAuliffe, J.D., et al.: Nonparametric empirical Bayes for the Dirichlet process mixture model. Stat. Comput. **16**(1), 5–14 (2006)
23. Muchene, L., Safari, W.: Two-stage topic modelling of scientific publications: a case study of University of Nairobi, Kenya. Plos One **16**(1), e0243208 (2021)
24. Newman, D., Asuncion, A.U., Smyth, P., Welling, M.: Distributed inference for latent Dirichlet allocation. In: Advances in Neural Information Processing Systems 20, Proceedings of the Twenty-First Annual Conference on Neural Information Processing Systems, Vancouver, British Columbia, Canada, 3–6 December 2007, pp. 1081–1088 (2007)
25. Paisley, J.: A simple proof of the stick-breaking construction of the Dirichlet process (2010)
26. Paisley, J.W., Carin, L.: Hidden Markov models with stick-breaking priors. IEEE Trans. Signal Process. **57**(10), 3905–3917 (2009)
27. Papaspiliopoulos, O., Roberts, G.O.: Retrospective Markov chain monte Carlo methods for Dirichlet process hierarchical models. Biometrika **95**(1), 169–186 (2008)
28. Porteous, I., Newman, D., Ihler, A.T., Asuncion, A.U., Smyth, P., Welling, M.: Fast collapsed Gibbs sampling for latent Dirichlet allocation. In: Proceedings of the 14th ACM SIGKDD International Conference on Knowledge Discovery and Data Mining, Las Vegas, Nevada, USA, 24–27 August 2008, pp. 569–577 (2008)
29. Postman, M., Huchra, J.P., Geller, M.J.: Probes of large-scale structure in the corona borealis region. Astron. J. **92**, 1238–1247 (1986)

30. Ramage, D., Manning, C.D., Dumais, S.T.: Partially labeled topic models for interpretable text mining. In: Proceedings of the 17th ACM SIGKDD International Conference on Knowledge Discovery and Data Mining, San Diego, CA, USA, 21–24 August 2011, pp. 457–465 (2011)
31. Serviansky, H., et al.: Set2Graph: learning graphs from sets. In: Advances in Neural Information Processing Systems, vol. 33 (2020)
32. Shi, Y., Laud, P., Neuner, J.: A dependent Dirichlet process model for survival data with competing risks. Lifetime Data Anal., 1–21 (2020)
33. Teh, Y.W.: A hierarchical Bayesian language model based on Pitman-YOR processes. In: ACL 2006, 21st International Conference on Computational Linguistics and 44th Annual Meeting of the Association for Computational Linguistics, Proceedings of the Conference, Sydney, Australia, 17–21 July 2006 (2006)
34. Teh, Y.W., Görür, D., Ghahramani, Z.: Stick-breaking construction for the Indian buffet process. In: Proceedings of the Eleventh International Conference on Artificial Intelligence and Statistics, AISTATS 2007, San Juan, Puerto Rico, 21–24 March 2007, pp. 556–563 (2007)
35. Teh, Y.W., Jordan, M.I., Beal, M.J., Blei, D.M.: Hierarchical Dirichlet processes. J. Am. Stat. Assoc. **101**(476), 1566–1581 (2006)
36. Teh, Y.W., Kurihara, K., Welling, M.: Collapsed variational inference for HDP. In: Advances in Neural Information Processing Systems 20, Proceedings of the Twenty-First Annual Conference on Neural Information Processing Systems, Vancouver, British Columbia, Canada, 3–6 December 2007, pp. 1481–1488 (2007)
37. Thibaux, R., Jordan, M.I.: Hierarchical beta processes and the Indian buffet process. In: Proceedings of the Eleventh International Conference on Artificial Intelligence and Statistics, AISTATS 2007, San Juan, Puerto Rico, 21–24 March 2007, pp. 564–571 (2007)
38. Wallach, H.M.: Structured topic models for language. Ph.D. thesis, University of Cambridge Cambridge, UK (2008)
39. Wang, Y., Bai, H., Stanton, M., Chen, W.-Y., Chang, E.Y.: PLDA: parallel latent Dirichlet allocation for large-scale applications. In: Goldberg, A.V., Zhou, Y. (eds.) AAIM 2009. LNCS, vol. 5564, pp. 301–314. Springer, Heidelberg (2009). https://doi.org/10.1007/978-3-642-02158-9_26
40. Williamson, S., Wang, C., Heller, K.A., Blei, D.M.: The IBP compound Dirichlet process and its application to focused topic modeling. In: ICML (2010)
41. Wood, J., et al.: Source-LDA: enhancing probabilistic topic models using prior knowledge sources. In: 33rd IEEE International Conference on Data Engineering (2016)

CoMSum and SIBERT: A Dataset and Neural Model for Query-Based Multi-document Summarization

Sayali Kulkarni$^{(\boxtimes)}$, Sheide Chammas, Wan Zhu, Fei Sha, and Eugene Ie

Google Research, Mountain View, USA
{sayali,sheide,fsha,eugeneie}@google.com

Abstract. Document summarization compress source document(s) into succinct and information-preserving text. A variant of this is query-based multi-document summarization (qMDS) that targets summaries to providing specific informational needs, contextualized to the query. However, the progress in this is hindered by limited availability to large-scale datasets. In this work, we make two contributions. First, we propose an approach for automatically generated dataset for both extractive and abstractive summaries and release a version publicly. Second, we design a neural model SIBERT for extractive summarization that exploits the hierarchical nature of the input. It also infuses queries to extract query-specific summaries. We evaluate this model on CoMSum dataset showing significant improvement in performance. This should provide a baseline and enable using CoMSum for future research on qMDS.

Keywords: Extractive summarization · Abstractive summarization · Neural models · Transformers · Summarization dataset

1 Introduction

Document summarization is an important task in natural language processing and has drawn a lot of research interests in recent years, instigated by the rise and success of neural-based techniques [34, 36, 41, 44].

Among all the variants, multi-document summarization (MDS) is especially challenging [9, 12, 13, 19, 22, 24, 26, 33, 37]. First, large-scale datasets for the task are lacking as high-quality human curated summaries are costly to collect. Human generated summaries vary greatly due to subjective comprehension and interpretation of the source texts, and judgement on what should be included in the summaries. Using news headlines or identifying how-to instructions have been used to obtain pre-annotated data from established web publishers for single document summarization (SDS) [15, 20]. However such (semi)automated techniques have not been extended for building large-scale MDS datasets with demonstrated wide adoption.

S. Kulkarni, S. Chammas and W. Zhu—Equal contribution.

J. Lladós et al. (Eds.): ICDAR 2021, LNCS 12822, pp. 84–98, 2021.
https://doi.org/10.1007/978-3-030-86331-9_6

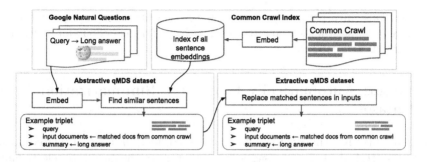

Fig. 1. Automatically generating abstractive or extractive query based multi-document summarization examples. The created dataset CoMSum will be released to public.

Secondly, many modern neural models are built on attention-based transformers, whose computational complexity is quadratic in the number of input tokens. For MDS, the input size easily exceeds the capability of standard models. Transformers that can deal with large input sizes are being proposed but their utility for MDS has not been thoroughly examined (e.g., [3,27]).

In this paper, we study a variant of MDS called query-based MDS (qMDS) which has crucial applications in augmenting information retrieval (IR) experiences [2,6,14,19,23,30]. Text documents are typically multi-faceted and users are often interested in identifying summaries specific to their area of interest. For example, suppose a user is interested in car reliability and cars from certain manufacturers, then an effective IR system could consolidate car reviews from across the web and provide concise summaries relating to the reliability of those cars of interest. Similar to MDS, qMDS also suffers from the lack of large-scale annotated data. Two recent datasets, WikiSum and Multi-News, have been designed to automatically harvest documents for MDS by crawling hyperlinks on Wikipedia and newser.com pages respectively [12,24]. But they target significantly longer (whole page) summaries that are not conditioned on contexts.

Our contributions in this paper are two-fold. We first introduce a scalable approach for machine generated qMDS datasets, with knobs to control generated summaries along dimensions such as document diversity and degree of relevance of the target summary. The main idea is to use an information retrieval engine to generate short summaries from top ranked search results. We use the cleaned Common Crawl (CC) corpus [32] to source relevant web documents that are diverse and multi-faceted for generating Natural Questions (NQ) (long-form) answers [21]. Figure 1 illustrates the overall procedure. We create an instance of such a dataset CoMSum containing 8,162 examples. Each example contains an average of 6 source documents to be synthesized into a long-form answer. CoMSum will be released after review period.

Our second contribution is to present the performance of several baseline methods as well as a new hierarchical summarization model on the dataset. Experimental results show that the proposed SIBERT is superior to others as it models multiple documents hierarchically to infuse information from all of them.

The model also incorporates queries as additional contexts to yield query-specific summaries. These studies demonstrate the utility of using CoMSum for developing new methods for qMDS.

2   CoMSum: Automatically Generated Summarization Dataset

The qMDS task is formalized as follows. Given a query q, a set of related documents $R = \{r_i\}$, document passages $\{r_{i,j}\}$ relevant to the query are selected and synthesized into a summary s. s should be fluent natural language text that covers the information in $\{r_{i,j}\}$ rather than just a concatenation of relevant spans. We propose an automated approach to generating large datasets for the qMDS task for training and evaluating both abstractive and extractive methods. We use Google's Natural Questions (NQ) and Common Crawl (CC) and release the instance of CoMSum[1] with the configurations discussed in this section. But the methodology is general enough to be extended to any other question answering dataset (containing answers that span multiple sentences) and web corpora (to serve as the domain for retrieval).

2.1   Overview

Suppose a long answer to a question q is comprised of n sentences $a = \{l_1, ..., l_n\}$ and a document corpus $D = \{d_i\}$ consisting of sentences $\{d_{ij}\}$.

We use the Universal Sentence Encoder [4] to encode sentences $\phi(l_k)$ and $\phi(d_{ij})$. The dot product $s_{kij} = \langle \phi(l_k), \phi(d_{ij}) \rangle$ gives the semantically similar sentences to the summary. This is the partial result set for the answer sentences l_k: $R_k = \{(d_i, s_{kij}) : \theta_U > s_{kij} > \theta_L\}$. The sets for all sentences $\{R_k\}$ are then combined into a document-level score specific to q: $\psi(d_i) = \sum_{k,j} s_{k,i,j}$ to yield document-level scores to get result set $R = \{(d_i, \psi(d_i))\}$. We then pick top-K ranked documents as the input.

While we have made specific choices for ϕ, $s_{k,i,j}$, ψ, they can be customized and tuned to construct result sets R with higher/lower diversity, tighter/looser topicality match, or number of documents retrieved. For instance, by tuning θ_U and θ_L, we can vary the semantic relevance of sentences admitted to the result set R. Lowering θ_U ensures any exact matches are dropped while raising θ_L increases semantic relevance. This gives the qMDS abstractive summarization examples (q, a, R).

For extractive version, we simply replace the sentences in R with exactly matched ones in the long answer. Table 1 gives one such abstractive-extractive duo.

---

[1] https://github.com/google-research-datasets/aquamuse/tree/main/v3.

Table 1. Exemplary documents for abstractive and extractive summarization

Query	How long has there been land plants on earth
Long answer (Target summary)	Evidence for the appearance of the first land plants occurs in the Ordovician, around 450 million years ago, in the form of fossil spores. Land plants began to diversify in the Late Silurian, from around 430 million years ago, ...
Document for abstractive summarization	Plants originally invaded the terrestrial environment in the Ordovician Period ... That's when we have evidence of the first land plant spores from the fossil record. While terrestrial plant evolution would ...
Document for extractive summarization	Plants originally invaded the terrestrial environment in the Ordovician Period ... Evidence for the appearance of the first land plants occurs in the Ordovician, around 450 million years ago, in the form of fossil spores.

2.2 Creating Contextual Multi-document Summarization (CoMSum)

Sources of Target Summaries and Documents. We use Google's Natural Questions (NQ) as the source of queries and target summaries. NQ is an open-domain question answering dataset containing 307,373 train, 7,830 dev and 7,842 eval examples respectively (version 1.0) [21]. Each example is a Google search query (from real users) paired with a crowd sourced short answer (one or more entities) and/or a long answer span (often a paragraph) from a Wikipedia page. Queries annotated with *only* a long answer span serve as summarization targets since these cannot be addressed tersely by entity names (e.g. *"Who lives in the Imperial Palace in Tokyo?"*) or a boolean. These queries result in open-ended and complex topics answers (e.g., *"What does the word China mean in Chinese?"*).

We use a pre-processed and cleaned version of the English CC corpus called the Colossal Clean Crawled Corpus as the source of our input documents [32]. It contains 355M web pages in total. Using TFHub (https://tfhub.dev/google/universal-sentence-encoder/4) we compute Universal Sentence Embeddings (which are approximately normalized) for sentences tokenized from both NQ and CC data sources. An exhaustive all pairwise comparison is performed using Apache Beam (https://beam.apache.org/). The sentences from the NQ long answers are matched with the CC corpus using efficient nearest neighbor searches over sentence embeddings indexed by space partitioning trees [25].

Selecting Sentences. θ_U and θ_L control the semantic relevance of the selected sentences matched between CC and NQ. we filter out sentences with matching scores below $\theta_L = 0.75$. To avoid exact sentence matches from pages with near-Wikipedia duplicates, we also filter out sentences with scores above $\theta_U = 0.95$.

Summary Recall (SC). It is entirely possible that we cannot locate a match for every sentence in a long answer. The *Summary Recall* is the fraction of sentences in a long answer with matching CC document sentences. A $SC = 1.0$

guarantees a summary can be generated from the documents. In our specific dataset instance, we restrict SC to 0.75. Though this may seem like a handicap, we observed in experiments that the input documents have enough information to reconstruct the target summary fully and can be used to test the language generation capabilities of qMDS approaches.

Top-K is analogous to the number of top results returned by web search and controls the degree of support as well as diversity across multiple documents for a qMDS task. We evaluated the quality of the matched documents as ranked by their match scores. We use $K = 7$ as we found document quality degrades after that (note that this depends on how θ_L, θ_U and SC are set). Since each sentence, and hence the document containing it, is selected independently of each other, we could have some overlap within the documents the top-K selected documents. This would be effective for models trained on this dataset for support when generating a strong summary.

2.3   Quality Evaluation of CoMSum

In this section, we carefully assess the quality of the automatically generated qMDS examples along several axes: correctness of matched documents, fluency of machine edited inputs for extractive summaries, and the overall quality. All our human evaluation tasks are based on human rater pools consisting of fluent English speakers. The annotation tasks are *discriminative* in nature (e.g., judging semantic match), which are cheaper to source and easier to validate through replication than generative annotation tasks (e.g., open-ended text generation).

Basic Statistics. Using the aforementioned procedure, we construct 8,162 examples that are split into training, development and testing sets. These examples are drawn from 52,702 documents from CC. 6,925 queries have 7 documents per query, and the rest have 1–6 documents per query, distributed roughly evenly. Table 2 compares our publicly available dataset to other commonly used SDS/MDS datasets. CNN/DM is for abstractive single-document summarization[29], while Multi-News is a large-scale MDS dataset [12]. While Multi-News has more summarization examples, our dataset includes query contexts and covers more documents, sentences and tokens per summary. Recently, a qMDS dataset derived from Debatepedia (http://www.debatepedia.org) has been introduced [30]. Their input documents are much shorter than CoMSum.

Correctness. We first evaluate how relevant the target summaries are to the retrieved documents. We use BLEU scores to measure the lexical overlapping. Since our retrieval is based on sentence-level matching, it does not guarantee a high recall of the N-grams in the summary.

   A perfect BLEU precision score implies that every N-gram in the summary can be mapped to a distinct N-gram in the source. Figure 2 shows the histogram of this overlap measure. A large number (>60%) of bigrams has BLEU scores

Table 2. A comparison of several summarization datasets

Dataset	# queries	Data split			Summary		Inputs		Per-input doc	
		Train	dev	Test	# w	# s	# w	# s	# w	# s
CoMSum	8162	6599	714	849	102	3.6	10734.4	447.5	1662.5	69.3
Debatepedia	13719	12000	719	1000	11.2	1	75.1	5.1	75	5.1
Multi-News	NA	45K	5622	5622	263.7	10.0	2103.5	82.7	489.2	23.4
CNN/DM	NA	287K	13K	11K	56.2	3.7	810.6	39.8	NA	NA

w: the number of words, # s: the number of sentences

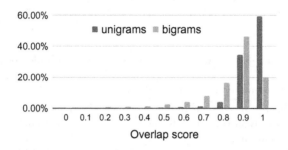

Fig. 2. Histogram of lexical overlap (in BLEU score) between the summary and input documents.

at most 0.8. Since we have new unseen bigrams, we evaluate the factual and semantic correctness of the CC documents that were matched with the NQ long answers. We focus on the abstractive setup. For extractive summarization, this is a less concern as the sentences in the answers are in the documents.

We presented the human raters a Wikipedia paragraph and a matched sentence. They were asked to rate "+1" if the CC sentence matched some of the content of the Wikipedia paragraph. Raters were instructed not to rely on external knowledge. Numerical facts were subjectively evaluated, e.g., *4B years* is close to *4.5B years*, but *3 electrons* and *4 electrons* is not. We rated a sample of 987 examples corresponding to 152 queries, each with a replication of 3 to account for subjectivity. We found that 99.7% examples were marked as relevant by majority.

Fluency. The extractive dataset is created by replacing sentences from the CC doc with the matched sentence in Wikipedia long answer *a*. This, however, may distort the overall flow of the original CC passage. This evaluation task ensures that the fluency is not harmed.

For this, we measured the perplexity of the paragraphs with replaced sentences using a language model[2]. The mean perplexity increased slightly from 80 to 82 after replacement. This small increase is expected since a foreign sentence was inserted, but the difference is small enough proving that the fluency is preserved.

---

[2] https://tfhub.dev/google/wiki40b-lm-en/1.

Overall Quality. We evaluate the overall quality of 60 random qMDS samples (q, R, a) along three dimensions—referential clarity, focus and the coherence of the summary—adapted from DUC2007 task[3]. Since the summary is a Wikipedia passage, grammatical correctness and redundancy need not be evaluated.

Table 3. Clarity, focus and coherence ratings on a 5-point scale (5 being the best. 60 examples have been evaluated.)

Scores	5	4	3	2	1		5	4	3	2	1		5	4	3	2	1
Clarity	59	–	–	–	1	Focus	50	3	–	4	3	Coherence	57	1	–	1	1

Table 4. Multi-faceted query with summary from multiple documents

Query: what is a dream and why do we dream

Source1: Dreams are successions of images, ideas, emotions and sensations occurring involuntarily in the mind during certain stages of sleep. The content and purpose of dreams are not yet understood, though they have been a topic of speculation and interest throughout recorded history. The scientific study of dreams is known as oneirology

Source2: A dream is a succession of images ideas emotions and sensations that usually occur involuntarily in the mind during certain stages of sleep. The content and purpose of dreams are not fully understood although they have been a topic of scientific philosophical and religious interest throughout recorded history. ...

Summary: A dream is a succession of images, ideas, emotions, and sensations that usually occur involuntarily in the mind during certain stages of sleep. The content and purpose of dreams are not fully understood, though they have been a topic of scientific speculation, as well as a subject of philosophical and religious interest, throughout recorded history. Dream interpretation is the attempt at drawing meaning from dreams and searching for an underlying message. The scientific study of dreams is called oneirology

Each sample was rated by 3 raters. They were instructed to consider the query q when evaluating the focus of the summary rather than just a generic summary that can be generated from the set of input documents. Ratings were on a 5-point scale, with 5 being "very good" and 1 being "very poor". Results in Table 3 shows the inter-annotator agreement indicating that the majority of scores are "(very) good" (4 or 5).

Qualitative Examples. We illustrate two specific challenging aspects of the qMDS task. The example in Table 4 demonstrates how a summary can cover multiple facets of a single query that can be sourced from multiple input documents. The example in Table 5 shows how the query context may require summarization models to attend to specific portions of the source documents.

---

[3] https://duc.nist.gov/duc2007/quality-questions.txt.

3 SIBERT: A Use of CoMSum for Extractive Summarization

We demonstrate the utility of CoMSum in the context of multi-document extractive summarization. In this task, the learning algorithm learns to select which sentences in the input documents should be included in the summary. The learning problem is often cast as a binary classification problem on each sentence.

Table 5. Contextual summarization requires attention to the middle of documents

Query: what do you mean by 32 bit operating system
Summary: In computer architecture, 32-bit integers, memory addresses, or other data units are those that are 32 bits (4 octets) wide. Also, 32-bit CPU and ALU architectures are those that are based on registers, address buses, or data buses of that size. 32-bit microcomputers are computers in which 32-bit microprocessors are the norm.
Source1: *Paragraph#1 is about what an operating system is. Paragraph#2 is about hardware. Paragraph#3 is about types of operating systems.* In an operating system, 32-bit integers, memory addresses, or other data units are those that are 32 bits (4 octets) wide. Also, 32-bit CPU and ALU architectures are those that are based on registers, address buses, or data buses of that size. 32-bit micro computers are computers in which 32-bit microprocessors are the norm. *Despite this, such processors could be labeled "32-bit," since they still had 32-bit registers...*   ***other sources...***

In addition to providing the performance of several baseline methods, we also propose a new model, **Stacked Infusion BERT** (SIBERT), for this task. The comparison of SIBERT to other methods highlights the challenges (and the potential) of query-based MDS: it is essential to model relationships among (many) input documents, with respect to the query as contexts.

3.1 The SIBERT Model

The proposed SIBERT model is based on the HIBERT model [45]. The main hypothesis is to exploit the hierarchy in the input, which consists of multiple documents, which are further composed of sentences that are made of word tokens. We design a model that encodes dependencies progressively.

Let INPUT $= \{d_1, d_2, \ldots, d_M\}$ be M documents and each document $d_i = \{s_{i1}, s_{i2}, \ldots, s_{iN}\}$ be N sentences. And each sentence s_{ij} has L words (or word-pieces): $\{w_{ij0}, w_{ij1}, w_{ij2}, \ldots, w_{ijL}\}$. Note the special token $w_{ij0} = $ [cls].

HIBERT is an extractive summarizer for single-document summarization. It first uses a transformer to encode each sentence independently and then uses their pooled representations to construct a second level sequence. This gets encoded by a second transformer.

We first transform the tokens at the lowest level into embeddings:

$$e_{ijl} \leftarrow \texttt{lookup}(w_{ijl}) + p_l \tag{1}$$

where $\texttt{lookup}$ is a lookup table, storing a vector embedding for each token in the dictionary (including the [cls] token). p_l is a position embedding vector to encode the sequential order information of the l-th token.

HIBERT then uses 8 Transformer layers to transform each sentence separately

$$\{h_{ij0}, h_{ij1}, \ldots, h_{ijL}\} \leftarrow \texttt{Transformer}^{(1)}(e_{ij0}, e_{ij1}, \ldots, e_{ijL}) \tag{2}$$

where the $\texttt{Transformer}^{(1)}$ denotes 8 repeated applications of self-attention and fully connected layers [39]. h_{ij0} corresponds to the [cls] token of the sentence.

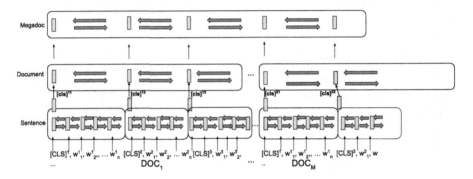

Fig. 3. SIBERT: words and sentences are encoded hierarchically from *multiple* documents.

A second Transformer of 4 Transformer layers follows,

$$h_{i0}^{(2)} = \{h_{i10}^{(2)}, h_{i20}^{(2)}, \ldots, h_{iN0}^{(2)}\} \leftarrow$$
$$\texttt{Transformer}^{(2)}(h_{i10} + p_1, h_{i20} + p_2, \ldots, h_{iN0} + p_N) \tag{3}$$

where the superscript $^{(2)}$ indicates the second Transformer's transformation and resulting embeddings. Note this second Transformer operates on the *sequence of sentence embedding* h_{ij0}. Moreover, the position embeddings p_j is now added to the j-th sentence's embedding for its position in the sequence. The final embeddings of all the sentences in d_i are thus $h_{i0}^{(2)}$.

SIBERT extends HIBERT in two ways. A straightforward use of HIBERT for qMDS is to ignore the query and concatenate all documents together as if it were a single one. In other words, the INPUT has $M = 1$ "mega"-document: a sentence in document d_i infuses with other sentences from d_j where j does not have to be i. On the other end, SIBERT aggregates sentences from all documents *after* they are fused with other sentences in the same document. This creates a hierarchy of information fusion with progressively broader granularity.

Specifically, SIBERT introduces another infusion stage:

$$h_0^{(3)} = \{h_{10}^{(3)}, h_{20}^{(3)}, \ldots, h_{M0}^{(3)}\} \leftarrow \texttt{Transformer}^{(3)}(h_{10}^{(2)}, h_{20}^{(2)}, \ldots, h_{M0}^{(2)}), \quad (4)$$

after each document is processed by the first two Transformers *independently*. In this final stage, all the sentences from all the documents are *flattened* and treated collectively as a mega-doc. The 4 Transformation layers transform all the sentences while allowing them to attend to others from all the documents[4]. Please refer to Fig. 3 for an illustration of the model architecture.

The second extension of SIBERT incorporates queries as additional contexts. This is achieved by encoding the query (sentence) $q = \{[\texttt{cls}], q_1, q_2, \ldots, q_K\}$ with the Transformer used to encode the lowest-level tokens in the documents,

$$z = (z_0, z_1, z_2, \ldots, z_K) \leftarrow \texttt{Transformer}^{(1)}([\texttt{cls}], q_1, q_2, \ldots, q_K) \quad (5)$$

For notation simplicity, we have omitted the steps of lookup and adding position embeddings. We fuse this query representation with the sentence representations $h_0^{(3)}$ (Eq. (4)) using a multi-heand attention module with residual connections:

$$h_0^z = \texttt{MultiHeadAttn}(z, h_{10}^{(3)}, h_{20}^{(3)}, \ldots, h_{M0}^{(3)}) \quad (6)$$

$$h_0 = \texttt{Normalize}(h_0^z + h_0^{(3)}) \quad (7)$$

Note that while the query representation z does not need to be updated, all the sentence embeddings in $h_0^{(3)}$ are updated with respect to z, resulting in a new set of embeddings h_0^z to be combined and normalized.

The training of SIBERT is similar to HIBERT for extractive summarization. Each of the sentence in h_0 has a binary label whether the sentence is selected for the target summary. A cross-entropy loss (after adding a linear projection layer and a sigmoid nonlinearity) is then used to train the model end to end.

Variants of SIBERT. There are several ways to incorporate queries as additional contexts. Cross-attention to the queries can occur at 4 levels: part of Transformer[1] (equivalently, appending the query to every sentence), part of Transformer$^{(2)}$, part of Transformer$^{(3)}$, or the outputs of Transformer$^{(3)}$. We observed that fusing at the output layer is optimal on the holdout set and hence use this configuration for our experiments (Eq. (6)).

---

[4] For notation simplicity, we omit the step adding position embeddings to all the sentences in the $\{h_{i0}^{(2)}\}_{i=1}^M$ according to their positions in the mega-doc—the embeddings are shared with the ones in the previous two Transformers.

3.2   Experiments

Implementation Details. To fit our training on a single GPU, the minibatch size is restricted to a single example. The total number of sentences in the input is restricted to 800 and the maximum sentence length to 30 tokens. The maximum number of the documents is 7 and each document is at most 800 sentences. We used an Adam optimizer with a learning rate of $1e^{-5}$. All weights except for the wordpiece embeddings were randomly initialized with a truncated normal having a standard deviation of 0.2. We use pretrained wordpiece embeddings with a vocabulary size of 30,522 that were allowed to be fine-tuned during training. We used a hidden dimension of 768 for all the embeddings (token, sentence, and position). The setup for all the Transformer layers is similar to those in the original BERT$_{\text{BASE}}$ [8], except those noted in the description of the SIBERT and the following: 12 attention heads are used for the self-attention operations with intermediate size of 1024. The activation function used throughout was the Gaussian Error Linear Unit (GELU).

Table 6. Extractive summarization results on CoMSum

Model	ROUGE-1	ROUGE-2	ROUGE-L
HIBERT	37.33	20.93	26.77
query HIBERT	37.85	21.40	27.11
TextRank	39.76	32.20	35.57
Biased TextRank	40.16	24.54	31.07
SIBERT w/o query	41.29	25.43	30.29
SIBERT	41.71	26.15	30.79

Alternative Methods. We compare SIBERT to HIBERT as a baseline method where we aggregate all documents as a single mega-document. We also extend HIBERT by appending the query to the mega-document so it is more suitable for query-based MDS.

We also compare to TextRank, which is an unsupervised approach for extractive summarization [28]. It builds a weighted graph of sentences from pairwise similarity and selects the top ranked sentences as summary. Biased TextRank is a more recent approach that extends TextRank by biasing the weights with the query [19]. In this study, we experiment with both approaches. We use the same TFHub Universal Sentence Encoder from creating CoMSum to generate similarity among the query and the sentences. We choose top 4 sentences as predicted summaries. We adjust the filter threshold in the bias weights for the Biased TextRank approach on the dev set data—0.7 achieved the best results.

Main Results. Table 6 shows the performance of various methods on the CoMSum dataset. We evaluate with BLEU scores. For all 3 methods, adding query generally improves the ROUGE scores; the exceptions are the Biased TextRank on ROUGE-2 and ROUGE-L where the scores are reduced.

Contrasting the two neural-based models HIBERT and SIBERT, we observe that adding an additional Transformer layer improves the ROUGE scores significantly: from $37.33 \rightarrow 41.29$ and $37.85 \rightarrow 41.71$ for ROUGE-1, for example. This suggests that hierarchical modeling for multi-document is beneficial. While SIBERT aggregates sentences from all documents, more refined modeling could bring further improvement.

Human Evaluation. We manually evaluated 44 examples, each of which was rated by 3 raters. The raters were presented with the query and the predicted summary using SIBERT model trained with and without the query context. We found in 34% cases, the generated summary was rated as "being similar" in both cases. 62.1% had a better (more relevant) summary using SIBERT with the query than the one without it. This suggests that the query context plays an important role in the modeling to yield more accurately the target summaries.

4   Related Work

Query-based summarization can be both extractive [5, 6, 23, 31, 35, 40, 42] or abstractive [2, 14, 17, 30]. Earlier studies were often extractive and relied on manually selected and curated datasets such as DUC2005 and DUC2006 [5] and recent ones for abstractive summaries [12]. While the query-based summarization does not get much limelight, there has been some work in extractive [19] and abstractive [11] summaries that take such focus into account.

Previous approaches have also explored creating QA datasets for short answers using news articles from CNN/DM [16], HotpotQA [43], TriviaQA [18], SearchQA [10], online debates with summarizing arguments and debating topics as queries [30], and community question answering websites [7]. BIOASQ released a domain-specific dataset based on biomedical scientific articles with their corresponding short user-understandable summaries [38]. Some of them involve extracting text spans of words as answers from multiple text passages. However, our work focuses on longer non-factoid answers independent of the domain. The MS Marco has 1M question-answer-context triplets where the answers are manually created using the top-10 passages from Bing's search queries and is close to our work in spirit [1]. But CoMSum uses Wikipedia passages as summaries thus bypassing any manual text generation. Other recent large-scale datasets for MDS over long input documents with much longer summaries unlike short concise answers like CoMSum are another flavor in this field [12, 20, 24].

5   Conclusion

We describe an automatic approach to generate examples for learning query-based multi-document summarization. The approach leverages the publicly available language resources, and does not require additional human annotation. Using this approach, we have created CoMSum, a large-scale dataset for qMDS.

We study its qualities and utility in developing learning methods for summarization. In particular, we propose a new method SIBERT for query-based multi-document extractive summarization. Through empirical studies, we have found that to perform well, models taking advantage of the query as additional contexts and the hierarchy in the multi-document input perform better. This is consistent with researchers' intuition about the task, thus supporting the utility of using CoMSum for methodology research.

References

1. Bajaj, P., et al.: MS MARCO: a human generated machine reading comprehension dataset (2018)
2. Baumel, T., Eyal, M., Elhadad, M.: Query focused abstractive summarization: incorporating query relevance, multi-document coverage, and summary length constraints into seq2seq models. CoRR abs/1801.07704 (2018). http://arxiv.org/abs/1801.07704
3. Beltagy, I., Peters, M.E., Cohan, A.: Longformer: the long-document transformer (2020)
4. Cer, D., et al.: Universal sentence encoder. CoRR abs/1803.11175 (2018). http://arxiv.org/abs/1803.11175
5. Dang, H.T.: Duc 2005: evaluation of question-focused summarization systems. In: Proceedings of the Workshop on Task-Focused Summarization and Question Answering, SumQA 2006, pp. 48–55. Association for Computational Linguistics, USA (2006)
6. Daumé III, H., Marcu, D.: Bayesian query-focused summarization. In: Proceedings of the 21st International Conference on Computational Linguistics and 44th Annual Meeting of the Association for Computational Linguistics, pp. 305–312. Association for Computational Linguistics, Sydney (July 2006). https://www.aclweb.org/anthology/P06-1039
7. Deng, Y., et al.: Joint learning of answer selection and answer summary generation in community question answering. In: AAAI (2020)
8. Devlin, J., Chang, M.W., Lee, K., Toutanova, K.: Bert: pre-training of deep bidirectional transformers for language understanding (2019)
9. Diego Antognini, B.F.: Gamewikisum: a novel large multi-document summarization dataset. LREC (2020). https://arxiv.org/abs/2002.06851
10. Dunn, M., Sagun, L., Higgins, M., Guney, V.U., Cirik, V., Cho, K.: SearchQA: a new q&a dataset augmented with context from a search engine (2017)
11. Egonmwan, E., Castelli, V., Sultan, M.A.: Cross-task knowledge transfer for query-based text summarization. In: Proceedings of the 2nd Workshop on Machine Reading for Question Answering, pp. 72–77. Association for Computational Linguistics, Hong Kong (November 2019). https://www.aclweb.org/anthology/D19-5810
12. Fabbri, A.R., Li, I., She, T., Li, S., Radev, D.R.: Multi-news: a large-scale multi-document summarization dataset and abstractive hierarchical model. ACL (2019)
13. Ghalandari, D.G., Hokamp, C., Pham, N.T., Glover, J., Ifrim, G.: A large-scale multi-document summarization dataset from the Wikipedia current events portal. ACL (2020). https://arxiv.org/abs/2005.10070
14. Hasselqvist, J., Helmertz, N., Kågebäck, M.: Query-based abstractive summarization using neural networks. CoRR abs/1712.06100 (2017). http://arxiv.org/abs/1712.06100

15. Hermann, K.M., et al.: Teaching machines to read and comprehend. In: Advances in Neural Information Processing Systems, pp. 1693–1701 (2015)
16. Hermann, K.M., et al.: Teaching machines to read and comprehend. CoRR abs/1506.03340 (2015). http://arxiv.org/abs/1506.03340
17. Ishigaki, T., Huang, H.H., Takamura, H., Chen, H.H., Okumura, M.: Neural query-biased abstractive summarization using copying mechanism. Adv. Inf. Retr. **12036**, 174–181 (2020)
18. Joshi, M., Choi, E., Weld, D., Zettlemoyer, L.: TriviaQA: a large scale distantly supervised challenge dataset for reading comprehension, pp. 1601–1611. Association for Computational Linguistics, Vancouver (July 2017). https://www.aclweb.org/anthology/P17-1147
19. Kazemi, A., Prez-Rosas, V., Mihalcea, R.: Biased TextRank: unsupervised graph-based content extraction (2020)
20. Koupaee, M., Wang, W.Y.: Wikihow: a large scale text summarization dataset. CoRR abs/1810.09305 (2018). http://arxiv.org/abs/1810.09305
21. Kwiatkowski, T., et al.: Natural questions: a benchmark for question answering research. Trans. Assoc. Comput. Linguist. **7**, 453–466 (2019)
22. Li, W., Xiao, X., Liu, J., Wu, H., Wang, H., Du, J.: Leveraging graph to improve abstractive multi-document summarization. ACL (2020). https://arxiv.org/abs/2005.10043
23. Litvak, M., Vanetik, N.: Query-based summarization using MDL principle. In: Proceedings of the MultiLing 2017 Workshop on Summarization and Summary Evaluation Across Source Types and Genres, pp. 22–31. Association for Computational Linguistics, Valencia (April 2017). https://www.aclweb.org/anthology/W17-1004
24. Liu, P.J., et al.: Generating Wikipedia by summarizing long sequences. CoRR abs/1801.10198 (2018). http://arxiv.org/abs/1801.10198
25. Liu, T., Moore, A.W., Gray, A.G., Yang, K.: An investigation of practical approximate nearest neighbor algorithms. In: NIPS (2004)
26. Liu, Y., Lapata, M.: Hierarchical transformers for multi-document summarization. ACL (2019). arXiv:1905.13164
27. Zaheer, M., et al.: Big bird: transformers for longer sequences. In: NeurIPS (2020). https://arxiv.org/abs/2007.14062
28. Mihalcea, R., Tarau, P.: TextRank: bringing order into text. In: Proceedings of the 2004 Conference on Empirical Methods in Natural Language Processing, pp. 404–411. Association for Computational Linguistics, Barcelona (July 2004). https://www.aclweb.org/anthology/W04-3252
29. Nallapati, R., Zhou, B., Gulcehre, C., Xiang, B., et al.: Abstractive text summarization using sequence-to-sequence RNNs and beyond. Computational Natural Language Learning (2016)
30. Nema, P., Khapra, M.M., Laha, A., Ravindran, B.: Diversity driven attention model for query-based abstractive summarization. In: Proceedings of the 55th Annual Meeting of the Association for Computational Linguistics, ACL 2017, Vancouver, Canada, July 30–August 4, vol. 1: Long Papers, pp. 1063–1072. Association for Computational Linguistics (2017). https://doi.org/10.18653/v1/P17-1098
31. Otterbacher, J., Erkan, G., Radev, D.R.: Biased LexRank: passage retrieval using random walks with question-based priors. Inf. Process. Manag. **45**(1), 42–54 (2009). http://www.sciencedirect.com/science/article/pii/S0306457308000666
32. Raffel, C., et al.: Exploring the limits of transfer learning with a unified text-to-text transformer. arXiv e-prints (2019)

33. Rossiello, G., Basile, P., Semeraro, G.: Centroid-based text summarization through compositionality of word embeddings. ACL (2017). https://www.aclweb.org/anthology/W17-1003
34. Rush, A.M., Chopra, S., Weston, J.: A neural attention model for abstractive sentence summarization. In: EMNLP (2015)
35. Schilder, F., Kondadadi, R.: Fastsum: fast and accurate query-based multi-document summarization. In: Proceedings of the 46th Annual Meeting of the Association for Computational Linguistics on Human Language Technologies: Short Papers, HLT-Short 2008, pp. 205–208. Association for Computational Linguistics, USA (2008)
36. See, A., Liu, P.J., Manning, C.D.: Get to the point: summarization with pointer-generator networks. CoRR abs/1704.04368 (2017). http://arxiv.org/abs/1704.04368
37. Su, D., Xu, Y., Yu, T., Siddique, F.B., Barezi, E.J., Fung, P.: Caire-covid: a question answering and query-focused multi-document summarization system for covid-19 scholarly information management. In: EMNLP2020 NLP-COVID Workshop (2020). https://arxiv.org/abs/2005.03975
38. Tsatsaronis, G., Balikas, G., Malakasiotis, P., et al.: An overview of the BIOASQ large-scale biomedical semantic indexing and question answering competition. BMC Bioinform. **16**, 1–28 (2015). https://doi.org/10.1186/s12859-015-0564-6
39. Vaswani, A., et al.: Attention is all you need (2017)
40. Wang, H., et al.: Self-supervised learning for contextualized extractive summarization. In: Proceedings of the 57th Annual Meeting of the Association for Computational Linguistics, pp. 2221–2227. Association for Computational Linguistics, Florence (July 2019). https://www.aclweb.org/anthology/P19-1214
41. Wang, L., Yao, J., Tao, Y., Zhong, L., Liu, W., Du, Q.: A reinforced topic-aware convolutional sequence-to-sequence model for abstractive text summarization. CoRR abs/1805.03616 (2018). http://arxiv.org/abs/1805.03616
42. Wang, L., Raghavan, H., Castelli, V., Florian, R., Cardie, C.: A sentence compression based framework to query-focused multi-document summarization. CoRR abs/1606.07548 (2016). http://arxiv.org/abs/1606.07548
43. Yang, Z., et al.: HotpotQA: a dataset for diverse, explainable multi-hop question answering. In: Proceedings of the 2018 Conference on Empirical Methods in Natural Language Processing, pp. 2369–2380. Association for Computational Linguistics, Brussels (October–November 2018). https://www.aclweb.org/anthology/D18-1259
44. Zhang, J., Zhao, Y., Saleh, M., Liu, P.J.: Pegasus: pre-training with extracted gap-sentences for abstractive summarization. ArXiv abs/1912.08777 (2019)
45. Zhang, X., Wei, F., Zhou, M.: HIBERT: document level pre-training of hierarchical bidirectional transformers for document summarization. CoRR abs/1905.06566 (2019). http://arxiv.org/abs/1905.06566

RTNet: An End-to-End Method for Handwritten Text Image Translation

Tonghua Su$^{(\boxtimes)}$ (ID), Shuchen Liu, and Shengjie Zhou

School of Software, Harbin Institute of Technology, Harbin, China
`thsu@hit.edu.cn`

Abstract. Text image recognition and translation have a wide range of applications. It is straightforward to work out a two-stage approach: first perform the text recognition, then translate the text to target language. The handwritten text recognition model and the machine translation model are trained separately. Any transcription error may degrade the translation quality. This paper proposes an end-to-end leaning architecture that directly translates English handwritten text in images into Chinese. The handwriting recognition task and translation task are combined in a unified deep learning model. Firstly we conduct a visual encoding, next bridge the semantic gaps using a feature transformer and finally present a textual decoder to generate the target sentence. To train the model effectively, we use transfer learning to improve the generalization of the model under low-resource conditions. The experiments are carried out to compare our method to the traditional two-stage one. The results indicate that the performance of end-to-end model greatly improved as the amount of training data increases. Furthermore, when larger amount of training data is available, the end-to-end model is more advantageous.

Keywords: Machine translation · Text recognition · Image text translation · Handwritten text · End-to-End

1 Introduction

Text image translation can be regarded as the combination of text recognition and machine translation, which has a wide range of applications. It can be used for the electronization of paper documents, such as books, newspapers, and magazines. What's more, scanning paper text with a mobile phone camera for recognition and translation helps people read foreign manuscripts, literature, letters, etc. This technology could also make it easier for visitors to read menus and signs in foreign languages. For example, with Google Translate [1], the camera-captured text image can be directly translated into the selected target language.

There has been a boom in independent research on text recognition and machine translation, but the idea of creating one system that combines both areas together, is still a challenging task. A traditional solution for text image translation is a two-stage approach of first performing the text recognition, then translating the text to target language [2–4]. However, the two-stage approach exists some drawbacks: Any minor transcription

J. Lladós et al. (Eds.): ICDAR 2021, LNCS 12822, pp. 99–113, 2021.
https://doi.org/10.1007/978-3-030-86331-9_7

Fig. 1. Illustration of a translation error in the two-stage approach. Due to the incorrect recognition of the second character, 'Jelly color' was translated into 'Fruit practice color'.

error may affect the semantics of the text, thereby degrading the translation quality (see Fig. 1).

This paper proposes an end-to-end leaning architecture that integrates Optical Character Recognition (OCR) [5] technology and Machine Translation (MT) [6] technology into a unified deep learning model, and directly translates English handwritten text lines in images into Chinese sentences. The key points are as follows: data augmentation in case of insufficient training data; application of transfer learning to improve the generalization of the model under low-resource conditions; how to design the model structure and loss function to ensure the quality of translation.

Our model architecture is based on CNN and Transformer [7]. In the model, the CNN part is mainly used to extract two-dimensional visual features of the image and convert them into one-dimensional feature sequences, and the Transformer part is used for modeling feature sequence to target sequence. Firstly, we conduct a visual encoding, next bridge the semantic gaps using a feature transformer and finally present a textual decoder to generate the target sentence. We apply the multi-task learning paradigm, with text image translation as the main task, and recognition as the auxiliary task. We perform experiments to verify the proposed architecture. The results reveal that there is a significant improvement in translation quality.

The rest of the paper is organized as follows. Section 2 surveys related work research; Sect. 3 presents pre-training models for text recognition and translation and the end-to-end text image translation method; Sect. 4 introduces the data sets, implementation process, and result analysis. Section 5 is the conclusion of this paper.

2   Related Works

2.1   Machine Translation

Machine translation problem can be addressed by the Encoder-Decoder framework [8]. When the input sequence is long, the context vector cannot represent the information of the entire sequence. To this end, researchers introduced the attention model [9], which generates a separate context vector when predicting each word of the output sequence. It can be trained in parallel, reducing the time cost of training while maintaining recognition accuracy.

Transformer is state-of-the-art model in the machine translation. The Transformer model [7] uses matrix multiplication to calculate the Attention matrix and output sequence, accelerating the training process. On the basis of Transformer, the Universal Transformer model [10] introduces an adaptive computing time mechanism, an enhanced version of Transformer (GRET) models sequence-level global information [11].

A hot topic related to Transformer research is to analyze the role of attention mechanism and the weight and connection of network [12, 13]. It's doubted that the Transformer network has learned too much redundant information and attention layers need pruning [14, 15]. Michel et al. demonstrated removing multiple heads from the trained Transformer model and reducing certain layers to one head will not significantly reduce test performance [15].

2.2 Text Line Recognition

Since the Encoder-Decoder framework was proposed, researchers have also explored it in text line recognition. Michael J proposed a handwritten text recognition method combining CNN and RNN based on attention mechanism [19], incorporating the location information and the content information.

A mainstream approach employs CNN + RNN + CTC architecture to solve the text line recognition task, called CRNN [20]. Inspired by the above research [21, 22], it is considered that applying the attention model to HTR may improve the recognition performance. Also it proves that fully convolutional model for sequence to sequence learning outperforms recurrent models on very large benchmark datasets at an order of magnitude faster speed [23]. However, the serial computing method of the RNN takes a lot of time in the training phase. Transformer [7] is composed of parallel computing structure to reduce training time consumption. Since text line recognition can be seen as a Seq2Seq task, we introduce Transformer into our OCR model.

2.3 Handwritten Text Line Translation

Handwritten text translation can be seen as a combination of Handwritten Text Recognition (HTR) and Machine Translation (MT). Researchers began to study the methods of detecting, recognizing and translating signage in camera pictures [4]. Some approaches accomplish the task of Arabic text image translation by combining the separate handwritten text recognition system and machine translation system [2, 3]. Other approaches subdivide the task of text image translation into multiple stages, and solve the problems of each stage one by one. In the research [24], the error correction module with Bigram is applied to correct OCR results. The experimental results show that the use of error correction module can greatly improve the final translation quality.

In addition to multi-stage text recognition and translation, researchers have proposed a more concise end-to-end image translation approach that can recognize printed and handwritten Arabic corpus documents and translate them into English [25]. This method adjusts the baseline MT system model to the problem of the in-domain model.

Since speech recognition and text line recognition have something in common, many approaches for speech recognition can also be used for text line recognition. Some researchers implemented a two-stage system to translate Arabic speech into English text [26]. Besides, some researchers have implemented an end-to-end model via related approaches for processing sequence-to-sequence task. It can directly input the speech sequence of the source language and output the speech sequence of the target language without the step of speech recognition [27]. Experiment was carried out on Spanish to English translation datasets. The results demonstrate that the proposed end-to-end method is slightly worse than the cascaded two-stage method (first recognized and then synthesized), indicating the end-to-end translation task is still challenging. The main reason is that if the end-to-end network is to be trained in a fully supervised manner, large-scale labeled data is required. End-to-end text line translation has extensive research prospects, but is extremely difficult due to the limit of dataset. Since there is no public dataset for image translation tasks, some researchers used bilingual subtitles for movies, and proposed two sequence to sequence methods, including a convolution based attention model and a BLSTM + CTC model, to solve the image translation problem [28].

At present, most methods for implementing handwritten text line translation are combined with OCR systems and MT systems. The output of OCR system is used as the input of the MT system to obtain the translation results. Our end-to-end recognition and translation model simplifies the above steps and improves the quality of translation.

3 Methodology

3.1 Architecture

A Transformer can be divided into an encoder and a decoder. We mainly focus on the role of decoder. The combination of encoder and decoder can be defined as a discriminative model. The condition of this discriminative model is the input of the encoder, and the category is the output of the decoder. Since the output of the decoder is a sequence, and the number of categories to be combined is countless, the decoder can be seen as a conditional generative model. With this view, we design an end-to-end handwritten text translation model.

Our end-to-end handwritten text translation architecture integrates OCR model and MT model into a unified deep learning model, consisting of an encoder in the OCR model, a decoder in the MT model and a feature converter. The input is English text images and the output is Chinese character sequences.

Since the decoder only needs a sequence of hidden layer vector to generate the sentence of the target language, we can also input the hidden layer sequences of the OCR encoder into the MT decoder. (Cf. Fig. 2).

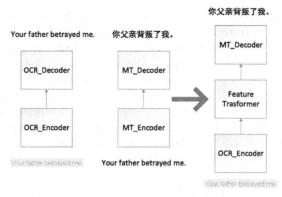

Fig. 2. End-to-end model architecture. We first pre-trained the OCR model to recognize English text images, the MT model to translate English sequences into Chinese sequences, and then recombined the encoder in the OCR model and the decoder in the MT model. As the features of sequences output by the OCR encoder and the MT encoder are inconsistent, we employ a feature transformer to map the distribution.

3.2 Theory

The two-stage approach for image translation is described as follows: let $p(y|f)$ be the probability of generating the target language text y when observing the input image f. The system finds the optimal target language text y^* such that:

$$y^* = \underset{y}{\operatorname{argmax}}\, p(y|f) \qquad (1)$$

Let $p(x|f)$ be the probability of recognition result x occurrence under the condition of the input image f. The text recognition system can be expressed as:

$$x^* = \underset{x}{\operatorname{argmax}}\, p(x|f) \qquad (2)$$

According to the total probability formula, the end-to-end text image translation model can be expressed as:

$$p(y|f) = \sum_x p(y|x,f)p(x|f) \qquad (3)$$

It is not feasible to enumerate all the recognition results x. The above formula can be approximated as:

$$p(y|f) = p(y|x^*,f)p(x^*|f) \qquad (4)$$

$$x^* = \underset{x}{\operatorname{argmax}}\, p(x|f) \qquad (5)$$

So the optimal translation sequence is expressed as:

$$y^* = \underset{y}{\operatorname{argmax}}\, p(y|f) \qquad (6)$$

In the two-stage approach, there are always errors in the above process even if the translation result y* is corrected through the language model. According to the total probability formula, the conditional probability of the output sentence y is obtained by summing up the conditional probability of all possible recognition results x. However the OCR system just greedily selects the optimal sentence in the recognition result as an approximate representation. It is possible to reduce the error by modeling $p(y|f)$ directly.

Suppose there is training data $<y^i, f^i>(i = 1, 2, ...N)$, of which there are N samples in total. $<y^i, f^i>$ represents the i-th sample, f^i represents a text image, y^i represents a translated sentence. yi is a sequence and can be expressed as $yi = \{y_1^i, y_2^i, ...y_T^i\}$. Then the conditional probability can be calculated as:

$$p\left(y^i, f^i\right) = p(y_1^i, y_2^i, ...y_T^i | f^i) \tag{7}$$

According to the probability multiplication chain rule, we know:

$$p(y_1^i, y_2^i, ...y_T^i | f^i) = p(y_1^i | f^i) p(y_2^i | y_1^i, f^i) ... p(y_T^i | y_1^i, y_2^i, ...y_{T-1}^i, f^i) \tag{8}$$

Assuming that the occurrence of each word in the sentence is independent of each other, the above formula is approximately expressed as:

$$p(y^i | f^i) = \prod_{t=1}^{T} p(y_t^i | f^i) \tag{9}$$

Suppose the probability of each word in the sentence satisfies the Markov property.

$$p(y_t^i | y_1^i, y_2^i, ...y_{t-1}^i, f^i) = p(y_t^i | y_{t-1}^i, f^i) \tag{10}$$

Then the sequence probability can be approximated as:

$$p(y^i | f^i) = p(y_1^i | f^i) \prod_{t=2}^{T} p(y_t^i | y_{t-1}^i, f^i) \tag{11}$$

If a neural network M is used to approximate the conditional probability, the input of M should be the text image f and translation result at time step $t - 1$, then M is represented as:

$$p(y_t^i | y_{t-1}^i, f^i) = M(y_{t-1}^i, f^i, \theta) \tag{12}$$

θ is the weight of the neural network. The likelihood function is derived from the known true tag sequence y^i.

$$L^i = \prod_{t=1}^{T} M\left(y_{t-1}^i, f^i, \theta\right) \tag{13}$$

The weight θ is estimated by maximum likelihood, that is, maximizing L^i, which is equivalent to minimizing the negative log-likelihood function $-\log L^i$:

$$-\log L^i = -\sum_{t=1}^{T} M\left(y_{t-1}^i, f^i, \theta\right) \tag{14}$$

The loss function of the entire training set is expressed as:

$$Loss = -\sum_{i=1}^{N} \log L^i \tag{15}$$

N represents the number of samples in the training set. The estimated weight can be written as:

$$\overline{\theta} = \underset{\theta}{argmax} \sum_{i=1}^{N} \sum_{t-1}^{T} M\left(y_{t-1}^i, f^i, \theta\right) \tag{16}$$

Since the neural network M is differentiable, the above loss function can be trained by a gradient descent algorithm.

We can model the mapping relationship from the original image to the target language using CNN and Transformer. Firstly, we use the text recognition dataset to pre-train the convolution module, and the machine translation data set to pre-train the machine translation module, and then use a small amount of end-to-end data (input the original image and output the target language sentence) to tune some parameters to obtain the end-to-end model.

3.3 Training Strategy

The detailed steps of training an end-to-end model are as follows:

Get an OCR Encoder and a MT Decoder. We pre-train an OCR model to acquire an OCR encoder that converts images into OCR hidden layer sequences, pre-train a MT model to get a MT encoder that converts source language sentences in the form of strings into MT hidden layer sequences, and a MT decoder that generates target lan-guage sentences from the hidden layer sequences.

Recombine the Network Structure. This is a critical part of the training strategy. The Objective is combine the OCR encoder and the MT decoder to achieve image text line translation. Since the output of the OCR encoder and the input of the MT decoder are feature vector sequence, we can feed the OCR encoder's output to the input of the MT decoder. However, the distribution of the OCR encoder's output feature sequence is not the same as the input of MT decoder, so a feature transformer is required.

Because the form of the input data has changed from an original string to an image, the MT encoder is discarded. OCR encoder and MT decoder are combined to implement end-to-end text recognition and translation. Considering that the OCR hidden layer sequences is different from that of the MT encoder, we apply some self-attention layers to re-extract features of OCR hidden layer sequences and convert them into features to be processed by the MT decoder. We call this part a feature transformer (see Fig. 3). A feature transformer several stacked multi-head self-attention layers. Assume the output of the OCR encoder is a feature map tensor of shape [N, C, H, W], where N is batch size, C is the channel size, H and W is height and width of feature map respectively. The feature transformer first reshape the tensor to a feature sequence of shape [N, T, M], where N is the batch size, T is the length of sequence, M is the size of a feature vector. We set T = W, Because on the image, the characters of a text line are arranged horizontally

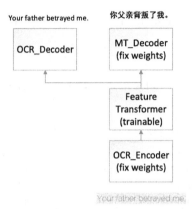

Fig. 3. Regulating network weights throSugh multi-task learning. We observe the recognition performance through the branch of the OCR decoder as an auxiliary task, and analyze the final model by the MT decoder. Only the weights inside the feature transformer are trainable. Note that the branch of the OCR decoder is only used in training, not in prediction.

in space, e.g. Each column corresponds to a character. And M = C * H, because we stack the feature map of different rows on each column. The reshaped feature sequence is then fed into several stacked multi-head self-attention layer. The output of these layers is a tensor of shape [N, T, K], where K equals to the feature vector size of the input of MT decoder.

Regulate the Network Weights on the Synthetic Training Dataset. The main issue is which weights to update. If adjusting the weights of the entire network, that is, OCR encoder, feature transformer, and MT decoder, it is easy to cause overfitting due to the lack of training data. Therefore we fix the weights of the OCR encoder and MT decoder.

The weights of the feature converter are randomly initialized and trained on the training data. (see Fig. 3) illustrates the process of regulating network weights through multi-task learning. We observe the recognition performance through the branch of the OCR decoder as an auxiliary task, and analyze the convergence result of the final model by the MT decoder. During training, the total loss is the sum of the OCR loss and the MT loss. Only the weights inside the feature transformer are trainable. Note that the branch of the OCR decoder is only used in training, not in prediction.

3.4   OCR Model

The OCR model combines a convolutional neural network and a Transformer architecture, so we called it CNN-Transformer (see Fig. 4). This model uses Cross Entropy loss function during training, and does not require forward and backward operations in CTC, further speeding up training. The decoder adopts the deep attention mechanism. Under the condition that the language samples are abundant, the decoder can learn the language model, which is beneficial for some text recognition applications.

The Hyperparameters of Conv-backbone in encoder are as follows: The maximum length of the label is 128, the maximum height of the picture is 64, the maximum width of the picture is 2048, the number of hidden layer units of each attention layer is 512, the number of attention layers of the encoder is 3, the number of attention layers of the decoder is 1, the number of headers in the attention layer is 8.

The shape of the input image is 64 × 2048 with channel 1, and the network structure is 32C3-MP2-64C3-MP2-64C3-64C3-MP2-96C3-96C3-MP2-512C3-MP2, where xCy represents a convolutional layer with kernel size of y × y and output channels of x, MPx represents a max pooling layer with kernel size of x × x and stride of x × x, followed by a reshape layer of 128 × 512.

3.5 MT Model

In order to integrate with the OCR model, the MT model applies the character-level machine translation method, which is to split each word in the input and output sentences into characters, and deploys the Transformer model to implement an English-Chinese translation system.

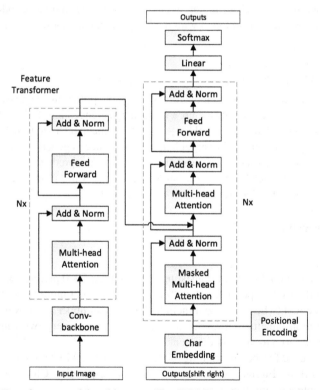

Fig. 4. CNN-Transformer model architecture. The CNN-Transformer model first extracts visual features from input images with a Conv-backbone, and converts images to feature sequences. The rest part of CNN-Transformer is consistent with the Transformer model in machine translation. Self-attention mechanism extracts global information from the input sequences. The decoder predicts each character with an autoregressive algorithm until the model finally outputs a <EOS> tag or the output sequences reach its maximum length.

Deep learning models cannot directly process strings, so we convert strings into integer ID sequences. Each English letter, Chinese character, and punctuation mark corresponds to a unique integer ID. The character set consists of all characters in the training set plus some special symbols. The special symbols are as follows: <go> indicates the beginning of the sequence, <eos> indicates the end of the sequence, <unk> represents an unknown character, for handling characters not in the character set, <pad> represents a padding blank.

The MT model is based on the standard Transformer network. Both the encoder and decoder have 6 layers. The encoder uses multi-head self-attention, and the decoder uses vanilla attention. Finally, the decoder uses a fully-connected layer to map the features of the hidden layer to a space of the character set size, and connected with a softmax layer to get the output probability distribution. To promote translation, the decoder's char embedding and the linear layer before softmax layer are shared.

Different from the standard Transformer network, the encoder and decoder reduce the number of nodes in the fully connected layer to 1/4 of the original. Other network hyperparameters are the same as the standard one. Since both the OCR model and the MT model are based on the Transformer architecture, they have a similar decoder structure. We replace the decoder part in Fig. 4 with the MT decoder so that we get the end-to-end model.

3.6 Autoregressive Decoding

The sequence alignment of OCR model and MT model applies autoregressive decoding mechanism. The encoder encodes source language sentences into hidden layer vector sequences, and then the decoder employs the attention mechanism to translate one character at a time step until the entire target language sentence is generated. It decodes one character in the predicted sequence once, and then inputs the predicted string prefix into the decoder when predicting the next character. The decoding process ends when the decoder outputs an <EOS> tag.

4   Experiments

4.1   Dataset

We pretrain OCR model on IAM consisting of more than 6000 pictures of handwritten text lines and their corresponding labels for training set, and more than 1800 pictures for test set. The text to be recognized includes 26 English upper and lower case letters, numbers, common punctuation marks, and spaces between every two English words. The dataset contains a total of 84 different symbols.

The dataset used for the pre-training of the MT model is AI-Challenger 2018, which contains 9.6 million English sentences and their corresponding Chinese translations. Most of the sentences are dialog data.

For training an end-to-end text image translation model, the input is an English text line picture and the output is a Chinese sentence, so we synthesize a new dataset. We use source language sentences in the parallel corpus of the dataset to synthesize pictures corresponding to the source language sentences, and save their corresponding translations. The specific scheme for generating a synthetic dataset is as follows:

Preprocess Source Language Sentences in Parallel Corpora to Generate Character Set. Put characters in each source language into character set **A** and remove duplicate characters.

Preprocess the Text Recognition Dataset and Generate a Picture Dictionary of Characters. For each character *a* in the character set **A**, find all the character pictures labeled *a* from the OCR dataset and put them into the character picture set.

Generate Pictures for Source Language Sentences. For each character in each source language sentence in the parallel corpus, randomly select a corresponding picture from the character picture set, and then splice the character picture with the picture of the generated sentence. After selecting picture for each character in the source language sentence, we stitch a complete text picture I.

Generate Data Samples. According to the target language sentence t and the text picture I corresponding to the source language sentence, generate a data sample <I, t>, where I is the text picture containing the source language sentence and t is the translated sentence. Add data samples to the dataset.

The end-to-end experiments were performed on the synthetic dataset containing 50,000 training samples.

Table 1. BLEU evaluation of machine translation

Model	AI-Challenger 2018 (BLEU)
Transformer-base	28.083256
Transformer-small	22.511538

Table 2. BLEU evaluation of End-to-End model

Synthetic data amount	Train BLEU	Test BLEU
50 K (end-to-end)	35.8459	3.1870
500 K (end-to-end)	23.8916	16.0736
50 K (two-stage, baseline)	30.7293	22.9472
500 K (two-stage, baseline)	18.9014	15.9567

4.2 Pre-training Results

The hyperparameter settings used in the text recognition and machine translation model during pre-training are as follows: the batch size is 64, the learning rate is 1e-4, the dropout rate is 0.1, L2 regularization coefficient is 1e-5, the ratio of random replacement of decoder input characters is 0.1, and the loss function is cross entropy. The training process uses the Adam optimizer.

Model evaluation adopts BLEU scores. The experimental results of machine translation tasks are shown in Table 1. We trained two Transformer models. The first is a standard Transformer, and the second is a simplified version that reduces the number of nodes in the fully connected layer to 1/4 of the original. The BLEU scores for both models are shown in the table. According to the data in the table, reducing the number of nodes in the fully connected layer has a great impact on the translation.

4.3 End-to-End Results

We use the two-stage approach as the baseline for our end-to-end approach: The OCR model recognizes the text on the image as a string, and inputs the recognition results into the MT model, which generates the target language sentences. OCR model and MT model are trained separately. The preprocessed text image is input into the OCR model and output strings. The strings are converted and input into the MT model to get the target language sentence.

The experiments were carried out on the synthetic dataset according to our end-to-end method and the two-stage baseline method. The experimental results are shown in Table 2.

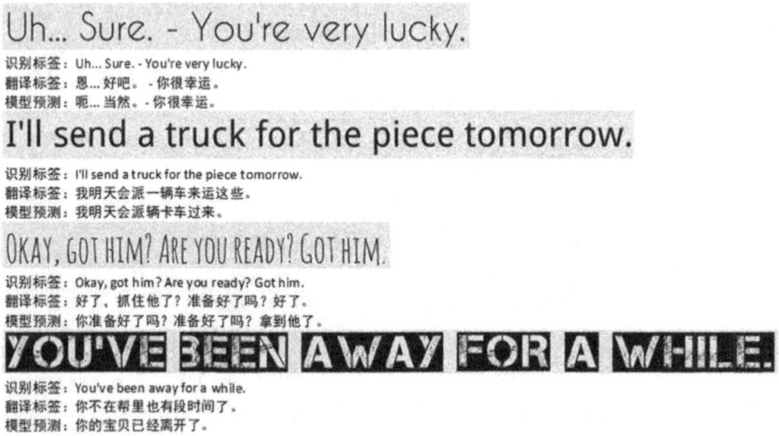

Fig. 5. Results on the end-to-end system. Machine translation provides relatively smooth and accurate translation.

With the increase of data volume, the BLEU scores of the two-stage method in the training set and test set will be decreased, indicating that the increment on data volume brings difficulties to the fitting of two-stage model. The experimental results indicate that the performance of the end-to-end model is related to the amount of training data. With 50,000 pieces of data, severe overfitting occurs in the end-to-end system. It shows that when data amount is insufficient, our end-to-end model performs strong in fitting and weak in generalization. Increasing the amount of data alleviates overfitting. With the 500,000 pieces of training data, although the performance of the end-to-end model is similar to the two-stage model, from the trend of fitting and generalization ability with the amount of data, the end-to-end model may be more suitable for large datasets. On the other hand, the end-to-end model can alleviate semantic confusion caused by transcription errors in the two-stage method in Fig. 1 to a certain extent. Figure 5 illustrates the test results on the end-to-end system. In the first case, 'Uh…Sure' are translated into different words, but their semantics are exactly the same. The second and third case also get complete sentences with correct semantics. In the fourth case, the model identifies the word "been" error as "baby", but the semantics of the predicted sentence is still smooth.

The following analyzes in principle why the end-to-end method can alleviate semantic confusion of the two-stage method: In the two-stage model, we obtain a sequence with maximum probability through decoding of the OCR model. The output of the neural network is a probability distribution sequence, but the model only choose the sequence with the highest probability, and some information will be lost in this process. In the end-to-end model, feature transformation is performed in the latent variable space, and the output of OCR model does not need to be decoded, thus avoiding loss of information.

Our model still exists limitations: since there is no ready-made dataset for training end-to-end image translation model, we adopt synthesis method to expand the dataset. In addition, the amount of synthetic data is limited, so large-scale experiments cannot be carried out yet. In order to enhance the generalization ability of the end-to-end model, the key is to improve the number of training samples labeled. We plan to explore more data expansion and augmentation method in future work.

5   Conclusions

In this paper, we implement the handwritten text recognition model and English-Chinese machine translation model based on the deep attention mechanism. This paper further explores the problem of combining the OCR model and MT model and proposes an end-to-end model for handwritten text translation. The end-to-end model is trained by transfer learning and compared with the two-stage model (baseline) on the synthesized dataset. The results reveals the end-to-end model will be greatly improved as the amount of training data increases. Furthermore, when large amount of training data is available, the end-to-end model is more advantageous.

Acknowledgment. This work was supported by the National Key Research and Development Program of China (No. 2020AAA0108003) and National Natural Science Foundation of China (No. 61673140 and 81671771).

References

1. Groves, M., Mundt, K.: Friend or foe? Google Translate in language for academic purposes. Engl. Specif. Purp. **37**, 112–121 (2015)

2. Dehghani, M., Gouws, S., Vinyals, O., et al.: Universal transformers. In: ICLR (2019)
3. Weng, R., Wei, H., Huang, S., et al.: GRET: Global Representation Enhanced Transformer. In: AAAI (2020)
4. Tang, G., Müller, M., Rios, A., et al.: Why self-attention? a targeted evaluation of neural machine translation architectures. In: EMNLP (2018)
5. Voita, E., Sennric, R., Titov, I.: The bottom-up evolution of representations in the transformer: a study with machine translation and language modeling objectives. In: Proceedings of the 2019 Conference on Empirical Methods in Natural Language Processing and the 9th International Joint Conference on Natural Language Processing (EMNLP-IJCNLP) (2019)
6. Voita, E., Talbot, D., Moiseev, F., et al.: Analyzing multi-head self-attention: specialized heads do the heavy lifting, the rest can be pruned. In: Proceedings of the 57th Annual Meeting of the Association for Computational Linguistics, pp. 19–1580. ACL, Italy (2019)
7. Michel, P., Levy, O., Neubig, G.: Are sixteen heads really better than one? In: Advances in Neural Information Processing Systems (2019)
8. Raganato, A., Scherrer, Y., Tiedemann, J.: Fixed encoder self-attention patterns in transformer-based machine translation. arXiv preprint (2020)
9. Poulos, J., Rafael, V.: "Attention networks for image-to-text". arXiv preprint, arXiv: 1712.04046 (2017)
10. Chorowski, J.K., Bahdanau, D., Serdyuk, D., et al.: Attention-based models for speech recognition. In: Advances in Neural Information Processing Systems (2015)
11. Michael, J., Labahn, R., Grüning, T., et al.: Evaluating sequence-to-sequence models for handwritten text recognition. In: ICDAR (2019)
12. Shi, B., Bai, X., Yao, C.: An end-to-end trainable neural network for image-based sequence recognition and its application to scene text recognition. IEEE Trans. Pattern Anal. Mach. Intell. **39**(11), 2298–2304 (2016)
13. Zhang, H., Wei, H., Bao, F., et al.: Segmentation-free printed traditional Mongolian OCR using sequence to sequence with attention model. In: 2017 14th IAPR International Conference on Document Analysis and Recognition (ICDAR). IEEE Computer Society (2017)
14. Lee, C.Y., Osindero, S.: Recursive recurrent nets with attention modeling for OCR in the wild. In: CVPR (2016)
15. Gehring, J., Auli, M., Grangier, D., et al.: Convolutional sequence to sequence learning. In: ICML, pp. 1243–1252 (2017)
16. Chang, Y., Chen, D., Zhang, Y., et al.: An image-based automatic Arabic translation system. Pattern Recogn. **42**(9), 2127–2134 (2009)
17. Cao, H., Chen, J., Devlin, J.: Document recognition and translation system for uncon-strained Arabic documents. In: International Conference on Pattern Recognition (2012)
18. Bender, O., Matusov, E., Hahn, S.: The RWTH Arabic-to-English spoken language trans-la-tion system. In: IEEE Workshop on Automatic Speech Recognition & Understanding (2008)
19. Jia, Y.: Direct speech-to-speech translation with a sequence-to-sequence model. In: Inters-peech (2019)
20. Morillot, O.: The UOB-Tele-com ParisTech Arabic Handwriting Recognition and Translation Systems for the OpenHart 2013 Competition (2013)
21. Alkhoury, I.: Arabic Text Recognition and Machine Translation. PhD Thesis (2015)
22. Yang, J., Gao, J., Zhang, Y.: Towards automatic sign translation. In: International Confe-rence on Human Language Technology Research (2003)
23. Patel, C., Patel, A., Patel, D.: Optical character recognition by open source OCR tool tesse-ract: a case study. Int. J. Comput. Appl. **55**(10), 50–56 (2012)
24. Ling, W., Trancoso, I., Dyer, C.: Character-based neural machine translation. In: ACL (2016)
25. Vaswani, A., et al.: Attention is all you need. In: Advances in Neural Information Processing Systems (2017)

26. Yonghui, W.: Google's neural machine translation system: Bridging the gap between human and machine translation. arXiv preprint (2016)
27. Bahdanau, D., Cho, K., Bengio, Y.: Neural Machine Translation by Jointly Learning to Align and Translate. arXiv preprint (2015)
28. Chen, Z., Yin, F., Zhang, X.Y., et al.: Cross-lingual text image recognition via multi-task sequence to sequence learning. In: International Conference on Pattern Recognition, pp. 3122–3129 (2020)

Multimedia Document Analysis

NTable: A Dataset for Camera-Based Table Detection

Ziyi Zhu[1], Liangcai Gao[1(✉)], Yibo Li[1,2], Yilun Huang[1], Lin Du[3], Ning Lu[3], and Xianfeng Wang[3]

[1] Institute of Computer Science and Technology, Peking University, Beijing, China
`gaoliangcai@pku.edu.cn`
[2] Center for Data Science of Peking University, Beijing, China
[3] Huawei AI Application Research Center, Beijing, China

Abstract. Comparing with raw textual data, information in tabular format is more compact and concise, and easier for comparison, retrieval, and understanding. Furthermore, there are many demands to detect and extract tables from photos in the era of Mobile Internet. However, most of the existing table detection methods are designed for scanned document images or Portable Document Format (PDF). And tables in the real world are seldom collected in the current mainstream table detection datasets. Therefore, we construct a dataset named NTable for camera-based table detection. NTable consists of a smaller-scale dataset NTable-ori, an augmented dataset NTable-cam, and a generated dataset NTable-gen. The experiments demonstrate deep learning methods trained on NTable improve the performance of spotting tables in the real world. We will release the dataset to support the development and evaluation of more advanced methods for table detection and other further applications in the future.

Keywords: NTable · Table detection · Target detection network

1 Introduction

Important information that relates to a specific topic is often organized in tabular format for convenient retrieval and comparison, it is obviously more structured and comprehensible in particular areas compared to natural language [31]. Therefore, tables are not only used in documents, but also prevalent in almost every aspect of daily life. For example, they appear commonly in notice boards, timetables, and bills. It brings a large demand for detecting tables for further information extraction, collection, and organization.

Due to various table layouts and complicated formats, table detection is a challenging task in the automatic processing of documents. Researches on table detection started decades ago. Table detection techniques based on heuristic rules and layout analysis were firstly used. Afterwards, object detection methods based on deep learning have been proposed, bringing significant performance improvement in table detection.

© Springer Nature Switzerland AG 2021
J. Lladós et al. (Eds.): ICDAR 2021, LNCS 12822, pp. 117–129, 2021.
https://doi.org/10.1007/978-3-030-86331-9_8

Although current models have made enormous progress on the task of table detection for PDFs and scanned images, tables in real-world photos are seldom collected in current mainstream table detection datasets. Therefore, existing methods still show an apparent gap for camera-based table detection. We can infer that camera-based table detection is neglected.

Most of the existing table detection methods are data-driven. Without corresponding data, it is difficult for them to detect tables in the real world. To solve this problem, we construct an entirely new dataset NTable and conduct a series of experiments on this dataset. This paper makes the following three contributions in camera-based table detection:

1) A new dataset, NTable, is proposed for camera-based table detection, which consists of a smaller-scale dataset NTable-ori, an augmented dataset NTable-cam, and a generated dataset NTable-gen.
2) We train and test different models on this dataset, and provide a series of benchmarks with different models.
3) The experimental results demonstrate the generated dataset NTable-gen can improve models' performance for camera-based table detection.

2 Related Work

2.1 Table Detection Dataset

Generally speaking, table detection is the task of localizing bounding boxes of tables in PDFs or images. There have been several datasets that can be used for this task. Table 1 shows the comparison between these datasets.

The UW-III dataset [1] consists of 1,600 corrected English documents images with manually annotated ground-truth of entity types and bounding boxes. It contains 215 table zones distributed over 110 pages.

The UNLV dataset [25] contains 2,889 pages of scanned documents images from a variety of sources. The ground-truth of manually marked zones with zone types are provided in text format. More detailed ground truth data is XML-encoded.

The Marmot dataset [3] contains 958 pages with tables. It is collected from Chinese e-Books, English conferences and journal papers crawled from the web. Marmot is built with "positive" and "negative" cases, which means it not only has pages containing at least one table, but also has pages with complex layout instead of tables.

ICDAR2013 [7] is from ICDAR 2013 Table Competition, it consists of 156 images collected from PDF documents and XML ground truth files. ICDAR2017 [5] is from ICDAR 2017 POD Competition, containing 2.4k scientific document images converted from PDFs.

DeepFigures [27] is designed for extracting scientific figures, but it also provides tables with location information. It contains 1.4m pages with tables collected from a large number of scientific documents from arXiv and PubMed.

ICDAR2019 [4] contains two subsets as Modern and Historical, a total of 1.6k images. The historical dataset contains a variety of tables such as handwritten accounting books, train timetables, stock exchange lists, prisoner lists, production census, etc. The modern dataset comes from different kinds of PDF documents such as scientific journals, financial statements, etc.

PubLayNet [32] contains 113k pages with tables. It generated annotations automatically by matching the XML representations and articles that are publicly available on PubMed Central™ Open Access (PMCOA).

TableBank [16] is a huge dataset for table detection which consists of 417k pages. The PDF pages in TableBank are obtained from Word documents and Latex documents. Annotations are extracted from the internal Office XML code and the generated PDF documents with a special command.

It can be seen that camera-based table data is scarce in the existing mainstream table detection dataset. This inspires us to propose a new dataset NTable for camera-based table detection.

Table 1. Datasets for table detection.

Dataset	Source	Tables number
UW-III	Scanned document images	215
UNLV	Scanned document images from magazines, newspapers, business Letter, annual report, etc.	427
Marmot	e-Books, conference and journal papers	958
ICDAR2013	PDF documents from the European Union and US Government	156
ICDAR2017	Scientific documents images (converted from PDF)	2417
DeepFigures	PDF scientific documents	1.4 m
ICDAR2019	The historical subset: accounting books, lists, timetables, etc. The modern subset: PDF documents such as scientific journals, financial statements, etc.	1.6 k
PubLayNet	PDF scientific documents	113 k
TableBank	PDF pages from Word documents in '.docx' format from the internet and Latex documents with Latex source code from 2014 to 2018 in arXiv	417 k

2.2 Table Detection

Pyreddy et al. [22] proposed an approach of detecting tables using heuristics. Kieninger et al. [14] presented T-Recs System that uses a bottom-up analysis approach to locate logical text block units on document images. Yildiz et al. [30] proposed the pdf2table system, which starts with text lines detection and finally merges them into tables. Kasar et al. [13] also extracted intersections of the horizontal and vertical lines and passed this information into SVM. Based on graph representation of layout regions, Koci et al. [15] proposed Remove and

Conquer (RAC) that implements a list of curated rules for table recognition. Melinda et al. [21] introduced parameter-free table detection methods separately for closed and open tables based on multi-gaussian analysis.

With the development of deep learning and object detection methods, significant performance improvements have been made in the table detection task. Hao et al. [8] presented a deep learning based approach for table detection. This system computes region proposals through predefined rules and uses CNN to detect whether a certain region proposal belongs to table region or not. Schreiber et al. [24] and Gilani et al. [6] proposed an approach using Faster R-CNN model and adding some adaptive adjustments. He et al. [9] used FCN with an end verification network to determine whether the segmented region is a table or not. Li et al. [17] used layout analysis to get candidate regions, then applied CRF and CNN to classify them. Siddiqui et al. [26] proposed a deep deformable CNN model, which is a combination of deformable CNN with faster R-CNN/FPN. Li et al. [18] proposed a feature generator model similar to GAN to improve the performance of less-ruled table detection. Huang et al. [11] presented a YOLOv3-based method and introduced some adaptive adjustments including an anchor optimization strategy and post-processing methods. Riba et al. [23] proposed a graph-based framework for table detection in document images, which makes use of Graph Neural Networks (GNNs) to describe and process local structural information of tables. To improve the precision of table boundary locating, Sun et al. [28] proposed Faster R-CNN based table detection combining corner locating method.

3   The NTable Dataset

The main motivation of developing the NTable dataset is to fill the gap between regular table pictures and tables that exist in the real world, which may facilitate a new research direction of camera-based table detection. We will introduce a series of details about building this dataset.

3.1   NTable-ori and NTable-cam

Source Acquisition. NTable-ori is made up of 2.1k+ images taken by different cameras and mobile phones. We collect images from abundant sources, such as digital screens, English/Chinese/Spanish/French textbooks (for high school or university), books and documents on psychology, mathematics, statistics, physics, physiology, ecology, art and computer, timetables, bulletin boards, price lists, accounting books, etc., trying to cover as many types as possible.

We provide two classification methods, one is based on the source, the other is based on the shape (see Fig. 1). According to the source, NTable-ori can be divided into textual, electronic and wild. Textual means that the image is taken from a paper document in a natural setting, electronic means that the image is taken from an electronic screen, and wild means that the image is a street view photograph with tables. According to the shape, NTable-ori can be divided into upright, oblique and distorted. Upright means the table is upright and almost rectangular, oblique means the table is tilted, but the border of it is still straight, and there is no distortion caused by perspective, and all the other tables fall into the distorted category. Table 2 counts the classification results.

Fig. 1. Samples images in each category and subcategory. The blue boxes and the pink boxes respectively represent the two categories 'source' and 'shape'. (Color figure online)

Table 2. Classification results of NTable-ori.

Category	Source			Shape		
Subcategory	Textual	Electronic	Wild	Upright	Oblique	Distorted
# of pages	1674	254	198	758	421	947

Data Augmentation. NTable-cam is augmented from NTable-ori. By changing rotation, brightness and contrast, original 2.1k+ images are expanded eightfold to 17k+ images. For brightness and contrast, we add jitter in the HSV color space to change the brightness and contrast within an appropriate range. For rotation, we first apply a random rotation of $[-5°, +5°]$ on the image, then using geometric methods to calculate the multiple of enlargement, so that the blank caused by rotation can be removed by cropping. Experiments show that these data augmentation methods improve data richness greatly.

3.2   NTable-gen

Limited by the data available, it is almost impossible to cover all real-world situations by manually folding or distorting document papers [20]. Therefore, models trained only on dataset that consists of a limited number of photos may not generalize well to various scenarios. To address the limitations of the current data, ulteriorly improve data richness, we add a synthetic dataset NTable-gen, which is used to simulate as much as possible the various deformation conditions.

Document Acquisition. We chose PubLayNet [32] as the original document images. There are 86950 pages with at least one table in PubLayNet's training set. We randomly select 8750 pages.

Distorted Images Generation. We use the same way to synthesize distorted images in 2D as DocUNet [20]. In a nutshell, generating a 2D distorted document image is to map the original image to a random 2D mesh. Therefore, an image can be warped into different distortions. This process can cover more different types of distortions in real-world situations, and is much faster compared to directly rendering distorted documents in 3D rendering engines.

The distortions in the real world can be included into two basic types: folds and curves, which usually appear together. The parameters α of folding and curving is respectively set to 0.8 and 2 the same as [19]. To guarantee the diversity of the background in natural scenes, we randomly chose images as backgrounds from the VOC2012 dataset [2].

3.3   Manually and Automatic Annotation

We manually annotate each table in an image with the biggest bounding box location information by a database and web-based tool for image annotation called Labelme [29]. When synthesizing a distorted document image (see Fig. 2), a mask of the original table area is also generated (without a background). Then connected component analysis is applied to the generated image of table mask, the location of pixel that belongs to the ith mask is (x_i, y_i). We calculate $(min(x_i), min(y_i))$, $(max(x_i), max(y_i))$ as the top left and the bottom right coordinates (X_i^{left}, Y_i^{left}), $(X_i^{right}, Y_i^{right})$ of the corresponding bounding box. Through these calculations, annotations in the same format with NTable-ori are obtained. With this information, we can construct annotations in any format, such as COCO, YOLO, etc. Figure 3 shows the visualized results of annotation.

In conclusion, the characteristics and difficulties of NTable can be summarized as the following notable points:

1) Image quality: The quality of different pictures varies sharply. Because tables can exist on a variety of materials (paper, digital screen, glass, plastic, etc.), all sorts of reflections, folds and shading will appear. What's more, watermarks, mosaic, perspective distortion and light will all change the ideal state of the tables.

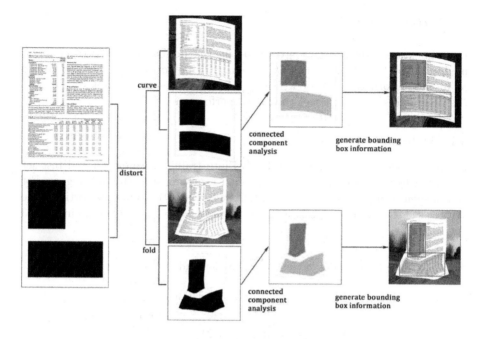

Fig. 2. Process of automatic annotation.

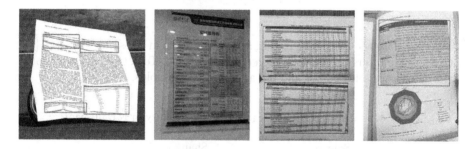

Fig. 3. Sample images with bounding boxes.

2) Table formats: Although the styles of tables in electronic documents are already diverse, they are still relatively rigid and monotonous compared to tables in real life. Tables completely without border, blank or half-filled tables, irregular (not rectangle) tables, tables that contain pictures in the cells, and hand-drawn tables are common in the real world.

3) Background components of tables: The background of the tables in previous datasets are usually white paper, it is indeed also a majority in real life. But there still exist color background and other complicated conditions, which cannot be included in simple text/figure/title/list.

4   Experiment

Experiments are designed to investigate 1) the performance of two common models, Mask-RCNN [10] and YOLOv5 [12]; 2) impact of the generated subset on model performance; 3) for which specific attributes of the tables in Ntable, the generated subset improves the model performance.

4.1   Data and Metrics

For NTable-cam, the samples are divided into training, validation and test set according to the ratio of 7:1:2 by stratified sampling in each subcategory. For NTable-gen, samples are randomly partitioned with the same proportion. The statistics of the training, validation, and test sets are depicted in detail in Table 3 and Table 4.

Table 3. Statistics of training, validation and test sets in NTable.

	Training	Validation	Test
NTable-cam	11904	3408	1696
NTable-gen	11984	3424	1712
Total	23888	6832	3408

We calculate the Precision, Recall, and F1-measure in the same way as in [11] with a specific Intersection over Union (IoU) threshold. To better compare model performance, we selected 0.6 and 0.8 as the IoU thresholds.

Table 4. Statistics of each category and subcategory in NTable-cam.

	Source			Shape		
	Textual	Electronic	Wild	Upright	Electronic	Wild
Train	7152	1336	3416	3072	2136	6696
Validation	944	200	552	464	304	928
Test	2064	288	1056	912	600	1896

4.2   Experimental Results and Analysis

During training, for the Mask-RCNN model, the input resolution is set to 600 × 800. The network is fine-tuned with an initial learning rate of 0.02, batch size is 32 on 4 GTX 1080Ti GPUs. The model is trained for 12 epochs.

For the YOLOv5 model, the input resolution is set to 640 × 640. The network is fine-tuned with an initial learning rate of 0.01, batch size is 8 on 4 GTX 1080Ti GPUs. The model is trained for 12 epochs.

NTable Baseline. The Mask-RCNN and YOLOv5 models are trained and tested on NTable-cam. Table 5 shows the experiment results.

Table 5. Experiment results of Mask-RCNN and YOLOv5 on NTable-cam.

Model	IoU	Precision	Recall	F1-measure
Mask-RCNN	0.6	0.901	0.921	0.911
	0.8	0.853	0.872	0.862
YOLOv5	0.6	0.901	0.971	0.935
	0.8	0.873	0.942	0.906

Impact of NTable-gen. In this part, we set up several training strategies by using TableBank and NTable, then test the YOLOv5 model on NTable-cam. The experimental results are listed in Table 6. Table 6 shows that if there is no photo data in the training set, tables cannot be detected as accurately as they are on other table detection datasets (see number 1 and 2). This illustrates the gap between the real world and scanned documents images. After training or fine tuning on NTable-cam, the YOLOv5 model shows performance improvements (see number 3 and 4), reaching a relatively good result. It also proves that even if there is a large electronic document dataset for pre-training, there is no obvious help to the camera-based data. The YOLOv5 model achieves the best performance when fine-tuned on both NTable-cam and NTable-gen (see number 5). Compared with number 3 and 4, when the IoU is 0.6, the F1-measure increases from 0.934/0.934 to 0.980, when the IoU is 0.8, the F1-measure increases from 0.906/0.907 to 0.972. This proves that better results can be obtained by adding a generated dataset to the fine-tuning process. As the best performance is obtained by TableBank, NTable-cam and NTable-gen together, while only using any part or both parts cannot achieve such good performance.

Table 6. Experiment results of training and fine tuning, results are reported on Ntable-cam using YOLOv5.

Number	Train	Fine tuning	IoU	Precision	Recall	F1-measure
1	TableBank	–	0.6	0.770	0.689	0.727
			0.8	0.657	0.588	0.621
2	Ntable-gen	–	0.6	0.707	0.578	0.636
			0.8	0.607	0.496	0.546
3	Ntable-cam	–	0.6	0.901	0.971	0.935
			0.8	0.873	0.942	0.906
4	TableBank	Ntable-cam	0.6	0.902	0.969	0.934
			0.8	0.875	0.941	0.907
5	TableBank	Ntable-cam and Ntable-gen	0.6	**0.964**	**0.997**	**0.980**
			0.8	**0.956**	**0.988**	**0.972**

Table 7. Experiment results (F1-measure) of YOLOv5 between number 4 and number 5 in Table 6.

Train	Fine tuning	IoU	Shape			Source		
			Distorted	Oblique	Upright	Textual	Electronic	Wild
TableBank	Ntable-cam	0.6	0.939	0.920	0.936	0.944	0.918	0.921
		0.8	0.925	0.898	0.873	0.934	0.830	0.876
TableBank	Ntable-cam and Ntable-gen	0.6	0.977	0.991	0.980	0.979	0.988	0.979
		0.8	0.972	0.988	0.961	0.975	0.952	0.972

Performance Analysis. Table 7 compares experiment results (F1-measure) of YOLOv5 between different fine-tune dataset. If the IoU threshold is 0.6, then after fine tuning, the performance improvement of the YOLOv5 model is most obvious on oblique and electronic. If the IoU threshold is 0.8, the performance improvement improves most on oblique, upright, wild and electronic. All the categories mentioned above are relatively small in data volume, which indicates that NTable-gen increases data richness, and is an effective supplement to NTable. Figure 4 provides two samples from number 4 and 5 in Table 6, illustrating that NTable-gen can be thought of as the transitional data from TableBank to NTable-cam. By adding NTable-gen to the fine-tune process, the YOLOv5 model successfully avoids errors caused by the light or surrounding content.

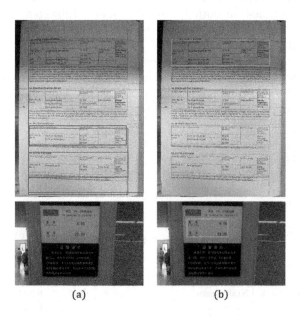

(a)   (b)

Fig. 4. Result samples of YOLOv5. (a) represents fine tuning on NTable-cam, (b) represents fine tuning on NTable-cam and NTable-gen.

5   Conclusion

To supplement the existing dataset of table detection and support the research of camera-based table detection, we collect and generate the NTable dataset, which consists of camera-shot and generated data. We use the Mask-RCNN and YOLOv5 model as the baseline. Experiments prove that the generated subset NTable-gen can improve performance of models on camera-based table detection. By evaluating the model accuracy on different categories, we further analyze on which subsets, the existing methods perform worse, and how NTable-gen contribute to improving model performance.

In the future, the dataset will be enlarged to cover more type of tables photos, such as bigger tilt angle and richer data sources, making the subsets as balanced as possible. We will release the dataset to support the development and evaluation of more advanced methods for table detection and other further applications, such as camera-based table structure recognition.

Acknowledgment. This work is supported by the projects of National Key R&D Program of China (2019YFB1406303) and National Natural Science Foundation of China (No. 61876003), which is also a research achievement of Key Laboratory of Science, Technology and Standard in Press Industry (Key Laboratory of Intelligent Press Media Technology).

References

1. Chen, S., Jaisimha, M., Ha, J., Haralick, R., Phillips, I.: Reference manual for UW English document image database i. University of Washington (1993)
2. Everingham, M., Van Gool, L., Williams, C.K.I., Winn, J., Zisserman, A.: The PASCAL Visual Object Classes Challenge 2012 (VOC2012) Results. http://www.pascal-network.org/challenges/VOC/voc2012/workshop/index.html
3. Fang, J., Tao, X., Tang, Z., Qiu, R., Liu, Y.: Dataset, ground-truth and performance metrics for table detection evaluation. In: 10th IAPR International Workshop on Document Analysis Systems, DAS 2012, pp. 445–449 (2012)
4. Gao, L., et al.: ICDAR 2019 competition on table detection and recognition (CTDAR). In: 2019 International Conference on Document Analysis and Recognition, ICDAR 2019, pp. 1510–1515 (2019)
5. Gao, L., Yi, X., Jiang, Z., Hao, L., Tang, Z.: ICDAR2017 competition on page object detection. In: 14th IAPR International Conference on Document Analysis and Recognition, ICDAR 2017, pp. 1417–1422. IEEE (2017)
6. Gilani, A., Qasim, S.R., Malik, M.I., Shafait, F.: Table detection using deep learning. In: 14th IAPR International Conference on Document Analysis and Recognition, ICDAR 2017, pp. 771–776 (2017)
7. Göbel, M.C., Hassan, T., Oro, E., Orsi, G.: ICDAR 2013 table competition. In: 12th International Conference on Document Analysis and Recognition, ICDAR 2013, pp. 1449–1453 (2013)
8. Hao, L., Gao, L., Yi, X., Tang, Z.: A table detection method for pdf documents based on convolutional neural networks. In: 2016 12th IAPR Workshop on Document Analysis Systems (DAS), pp. 287–292 (2016). https://doi.org/10.1109/DAS.2016.23

9. He, D., Cohen, S., Price, B.L., Kifer, D., Giles, C.L.: Multi-scale multi-task FCN for semantic page segmentation and table detection. In: 14th IAPR International Conference on Document Analysis and Recognition, ICDAR 2017, pp. 254–261 (2017)

10. He, K., Gkioxari, G., Dollár, P., Girshick, R.B.: Mask R-CNN. In: IEEE International Conference on Computer Vision, ICCV 2017, pp. 2980–2988 (2017)

11. Huang, Y., et al.: A yolo-based table detection method. In: 2019 International Conference on Document Analysis and Recognition, ICDAR 2019, pp. 813–818 (2019)

12. Jocher, G., Stoken, A., Borovec, J., et. al.: ultralytics/yolov5: v3.1 - Bug Fixes and Performance Improvements (October 2020). https://doi.org/10.5281/zenodo.4154370

13. Kasar, T., Barlas, P., Adam, S., Chatelain, C., Paquet, T.: Learning to detect tables in scanned document images using line information. In: 12th International Conference on Document Analysis and Recognition, ICDAR 2013, pp. 1185–1189 (2013)

14. Kieninger, T.: Table structure recognition based on robust block segmentation. In: Document Recognition V, 1998. SPIE Proceedings, vol. 3305, pp. 22–32 (1998)

15. Koci, E., Thiele, M., Lehner, W., Romero, O.: Table recognition in spreadsheets via a graph representation. In: 13th IAPR International Workshop on Document Analysis Systems, DAS 2018, pp. 139–144 (2018)

16. Li, M., Cui, L., Huang, S., Wei, F., Zhou, M., Li, Z.: TableBank: table benchmark for image-based table detection and recognition. CoRR abs/1903.01949 (2019)

17. Li, X., Yin, F., Liu, C.: Page object detection from PDF document images by deep structured prediction and supervised clustering. In: 24th International Conference on Pattern Recognition, ICPR 2018, pp. 3627–3632 (2018). https://doi.org/10.1109/ICPR.2018.8546073

18. Li, Y., Gao, L., Tang, Z., Yan, Q., Huang, Y.: A GAN-based feature generator for table detection. In: 2019 International Conference on Document Analysis and Recognition, ICDAR 2019, pp. 763–768 (2019)

19. Liu, X., Meng, G., Fan, B., Xiang, S., Pan, C.: Geometric rectification of document images using adversarial gated unwarping network. Pattern Recognit. **108**, 107576 (2020)

20. Ma, K., Shu, Z., Bai, X., Wang, J., Samaras, D.: DocUNet: document image unwarping via a stacked u-net. In: 2018 IEEE Conference on Computer Vision and Pattern Recognition, CVPR 2018, pp. 4700–4709 (2018)

21. Melinda, L., Bhagvati, C.: Parameter-free table detection method. In: 2019 International Conference on Document Analysis and Recognition, ICDAR 2019, pp. 454–460 (2019)

22. Pyreddy, P., Croft, W.B.: TINTIN: a system for retrieval in text tables. In: Proceedings of the 2nd ACM International Conference on Digital Libraries, pp. 193–200 (1997)

23. Riba, P., Dutta, A., Goldmann, L., Fornés, A., Terrades, O.R., Lladós, J.: Table detection in invoice documents by graph neural networks. In: 2019 International Conference on Document Analysis and Recognition, ICDAR 2019, pp. 122–127 (2019)

24. Schreiber, S., Agne, S., Wolf, I., Dengel, A., Ahmed, S.: Deepdesrt: deep learning for detection and structure recognition of tables in document images. In: 2017 14th IAPR International Conference on Document Analysis and Recognition (ICDAR), vol. 01, pp. 1162–1167 (2017). https://doi.org/10.1109/ICDAR.2017.192

25. Shahab, A., Shafait, F., Kieninger, T., Dengel, A.: An open approach towards the benchmarking of table structure recognition systems. In: The Ninth IAPR International Workshop on Document Analysis Systems, DAS 2010, pp. 113–120. ACM International Conference Proceeding Series (2010)
26. Siddiqui, S.A., Malik, M.I., Agne, S., Dengel, A., Ahmed, S.: Decnt: deep deformable CNN for table detection. IEEE Access 6, 74151–74161 (2018). https://doi.org/10.1109/ACCESS.2018.2880211
27. Siegel, N., Lourie, N., Power, R., Ammar, W.: Extracting scientific figures with distantly supervised neural networks. CoRR abs/1804.02445 (2018)
28. Sun, N., Zhu, Y., Hu, X.: Faster R-CNN based table detection combining corner locating. In: 2019 International Conference on Document Analysis and Recognition (ICDAR), pp. 1314–1319 (2019). https://doi.org/10.1109/ICDAR.2019.00212
29. Wada, K.: Labelme: image polygonal annotation with Python (2016). https://github.com/wkentaro/labelme
30. Yildiz, B., Kaiser, K., Miksch, S.: pdf2table: a method to extract table information from PDF files. In: Proceedings of the 2nd Indian International Conference on Artificial Intelligence, pp. 1773–1785 (2005)
31. Zhong, X., ShafieiBavani, E., Jimeno-Yepes, A.: Image-based table recognition: data, model, and evaluation. CoRR abs/1911.10683 (2019)
32. Zhong, X., Tang, J., Jimeno-Yepes, A.: PubLayNet: largest dataset ever for document layout analysis. CoRR abs/1908.07836 (2019)

Label Selection Algorithm Based on Boolean Interpolative Decomposition with Sequential Backward Selection for Multi-label Classification

Tianqi Ji, Jun Li, and Jianhua Xu[✉]

School of Computer and Electronic Information, School of Artificial Intelligence,
Nanjing Normal University, Nanjing 210023, Jiangsu, China
182202009@stu.njnu.edu.cn, {lijuncst,xujianhua}@njnu.edu.cn

Abstract. In multi-label classification, an instance may be associated with multiple labels simultaneously and thus the class labels are correlated rather than exclusive one another. As various applications emerge, besides large instance size and high feature dimensionality, the dimensionality of label space also grows quickly, which would increase computational costs and even deteriorate classification performance. To this end, dimensionality reduction strategy is applied to label space via exploiting label correlation information, resulting in label embedding and label selection techniques. Compared with a lot of label embedding work, less attention has been paid to label selection research due to its difficulty. Therefore, it is a challenging task to design more effective label selection techniques for multi-label classification. Boolean matrix decomposition (BMD) finds two low-rank binary matrix Boolean multiplication to approximate the original binary matrix. Further, Boolean interpolative decomposition (BID) version specially forces the left low-rank matrix to be a column subset of original ones, which implies to choose some informative binary labels for multi-label classification. Since BID is an NP-hard problem, it is necessary to find out a more effective heuristic solution method. In this paper, after executing exact BMD which achieves an exact approximation via removing a few uninformative labels, we apply sequential backward selection (SBS) strategy to delete some less informative labels one by one, to detect a fixed-size column subset. Our work builds a novel label selection algorithm based on BID with SBS. This proposed method is experimentally verified through six benchmark data sets with more than 100 labels, according to two performance metrics (precision@n and discounted gain@n, $n = 1$, 3 and 5) for high-dimensional label situation.

Keywords: Multi-label classification · Label selection · Boolean matrix decomposition · Interpolative decomposition · Sequential backward selection.

Supported by the Natural Science Foundation of China (NSFC) under grants 62076134 and 62173186.

J. Lladós et al. (Eds.): ICDAR 2021, LNCS 12822, pp. 130–144, 2021.
https://doi.org/10.1007/978-3-030-86331-9_9

1   Introduction

Traditional supervised learning mainly deals with a single-label (binary or multi-class) classification problem where each instance only has one of predefined labels [8]. However, for some real-world applications, it is possible to assign several labels to an instance at the same time, resulting in multi-label classification paradigm [13,19], whose speciality is that the classes are correlated, rather than exclusive in single-label case. Nowadays, its main application domains cover text categorization [16], image annotation [6], bioinformatics [11] and recommendation system [31].

As many new applications occur continuously, besides traditional large instance size and high dimensional feature space, a pivotal challenge in multi-label classification is its high-dimensional label space [25], which increases training consumption and even deteriorates classification performance for traditional multi-label classifiers. In order to face this special situation, considerable research efforts have been devoted to label space dimensionality reduction (LSDR), which is mainly to exploit correlation information between labels to achieve a low-dimensional label space for reducing training costs, and even enhancing classification performance.

Via LSDR, the multi-label classification problem generally is coped with by slightly complicated training and testing procedures. Its training stage includes: (a) to design a compressed operator (or procedure) and its corresponding recovery one, (b) to reduce the dimensionality of label space by the compressed operator, and (c) to train a classifier or regressor in the low-dimensional label space. In its testing phase, a low label vector is predicted by classifier or regressor, and then is recovered back to the original high-dimensional label space via the recovery operator. Therefore, the key work in LSDR is how to design compressed and recovery operators. Generally, existing LSDR algorithms could be divided into two categories: label embedding and label selection [25,30], similarly to feature extraction and selection in feature space dimensionality reduction.

Label embedding methods reduce the binary high-dimensional label space into a real or binary low-dimensional one, which implies that the original high-dimensional classification problem is converted into a low-dimensional regression or classification problem.

The first real embedding approach (ML-CS) [14] is based on compressed sensing theory [23]. This method assumes that the label vectors are sparse, and then are converted into low-dimensional real vectors via a linear random projection matrix extremely efficiently. In its recovery stage, it is needed to solve a complicated L1 minimization problem to rebuild the high-dimensional label vectors, resulting in an extremely high computational complexity. In PLST [27], principal component analysis (PCA) is executed on binary label matrix, which induces effective and efficient compressed and recovery matrices with several top-ranked eigenvectors. In NMF [9], nonnegative matrix factorization [10] is directly applied to achieve two low-rank nonnegative matrices as low-dimensional label matrix and recovery matrix, respectively. Considering binary label matrix situation, another strategy is to obtain a low-dimensional binary label space. The representative method is MLC-BMaD [29], which is based on Boolean matrix decomposition (BMD) [2,20] to decompose the original binary matrix into two low-rank binary matrix multiplication, whose left and right low-rank matrices are regarded as low-dimensional label and recovery matrices, respectively. It is worth noting that BMD is NP-hard [26], its solution techniques are all based on proper heuristic approaches. These label embedding methods mainly utilize label data from multi-label problem only.

It is also recognized that LSDR methods could be improved via introducing feature information. Thus, canonical correlation analysis (CCA) and Hilbert-Schmidt independence criterion are used to describe relevance between label data and feature ones. In [5] HSIC is directly applied to build an eigenvalue problem to achieve both projection and recovery matrices. At first, an orthonormal CCA (OCCA) is proposed in [27], which then is combined with PCA to build conditional PLST (i.e., CPLST). On the other hand, FaIE [17] combines PCA with CCA to obtain low-dimensional label matrix directly and recovery matrix. Apart from PCA and HSIC, via adding a local covariance matrix for each instance with its several nearest neighbour instances, ML-mLV [28] defines a trace ratio optimization problem to find out projection and recovery matrices.

The aforementioned label embedding methods have many successful applications, but their major limitation is that the transformed labels would be lack of original label real-world meanings.

In order to preserve label physical meanings, label selection methods are to choose an informative label subset, so that those unselected labels can be recovered effectively. In [1], a multiple output prediction landmark selection method based on group-sparse learning is proposed, which is realized by solving an L1 and L1,2 regularized least square regression problem. However, this method needs an additional procedure to estimate conditional probabilities for Bayesian rule to build its label recovery way. In ML-CSSP [4], the label selection is regarded as a column subset selection problem (CSSP), which is NP-complete [24] and is solved via a randomized sampling method. Its advantage is that the recovery way is obtained directly and its disadvantage is that two high correlated labels are still selected at the same time. Sun et al. [26] proposed a special BMD algorithm to approximate the original matrix exactly (EBMD), where the left low-rank matrix is a column subset of original matrix and the right low-rank right matrix comes from a binary correlation matrix of original matrix. This decomposition implements so-called BCX in [21] or interpolative decomposition (ID) in [12], where the latter is regarded as a specific term used in this study for our convenient statement. In [18], this EBMD is directly applied to label selection for multi-label classification (MLC-EBMD) to remove a few uninformative labels. However, when selecting fewer labels, although this method could be slightly modified to rank those remained labels using the number of "1" components from each label and its corresponding recovery vector in descending order, its solution is not optimal in principle. This inspires us to find out a more effective method to sort those remained labels to realize a better label selection algorithm for multi-label classification.

In this study, in order to detect most informative labels of fixed size, we remove a few uninformative labels via EBMD [26] and then delete some less informative labels using sequential backward selection (SBS) strategy widely used in feature selection field [8], which builds a novel label selection algorithm based on BID with SBS. The detailed experiments on six benchmark data sets with more than 100 labels illustrate our proposed method is more effective than three state-of-the-art techniques (MLC-BMaD [29], ML-CSSP [4] and MLC-EBMD [18]) according to two metrics (precision@n and discounted gain@n, $n =$ 1, 3 and 5) for high-dimensional label applications.

Our contributions of this paper are summarized into three folds: (a) A more effective BID is realized via cascading EBMD and SBS; (b) A new label selection algorithm is proposed based on BID with SBS for multi-label classification, (c) our proposed method is validated by six benchmark data sets experimentally.

This paper is organized as follows. Section 2 provides some preliminaries for this paper. In Sects. 3 and 4, we introduce and validate our proposed label selection methods, respectively, in detail. Finally, this paper ends with some conclusions.

2 Preliminaries

Suppose there is a training multi-label data set of size N as follows

$$\{(\mathbf{x}_1, \mathbf{y}_1), \ldots, (\mathbf{x}_i, \mathbf{y}_i), \ldots (\mathbf{x}_N, \mathbf{y}_N)\} \tag{1}$$

where, for the i-th instance, the $\mathbf{x}_i \in R^D$ and $\mathbf{y}_i \in \{0,1\}^C$ denote its D-dimensional feature column vector and C-dimensional binary label column vector, and N, D and C are the numbers of instances, features and labels, respectively. Additionally, let $\mathbf{z}_i \in \{0,1\}^c$ be the low-dimensional label vector of $\mathbf{y}_i$ after label selection operation, where c is the number of selected labels ($c < C$). For the simplification of mathematical formulations, we also define three matrices: feature matrix $\mathbf{X}$, original label matrix $\mathbf{Y}$ and low-dimensional label matrix $\mathbf{Z}$, i.e.,

$$\begin{aligned}
\mathbf{X} &= [\mathbf{x}_1, ..., \mathbf{x}_i, ..., \mathbf{x}_N]^T \in R^{N \times D} \\
\mathbf{Y} &= [\mathbf{y}_1, ..., \mathbf{y}_i, ..., \mathbf{y}_N]^T = [\mathbf{y}^1, ..., \mathbf{y}^j, ..., \mathbf{y}^C] \in \{0,1\}^{N \times C} \\
\mathbf{Z} &= [\mathbf{z}_1, ..., \mathbf{z}_i, ..., \mathbf{z}_N]^T = [\mathbf{z}^1, ..., \mathbf{z}^j, ..., \mathbf{z}^c] \in \{0,1\}^{N \times c}
\end{aligned} \tag{2}$$

where the column vectors $\mathbf{y}^j$ and $\mathbf{z}^j$ indicate the j-th label in $\mathbf{Y}$ and $\mathbf{Z}$, respectively.

Boolean matrix decomposition (BMD) is to decompose the original label matrix $\mathbf{Y}$ into two low-rank binary matrix ($\mathbf{Z}$ and $\mathbf{B}$) multiplication, i.e.,

$$\mathbf{Y} \approx \mathbf{Z} \circ \mathbf{B} \tag{3}$$

where the operator "$\circ$" implies the Boolean multiplication (specially $1 + 1 = 1$), and $\mathbf{B} \in \{0,1\}^{c \times C}$ denotes a binary matrix that reflects the correlation between labels, which acts as a label recovery matrix in this study. Generally, BMD is implemented via minimizing the following approximated error

$$E(\mathbf{Y}, \mathbf{Z} \circ \mathbf{B}) = \|\mathbf{Y} - \mathbf{Z} \circ \mathbf{B}\|_0 \tag{4}$$

where the operation $\|\cdot\|_0$ represents the zero-norm to calculate the number of non-zero elements of binary matrix or vector. It is noted that BMD is an NP-hard problem stated in [26] according to Orlin's work in [22]. Therefore BMD is solved by various heuristic approaches [2].

If $\|\mathbf{Y} - \mathbf{Z} \circ \mathbf{B}\|_0 = 0$ holds true, such a decomposition (3) is named as exact BMD (i.e., EBMD) [3,26], where the smallest possible c is called as the Boolean rank.

When the matrix $\mathbf{Z}$ consists of some columns of matrix $\mathbf{Y}$, BMD (3) is specially referred to as BCX in [21]. For real matrices $\mathbf{Y}, \mathbf{Z}$ and $\mathbf{B}$, $\mathbf{Y} = \mathbf{Z}\mathbf{B}$ is regarded as interpolative decomposition (simply ID or CX [12]). In this case, BCX is replaced by Boolean interpolative decomposition (BID) in this study for our concise statement.

For multi-label classification, when BID is executed on $\mathbf{Y}$, we could obtain the label selection matrix $\mathbf{Z}$ and its corresponding recovery matrix $\mathbf{B}$. After that, we still need to deal with a multi-label classification problem with fewer labels. In this case, the training procedure includes the following steps:

(i) To input feature matrix $\mathbf{X}$ and high-dimensional label matrix $\mathbf{Y}$.
(ii) To execute BID $\mathbf{Y} \approx \mathbf{Z} \circ \mathbf{B}$ to obtain low-rank binary label matrix $\mathbf{Z}$ and binary recovery matrix $\mathbf{B}$, given a fixed c.
(iii) To train a multi-label classifier $f(\mathbf{x}) : R^D \rightarrow \{0,1\}^c$ using $\mathbf{X}$ and $\mathbf{Z}$.

In the testing stage, for a testing instance $\mathbf{x}$, we sequentially execute the following two steps:

(i) To predict the low-dimensional binary label vector $\mathbf{z}$ for $\mathbf{x}$ via multi-label classifier $\mathbf{z} = f(\mathbf{x})$.
(ii) To recover its high-dimensional label vector $\mathbf{y}$ via $\mathbf{y} = \mathbf{B}^T \circ \mathbf{z}$.

In the next section, we will propose an effective method to achieve such two low-rank binary matrices $\mathbf{Z}$ and $\mathbf{B}$ for a pre-specific c.

3   A Label Selection Algorithm Based on Boolean Interpolative Decomposition with Sequential Backward Selection

In this section, we introduce a novel label selection algorithm based on Boolean interpolative decomposition with sequential backward selection (LS-BaBID), which includes two key phases: exact Boolean matrix decomposition to remove some uninformative labels and sequential backward selection to delete some less informative labels gradually.

3.1   Uninformative Label Reduction via Exact Boolean Matrix Decomposition

To implement a heuristic Boolean interpolative decomposition technique, Sun et al. [26] proposed an exact Boolean matrix decomposition (EBMD), which applies a remove-smallest heuristic strategy to find out smallest $\bar{C}$ columns of $\mathbf{Y}$ to form $\mathbf{Z}$ for satisfying $\mathbf{Y} = \mathbf{Z} \circ \mathbf{B}$ exactly. From label selection viewpoint, this method essentially removes some uninformative labels without any contribution to the original matrix $\mathbf{Y}$. Here we concisely restate this EBMD as follows.

According to the original label matrix $\mathbf{Y}$, an entire recovery matrix $\mathbf{B}$ of size $C \times C$ is estimated

$$\mathbf{B} = [\mathbf{b}_1^T, ..., \mathbf{b}_i^T, ..., \mathbf{b}_C^T]^T = \overline{\mathbf{Y}^T \circ \overline{\mathbf{Y}}} \tag{5}$$

where "$-$" is the "NOT" logical operation and $\mathbf{b}_i \in \{0,1\}^C$ is the i-th row of $\mathbf{B}$. After let $\mathbf{Z} = \mathbf{Y}$, for each pair of $(\mathbf{z}^j, \mathbf{b}_j)$ $(j = 1, ..., C)$, its sub-matrix $\mathbf{T}_j$ of size $N \times C$ is defined as

$$\mathbf{T}_j = \mathbf{z}^j \circ \mathbf{b}_j = \mathbf{z}^j \mathbf{b}_j \tag{6}$$

and a total matrix is constructed as

$$\mathbf{T} = \mathbf{ZB} = \sum_{j=1}^{C} \mathbf{T}_j = \sum_{j=1}^{C} \mathbf{z}^j \circ \mathbf{b}_j. \tag{7}$$

To characterize the contribution of each $\mathbf{T}_j$ to $\mathbf{T}$, the number of "1" components is calculated

$$\sigma_j = \|\mathbf{T}_j\|_0 = \|\mathbf{z}^j \circ \mathbf{b}_j\|_0 = \|\mathbf{z}^j\|_0 \|\mathbf{b}_j\|_0, j = 1, ..., C. \tag{8}$$

The larger σ_j of $\mathbf{T}_j$ is, the higher its contribution to $\mathbf{T}$ is. After sorting $\sigma_j (j = 1, ..., C)$ in descending order, correspondingly, the columns in $\mathbf{Z}$ and the rows in $\mathbf{B}$ are rearranged.

Next, each label is checked sequentially to determine whether this label could be removed from $\mathbf{Z}$ or not. For the j-th label $\mathbf{z}^j$, if the following relation holds

$$\|\mathbf{T}\|_0 = \|\mathbf{T} - \mathbf{T}_j\|_0 \tag{9}$$

$\mathbf{z}^j$ and $\mathbf{b}_j$ are deleted from $\mathbf{Z}$ and $\mathbf{B}$, respectively and then let

$$\mathbf{T} = \mathbf{T} - \mathbf{T}_j. \tag{10}$$

Finally, the above procedure is summarized in Algorithm 1. In [18], this algorithm is directly used to construct a label selection method (MLC-EBMD) to remove a few uninformative labels only. In order to select c labels (i.e., $c < \bar{C}$), we detect the first c columns from $\mathbf{Z}$ and the first c rows from $\mathbf{B}$ as the final $\mathbf{Z}_{N \times c}$ and $\mathbf{B}_{c \times C}$. Such a slightly modified version is still referred to as MLC-EBMD in this study.

Algorithm 1. Uninformative label reduction via EBMD

Input: Label matrix $\mathbf{Y}_{N \times C}$

Process:

1: Calculate two matrices $\mathbf{B}$ and $\mathbf{T}$ using (5) and (7), respectively.
2: Estimate $\sigma_j (j = 1, ..., C)$ via (8) and sort them in descending order.
3: Rearrange $\mathbf{Z}$ and $\mathbf{B}$, correspondingly.
4: Let $\bar{C} = C$.
5: **for** $j = 1$ to $\bar{C}$ **do**
6:     Calculate $\mathbf{T}_j$ according to (6).
7:     If $\|\mathbf{T}\|_0 = \|\mathbf{T} - \mathbf{T}_j\|_0$ then
8:         Remove the j-th column from $\mathbf{Z}$, and the j-th row from $\mathbf{B}$.
9:         Set $\mathbf{T} = \mathbf{T} - \mathbf{T}_j, \bar{C} = \bar{C} - 1$ and $j = j - 1$.
10: **end for**

Output: Label matrix: $\mathbf{Z}_{N \times \bar{C}}$ and recovery matrix: $\mathbf{B}_{\bar{C} \times C}$.

3.2 Less Informative Label Reduction via Sequential Backward Selection

In the aforementioned Algorithm 1, the $(C - \bar{C})$ uninformative labels could be deleted. For the $\bar{C}$ remained labels (generally, $c < \bar{C} < C$), we only could obtain a simple ranking from $\sigma_j (j = 1, ..., \bar{C})$ to show their significance. In this subsection, we will choose the most informative c labels from those $\bar{C}$ pre-selected labels from Algorithm 1 via sequential backward selection (SBS) strategy [8].

The SBS is a widely-used greedy search strategy in pattern recognition, data mining and machine learning. Starting from all labels, we remove some less informative labels one by one, until the number of pre-fixed labels are reached. In this paper, we will remove $\bar{C} - c$ less informative labels via such a strategy.

Now assume that given a fixed k, its corresponding label and recovery matrices are denoted by $\mathbf{Z}_{N \times k}$ and $\mathbf{B}_{k \times C}$, from which we define a relative reconstruction error (RRE) as follows

$$\text{RRE}(\mathbf{Y}, \mathbf{Z} \circ \mathbf{B}) = \frac{\|\mathbf{Y} - \mathbf{Z} \circ \mathbf{B}\|_0}{\|\mathbf{Y}\|_0}. \tag{11}$$

When the m-th column in $\mathbf{Z}$ and m-th row of $\mathbf{B}$ are removed to obtain $\mathbf{Z}^{\bar{m}}$ and $\mathbf{B}_{\bar{m}}$, the corresponding RRE is defined as,

$$\text{RRE}(m) = \text{RRE}(\mathbf{Y}, \mathbf{Z}^{\bar{m}} \circ \mathbf{B}_{\bar{m}}) = \frac{\|\mathbf{Y} - (\mathbf{Z}^{\bar{m}} \circ \mathbf{B}_{\bar{m}})\|_0}{\|\mathbf{Y}\|_0}. \tag{12}$$

For $m = 1, ..., k$, we calculate their corresponding $RRE(m)$, and then find out an optimal m^{opt}, such that

$$m^{opt} = \text{argmin}\{\text{RRE}(m) | m = 1, ..., k\}. \tag{13}$$

Let $\mathbf{Z} = \mathbf{Z}^{\bar{m}^{opt}}$, $\mathbf{B} = \mathbf{B}^{\bar{m}^{opt}}$ and $k = k - 1$. This procedure is executed until $k = c$. We summarize such a less informative label reduction method in Algorithm 2, which finally implements a new heuristic technique for BID.

Algorithm 2. Less Informative Label Reduction via SBS

Input: Label matrix $\mathbf{Z}_{N \times \bar{C}}$, matrix $\mathbf{B}_{\bar{C} \times C}$ from **Algorithm-1**, and the number of selected labels: c.

Process:

1: Let $k = \bar{C}$.
2: **for** $i = c$ to $\bar{C}$ **do**
3:     **for** $m = 1$ to k **do**
4:         Calculate $RRE(m)$ using (12).
5:     **endfor**
6:     Find out the optimal label m^{opt} to minimize $RRE(m)(m = 1, ..., k)$.
7:     Remove the m^{opt}-th column from $\mathbf{Z}$,
8:     Delete the m^{opt}-th row from $\mathbf{B}$,
9:     Set $k = k - 1$.
10: **end for**

Output: Label matrix $\mathbf{Z}_{N \times c}$, and recovery matrix $\mathbf{B}_{c \times C}$.

Algorithm 3. Label selection algorithm: LS-BaBID

Input: label matrix $\mathbf{Y}_{N \times C}$ and the number of selected labels c.

1: Execute **Algorithm-1** to achieve two $\bar{C}$-rank matrices $\mathbf{Z}_{N \times \bar{C}}$ and $\mathbf{B}_{\bar{C} \times C}$.
2: **if** $\bar{C} > c$ **do**
3:     Execute **Algorithm-2** to obtain two c-rank matrices $\mathbf{Z}_{N \times c}$ and $\mathbf{B}_{c \times C}$.
4: **endif**

Output: Label matrix: $\mathbf{Z}_{N \times c}$ and recovery matrix: $\mathbf{B}_{c \times C}$.

3.3   Label Selection Algorithm Based on BID with SBS

According to the aforementioned sub-section work, we propose a label selection algorithm for multi-label classification, which consists of two stages. The one is uninformative label reduction phase listed in Algorithm 1, and the other is less informative label reduction process listed in Algorithm 2. Finally, our entire label selection algorithm is summarized in Algorithm 3, which is referred to as label selection algorithm based on Boolean Interpolative decomposition (BID) with sequential backward selection (SBS), simply LS-BaBID.

4   Experiments

In this section, we experimentally evaluate the proposed LS-BaBID on six benchmark multi-label data sets, via comparing it with three existing methods: MLC-EBMD [18], MLC-BMaD [29] and ML-CSSP [4].

4.1   Six Benchmark Data Sets and Two Evaluation Metrics

All of our experiments are based on six benchmark multi-label data sets (Mediamill, Bibtex, Corel16k-s2, CAL500, Bookmarks and Chess) downloaded from

Table 1. Statistics of six experimented multi-label data sets.

Dataset	Domain	#Train	#Test	#Features	#Labels	#Cardinality
Mediamill	Video	30993	12914	120	101	4.376
Bibtex	Text	4880	2515	1836	159	2.402
Corel16k-s2	Image	5241	1783	500	164	3.136
CAL500	Music	300	202	68	174	26.044
Bookmarks	Text	57985	29871	2150	208	2.028
Chess	Text	1508	168	585	227	2.418

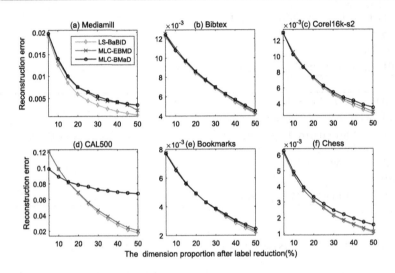

Fig. 1. Relative reconstruction error of three BMD-based methods on six data sets.

Mulan[1] and Cometa[2]. These data sets are listed in Table 1 according to the number of original labels in ascending order, whose more statistics also are provided, such as application domain, the number of instances in their training and testing subsets, the dimensionality of features and labels, and the label cardinality. It is worth noting that the number of labels is more than 100 for all six data sets and three data sets belong to text categorization applications.

The traditional multi-label classification evaluation metrics are designed for low-dimensional label space [13], which are not suitable for large-scale label one [15]. In our experiments, we utilize two new evaluation metrics [15]: precision@n, and (DisCounted Gain) DCG@$n(n = 1, 2, 3, ...)$.

---

[1] http://mulan.sourceforge.net/datasets-mlc.html.

[2] https://cometa.ml.

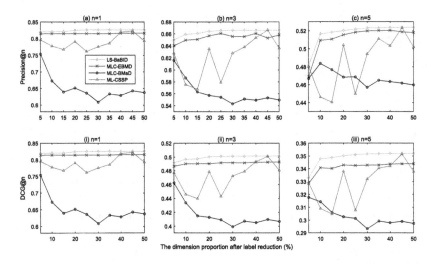

Fig. 2. Two metrics (at n = 1, 3 and 5) from four methods on Mediamill.

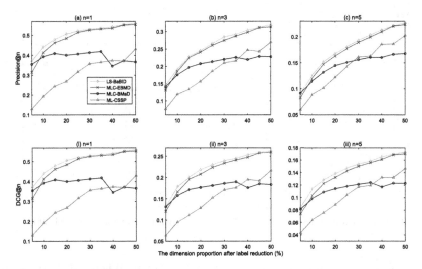

Fig. 3. Two metrics (at n = 1, 3 and 5) from four methods on Bibtex.

For a testing instance $\mathbf{x}$, its ground label vector is $\mathbf{y} = [y_1, \ldots, y_i, \ldots, y_C]^T \in \{0,1\}^C$ and predicted function values $\hat{\mathbf{y}} = [\hat{y}_1, \ldots, \hat{y}_i, \ldots, \hat{y}_C]^T \in \{0,1\}^C$, and then such two metrics are defined as follows:

$$
\begin{aligned}
\text{Precision@}n &= \frac{1}{C} \sum_{i \in rank_n(\hat{\mathbf{y}})} y_i \\
\text{DCG@}n &= \frac{1}{C} \sum_{i \in rank_n(\hat{\mathbf{y}})} \frac{y_i}{\log_2(i+1)}.
\end{aligned}
\tag{14}
$$

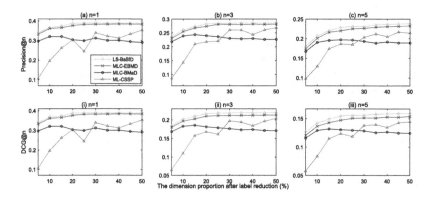

Fig. 4. Two metrics (at n = 1, 3 and 5) from four methods on Corel16k-s2.

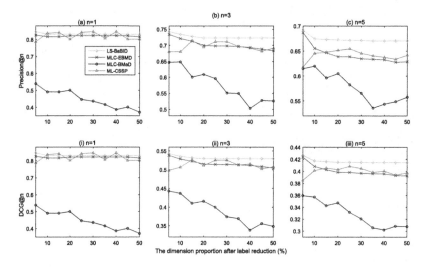

Fig. 5. Two metrics (at n = 1, 3 and 5) from four methods on CAL500.

where $rank_n(\hat{\mathbf{y}})$ returns the top n label indexes of $\hat{\mathbf{y}}$. When $n = 1$, these two metrics are the same. Finally, their average values are calculated via averaging them over all testing instances. Additionally, the higher these two metric values are, the better the label selection techniques perform.

4.2   Experimental Settings

In our experiments, we evaluate LS-BaBID and three existing techniques (MLC-EBMD, MLC-BMaD and ML-CSSP) via training versus testing mode. To this end, we choose random forest (RF) as our base classifier, in which the number of trees is set to 100. In order to investigate how the number of reduced labels

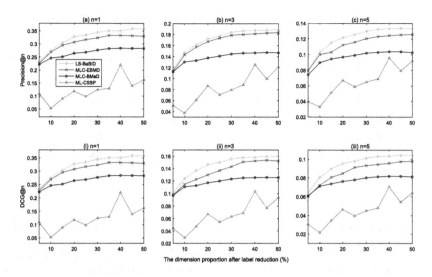

Fig. 6. Two metrics (at n = 1, 3 and 5) from four methods on Bookmarks.

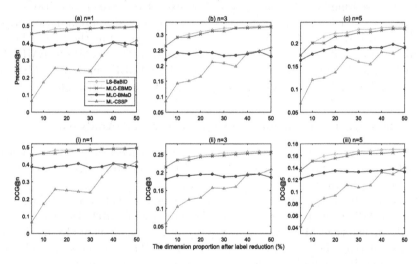

Fig. 7. Two metrics (at n = 1, 3 and 5) from four methods on Chess.

(i.e., c) would affect reconstruction error and classification performance, the dimension proposition after label reduction (i.e., c/C) is to set as 5% to 50% of the original label size (C) with the step of 5%. In particular, the threshold for constructing the label association matrix in MLC-BMaD is tuned to be 0.7 (rather than the default 0.5) for achieving a satisfactory comparison performance. For two metrics (14), we set n = 1, 3, and 5.

Table 2. The number of wins for each method and metric across six data sets.

Metric	ML-CSSP	MLC-BMaD	MLC-EBMD	LS-BaBID
Precision@1	5	0	5	**50**
Precision@3	1	0	3	**56**
Precision@5	0	0	2	**58**
DCG@1	5	0	5	**50**
DCG@3	0	0	2	**58**
DCG@5	0	0	2	**58**
Total wins	11	0	19	**330**

4.3 Reconstruction Error Analysis for Three BMD-Type Methods

At first, for three BMD-type methods (except ML-CSSP), we analyze their relative reconstruction errors defined in (11) as a function of different label dimension propositions as shown in Fig. 1, which is to illustrate the approximation ability for BMD-based methods. It is observed that, except for three propositions (5%, 10% and 15%) for CAL500, our LS-BaBID achieves lowest approximation errors for all propositions, which demonstrates that our proposed method has a better approximation ability than other two BMD-based methods (MLC-EBMD [18] and MLC-BMaD [29]). Specially, our proposed LS-BaBID indeed could select some more informative labels than MLC-EBMD does experimentally.

4.4 Classification Performance Comparison and Analysis

In this sub-section, we investigate two classification metrics (i.e., precision@n and DCG@n) as a function of the different dimension proportions after label reduction, as shown in Figs. 2, 3, 4, 5, 6 and 7.

It can be obviously found out from these six figures that two BID-type algorithms, including our proposed LS-BaBID and MLC-EBMD [18], have a better performance than ML-CSSP and MLC-BMaD, which benefits from selecting more informative labels via BID. Although ML-CSSP also belongs to label selection category, its random selection strategy affects its performance.

For the sake of evaluating four compared approaches thoroughly, the "win" index in [7] is applied in our comparison, which represents how many times each technique reaches the best metric values for all data sets and all dimension proportions of reduced labels, as shown in Table 2. ML-CSSP, MLC-BMaD and MLC-EBMD win 11, 0 and 19 times, respectively. Our LS-BaBID achieves the best values of 330 times for all metrics across six data sets, which is greatly than the number of wins from the total summation (30) of other three methods.

In short, our aforementioned detailed experiments show that our proposed method LS-BaBID is more effective experimentally, compared with three existing methods (MLC-EBMD, MLC-BMaD and ML-CSSP).

5   Conclusions

In this paper, we proposed a novel label selection algorithm for multi-label classification applications with many labels, to preserve selected label physical meanings effectively. Our method consists of two procedures, in which the one is to remove some uninformative labels via exact Boolean matrix decomposition, and the other is to delete some less informative labels one by one via sequential backward selection strategy, to achieve a fixed number of labels. This method is referred to as label selection algorithm based Boolean interpolative decomposition with sequential backward selection. Our detailed experiments demonstrate that our proposed method is superior to three existing techniques with low-dimensional binary labels.

In future, we will execute more experiments on more benchmark data sets to validate multi-label classification performance and speed up our computational procedure for our proposed method.

References

1. Balasubramanian, K., Lebanon, G.: The landmark selection method for multiple output prediction. In: ICML, pp. 283–290 (2012)
2. Belohlavek, R., Outrata, J., Trnecka, M.: Toward quality assessment of Boolean matrix factorizations. Inf. Sci. **459**, 71–85 (2018)
3. Belohlavek, R., Trnecka, M.: A new algorithm for Boolean matrix factorization which admits overcovering. Discret. Appl. Math. **249**, 36–52 (2018)
4. Bi, W., Kwok, J.: Efficient multi-label classification with many labels. In: ICML, pp. 405–413 (2013)
5. Cao, L., Xu, J.: A label compression coding approach through maximizing dependance between features and labels for multi-label classification. In: IJCNN, pp. 1–8 (2015)
6. Chen, Z.M., Wei, X.S., Wang, P., Guo, Y.: Multi-label image recognition with graph convolutional networks. In: CVPR, pp. 5177–5186 (2019)
7. Demsar, J.: Statistical comparisons of classifiers over multiple data sets. J. Mach. Learn. Res. **7**, 1–30 (2006)
8. Duda, R.O., Hart, P.E., Stork, D.G.: Pattern Classification, 2nd edn. Wiley, New York (2001)
9. Firouzi, M., Karimian, M., Baghshah, M.S.: NMF-based label space factorization for multi-label classification. In: ICMLA, pp. 297–303 (2017)
10. Gillis, N.: Nonnegative Matrix Factorization. Society for Industrial and Applied Mathematics, Philadelphia (2021)
11. Guo, Y., Chung, F.L., Li, G., Zhang, L.: Multi-label bioinformatics data classification with ensemble embedded feature selection. IEEE Access **7**, 103863–103875 (2019)
12. Halko, N., Martinsson, P., Tropp, J.A.: Finding structure with randomness: probabilistic algorithms for constructing approximate matrix decompositions. SIAM Rev. **53**(2), 217–288 (2011)
13. Herrera, F., Charte, F., Rivera, A.J., del Jesus, M.J.: Multilabel Classification Problem Analysis, Metrics and Techniques. Springer, Switzerland (2016). https://doi.org/10.1007/978-3-319-41111-8

14. Hsu, D.J., Kakade, S.M., Langford, J., Zhang, T.: Multi-label prediction via compressed sensing. In: NIPS, pp. 772–780 (2009)
15. Jain, H., Prabhu, Y., Varma, M.: Extreme multi-label loss functions for recommendation, tagging, ranking & other missing label applications. In: SIGKDD, pp. 935–944 (2016)
16. Lee, J., Yu, I., Park, J., Kim, D.W.: Memetic feature selection for multilabel text categorization using label frequency difference. Inf. Sci. **485**, 263–280 (2019)
17. Lin, Z., Ding, G., Hu, M., Wang, J.: Multi-label classification via feature-aware implicit label space encoding. In: ICML, pp. 325–333 (2014)
18. Liu, L., Tang, L.: Boolean matrix decomposition for label space dimension reduction: method, framework and applications. In: CISAT, p. 052061 (2019)
19. Liu, W., Shen, X., Wang, H., Tsang, I.W.: The emerging trends of multi-label learning. arXiv2011.11197v2 (December 2020)
20. Miettinen, P., Neumann, S.: Recent developments in Boolean matrix factorization. In: IJCAI, pp. 4922–4928 (2020)
21. Miettinen, P.: The Boolean column and column-row matrix decompositions. Data Min. Knowl. Discov. **17**(1), 39–56 (2008)
22. Orlin, J.: Contentment in graph theory: covering graphs with cliques. Indag. Math. **80**(5), 406–424 (1977)
23. Qaisar, S., Bilal, R.M., Iqbal, W., Naureen, M., Lee, S.: Compressive sensing: from theory to applications, a survey. J. Commun. Netw. **15**(5), 443–456 (2013)
24. Shitov, Y.: Column subset selection is NP-complete. Linear Algebra Appl. **610**, 52–58 (2021)
25. Siblini, W., Kuntz, P., Meyer, F.: A review on dimensionality reduction for multi-label classification. IEEE Trans. Knowl. Data Eng. **33**(3), 839–857 (2021)
26. Sun, Y., Ye, S., Sun, Y., Kameda, T.: Exact and approximate Boolean matrix decomposition with column-use condition. Int. J. Data Sci. Anal. **1**(3–4), 199–214 (2016)
27. Tai, F., Lin, H.T.: Multilabel classification with principal label space transformation. Neural Comput. **24**(9), 2508–2542 (2012)
28. Wang, X., Li, J., Xu, J.: A label embedding method for multi-label classification via exploiting local label correlations. In: Gedeon, T., Wong, K.W., Lee, M. (eds.) ICONIP 2019, Part V. CCIS, vol. 1143, pp. 168–180. Springer, Cham (2019). https://doi.org/10.1007/978-3-030-36802-9_19
29. Wicker, J., Pfahringer, B., Kramer, S.: Multi-label classification using Boolean matrix decomposition. In: SAC, pp. 179–186 (2012)
30. Xu, J., Mao, Z.H.: Multilabel feature extraction algorithm via maximizing approximated and symmetrized normalized cross-covariance operator. IEEE Trans. Cybern. **51**(7), 3510–3523 (2021). https://doi.org/10.1109/TCYB.2019.2909779
31. Zhang, D., Zhao, S., Duan, Z., Chen, J., Zhang, Y., Tang, J.: A multi-label classification method using a hierarchical and transparent representation for paper-reviewer recommendation. ACM Trans. Inf. Syst. **38**(1), 1–20 (2020)

GSSF: A Generative Sequence Similarity Function Based on a Seq2Seq Model for Clustering Online Handwritten Mathematical Answers

Huy Quang Ung[ID], Cuong Tuan Nguyen[(✉)] [ID], Hung Tuan Nguyen[ID],
and Masaki Nakagawa[ID]

Tokyo University of Agriculture and Technology, Tokyo, Japan
fx4102@go.tuat.ac.jp, nakagawa@cc.tuat.ac.jp

Abstract. Toward a computer-assisted marking for descriptive math questions, this paper presents clustering of online handwritten mathematical expressions (OnHMEs) to help human markers to mark them efficiently and reliably. We propose a generative sequence similarity function for computing a similarity score of two OnHMEs based on a sequence-to-sequence OnHME recognizer. Each OnHME is represented by a similarity-based representation (SbR) vector. The SbR matrix is inputted to the k-means algorithm for clustering OnHMEs. Experiments are conducted on an answer dataset (Dset_Mix) of 200 OnHMEs mixed of real patterns and synthesized patterns for each of 10 questions and a real online handwritten mathematical answer dataset of 122 student answers at most for each of 15 questions (NIER_CBT). The best clustering results achieved around 0.916 and 0.915 for purity, and around 0.556 and 0.702 for the marking cost on Dset_Mix and NIER_CBT, respectively. Our method currently outperforms the previous methods for clustering HMEs.

Keywords: Clustering · Handwritten mathematical expressions · Computer-assisted marking · Similarity function · Online patterns

1 Introduction

In 2020, the widespread of the SARS-CoV-2 (COVID-19) has a strong impact on education over the world, which caused almost schools and universities to be temporally closed. Many educational organizers resume the learners' studies via online platforms in response to significant demand. The adoption of online learning might continue persisting in post-pandemic, and the shift would impact the worldwide education market. In this context, the self-learning and e-testing applications would be necessary options for students in the near future.

Nowadays, touch-based and pen-based devices are widely used as learning media. Learners use them to read textbooks, study online courses and do exercise. Moreover, those devices are suitable for learners to quickly input mathematical expressions (MEs),

© Springer Nature Switzerland AG 2021
J. Lladós et al. (Eds.): ICDAR 2021, LNCS 12822, pp. 145–159, 2021.
https://doi.org/10.1007/978-3-030-86331-9_10

which seems better than using common editors such as Microsoft Equation Editor, MathType, and LaTeX.

Over the last two decades, the study of online handwritten mathematical expressions (OnHMEs) recognition has been actively carried out due to demands for its application on tablets. Competitions on recognizing OnHMEs have been ongoing under the series of CROHME with better recognition performance [1]. In this context, many e-learning interfaces based on pen-based devices have been studied and adopted in practical applications [2–4]. If learners can verify and confirm the recognition results of their answers, we can incorporate the handwritten mathematical expression (HME) recognition into self-learning and e-testing applications. Although learners have to do additional work, they can receive immediate feedback.

HME recognition can also be used for marking. Automatically marking answers by comparing the recognition result of an HME answer with the correct answer is one solution. However, there remain the following problems as mentioned in [5]. Firstly, it is complex to mark partially correct answers. Secondly, there may be several correct answers as well as some different but equivalent notations for an answer (e.g., "a + b" and "b + a"). Therefore, it is not practical to pre-define all possible cases. Thirdly, it requires examiners or examinees to confirm the automatic marking since the recognition results are not entirely perfect. However, large-scale examinations (e.g., national-wide qualification examinations) do not often provide opportunities for the examinees to confirm the marking, so examiners or someone else must confirm the marking.

Computer-assisted marking is an alternative approach for marking HME answers. Instead of recognizing, answers are clustered into groups of similar answers then marked by human markers. Ideally, when the answers in each group are the same, it takes only one action to mark all the answers for each group. It reduces the huge amount of marking efforts and reduces the marking errors. Since examinees make the final marking, their anxieties will also decrease.

Although clustering HMEs is promising for computer-assisted marking, the approach encounters two main challenges. Firstly, it is challenging to extract the features that represent the two-dimensional structure of HMEs. The bag-of-features approach [5, 6] combined low-level features such as directional features and high-level features such as bag-of-symbols, bag-of-relations and bag-of-position. These features, however, rather capture the local features instead of the global structure of HMEs. Secondly, we need a suitable similarity/dissimilarity measurement between the feature representations of two HMEs, in which the previous research utilized Euclidean distance [5, 6] or cosine distance [7]. However, Euclidean distance and p-norm seem ineffective in a high-dimensional space because of the concentration phenomenon, in which all pairwise distances among data-points seem to be very similar [8]. These approaches are also limited to measure the fixed-size features.

This paper proposes a method that utilizes a generative sequence similarity function (GSSF) and a data-driven representation for each OnHME. GSSF is formed by high-level sequential features, probability terms of the output sequence generated from a sequence-to-sequence (Seq2Seq) OnHME recognizer. The sequential features are dynamic, and they could represent the global structure of OnHME. Each OnHME is then represented by a vector of similarity scores with other OnHMEs, namely similarity-based representation

(SbR). SbR allows controlling the dimensionality of the feature space to reduce the influence of the concentration phenomenon. Finally, we input the SbR matrix into a clustering algorithm such as k-means to obtain the clusters of OnHMEs.

The rest of the paper is organized as follows. Section 2 briefly presents related research. Section 3 describes our method in detail. Section 4 focuses on problems related to the cost of clustering-based marking. Section 5 presents our experiments for evaluating the proposed method. Finally, Sect. 6 concludes our work and discusses future works.

2 Related Works

This section presents previous studies of computer-assisted marking and Seq2Seq-based OnHME recognition.

2.1 Computer-Assisted Marking

There are several past studies on clustering offline (bitmap image) HMEs (OfHMEs). Khuong et al. [5] also combined low-level features and high-level features (bag-of-symbols, bag-of-relations, and bag-of-positions) to represent each OfHME. However, those high-level features are formed by recognizing offline isolated symbols from connected components in an OfHME along with pre-defined heuristic rules. Hence, there are problems related to segmentation and determination of spatial relationships. Nguyen et al. [7] presented features based on hierarchical spatial classification using a convolutional neural network (CNN). Their CNN model is trained to localize and classify symbols in each OfHME using weakly supervised learning. Then, spatial pooling is applied to extract those features from the class activation maps of the CNN model. The authors also proposed a pairwise dissimilarity representation for each OfHME by computing the cosine distance between two OfHMEs.

For essay assessment [9–11] and handwritten essay scoring [12], extensive research has been conducted. Basu et al. also proposed a method for clustering English short answers [13]. They formed a similarity metric by computing a distance between two answers using logistic regression. Then, they utilized a modified k-Medoids and a latent Dirichlet allocation algorithm for forming clusters of answers. As an extended work of [13], Brooks et al. designed a cluster-based interface and demonstrated its effectiveness [14].

There are deep neural network-based methods for clustering sequential data such as OnHMEs. Several methods aim to embed sequential data into feature vectors based on the reconstruction loss and the clustering loss [24, 25]. Another approach is to compute the pairwise similarity/dissimilarity instead of embedded features [26]. However, those methods without information about symbols and relations encounter difficulty for clustering OnHMEs since there are infinite combinations of symbols and relations to form mathematical expression (MEs). Nguyen et al. [7] showed that metric learning methods do not work well compared to CNN-based spatial classification features for clustering offline HMEs.

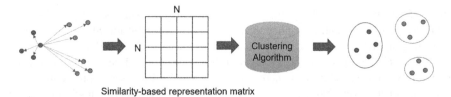

Fig. 1. Clustering process of our proposed method

2.2 Seq2seq OnHME Recognizer

Seq2Seq models for recognizing OnHMEs are deep learning models that directly convert an input sequence into an output sequence. A Seq2Seq model consists of two main components, an encoder and a decoder. The encoder, an LSTM or a BLSTM network, accepts a time series input of arbitrary length and encodes information from the input into a hidden state vector. The decoder, commonly an LSTM network, generates a sequence corresponding to the input sequence.

Zhang et al. [15] proposed a track, attend and parse (TAP) architecture, which parses an OnHME into a LaTeX sequence by tracking a sequence of input points. The encoder or the tracker stacks several layers of bidirectional Gated Recurrent Units (GRUs) to get the high-level representation. The decoder or the parser is composed of unidirectional GRUs combined with a hybrid attention mechanism and a GRU-based language model. The hybrid attention consists of two attentions: coverage based spatial attention and temporal attention. Hong et al. [16] improved TAP by adding residual connections in the encoder to strengthen the feature extraction and jointly using a transition probability matrix in the decoder to learn the long-term dependencies of mathematical symbols.

3 Proposed Method

Cluster analysis is useful for exploratory analysis by partitioning unlabeled samples into meaningful groups. For this problem, we traditionally extract useful features from each sample and pass them to a clustering algorithm such as k-means to partition them into groups. In this paper, we utilize a type of data-driven representation for each sample. We represent each sample by pairwise similarities between it and other samples. Then, we form a similarity-based representation (SbR) matrix. The SbR matrix is inputted to a clustering algorithm to obtain clusters of OnHMEs. The overall process of our proposed method is shown in Fig. 1. This section presents our proposed similarity function (SF), then describes the SbR.

3.1 Generative Sequence Similarity Function

Our proposed SF gives a similarity score between two OnHMEs based on a Seq2Seq OnHME recognizer. We expect that the similarity score of two OnHMEs containing the same ME is significantly higher than those with different MEs.

A standard Seq2Seq model, as shown in Fig. 2, consists of two parts: an encoder that receives an input sequence to represent high-level features and a decoder that sequentially generates an output sequence from the encoded features and the previous prediction. Given two input OnHMEs denoted as S_1 and S_2, the recognizer generates LaTeX sequences of $\{y_i^1\}_{i=\overline{1,N}}$ and $\{y_j^2\}_{j=\overline{1,M}}$, where y_i^1 and y_j^2 are symbol classes in the vocabulary, and N and M are the lengths of the output sequences.

A simple idea to form the similarity score of two OnHMEs is to calculate the edit distance of the two output sequences $\{y_i^1\}_i$ and $\{y_j^2\}_j$. However, this method might not be effective since the edit distance only utilizes the differences in terms of recognized symbol classes, but the probabilities of recognized symbols seem more important than the symbol classes. Our proposed SF utilizes terms of probabilities of recognized symbol classes instead of the symbol classes.

Another difficulty of directly comparing two output sequences or generated probabilities is that their lengths are variant. The proposed SF uses the symbol predictions of an OnHME to input into the decoder of another OnHME for computing the terms of probabilities. Those probabilities are formed on the output sequence of one of those two OnHMEs. Hence, the SF is not influenced by the size-variant problem.

Our SF consists of two main components: the similarity score of S_1 compared to S_2 and the one of S_2 compared to S_1 denoted as $F(S_1|S_2)$ and $F(S_2|S_1)$, respectively. We firstly define $F(S_1|S_2)$ as follows:

$$F(S_1|S_2) = \sum_{i=1}^{N}\left(\log\left(P\left(y_i^1|S_2, y_{i-1}^1\right)\right) - \log\left(P\left(y_i^1|S_1, y_{i-1}^1\right)\right)\right) \quad (1)$$

where $P(x|y, z)$ is the probability of x given y and z. An illustration of computing $F(S_1|S_2)$ is shown in Fig. 3. The predicted symbol y_i^1 at the i-th time step is inputted to the $(i + 1)$-th time step of the S_2 decoder for computing the probability $P\left(y_i^1|S_2, y_{i-1}^1\right)$. Similarly, we define $F(S_2|S_1)$ as follows:

$$F(S_2|S_1) = \sum_{j=1}^{M}\left(\log\left(P\left(y_j^2|S_1, y_{j-1}^2\right)\right) - \log\left(P\left(y_j^2|S_2, y_{j-1}^2\right)\right)\right) \quad (2)$$

The F function is not appropriate for the clustering algorithms such as k-means because it is asymmetrical. Thus, we compute an average of $F(S_1|S_2)$ and $F(S_2|S_1)$ that is symmetrical measurement. We name it as the Generative Sequence Similarity Function (GSSF), which is computed as follows:

$$\text{GSSF}(S_1, S_2) = \frac{F(S_1|S_2) + F(S_2|S_1)}{2} \quad (3)$$

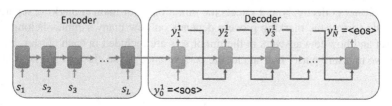

Fig. 2. A standard Seq2Seq model

Assume that the Seq2Seq recognizer are well recognized S_1 and S_2. GSSF has some properties as follows:

- GSSF(S_1, S_1) equals to zero if and only if $F(S_1|S_1)$ equals to zero.
- GSSF(S_1, S_2) is approximately zero if S_1 and S_2 denote the same ME. In this case, $F(S_1|S_2)$ and $F(S_2|S_1)$ are both around zero.
- GSSF(S_1, S_2) is negative if S_1 and S_2 denote two different MEs. In this case, both $F(S_1|S_2)$ and $F(S_2|S_1)$ are much lower than zero.
- GSSF is a symmetric function.

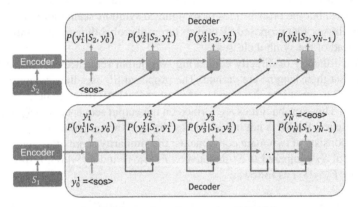

Fig. 3. Illustration of computing $F(S_1|S_2)$.

3.2 Similarity-Based Representation

Given N OnHMEs $\{X_1, X_2, \ldots, X_N\}$, SbR of X_i is formed by a pre-defined pairwise SF:

$$SbR(X_i) = [SF(X_i, X_1), \ldots, SF(X_i, X_i), \ldots, SF(X_i, X_N)] \tag{4}$$

The values of $SbR(X_i)$ is normalized into $[0, 1]$ before inputting to the clustering algorithm.

4 Clustering-Based Measurement

In clustering-based marking systems, a human marker marks the major set of answers for each cluster collectively and selects the minor ones for manual marking separately. Hence, the cost of the marking process depends on how many samples belong to the major set and how few answers in the minor sets are included in each cluster. For this reason, we measure purity to evaluate the performance of the clustering task as shown in Eq. (5):

$$Purity(G, C) = \frac{1}{H} \sum_{k=1}^{K} \max_{1 \leq i \leq J} |g_k \cap c_i| \tag{5}$$

where H is the number of samples, J is the number of categories (the right number of clusters), $C = \{c_1, c_2, \ldots, c_J\}$ is a set of categories, K is the number of clusters, and $G = \{g_1, g_2, \ldots, g_K\}$ is a set of obtained clusters.

However, high purity is easy to achieve when the number of clusters is large. For example, when K is equal to H, we obtain a perfect purity of 1.0. Hence, we set the number of clusters as the number of categories to evaluate in our experiments.

The purity alone does not show the quality of clustering in the clustering-based marking systems. We employ a cost function presented in Khuong et al. [5], reflecting a scenario of verifying and marking answers in the clustering-based marking systems. For each cluster, the verifying task is to find a major set of answers by filtering minor answers, while the marking task is to compare the major set and minor answers with the correct answer or the partially correct answers. The marking cost (MC) is composed of the verifying time C_{ver} and the marking time C_{mark} as shown in Eq. (6):

$$f(G, C) = \sum_{i=1}^{K} cost(g_i, C) = \sum_{i=1}^{K} (C_{ver}(g_i, C) + C_{mark}(g_i, C))$$

$$= \sum_{i=1}^{K} (|g_i| \times \alpha T + (1 + |g_i| - |M_i|) \times T) \qquad (6)$$

where T is the time unit to mark an answer. There exists a real number $\alpha (0 < \alpha \leq 1)$ so that the verifying cost of an answer is αT. $C_{ver}(g_i, C) = |g_i| \times \alpha T$ is the cost of verifying all answers in the cluster g_i. $C_{mark}(g_i, C) = (1 + |g_i| - |M_i|) \times T$ is the cost of marking the major set of answers M_i and all minor answers in the cluster g_i. For simplicity, we assume $\alpha = 1$, implying that the verification time is the same as the marking time T. We normalize Eq. (6) into [0, 1], and obtain Eq. (7) as follows:

$$MC(G, C) = \frac{f(G, C)}{2NT} = \frac{K}{2H} + \left(1 - \frac{1}{2}Purity(G, C)\right) \qquad (7)$$

MC equals 1 in the worst case if the number of clusters equals the number of answers. It implies that MC approaches the cost of marking all answers one-by-one.

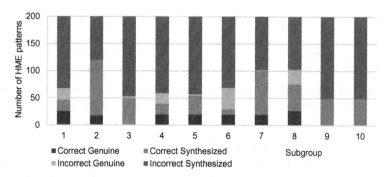

Fig. 4. Details of each subgroup in Dset_Mix.

$$\frac{tan\alpha - tan\beta}{1 + tan\alpha\cdot tan\beta} \qquad \frac{tan\alpha - tan\beta}{1 + tan\alpha\cdot tan\beta} \qquad \frac{1 + tan\alpha}{1 - tan\beta} \qquad \frac{1 + tan\beta}{1 - tan\alpha} \qquad \frac{1}{1} + \frac{tan\beta}{1 + tan\alpha} \qquad \frac{2 tan\alpha}{1 - tan^2\alpha}$$

Correct answer (genuine OnH-ME patterns) | Correct answer (synthesized) | Incorrect answer (synthesized) | Incorrect answer (synthesized) | Incorrect answer (synthesized) | Incorrect answer (synthesized)

Fig. 5. Samples in subgroup 8 of Dset_Mix.

5   Experiments

This section presents the evaluation of our proposed SF on two answer datasets, i.e., Dset_Mix and NIER_CBT, by using the TAP recognizer proposed in [15] without the language model. We name this modified recognizer as MTAP.

5.1   Datasets

We use two datasets of OnHMEs to evaluate our proposed method. The first dataset, named Dset_Mix [17], consists of mixed patterns of real (genuine) OnHMEs from CROHME 2016 and synthesized patterns. The synthesized patterns are generated according to LaTeX sequences and isolated handwritten symbol patterns from CROHME 2016. Dset_Mix consists of ten subgroups, corresponding to ten questions. Each subgroup consists of 200 OnHMEs, a mixture of genuine patterns and synthesized patterns. The synthesized OnHMEs were generated according to the method proposed in [18]. The sample size for each question is set according to the number of students in each grade of common schools. The details of each subgroup are shown in Fig. 4. Each subgroup contains a few correct answers and several incorrect answers. Figure 5 shows some samples in subgroup 8 of Dset_Mix.

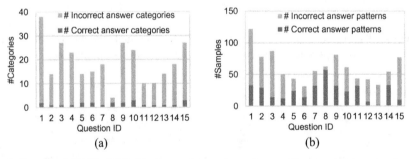

Fig. 6. Details of NIER_CBT. (a) shows the number of correct\incorrect categories in each question, and (b) shows the number of correct\incorrect patterns in each question.

The second dataset, named NIER_CBT, is a real answer dataset collected by a collaboration with National Institute for Educational Policy Research (NIER) in Tokyo, Japan. NIER carried out math tests for 256 participants, consisting of 249 students of grade 11 and 7 students of grade 7 at nine high schools. The participants answered a set of 5 questions within 50 min, then wrote their results on iPad by using an Apple pen and a developed tool with OnHMEs captured. The details of the collection are presented in [19]. There are three sets of questions to obtain 934 answers for 15 questions. Since our OnHME recognizer was trained for 101 common math symbols that appeared in the dataset of CROHME [1], we removed 15 OnHMEs that contain out-of-vocab symbols. The number of correct/incorrect answer categories and the number of correct/incorrect answer patterns in NIER_CBT are shown in Fig. 6.

5.2 OnHME Recognizer

The overview of MTAP is shown in Fig. 7. It includes three main parts: the feature extraction, the encoder, and the decoder.

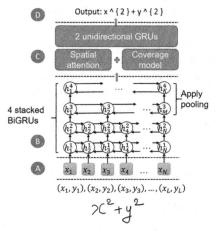

Fig. 7. Overview of TAP consisting of point-based features as the input (A), the encoder part (B), the decoder part (C), and the output (D).

Trajectory Feature Extraction. We utilized the set of point-based features used in [15]. An OnHME is a sequence of trajectory points of pen-tip movements. We denote the sequence of L points as $\{X_1, X_2, X_3, \ldots, X_L\}$ with $X_i = (x_i, x_i, s_i)$ where (x_i, y_i) are the coordination of each point and s_i is the corresponding stroke index of the i-th point. We store the sequence $\{X_i\}$ in the order of writing process. Before extracting the features, we firstly interpolate and normalize the original coordinates accoding to [20]. They are necessary to deal with non-uniform sampling in terms of writing speed and the size variations of the coordinate by using different devices to collect patterns. For each point, we extract an 8-dimensional feature vector as follows:

$$\left[x_i, y_i, dx_i, dy_i, d'x_i, d'y_i, \delta(s_i = s_{i+1}), \delta(s_i \neq s_{i+1})\right] \tag{8}$$

where $dx_i = x_{i+1} - x_i, dy_i = y_{i+1} - y_i, d'x_i = x_{i+2} - x_i, d'y_i = y_{i+2} - y_i$ and $\delta(\cdot) = 1$ when the conditional expression is true or otherwise zero, which presents the state of the pen (down/up).

Encoder. The encoder of MTAP is a combination of 4 stacked bidirectional GRUs [21] (BiGRUs) with a pooling operator, as shown in Fig. 7. Stacking multiple BiGRU layers could make the model learn high-level representation from the input. The input sequence of an upper BiGRU layer is the sequence of the hidden state of its lower BiGRU layer. Each layer has 250 forward and 250 backward units. Since the encoded features of two adjacent points are slightly different, the pooling layers are applied to reduce the complexity of the model and make the decoder part easier to parse with a fewer number of hidden states of the encoder. The pooling operator applied on the 2 top BiGRUs layers is to drop the even time steps of the lower BiGRU layer outputs and receive the odd outputs as the inputs. The hidden states outputted from the 4th layer are inputted to the decoder.

Decoder. The MTAP decoder receives the hidden states $\{h_i\}_{i=\overline{1,K}}$ from the encoder and generates a corresponding LaTeX sequence of the input traces. The decoder consists of a word embedding layer of 256 dimensions, two layers of unidirectional GRU with 256 forward units, spatial attention, and a coverage model. The spatial attention points out the suitable local region in $\{h_i\}$ to attend for generating the next LaTeX symbol by assigning higher weights to a corresponding local annotation vector $\{a_i\}_{i=\overline{1,K}}$. The coverage model is a 1-dimensional convolutional layer to indicate whether a local region in $\{h_i\}$ has been attended to the generation. The model is trained with an attention guider using the oracle alignment information from the training OnHMEs to force the attention mechanism to learn well.

Table 1. ExpRate and CER of MTAP.

Dataset	MTAP		TAP [15]	
	ExpRate (%)	CER (%)	ExpRate (%)	CER (%)
CROHME 2014 testing set	48.88	14.54	50.41	13.39
Dset_Mix	34.62	17.51	–	–
NIER_CBT	57.89	18.57	–	–

Training and Testing. We trained MTAP by the training data of CROHME 2016. We removed genuine OnHME patterns in Dset_Mix for fair evaluation, because they are in the training data set. The optimizer and hyperparameters are the same as in [15]. Then, we measured the expression rate (ExpRate) and the character error rate (CER) on the testing data of CROHME 2014, Dset_Mix, and NIER_CBT, as shown in Table 1. The recognition rate of MTAP is 1.53% points lower than the original TAP model on the CROHME 2014 testing set.

5.3 Experimental Settings

We utilized the k-means algorithm and the complete linkage (CL) method for the clustering task. For k-means, we applied the Euclidean distance and initialized centroids using k-means++ [22], a popular variant of the k-means algorithm that tries to spread out initial centroids. To evaluate the proposed features, we set the number of clusters as the number of categories in our experiments.

5.4 Evaluation

In this section, we compare the proposed method with the previous methods. Moreover, we conduct experiments to evaluate our proposed SF.

Comparison with Other Methods. So far, clustering OnHMEs is not shown on the common dataset. We re-implemented the online bag-of-features proposed by Ung et al. [6] denoted as *M1*) to compare with our proposed method. *M1* used an OnHME recognizer outputting a symbol relation tree instead of a LaTeX sequence, so that we used the OnHME recognizer proposed in [23] combined with an n-gram language model to extract the online bag-of-features. This recognizer achieves the expression rate of 51.70% on the CROHME 2014 testing set, which is 2.82% points better than MTAP. In addition, we compared with the method using the edit distance, which is to compute the dissimilarity between two LaTeX sequences outputted from MTAP, denoted as *M2*. SbRs produced by the edit distance are inputted to the k-means algorithm.

We carried out several experiments to evaluate the effectiveness of GSSF and SbR on the representation. Firstly, we directly used the absolute of GSSF as the distance function to input into CL, denoted as *M3*. Secondly, we used the SbR matrix produced by MTAP and GSSF to input into CL by using the Euclidean distance, denoted as *M4*. Thirdly, we used the SbR matrix produced by MTAP and GSSF to input into the k-means algorithm, denoted as *M5*.

We also compared our proposed method with previous methods for clustering OfHMEs, which consists of the offline bag-of-features proposed by Khuong et al. [5] (denoted as M6) and the CNN-based features proposed by Nguyen et al. [7] (denoted as M7). OfHMEs are converted from OnHMEs. We used a symbol classifier to extract the offline bag-of-features in M6. We also trained a CNN model to extract spatial classification features for M7. Those models in M1, M6, and M7 were trained in the same dataset with MTAP.

Table 2 shows that our proposed GSSF combined with MTAP, i.e., M3, M4, and M5, outperforms M1, M2, and M7 in purity and MC on both Dset_Mix and NIER_CBT. M3 and M4 have lower performance than M6 on Dset_Mix. M5 yields the best performance on Dset_Mix, while M3 performs best on NIER_CBT. However, M5 achieves a high purity on both datasets. Regarding MC, M5 achieves the marking cost of around 0.556 and 0.702 in Dset_Mix and NIER_CBT. Consequently, the marking cost is reduced by 0.444 and 0.298 than manual marking.

There are some discussions based on the results of M3, M4, and M5. Firstly, our GSSF without SbR works well when using the CL method on NIER_CBT. Secondly, GSSF combined with SbR achieves more stable performance when using the k-means algorithm than the CL method with the Euclidean distance.

Evaluations on Our Similarity Function. We conducted experiments on forming our proposed GSSF. According to Eq. (3), GSSF is formed by taking an average of two

Table 2. Comparisons with other methods of clustering HMEs. Values are presented in form of "average value (standard deviation)"

HME type	Name	Features	Clustering algorithm	Dset_Mix		NIER_CBT	
				Purity	*MC*	Purity	*MC*
OnHME	Ung et al. (*M1*)	Online bag-of-features	*k*-means	0.764 (0.11)	0.629 (0.09)	0.895 (0.05)	0.712 (0.06)
	Ours (*M2*)	MTAP + Edit distance + SbR	*k*-means	0.638 (0.12)	0.700 (0.05)	0.867 (0.05)	0.725 (0.07)
	Ours (*M3*)	MTAP + GSSF	CL (GSSF)	0.806 (0.10)	0.611 (0.05)	**0.921 (0.05)**	0.698 (0.06)
	Ours (*M4*)	MTAP + GSSF + SbR	CL (Euclidean)	0.775 (0.15)	0.633 (0.07)	**0.920 (0.04)**	0.700 (0.06)
	Ours (*M5*)	**MTAP + GSSF + SbR**	*k*-means	**0.916 (0.05)**	**0.556 (0.03)**	**0.915 (0.03)**	**0.702 (0.07)**
OfHME	Khuong et al. (*M6*)	Offline bag-of-features	*k*-means	0.841 (0.15)	0.595 (0.07)	0.834 (0.07)	0.739 (0.08)
	Nguyen et al. (*M7*)	CNN-based features	*k*-means	0.723 (0.16)	0.653 (0.08)	0.829 (0.06)	0.744 (0.06)

Table 3. Comparisons with other variant of SFs

Method	Purity	
	Dset_Mix	NIER_CBT
Asymmetric_GSSF	0.857 ± 0.07	0.907 ± 0.05
Min_GSSF	0.861 ± 0.08	0.907 ± 0.04
Max_GSSF	0.909 ± 0.08	0.913 ± 0.04
GSSF	**0.916 ± 0.05**	**0.915 ± 0.03**

similarity components $F(S_1|S_2)$ and $F(S_2|S_1)$ since we aim to make sf symmetric. We compare GSSF with Three possible variants as follows:

Evaluations on our Similarity Function. We conducted experiments on forming our proposed GSSF. According to Eq. (3), GSSF is formed by taking an average of two similarity components $F(S_1|S_2)$ and $F(S_2|S_1)$ since we aim to make SF symmetric. We compare GSSF with three possible variants as follows:

- We directly use function $F(x|y)$ as SF so that SbR of X_i in Eq. (4) is as $[F(X_i|X_1), \ldots, F(X_i|X_i), \ldots, F(X_i|X_N)]$. Since F is not a symmetric function, we denote this SF as Asymmetric_GSSF.

- We define two SFs by getting the minimum and maximum value between $F(S_1|S_2)$ and $F(S_2|S_1)$ instead of taking the average of them. We denote them as Min_GSSF and Max_GSSF, respectively.

Table 3 presents the performance of our proposed SF with Asymmetric_GSSF, Min_GSSF, and Max_GSSF in terms of purity by using the k-means algorithm. Our GSSF performs better than the other SFs on Dset_Mix and NIER_CBT, which implies that taking the average of two similarity components is better than using them directly or taking the minimum or maximum value between them. However, Max_GSSF yields comparable results with GSSF.

5.5 Visualizing Similarity-Based Representation Matrix

This section shows the visualization of the SbR matrix to see how this representation discriminates for clustering. Figure 8 presents the SbR matrix of the subgroup 3 and 8 in Dset_Mix. OnHMEs belonging to the same class are placed together in both dimensions. For subgroup 3, its categories are significantly distinct. We can see that SbR well represents for OnHMEs in the same category. The similarity scores among intra-category OnHMEs almost near 0, and they are much higher than those among inter-category OnHMEs. On the other hand, some categories in subgroup 8 are slightly different, such as categories C_3 and C_5 or category C_4 and C_6, and SbR among them is not so different. Purity on subgroups 3 and 8 are 0.984 and 0.832, respectively.

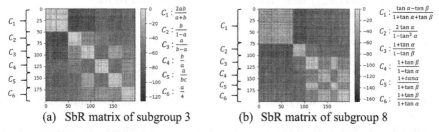

(a) SbR matrix of subgroup 3   (b) SbR matrix of subgroup 8

Fig. 8. Visualization of the SbR matrix of the subgroup 3 and 8 before normalizing them into [0, 1].

6   Conclusion and Future Works

This paper presented a similarity-based representation (SbR) and a generative sequence similarity function (GSSF) based on the Seq2Seq recognizer for clustering to provide computer-assisted marking for handwritten mathematics answers OnHMEs. The SbR matrix is then inputted to the k-means algorithm by setting the number of clusters as the number of categories. We achieved around 0.916 and 0.915 for purity and around

0.556 and 0.702 for the marking cost on the two answer datasets, Dset_Mix and NIER, respectively. Our method outperforms other methods on clustering HMEs.

There are several remaining works. We need to study the method to estimate the correct number of clusters and consider mini-batch clustering for larger answer sets. User interface for markers is also another problem to study.

Acknowledgement. This research is being partially supported by the grant-in-aid for scientific research (A) 19H01117 and that for Early Career Research 18K18068.

References

1. Mahdavi, M., Zanibbi, R., Mouchere, H., Viard-Gaudin, C., Garain, U.: CROHME+TFD: competition on recognition of handwritten mathematical expressions and typeset formula detection. In: Proceedings of International Conference on Document Analysis and Recognition, pp. 1533–1538 (2019)
2. LaViola, J.J., Zeleznik, R.C.: MathPad2: a system for the creation and exploration of mathematical sketches. ACM Trans. Graph. **23**, 432–440 (2004)
3. Chan, K.F., Yeung, D.Y.: PenCalc: a novel application of on-line mathematical expression recognition technology. In: Proceedings of International Conference on Document Analysis and Recognition, pp. 774–778 (2001)
4. O'Connell, T., Li, C., Miller, T.S., Zeleznik, R.C., LaViola, J.J.: A usability evaluation of AlgoSketch: a pen-based application for mathematics. In: Proceedings of Eurographics Symposium on Sketch-Based Interfaces Model, pp. 149–157 (2009)
5. Khuong, V.T.M., Phan, K.M., Ung, H.Q., Nguyen, C.T., Nakagawa, M.: Clustering of handwritten mathematical expressions for computer-assisted marking. IEICE Trans. Inf. Syst. **E104.D**, 275–284 (2021). https://doi.org/10.1587/transinf.2020EDP7087
6. Ung, H.Q., Khuong, V.T.M., Le, A.D., Nguyen, C.T., Nakagawa, M.: Bag-of-features for clustering online handwritten mathematical expressions. In: Proceedings of International Conference on Pattern Recognition and Artificial Intelligence, pp. 127–132 (2018)
7. Nguyen, C.T., Khuong, V.T.M., Nguyen, H.T., Nakagawa, M.: CNN based spatial classification features for clustering offline handwritten mathematical expressions. Pattern Recognit. Lett. **131**, 113–120 (2020)
8. François, D., Wertz, V., Verieysen, M.: The concentration of fractional distances. IEEE Trans. Knowl. Data Eng. **19**, 873–886 (2007)
9. Cummins, R., Zhang, M., Briscoe, T.: Constrained multi-task learning for automated essay scoring. In: Proceedings of Annual Meeting Association and Computing Linguistics, pp. 789–799 (2016)
10. Salvatore, V., Francesca, N., Alessandro, C.: An Overview of current research on automated essay grading. J. Inf. Technol. Educ. Res. **2**, 319–330 (2003)
11. Ishioka, T., Kameda, M.: Automated Japanese essay scoring system: jess. In: Proceedings of International Workshop Database Expert Systema and Applications, pp. 4–8 (2004)
12. Srihari, S., Collins, J., Srihari, R., Srinivasan, H., Shetty, S., Brutt-Griffler, J.: Automatic scoring of short handwritten essays in reading comprehension tests. Artif. Intell. **172**, 300–324 (2008)
13. Basu, S., Jacobs, C., Vanderwende, L.: Powergrading: a clustering approach to amplify human effort for short answer grading. Trans. Assoc. Comput. Linguist. **1**, 391–402 (2013)
14. Brooks, M., Basu, S., Jacobs, C., Vanderwende, L.: Divide and correct: using clusters to grade short answers at scale. In: Proceedings of ACM Conference on Learning @ Scale, pp. 89–98 (2014)

15. Zhang, J., Du, J., Dai, L.: Track, attend, and parse (TAP): an end-to-end framework for online handwritten mathematical expression recognition. IEEE Trans. Multimed. **21**, 221–233 (2019)
16. Hong, Z., You, N., Tan, J., Bi, N.: Residual BiRNN based Seq2Seq model with transition probability matrix for online handwritten mathematical expression recognition. In: Proceedings of the International Conference on Document Analysis and Recognition, ICDAR, pp. 635–640 (2019). https://doi.org/10.1109/ICDAR.2019.00107
17. Khuong, V.T.M.: A Synthetic Dataset for Clustering Handwritten Math Expression TUAT (Dset_Mix) - TC-11. http://tc11.cvc.uab.es/datasets/Dset_Mix_1
18. Phan, K.M., Khuong, V.T.M., Ung, H.Q., Nakagawa, M.: Generating synthetic handwritten mathematical expressions from a LaTeX sequence or a MathML script. In: Proceedings of International Conference on Document Analysis and Recognition, pp. 922–927 (2020)
19. Yasuno, F., Nishimura, K., Negami, S., Namikawa, Y.: Development of mathematics items with dynamic objects for computer-based testing using tablet PC. Int. J. Technol. Math. Educ. **26**, 131–137 (2019)
20. Zhang, X.Y., Yin, F., Zhang, Y.M., Liu, C.L., Bengio, Y.: Drawing and recognizing Chinese characters with recurrent neural network. IEEE Trans. Pattern Anal. Mach. Intell. **40**, 849–862 (2018)
21. Cho, K., et al.: Learning phrase representations using RNN encoder-decoder for statistical machine translation. In: Conference on Empirical Methods in Natural Language Processing 2014, pp. 1724–1734 (2014).https://doi.org/10.3115/v1/d14-1179
22. Arthur, D., Vassilvitskii, S.: k-means++: the advantages of careful seeding. In: Proceedings of Annual ACM-SIAM Symposium on Discrete Algorithms, pp. 1027–1035 (2007)
23. Nguyen, C.T., Truong, T.N., Ung, H.Q., Nakagawa, M.: Online handwritten mathematical symbol segmentation and recognition with bidirectional context. In: Proceedings of International Conference on Frontiers Handwriting Recognition, pp. 355–360 (2020)
24. Ienco, D., Interdonato, R.: Deep multivariate time series embedding clustering via attentive-gated autoencoder. In: Lauw, H.W., Wong, R.-W., Ntoulas, A., Lim, E.-P., Ng, S.-K., Pan, S.J. (eds.) PAKDD 2020. LNCS (LNAI), vol. 12084, pp. 318–329. Springer, Cham (2020). https://doi.org/10.1007/978-3-030-47426-3_25
25. Ma, Q., Zheng, J., Li, S., Cottrell, G.W.: Learning representations for time series clustering. In: Advances in Neural Information Processing Systems, pp. 3776–3786 (2019)
26. Rao, S.J., Wang, Y., Cottrell, G.: A deep siamese neural network learns the human-perceived similarity structure of facial expressions without explicit categories. In: Proceedings of the 38th Annual Conference of the Cognitive Science Society, pp. 217–222 (2016)

C2VNet: A Deep Learning Framework Towards Comic Strip to Audio-Visual Scene Synthesis

Vaibhavi Gupta, Vinay Detani, Vivek Khokar,
and Chiranjoy Chattopadhyay$^{(\boxtimes)}$ (ORCID)

Indian Institute of Technology Jodhpur, Jodhpur, India
{gupta.33,detani.1,khokar.1,chiranjoy}@iitj.ac.in

Abstract. Advances in technology have propelled the growth of methods and methodologies that can create the desired multimedia content. "Automatic image synthesis" is one such instance that has earned immense importance among researchers. In contrast, audio-video scene synthesis, especially from document images, remains challenging and less investigated. To bridge this gap, we propose a novel framework, Comic-to-Video Network (C2VNet), which evolves panel-by-panel in a comic strip and eventually creates a full-length video (with audio) of a digitized or born-digital storybook. This step-by-step video synthesis process enables the creation of a high-resolution video. The proposed work's primary contributions are; (1) a novel end-to-end comic strip to audio-video scene synthesis framework, (2) an improved panel and text balloon segmentation technique, and (3) a dataset of a digitized comic storybook in the English language with complete annotation and binary masks of the text balloon. Qualitative and quantitative experimental results demonstrate the effectiveness of the proposed C2VNet framework for automatic audio-visual scene synthesis.

Keywords: Comic strip · Video synthesis · Deep learning · Segmentation · Multimedia

1 Introduction

Comics are loved widely across the world. People of all age groups read and appreciate them. As digitization increases globally, so are the need to digitize the cultural heritage of which comics beholds a crucial position. It gives a graphical view of society and culture. In the era, people get inclined towards the multimedia aspect of digital documents, so is the case with comics which is a mixture of graphics and texts. The comic book video we see today is being manually created by graphic and video-making tools or software. However, to the best of our knowledge, there has been no work done for the automatic audio-visual synthesis of a comic book. Therefore, we propose a framework Comic-to-Video Network (C2VNet), which evolves panel-by-panel in a comic strip and eventually creates a full-length video (with audio) from a digitized or born-digital storybook. The

© Springer Nature Switzerland AG 2021
J. Lladós et al. (Eds.): ICDAR 2021, LNCS 12822, pp. 160–175, 2021.
https://doi.org/10.1007/978-3-030-86331-9_11

problem statement that we attempt to solve in this paper is *"Given a PDF of a comic book, C2VNet automatically generates a video with enhanced multimedia features like audio, subtitles, and animation"*.

Comics consists of different complex elements like panels, speech balloons, narration boxes, characters, and texts. A story or narration emerges from the combination of these elements, which makes reading engaging. Comic research mainly started with reading digital comics on cellular phones [29]. To study the page structure, in [27], a density gradient approach was proposed. In [28], an automated electronic comic adaptation technique was proposed for reading in mobile devices and separating the connected frames. Rigaud [23] simultaneously segmented frames and text area by classifying RoI in 3 categories "noise", "text", and "frame" by contours and clustering. In [18], for manga comics, a three steps approach was proposed. Deep learning frameworks gave a boost to comic image analysis research. A deep learning model was proposed in [13] to find a connection between different panels forming coherent stories. Ogawa et al. [17] proposed a CNN-based model for highly overlapped object problems using anchor boxes.

Speech balloon and narration box segmentation is another area of research in comics analysis. In [10], a deep neural network model was proposed for this task. In [6], extracted the texts in manga comics by following a step-wise process of extracting panels, detecting speech balloons, extracting text blobs with morphological operations, and recognizing text using OCR. In another work, [19] also used OCR for extracting texts and converting text to speech for people with special needs. A morphological approach was proposed in [12].

To do the open and closed comic balloon localization, [22] used active contouring techniques. For text region extraction [7] used two deep learning techniques and classify text regions using SVM. In their further work to recognize typewritten text [24], use OCR Tesseract and ABBYY FineReader and recognize handwritten text. They trained OCR using OCRopus (an OCR system). Later, [21] compared the performances of two systems for pre-trained OCR and other segmentation-free approaches. Many deep learning models using CNN like [30] are widely used to detect text regions from digital documents or images.

Though there has been much work in the analysis of comics elements, comic audio-visual synthesis is less investigated. Comics videos that we see are manually created using graphics and video-making software. The advent of newer technology has changed the habit of viewers' experience and demand. Comics enthusiasts look for new ways to read and explore the comics world. There is the massive popularity of anime, which are Japanese animation based on Manga comics. There is some work in the field of text recognition and converting text to speech. An interactive comic book is created by Andre Bergs [9] named "Protanopia", which has animated images. It is available for android and iOS, which is created using animation software.

To eliminate the manual work, we propose a novel framework, Comic-to-Video Network (C2VNet), which evolves panel-by-panel in a comic strip and ultimately creates a full-length video of a digitized or born-digital storybook. This step-by-step video synthesis process enables the creation of a high-resolution video. The significant contributions of the paper include the following:

Table 1. Details of the number of assets in various categories in IMCDB.

Comic pages	Panels	Text files	Masks	Stories
2,303	9,212	18,424	2,303	331

1. A novel end-to-end comic strip to audio-video scene synthesis framework
2. An improved panel and text balloon segmentation technique
3. A dataset of digitized comic storybooks in the English language with complete annotation and binary masks of the text balloon is proposed.

The rest of the paper is organized as: Sect. 2 presents the details of the proposed dataset. Section 3 presents the proposed C2VNet framework. Section 4 refers to the results and discussions and the paper concludes with Sect. 5.

2   IMCDB: Proposed Indian Mythological Comic Dataset

Machine learning models require training datasets, and applying them to such diverse semistructured comics requires vast datasets. There are several publicly available datasets [1,5,11,13,14]. Indian comics have a wide variety of genres like mythology based on Puranas and The epics; history consists of comics of ancient, medieval, and modern India, Literature, which has comics of folktales, fables, legends, and lore including animal stories and humorous comics. Since there is no Indian Comic dataset available for comic analysis, we have created our dataset IMCDB (Indian Mythological Comic Database) by collecting comic books from archiving websites where these comics are digitally available. We have made ground truth annotations for each panel in pages and ground truth text files for each narration box and speech balloon within a panel. Additionally, ground truth binary masks of speech balloons and narration box for each page. The dataset is available at [2].

IMCDB consists of comics of "Amar Chitra Katha" and "Tinkle Digest". Amar Chitra Katha was founded in 1967 to visualize the Indian culture of great Indian epics, mythology, history, and folktales. Tinkle Digest also contains comics with famous characters like "Suppandi" and "Shikari shambhu", which gain popularity among all age groups.

Comics consists of different complex elements like panels, speech balloons, narration boxes, characters, and texts. A story or narration emerges from the combination of these elements, which makes reading engaging. Different comic book elements in the proposed dataset presented in Fig. 1. There are four specific elements of comic books in the dataset which are comic images, panels, speech balloon masks and text files. All the comics in IMCDB are written in the English language, so the reading order is from left to right, hence the order of panels, speech balloons, and texts. Table 1 summarizes the contents of IMCDB dataset.

Comic Images: It contains the comic book PDF pages in ".png" format.

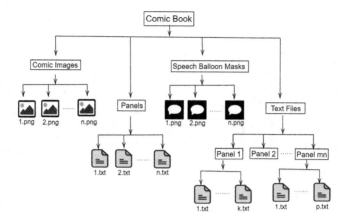

Fig. 1. An illustration of the hierarchy of different assets in IMCDB.

Panels: For ground truths of the panel, graphical image annotation tool "Labe-lImg" was utilized to create YOLOv3 compatible files. The panels description in "Yolov3" format for each comic page is present in ".txt" files.

Speech Balloon Masks: The speech balloons in IMCDB are of varying shape and size, as well as there are overlaps. We created binary masks as ground truth for all the speech balloons and narration boxes. For each comic book page there is a speech balloon masks in ".png" format.

Texts: The source text is all handwritten in uppercase with varying styles. For ground truth, we provided ".txt" files corresponding to each narration box/speech balloon within a panel of a comic book page.

3   C2VNet Architecture

This section presents the step-by-step approach followed in C2VNet, shown by Fig. 2, to develop interactive comic visuals that evolve panel-by-panel in a comic strip and ultimately creates a full-length video of a digitized or born-digital storybook. It involves four main steps; (1) Preprocessing, (2) Extraction of comic elements, (3) Video Asset creation, and (4) Video composition and animation. In the following sub-section, we describe each framework module in detail.

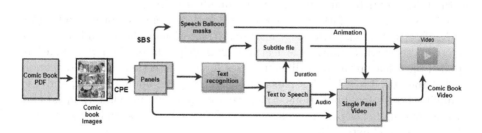

Fig. 2. A block diagram of the proposed C2VNet framework.

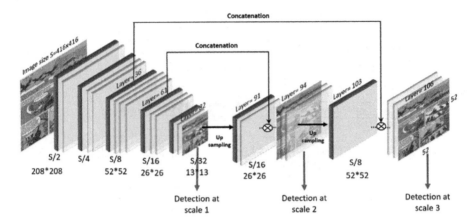

Fig. 3. An illustration of the proposed CPENet model for panel extraction.

3.1 Preprocessing

A digital comic book is composed of pages containing elements like panels, speech balloons, etc. Comic PDFs are first preprocessed by extracting each page as an image with a dpi parameter of 300. IMCDB's comic panels are rectangular. However, due to the digitization of old comics, some comics do not have correctly defined borders of the panel. Therefore, before testing, we have introduced borders to the comic page and remove noise from the images.

3.2 Extraction of Comic Elements

Comic Panel Extraction (CPE): In C2VNet, the comic panel extraction (CPE) task (see Fig. 2) is achieved by CPENet (see Fig. 3), which is adapted from YOLOv3 [20]. We first divide the input image into $S \times S$ grid cells and predicts bounding boxes for each cell. Each bounding box contains five predictions (x, y, w, h, C). Here x and y are center coordinates, while w and h are the width and height of the bounding box. All these coordinates are relative to image size. The confidence score (C), for each bounding box, shows how confident a model is to have a class label in that bounding box and is defined as $C = P(O) \times B$. Here, $P(O)$ is the probability of object present in the image, and B is defined as the IOU (predicted and ground-truth bounding box).

CPENet has two darknet-53 models stacked together to form a 106 layer CNN. The first model acts as a feature extractor, and the second acts as a detector. To extract multi-scale features at multiple layers, a 1×1 kernel is applied on a feature map of different sizes for the detection. Kernel detection is of shape $(1 \times 1 \times (B \times (5 \times class)))$, where B is the bounding box, and 5 is bounding boxe's five predictions. Since we are detecting one class (panels) and three bounding boxes per grid cell, our kernel shape becomes $= (1 \times 1 \times 18)$. We discuss the panel extraction in detail in Sect. 4.1.

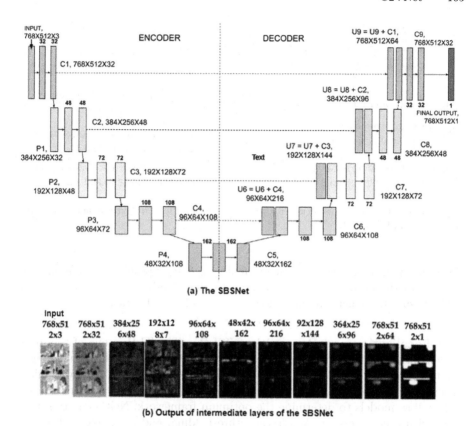

(a) The SBSNet

(b) Output of intermediate layers of the SBSNet

Fig. 4. An illustration of the proposed SBSNet and intermediate results.

Speech Balloon Segmentation (SBS): C2VNet has adapted the U-Net [25] architecture for the speech balloon segmentation (SBS) task (see Fig. 2). The architecture (SBSNet) is divided into two sub-networks (see Fig. 4a), encoding and decoding. An encoder network is used to capture the context in an image. Each block in this section applies two 3×3 convolutional layers followed by one 2×2 max-pooling down-scaling to encode the input image to extract features at multiple levels. The number of feature maps doubles in the standard UNET with a unit increase in encoding level. Therefore the architecture learns "What" information in the image, forgetting the "Where" information at the same time.

The decoder network is the second half of this architecture. We apply transposed convolution along with regular convolutions in this expansion path of the decoder network. This network aims to semantically project the discriminative feature (lower resolution) learned by the encoder onto the pixel space (higher resolution) to achieve a dense classification. The network here learns the

(a) |) WISH I COULP PO J] SOMETHING ≪++

(b) | WISH | COULP PO ISOMETHING + ANS
THING +=* WHICH WOULD MAKE EVERYONE'+'

(c) I WISH I COULD DO ISOMETHING +++
ANYTHING!" WHICH WOULD MAKE EVERYONE:

(a) | SHALL REWARP 40U HANPSOMELY
IF 4OU HELP 4S FINE HIM,

(b) I SHALL REWARD 4OU HANDSOMELY
IF 4OU CAN HELP US FINE HIM,

(c) I SHALL REWARD YOU HANDSOMELY
IF YOU CAN HELP US FINE HIM.

Fig. 5. Output of text detection by (a) Pytesseract [3], (b) Pytesseract after pre-processing, and (c) DTR.

forgotten "Where" information by gradually applying upsampling. The number of levels in the encoder network is equal to the decoder network's number of levels. Figure 4b depicts the intermediate result of the SBS task.

Text Localization and Recognition (TLR): The TLR is one of the most challenging tasks due to variations in fonts and writing styles. The comic images also become noisy and distorted due to digitization and aging. We analyze different existing models for this task. We have experimented with Pytesseract, deep and shallow models to determine the optimum framework. Neither Pytesseract (even with pre-processing like filtering, thresholding, and morphological operation) nor the shallow models yield satisfactory results. Figure 5 depicts qualitative results of the TLR task.

For TLR, we adapted a deep learning approach (henceforth referred as Deep Text Recognizer (DTR)). We applied two deep models [8, 26] as part of DTR. The CRAFT model [8] localizes the text while the Convolution recurrent neural network (CRNN) [26] model recognizes it. Text Recognition based on CRNN, which process image in 3 layers. First convolutional layer process images and extract the feature sequence. The recurrent network predicts each feature sequence; the recurrent network has bidirectional LSTM (Long short-term memory), combining forwarding and backward LSTMS. The output of LSTMs goes into the transcription network, which translates the prediction of LSTM into word labels. The output shown in Fig. 5 depicts that the proposed DTR model is performing the best as compared to the other approaches.

3.3   Video Asset Creation

Video assets are various components like audio, frame timings, transitions, and subtitles, which are essential components for creating a real video.

Text to Speech Conversion: To convert recognized text from each panel to speech (refer to Fig. 2), we have incorporated the gTTs (Google Text to Speech engine) [4] into C2VNet. The .mp3 file generated by gTTs is embedded into the single panel video as an audio source. The audio duration of each mp3 file is taken as one of the input by the subtitle file module.

Frame Duration Prediction: This module's main motive is to predict the subtitle file's frame duration when given a word count. A linear regression model is the best fit to find the relationship between these two variables. The dataset used is the English subtitle file of the movie "Avengers". Data preprocessing involves removing punctuation and symbols that are not counted as words or converted to speech. Dataset consists of 2140 instances for word count and frame duration, split into a 3:2 ratio for training and testing.

$$\text{Predicted frame duration} = \theta_1(\text{word count}) + \theta_2 \tag{1}$$

where, the parameters $\theta_1 = 1.199661$ and $\theta_2 = 0.134098$. To evaluate the model (1), error is calculated as Mean Square error which comes out 0.29488. The model, combined with text to speech module predicts frame duration for IMCDB recognised.

Subtitle File: To generate a subtitle (.srt) file of a complete comic book video as shown in Fig. 2, we have considered two inputs. The first one is the text file generated by the text recognition module of each panel. The second one

---

Algorithm 1. Animation creation algorithm

Input: P, M, t
Output: V

1: **procedure** SYNTHESZIEANIMATION(P, M, t) ▷ Animation synthesis
2:   $Contour \leftarrow$ Contour on mask   ▷ Using M
3:   Draw Contour on original panel image
4:   $A \leftarrow Area(Contour)$
5:   **if** $A > 0.02 \times M$ **then**
6:     $V_C \leftarrow Contour$ ▷ Valid Contour
7:     $S_C \leftarrow SortContour(V_C)$ ▷ Sort contours with respect to their position in the page
8:   **while** $C \leftarrow S_C$ **do**   ▷ For each contour
9:     $L \leftarrow ArcLength(C)$
10:     $X = approxPolyDp(C, L)$
11:     **if** $Length(X) > 6$ **then**
12:       $DrawContour(C, Red)$ ▷ Narration Box
13:     **else**
14:       $DrawContour(C, Yellow)$ ▷ Speech Balloon
15:   $FPS \leftarrow 10$ ▷ Video synthesis parameter
16:   $Frame_{Loop} \leftarrow t \times 5$ ▷ Video synthesis parameter
17:   $V \leftarrow CreateVideo(FFMPEG, P, C)$
18:   **return** V ▷ The Synthesized video is V

---

is the minimum of the audio duration from text to speech module and the regression prediction model. We have iteratively updated the subtitle file corresponding to each panel. The subtitle file is consisting of textual description representing the frame number, duration and the text spoken as part of narration box/speech balloon, i.e. the voice over. For each frame these information is grouped into three consecutive lines in the .srt file. If there are n panels in comic book, then $3n$ number of lines will be there in a subtitle file of that comic book video. The details of the lines are as follows: first line mentions the frame number, second line mentions duration for subtitle and the third line mentions the text spoken. The frame numbering starts from 1 and the duration is of format "hours:minutes:seconds:milliseconds" with the milliseconds are rounded to 3 decimal places.

3.4   Video Composition

Animation: To have an animation effect for every frame of the individual panel video, we loop two images, the original panel image and another is the same panel image but with the highlighted contour of speech balloons. Algorithm 1 shows the steps of applying animation to the extracted panels. The input to the algorithm are panel image from CPE (P), speech balloon mask from SBS (M) and audio duration for panel (t), and the output is the video (V) with animation effect. Line no 8 initiates the animation process by collating all the components extracted from line 2 till this point. The thresholds for valid contour selection (line no 5) and speech balloon segmentation (line no 11) are empirically determined. Figure 6 depicts the results of the proposed animation creation algorithm on two different input panels. The images on the left (Fig. 6(a)) are the extracted panels, while the right ones (Fig. 6(b)) shows the output. Different colors are used to highlight the narration box ("Red") and the speech balloon ("Yellow").

Individual Panel Video Container: To make the video of the comic, we have utilized the "FFmpeg". It is an open-source tool used for video and audio processing. This module takes the original panel, highlighted panel image, and audio duration as input and generates a video sequence of the panel with a frame rate as per its audio duration (refer to Fig. 2). Audio of respective text files (following speech balloon ordering from left to right) for individual panels then embeds into the video. In this way, the video container of a single panel is built.

Comic Book Video: After completing all video containers of individual panels, this module, using FFmpeg, combines all the separate panel video containers to form a video with audio, subtitles, and animation (see Fig. 2).

4   Experiments and Results

The implementation prototype is developed in Python, using a standard Python library, while video processing is performed by "FFmpeg" in Python. Image processing relies on the open-source library "OpenCV" which provides standard tools for thresholding, masking, contouring, and resizing the image.

(a) Input   (b) Output

Fig. 6. An illustration of detection and classification of speech balloon and narration boxes in two different panels.

4.1   Training and Testing

This section describes the training and testing methodologies being followed in this work for the individual tasks, i.e. CPE and SBS.

CPE Task: For the CPE task by CPENet (Fig. 3 in Sect. 3.2), first we resize all the comic pages to 416×416 while maintaining the aspect ratio. The first detection is 82nd layer with a down-sampling image by 32. Our detection feature map here becomes $(13 \times 13 \times 18)$. Similarly, the second detection is at the 92nd layer, where down-sampling is by 16, and the detection feature map is $(26 \times 26 \times 18)$. The 3rd detection is at the last 106th layer with down-sampling by eight, and the feature map is $(52 \times 52 \times 18)$. For every grid cell, 3 per scale, i.e., 9 anchor boxes, are assigned. The box having maximum overlapping with ground truth is chosen. The Sigmoid classifier is used for object class prediction and confidence score. For training, we leverage transfer learning by using pre-trained weights of ImageNet on VGG16. We kept the learning rate = 0.001 and used a batch size of 64 with an IOU of 0.75. We trained using 414 comic page images of different size rectangular panels. The preprocessing stage discussed in Sect. 3.1 improves CPENet's performance.

SBS Task: We have used 1502 comic book pages and taken a training:validation: testing ratio of 70:10:20. The entire dataset is shuffled randomly to ensure that the same comic book is not used together in a single batch, resulting in overfitting the model. Resizing of pages $(768 \times 512 \times 3)$ and binary masks $(768 \times 512 \times 1)$ is also performed. Adam optimizer [15] is used with a learning rate of 0.001. The training runs for 25 epochs with a batch size of 16 and a callback function with the patience of 2. A Sigmoid activation layer is added to

(a)   (b)

Fig. 7. Representing panel detection by model, where panel in red is the ground truth and dashed blue line is model prediction.

give the output pixel value between 0.0 and 1.0. We have empirically determined a threshold value of 0.5, greater than which a pixel is classified speech balloon (or narrative text), otherwise non-speech balloon region.

4.2   Results

We could not compare the overall framework C2VNet for audio-visual synthesis due to the unavailability of work. However, in this sub-section, we present the results and analysis of the intermediate stages.

CPE Task: Figure 7 shows correct panel detection by CPENet model. The figure represents panel detection by model, where panel in red is the groundtruth and dashed blue line is model prediction. Object detection accuracy is calculated using Precision and Recall.

$$Precision = \frac{TP}{TP + FP} \quad (2) \qquad Recall = \frac{TP}{TP + FN} \quad (3)$$

where TP = True positive, FP = False Positive, FN = False negative.

Table 2. Quantitative comparison between CPENet and Nhu et al. [16].

Method	By [16]	CPENet
Accuracy	0.952	0.986
IOU score	0.944	0.977
Recall	0.990	0.997
Precision	0.952	0.978

(a) Training Performance

Method	By [16]	CPENet
Accuracy	0.932	0.979
IOU score	0.937	0.961
Recall	0.988	0.976
Precision	0.948	0.975

(b) Testing Performance

Precision and Recall are calculated using Intersection over Union (IOU), in which we define an IOU threshold and prediction with IOU > threshold are true positive and otherwise false positive. For the accuracy calculation, we took IOU threshold as 0.5. To evaluate our model's performance, we plotted a precision and recall curve that varies with each input. We used all points interpolation and calculate the area under the curve, which gives accuracy. Test dataset consists of 189 comics with ground truths containing 5 attributes (*class*, x, y, w, h) and prediction contains 6 attributes (*class*, $confidence score$, x, y, w, h).

Comparative Analysis on CPE Task: To evaluate the model performance, we compared CPENet with the state-of-the-art model [16] published recently. We computed the results of the state-of-the-art on IMCDB by integrating available source codes. Table 2a shows the training, and Table 2b shows the testing comparison on the IMCDB dataset. The method [16] has a 1.2% better recall than our CPENet model, however, our model has 2.7% better precision and yielded a highly accurate panel extraction which is 4.5% more than [16]. This means that our proposed model CPENet can predict relevant panels over [16] as a better precision value is essential for proper video synthesis.

SBS Task: Figure 8 depicts the proposed model's qualitative results for the SBS task. It can be observed that SBSNet is able to achieve high accuracy for the given task as compared to the ground truth. Four metrics, i.e., DICE, IOU-score, Recall, and Accuracy, are followed for quantitatively evaluating the SBS task. The dice coefficient (F1 score) is two times the area overlap divided by the total number of pixels in both images. IOU-score is the ratio of logical-and and logical-or of the predicted and true output.

$$DICE = \frac{2 * TP}{TP + FP + FN} \quad (4) \qquad IOU = \frac{Logical\,And}{logical\,Or} \quad (5)$$

where TP = True positive, FP = False positive, FN = False negative.

Ablation Study: We have experimented with various filters for the network layers and reported in Table 3. The model column depicts the filters in the network

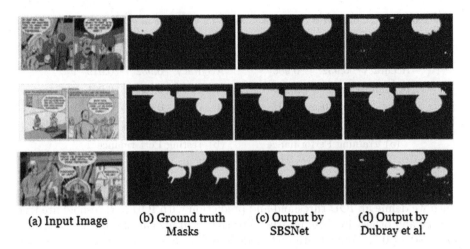

(a) Input Image   (b) Ground truth   (c) Output by   (d) Output by
                        Masks   SBSNet   Dubray et al.

Fig. 8. Qualitative comparison between Dubray et al. [10] and SBSNet.

model. In the first model, the number of filters in the first layer is 16, which doubles in the subsequent layer to 32, 64, 128 and 256. Total trainable parameters in this network is $1,941,105$. The last four columns are of the performance measures that help us to determine the best model. The performance of the model in the first and last row is approximately the same. The last one has slightly lesser accuracy, however, better in IOU-score and Recall. Still if both are considered same in the performance, the last one $582k$, i.e., $(1,941,105 - 1,358,485 = 582,620)$ lesser trainable parameters than the first one. This means that the last model is giving us the same result with lesser resources. The second and third models are also performing well. However, the second lags in recall and the third one in IOUs-score compared to the last (4th) one. Hence, we have considered the fourth model in C2VNet. The performance (accuracy, IOU score, and recall) level slightly drops (except for the precision, which increases) when we move over to testing w.r.t training and validation. We got a 98.8% accuracy along with a 0.913 IOU score, a 0.969 recall value, and a precision of 0.967 when we tested our model on IMCDB. During our model's training on IMCDB, the scores were $0.993, 0.948, 0.972,$ and 0.965 for accuracy, IOU score, recall, and precision,

Table 3. Ablation study: model comparison.

Model	Trainable parameters	DICE	IOU	Recall	Accuracy
(16->32->64->128->256)	1,941,105	0.97089	0.94343	0.95633	0.99269
(8->16->32->64->128)	485,817	0.96285	0.92837	0.93744	0.99078
(16->24->36->54->81)	340,252	0.96810	0.93818	0.96274	0.99191
(32->48->72->108->162)	1,358,485	0.96927	0.94039	0.97185	0.99214

Table 4. Quantitative performance of SBSNet on IMCDB.

Result/Data	Training	Validation	Testing
Accuracy	0.993	0.989	0.988
IOU score	0.948	0.918	0.913
Recall	0.972	0.974	0.969
Precision	0.965	0.961	0.967

respectively. Table 4 presents the performance of the 4th model over the training, validation, and testing set of the IMCDB.

Comparative Analysis on SBS Task. Table 5 shows the comparative study of the SBSNet with [10]. The comparison reveals that SBSNet is lighter, faster, and more accurate than [10]. The total number of parameters was reduced more than 13 times. The model given by [10] was trained and tested on IMCDB. The training time of SBSNet was reduced about 16 times, while testing time was reduced by 19.5 s. SBSNet took 13.4 s when tested on over 300 comic book pages. The size of our model is 16Mb in contrast to 216Mb by [10]. SBSNet also gives a comparable result. The accuracy, IOU score, recall, and precision are in proportion with that of [10] as seen in Table 5a.

Results of Video Synthesis: The proposed C2VNet framework generates the output as an interactive comic book video with enhanced multimedia features like audio, subtitles, and animation, where the input is given as a digitized or born digital comic book. Figure 9 depicts a few resultant video frames created from one such comic book. At the time $t = 0$ second, the first extracted panel of comic storybooks is made as the first frame of the video followed by its highlighted contour frame. Following this, all single panel video frames are combined with their respective durations forming a complete video of the comic storybook. As indicated in Fig. 9 as dummy signal waveform, the audio information, generated using the technique described in Sect. 3.3 is also embedded into the video frames. The reader has the option to toggle between, enable or disable the subtitle of the video, which is also part of the video file.

Table 5. Quantitative comparison between SBSNet and Dubray et al. [10].

Result/Model	**By [10]**	**Proposed**
Accuracy	0.984	0.988
IOU score	0.877	0.913
Recall	0.920	0.969
Precision	0.949	0.967

(a) Performance

Result/Model	**By [10]**	**Proposed**
Parameters	18,849,033	1,358,485
Training time	6781 sec	431 sec
Testing time	32.9 sec	13.4 sec
Size of output	216 Mb	16 Mb

(b) Network model

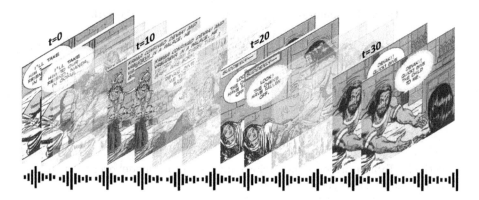

Fig. 9. An illustration of an output video generated by C2VNet.

5   Conclusion

This paper proposed a framework *"Comic-to-Video Network (C2VNet)"*, which evolves panel-by-panel in a comic strip and eventually creates a full-length video of a digitized or born-digital storybook embedded with multimedia features like audio, subtitle, and animation. To support our work, we proposed a dataset named *"IMCDB: Indian Mythological Comic Dataset of digitized Indian comic storybook"* in the English language with complete annotation for panels, binary masks of the text balloon and text files for each speech balloon and narration box within a panel and will make it publicly available. Our panel extraction model *"CPENet"* shows more than 97% accuracy, and the speech balloon segmentation model *"SBSNet"* gives 98% accuracy with the reduced number of parameters, and both performed better than state-of-art models. C2VNet is a first step towards the big future of automatic multimedia creation of comic books to bring new comic reading experiences.

References

1. Digital comic museum. https://digitalcomicmuseum.com/
2. IMCDB dataset. https://github.com/gesstalt/IMCDB
3. Pytesseract. https://pypi.org/project/pytesseract/
4. Text-to-speech. https://cloud.google.com/text-to-speech
5. Aizawa, K., et al.: Building a manga dataset "manga109" with annotations for multimedia applications. IEEE Multimedia **2**, 8–18 (2020)
6. Arai, K., Tolle, H., Arai, K., Tolle, H.: Method for real time text extraction of digital manga comic, pp. 669–676 (2011)
7. Aramaki, Y., Matsui, Y., Yamasaki, T., Aizawa, K.: Text detection in manga by combining connected-component-based and region-based classifications. In: ICIP, pp. 2901–2905 (2016)
8. Baek, Y., Lee, B., Han, D., Yun, S., Lee, H.: Character region awareness for text detection. In: CVPR, pp. 9357–9366 (2019)

9. Bergs, A.: Protanopia, a revolutionary digital comic for iPhone and iPad (2018). http://andrebergs.com/protanopia
10. Dubray, D., Laubrock, J.: Deep CNN-based speech balloon detection and segmentation for comic books. In: ICDAR, pp. 1237–1243 (2019)
11. Guérin, C., et al.: eBDtheque: a representative database of comics. In: ICDAR, pp. 1145–1149, August 2013
12. Ho, A.K.N., Burie, J., Ogier, J.: Panel and speech balloon extraction from comic books. In: DAS, pp. 424–428 (2012)
13. Iyyer, M., et al.: The amazing mysteries of the gutter: Drawing inferences between panels in comic book narratives. In: CVPR, pp. 6478–6487 (2017)
14. Khan, F.S., Anwer, R.M., van de Weijer, J., Bagdanov, A.D., Vanrell, M., Lopez, A.M.: Color attributes for object detection. In: CVPR, pp. 3306–3313 (2012)
15. Kingma, D.P., Ba, J.: Adam: a method for stochastic optimization. In: ICLR (2015). http://arxiv.org/abs/1412.6980
16. Nguyen Nhu, V., Rigaud, C., Burie, J.: What do we expect from comic panel extraction? In: ICDARW, pp. 44–49 (2019)
17. Ogawa, T., Otsubo, A., Narita, R., Matsui, Y., Yamasaki, T., Aizawa, K.: Object detection for comics using manga109 annotations. CoRR (2018)
18. Pang, X., Cao, Y., Lau, R.W., Chan, A.B.: A robust panel extraction method for manga. In: ACM MM, pp. 1125–1128 (2014)
19. Ponsard, C., Ramdoyal, R., Dziamski, D.: An OCR-enabled digital comic books viewer. In: ICCHP, pp. 471–478 (2012)
20. Redmon, J., Farhadi, A.: Yolov3: an incremental improvement. arXiv (2018)
21. Rigaud, C., Burie, J., Ogier, J.: Segmentation-free speech text recognition for comic books. In: ICDAR, pp. 29–34 (2017)
22. Rigaud, C., Burie, J., Ogier, J., Karatzas, D., Van De Weijer, J.: An active contour model for speech balloon detection in comics. In: ICDAR, pp. 1240–1244 (2013)
23. Rigaud, C.: Segmentation and indexation of complex objects in comic book images. ELCVIA (2014)
24. Rigaud, C., Pal, S., Burie, J.C., Ogier, J.M.: Toward speech text recognition for comic books. In: MANPU, pp. 1–6 (2016)
25. Ronneberger, O., Fischer, P., Brox, T.: U-net: convolutional networks for biomedical image segmentation. In: MICCAI, pp. 234–241 (2015)
26. Shi, B., Bai, X., Yao, C.: An end-to-end trainable neural network for image-based sequence recognition and its application to scene text recognition. TPAMI **39**, 2298–2304 (2017)
27. Tanaka, T., Shoji, K., Toyama, F., Miyamichi, J.: Layout analysis of tree-structured scene frames in comic images. In: IJCAI, pp. 2885–2890 (2007)
28. Tolle, H., Arai, K.: Automatic e-comic content adaptation. IJUC **1**, 1–11 (2010)
29. Yamada, M., Budiarto, R., Endo, M., Miyazaki, S.: Comic image decomposition for reading comics on cellular phones. IEICE Trans. Inf. Syst. **87**, 1370–1376 (2004)
30. Zhou, X., et al.: EAST: an efficient and accurate scene text detector. In: CVPR, pp. 2642–2651 (2017)

LSTMVAEF: Vivid Layout via LSTM-Based Variational Autoencoder Framework

Jie He⬤, Xingjiao Wu⬤, Wenxin Hu, and Jing Yang$^{(\boxtimes)}$

East China Normal University, Shanghai, China
{51184506012,Xingjiao.Wu}@stu.ecnu.edu.cn, wxhu@cc.ecnu.edu.cn,
jyang@cs.ecnu.edu.cn

Abstract. The lack of training data is still a challenge in the Document Layout Analysis task (DLA). Synthetic data is an effective way to tackle this challenge. In this paper, we propose an LSTM-based Variational Autoencoder framework (LSTMVAF) to synthesize layouts for DLA. Compared with the previous method, our method can generate more complicated layouts and only need training data from DLA without extra annotation. We use LSTM models as basic models to learn the potential representing of class and position information of elements within a page. It is worth mentioning that we design a weight adaptation strategy to help model train faster. The experiment shows our model can generate more vivid layouts that only need a few real document pages.

Keywords: Document Layout Analysis · Variational Autoencoder · Document generation

1 Introduction

Document Layout Analysis (DLA) is an important visual basis for document semantic segmentation and understanding [25,30]. The datasets use now are not large and targeted enough, which makes the training of DLA models hard. Generating more realistic samples would be helpful for both domain-specific and common layout analysis models. People often generate documents with various page elements randomly arranged if more training data is needed (*e.g.* generate document layout based on LaTeX). These results always lack diversity and are not in line with human aesthetics, which means the following analysis models may not be able to learn the underlying relationship of the document elements to get better performance.

Our aim is to generate new layouts that have a more meaningful latent connection between page elements than random ones. Previous methods for image generation preferred a convolutional neural network (CNN) with generating pixels, which make generated images blurry and noisy. We hope to directly get clear layouts with each element on the page is shown as its category and location.

J. He and X. Wu—These authors contributed equally to this work.

© Springer Nature Switzerland AG 2021
J. Lladós et al. (Eds.): ICDAR 2021, LNCS 12822, pp. 176–189, 2021.
https://doi.org/10.1007/978-3-030-86331-9_12

The location here represented as a bounding box with coordinates of two diagonal vertices. Consider the arbitrariness of elements number within one page and the ability of the Long Short-Term Memory (LSTM) [10] model to deal with the unlimited data, we try to build our generation model with LSTM.

Considering the potential structural relationships within one page, it is possible to treat the layout generation problem as a sequence generation problem, which requires reasonable category order and location information. Nowadays there are two mainstream frameworks to generate new samples: Generative Adversarial Networks (GANs) [6] and Variational AutoEncoder (VAE) [15] models. They have been proved useful in generating sequential data in many tasks and we use a similar idea to orderly produce a reasonable layout.

We have tried to generate samples with category and bounding box information in an LSTM-based GAN. The results are not vivid enough and then we try to build an LSTM-based VAE framework to generate a high-quality layout.

We feed the elements information into the LSTM-based encoder and get a latent vector representing the whole page. This latent vector would be the input of the LSTM-based decoder at each time-step and get new layouts. The VAE framework is difficult to train and we design a weight adaptation strategy to adjust the loss weights during training. We generate new samples with our framework on different datasets. The results of different datasets show the ability of our model in generating samples on both common and specific domains. We use generated layouts to synthesize the document images and pre-train the DLA model. The results on different benchmarks show our generated data is helpful to the model training.

In summary, this paper makes the following contributions:

- We propose an LSTM-based Variational Autoencoder framework (LSTM-VAEF) to generate realistic layouts to augment the training data of DLA. It regards the layout generator as an indefinite length page element sequence generation task rather than an image generation task.
- We design the adaptive loss weight to help our framework train.
- The experiments on different datasets show our framework can generate new layouts with only a few real data and the generated ones are more vivid than randomly arranged. These generated data are helpful to train the DLA model.

2   Related Work

Here are many applications of GANs or VAEs to generate data for different purposes. For example, many methods of GANs are proposed to generate different images, including generating new samples similar to original ones [1,8,20], generating new images with a different perspective from the original image [11], completing the missing part of the original image [27], producing high-resolution images [24], changing the style of images [12,31] and so on. GANs also have many applications to deal with or generate sequence data. They generate single image from text description [17] or produce videos from content and motion information [22]. People use VAE because its architecture can be proven by mathematical principles and avoid the "posterior collapse" problem. People can feed a random vector followed by a specific distribution to the decoder part to generate new samples. There are also many applications to generate sequential [21, 23] or non-sequential data with VAE.

All models mentioned above are designed for specific tasks, but not for generating training data for DLA. We build our generation framework to generate new vivid layout information and then generate both document images and corresponding annotations for DLA training. This work could generate a large number of new training data and greatly reduce the work of human labeling.

Directly generate document images would be difficult to get layouts information, and the reverse is equally difficult. The Layout generation aims to generate reasonable layouts. Realistic layouts can generate many new images when choosing different real document elements to replace layout bounding boxes. That is also easy to get both document images and label information. Here are some works that try to generate layouts. [14] proposed a model named LayoutVAE to generate scene layouts. It uses some labels as inputs and generates the number of occurrences of each element with their bounding boxes. [16] proposed a common method to generate different kinds of layouts. It takes random graphic elements as inputs and trains a self-attention model. The generated label and geometric parameters will produce a realistic layout with a differentiable wireframe rendering discriminator. [29] used a random vector to generate layouts, with the image, keywords, and high-level attributes as additional information. An encoder will encode the real layout into a vector and the discriminator learn to distinguish whether the layout-vector pair is real or fake. [19] present a REcursive Autoencoders for Document layout generation (READ). It extracts a structure decomposition from a single document and trains a VAE. A new document can be generated from a random vector.

These works may not propose to generate document layout or need to collect other information before training the generate model. We learned from these works to treat the layout generation problem as a sequence generation problem. We build a VAE with LSTM modules, which can be trained only with data of DLA task. We produce more new layouts with unlimited length and help to augment training data.

3   Framework

This section will introduce our method from two aspects: the structure of our LSTM-based VAE layout generation model and the adaptive loss we designed for training.

3.1   LSTM-Based VAE Framework

The Long Short-Term Memory (LSTM) [5] is an extension of the Recurrent Neural Networks (RNNs, [4]). Researchers use LSTM-based model in many tasks

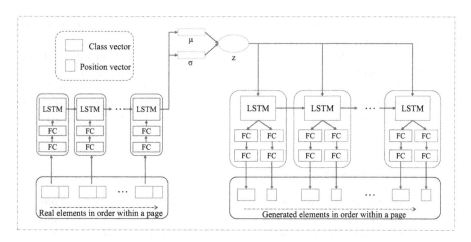

Fig. 1. The framework of layout generation. The encoder is an LSTM-based model, which gets element information as input and generates a latent vector. The decoder gets this vector and output the category and coordinates values at each time step.

such as classification [28], different kinds of object recognition [7] and data generation [13]. We use LSTM to encode the layout information with a variable number of elements into a fixed-size vector and vice versa.

Because of the latent order of document elements and the uncertainty of the number of elements in one page, we design an LSTM-based VAE framework to produce layouts. When we get a set of elements of one page from real samples, the class label and element position has been determined. We use one-hot encoding to represent the element class and four variables to represent the element position. The final element is represented as follows:

$$e_i^n = (c_i^n, p_i^n) \tag{1}$$

$$p_i^n = (x_i^l, y_i^t, x_i^r, y_i^b) \tag{2}$$

The e_i^n means the $i-th$ element in $n-th$ page, c_i^n is the one-hot encoding of $i-th$ element's class. p_i^n is the position of $i-th$ element and four variables are left, top, right and buttom coordinates respectively. Finally one page (or one sample) should be $(e_0^n, \cdots, e_i^n)$ for both real and synthesis data. The coordinate values will be normalized to a range of 0-1, and elements in every page will be reordered by position from top to bottom, left to right. When elements overlap, they will be reordered from bottom to top.

Considering the sequence and length uncertainty of data, we build our network with LSTM to get the features representing the whole page. As shown in Fig. 1, We first use two fully-connected (FC) layers to expand the dimension of element information and then feed it into an LSTM model to encode elements into a latent vector of fixed size. This latent vector would be the input of each time step in the LSTM-based decoder. At each state, the LSTM outputs the features to generate one element with FC layers and all outputs generate an elements sequence.

The FC layers of the encoder are $(10, 64)$, $(64, 128)$. (in, out) means the input and output dimension of each layer. The LSTMs of the encoder and decoder have depth 2 and 512 hidden units. The final output of this LSTM model will be fed into different FC layers with $(512, 20)$, $(20, 10)$ to build the μ and σ vector. The layers of coordinates and class generation are separate. The Class generation is: $(512, 64)$, $(64, 6)$. The position generation is: $(512, 64)$, $(64, 4)$, Sigmoid.

3.2   Loss Function Design

The training of VAE is to narrow the distance between the original sample and generated sample, at the same time the hidden vector z should try to follow a specific distribution, which we use Gaussian distribution there. The training loss is as follows:

$$Loss_{VAE} = \lambda Loss_{KL} + Loss_P + Loss_C \qquad (3)$$

The first term is the KL-divergence loss. It is used to measure the distence between two distributions $\mathcal{N}_1 \sim (\mu_1, \sigma_1^2)$ and $\mathcal{N}_2 \sim (\mu_2, \sigma_2^2)$:

$$KL(\mathcal{N}_1 \| \mathcal{N}_2) = log(\frac{\sigma_2}{\sigma_1}) + \frac{\sigma_1^2 + (\mu_1 - \mu_2)^2}{2\sigma_2^2} - \frac{1}{2} \qquad (4)$$

Here we use it to measure the distance between latent distribution $\mathcal{N}_z \sim (\mu, \sigma^2)$ and Gaussion distribution $\mathcal{N} \sim (0, 1)$:

$$Loss_{KL} = -log(\sigma) + \frac{\sigma^2 + \mu^2}{2} - \frac{1}{2} \qquad (5)$$

Due to the difficultly of training the VAE network, we add a weight λ to control the $Loss_{KL}$. The λ will be set 0 at the beginning of training and the model will focus on generating samples as similar as it can. As training progresses, we increase the λ gradually to force the encoder better compress information. The λ will raise to 1 and no longer increase:

$$\lambda = min(1, \alpha \times epoch) \qquad (6)$$

Here the *epoch* means our training progress and we set α as $1e^{-4}$.

The second term is the relative error of element box coordinates. Due to the uncertain length of samples, there are fewer behind elements and the training will be insufficient. We set a weight to make the relative errors of the latter elements have more influence on updates. The loss is as follows:

$$Loss_P = \frac{1}{n} \sum_{i=1}^{n} w \times (p_i - \widetilde{p}_i)^2 \qquad (7)$$

$$w = 1 + \beta \times i \qquad (8)$$

Here the p_i is the real element coordinates and $\tilde{p}_i$ is the generated element coordinates. The n is the number of elements within a page and w is the weight of element errors, and we set β as 0.05.

The third term is a categorical cross-entropy loss. The latter elements also have greater weight:

$$Loss_C = \lambda_c \sum_{i=1}^{n} w \times H(\tilde{c}_i, c_i) \tag{9}$$

$$H(\tilde{c}_i, c_i) = -\sum_{i=1}^{n_c} c_i log(softmax(\tilde{c}_i)) \tag{10}$$

where n_c is the number of categories, $\tilde{c}^i$ is the generated class vector and c_i is the corresponding real category. Considering the numerical gap between $Loss_P$ and $Loss_C$, we add the λ_c to zoom the $Loss_C$. Here we set λ_c as 100.

4 Experiments

4.1 Details for Generation

Details for Training. We build the framework on Pytorch. Here we trained on different datasets and all of them have the same categories: Text, Section, List, Figure, Table, and Caption. When the number of elements within a page is len, the size of the training sample is $10 \times len$, including $6 \times len$ class and $4 \times len$ position information. We set the latent vector size as 10. We use Adam to update the parameters with an initial learning rate of 5×10^{-4}. When testing the generate samples all with the length of 20 and cut the generation with an overlap rate.

Details About Datasets: We conducted experiments on several datasets to prove the generation ability of our framework, including one-column pages from the DSSE-200 dataset, 96 pages we got and labeled from conference articles, and 246 pages we collected from magazines.

One Column from DSSE-200: The one-column pages in the DSSE-200 dataset mainly come from magazines and books. We select 123 pages and there are 1 to 52 elements within a page and have an average of 9.78.

Samples from Conference: We get 16 papers, 96 pages from one conference, and label the elements into 6 classes. They come from the same conference and should have higher consistency in style. There are 4 to 21 elements within one page and with an average of 12.54.

Samples from Magazines: We collect 246 pages from magazines. They come from different issues of the same magazine and have different layout styles from the papers. We split 80 pages for training our model and 166 pages for the test. There are 2 to 26 elements within one page in training data and with an average of 10.19. In test data, there are 1 to 62 elements within one page and with an average of 13.45.

4.2 Layout Generation Results

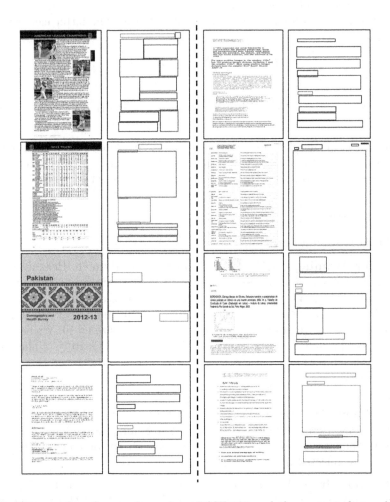

Fig. 2. The samples of one column from DSSE-200 and the generated ones by our method. To the left of the dotted line are the real pages with corresponding layouts, and to the right are the generated layouts and generated pages. Color legend: Text, Section, List, **Figure**, Table and Caption.

In Fig. 2 we visualize some one-column samples from DSSE-200 and generated by our framework. The results prove that we can generate new layouts based on small datasets and really learn some meaningful layout styles.

In Fig. 3 we show the samples generated with conference papers. The papers from the same conference follow the same detailed rules. They are highly similar in the conference but are significantly different from samples outside.

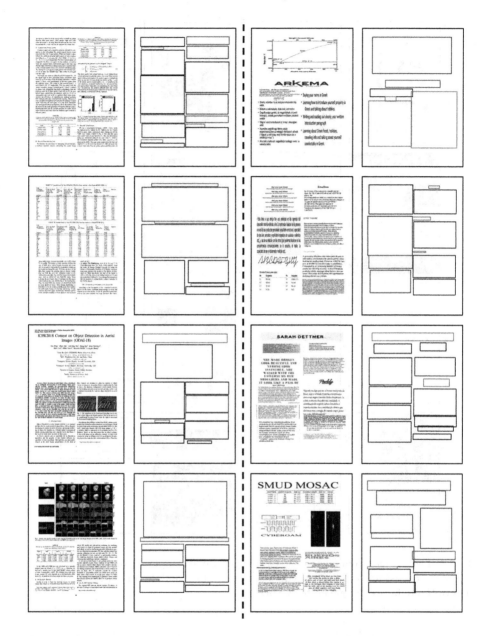

Fig. 3. The samples from conference papers and the generated ones by our method. To the left of the dotted line are the real pages with corresponding layouts, and to the right are the generated layouts and generated pages.

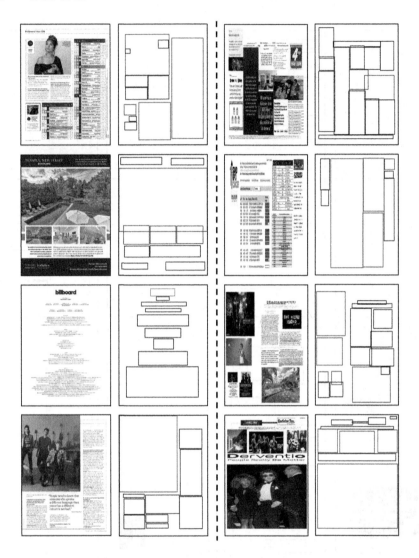

Fig. 4. The samples from magazines and the generated samples by our method. To the left of the dotted line are the real pages from maganizes and its corresponding layouts, and to the right are our generated layouts.

In Fig. 4 we show the samples from magazines and generated by our framework. The magazine has a different layout style from papers but our method also gets good layouts which are hard to imitate by the traditional method.

These experiments show our model can generate layouts that follow the special rules of a magazine or a conference paper. It will be very helpful when we are lacking domain-specific training data in DLA.

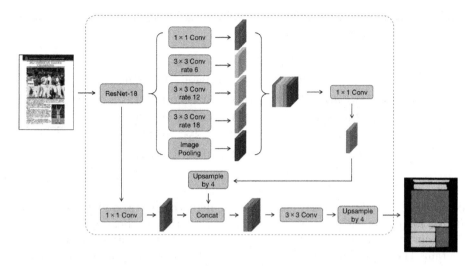

Fig. 5. The structure of our DLA base model. It is based on DeepLabv3+ and the backbone was replaced by ResNet-18 [9] to alleviate the overfitting.

4.3 Pre-training for Document Layout Analysis

We use the generated layouts above to synthetic document images to pre-train the DLA model, and then fine-tune it on different DLA benchmarks. The results show our generated data can help improve the ability of the DLA model to recognize the document better.

Document Generation. The steps to generate document images are as follows:

First, we use the conference papers and our annotations to cut element images. These images are just used to get similar document element images on Google Images. The search results are filtered by the human to make sure all pictures are clear enough and the categories are correct.

Second, we use searched images and the generated layouts data to synthetic document images. The layout data generated from our model are all 20 elements in order and we will stop the generation with an overlap rate. The element information can be directly converted to the ground truth for training the DLA model.

Experiments for DLA. We use the DeepLabv3+ [2] to be our DLA model and replace the backbone with ResNet-18 [9] to complete the experiment. The model structure is shown in Fig. 5. We used Adam to update the parameters in our DLA model with an initial learning rate of 1×10^{-3} and trained 300 epochs.

We collect two kinds of document samples generated by our method to pre-train our base model and fine-tune it on several datasets: DSSE-200 [26], CS-150 [3] and our magazine dataset. The DSSE-200 is a dataset consist of different

Table 1. The results of DSSE-200 and CS-150 datasets, which are trained on same model with or without our generated samples.

	DSSE-200		CS-150	
	Base	Pre-train	Base	Pre-train
Accuracy	0.8776	0.8938	0.9888	0.9893
Precision	0.8223	0.8700	0.9259	0.9323
Recall	0.8290	0.8531	0.8547	0.8746
F_1	0.8256	0.8614	0.8889	0.9025

documents, including books, magazines, conference papers, and PowerPoints. It has 200 document pages in total and we split the benchmark randomly to finish our experiment. The CS-150 is a dataset with 150 scientific papers, which contains 1175 two-column pages and has a similar layout style to our conference pages. For DSSE-200 we use 60 pages to train and 140 to test, and for CS-150 we use 235 for training and 940 for testing. We generate about 700 samples whose generation model is trained on 96 conference pages. For DSSE-200 the DLA model would classify the document pixel into 4 classes: Text, Figure, Table, and Background. For CS-150 the DLA model would classify the document pixel into 3 classes: Figure, Table, and Others. We would convert our generated samples into these classes to pre-train the DLA model. The magazine data is collected by us and we will only public the samples on GitHub[1]. We collect 246 pages from the same magazine and split 80 to train, 166 to test. We use 80 pages to train our generation model and get 2275 samples to pre-train the DLA model.

Results for DLA. We follow the work [18] to evaluate the results of DLA. The DLA model would predict the class of every pixel in the document image and we can get a confusion matrix M. When M_{ij} means the number of pixels that were predicted as class j with real class i and num means the number of classes, the evaluation methods are designed as follows:

Accuracy: The ratio of the true predicted pixels and the total number of pixels.

$$Accuracy = \frac{\sum_{i=1}^{num} M_{ii}}{\sum_{i=1}^{num} \sum_{j=1}^{num} M_{ij}} \qquad (11)$$

Precision: Firstly, for each category, we calculate the ratio of the true predicted pixels and the total number of pixels predicted to this category. Secondly, we get an average of all categories.

$$Precision = \frac{1}{num} \sum_{i=1}^{num} \frac{M_{ii}}{\sum_{j=1}^{num} M_{ji}} \qquad (12)$$

---

[1] https://github.com/Pandooora/LSTMVAEFD.

Table 2. The results of magazine data, which are trained on same DLA model with only real data, with LaTeX based generated samples or with our generated samples.

	Base	Augmentation	LaTeX	Pre-train
Accuracy	0.7443	0.7482	0.7436	0.7962
Precision	0.8303	0.8251	0.8163	0.8552
Recall	0.7027	0.7289	0.7702	0.8054
F_1	0.7612	0.7740	0.7926	0.8296

Recall: Firstly, for each category, we calculate the ratio of the true predicted pixels and the total number of pixels which belong to this category. Secondly, we get an average of all categories.

$$Recall = \frac{1}{num} \sum_{i=1}^{num} \frac{M_{ii}}{\sum_{j=1}^{num} M_{ij}} \tag{13}$$

F_1: F-measure is a score to combine Precision and Recall. Here we use F_1 to balance these two accuracies.

$$F_1 = \frac{2 \times Precision \times Recall}{Precision + Recall} \tag{14}$$

We compare the results about conference paper layouts in Table 1 and the magazine layouts in Table 2.

The **base** in two tables means we only use the real data to train the model. The **Pre-train** means we use our generated data to pre-train the same model first and then use the data from the dataset to fine-tune.

The results show our generated data can help improve the performance of the DLA model. For Table 1, two datasets both have some conference pages, which is similar to our generated ones in a layout style. It shows our method to augment the training set is useful. The results in Table 2 also prove this point.

We conducted additional comparative experiments on the magazine dataset. We use different methods to increase the training data and keep the total amount of training consistent to explore the role of our generation method.

The **Augmentation** means we use the traditional image augmentation to increase training data for DLA. For Table 2 we randomly decide whether to flip, how much to zoom the image and keep the total number of training images for each epoch at 2000 to train the DLA model. The **LaTeX** means we randomly generate one or two-column pages by LaTeX to pre-train the same DLA model and then use the real data to fine-tune. For Table 2 we generate 2000 pages so that we can compare the usage of LaTeX generated samples and our generated samples.

For Table 2, the magazine layout style is complex and hard to be generated by a simple rules-based LaTeX method. Our generation can provide more vivid layouts and provide greater help to the DLA model. And we also compare the results with image augmentation. After using the method in this paper to amplify the training set, the DLA performance improvement is higher than that of traditional image augmentation under the same training sample size. The results show that the method in this paper can provide useful information while increasing the training data.

5 Conclusions

In this paper, we proposed a method to generate new layouts based on small existing datasets for Document Layout Analysis. We treat the elements within a page as a sequence that appears from top to bottom and from left to right. The categories and positions of elements within one page will potentially influence each other and we try to model this latent influence with Variational Autoencoder and generate new layouts. We design the framework with LSTMs and FC layers. The experiment shows the approach can generate domain-specific layouts with just little training samples, and our generated images can help the DLA model training.

Acknowledgement. This work was supported in part by the Fundamental Research Funds for the Central Universities, the 2020 East China Normal University Outstanding Doctoral Students Academic Innovation Ability Improvement Project (YBNLTS2020-042), and the computation is performed in ECNU Multifunctional Platform for Innovation (001).

References

1. Arjovsky, M., Chintala, S., Bottou, L.: Wasserstein generative adversarial networks. In: International Conference on Machine Learning, pp. 214–223 (2017)
2. Chen, L.-C., Zhu, Y., Papandreou, G., Schroff, F., Adam, H.: Encoder-decoder with atrous separable convolution for semantic image segmentation. In: Ferrari, V., Hebert, M., Sminchisescu, C., Weiss, Y. (eds.) ECCV 2018. LNCS, vol. 11211, pp. 833–851. Springer, Cham (2018). https://doi.org/10.1007/978-3-030-01234-2_49
3. Clark, C.A., Divvala, S.: Looking beyond text: extracting figures, tables and captions from computer science papers. In: AAAI (2015)
4. Elman, J.L.: Finding structure in time. Cogn. Sci. **14**(2), 179–211 (1990)
5. Gers, F.A., Schmidhuber, J., Cummins, F.: Learning to forget: continual prediction with LSTM. Neural Comput. **12**(10), 2451–2471 (1999)
6. Goodfellow, I.J., et al.: Generative adversarial nets. In: NIPS (2014)
7. Graves, A., Jaitly, N., Mohamed, A.R.: Hybrid speech recognition with deep bidirectional LSTM. In: IEEE Workshop on Automatic Speech Recognition and Understanding, pp. 273–278. IEEE (2013)
8. Gulrajani, I., Ahmed, F., Arjovsky, M., Dumoulin, V., Courville, A.C.: Improved training of Wasserstein GANs. In: NIPS, pp. 5767–5777 (2017)
9. He, K., Zhang, X., Ren, S., Sun, J.: Deep residual learning for image recognition. In: CVPR, pp. 770–778 (2016)
10. Hochreiter, S., Schmidhuber, J.: Long short-term memory. Neural Comput. **9**(8), 1735–1780 (1997)

11. Huang, R., Zhang, S., Li, T., He, R.: Beyond face rotation: global and local perception GAN for photorealistic and identity preserving frontal view synthesis. In: ICCV, pp. 2439–2448 (2017)
12. Isola, P., Zhu, J.Y., Zhou, T., Efros, A.A.: Image-to-image translation with conditional adversarial networks. In: CVPR, pp. 1125–1134 (2017)
13. Jia, X., Gavves, E., Fernando, B., Tuytelaars, T.: Guiding the long-short term memory model for image caption generation. In: ICCV, pp. 2407–2415 (2015)
14. Jyothi, A.A., Durand, T., He, J., Sigal, L., Mori, G.: Layoutvae: stochastic scene layout generation from a label set. In: ICCV, pp. 9895–9904 (2019)
15. Kingma, D.P., Welling, M.: Auto-encoding variational bayes. In: ICLR (2014)
16. Li, J., Yang, J., Hertzmann, A., Zhang, J., Xu, T.: Layoutgan: generating graphic layouts with wireframe discriminators. In: ICLR (2019)
17. Li, W., et al.: Object-driven text-to-image synthesis via adversarial training. In: CVPR, pp. 12174–12182 (2019)
18. Mehri, M., Nayef, N., Héroux, P., Gomez-Krämer, P., Mullot, R.: Learning texture features for enhancement and segmentation of historical document images. In: ICDAR, pp. 47–54 (2015)
19. Patil, A.G., Ben-Eliezer, O., Perel, O., Averbuch-Elor, H.: READ: recursive autoencoders for document layout generation. In: Proceedings of IEEE Conference on Computer Vision and Pattern Recognition Workshops, pp. 2316–2325 (2020)
20. Radford, A., Metz, L., Chintala, S.: Unsupervised representation learning with deep convolutional generative adversarial networks. In: ICLR (2015)
21. Roberts, A., Engel, J., Raffel, C., Hawthorne, C., Eck, D.: A hierarchical latent vector model for learning long-term structure in music. In: ICML (2018)
22. Tulyakov, S., Liu, M.Y., Yang, X., Kautz, J.: Mocogan: decomposing motion and content for video generation. In: CVPR, pp. 1526–1535 (2018)
23. Wang, T., Wan, X.: T-CVAE: transformer-based conditioned variational autoencoder for story completion. In: IJCAI, pp. 5233–5239 (2019)
24. Wang, X., et al.: ESRGAN: enhanced super-resolution generative adversarial networks. In: Leal-Taixé, L., Roth, S. (eds.) ECCV 2018. LNCS, vol. 11133, pp. 63–79. Springer, Cham (2019). https://doi.org/10.1007/978-3-030-11021-5_5
25. Wu, X., Hu, Z., Du, X., Yang, J., He, L.: Document layout analysis via dynamic residual feature fusion. In: ICME (2021)
26. Yang, X., Yumer, E., Asente, P., Kraley, M., Kifer, D., Lee Giles, C.: Learning to extract semantic structure from documents using multimodal fully convolutional neural networks. In: CVPR, pp. 5315–5324 (2017)
27. Yeh, R.A., Chen, C., Yian Lim, T., Schwing, A.G., Hasegawa-Johnson, M., Do, M.N.: Semantic image inpainting with deep generative models. In: CVPR, pp. 5485–5493 (2017)
28. Yue-Hei Ng, J., Hausknecht, M., Vijayanarasimhan, S., Vinyals, O., Monga, R., Toderici, G.: Beyond short snippets: deep networks for video classification. In: CVPR, pp. 4694–4702 (2015)
29. Zheng, X., Qiao, X., Cao, Y., Lau, R.W.: Content-aware generative modeling of graphic design layouts. ACM Trans. Graph. **38**(4), 1–15 (2019)
30. Zheng, Y., Kong, S., Zhu, W., Ye, H.: Scalable document image information extraction with application to domain-specific analysis. In: IEEE International Conference on Big Data (2019)
31. Zhu, J.Y., Park, T., Isola, P., Efros, A.A.: Unpaired image-to-image translation using cycle-consistent adversarial networks. In: ICCV, pp. 2223–2232 (2017)

Mobile Text Recognition

HCRNN: A Novel Architecture for Fast Online Handwritten Stroke Classification

Andrii Grygoriev[1]([✉]) [iD], Illya Degtyarenko[1] [iD], Ivan Deriuga[1] [iD],
Serhii Polotskyi[1] [iD], Volodymyr Melnyk[1] [iD], Dmytro Zakharchuk[1,2] [iD],
and Olga Radyvonenko[1] [iD]

[1] Samsung Research, Kyiv, Ukraine
{a.grigoryev,i.degtyarenk,i.deriuga,s.polotskyi,v.melnyk,
o.radyvonenk}@samsung.com
[2] Taras Shevchenko National University of Kyiv, Kyiv, Ukraine
d.zakharchuk@samsung.com

Abstract. Stroke classification is an essential task for applications with free-form handwriting input. Implementation of this type of application for mobile devices places stringent requirements on different aspects of embedded machine learning models, which results in finding a trade-off between model performance and model complexity. In this work, a novel hierarchical deep neural network (HDNN) architecture with high computational efficiency is proposed. It is adopted for handwritten document processing and particularly for multi-class stroke classification. The architecture uses a stack of 1D convolutional neural networks (CNN) on the lower (point) hierarchical level and a stack of recurrent neural networks (RNN) on the upper (stroke) level. The novel fragment pooling techniques for feature transition between hierarchical levels are presented. On-device implementation of the proposed architecture establishes new state-of-the-art results in the multi-class handwritten document processing with a classification accuracy of 97.58% on the IAMonDo dataset. Our method is also more efficient in both processing time and memory consumption than the previous state-of-the-art RNN-based stroke classifier.

Keywords: Handwriting input · Stroke classification · Hierarchical deep neural networks · Convolutional neural network · Recurrent neural network · Pooling · Mobile computing

1 Introduction

Over the past few years, digital note-taking has become available on a wide range of devices, providing users with a natural writing or drawing experience supported with AI features, a trend that is likely to intensify in the future [1]. The main reason lies not only in the continuous technical improvement of devices with touch input, smart pens, smart writing surfaces, but also in the technological breakthrough in handwriting recognition technologies through the implementation of deep learning approaches. Applying them to the online handwritten

© Springer Nature Switzerland AG 2021
J. Lladós et al. (Eds.): ICDAR 2021, LNCS 12822, pp. 193–208, 2021.
https://doi.org/10.1007/978-3-030-86331-9_13

text recognition has shown outperforming results [10], while the classification of more complex online structures for free-form handwritten input still requires more research [20].

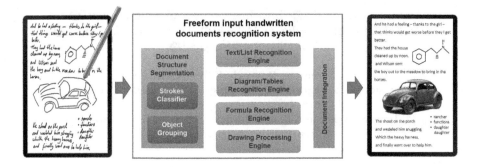

Fig. 1. Structure of free-form handwriting recognition system

Free-form handwritten input and recognition suppose processing of handwritten documents with mixed content, including text blocks, diagrams, tables, formulas, etc. [28]. Each type of content demands on a specialized recognition engine (Fig. 1). Thus, for accurate recognition, the document has to be contextually separated into objects and every separated object has to be dispatched to the corresponding specialized recognizer. For solving this problem, we use document structure segmentation (DSS), which aimed at performing the following steps:

– handwritten stroke classification;
– handwritten stroke grouping into the objects such as text blocks, tables, diagrams, etc.

To ensure a comfortable user experience, these procedures should be performed with imperceptible latency and subject to limited memory consumption. This paper focuses on speeding-up and improving the accuracy of stroke classification as a part of DSS for on-device online handwritten document recognition.

The paper describes an original stroke classification approach based on HDNN architecture for online handwritten documents with multi-class content. The combination of hierarchical architecture with modern deep learning techniques allows extracting information effectively from the documents, taking into account all the features of handwritten data. This work presents a novel handwritten stroke classifier structure grounded on hierarchical convolutional and recurrent neural networks (HCRNN). The key element of the proposed hierarchical architecture is the fragment pooling layer, which allows optimizing both model accuracy and latency.

The proposed approach was evaluated on the benchmark IAMonDo dataset [12] and has shown new state-of-the-art results for the task of multi-class handwritten document processing. Experimental studies were conducted for the five-class classification.

The structure of the paper is as follows: after this introduction, Sect. 2 presents related work analysis. Section 3 introduces the background of online document structure and describes the proposed HCRNN architecture for handwritten stroke classification. The novel fragment pooling layers are described in Sect. 4. Section 5 includes details about the dataset, the evaluation process, and the results of the ablation study.

2   Related Work

Stroke classification in handwritten documents is still a challenging problem, even though it has been studied since the early 2000s. The first studies were focused on the handcrafted features [2,22], where the authors used their own datasets to evaluate the results. One of the current benchmark datasets was presented in [12].

There are two modes in document stroke processing during classification. The first one considers and classifies all the document strokes together [7,25]. We call it batch mode. The second one considers and classifies stroke subsets (groups) [6,11,19,22]. This approach is referred as edit mode. Handwritten strokes were classified using conditional random fields and Markov random fields in [7]. In [8,11,14,19] classification was conducted using different types of RNNs. In general, conventional RNNs do not handle context-based problems very well. Contextual stroke classification with graph attention network was proposed in [25] and showed promising results for the batch mode. State-of-the-art result for edit mode was obtained in [19]. Also [19] and [4] are the first researches where classification was performed on the point level and aggregated on the stroke one further. But RNNs have a known peculiarity that should be taken into account—latency both in terms of training [13] and in terms of inference [9]. Hence there is a need for more research to find ways to optimize existing architectures and develop new ones.

Hierarchical NNs have proved to be very powerful for solving different problems related to sequential data. The learning of both hierarchical and temporal representation from data has been one of the enduring problems of DNNs [9]. Multi-level RNN [3] was seen as a promising approach to solving this problem. Since the hierarchical architecture of NNs mirrors the hierarchical structure of documents and can help to learn it effectively, the HDNNs are appropriate for document analysis. They demonstrated a plausible performance in natural language processing tasks for document content analysis, for example, text/sentence classification [23,24] where character, word and sentence levels were represented by different layers of NN.

Most frequently RNNs are applied to get information of each sentence while learning word representations, CNNs applied to obtain the contextual information of the whole document [16]. The approach proposed in [5] demonstrated 97.25% accuracy for multi-class classification using hierarchical RNN that is better on 2.61% than in [25]. The combination of CNN and RNN has proved its effectiveness for solving problems with sequential data [15,17,18]. This combination has shown promising results in [4]. But there were limitations like an equal length of all the document strokes. In order to improve the performance of CNN we paid attention to the pooling [26,27]. A comprehensive overview of pooling techniques and their contribution to computer vision tasks is given in [21]. This paper evolves the idea of hierarchical DNN [5] by involving CNN and novel fragment pooling techniques.

3   HDNN for Handwritten Stroke Classification

3.1   Background

The online handwritten documents have a cogent hierarchical structure. There are three main levels here (Fig. 2).

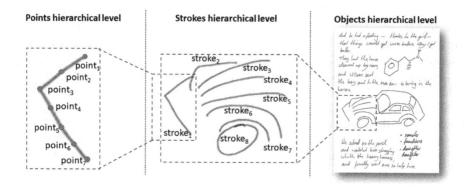

Fig. 2. Hierarchy of online handwritten document

This hierarchy must be taken into account at the design stage of the document recognition system and its components. The data at each level has its own particular features and implies a certain processing procedure. We propose to use HDNN for the semantic stroke segmentation procedure. This procedure uses features of point and stroke levels to provide information about the structure of a handwritten document to the upper (object) hierarchical levels.

In [5], the HDNN structure has been used for handwritten stroke classification where point and stroke levels are handled by different RNNs. The main element of such an architecture is, so-called, fragment pooling. It is a procedure that compresses knowledge from the first RNN and transfers it to the second one. The term *fragment* corresponds to a handwritten stroke in a stroke classification problem. In this work, we consider a more general structure. Instead of two RNNs and one type of pooling, we explore CNN and RNN on the point and stroke hierarchical levels respectively, that are connected with different types of fragment pooling. A detailed description of this architecture is given below.

3.2 HCRNN Architecture

The main idea of HCRNN architecture is to handle the combination of CNN layers at the point level for extracting features based on information from neighbouring points (local context) and RNN layers at the stroke level for taking into account global context from wide range stroke neighbours (Fig. 3).

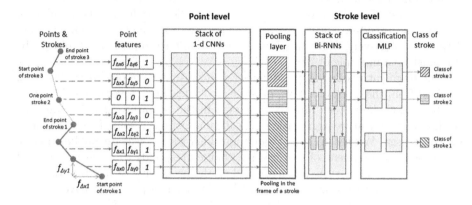

Fig. 3. HCRNN architecture and features

We parameterized the complexity of HCRNN architecture by an integer parameter N_F. It defines the number of hidden features, produced at every layer of HCRNN. The model parameterizing makes it scalable and allows us to explore the impact of complexity on performance and memory consumption. The experimental selection of parameter N_F is presented in Sect. 5.

As an input, it uses the vectorized trajectory of the digital pen movement that reproduces the way the document was drawn by the user. Each element of the trajectory is encoded with the following features:

- f_{Δ_x} is the difference between two neighbour points by x coordinate;
- f_{Δ_y} is the difference between two neighbour points by y coordinate;
- p is pen-up/pen-down: a boolean variable indicating whether or not the pen-tip touches the screen (pen-down position is indicated by 1, and pen-up by 0).

At the point level, we extract features based on information from neighbouring points (local context). We assume that the CNN extracts some specific patterns, which can help to detect stroke type with respect to the local and a global context, which will be extracted at later stages. The choice of these patterns during training is determined at the later stages, and by 'specificity' we mean distinctive local behaviour of the stroke curve, which in combination with global-context can be used to determine the stroke type. On this purpose, we use a cascade of three consequent 1D-convolutional layers. Each layer contains N_F filters with an aperture length of 3 and ReLU activation function.

The choice of convolutional layers at point level was made due to the simplicity of convolutions compared to recurrent layers used in HRNN. By *'simplicity'* we mean the absence of heavy non-linear mathematical functions and the ability to parallelize calculation both at the training stage and at the inference stage. It should be noted that simplicity is important at this stage because the number of points is many times higher than the number of strokes. In addition, to further simplifying of the architecture at the 2nd and 3rd layers, we used grouping with a division into 5 and 4 groups, respectively. It allows reducing the number of weights and speed-up matrix calculations on heavy 2nd and 3rd layers in 5 and 4 times, respectively. Since the division into 5 and 4 groups is used, obviously, N_F must be a multiple of 20.

The fragment pooling layer provides a transformation of point-level features, received from the stack of CNNs, into the stroke features. Several possible pooling techniques have been already discussed in the literature [26,27]. Standard pooling implies feature aggregation or feature selection in some fixed-size frame of feature set. We use fragment pooling which means feature pooling in the frame of some fragment of features set that corresponds to some hierarchical entities with variable size. When applied to the processing of handwritten documents, these entities are strokes and handwritten objects. We have studied several options for fragment pooling techniques (Sects. 4 and 5).

After the pooling layer, there is a sequence of N_F–dimensional vectors describing each stroke of the document. We assume that this representation can describe the shape of every stroke focusing on information about the presence of special fragments, their relative quantities and the combination of these fragments in stroke trace and, possibly, the stroke's direct neighbours. But of course, stroke classification is highly dependent on the document context. To sum up this context, we use two recurrent layers before the classification. To this end, we employ two bidirectional GRU RNNs with N_F hidden elements in each direction to obtain contextual information sufficient for stroke classification.

At the classification stage, the strokes are categorized based on N_F features received from the stroke level. Here we apply a multi-layer perceptron (MLP) classifier for each stroke, represented by two biased fully-connected layers. The first layer has N_F outputs and ReLU activation function. The second one has five outputs, corresponding to the number of stroke classes, and a soft-max activation function.

4   Fragment Pooling

The fragment pooling technique implies the calculation of higher-level features through the aggregation of lower-level features over corresponding fragments. Regarding the pooling procedure, we make the feature transition from the lower hierarchical level to the higher one, keeping the number of feature vectors in accordance with the number of elements at each level. It allows reducing the amount of processing data without sufficient information loss. As a result, we also reduce neural network training and inference time.

Proposed in [5] edge fragment pooling is the first method of fragment pooling designed for bidirectional RNNs and cannot be implemented for other types of layers, for example, MLP, CNN, graph-based layers, etc. To cope with this limitation, we propose the adoption of max and average pooling, moreover we designed novel techniques of stochastic and mixed pooling for fragment aggregation case. Fragment average and fragment max pooling are used during model training and the inference. The rest is used only on the training phase in combination with max and average pooling techniques at the inference phase. The mixed fragment pooling is a superstructure over the other ones. This imposes an additional regularization effect due to the random switching between different pooling techniques.

Formally, all the proposed layers work with the following input data:

- sequence $(\mathbf{x_1}, \mathbf{x_2}, \ldots, \mathbf{x_n})$ of length n containing lower-level entities, represented by k–dimensional real valued feature vectors $\mathbf{x_i} = (x_{i1}, x_{i2}, \ldots, x_{ik})$.
- sequences $(s_1, s_2, \ldots, s_m)$ of length m and $(e_1, e_2, \ldots, e_m)$, containing information about the starting and final index of every sequence fragment. Indexes s_i, e_i must fit the following conditions:

$$1 \leq s_i \leq e_i \leq n, \forall i \in 1, \ldots, m;$$

$$\forall i, j \in 1, \ldots, m, i \neq j : [s_i, e_i] \cap [s_j, e_j] = \varnothing.$$

Corresponding to the given lower level input sequence and fragment locations, the layers aggregate data over the fragments and generate a higher level sequence $(\mathbf{y_1}, \mathbf{y_2}, \ldots, \mathbf{y_m})$, $\mathbf{y_i} = (y_{i1}, y_{i2}, \ldots, y_{ik})$.

As input for backpropagation, we get a sequence of errors $(\delta_1, \delta_2, \ldots, \delta_m)$ from higher level, where $\delta_i = (\delta_{i1}, \delta_{i2}, \ldots, \delta_{ik})$. Since all the presented layers have no parameters, for training we only need to implement error backpropagation.

4.1   Fragment Max Pooling

Here we use max function for aggregating features over every fragment of the document. This layer makes an embedding representation of the corresponding fragments. Fragment aggregation with max can be interpreted as the fuzzy existential quantifier. In fact, this pooling technique captures the presence of a certain activation peak in the current fragment:

$$y_{ij} = \max_{p \in \{s_i, \ldots, e_i\}} x_{pj} \tag{1}$$

The error backpropagation is transmitted only over links, which were used in forward propagation, i.e. having maximum activation on a fragment:

$$\frac{\partial E}{\partial x_{ij}} = \begin{cases} \delta_{pj}, & \text{if } \exists p: \ i \in \{s_p, \ldots, e_p\} \text{ and } x_{ij} = y_{pj}; \\ 0, & \text{otherwise.} \end{cases} \tag{2}$$

4.2   Fragment Average Pooling

This layer is similar to the fragment max pooling, but here we use mean aggregation instead of the maximum. Unlike the max – pooling. Here not only the presence of a feature within a fragment is captured, but also is the amount of this feature within a fragment is estimated indirectly. In addition, the aggregation is smoother, since all the fragment elements take part in the layer inference:

$$y_{ij} = \frac{1}{e_i - s_i + 1} \sum_{p=s_i}^{e_i} x_{pj}. \tag{3}$$

During backpropagation, the error spreads evenly across all low-level elements, in contrast to max – pooling where error is fed only at one fragment element:

$$\frac{\partial E}{\partial x_{ij}} = \begin{cases} \dfrac{\delta_{pj}}{e_p - s_p + 1}, & if \ \exists p : \ i \in \{s_p, \dots, e_p\}; \\ 0, \ otherwise. \end{cases} \tag{4}$$

4.3   Fragment Stochastic Pooling

Fragment stochastic pooling is grounded on the approach presented in [27]. This technique can be interpreted as a generalization of max-pooling for stochastic cases. In [27] authors randomly select pooling region element with a probability proportional to the element's activation.

In our approach we calculate stochastic pooling in two steps:

1. for every fragment $[s_i, e_i]$ we randomly, with discrete uniform distribution $U\{s_i, e_i\}$, choose index ξ_i;
2. transfer feature vector at index ξ_i as an output features for i-th fragment: $y_{ij} = x_{\xi_i j}$.

During backpropagation, the error propagates to the element, which was selected for forward-propagation, similar to max – pooling:

$$\frac{\partial E}{\partial x_{ij}} = \begin{cases} \delta_{pj}, & if \ \exists p : i \in \{s_p, \dots, e_p\} \ and \ i = \xi_p, \\ 0, otherwise. \end{cases} \tag{5}$$

As a low-level element was selected with discrete uniform distribution, so the expected output of fragment stochastic pooling is the same as output for fragment average pooling. So, our variant of stochastic pooling can be interpreted as a generalization of fragment average pooling. Thus, at the inference phase, we use the average pooling as its deterministic version.

4.4   Mixed Fragment Pooling

Classical max and average pooling methods have their own drawbacks [26, 27]. To minimize these drawbacks the Mixed pooling technique was implemented [26].

It randomly employs the local max pooling and average pooling methods during training CNNs.

By analogy, we have formed Mixed fragment pooling. This method implies random selection between fragment max, average and stochastic pooling while model training. From the model inference view, we have studied the implementation of two different configurations:

- with average pooling: HCRNN [Mix avg];
- with max pooling: HCRNN [Mix max].

For both model configurations, we use fixed probabilities for pooling technique selection at the training stage. These probabilities were selected having based on the following considerations:

- a half of training time the same pooling as on inference side is used (to ensure the progressive changes in weights towards the local optimum);
- alternative deterministic pooling has the probability of 0.3 (it plays an auxiliary role in moving to the optimum);
- fragment stochastic pooling has the probability of 0.2 (it is used for the regularization and avoiding the model being trapped into local minima).

All HCRNN configurations studied in this work are shown in Fig. 4.

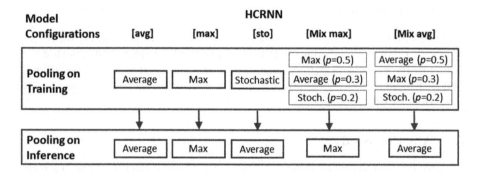

Fig. 4. HCRNN models configuration

5 Experimental Results

5.1 Dataset Description

All the configurations of proposed HCRNN architecture were evaluated on the benchmark IAMonDo dataset [12]. The dataset contains 941 online handwritten documents with different handwritten objects: text, drawings, diagrams, formulas, tables, lists and markings). All the strokes are marked and have time-stamps. It allows using this dataset for the evaluation of DSS algorithms and models. We use well-established dataset division on the training/validation/test for five-class classification [5, 25].

5.2   Evaluation Procedure

For the proposed approaches evaluation, the machine-learning models were implemented using PyTorch framework. We apply Adam optimizer with a learning rate of 0.001. As a loss function, we used MSE (mean squared error). The fragment pooling layers, described in the previous section, had been implemented as custom modules for PyTorch on CUDA C language. The models were trained for 2000 epochs with a batch size of 20 documents.

At the data preparation stage, The Ramer-Douglas-Peucker (RDP) decimating algorithm with $\varepsilon = 1.0$ was applied for every stroke. Having decimated, the average number of points per document reduced in 2.25 times. For each described above HCRNN configuration, two models were evaluated: model trained on raw train dataset, and model trained on augmented dataset.

The data augmentation was implemented on-the-fly before training batch formation. The following transformations were applied for each document:

- rotation of the whole document by a random angle, chosen according to the normal distribution with $\mu = 0$ and $\sigma = \pi/6$;
- partition and permutation transform. This procedure supposes a random split of a document into n sequence parts (n obeys the uniform distribution $U\{1, 6\}$). Finally, the random shuffle procedure is implemented for the given sequence parts.

To evaluate models, we used the stroke classification accuracy and other standard metrics [25]. We have run each experiment 10 times and reported the results of the classifier's performance (mean accuracy and standard deviation). The best model for every experiment was chosen by the highest accuracy on the validation set during 2000 epochs.

5.3   Ablation Study

We have conducted several experiments to estimate an influence of pooling techniques and parameter N_F on the HCRNN models performance. The first experiment was aimed at the investigation of the proposed fragment pooling techniques. We use the same HCRNN architecture ($N_F = 40$) with different pooling layers. Also we estimated the influence of the proposed hierarchical structure on the model accuracy and performance. On this purpose, we implemented CRNN model, which differs from HCRNN by the position of fragment pooling in neural network architecture. In CRNN we moved fragment mix average pooling to the end of the architecture. Though this model does not contain a fragment pooling layer between CNN and RNN sub-networks, but the transition to the stroke level is implemented as an aggregation of point-wise classifier results.

The test results for all the model configurations are shown in Table 1. The best model, trained on the raw data, was achieved using fragment Mix max pooling layer. Adding of the augmented data allows increasing accuracy on ~1%. For augmented train data, the model with fragment Mix average pooling layer has the best performance. Generally, the models with fragment Mix polling show 0.5% better accuracy comparing to other models. CRNN model has the worst accuracy in the case of training on raw data and the second place from the end in the case of training on augmented data. In addition, the CRNN model trained four times slower than hierarchical models.

Table 1. Performance of the models with different pooling layers

Architecture [pooling technique]	Accuracy, % models trained with raw data	Accuracy, % models trained with augmented data
HCRNN [avg]	95.76 ± 0.35	97.21 ± 0.18
HCRNN [max]	95.78 ± 0.30	96.99 ± 0.12
HCRNN [sto]	95.70 ± 0.29	96.44 ± 0.25
HCRNN [Mix avg]	96.23 ± 0.28	$\mathbf{97.42 \pm 0.16}$
HCRNN [Mix max]	$\mathbf{96.39 \pm 0.23}$	97.32 ± 0.24
CRNN [Mix avg]	95.17 ± 0.41	96.68 ± 0.31

In the frame of the second experiment, we have studied the influence of parameter N_F on HCRNN model performance. Here we use HCRNN architecture with fragment Mix average pooling, which shows the best accuracy in case of training on augmented data. There are five N_F values in range $[20, 100]$ with step 20 were studied. The results of classifiers performance are presented in Table 2 and Fig. 5.

Table 2. Comparison of HCRNN [Mix avg] models with different complexity

Parameter N_F	Accuracy, %	Number of weights	Latency, ms per document
20	96.35 ± 0.32	14125	10.9
40	97.42 ± 0.16	54925	23.2
60	$\mathbf{97.58 \pm 0.21}$	122405	45.1
80	97.53 ± 0.22	216565	77.7
100	97.55 ± 0.19	337405	120.6

Also they contain information about the size and latency for each model configuration. These are very important parameters for on-device solutions. The average latency was estimated by demo applications for on-device performance testing. The test procedures were conducted ten times for each model in the same condition on the Samsung Note 20 device. Algorithm implemented in C++

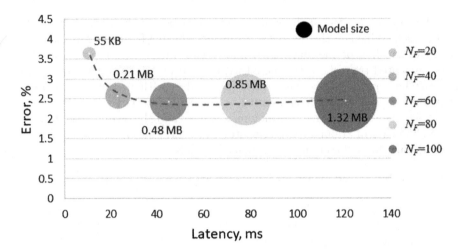

Fig. 5. Comparison of HCRNN models with different complexity

for CPU in single-threaded mode. The average latency for one IAMonDo document is listed in Table 2. The implementation of hierarchical architecture with fragment pooling layer allows reducing latency four times comparing to CRNN architecture (from 94.6 ms to 23.2 ms).

The best accuracy was obtained by the HCRNN model with fragment Mix average polling for the parameter $N_F = 60$. Further increase of parameter N_F leads to rising of the model size and latency without any gain in accuracy. The confusion matrix for the best model from the series of 10 experiments is presented in Table 3. The accuracy of this model is 97.92%. The highest precision got the *Text* class and the *List* class got the lowest one. The lower precision of *Lists* can be explained by fewer presents of this class in the train data and confusion with *Text* and *Tables*. The examples of inaccuracies of the classification are shown in Fig. 6.

Table 3. Confusion matrix of best HCRNN model

	Drawing	Text	Table	List	Math
Drawing	**97.52%**	0.59%	0.86%	0.15%	0.86%
Text	0.18%	**99.12%**	0.20%	0.44%	0.07%
Table	0.68%	0.81%	**97.91%**	0.41%	0.19%
List	2.60%	4.82%	4.65%	**87.61%**	0.32%
Math	3.47%	0.64%	1.01%	0.03%	**94.85%**

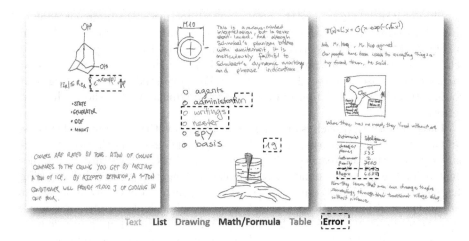

Text   **List   Drawing   Math/Formula   Table**   Error

Fig. 6. Examples of inaccuracies in the classification

The presented HCRNN model has established a novel state-of-the-art accuracy for the task of multi-class handwritten stroke classification (Table 4). Comparing to the previous state-of-the-art, the classification accuracy was increased on 0.33% up to **97.58%**. Also, this HCRNN model has twice smaller latency: 45.1 ms vs 86.2 ms [5].

Table 4. Performance characteristic of the best model

Method	Description	Accuracy, %
A. Delaye [7]	Hierarchical CRF with tree structure	93.46
Jun-Yu Ye [25]	Graph attention networks	94.64 ± 0.29
HRNN [5]	Hierarchical RNN+RNN	97.25 ± 0.25
HCRNN	Hierarchical CNN+RNN	**97.58 ± 0.21**

6   Conclusion and Outlook

This work presents a novel approach for highly efficient multi-class handwritten stroke classification. The proposed solution uses the HDNN architecture, which is formed by taking into account the hierarchical structure of handwritten documents. The point and stroke levels are distinguished in the classifier and the special fragment pooling techniques are proposed to align these levels. The introduced HCRNN architecture uses the stack of 1D CNNs on the point level and the stack of RNNs on the stroke level. Such an approach allows extracting information effectively regarding the local and global context of a handwritten document.

The substantial element of the proposed architecture is the fragment pooling layer. This layer supposes that the procedure of feature aggregation or selection is held in the frame of the particular level entity. The fragment pooling allows getting a higher computational efficiency and lower memory consumption comparing to the case of direct level stacking. We have studied several fragment pooling techniques and found out that mixed fragment polling shows the best performance.

The presented approach is aimed at classification with any number of classes, but the evaluation was carried out for the case of 5-classes due to the need to compare with existing approaches. The presented HCRNN architecture was implemented for the five-class handwritten stroke classification and evaluated using the benchmark IAMonDo dataset. The work establishes a novel state-of-the-art rate in the multi-class stroke classification for online handwritten documents with the accuracy **97.58% ± 0.21**. In addition, the proposed approach demonstrates high computing efficiency, which results in low memory consumption and high document processing speed.

The presented results provide the background for high-performance mobile applications with free-form handwriting input. These results can be used not only for handwritten stroke classification but also for handwritten object grouping and recognition in the frame of DSS. The principles of the hierarchical architecture supplemented by the proposed fragment pooling techniques can be applied for other problems of hierarchical data processing and recognition, for example, text, speech, industrial data, etc.

References

1. Artz, B., Johnson, M., Robson, D., Taengnoi, S.: Taking notes in the digital age: evidence from classroom random control trials. J. Econ. Educ. **51**(2), 103–115 (2020)
2. Bishop, C.M., Svensen, M., Hinton, G.E.: Distinguishing text from graphics in online handwritten ink. In: Proceedings of International Workshop on Frontiers in Handwriting Recognition, pp. 142–147 (2004)
3. Chung, J., Ahn, S., Bengio, Y.: Hierarchical multiscale recurrent neural networks. In: Proceedings of ICLR (2017)
4. Darvishzadeh, A., et al.: CNN-BLSTM-CRF network for semantic labeling of students' online handwritten assignments. In: Proceedings of International Conference on Document Analysis and Recognition, pp. 1035–1040 (2019)
5. Degtyarenko, I., et al.: Hierarchical recurrent neural network for handwritten strokes classification. In: ICASSP 2021–2021 IEEE International Conference on Acoustics, Speech and Signal Processing (ICASSP), pp. 2865–2869 (2021)
6. Degtyarenko, I., Radyvonenko, O., Bokhan, K., Khomenko, V.: Text/shape classifier for mobile applications with handwriting input. Int. J. Doc. Anal. Recognit. **19**(4), 369–379 (2016)
7. Delaye, A., Liu, C.L.: Contextual text/non-text stroke classification in online handwritten notes with conditional random fields. Pattern Recogn. **47**(3), 959–968 (2014)

8. Dengel, A., Otte, S., Liwicki, M.: Local feature based online mode detection with recurrent neural networks. In: Proceedings of International Conference on Frontiers in Handwriting Recognition, pp. 533–537 (2012)
9. El Hihi, S., Bengio, Y.: Hierarchical recurrent neural networks for long-term dependencies. In: Proceedings of NIPS (1995)
10. Gonnet, P., Deselaers, T.: Indylstms: independently recurrent LSTMs. In: Proceedings of ICASSP, pp. 3352–3356 (2020)
11. Indermühle, E., Frinken, V., Bunke, H.: Mode detection in online handwritten documents using BLSTM neural networks. In: Proceedings of International Conference on Frontiers in Handwriting Recognition, pp. 302–307 (2012)
12. Indermühle, E., Liwicki, M., Bunke, H.: IAMonDo-database: an online handwritten document database with non-uniform contents. In: Proceedings of International Workshop on Document Analysis Systems, pp. 97–104 (2010)
13. Khomenko, V., Shyshkov, O., Radyvonenko, O., Bokhan, K.: Accelerating recurrent neural network training using sequence bucketing and multi-GPU data parallelization. In: Proceedings of IEEE DSMP, pp. 100–103 (2016)
14. Khomenko, V., Volkoviy, A., Degtyarenko, I., Radyvonenko, O.: Handwriting text/non-text classification on mobile device. In: Proceedings of International Conference on Artificial Intelligence and Pattern Recognition, pp. 42–49 (2017)
15. Lee, K., Kim, J., Kim, J., Hur, K., Kim, H.: Stacked convolutional bidirectional LSTM recurrent neural network for bearing anomaly detection in rotating machinery diagnostics. In: Proceedings of IEEE International Conference on Knowledge Innovation and Invention, pp. 98–101 (2018)
16. Lei, L., Lu, J., Ruan, S.: Hierarchical recurrent and convolutional neural network based on attention for Chinese document classification. In: Proceedings of IEEE Chinese Control And Decision Conference, pp. 809–814 (2019)
17. Liu, F., et al.: An attention-based hybrid LSTM-CNN model for arrhythmias classification. In: Proceedings of International Joint Conference on Neural Networks, pp. 1–8 (2019)
18. Mousavi, S., Weiqiang, Z., Sheng, Y., Beroza, G.: CRED: a deep residual network of convolutional and recurrent units for earthquake signal detection. Sci. Rep. **9**, 1–14 (2019)
19. Polotskyi, S., Deriuga, I., Ignatova, T., Melnyk, V., Azarov, H.: Improving online handwriting text/non-text classification accuracy under condition of stroke context absence. In: Rojas, I., Joya, G., Catala, A. (eds.) IWANN 2019. LNCS, vol. 11506, pp. 210–221. Springer, Cham (2019). https://doi.org/10.1007/978-3-030-20521-8_18
20. Polotskyi, S., Radyvonenko, O., Degtyarenko, I., Deriuga, I.: Spatio-temporal clustering for grouping in online handwriting document layout analysis with GRU-RNN. In: Proceedings of International Conference on Frontiers in Handwriting Recognition, pp. 276–281 (2020)
21. Saeedan, F., Weber, N., Goesele, M., Roth, S.: Detail-preserving pooling in deep networks. In: Proceedings of International Conference on Computer Vision and Pattern Recognition, pp. 9108–9116 (2018)
22. Willems, D., Rossignol, S., Vuurpijl, L.: Features for mode detection in natural online pen input. In: Proceedings of Conference of the International Graphonomics Society, pp. 113–117 (2005)
23. Yang, J., et al.: A hierarchical deep convolutional neural network and gated recurrent unit framework for structural damage detection. Inf. Sci. **540**, 117–130 (2020)

24. Yang, Z., Yang, D., Dyer, C., He, X., Smola, A., Hovy, E.: Hierarchical attention networks for document classification. In: Proceedings of NAACL-HLT, pp. 1480–1489 (2016)
25. Ye, J., Zhang, Y., Yang, Q., Liu, C.: Contextual stroke classification in online handwritten documents with graph attention networks. In: Proceedings of the International Conference on Document Analysis and Recognition, pp. 993–998 (2019)
26. Yu, D., Wang, H., Chen, P., Wei, Z.: Mixed pooling for convolutional neural networks. In: Rough Sets and Knowledge Technology, pp. 364–375 (2014)
27. Zeiler, M., Fergus, R.: Stochastic pooling for regularization of deep convolutional neural networks. In: Proceedings of ICLR (2013)
28. Zhelezniakov, D., Zaytsev, V., Radyvonenko, O., Yakishyn, Y.: InteractivePaper: minimalism in document editing UI through the handwriting prism. In: Proceedings of ACM Symposium on UIST, pp. 13–15 (2019)

RFDoc: Memory Efficient Local Descriptors for ID Documents Localization and Classification

Daniil Matalov[1,2(✉)] 📷, Elena Limonova[1,2] 📷, Natalya Skoryukina[1,2] 📷, and Vladimir V. Arlazarov[1,3] 📷

[1] Smart Engines Service LLC, Moscow, Russia
{d.matalov,limonova,skleppy.inc,vva}@smartengines.com
[2] Federal Research Center Computer Science and Control RAS, Moscow, Russia
[3] Institute for Information Transmission Problems (Kharkevich Institute) RAS, Moscow, Russia

Abstract. Majority of recent papers in the field of image matching introduces universal descriptors for arbitrary keypoints matching. In this paper we propose a data-driven approach to building a descriptor for matching local keypoints from identity documents in the context of simultaneous ID document localization and classification on mobile devices. In the first stage, we train features robust to lighting changes. In the second stage, we select the most best-performing and discriminant features and forms a set of classifiers which we called RFDoc descriptor. To address the problem of limited computing resources the proposed descriptor is binary rather than a real-valued one. To perform experiments we prepared a dataset of aligned patches from a subset of identity document types presented in MIDV datasets and made it public. RFDoc descriptor showed similar performance in complex document detection and classification on the test part of the MIDV-500 dataset to the state-of-the-art BEBLID-512 descriptor, which is more than 2.6 times less memory efficient than RFDoc. On a more complex MIDV-2019 dataset RFDoc showed 21% fewer classification errors.

Keywords: Local features learning · Document classification · Identity documents recognition · Keypoint descriptors

1 Introduction

ID document recognition systems are already deeply integrated into human activity, and the pace of integration is only increasing [17]. Almost every person faces document recognition algorithms in banking, sharing economy, border

This work was partially financially supported by Russian Foundation for Basic Research, projects 19-29-09066 and 18-29-26035.

J. Lladós et al. (Eds.): ICDAR 2021, LNCS 12822, pp. 209–224, 2021.
https://doi.org/10.1007/978-3-030-86331-9_14

crossing, hospitality, medicine, insurance, and other areas that require authentication by the ID document.

The very first fundamental document recognition problems are image document detection and classification [9]. Document image classification is an assignment of the document image to one of several categories based on its contents. Document localization in a digital image usually consists of finding a document quadrangle whose coordinates are the coordinates of the document's corners in the image basis. A high-quality solution to these problems plays a significant role in a document recognition process. Sometimes, document type misclassification can be fatal for the whole ID document recognition process. Specific knowledge about the document type allows defining optimal parameters for document image processing to achieve high recognition accuracy. This also may decrease the size of the input data for character recognition algorithms and affect the required amount of computing resources. For example, modern text-in-the-wild recognition algorithms, which are supposed to recognize all text characters regardless of their font size and semantics, require high-end computational performance [20, 36].

There are several document location approaches based on extraction and analysis of the document boundaries, lines, segments, and corners [25, 37, 41]. These methods perform testing a set of hypotheses about the consistency of the extracted geometric primitives with the distortion model of the document rectangle.

Some modern deep learning-based approaches are used for document classification and retrieval [15, 19]. However, training high accuracy algorithm requires tens of thousands of training samples, which is complicated to obtain for regulatory and privacy reasons. Such algorithms also contain millions of parameters that significantly affect computation time and memory consumption including RAM and persistent storage.

ID documents often have a fixed geometry and a set of predefined elements such as static text filling, logos, and fixed background. Paper [6] states that visual-based approaches are the best suited for ID document recognition. The statement is supported by experimental results of the proposed template matching-based framework for ID document detection and classification. A block diagram summarizing the main blocks of the approach is shown in Fig. 1. The approach has the following properties:

- Robustness to capturing conditions. This property essentially depends on the robustness of keypoints extraction and descriptors computing algorithms;
- There is no need for a large training dataset. The model of a specific document type may be built from a single example;
- The algorithm exploits a model of projecting a 2D plane from a 3D world to a 2D plane of the digital image, which leads to high precision localization;
- The approach is extensible to a large number of various document types;
- Every algorithm (keypoint extraction, descriptor computation, matching, etc.) in the approach can be replaced or improved independently of each other;

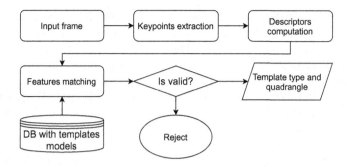

Fig. 1. Template matching with local feature descriptors

– The method's design is extensible and flexible. There exist several papers [29,31] proposing the usage of lines and various rejection rules for RANSAC-provided hypotheses.

One of the most important advantages of the approach is that it demonstrates industrial precision and performance on such a widespread class of devices as smartphones [9,29]. However, with the development of such industrial systems, the number of supported documents increases along with the number of indexed document template models, which affects the algorithm's memory consumption and speed. Paper [31] shows that usage of the BinBoost [38] descriptors instead of SURF [8] allowed to significantly reduce memory consumption without noticeable changes in classification quality.

In this paper, we focus on constructing a fast algorithm for constructing a memory-efficient, highly discriminative keypoint descriptor for ID document detection and classification with template matching approach.

2   Previous Work

SIFT [23] is the most known and well-studied algorithm for local image feature detection and description. It was inspired by the properties of neurons from the inferior temporal cortex, which are responsible for arbitrary object recognition in primate vision, and uses handcrafted gradient-based features. According to paper [12], the SIFT descriptors are still among the most widely used description algorithms producing the best accuracy for text document matching. However, SIFT description of a single keypoint is a 128 dimensioned real-valued vector. In the case of 32-bits floating-point values, it leads to the requirement of 4096 bits of memory for a single local feature point representation. In addition, the SIFT algorithm requires large computing resources and special GPU implementations to ensure real-time performance [4].

There are many universal descriptors proposed to address the problem of intense calculations and dimensionality of the SIFT, for example, SURF [8]. To increase the calculation speed and to reduce memory consumption, many binary descriptors, such as BRIEF [11], BRISK [22], ORB [26], RFD [16] were proposed. They allow to calculate Hamming distance with efficient hardware instruction and enhance memory usage efficiency. The main problem with these descriptors is that they show significantly lower accuracy than real-valued descriptors [30,34].

There are many supervised learning-based local keypoints description algorithms proposed in recent time. Such descriptors as BinBoost [38], BEBLID [33] use modification of AdaBoost algorithm [27] to train compact image representations. Deep learning is also widely used to train descriptors. Some of them use classical approach with L_2 loss minimization [28,32,35], while the most modern approaches propose training Siamese networks [18,21] using triplet loss optimization [7,21].

Several papers exist that propose special document-oriented descriptors. Paper [14] introduces a descriptor based on the usage of geometrical constraints between pairs of nearest points around a keypoint. The descriptor directly addresses the problem of robustness of the keypoint neighborhood description under projective and affine distortions. However, the algorithm considers centroids of each word connected component as a keypoint, which requires an algorithm of connected component extraction with high quality and robustness to local lightning. The further modifications of the SRIF method [14] resulted in a SSKRIF [13] descriptor, which supports such classical local keypoints as SURF, SIFT, etc. The paper [13] states that the stability of the keypoints extraction directly affects the algorithm performance.

3   A Dataset of Aligned Patches from ID Document Images

There exist several datasets for benchmarking arbitrary patch verification problems, however, there are no datasets of patches extracted from ID documents. To perform experiments we prepared a dataset of patches from ID documents using a subset of document types from well-known MIDV-500 [5] and MIDV-2019 [10] datasets. The MIDV-500 [5] dataset contains video clips of 50 different document types captured with a smartphone camera. The dataset addresses such problems of mobile camera-based document recognition as complex background, projective distortions, highlighting, and other hand-held capturing caused distortions. The MIDV-2019 [10] is an extension, which includes the documents captured under extreme projective distortions and low-lightning in 4K resolution. Each video-clip in the mentioned datasets consists of 30 frames and corresponding ground-truths including coordinates of the document corners. We randomly selected 10 document types (chn.id, deu.passport.new, dza.passport, esp.id.new, fin.drvlic, grc.passport, hun.passport, irn.drvlic, prt.id, usa.passportcard) from each of the mentioned MIDV datasets. For each video clip we randomly selected no more than 7 frames rejecting those frames with document corners falling out of the frame for more than 10% of the average diagonal length of a document. The procedure for extracting matching patches from a single frame is described in Algorithm 1.

Algorithm 1: Retrieving a set of matching document patches from a single frame

---

Input: F – an input frame, Q – a document quadrangle in the input frame basis, N_b – a document quadrangle in the normalized frame basis, R – a set of rectangles in N_b basis;

Output: A set of pairs of image patches Y

Compute transformation matrix $H : Q \rightarrow N_b$;

Compute scale $s \leftarrow \frac{A_{N_b}}{A_Q}$ as a ratio of areas N_b and Q;

Compute normalized document image D applying H to F;

Run YACIPE [24] algorithm to D and get a set of keypoints $K \leftarrow \{(x, y, size)\} \in D$;

Remove points from a set K that do not lie inside at least one rectangle from R;

foreach *keypoint* $k_i = (x, y, size) \in K$ **do**

$\quad$ $k_i' \leftarrow (x_i', y_i', size_i')$, where $(x_i', y_i') \leftarrow H^{-1}(x_i, y_i)$, $size' \leftarrow size \cdot s$;

$\quad$ Extract aligned image patches (P_i, P_i') for keypoints k_i and k_i' from the original frame and the normalized one;

$\quad$ $Y \leftarrow Y \cup (P_i, P_i', l_i)$, where $l_i = 1$;

end

The parameters N_b and R were document type specific. Quadrangle N_b defined document coordinates in a fixed-size basis where a document occupied the entire image plane. The size of the basis was computed according to the physical sizes of the document type to ensure that 1 physical millimeter occupied 10 pixels. For example, the physical width and height of the alb.id template are 85.6 and 54 mm and the size of the N_b basis is (856, 540). The set of rectangles R specified document regions containing personal data (red rectangles in Fig. 2). The filtering of points by rectangles in Algorithm 1 is done to avoid potential

Fig. 2. Regions of static document type dependent elements (green) and personal data (red) (Color figure online)

overfitting. We proceed from the assumption that the document model in template matching approach is built from the descriptions of the points of the static elements of the document (green rectangles in Fig. 2). To complete the dataset of aligned patches with negative pairs we used a well-known Brown dataset [40]. A single negative pair was obtained by random sampling a patch from Y and a patch from sets of 500k patches in the Brown dataset. All the patches from Y that were used in negative pair construction were removed from the matching part of the dataset. The retrieved dataset of patches were divided into 3 sets: 75k pairs for features training, 75k for features selection and 350k for test. All the sets are balanced (50% of matching and 50% of dismatching pairs) and does not intersect. Examples of pairs of patches from the collected dataset presented in Fig. 3. The dataset is available for download at ftp://smartengines.com/rfdoc. The structure of the dataset and the way of indexing patches is similar to that used in the Brown dataset [40].

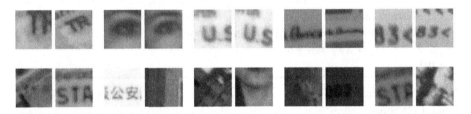

Fig. 3. Examples of matching (top row) and non-matching (bottom row) pairs from the dataset

4   Proposed Image Descriptor

The descriptor that we propose is named RFDoc since the training procedure is very similar to RFD [16].

4.1   Image Patch Descriptor Construction

RFDoc descriptor of the image patch is a response vector of a set of binary classifiers. Firstly, the feature space is constructed. Input image patch P is mapped into 8 images, each produced with a bilinear soft assignment of image gradient orientation. GM_0 and GM_6 in second column in Fig. 4 shows the magnitude of gradients which orientations mostly drop to sectors with $[0; \frac{\pi}{4}]$ and $[\frac{6\pi}{4}; \frac{7\pi}{4}]$ relatively. Unlike many other gradient orientation based-features, we use the L_1 norm of gradient magnitude in our implementation to increase robustness, computation speed and reduce memory consumption. A binary classifier h in RFDoc is defined with rectangular region R, gradient map index or orientation map c

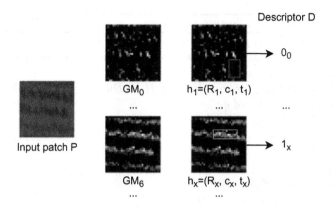

Fig. 4. Image patch description

and threshold t. The classification rule of a single classifier for an input match P is implemented as follows:

$$h(P) = \begin{cases} 1, & \text{if } g(P) \geq t \\ 0, & \text{otherwise} \end{cases}, \tag{1}$$

where

$$g(P) = \frac{\sum_{(x,y) \in R} GM_c(x,y)}{\sum_{i=0}^{7} \sum_{(x,y) \in R} GM_i(x,y)}. \tag{2}$$

The values of the function g are efficiently computed using the concept of integral images [39].

4.2   Training and Features Selection

Features training and selection are based on the idea that Hamming distances between supposedly matching pairs are significantly less than the distances between dissimilar pairs of patches. The training process is organized in two steps: 1) independent training of a set of classifiers of the type h setting thresholds leading to the best accuracy [16] and 2) selecting the most accurate features considering the potentially excessive correlation of the classifiers' responses. Unlike the RFD [16] feature selection algorithm, we use a different way of calculating the feature correlation and limit the number of resultant features. Given M labeled pairs of patches (P_i^1, P_i^2, l_i), where $l_i \in \{1, -1\}$ indicates matching and non-matching pairs, a feature selection is performed according to Algorithm 2. The *corr* value models the frequency of coincidence of the responses of the classifiers on the corresponding patches and the values is normalized to $[0, 1]$.

Algorithm 2: Feature selection algorithm

Input: Set of labeled pairs of patches M, set of classifiers H, maximum
number of features to select N, correlation threshold t_c

Output: Set of selected features S

Compute accuracy q_i for each h_i from H on the set of patches M and sort
H in descending order of q_i;

$S \leftarrow S \cup \{h_0\}, j \leftarrow 1$;

while $|S| \leq N$ *and* $j < |H|$ **do**

    foreach $j \in \{1, ..., |H|\}$ **do**

        $g_{ik} \leftarrow true$;

        foreach $k \in \{1, ..., |S|\}$ **do**

            $nm_{jk} \leftarrow 0, nd_{jk} \leftarrow 0$;

            foreach $m \in \{1, ..., |M|\}$ **do**

                if $h_j(P_m^1) = h_k(P_m^1)$ **then**

                    | $nm_{jk} \leftarrow nm_{jk} + 1$

                else

                    | $nd_{jk} \leftarrow nd_{jk} + 1$

                end

                if $h_j(P_m^2) = h_k(P_m^2)$ **then**

                    | $nm_{jk} \leftarrow nm_{jk} + 1$

                else

                    | $nd_{jk} \leftarrow nd_{jk} + 1$

                end

            end

            $corr_{jk} \leftarrow (nm_{jk} - nd_{jk})/(4 \cdot |M|) + 0.5$;

            if $corr_{jk} \geq t_c$ **then**

                $g_{ik} \leftarrow false$;

                break;

            end

        end

        if $g_{ik} = true$ **then**

            | $S \leftarrow S \cup \{h_j\}$;

        end

    end

end

---

5   Experimental Evaluation

5.1   Training and Features Selection

In our RFDoc implementation we built a descriptor which took a square image with size of 32 pixels. Millions of rectangles R may be generated in this basis, however a huge number of them is usually highly correlated [16]. In our experiments, we generated rectangles with sides greater or equal to 3 pixels and with area less or equal to 256. Threshold fitting was performed on a set of patches from ID documents (see in Sect. 3). To ensure compactness and high speed of comparison of a pair of descriptors (see in Sect. 5.5) we fixed the maximum number of the selected features N to 192. In the feature selection algorithm, we fixed the t_c to 0.7, since smaller values of t_c did not allow to select 192 features, but greater values led to less discriminant descriptor and showed a slightly worse performance.

5.2   Performance in ID Document Patches Verification

To explore the quality of the RFDoc descriptor we used a separate set of 350k pairs of patches from documents. Unfortunately, to our knowledge, there was no publicly available authorized RFD [16] implementation and we were unable to explore RFD performance in our test configurations. All experimental results were obtained using the OpenCV implementation of version 4.5.1. According to the ROC curve presented in Fig. 5 RFDoc descriptor outperformed state-of-the-art binary descriptors including the most efficient binary descriptor for arbitrary image matching BEBLID-512 [33].

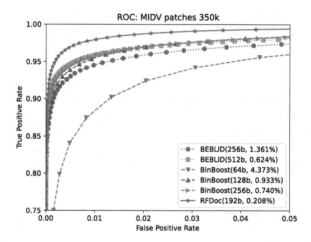

Fig. 5. Comparison of our RFDoc descriptor to the state-of-the-art binary descriptors. The amount of bits required to store a keypoint representation and error rate to provide 95% True Positive Rate are stated in the legend

5.3   Complex Camera-Based Document Recognition

To explore the performance of the RFDoc descriptor in complex ID document location and classification we evaluated the approach proposed in [6] on the rest 40 document types from the MIDV datasets that did not participate in the patch verification setup. Unlike [6] we used a bruteforce matching instead of FLANN to achieve reproducible results and to exclude potential collisions caused by approximate nearest neighbors search, and we did not used restrictions

Table 1. Experimental results on MIDV-500 dataset

Rank	Descriptor	Classification	Localization	Bits
1	BEBLID-512 [33]	**93.508**	**85.226**	512
2	*RFDoc*	93.458	85.128	192
3	BEBLID-256 [33]	92.783	84.072	256
4	BinBoost-256 [38]	91.116	81.132	256
5	SURF [8]	91.241	82.783	2048
6	BinBoost-128 [38]	85.958	73.588	128

Table 2. Experimental results on MIDV-2019 dataset

Rank	Descriptor	Classification	Localization	Bits
1	*RFDoc*	**88.875**	**75.535**	192
2	BEBLID-512 [33]	85.854	75.001	512
3	BEBLID-256 [33]	83.833	72.368	256
4	BinBoost-256 [38]	79.916	63.074	256
5	SURF [8]	75.666	61.542	2048
6	BinBoost-128 [38]	68.791	50.262	128

since they were too strict for MIDV datasets. All the parameters related to ideal template images, keypoint extraction and RANSAC were fixed according to [29]. The algorithm of computing detection performance is based on the calculation of the maximal deviation of the computed document corner coordinates divided by the length of the shortest document boundary side [29]. We used the SURF [8] algorithm for keypoints extraction. According to experimental results on the MIDV-500 dataset presented in Table 1, the RFDoc descriptor showed similar document classification and localization performance to BEBLID-512 being 2.65 times more memory efficient. On a more challenging MIDV-2019 dataset, which includes extreme projective distortions under low-lightning, RFDoc descriptor outperformed BEBLID-512 and showed 21% fewer classification errors (Table 2).

5.4   Memory Consumption

A required amount of memory to index keypoint descriptors with OpenCV brute force matcher for 50 document types presented in MIDV-500 dataset is showed in Table 3. The RFDoc descriptor that showed close and better performance on in our experiments is more memory efficient than BEBLID-512. If we scale the number of document types supposed to be detected to a number of supporting documents by industrial recognition systems, for example, 2000 document types, a difference in memory consumption will become critical.

Table 3. Required amount of memory to index descriptors of MIDV-500 templates

Descriptor	Bits	Memory (unpacked), MB	Memory (zip), MB
SURF [8]	2048	82.0	23.6
BEBLID [33]	256	11.5	2.9
	512	22.7	5.9
BinBoost [38]	128	5.8	1.4
	256	11.5	2.9
RFDoc	192	8.6	2.2

5.5 Matching Speed Comparison

It is rather difficult to reliably assess the impact of the type of descriptors on the matching speed. Things like the complexity of the descriptor computing algorithm, descriptor comparison metric, and the descriptor's data type, the number of document templates in the database, and the indexing algorithm directly affect the matching speed. We do not compare the overall matching time of different types of descriptors implemented with different degrees of code optimization. Instead, we evaluate the influence of different types of binary descriptors on matching speed in terms of the number of CPU instructions required to calculate the distance between a pair of descriptors.

One of the most important advantages of the binary descriptors above integer and real-valued ones is a usage of Hamming distance as a measure of similarity of a pair of descriptors. The Hamming distance computation can be implemented with bitwise XOR and population count instructions, usually included in the instruction set of modern processors.

The main application for our binary descriptors is edge computing, that is, the inference devices are smartphones, embedded systems, etc. They are mainly built as a system on a chip (SoC). Most modern industrial SoC's implement x86-64, ARM, or MIPS architectures. These architectures provide hardware instruction for the population count (popcnt) with a varying input width and processed width (see Table 4) [1–3]. Processed width is the size of the element inside an input data chunk for which the popcnt is computed. The principle difference between them is that basic x86-64 (since SSE4a on Intel and TBA on AMD) supports popcnt via single hardware instruction, while ARM, MIPS introduce it as a part of SIMD extensions. Population count is also supported on x86-64 via AVX-512 vector extension on modern CPUs. We propose the following model to compare the speed of calculating the distance between different binary descriptors. Algorithm 3 of Hamming distance computing using SIMD vector extension is presented. On x86-64 without AVX-512 we have $P = L$ and simply sequentially process 64-bit data chunks and accumulate total Hamming distance.

To obtain accurate performance of the algorithm we should consider the number of arithmetic logic units (ALUs) able to perform each operation and its latency. Such an estimate should be determined for each specific processor,

---

Algorithm 3: Hamming distance computation using SIMD extension

---

Input: Descriptors $d0$, $d1$ of length D bit, L – SIMD vector length in bits, P – processed by popcnt width in bits, padd is an operation of pairwise addition of neighbouring elements in a vector

Output: Hamming distance H between $d0$ and $d1$

$h \leftarrow 0$;

foreach $i \in \{0, ..., D/L - 1\}$ **do**

　　$h_{[iL+L-1:iL]} \leftarrow h_{[iL+L-1:iL]} + \texttt{popcnt}(XOR(d0_{[iL+L-1:iL]}, d1_{[iL+L-1:iL]}))$;

end

foreach $i \in \{0, ..., \log_2(L/P) - 1\}$ **do**

　　$h \leftarrow padd_i(h)$;

end

$H \leftarrow h$;

---

which is outside the scope of this work. Instead, we estimate the total number of instructions, which corresponds to the case with one ALU and the same execution time for all instructions. The number of instructions that process a fixed-size bit subset to calculate the Hamming distance for different architectures and binary descriptors is shown in Table 5.

Hamming distance computation between a pair of RFDoc descriptors is more than two times faster than the closest in terms of quality BEBLID-512 descriptor, and only slower than poor-performed BinBoost-128.

Table 4. Hardware support of population count instruction

Architecture	Instruction	Input width, bit	Processed width, bits
x86-64	POPCNT	32/64	32/64
x86-64 (AVX-512)	VPOPCNTW	128/256/512	16/32/64
ARM	VCNT	64/128	8
MIPS	PCNT	128	8/16/32/64

Table 5. A number instructions to compute Hamming distance for various binary descriptors on the most common SoC architectures in xor/popcnt/add format (less is better)

Descriptor	Bits	x86-64		ARM		MIPS	
		x/y/z	Total	x/y/z	Total	x/y/z	Total
BEBLID [33]	256	4/4/3	**11**	2/2/5	**9**	2/2/2	**6**
	512	8/8/7	**23**	4/4/7	**15**	4/4/4	**12**
BinBoost [38]	128	2/2/1	**5**	1/1/4	**6**	1/1/1	**3**
	256	4/4/3	**11**	2/2/5	**9**	2/2/2	**6**
RFDoc	192	3/3/2	**8**	2/2/5	**9**	2/2/2	**6**

5.6   Conclusion

In this paper we proposed a memory efficient local keypoint descriptor for ID documents localization and classification. The problem-specific dataset of aligned patches extracted from camera captured ID documents was created published. The proposed RFDoc descriptor showed 3 times fewer errors on the TPR95% metric than the state-of-the-art BEBLID-512 descriptor in the ID document patches verification. In complex camera-based ID document localization and classification on the MIDV-500 dataset, the proposed descriptor slightly loses only to BEBLID-512 showing 5% more classification errors being 2.65 times more memory efficient. On more challenging dataset for document recognition MIDV-2019 which includes more projective distortions and low-lightning, the proposed descriptor demonstrated the best performance with 21% fewer errors than the closest BEBLID-512 descriptor.

For future work extension, we plan to explore the possibility of decreasing memory footprint even more by reducing the total amount of features in the descriptor and to investigate the quality characteristics of the RFDoc descriptor in other tasks.

References

1. ARM NEON documentation. https://developer.arm.com/architectures/instruction-sets/simd-isas/neon/intrinsics
2. Intel Intrinsics Guide. https://software.intel.com/sites/landingpage/Intrinsics Guide/
3. MIPS SIMD documentation. https://www.mips.com/products/architectures/ase/simd
4. Acharya, K.A., Babu, R.V., Vadhiyar, S.S.: A real-time implementation of SIFT using GPU. J. Real-Time Image Process. **14**(2), 267–277 (2018)
5. Arlazarov, V., Bulatov, K., Chernov, T., Arlazarov, V.: MIDV-500: a dataset for identity document analysis and recognition on mobile devices in video stream. Comput. Opt. **43**(5), 818–824 (2019). https://doi.org/10.18287/2412-6179-2019-43-5-818-824
6. Awal, A.M., Ghanmi, N., Sicre, R., Furon, T.: Complex document classification and localization application on identity document images. In: 2017 14th IAPR International Conference on Document Analysis and Recognition (ICDAR). IEEE (November 2017). https://doi.org/10.1109/icdar.2017.77
7. Balntas, V., Johns, E., Tang, L., Mikolajczyk, K.: PN-Net: conjoined triple deep network for learning local image descriptors. arXiv preprint arXiv:1601.05030 (2016)
8. Bay, H., Tuytelaars, T., Van Gool, L.: SURF: speeded up robust features. In: Leonardis, A., Bischof, H., Pinz, A. (eds.) ECCV 2006, Part I. LNCS, vol. 3951, pp. 404–417. Springer, Heidelberg (2006). https://doi.org/10.1007/11744023_32
9. Bulatov, K., Arlazarov, V.V., Chernov, T., Slavin, O., Nikolaev, D.: Smart IDReader: document recognition in video stream. In: 2017 14th IAPR International Conference on Document Analysis and Recognition (ICDAR), vol. 06, pp. 39–44 (2017). https://doi.org/10.1109/ICDAR.2017.347

10. Bulatov, K., Matalov, D., Arlazarov, V.V.: MIDV-2019: challenges of the modern mobile-based document OCR. In: Twelfth International Conference on Machine Vision (ICMV 2019). SPIE (January 2020). https://doi.org/10.1117/12.2558438

11. Calonder, M., Lepetit, V., Strecha, C., Fua, P.: BRIEF: binary robust independent elementary features. In: Daniilidis, K., Maragos, P., Paragios, N. (eds.) ECCV 2010, Part IV. LNCS, vol. 6314, pp. 778–792. Springer, Heidelberg (2010). https://doi.org/10.1007/978-3-642-15561-1_56

12. Dang, B., Coustaty, M., Luqman, M., Ogier, J.M.: A comparison of local features for camera-based document image retrieval and spotting. Int. J. Doc. Anal. Recognit. (IJDAR) **22**, 247–263 (2019). https://doi.org/10.1007/s10032-019-00329-w

13. Dang, Q.B., Coustaty, M., Luqman, M.M., Ogier, J.M., Tran, C.D.: SSKSRIF: scale and rotation invariant features based on spatial space of keypoints for camera-based information spotting. In: 2018 International Conference on Content-Based Multimedia Indexing (CBMI), pp. 1–6 (2018). https://doi.org/10.1109/CBMI.2018.8516532

14. Dang, Q.B., Luqman, M.M., Coustaty, M., Tran, C.D., Ogier, J.M.: SRIF: scale and rotation invariant features for camera-based document image retrieval. In: 2015 13th International Conference on Document Analysis and Recognition (ICDAR), pp. 601–605 (2015). https://doi.org/10.1109/ICDAR.2015.7333832

15. Das, A., Roy, S., Bhattacharya, U., Parui, S.K.: Document image classification with intra-domain transfer learning and stacked generalization of deep convolutional neural networks. In: 2018 24th International Conference on Pattern Recognition (ICPR), pp. 3180–3185. IEEE (2018)

16. Fan, B., Kong, Q., Trzcinski, T., Wang, Z., Pan, C., Fua, P.: Receptive fields selection for binary feature description. IEEE Trans. Image Process. **23**(6), 2583–2595 (2014). https://doi.org/10.1109/TIP.2014.2317981

17. Goode, A.: Digital identity: solving the problem of trust. Biom. Technol. Today **2019**(10), 5–8 (2019)

18. Han, X., Leung, T., Jia, Y., Sukthankar, R., Berg, A.C.: Matchnet: unifying feature and metric learning for patch-based matching. In: Proceedings of the IEEE Conference on Computer Vision and Pattern Recognition, pp. 3279–3286 (2015)

19. Harley, A.W., Ufkes, A., Derpanis, K.G.: Evaluation of deep convolutional nets for document image classification and retrieval. In: 2015 13th International Conference on Document Analysis and Recognition (ICDAR). IEEE (August 2015). https://doi.org/10.1109/icdar.2015.7333910

20. Jaderberg, M., Simonyan, K., Vedaldi, A., Zisserman, A.: Reading text in the wild with convolutional neural networks. Int. J. Comput. Vis. **116**(1), 1–20 (2016)

21. Kumar BG, V., Carneiro, G., Reid, I.: Learning local image descriptors with deep Siamese and triplet convolutional networks by minimising global loss functions. In: Proceedings of the IEEE Conference on Computer Vision and Pattern Recognition (CVPR) (June 2016)

22. Leutenegger, S., Chli, M., Siegwart, R.Y.: Brisk: binary robust invariant scalable keypoints. In: 2011 International Conference on Computer Vision, pp. 2548–2555. IEEE (2011)

23. Lowe, D.G.: Object recognition from local scale-invariant features. In: Proceedings of the Seventh IEEE International Conference on Computer Vision, vol. 2, pp. 1150–1157 (1999). https://doi.org/10.1109/ICCV.1999.790410

24. Lukoyanov, A., Nikolaev, D., Konovalenko, I.: Modification of YAPE keypoint detection algorithm for wide local contrast range images. In: Tenth International Conference on Machine Vision (ICMV 2017), vol. 10696, pp. 305–312. International Society for Optics and Photonics, SPIE (2018). https://doi.org/10.1117/12.2310243

25. Puybareau, É., Géraud, T.: Real-time document detection in smartphone videos. In: 2018 25th IEEE International Conference on Image Processing (ICIP), pp. 1498–1502. IEEE (2018)

26. Rublee, E., Rabaud, V., Konolige, K., Bradski, G.: ORB: an efficient alternative to SIFT or SURF. In: 2011 International Conference on Computer Vision, pp. 2564–2571. IEEE (2011)

27. Schapire, R.E., Singer, Y.: Improved boosting algorithms using confidence-rated predictions. Mach. Learn. **37**(3), 297–336 (1999)

28. Simo-Serra, E., Trulls, E., Ferraz, L., Kokkinos, I., Fua, P., Moreno-Noguer, F.: Discriminative learning of deep convolutional feature point descriptors. In: Proceedings of the IEEE International Conference on Computer Vision, pp. 118–126 (2015)

29. Skoryukina, N., Arlazarov, V., Nikolaev, D.: Fast method of ID documents location and type identification for mobile and server application. In: 2019 International Conference on Document Analysis and Recognition (ICDAR). IEEE (September 2019). https://doi.org/10.1109/icdar.2019.00141

30. Skoryukina, N., Arlazarov, V.V., Milovzorov, A.: Memory consumption reduction for identity document classification with local and global features combination. In: Thirteenth International Conference on Machine Vision, vol. 11605, p. 116051G. International Society for Optics and Photonics (2021). https://doi.org/10.1117/12.2587033

31. Skoryukina, N., Faradjev, I., Bulatov, K., Arlazarov, V.V.: Impact of geometrical restrictions in RANSAC sampling on the id document classification. In: Twelfth International Conference on Machine Vision (ICMV 2019), vol. 11433, p. 1143306. International Society for Optics and Photonics (2020). https://doi.org/10.1117/12.2559306

32. Stankevièius, G., Matuzevièius, D., et al.: Deep neural network-based feature descriptor for retinal image registration. In: 2018 IEEE 6th Workshop on Advances in Information, Electronic and Electrical Engineering (AIEEE), pp. 1–4. IEEE (2018). https://doi.org/10.1109/AIEEE.2018.8592033

33. Suárez, I., Sfeir, G., Buenaposada, J.M., Baumela, L.: BEBLID: boosted efficient binary local image descriptor. Pattern Recognit. Lett. **133**, 366–372 (2020). https://doi.org/10.1016/j.patrec.2020.04.005

34. Tareen, S.A.K., Saleem, Z.: A comparative analysis of SIFT, SURF, kaze, akaze, ORB, and BRISK. In: 2018 International Conference on Computing, Mathematics and Engineering Technologies (iCoMET), pp. 1–10. IEEE (2018)

35. Tian, Y., Fan, B., Wu, F.: L2-net: deep learning of discriminative patch descriptor in Euclidean space. In: Proceedings of the IEEE Conference on Computer Vision and Pattern Recognition (CVPR) (July 2017)

36. Tong, G., Li, Y., Gao, H., Chen, H., Wang, H., Yang, X.: MA-CRNN: a multi-scale attention CRNN for Chinese text line recognition in natural scenes. Int. J. Doc. Anal. Recognit. (IJDAR) **23**(2), 103–114 (2019). https://doi.org/10.1007/s10032-019-00348-7

37. Tropin, D., Konovalenko, I., Skoryukina, N., Nikolaev, D., Arlazarov, V.V.: Improved algorithm of ID card detection by a priori knowledge of the document aspect ratio. In: Thirteenth International Conference on Machine Vision. SPIE (January 2021). https://doi.org/10.1117/12.2587029

38. Trzcinski, T., Christoudias, M., Lepetit, V.: Learning image descriptors with boosting. IEEE Trans. Pattern Anal. Mach. Intell. **37**(3), 597–610 (2015). https://doi.org/10.1109/tpami.2014.2343961

39. Viola, P., Jones, M.: Rapid object detection using a boosted cascade of simple features. In: Proceedings of the 2001 IEEE Computer Society Conference on Computer Vision and Pattern Recognition, CVPR 2001, vol. 1, pp. I-I. IEEE (2001)

40. Winder, S.A.J., Brown, M.: Learning local image descriptors. In: 2007 IEEE Conference on Computer Vision and Pattern Recognition. IEEE (June 2007). https://doi.org/10.1109/cvpr.2007.382971

41. Zhu, A., Zhang, C., Li, Z., Xiong, S.: Coarse-to-fine document localization in natural scene image with regional attention and recursive corner refinement. Int. J. Doc. Anal. Recognit. (IJDAR) **22**(3), 351–360 (2019)

Dynamic Receptive Field Adaptation for Attention-Based Text Recognition

Haibo Qin⬤, Chun Yang⬤, Xiaobin Zhu⬤, and Xucheng Yin$^{(\boxtimes)}$⬤

University of Science and Technology Beijing, Beijing, China
{chunyang,zhuxiaobin,xuchengyin}@ustb.edu.cn

Abstract. Existing attention-based recognition methods generally assume that the character scale and spacing in the same text instance are basically consistent. However, this hypothesis not always hold in the context of complex scene images. In this study, we propose an innovative dynamic receptive field adaption (DRA) mechanism for recognizing scene text robustly. Our DRA introduces different levels of receptive field features for classifying character and designs a novel way to explore historical attention information when calculating attention map. In this way, our method can adaptively adjust receptive field according to the variations of character scale and spacing in a scene text. Hence, our DRA mechanism can generate more informative features for recognition than traditional attention-based mechanisms. Notablely, our DRA mechanism can be easily generalized to off-the-shelf attention-based methods in text recognition to improve their performances. Extensive experiments on various public available benchmarks, including the IIIT-5K, SVT, SVTP, CUTE80, and ICDAR datasets, indicate the effectiveness and robustness of our method against the state-of-the art methods.

Keywords: Dynamic receptive field · Attention mechanism · Text recognition · Deep learning

1 Introduction

Scene images contain rich and informative texts which are beneficial for many applications, such as image retrieval, image content understanding, intelligent driving, and visual question answering. Although a multiplicity of solutions have been proposed for text recognition, scene text recognition is still a challenging task due to complex background, irregular arrangement, and other factors relied in scene image.

With the rapid progress of deep learning, text recognition solutions have achieved promising performance improvements. Among these solutions, encoder-decoder framework for attention-based recognizers gain popularity for effectively concentrating on informative areas. The encoder generally consists of an CNN sub-network for feature extraction and an RNN sub-network for modeling sequence information. The decoder generally contains an LSTM or GRU, an attention calculation module and a classification prediction layer. In the existing

© Springer Nature Switzerland AG 2021
J. Lladós et al. (Eds.): ICDAR 2021, LNCS 12822, pp. 225–239, 2021.
https://doi.org/10.1007/978-3-030-86331-9_15

Fig. 1. Examples of inconsistent character aspect ratios. In the same text example, the characters with red boxes have different ratios of width to height. For instance, the aspect ratios of character 'S' and character 'i' in the fourth picture of the second column are quite different. (Color figure online)

attention-based methods [2,3,13,23,28], the hidden state of LSTM in decoder contains the absolute position information of the decoding step. These methods perform well in scenarios with the same character width-height ratio and uniform character arrangement. However, for scene images, due to various factors such as camera angle and font style, the character width-height ratio are often not consistent. In addition, the character spacing may be varied. Figure 1 are some representative text cases from natural scenes. Besides, for those extremely curved images, some characters will be stretched or compressed after TPS rectification, resulting in changes in character scale and aspect ratio. From our key observation and experiments, keeping the same receptive field in recognizing texts on these images may greatly degrade the performance.

To address the above-mentioned problem, we propose a novel mechanism named Dynamic Receptive Field Adaption (DRA). To be concrete, we introduce a dynamic feature fusion module to ingeniously fuse the low and high receptive field information during decoding. Our method can automatically select the most predictive feature from high and low-level features when predicting a character. For example, in the sub-image in the first column and second row of Fig. 1, the character 'r' requires a smaller receptive field than the character 'S'. The model may predict 'n' in this step if just using high-level information. In addition, to adapt to the varied spacing arrangement of the character, we introduce historical attention maps as prior information to guide the accurate computation of current character regions. In this way, our DRA can improve recognition performance on text instances with inconsistent character scale and spacing. In conclusion, our primary contributions are three-fold:

- We found that attention-based recognizer requires different receptive fields for each character under the inconsistent character scale and spacing, and propose a framework to solve this problem.
- We propose a novel mechanism named DRA to improve the performance of attention-based recognizer on images with inconsistent character scale and spacing.
- Extensive experiments on several public scene text benchmarks demonstrate the superiority of our method.

2 Related Work

In recent years, several methods have been proposed for scene text recognition. Existing methods are mainly divided into two categories, i.e., segmentation-based methods and sequence-based methods.

2.1 Segmentation-Based Approaches

The segmentation-based methods try to locate each character's position in a text instance image, apply a character classifier to recognize individual character, and finally integrate the recognition results to obtain a complete text sequence. Wang et al. [26] introduced a multi-scale sliding window to extract HOG features and used a trained classifier for character classification. Character segmentation is extremely challenging owing to the complex background and irregular arrangement in natural scene images, while the classification effect is largely dependent on the result of character segmentation. Character segmentation by traditional approaches often shows poor performance. Therefore, some researchers completed character segmentation and character classification by deep learning methods. Liao et al. [14] applied a fully convolutional neural network to locate and classify individual characters and combine them according to certain rules. Recently, Wan et al. [25] proposed a method based on semantic segmentation, including two branch. The classification branch is used for character classification; the geometric branch is used to define characters' order and position. All of the above methods require the accurate segmentation of characters. They cannot model the context information other than a single character, leading to poor word-level recognition results.

2.2 Sequence-Based Approaches

The sequence-based method directly maps the entire text instance image to the target string sequence through the encoder-decoder framework. According to the decoder, such methods can be roughly divided into two categories: Connectionist Temporal Classification (CTC) [5] and Attention mechanism. CTC-based decoding approaches generally use CNN for image feature extraction and employ RNN for contextual information modeling, use CTC loss to train the model end-to-end without any alignment between the input sequence and the output. The encoder-decoder with attention framework was first proposed for machine translation [1]. Later, it was introduced into scene text recognition to align the character in the output sequence with local image regions. With the development of deep learning, more and more new methods were proposed to improve the performance of this framework. Due to the inherent decoding mechanism based on attention, this framework has the phenomenon of attention drift. To alleviate this problem, Cheng *et al.* [3] proposed a focus attention network (FAN). The above works assume that the text is horizontal and can't handle irregularities such as curvature, affine, etc. Shi *et al.* [24] combined the Spatial Transformer Network (STN) [8] to rectify the irregular text images to obtain horizontal text and then process them through the recognition network. In addition, iterative rectification [29] and fine-grained geometric constraints [27] have been used to achieve better rectification effect. Instead of rectifying images, Li *et al.* [13] introduced a two-dimension attention mechanism to realize alignment in 2D space. SATRN [12] replaced RNN with transformer's decoder, introduced a long-range dependency, and achieved better results. The research on multi-scale characters recognition mainly at mathematical formula recognition. Zhang *et al.* [30] obtained multi-scale features by performing the attention calculation process on the multi-level feature maps.

3   Method

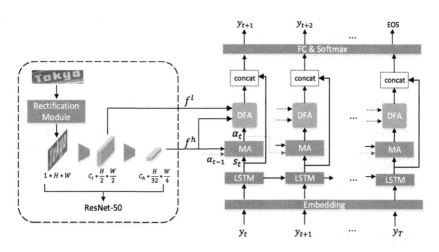

Fig. 2. Framework of our method. The pipeline of our method includes two parts: a feature encoder and a decoder. In the decoder, we proposes two modules, dynamic features adaption (DFA) and memory attention (MA), to realize the dynamic receptive field mechanism.

As shown in Fig. 2, our method mainly consists of two component, i.e.: an encoder for extracting visual information from images and a cyclic decoder based on attention. During the decoding stage, features from different levels will be dynamically merged in DFA, constructing more informative features. In addition, we propose an attention method with historical information named MA, which makes the alignment effect better.

3.1 Feature Encoder

Before the feature extraction stage, we use a rectification network to rectify images so that the method can get better performance on irregular text images. Then we use a backbone network based on ResNet to extract the visual information of an image. The dimension of the last feature map is $C_h \times \frac{H}{32} \times \frac{W}{4}$, where W and H mean the width and height of input image, C_h means the channel number. In the case of 1D attention, after several downsampling layers, the feature map becomes a feature sequence with a length N. As a sequence prediction task, the contextual information among vectors in a sequence is also important. Therefore, we use a two-layer bidirectional LSTM to enlarge the feature context. The output of the encoder is represented as follows,

$$F = [f_1, f_2, ..., f_N], N = \frac{W}{4}. \tag{1}$$

3.2 Recurrent Attention Decoder

The sequence-to-sequence model translates the feature sequence $(f_1, f_2, ..., f_N)$ into a character sequence $(\hat{y}_1, \hat{y}_2, ..., \hat{y}_T)$. To align the two sequences, we build a decoder based on attention. At the t-th step, the decoder generates an output y_t as

$$y_t = argmax(W_o[s_t; g_t] + b_o), \tag{2}$$

where W_o is trainable weights and s_t is the hidden state of LSTM at step t, computed by

$$s_t = LSTM([f_{embed}(y_{t-1}); g_{t-1}], s_{t-1}), \tag{3}$$

and g_t is the weight sum of sequential feature vectors $(f_1, f_2, ..., f_N)$. Here we adapt a 1D attention mechanism and g_t is expressed as

$$g_t = \sum_{j=1}^{N} \alpha_{t,j} f_j. \tag{4}$$

We obtain two g_t vectors on feature maps of different levels and then perform adaptive fusion in our approach. This part of the details will be elaborated in Sect. 3.3 DFA. N represents the length of feature sequence, and $\alpha_t \in R^N$ is the vector of attention weight, the calculation process is as follows

$$\alpha_{t,j} = \frac{exp(e_{t,j})}{\sum_{i=1}^{N} exp(e_{t,i})}, \tag{5}$$

where $e_{t,j}$ represents the relevance between the t-th character and the j-th feature region and is often expressed as follows

$$e_{t,j} = s_t^T f_j. \tag{6}$$

We adopt the dot product attention implementation method. The performance is similar to additive attention, the calculation speed is faster owing to the parallel nature of the matrix. Simultaneously, in our method, f_j is not only the feature output by the encoder but also contains the attention information at the last moment when calculating the attention map. It will be described in detail in Sect. 3.4 MA.

The Decoder cyclically predicts the target character until it generates a special EOS token. The loss function of the entire model is

$$L_{Att} = -\sum_t lnP(\hat{y}_t|I, \theta), \tag{7}$$

where $\hat{y}_t$ is the t-th character in ground-truth, I is the input image and θ is the parameters of whole network.

3.3 Dynamic Feature Adaption

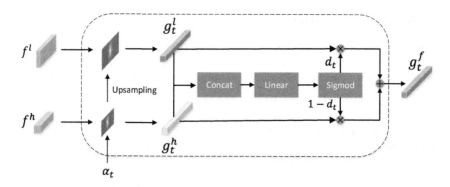

Fig. 3. Details of dynamic feature adaption. f^l and f^h are the feature maps of low-level and high-level while the α_t is the attention map at current step. The module takes f^l, f^h and α_t as inputs and output one adaptively fused feature vector g_t^f.

The existing attention-based methods generally calculate the attention weights on the last encoder layer's feature map, owing to its sizeable receptive field. A larger receptive field can perceive more accurate character area. Then these methods usually multiply attention weights by the uppermost feature map to compute a vector related to the current character. The receptive field of the uppermost feature map is fixed, which is not robust to scenes with inconsistent character scale and spacing. We introduce a dynamic feature adaptive module to choose different receptive fields' features according to the character ratio and spacing to handle this problem. The Dynamic Feature Adaption Strategy (DFA) is illustrated in Fig. 3, consisting of two main steps: 1) compute relevant vectors; 2) dynamically fuse relevant vectors.

Computing Relevant Vectors. As shown in Eqs. (5) and (6), we compute the attention weights on the high-level feature sequence and get an attention vector related to the character. At the same time, we weight the low-level feature sequence with this attention weight to obtain a relevant vector containing more detailed information. In this process, due to the different dimensions of high-level and low-level features sequence, we need upsample the attention map. After the encoding process, the size of high-level feature map becomes one in the height dimension, which is $C_h \times 1 \times W_h$. The high-level sequence is represented as follows,

$$F_h = [f_1, f_2, ..., f_{W_h}] \tag{8}$$

Considering of the trade-off between receptive field and detailed information, we select the output of the first block of ResNet-50 (five blocks in total) as our low-level feature map. The dimension of the low-level feature map is $C_l \times H_l \times W_l$, we reshape it as $C_l H_l \times 1 \times W_l$ to adopt 1D attention mechanism, and the low-level sequence is represented as

$$F_l = [f_1, f_2, ..., f_{W_l}], \tag{9}$$

where the W_l is twice as many as W_h. Then we will obtain two relevant vectors g_t^h and g_t^l by

$$g_t^h = \sum_{j=1}^{W_h} \alpha_{t,j} f_j^h, \tag{10}$$

$$g_t^l = \sum_{j=1}^{W_l} Upsample(\alpha_{t,j}) f_j^l, \tag{11}$$

Here, we perform linear interpolation on the 1D attention map to obtain the upsampled attention map in Eq. (11). The upsampling process preserves the distribution information of attention map and ensures that the dimension of upsampled map is consistent with low-level feature map.

Dynamically Fusing Relevant Vectors. As shown in Fig. 3, we propose to dynamically fuse the low-level relevant vector g_t^l and the high-level relevant vector g_t^h at each time step t. Inspired by the gated unit, we introduce a dynamic weight to determine the contributions of different parts of feature, named attention gate and can be formulated as

$$d_t = \sigma(\mathbf{W_d}[g_t^h, g_t^l]) \tag{12}$$

$$g_t^f = d_t \odot g_t^l + (1 - d_t) \odot g_t^h \qquad (13)$$

where $\mathbf{W_d}$ is trainable parameters, σ is the Sigmoid function and the $\odot$ represents the element wise multiplication, g_t^f is the correlation vector used for character classification.

3.4 Memory Attention

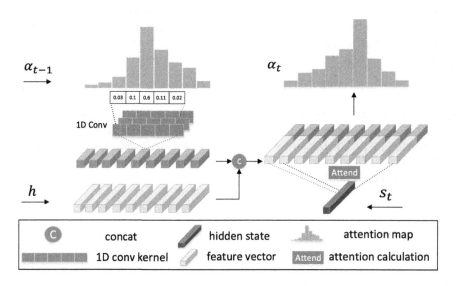

Fig. 4. Detailed structure of MA

To make the attention alignment effect better and obtain more informative features, we introduce historical attention map as prior information when calculating the attention weight. In particular, we use historical attention map in a novel way, and we named it Memory Attention. First, we perform feature extraction on the attention map through 1D convolution to obtain the approximate regional location information and character scale information of the last moment p_t, expressed by

$$p_t = Conv1D(\alpha_{t-1}) \qquad (14)$$

Then, the extracted distribution features and the encoder's feature sequence are concatenated in the channel dimension to get f in Eq. (6).

$$f = [h; p_t] \qquad (15)$$

Finally, The hidden state s_t at the current moment performs attention computation with the combined features. The specific process is shown in Fig. 4.

4   Experiments

In this section, we conduct extensive experiments to validate the proposed DRA mechanism on several general recognition benchmarks. First, we introduce public datasets for training and evaluation. Then, the implementation details of our entire method are described. Finally, we perform ablation studies to analyze each module's performance and compared it with state-of-the-art methods on several benchmarks.

4.1   Datasets

Our model is trained on two synthetic datasets without finetuning on other datasets and tested on 6 standard datasets to evaluate its recognition performance.

Synth90K [7] contains 8 million synthetic images of cropped words generated from a set of 90K common English words. Words are rendered onto natural images with random transformations and effects. Every image in this dataset has a corresponding ground-truth, and all of them are used for training.

SynthText [6] contains 6 million synthetic images of cropped word for training. This dataset is mainly for text detection and contains complete images with text, therefore we crop word images with the ground truth word bounding boxes.

IIIT 5K-Words (IIIT5K) [18] contains 3,000 cropped word images for testing that are collected from the Internet. It also provides a 50-word lexicon and a 1k-word lexicon.

ICDAR2013 (IC13) [10] contains 1,015 testing images. This dataset is filtered by removing words that have non-alphanumeric characters. No lexicon is provided.

ICDAR2015 (IC15) [9] contains 2,077 cropped images. A large number of this datasets is blur irregular text images.

Street View Text (SVT) [26] contains 647 images of cropped word images that are collected from street view. A 50-word lexicon is provided for each image.

SVT-Perspective (SVTP) [19] contains 639 cropped images for testing. Most of the images are heavily distorted.

CUTE80 (CUTE) [21] focus on irregular text images, containing 80 high-resolution images captured in natural scenes. This dataset consists of 288 cropped realistic images for testing, and there is no lexicon provided.

4.2   Implementation Details

The proposed framework is implemented by using PyTorch and trained end-to-end on the Synth90K and SynthText. The training images are all from these two datasets without any data augmentation and selection. The final classification output layer has a dimension of 97, contains digits, uppercase letters, lowercase letters and 32 ASCII punctuation marks, EOS symbol, padding symbol, and unknown symbol.

Images are resized to 64 × 256 without keeping ratio before input to rectification network, and 32 × 100 image is sampled to the recognition network after rectification process. ADADELTA was chosen as the optimizer to minimize the entire model's objective function end-to-end. The model is trained with a batch size of 512. During the training, the learning rate is initiated from 1 and is decayed to 0.1 and 0.01 respectively at the 4th epoch and 5th epoch. Moreover, all the experiments are accomplished on 2 NVIDIA TITAN X Pascal 12 GB GPU.

4.3   Ablation Study

Exploration of the Two Modules. We choose ResNet50 Network as our backbone, which is similar to the ASTER model. Our basemodel's structure shown in Fig. 2, with the DFA and MA modules removed from the decoder. The output of the last encoder layer is a 25 × 512 vector sequence for each image. By introducing historical attention information, our model has been improved on several datasets. The experimental result in Table 1. The improvement effect is most evident on several irregular text datasets, respectively improved by 2.0% on CUTE, 3.1% on IC15, and 6.2% on SVTP. It shows that irregular text arrangement is more changeable, so the introduction of attention map at historical moments can make the character area that the model focuses on more accurate. As some attention maps shown in Fig. 5, the obtained attention maps can pay more attention to the visual areas of corresponding characters correctly under the guidance of previous attention information. By dynamically fusing different receptive field features, our model have also been greatly improved on most datasets. The combination of the two modules has achieved the best performance on several datasets. Compared with the basemodel without using previous prior information and multi-level feature, DRA improves the accuracy by 0.9% on IIIT5K, 3.7% on IC15, 4.8% on SVTP, 3.7% on CUTE. SVTP dataset contains many fuzzy and distorted text images, and our DFA module mainly focuses on the characters with different scales. We observe that our method achieves poor performance on extremely blurry images with the setting of two modules. According to our analysis, this phenomenon on these images may be due to the noise informative features introduced by the low-level feature map.

Table 1. Performance comparisons between different module

Methods	MA	DFA	IIIT5K	SVT	IC13	IC15	SVTP	CUTE
Basemodel			93.3	88.9	90.4	76.5	77.8	79.6
Basemodel	√		93.7	89.6	**92.3**	79.6	**84.0**	81.6
Basemodel		√	93.8	89.4	91.6	78.5	82.0	82.5
DRA	√	√	**94.2**	**90.1**	91.8	**80.2**	82.6	**83.3**

Exploration of Feature Fusion Methods. There are many variants of feature fusion methods. In this paper, we introduce a novel dynamic fusion method

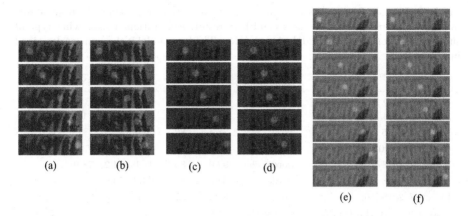

Fig. 5. Examples of attention visualization. (a), (c), (e) are the attention maps of basemodel and (b), (d), (f) are the attention maps of basemodel with MA.

Table 2. Performance comparisons between different fusion methods

Fusion methods	IIIT5K	SVT	IC13	IC15	SVTP	CUTE
Concat	93.4	88.1	90.8	75.8	78.1	79.2
Add	94.0	87.8	91.7	78.5	80.8	82.6
Vector-level fusion	93.1	89.2	**92.4**	77.7	80.6	79.9
Attention gate	**94.2**	**90.1**	91.8	**80.2**	**82.6**	**83.3**

to achieve element-level feature fusion named attention gate. In this section, we conduct experiments on the basemodel containing MA to verify the effectiveness of our proposed dynamic fusion methods. Some conventional fusion methods such as concatenate and add are involved in the comparison. Also, we have verified the effect of vector-level fusion. Vector-level fusion means that each channel is not considered separately but only two weights are calculated for two feature vectors. The experimental results in Table 2 indicate that element-level adaptive fusion is more robust and more applicable.

4.4 Comparisons with State-of-the-Arts

We also compare our methods with previous excellent methods on several benchmarks. The results are shown in Table 3. Our proposed model achieves the best results on two irregular text datasets and achieves second results on two regular benchmarks. On several low-quality datasets such as SVT and SVTP, they contain a lot of low resolution, blur, and distortion word images, thus more suitable receptive fields, and richer detailed information help to classify characters more accurately. For those irregular text datasets, the ratio and scale of the characters in the same text instance are quite different. The receptive field can effectively adapt to character scale changes with DRA mechanism.

Table 3. Lexicon-free results on public benchmarks. Bold represents the best performance. Underline represents the second best result. Annotations means what type of supervision information is used during training. The basemodel is the same as described in Sect. 4.3

Methods	Annotations	IIIT5K	SVT	IC13	IC15	SVTP	CUTE
Shi *et al.* 2016 [22] (CTC)	word	81.2	82.7	89.6	-	-	-
Shi *et al.* 2016 [23]	word	81.9	81.9	88.6	-	71.8	59.2
Lee *et al.* 2016 [11]	word	78.4	80.7	90.0	-	-	-
Cheng *et al.* 2018 [4] (AON)	word	87.0	82.8	-	68.2	73.0	76.8
Liu *et al.* 2018 [15]	word& char	92.0	85.5	91.1	74.2	78.9	-
Bai *et al.* 2018 [2]	word& char	88.3	87.5	**94.4**	73.9	-	-
Liu *et al.* 2018 [16]	word	89.4	87.1	<u>94.0</u>	-	73.9	62.5
Liao *et al.* 2019 [14] (FCN)	word& char	91.9	86.4	91.5	-	-	79.9
Zhan *et al.* 2019 [29] (ESIR)	word	93.3	**90.2**	91.3	76.9	79.6	83.3
Li *et al.* 2019 [13] (SAR)	word	91.5	84.5	91.0	69.2	76.4	83.3
Luo *et al.* 2019 [17] (MORAN)	word	91.2	88.3	92.4	68.8	76.1	77.4
Yang *et al.* 2019 [27] (ScRN)	word& char	**94.4**	88.9	93.9	78.7	80.8	**87.5**
Shi *et al.* 2018 [24] (ASTER)	word	93.4	89.5	91.8	76.1	78.5	79.5
Qiao *et al.* 2020 [20] (SEED)	word	93.8	89.6	92.8	<u>80.0</u>	<u>81.4</u>	<u>83.6</u>
Basemodel	word	93.3	88.9	90.4	76.5	77.8	79.6
DRA	word	<u>94.2</u>	<u>90.1</u>	91.8	**80.2**	**82.6**	83.3

The encoder part of our method is the same as the ASTER [24] model, while the decoder part has an additional dynamic fusion features module and a memory attention module compared with the conventional attention-based decoding method. As can be seen from the Table 3, the basemodel have a similar performance with ASTER, while our DRA has a improvement in multiple datasets than the ASTER model. Quantitatively speaking, it is improved by 0.8% on IIIT5K, 0.6% on SVT, 4.1% on IC15, 4.1% on SVTP, 3.8% on CUTE. Compared with other attention-based methods, such as SAR [13] etc., our DRA also achieve better performance on several datasets. Many recent studies have introduced semantic information to assist the recognition task. The introduction of semantic information can significantly improve the performance of recognition. Compared to SEED [20], using global semantic information as auxiliary information, our model only uses visual information, achieving similar or even better performance on some benchmarks.

It can be noticed that our method still doesn't achieve the best results on several datasets, but it is very flexible and can be inserted into most existing attention-based methods to improve the visual feature representation and get better performance.

4.5   Visual Illustrations

Figure 6 shows the result of performance visualization. As can be seen, for the images in the first column, the scale and spacing of characters are basically

Fig. 6. Example of inconsistent aspect ratio images and recognition results. In the recognition result of each picture, the first row is the result recognized by ASTER and the second and third row is the result obtained by our DRA.

consistent, ASTER model and our DRA can both recognize these pictures correctly. However, for those pictures with inconsistent character scales and spacing, showing in the second and third column, our model can adapt to changes in character receptive fields compared to conventional attention models. For example, in the second picture of the second column, since the character 'r' is close to the character 'i', conventional attention-based recognizer will recognize these two characters as 'n' without considering the change in the ratio of character aspect. In the third picture of the first column, due to local affine on the right size of the image, the two 'L' are extremely close, it need a smaller receptive field for recognizing the character correctly. If the receptive field information is not adapted, it will lead to missed recognition. DFA module can ensure that appropriate receptive field information is selected when decoding. At the same time, guided by historical attention map information, the model can be aware of historical location when calculating the area of interest. Benefit from this, the model has the ability to adapt to the changes of character spacing.

5   Conclusion

In this work, we propose a dynamic receptive field adaptive (DRA) mechanism for attention-based scene text recognition. Our DRA can adapt features and receptive fields for characters of different scales in the same text instance. More informative features are obtained by aggregating multi-level features. In addition, the previous attention information as a priori information can make our attention computation more accurate, resulting in better discriminative features. Our DRA is flexible and can be generalized to other attention-based methods to improve their performance.

Acknowledgement. This work was partly supported by National Key Research and Development Program of China (2019YFB1405900), National Natural Science Foundation of China (62006018), and the Fundamental Research Funds for the Central Universities and USTB-NTUT Joint Research Program.

References

1. Bahdanau, D., Cho, K., Bengio, Y.: Neural machine translation by jointly learning to align and translate. arXiv preprint arXiv:1409.0473 (2014)
2. Bai, F., Cheng, Z., Niu, Y., Pu, S., Zhou, S.: Edit probability for scene text recognition. In: CVPR, pp. 1508–1516 (2018)
3. Cheng, Z., Bai, F., Xu, Y., Zheng, G., Pu, S., Zhou, S.: Focusing attention: towards accurate text recognition in natural images. In: ICCV, pp. 5076–5084 (2017)
4. Cheng, Z., Xu, Y., Bai, F., Niu, Y., Pu, S., Zhou, S.: AON: towards arbitrarily-oriented text recognition. In: CVPR, pp. 5571–5579 (2018)
5. Graves, A., Fernández, S., Gomez, F., Schmidhuber, J.: Connectionist temporal classification: labelling unsegmented sequence data with recurrent neural networks. In: ICML, pp. 369–376 (2006)
6. Gupta, A., Vedaldi, A., Zisserman, A.: Synthetic data for text localisation in natural images. In: CVPR, pp. 2315–2324 (2016)
7. Jaderberg, M., Simonyan, K., Vedaldi, A., Zisserman, A.: Synthetic data and artificial neural networks for natural scene text recognition. arXiv preprint arXiv:1406.2227 (2014)
8. Jaderberg, M., Simonyan, K., Zisserman, A., Kavukcuoglu, K.: Spatial transformer networks. arXiv preprint arXiv:1506.02025 (2015)
9. Karatzas, D., et al.: ICDAR 2015 competition on robust reading. In: ICDAR, pp. 1156–1160. IEEE (2015)
10. Karatzas, D., et al.: ICDAR 2013 robust reading competition. In: ICDAR, pp. 1484–1493. IEEE (2013)
11. Lee, C.Y., Osindero, S.: Recursive recurrent nets with attention modeling for OCR in the wild. In: CVPR, pp. 2231–2239 (2016)
12. Lee, J., Park, S., Baek, J., Oh, S.J., Kim, S., Lee, H.: On recognizing texts of arbitrary shapes with 2D self-attention. In: CVPR Workshops, pp. 546–547 (2020)
13. Li, H., Wang, P., Shen, C., Zhang, G.: Show, attend and read: A simple and strong baseline for irregular text recognition. In: AAAI, pp. 8610–8617 (2019)
14. Liao, M., et al.: Scene text recognition from two-dimensional perspective. In: AAAI, pp. 8714–8721 (2019)
15. Liu, W., Chen, C., Wong, K.Y.: Char-Net: a character-aware neural network for distorted scene text recognition. In: AAAI (2018)
16. Liu, Y., Wang, Z., Jin, H., Wassell, I.: Synthetically supervised feature learning for scene text recognition. In: Ferrari, V., Hebert, M., Sminchisescu, C., Weiss, Y. (eds.) ECCV 2018. LNCS, vol. 11209, pp. 449–465. Springer, Cham (2018). https://doi.org/10.1007/978-3-030-01228-1_27
17. Luo, C., Jin, L., Sun, Z.: Moran: a multi-object rectified attention network for scene text recognition. Pattern Recogn. 90, 109–118 (2019)
18. Mishra, A., Alahari, K., Jawahar, C.: Scene text recognition using higher order language priors. In: BMVC. BMVA (2012)
19. Phan, T.Q., Shivakumara, P., Tian, S., Tan, C.L.: Recognizing text with perspective distortion in natural scenes. In: ICCV, pp. 569–576 (2013)
20. Qiao, Z., Zhou, Y., Yang, D., Zhou, Y., Wang, W.: Seed: semantics enhanced encoder-decoder framework for scene text recognition. In: CVPR, pp. 13528–13537 (2020)
21. Risnumawan, A., Shivakumara, P., Chan, C.S., Tan, C.L.: A robust arbitrary text detection system for natural scene images. Expert Syst. Appl. 41(18), 8027–8048 (2014)

22. Shi, B., Bai, X., Yao, C.: An end-to-end trainable neural network for image-based sequence recognition and its application to scene text recognition. TPAMI **39**(11), 2298–2304 (2016)
23. Shi, B., Wang, X., Lyu, P., Yao, C., Bai, X.: Robust scene text recognition with automatic rectification. In: CVPR, pp. 4168–4176 (2016)
24. Shi, B., Yang, M., Wang, X., Lyu, P., Yao, C., Bai, X.: ASTER: an attentional scene text recognizer with flexible rectification. TPAMI **41**(9), 2035–2048 (2018)
25. Wan, Z., He, M., Chen, H., Bai, X., Yao, C.: TextScanner: reading characters in order for robust scene text recognition. In: AAAI, pp. 12120–12127 (2020)
26. Wang, K., Babenko, B., Belongie, S.: End-to-end scene text recognition. In: ICCV, pp. 1457–1464. IEEE (2011)
27. Yang, M., et al.: Symmetry-constrained rectification network for scene text recognition. In: ICCV, pp. 9147–9156 (2019)
28. Yue, X., Kuang, Z., Lin, C., Sun, H., Zhang, W.: RobustScanner: dynamically enhancing positional clues for robust text recognition. In: Vedaldi, A., Bischof, H., Brox, T., Frahm, J.-M. (eds.) ECCV 2020. LNCS, vol. 12364, pp. 135–151. Springer, Cham (2020). https://doi.org/10.1007/978-3-030-58529-7_9
29. Zhan, F., Lu, S.: ESIR: end-to-end scene text recognition via iterative image rectification. In: CVPR, pp. 2059–2068 (2019)
30. Zhang, J., Du, J., Dai, L.: Multi-scale attention with dense encoder for handwritten mathematical expression recognition. In: ICPR, pp. 2245–2250. IEEE (2018)

Context-Free TextSpotter for Real-Time and Mobile End-to-End Text Detection and Recognition

Ryota Yoshihashi[✉], Tomohiro Tanaka, Kenji Doi, Takumi Fujino, and Naoaki Yamashita

Yahoo Japan Corporation, Tokyo, Japan
ryoshiha@yahoo-corp.jp

Abstract. In the deployment of scene-text spotting systems on mobile platforms, lightweight models with low computation are preferable. In concept, end-to-end (E2E) text spotting is suitable for such purposes because it performs text detection and recognition in a single model. However, current state-of-the-art E2E methods rely on heavy feature extractors, recurrent sequence modellings, and complex shape aligners to pursue accuracy, which means their computations are still heavy. We explore the opposite direction: *How far can we go without bells and whistles in E2E text spotting?* To this end, we propose a text-spotting method that consists of simple convolutions and a few post-processes, named *Context-Free TextSpotter*. Experiments using standard benchmarks show that Context-Free TextSpotter achieves real-time text spotting on a GPU with only three million parameters, which is the smallest and fastest among existing deep text spotters, with an acceptable transcription quality degradation compared to heavier ones. Further, we demonstrate that our text spotter can run on a smartphone with affordable latency, which is valuable for building stand-alone OCR applications.

Keywords: Scene text spotting · Mobile text recognition · Scene text detection and recognition

1 Introduction

Scene text spotting is a fundamental task with a variety of applications such as image-base translation, life logging, or industrial automation. To make the useful applications more conveniently accessible, deploying text-spotting systems on mobile devices has shown promise, as we can see in the recent spread of smartphones. If text spotting runs client-side on mobile devices, users can enjoy the advantages of edge computing such as availability outside of the communication range, saving of packet-communication fees consumed by uploading images, and fewer concerns about privacy violation by leakage of uploaded images.

However, while recent deep text-detection and recognition methods have become highly accurate, they are now so heavy that they can not be easily

© Springer Nature Switzerland AG 2021
J. Lladós et al. (Eds.): ICDAR 2021, LNCS 12822, pp. 240–257, 2021.
https://doi.org/10.1007/978-3-030-86331-9_16

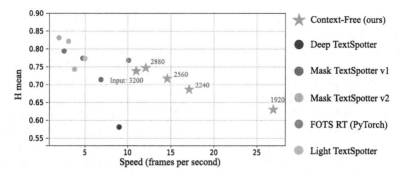

Fig. 1. Recognition quality vs. speed in scene text spotters, evaluated with a GPU on ICDAR2015 incidental scene text benchmark with Strong lexicon. The proposed Context-Free TextSpotter runs faster than existing ones with similar accuracy, and enables near-real-time text spotting with a certain accuracy degradation.

deployed on mobile devices. A *heavy* model may mean that its inference is computationally expensive, or that it has many parameters and its file size is large, but either is unfavorable for mobile users. Mobile CPUs and GPUs are generally powerless compared to the ones on servers, so the inference latency of slow models may become intolerable. The large file sizes of models are also problematic, because they consume a large amount of memory and traffic, both of which are typically limited on smartphones. For example, the Google Play app store has limited the maximum app size to 100 MB so that developers are cognizant of this problem, but apps that are equipped with large deep recognition models may easily exceed the limit without careful compression.

The above mentioned problems motivated us to develop a lightweight text spotting method for mobiles. For this purpose, the recently popular end-to-end (E2E) text spotting shows promise. E2E recognition, which performs joint learning of modules (e.g., text detection and recognition) on a shared feature representation, has been increasingly used in deep learning to enhance the efficiency and effectiveness of models, and already much research effort has been dedicated to E2E text spotters [4,20,28,29,32]. Nevertheless, most of the existing E2E text spotters are not speed-oriented; they typically run around five to eight frames per second (FPS) on high-end GPUs, which is not feasible for weaker mobile computation power. The reasons for this heaviness lies in the complex design of modern text spotters. Typically, a deep text spotter consists of 1) backbone networks as a feature extractor, 2) text-detection heads with RoI pooling, and 3) text-recognition heads. Each of these components has potential bottlenecks of computation; existing text spotters 1) adopt heavy backbones such as VGG-16 or ResNet50 [4], 2) often utilize RoI transformers customized for text recognition to geometrically align curved texts [32,33], and 3) the text-recognition head, which runs per text box, has unshared convolution and recurrent layers, which are necessary for the sequence modelling of word spells [20,32].

Then, how can we make an E2E text spotter light enough for mobile usage? We argue that the answer is an E2E text spotter with as few additional modules to convolution as possible. Since convolution is the simplest building block of deep visual systems and the de facto standard operation for which well-optimized kernels are offered in many environments, it has an intrinsic advantage in efficiency. Here, we propose *Context-Free TextSpotter*, an E2E text spotting method without bells and whistles. It simply consists of lightweight convolution layers and a few post-processes to extract text polygons and character labels from the convolutional features. This text spotter is *context-free* in the sense that it does not rely on the linguistic context conventionally provided by per-word-box sequence modelling through LSTMs, or the spatial context provided by geometric RoI transformations and geometry-aware pooling modules. These simplifications may not seem straightforward, as text spotters usually needs sequence-to-sequence mapping to predict arbitrary-length texts from visual features. Our idea to alleviate this complexity is to decompose a text spotter into a character spotter and a text-box detector that work in parallel. In our framework, a character spotting module spots points in convolutional features that are readable as a character and then classifies the point features to give character labels to the points. Later text boxes and point-wise character labels are merged by a simple rule to construct word-recognition results. Intuitively, characters are less curved in scenes than whole words are, and character recognition is easier to tackle without geometric operations. Sequence modelling becomes unnecessary in character classification, in contrast to the box-to-word recognition utilized in other methods. Further, by introducing weakly supervised learning, our method does not need character-level annotations of real images to train the character-spotting part in our method.

In experiments, we found that Context-Free TextSpotter worked surprisingly well for its simplicity on standard benchmarks of scene text spotting. It achieved a word-spotting Hmean of 84 with 25 FPS on the ICDAR 2013focused text dataset, and an Hmean of 74 with 12 FPS on the ICDAR2015 incidental text dataset in the Strong-lexicon protocol. Compared to existing text spotters, it is around three times faster than typical recent text spotters, while its recognition degradation is around five to ten percent-points. Another advantage is the flexibility of the model: it can control the accuracy-speed balance simply by scaling input-image resolution within a single trained model, as shown in Fig. 1. Context-Free TextSpotter ran the fastest among deep text spotters that reported their inference speed, and thus it is useful to extrapolate the current accuracy-speed trade-off curve into untrodden speed-oriented areas. Finally, we demonstrate that our text spotter can run on a smartphone with affordable latency, which is valuable for building stand-alone OCR applications.

Our contributions summarize as follows: 1) we design a novel simple text-spotting framework called Context-Free TextSpotter. 2) We develop techniques useful for enhancing efficiency and effectiveness of our text spotter, including the linear point-wise character decoding and the hybrid approach for character spotting, which are described in Sec. 3.3) In experiments, we confirm that our text spotter runs the fastest among existing deep text spotters, and able to be deploy in iPhone to run with acceptable latency.

2   Related Work

2.1   Text Detection and Recognition

Conventionally, text detection and recognition are regarded as related but separate tasks, and most studies have focused on either one or the other. In text recognition, while classical methods use character classifiers in a sliding-window manner [48], more modern plain methods consist of convolutional feature extractors and recurrent-network-based text decoders [42]. Among recurrent architecture, bidirectional long short-term memories (BiLSTM [14]) is the most popular choice. More recently, self-attention-based transformers, which have been successful in NLP, are intensively examined [54]. Connectionist temporal classification (CTC) [13] is used as an objective function in sequence prediction, and later attention-based text decoding has been intensively researched as a more powerful alternative [6]. For not-well-cropped or curved texts, learning-based image rectification has been proposed [43]. These techniques are also useful in text spotting to design recognition branches.

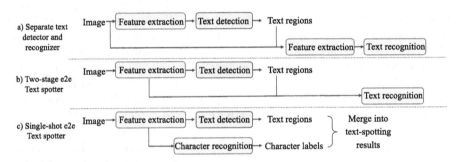

Fig. 2. Three typical text-spotting pipelines. a) Non-E2E separate text detectors and recognizers, b) two-stage E2E text spotters, and c) single-shot text spotters. We use c) for the highest parallelizability.

Deep text detectors are roughly categorized into two types: box-based detectors and segment-based ones. Box-based detectors typically estimate text box coordinates by regression from learned features. Such techniques have been intensively studied in object detection and they are easily generalized to text detection. However, the box-based detectors often have difficulty in localizing curved texts accurately, unless any curve handling mechanism is adopted. TextBoxes [30] is directly inspired by an object detector SSD [31], which regress rectangles from feature maps. SSTD [19] exploits text attention modules that are supervised by text-mask annotations. EAST [52] treats oriented texts by utilizing regression for box angles. CTPN [44] and RTN [53] similarly utilize strip-shaped boxes and connect them to represent curved text regions.

Segment-based detectors estimate foreground masks in a pixel-labeling manner, and extract their connected components as text instances. While they are naturally able to handle oriented and curved texts, they sometimes mis-connect multiple text lines when the lines are close together. For example, fully convolutional networks [34] can be applied to text detection with text/non-text mask prediction and some post processing of center-line detection and orientation estimation [51]. PixelLink [11] enhances the separability of text masks by introducing 8-neighbor pixel-wise connectivity prediction in addition to text/non-text masks. TextSnake [35] adopts a disk-chain representation that is predicted by text-line masks and radius maps. CRAFT [2] enhances learning effectiveness by modeling character-region awareness and between-character affinity using a sophisticated supervisory-mask generation scheme, which we also adopt in our method.

2.2   End-to-End Text Recognition

End-to-end (E2E) learning refers to a class of learning methods that can link inputs to outputs with single learnable pipelines. It has been successfully applied for text recognition and detection. While the earliest method used a sliding-window character detector and word construction through a pictorial structure [46], most of the modern E2E text recognizers have been based on deep CNNs. The majority of existing E2E text spotters are based on two-stage framework (Fig. 2 b). Deep TextSpotter [4] is the first deep E2E text spotter, that utilizes YOLOv2 as a region-proposal network (RPN) and cascades recognition branches of convolution + CTC after it. FOTS [32] adopts a similar framework of RPN and recognition branches, but improved the oriented text recognition by adding an RoI Rotate operation. Mask TextSpotter [28,29,36] has an instance-mask-based recognition head, that does not rely on CTC. The recently proposed CRAFTS [3] is a segment-based E2E text spotter featuring enhanced recognition by sharing character attention with a CRAFT-based detection branch.

Single-shot text spotting (Fig. 2 c), in contrast, has not been extensively researched up to now. CharNet [50] utilizes parallel text-box detection and character semantic segmentation. Despite of its conceptual simplicity, it is not faster than two-stage spotters due to its heavy backbone and dense character labeling. MANGO [39] exploits dense mask-based attention instead to eliminate the necessity for RoI operation. However, its character decoder relies on attention-based BiLSTM, and iterative inference is still needed, which might create a computational bottleneck in long texts. Here, our contribution is to show that a single-shot spotter actually runs faster than existing two-stage ones if proper simplification is done.

2.3   Real-Time Image Recognition

The large computation cost of deep nets is an issue in many fields, and various studies have explored general techniques for reducing computation cost, while others focus on designing lightweight models for a specific tasks. Some of examples of the former are pruning [18], quantization [9,23], and lighter operators [5,17]. While these are applicable to general models, it is not clear whether the loss of computational precision they cause is within a tolerable range in text localization, which require high exactness for correct reading. For the latter, lightweight models are seen in many vision tasks such as image classification, object detection, and segmentation. However, lightweight text spotting is much less studied, seemingly due to the complexity of the task, although there has been some research on mobile text detection [8, 10, 12, 24]. We are aware of only one previous work: Light TextSpotter [15], which is a slimmed down mask-based text spotter that features distillation and a ShuffleNet backbone. Although it is 40%-lighter than Mask TextSpotter, its inference speed is not as fast as ours due to its big mask-based recognition branch.

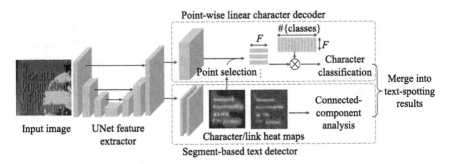

Fig. 3. Overview of proposed Context-Free TextSpotter.

3   Method

The design objective behind the proposed Context-Free Text Spotter is to pursue minimalism in text spotting. As a principle, deep E2E text spotters need to be able to detect text boxes in images and recognize their contents, while the detector and recognizer can share feature representations they exploit. Thus, a text spotter, at minimum, needs to have a feature extractor, a text-box detector, and a text recognizer.

With text spotters, the text recognizers tend to become the most complex part of the system, as text recognition from detected boxes needs to perform a sequence-to-sequence prediction, that is to relates an arbitrary-length feature sequence to an arbitrary-length text. Also, the recognizer's computation depends on the detector's output, which makes the pipeline less parallelizable (Fig. 2 b). We break down the text recognizer into a character spotter and classifier, where the character spotter pinpoints coordinates of characters within uncropped feature maps of whole images, and the character classifier classifies each of the spotted characters regardless of other characters around it or the text box to which it belongs (Fig. 2 c).

The overview of Context-Free Text Spotter is shown in Fig. 3. It consists of 1) a U-Net-based feature extractor, 2) heat-map-based character and text-box detectors, and 3) a character decoder called *the point-wise linear decoder*, which is specially tailored for our method. The later part of this section describes the details of each module.

3.1   Text-Box Detector and Character Spotter

To take advantage of shape flexibility and interpretability, we adopt segment-based methods for the character and text-box detection. More specifically, we roughly follow the CRAFT [2] text detector in the text-box localization procedure. Briefly, our text-box detector generates two heat maps: a region map and an affinity map. The region map is trained to activate strongly around the centers of characters, and the affinity map is trained to activate in areas between characters within single text boxes. In inference, connected components in the sum of the region map and affinity map are extracted as text instances.

For character spotting, we reuse the region map of the CRAFT-based text detector, and thus we do not need to add extra modules to our network. Further, we adopt point-based character spotting, instead of the more common box- or polygon-based character detection, to eliminate the necessity for box/polygon regression and RoI pooling operations.

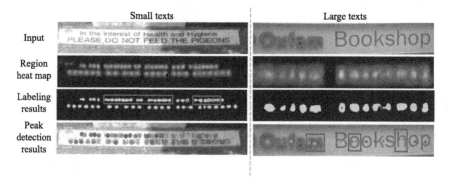

Fig. 4. Examples of small and large texts. Corresponding region heat maps (second row) show different characteristics that lead labeling-based approaches (third row) to failure (red boxes) in small texts, and peak-detection-based approaches (last row) to failure in large texts. Thus, we adopt a hybrid approach in character spotting. (Color figure online)

To select character points from the region map, we prefer simple image-processing-based techniques to learning-based ones in order to avoid unnecessary overheads. We consider two approaches: a labelling-based approach that performs grouping of heat maps, and a peak-detection based approach that picks up local maxima of the heat map. The former is adopted for spotting large characters, and the latter for spotting small characters. Here, our insight helpful to the point selector design is that the form of heat maps' activation to characters largely differs by the scales of the characters in input images. Examples of heat-map activation to large and small characters are shown in Fig. 4. For small characters, labels tend to fail to disconnect close characters, while they cause less duplicated detection of single characters. Apparently, such tendency comes from the difference of activation patterns in region heat maps; the heat-map activation to large texts reflects detailed shapes of characters that may cause multiple peaks in them, while that to small texts results in simpler single-peaked blobs.

For labelling-based spotting for large characters, we use connected-component analysis [49] to link separate characters. First we binarize region heat maps by a threshold, then link the foregrounds, and finally pick up centroids of each connected component. For peak-based spotting for small objects, we use local maxima of region heat maps, namely

$$P = \left\{ (x,y) \mid R(x,y) - \max_{(dx,dy)\in\mathcal{N}} (R(x + dx, y + dy)) = 0 \right\}, \qquad (1)$$

where R denotes the region heat map and $\mathcal{N}$ denotes 8-neighbors in the 2-D coordinate system. Later, the extracted points from the labelling-based spotter are used in character decoding if they are included in *large* text boxes, and the ones from peak-based spotting are used if included in *small* text boxes, where *large* and *small* are defined by whether the shorter side of the box is longer or shorter than a certain threshold.

3.2   Character Decoder

Our character decoder is called *the linear point-wise decoder*, which simply performs linear classification of feature vectors on detected character points. Given a set of points $P = [(x_1, y_1), (x_2, y_2), ..., (x_N, y_N)]$, a feature map $\boldsymbol{f}$ with F channels and size $H \times W$, and the number of classes (i.e., types of characters) C, the decoder is denoted as

$$\text{cls} = \text{Softmax}(\boldsymbol{f}[:,P]\boldsymbol{w} - \mathbb{1}\boldsymbol{b}),$$

$$= \text{Softmax} \left(\left(\begin{matrix} \boldsymbol{f}[:,y_1,x_1] \\ \boldsymbol{f}[:,y_2,x_2] \\ \vdots \\ \boldsymbol{f}[:,y_N,x_N] \end{matrix} \right) \left(\begin{matrix} w_1^1 & w_1^2 & \dots & w_1^C \\ w_2^1 & w_2^2 & \dots & w_2^C \\ \vdots & \vdots & \dots & \vdots \\ w_F^1 & w_F^2 & \dots & w_F^C \end{matrix} \right) - \left(\begin{matrix} b_1 & b_2 & \dots & b_C \\ b_1 & b_2 & \dots & b_C \\ \vdots & \vdots & \dots & \vdots \\ b_1 & b_2 & \dots & b_C \end{matrix} \right) \right), (2)$$

where $f[:,P]$ denotes the index operation that extracts feature vectors at the points P and stacks them into an (N,F)-shaped matrix. The parameters of linear transformation $(\boldsymbol{w}, \boldsymbol{b})$ are parts of the learnable parameters of the model, and are optimized by the back propagation during training jointly to the other network parameters, where $\boldsymbol{w}$ is an $F \times C$ matrix and $\boldsymbol{b} = [b_1, b_2, \dots, b_C]$ is a C-dimensional vector broadcast by $\mathbb{1} = [1, 1, \dots, 1]^T$ with length N. Finally, row-wise softmax is taken along the channel axis to form classification probabilities. Then cls becomes an (N, C)-shaped matrix, where its element at (i, j) encodes the probability that the i-th points is recognized as the j-th character.

After giving character probability to each point, we further filter out the points that are not confidently classified by applying a threshold. Finally, each character point is assigned to the text box that include it, and for each text box, the character points that are assigned to it are sorted by x coordinates and read left-to-right.

It is worth noting that the linear point-wise decoding is numerically equivalent to semantic-segmentation-style character decoding, except that it is computed sparsely. By regarding the classification weights and biases in Eq. 2 as a

1×1 convolution layer, it can be applied densely to the feature map, as shown in Fig. 5 a. Despite of its implementation simplicity, it suffers from its heavy output tensor. The output has a $(W \times H \times \#\{classes\})$-sized label space, where $W \times H$ is the desired output resolution and $\#\{classes\}$ is the number of types of characters we want to recognize. In text recognition, $\#\{classes\}$ is already large in the Latin script, which has over 94 alphanumerics and symbols in total, and may be even larger in other scripts (for example, there are over 2,000 types of Chinese characters). In practice, dense labeling of 94 characters from a $(400 \times 400 \times 32)$-sized feature map consumes around 60 MB memory and 0.48 G floating-point operations only by the output layer, regardless of how many letters are present in the image. If given 2,000 characters, the output layer needs 1.2 GB memory, which critically limits the training and inference efficiency. Thus, we prefer the point-wise alternative (Fig. 5 b) to maintain scalability and to avoid unnecessary computation when there are few letters in images.

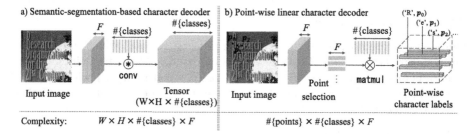

Fig. 5. Comparison between semantic-segmentation-based character decoding and our linear point-wise decoding. The linear point-wise decoding has advantages in space and time complexity when $\#\{points\}$ is small and $\#\{classes\}$ is large.

3.3   Feature Extractor

Most scene text spotters use VGG16 or ResNet50 as their backbone feature extractors, which are heavy for mobiles. Our choice for lightweight design is CSP-PeleeNet [45], which was originally designed for mobile object detection [47], and further lightened by adding cross-stage partial connection [45] without reducing ImageNet accuracy. The network follows the DenseNet [22] architecture but is substantially smaller thanks to having 32, 128, 256, 704 channels in its 1/1-, 1/4-, 1/8-, 1/16-scaled stage each, and a total of 21 stacked dense blocks.

On the CSPPeleeNet backbone, we construct U-Net [40] by adding upsampling modules for feature decoding. We put the output layer of the U-Net at the 1/4-scaled stage, which corresponds to the second block in the CSPPeleeNet.[1]

3.4   Training

Context-Free TextSpotter is end-to-end differentiable and trainable with ordinary gradient-based solvers. The training objective L is defined as

---

[1] Note that Fig. 3 does not precisely reflect the structure due to layout constraints.

$$L = L_{\text{det}} + \alpha L_{\text{rec}}, \tag{3}$$

$$L_{\text{det}} = \frac{1}{WH}\left(|R - R_{\text{gt}}|^2 + |A - A_{\text{gt}}|^2\right), \tag{4}$$

$$L_{\text{rec}} = \frac{1}{N}\sum_{(x_{\text{gt}},y_{\text{gt}},c_{\text{gt}})\in P_{\text{gt}}} \text{CE}\left(\text{Softmax}\left(\boldsymbol{f}[:,y_{\text{gt}},x_{\text{gt}}]\boldsymbol{w} - \boldsymbol{b}\right), c_{\text{gt}}\right), \tag{5}$$

where (W, H) denotes feature-map size, $R, A, R_{\text{gt}}, A_{\text{gt}}$ denote the predicted region and affinity map and their corresponding ground truths, and $(x_{\text{gt}}, y_{\text{gt}}, c_{\text{gt}}) \in P_{\text{gt}}$ denotes character-level ground truths that indicate locations and classes of the characters. CE denotes the cross-entropy classification loss, and α is a hyperparameter that balances the two losses.

This training objective needs character-level annotations, but many scene text recognition benchmarks only provide word-level ones. Therefore, we adopt weakly-supervised learning that exploits approximated character-level labels in the earlier training stage, and updates the labels by self-labeling in the later stage. In the earlier stage, for stability, we fix P_{gt} to the centers of *approximated character polygons*. Here, the approximated character polygons are calculated by equally dividing the word polygons into the word-length parts along their word lines. The ground truths of the region and affinity maps are generated using the same approximated character polygons following the same manner as the CRAFT [2] detector.

In the later stage, P_{gt} is updated by self-labeling: the word regions of training images are cropped by ground-truth polygon annotation, the network under training predicts heat maps from the cropped images, and spotted points in them are used as new P_{gt} combined with the ground-truth word transcriptions. Word instances whose number of spotted points and length of the ground-truth transcription are not equal are discarded because the difference suggests inaccurate self-labeling. We also exploit synthetic text images with character-level annotation by setting their P_{gt} at the character-polygon centers.

4 Experiments

To evaluate the effectiveness and efficiency of Context-Free TextSpotter, we conducted intensive experiments using scene-text recognition datasets. Further, we analyzed factors that control quality and speed, namely, choice of backbones and input-image resolution, and conducted ablative analyses of the modules. Finally, we deployed Context-Free TextSpotter as an iPhone application and measured the on-device inference speed to demonstrate the feasibility of mobile text spotting.

Datasets. We used ICDAR2013 [26] and ICDAR2015 [25] dataset for evaluation. ICDAR2013 thematizes focused scene texts and consists of 229 training and 233 testing images. ICDAR2015 thematizes incidental scene texts and consists of 1,000 training and 500 testing images. In the evaluation, we followed the competition's official protocol. For detection, we used DetEval for ICDAR2013 and IoU with threshold 0.5 for ICDAR2015, and calculated precision, recall, and their harmonic means (H means). For recognition, we used *end-to-end* and *word spotting* protocol, with provided *Strong* (100 words), *Weak* (1,000 words), and *Generic* (90,000 words) lexicons.

For training, we additionally used the SynthText [16] and ICDAR2019 MLT [37] datasets to supplement the relatively small training sets of ICDAR2013/2015. SynthText is a synthetic text dataset that provides 800K images and character-level annotations by polygons. ICDAR2019 MLT is a multilingual dataset with 10,000 training images. We replaced non-Latin scripts with *ignore* symbols so that we could use the dataset to train a Latin text spotter.

Implementation Details. Character-recognition feature channels F was set to 256. Input images were resized into 2,880 pixels for ICDAR2015 and 1,280 pixels for ICDAR2013 in the longer side keeping the aspect ratio. For training, we used Adam with initial learning rate 0.001, recognition-loss weight 0.01, and the batch size of five, where three are from real the and the rest are from SynthText training images. For evaluation, we used weighted edit distance [36] for lexicon matching. Our implementation was based on PyTorch-1.2.0 and ran on a virtual machine with an Intel Xeon Silver 4114 CPU, 16 GB RAM, and an NVIDIA Tesla V100 (VRAM16 GB × 2) GPU. All run times were measured with the batch size of one.

Table 1. Text detection and recognition results on ICDAR2015. **Bold** indicates the best performance and underline indicates the second best within lightweight text-spotting models. * indicates our re-measured or re-calculated numbers, while the others are excerpted from the literature.

Method	Detection			Word spotting			End-to-end			Prms	FPS
	P	R	H	S	W	G	S	W	G	(M)	
Standard models											
EAST [52]	83.5	73.4	78.2	–	–	–	–	–	–	–	13.2
Seglink [41]	73.1	76.8	75.0	–	–	–	–	–	–	–	8.9
CRAFT [2]	89.8	84.3	86.9	–	–	–	–	–	–	20*	5.1*
StradVision [25]	–	–	–	45.8	–	–	43.7	–	–	–	–
Deep TS [4]	–	–	–	58	53	51	54	51	47	–	9
Mask TS [20]	91.6	81.0	86.0	79.3	74.5	64.2	79.3	73.0	62.4	87*	2.6
FOTS [32]	91.0	85.1	87.9	84.6	79.3	63.2	81.0	75.9	60.8	34	7.5
Parallel Det-Rec [27]	83.7	96.1	89.5	89.0	84.4	68.8	85.3	80.6	65.8	–	3.7
CharNet [50]	89.9	91.9	90.9	–	–	–	83.1	79.1	69.1	89	0.9*
CRAFTS [3]	89.0	85.3	87.1	–	–	–	83.1	82.1	74.9	–	5.4
Lightweight models											
PeleeText [8]	85.1	72.3	78.2	–	–	–	–	–	–	10.3	11
PeleeText++ [7]	81.6	78.2	79.9	–	–	–	–	–	–	7.0	15
PeleeText++ MS [7]	87.5	76.6	81.7	–	–	–	–	–	–	7.0	3.6
Mask TS mini [20]	–	–	–	71.6	63.9	51.6	71.3	62.5	50.0	87*	6.9
FOTS RT [32]	85.9	79.8	82.7	76.7	69.2	53.5	**73.4**	**66.3**	51.4	28	10*
Light TS [15]	**94.5**	70.7	80.0	**77.2**	**70.9**	**65.2**	–	–	–	34*	4.8
Context-Free (ours)	88.4	77.1	**82.9**	74.3	67.1	54.6	70.2	63.4	52.4	**3.1**	**12**

Table 2. Text detection and recognition results on ICDAR2013.

Method	Detection			Word spotting			End-to-end			Prms	FPS
	P	R	H	S	W	G	S	W	G	(M)	
Standard models											
EAST [52]	92.6	82.6	87.3	–	–	–	–	–	–	–	13.2
Seglink [41]	87.7	83.0	85.3	–	–	–	–	–	–	–	20.6
CRAFT [2]	97.4	93.1	95.2	–	–	–	–	–	–	20*	10.6*
StradVision [25]	–	–	–	85.8	82.8	70.1	81.2	78.5	67.1	–	–
Deep TS [4]	–	–	–	92	89	81	89	86	77	–	10
Mask TS [20]	91.6	81.0	86.0	92.5	92.0	88.2	92.2	91.1	86.5	87*	4.8
FOTS [32]	–	–	88.3	92.7	90.7	83.5	88.8	87.1	80.8	34	13*
Parallel Det-Rec [27]	–	–	–	95.0	93.7	88.7	90.2	88.9	84.5	–	–
CRAFTS [3]	96.1	90.9	93.4	–	–	–	94.2	93.8	92.2	–	8.3
Lightweight models											
MobText [10]	88.3	66.6	76.0	–	–	–	–	–	–	9.2	–
PeleeText [8]	80.1	79.8	80.0	–	–	–	–	–	–	10.3	18
PeleeText++ [7]	87.0	73.5	79.7	–	–	–	–	–	–	7.0	23
PeleeText++ MS [7]	**92.4**	80.0	**85.7**	–	–	–	–	–	–	7.0	3.6
Context-Free (ours)	85.1	**83.4**	84.2	83.9	80.0	69.1	80.1	76.4	67.1	**3.1**	**25**

4.1   Results

Table 1 summarizes the detection and recognition results on ICDAR2015. Context-Free TextSpotter achieved an 82.9 detection H mean and 74.3 word-spotting H mean with the Strong lexicon with 3.8 million parameters and 12 FPS on a GPU. The inference speed and model size of Context-Free was the best among all compared text spotters.

For a detailed comparison, we separated the text spotters into two types: *standard models* and *lightweight models*. While it is difficult to introduce such separation due to the continuous nature of the size-performance balance in neural networks, we categorized 1) models that are claimed to be lightweight in their papers and 2) smaller or faster versions reported in papers whose focus were creating standard-sized models. The PeleeText [7,8] family is detection-only models based on Pelee [47], but the model sizes are larger than ours due to feature-pyramid-based bounding box regression. Mask TextSpotter mini (Mask TS mini) and FOTS Real Time (FOTS RT) are smaller versions reported in the corresponding literature, and are acquired by scaling input image size. However, their model sizes are large due to the full-size backbones, while the accuracies were comparable to ours. This suggests that simple minification of standard models is not optimal in repurposing for mobile or fast inference. On another note, FOTS RT was originally reported to run in 22.6 FPS in a customized Caffe with a TITAN-Xp GPU. Since it is closed-source, we re-measured a 3rd-party

Fig. 6. Text spotting results with Context-Free TextSpotter on ICDAR2013 and 2015. The lexicons are not used in this visualization. Red characters indicate wrong recognition and purple missed ones. Best viewed zoomed in digital. (Color figure online)

reimplementation in PyTorch[2], and the FPS was 10 when the batch size was one. We reported the latter number in the table in order to prioritize comparisons in the same environment. The other methods that we re-measured showed faster speed than in their original reports, seemingly due to our newer GPU. Light TextSpotter, which is based on lightened Mask TextSpotter, had an advantage in the evaluation with the Generic lexicon, but is much heavier than ours due to the instance-mask branches in the network.

Table 2 summarizes the detection and recognition results on ICDAR2013. ICDAR2013 contains focused texts with relatively large and clear appearance, which enables good recognition accuracy in smaller test-time input-image sizes than in ICDAR2015. While all text spotters ran faster, Context-Free was again the fastest among them. Light TextSpotter and the smaller versions of FOTS and Mask TextSpotter were not experimented in ICDAR2013 and thus are omitted here. This means we have no lightweight competitors in text spotting, but ours outperformed the lightweight text detectors while our model size was smaller and our inference was faster than theirs.

To analyze the effect of input-image sizes on inference, we collected the results with different input sizes and visualize them in Fig. 1. Here, input images were resized using bilinear interpolation keeping their aspect ratios before inference. We used a single model trained with cropped images 960 × 960 in size without retraining. The best results were obtained when the longer sides were 2,880 pixels, where the speed was 12 FPS. We also obtained the near-real-time speed

---

Table 3. Ablative analyses of modules and techniques we adopted.

	Training	E2E	Removing unreadables	Character spotting	Lexicon matching	Det H	WS S
a)	Synth		–	–	–	64.8	–
b1)	+ IC15		–	–	–	79.9	–
b2)	+ IC15	✓		Peak	Edit dist	80.2	–
c1)	+ IC15	✓	✓	Peak	Edit dist	81.0	70.5
c2)	+ IC15	✓	✓	Label	Edit dist	78.4	48.4
c3)	+ IC15	✓	✓	Hybrid	Edit dist	81.0	71.3
d)	+ IC15	✓	✓	Hybrid	Weighted ED	81.0	72.8
e)	+ IC15 & 19	✓	✓	Hybrid	Weighted ED	82.9	74.3

Table 4. Comparisons of lightweight backbones in our framework.

Backbone	IC13	IC15	FPS
MobileNetV3	67.1	50.1	11
GhostNet	74.6	63.3	9.7
CSPPeleeNet	83.9	74.3	12

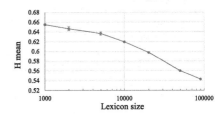

Fig. 7. Lexicon size vs. recognition H mean.

of 26 FPS with inputs 1,600 pixels long, even on the more difficult ICDAR2015 dataset with 62.8 H mean.

Table 3 summarizes the ablative analyses by removing the modules and techniques we adopted in the final model. Our observations are four-fold: First, real-image training is critical (a vs. b1). Second, end-to-end training itself did not much improve detection (b1 vs. b2), but removing boxes that were not validly read as texts (i.e., ones transcribed as empty strings by our recognizer) significantly boosted the detection score (c1). Third, the combination of peak- and labeling-based character spotting (called hybrid) was useful (c1–3). Fourth, weighted edit distance [28] and adding MLT training data [27], the known techniques in the literature, were also helpful for our model (d and e). Table 4 shows comparisons of the different backbones in implementing our method. We tested MobileNetV3-Large [21] and GhostNet [17] as recent fast backbones, and confirmed that CSPPeleeNet was the best among them.

Finally, as a potential drawback of context-free text recognition, we found that duplicated reading of a single characters and minor spelling errors sometimes appeared in outputs uncorrected, as seen in examples in Fig. 6 and the recognition metrics with the *Generic* lexicon. These may become a problem more when the target lexicon is large. Thus, we investigated the effect of lexicon sizes on text-spotting accuracy. We used *Weak* as a basic lexicon and inflated it by adding randomly sampled words from *Generic*. The trials were done five times and we plotted the average. The results are shown in Fig. 7, which shows the recognition degradation was relatively gentle when the lexicon size is smaller than 10,000. This suggests that Context-Free TextSpotter is to some extent robust against large vocabulary, despite it lacks language modelling.

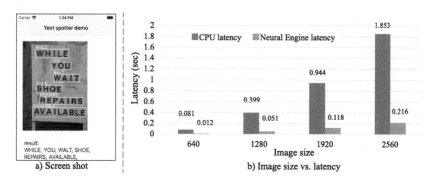

a) Screen shot   b) Image size vs. latency

Fig. 8. On-device Benchmarking with iPhone 11 Pro.

4.2   On-Device Benchmarking

To confirm the feasibility of deploying our text spotter on mobiles, we conducted on-device benchmarking. We used an iPhone 11 Pro with the Apple A13 CPU and 4GB memory, and ported our PyTorch model to the CoreML framework to run on it. We also implemented Swift-based post-processing (i.e., connected-component analysis and peak detection) to include their computation times in the benchmarking. With our ICDAR2013 setting (input size = 1, 280), the average inference latency was 399 msec with a CPU, and 51 msec with a Neural Engine [1] hardware accelerator. Usability studies suggest that latency within 100 msec makes "the user feel that the system is reacting instantaneously", and 1,000 msec is "the limit for the user's flow of thought to stay uninterrupted, even though the user will notice the delay" [38]. Our text spotter achieves the former if an accelerator is given, and the latter on a mobile CPU. More latencies with different input sizes are shown in Fig. 8.

We also tried to port existing text spotters. MaskTextSpotters [28,29] could not be ported to CoreML, due to their customized layers. While implementing the special text-spotting layers in the CoreML side would solve this, we also can say that simpler convolution-only models, like ours, have an advantage in portability. Another convolution-only model, CharNet [50] ran on iPhone but fairly slowly. With NeuralEngine, it took around 1.0 s to process 1,280-pixel-sized images, and could not process larger ones due to memory limits. With the CPU, it took 8.1 s for 1,280-pixel images, 18.2 s for 1,920, and ran out of memory for larger sizes.

5   Conclusion

In this work, we have proposed Context-Free TextSpotter, an end-to-end text spotting method without bells and whistles that enables real-time and mobile text spotting. We hope our method inspires the developers for OCR applications, and will serve as a stepping stone for researchers who want to prototype their ideas quickly.

Acknowledgements. We would like to thank Katsushi Yamashita, Daeju Kim, and members of the AI Strategy Office in SoftBank for helpful discussion.

References

1. Apple Developer Documentation: MLComputeUnits. https://developer.apple. com/documentation/coreml/mlcomputeunits. Accessed 28 Jan 2021
2. Baek, Y., Lee, B., Han, D., Yun, S., Lee, H.: Character region awareness for text detection. In: CVPR, pp. 9365–9374 (2019)
3. Baek, Y., et al.: Character region attention for text spotting. In: Vedaldi, A., Bischof, H., Brox, T., Frahm, J.-M. (eds.) ECCV 2020. LNCS, vol. 12374, pp. 504–521. Springer, Cham (2020). https://doi.org/10.1007/978-3-030-58526-6_30
4. Busta, M., Neumann, L., Matas, J.: Deep TextSpotter: an end-to-end trainable scene text localization and recognition framework. In: ICCV, pp. 2204–2212 (2017)
5. Chen, H., et al.: AdderNet: do we really need multiplications in deep learning? In: CVPR, pp. 1468–1477 (2020)
6. Cheng, Z., Bai, F., Xu, Y., Zheng, G., Pu, S., Zhou, S.: Focusing attention: towards accurate text recognition in natural images. In: ICCV, pp. 5076–5084 (2017)
7. Córdova, M., Pinto, A., Pedrini, H., Torres, R.D.S.: Pelee-Text++: a tiny neural network for scene text detection. IEEE Access (2020)
8. Córdova, M.A., Decker, L.G., Flores-Campana, J.L., dos Santos, A.A., Conceição, J.S.: Pelee-Text: a tiny convolutional neural network for multi-oriented scene text detection. In: ICMLA, pp. 400–405 (2019)
9. Courbariaux, M., Bengio, Y., David, J.P.: BinaryConnect: training deep neural networks with binary weights during propagations. In: NeurIPS (2015)
10. Decker1a, L.G.L., et al.: MobText: a compact method for scene text localization. In: VISAPP (2020)
11. Deng, D., Liu, H., Li, X., Cai, D.: PixelLink: detecting scene text via instance segmentation. In: AAAI, vol. 32 (2018)
12. Fu, K., Sun, L., Kang, X., Ren, F.: Text detection for natural scene based on MobileNet V2 and U-Net. In: ICMA, pp. 1560–1564 (2019)
13. Graves, A., Fernández, S., Gomez, F., Schmidhuber, J.: Connectionist temporal classification: labelling unsegmented sequence data with recurrent neural networks. In: NeurIPS, pp. 369–376 (2006)
14. Graves, A., Fernández, S., Schmidhuber, J.: Bidirectional LSTM networks for improved phoneme classification and recognition. In: Duch, W., Kacprzyk, J., Oja, E., Zadrożny, S. (eds.) ICANN 2005. LNCS, vol. 3697, pp. 799–804. Springer, Heidelberg (2005). https://doi.org/10.1007/11550907_126
15. Guan, J., Zhu, A.: Light Textspotter: an extreme light scene text spotter. In: Yang, H., Pasupa, K., Leung, A.C.-S., Kwok, J.T., Chan, J.H., King, I. (eds.) ICONIP 2020. CCIS, vol. 1332, pp. 434–441. Springer, Cham (2020). https://doi.org/10. 1007/978-3-030-63820-7_50

16. Gupta, A., Vedaldi, A., Zisserman, A.: Synthetic data for text localisation in natural images. In: CVPR, pp. 2315–2324 (2016)
17. Han, K., Wang, Y., Tian, Q., Guo, J., Xu, C., Xu, C.: GhostNet: more features from cheap operations. In: CVPR, pp. 1580–1589 (2020)
18. Han, S., Pool, J., Tran, J., Dally, W.J.: Learning both weights and connections for efficient neural networks. In: NeurIPS (2015)
19. He, P., Huang, W., He, T., Zhu, Q., Qiao, Y., Li, X.: Single shot text detector with regional attention. In: ICCV, pp. 3047–3055 (2017)
20. He, T., Tian, Z., Huang, W., Shen, C., Qiao, Y., Sun, C.: An end-to-end TextSpotter with explicit alignment and attention. In: CVPR, pp. 5020–5029 (2018)
21. Howard, A., et al.: Searching for MobileNetV3. In: ICCV (2019)
22. Huang, G., Liu, Z., Van Der Maaten, L., Weinberger, K.Q.: Densely connected convolutional networks. In: CVPR, pp. 4700–4708 (2017)
23. Hubara, I., Courbariaux, M., Soudry, D., El-Yaniv, R., Bengio, Y.: Quantized neural networks: training neural networks with low precision weights and activations. J. Mach. Learn. Res. 18(1), 6869–6898 (2017)
24. Jeon, M., Jeong, Y.S.: Compact and accurate scene text detector. Appl. Sci. 10(6), 2096 (2020)
25. Karatzas, D., et al.: ICDAR 2015 competition on robust reading. In: ICDAR, pp. 1156–1160 (2015)
26. Karatzas, D., et al.: ICDAR 2013 robust reading competition. In: ICDAR, pp. 1484–1493 (2013)
27. Li, J., Zhou, Z., Su, Z., Huang, S., Jin, L.: A new parallel detection-recognition approach for end-to-end scene text extraction. In: ICDAR, pp. 1358–1365 (2019)
28. Liao, M., Lyu, P., He, M., Yao, C., Wu, W., Bai, X.: Mask TextSpotter: an end-to-end trainable neural network for spotting text with arbitrary shapes. PAMI (2019)
29. Liao, M., Pang, G., Huang, J., Hassner, T., Bai, X.: Mask TextSpotter v3: segmentation proposal network for robust scene text spotting. In: Vedaldi, A., Bischof, H., Brox, T., Frahm, J.-M. (eds.) ECCV 2020. LNCS, vol. 12356, pp. 706–722. Springer, Cham (2020). https://doi.org/10.1007/978-3-030-58621-8_41
30. Liao, M., Shi, B., Bai, X.: TextBoxes++: a single-shot oriented scene text detector. Trans. Image Process. 27(8), 3676–3690 (2018)
31. Liu, W., et al.: SSD: single shot MultiBox detector. In: Leibe, B., Matas, J., Sebe, N., Welling, M. (eds.) ECCV 2016. LNCS, vol. 9905, pp. 21–37. Springer, Cham (2016). https://doi.org/10.1007/978-3-319-46448-0_2
32. Liu, X., Liang, D., Yan, S., Chen, D., Qiao, Y., Yan, J.: FOTS: fast oriented text spotting with a unified network. In: CVPR, pp. 5676–5685 (2018)
33. Liu, Y., Chen, H., Shen, C., He, T., Jin, L., Wang, L.: ABCNet: real-time scene text spotting with adaptive Bezier-curve network. In: CVPR, pp. 9809–9818 (2020)
34. Long, J., Shelhamer, E., Darrell, T.: Fully convolutional networks for semantic segmentation. In: CVPR, pp. 3431–3440 (2015)
35. Long, S., Ruan, J., Zhang, W., He, X., Wu, W., Yao, C.: TextSnake: a flexible representation for detecting text of arbitrary shapes. In: Ferrari, V., Hebert, M., Sminchisescu, C., Weiss, Y. (eds.) ECCV 2018. LNCS, vol. 11206, pp. 19–35. Springer, Cham (2018). https://doi.org/10.1007/978-3-030-01216-8_2
36. Lyu, P., Liao, M., Yao, C., Wu, W., Bai, X.: Mask TextSpotter: an end-to-end trainable neural network for spotting text with arbitrary shapes. In: Ferrari, V., Hebert, M., Sminchisescu, C., Weiss, Y. (eds.) Computer Vision – ECCV 2018. LNCS, vol. 11218, pp. 71–88. Springer, Cham (2018). https://doi.org/10.1007/978-3-030-01264-9_5

37. Nayef, N., et al.: ICDAR2019 robust reading challenge on multi-lingual scene text detection and recognition? RRC-MLT-2019. In: ICDAR (2019)
38. Nielsen, J.: Usability Engineering. Morgan Kaufmann (1994)
39. Qiao, L., et al.: Mango: a mask attention guided one-stage scene text spotter. In: AAAI (2021)
40. Ronneberger, O., Fischer, P., Brox, T.: U-net: convolutional networks for biomedical image segmentation. In: Navab, N., Hornegger, J., Wells, W.M., Frangi, A.F. (eds.) MICCAI 2015. LNCS, vol. 9351, pp. 234–241. Springer, Cham (2015). https://doi.org/10.1007/978-3-319-24574-4_28
41. Shi, B., Bai, X., Belongie, S.: Detecting oriented text in natural images by linking segments. In: CVPR, pp. 2550–2558 (2017)
42. Shi, B., Bai, X., Yao, C.: An end-to-end trainable neural network for image-based sequence recognition and its application to scene text recognition. PAMI **39**(11), 2298–2304 (2016)
43. Shi, B., Wang, X., Lyu, P., Yao, C., Bai, X.: Robust scene text recognition with automatic rectification. In: CVPR, pp. 4168–4176 (2016)
44. Tian, Z., Huang, W., He, T., He, P., Qiao, Yu.: Detecting text in natural image with connectionist text proposal network. In: Leibe, B., Matas, J., Sebe, N., Welling, M. (eds.) ECCV 2016. LNCS, vol. 9912, pp. 56–72. Springer, Cham (2016). https://doi.org/10.1007/978-3-319-46484-8_4
45. Wang, C.Y., Mark Liao, H.Y., Wu, Y.H., Chen, P.Y., Hsieh, J.W., Yeh, I.H.: CSP-Net: a new backbone that can enhance learning capability of CNN. In: International Conference on Computer Vision and Pattern Recognition (CVPR) Workshops, pp. 390–391 (2020)
46. Wang, K., Babenko, B., Belongie, S.: End-to-end scene text recognition. In: ICCV (2011)
47. Wang, R.J., Li, X., Ling, C.X.: Pelee: a real-time object detection system on mobile devices. NeurIPS **31**, 1963–1972 (2018)
48. Wang, T., Wu, D.J., Coates, A., Ng, A.Y.: End-to-end text recognition with convolutional neural networks. In: ICPR (2012)
49. Wu, K., Otoo, E., Suzuki, K.: Optimizing two-pass connected-component labeling algorithms. Pattern Anal. Appl. **12**(2), 117–135 (2009)
50. Xing, L., Tian, Z., Huang, W., Scott, M.R.: Convolutional character networks. In: ICCV, pp. 9126–9136 (2019)
51. Zhang, Z., Zhang, C., Shen, W., Yao, C., Liu, W., Bai, X.: Multi-oriented text detection with fully convolutional networks. In: CVPR, pp. 4159–4167 (2016)
52. Zhou, X., et al.: EAST: an efficient and accurate scene text detector. In: CVPR, pp. 5551–5560 (2017)
53. Zhu, X., et al.: Deep residual text detection network for scene text. In: ICDAR, vol. 1, pp. 807–812. IEEE (2017)
54. Zhu, Y., Wang, S., Huang, Z., Chen, K.: Text recognition in images based on transformer with hierarchical attention. In: ICIP, pp. 1945–1949. IEEE (2019)

MIDV-LAIT: A Challenging Dataset for Recognition of IDs with Perso-Arabic, Thai, and Indian Scripts

Yulia Chernyshova[1,2]($\boxtimes$) (ID), Ekaterina Emelianova[2,3] (ID),
Alexander Sheshkus[1,2] (ID), and Vladimir V. Arlazarov[1,2] (ID)

[1] FRC CSC RAS, Moscow, Russia
[2] Smart Engines Service LLC, Moscow, Russia
chernyshova@smartengines.com
[3] NUST MISIS, Moscow, Russia

Abstract. In this paper, we present a new dataset for identity documents (IDs) recognition called MIDV-LAIT. The main feature of the dataset is the textual fields in Perso-Arabic, Thai, and Indian scripts. Since open datasets with real IDs may not be published, we synthetically generated all the images and data. Even faces are generated and do not belong to any particular person. Recently some datasets have appeared for evaluation of the IDs detection, type identification, and recognition, but these datasets cover only Latin-based and Cyrillic-based languages. The proposed dataset is to fix this issue and make it easier to evaluate and compare various methods. As a baseline, we process all the textual field images in MIDV-LAIT with Tesseract OCR. The resulting recognition accuracy shows that the dataset is challenging and is of use for further researches.

Keywords: Dataset · Identity documents · Document analysis and recognition · Arabic script · Thai script · Indic scripts

1 Introduction

The processing and recognition of camera-captured images of various documents gain more popularity every year with the evolution of smartphones, tablets, and small digital cameras. Moreover, Know Your Customer (KYC) is an important and mandatory practice for any financial or governmental institution. To enhance the user experience, many organizations want to automatize the process necessary for KYC. A very time-consuming part of this process is entering the client's data from the photography of their identity document (ID). Recognition of an ID captured in the uncontrolled conditions is a complex iterative

This work was supported in part by RFBR according to the research projects 19-29-09075 and 19-29-09092.

Fig. 1. Cropped example images and text line images from MIDV-LAIT

process [12] employing a variety of image processing and machine learning methods. For modern recognition algorithms, public datasets are de facto the only possible comparison tool. Yet for ID images, GDPR [4] and other local laws prohibit the creation of datasets with real data. As a result, several datasets with synthetic and sample ID images appeared [9,13,21,32,38]. Out of them, three datasets MIDV-500 [9], MIDV-2019 [13], and BID [38] provide both the geometrical and textual ground truth. But BID contains only the Brazilian IDs, while both MIDV-500 and its addition MIDV-2019 contain 50 different IDs. Yet MIDV-500 has important drawbacks. Firstly, it contains only one document per type, e.g., one Chinese passport, or one Iranian driving license. Thus, it cannot be divided into independent training and test parts. Secondly, to create it the authors employed rather old smartphones (Apple iPhone 5 and Samsung Galaxy S3). Finally, MIDV-500 contains textual values only for the text fields printed with Latin and Cyrillic scripts and Western Arabic digits. These three drawbacks inspired us to create a new dataset called MIDV-LAIT that contains IDs with textual fields in Perso-Arabic, Thai, and Indian scripts. What is more, we created at least five unique documents per type and captured all the video clips with Apple iPhone X Pro. Figure 1 show the cropped example images (original images contain more background) and the example field images from MIDV-LAIT. The dataset is available at ftp://smartengines.com/midv-lait/.

2   Datasets for the Indian, Thai, and Perso-Arabic Scripts

To begin with, we want to describe some datasets with printed text for the Indian and Thai languages. In [29], the authors employ the UHTelPCC dataset with images of Telugu syllables. The authors of [28] propose a recognition system for documents printed in Kannada script and with Kannada, Sanskrit, Konkani, and Tulu languages printed in Kannada script. In [30], the authors present a dataset of scanned documents printed in Devanagari. The authors of [23] propose a system for classical Indian documents recognition written in the Sanskrit language in Devanagari script. To test their method, they use synthetically rendered text line images and real document images for three Sanskrit texts. The survey of Gujarati text recognition [26] mentions several printed Gujarati datasets. A Bangla dataset of camera-captured documents is described in paper [24], but, unfortunately, currently, it is unavailable. For Thai script, both the printed and handwritten datasets for National Software Contest exist [1]. All these datasets have several important drawbacks. Firstly, they contain either synthetically rendered or scanned images. Secondly, the majority of them contain binary images, some contain grayscale images with simple backgrounds. Finally, they usually do not cover all the syllables possible in the corresponding languages and scripts.

Speaking of printed Perso-Arabic datasets, a Yarmouk Arabic OCR dataset [22] contains scanned images of printed pages from Arabic Wikipedia. The APTID/MF database [25] contains synthetic images of Arabic text lines printed in 10 different fonts (all fonts represent the Naskh script). It is similar to the previously published APTI database [37] that is for the evaluation of screen-based OCR systems. An ALTID database [18] is a development of APTID/MF enriched with synthetic documents in Latin script and handwritten documents in both Arabic and Latin scripts. SmartATID [15] is the first dataset of camera-captured images of Arabic documents. The authors used Samsung Galaxy S6 and Apple iPhone 6S plus as the shooting devices. A simple dataset for Persian language text from [34] contains synthetic images of black text in white backgrounds. A dataset for Urdu ligatures recognition called Qaida [35] contains about 4M synthetically images in 256 different fonts. An UPTI database [36] consists of synthetically rendered images with Perso-Arabic Nastaliq script. Out of all listed Perso-Arabic datasets, only SmartATID contains camera-captured images, but its shooting conditions are still too controlled, e.g., for the out-of-focus blur, the document is 35 cm away from the camera and the focus is at the 22 cm distance. The other datasets contain binary or grayscale synthetically rendered images.

Recognition of handwritten text images is a popular task, and several datasets exist for all our scripts of interest. For example, PHDIndic_11 [31] is a dataset of scanned handwritten documents in 11 official Indian scripts. In MIDV-LAIT we cover the same scripts as they can be used in official documents and IDs. Yet we did not find any datasets for camera-captured handwritten text in Arabic, Thai, or Indian languages.

Besides the scanned documents' recognition, another popular task is the recognition of ancient manuscripts and historical inscriptions. The authors of [16] use a dataset of Kannada characters from the digitized estampages of stone inscriptions. This sample is similar to the famous MNIST dataset as it contains separate characters (or syllables) and is preprocessed. In 2017, a competition on recognition of early Indian printed documents was held as a part of ICDAR 2017. The target dataset consisted of scanned images of printed books in Bengali script containing such distortions typical for old books, for example, faded ink, ink bleed-through and show-through.

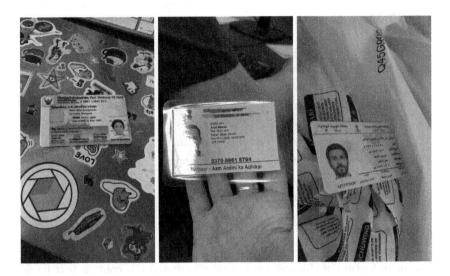

Fig. 2. Example images from MIDV-LAIT

The other problem in text recognition that cannot be skipped is the scene text recognition. In fact, the scene text recognition is very similar to camera-captured ID recognition as for both tasks the images can be highly distorted, contain complex backgrounds both outside and inside the target object, and a variety of fonts appear. EASTR-42k [7] is a dataset of 42K scene text images printed in both Arabic and Latin scripts. This dataset is one of the first datasets for Arabic scene text recognition and can be used as a benchmark for further research. We hope that our MIDV-LAIT can be of the same purpose for camera-captured ID recognition. The authors of [41] present a NI-VI dataset that contains images of Iraqi vehicle images for number plates recognition. An Urdu-Text dataset [8] provides images of Urdu text in natural scenes. A Thai scene text dataset recently appeared [39].

In our overview of the datasets for Arabic, Thai, and Indian languages we skipped a lot of datasets, as new small datasets for these languages tend to appear in every new paper. Yet we consider our overview rather consistent in terms of target problems of these datasets. A strong shortage of datasets with camera-captured documents printed in Arabic, Thai, or Indian scripts exists. Moreover, a dataset for the recognition of IDs with these scripts does not exist at all. In our view, it creates problems for a clear comparison of methods and approaches developed for this task.

3   MIDV-LAIT Dataset

In this paper, we present a dataset of synthetic identity documents images called MIDV-LAIT (Fig. 2).

Table 1. List of writing systems and languages in MIDV-LAIT

Writing system	Language	Writing system	Language
thai	Thai	bengali	Bengali
perso-arabic	Arabic	gujarati	Gujarati
perso-arabic	Persian	odia (oriya)	Odia (Oriya)
perso-arabic	Urdu	kannada	Kannada
perso-arabic	Urdu (Nastaliq)	malayalam	Malayalam
perso-arabic	Pushto	devanagari	Hindi
gurmukhi	Punjabi	devanagari	Marathi
tamil	Tamil	devanagari	Nepali
telugu	Telugu	latin	English

The main feature of this dataset is that all created IDs contain text fields in either Perso-Arabic script, or Thai script, or one of the Indian scripts. Thus, in the title MIDV-LAIT "L" stands for the Latin script (mainly, English language), "A" stands for the Perso-Arabic script, "I" stands for a variety of Indian scripts, and "T" is for the Thai one. The presented dataset contains images of the IDs in twelve writing systems apart from Latin. For Perso-Arabic script, we present lines printed with both Naskh and Nastaliq scripts. The full list of writing systems and languages is provided in Table 1. To be clear, we understand that Urdu and Urdu (Nastaliq) are the same in terms of language. But they are different in appearance and script, and many text line recognition methods that are effective for Naskh fail for Nastaliq. This situation is quite similar to the recognition of printed and handwritten texts - while the language itself can be the same, the methods would differ. Thus, we decided to separate the into two groups in all our experiments.

It is a well-known fact that public datasets of identity documents (ID) are forbidden by various laws on personal data protection [2,4]. To create our dataset we used sample document images that can be found in Wikipedia and are distributed under the Creative Commons license. The full list of links to the source images is provided along with the resulting dataset. As a rule, the IDs contain photos of their holders. We used the generated photos of the non-existent people from [3] in our synthesized documents. Figure 3 present the samples of the created IDs. Textual information was generated in several different ways depending on its type. Various numbers, dates and other similar fields was created randomly. The only restrictions were for dates to be adequate for the document, and for numbers to match the template. For fields with names and surnames we searched the internet for the web sites that provide the names for the specific country or language and then randomly chose the names and surnames from the list. As for the genders, we made an almost equal number of documents for both. In the case of addresses, professions, and so on we just made up some information. Machine-readable zones in the created ID documents are deliberately invalid as they contain incorrect check sums.

Fig. 3. Synthetic ideal IDs from MIDV-LAIT

```
{     "quad": [
          [512,  1466],
          [1810, 1463],
          [1794, 2351],
          [416,  2279]
      ]
}
```

Fig. 4. Per frame ground truth in JSON format

After we created the ideal images, we printed and laminated them. Then we captured one video clip per document using iPhone X Pro. The videos are captured in various conditions and backgrounds and also contain projective transform, highlights, low lightening, motion blur, and other distortions typical for mobile-captured video clips. Also, some of the videos were shot with daylight and some - with lamplight. Then we split the first two seconds of each video with 10 frames per second and acquired 20 images per document. As we created the images recently the dataset does not provide the distortions typical for years old ID documents, such as partial fading of printed information. The other distortion we could not cover is the structured highlights that appear because of antifraud protection measures.

For each frame, we provide a ground truth file in JSON format that contains the coordinates of four vertices of the document in the image (Fig. 4). For each created ID, we provide a ground truth file that contains a list of fields in this ID, e.g., name, date of birth. For each field, we provide its textual value, its script, and its rectangle on the ideal image (Fig. 5). The combination of two types of provided ground truth allows us to find all the fields in each frame.

In Fig. 6 we provide the examples of the Arabic field ground truth to show a fact that is likely to be of no importance for an Arabic-writing person, but tends to cause misunderstandings for others - the order of letters in Arabic text in two

```
{ "fields": [ //one field sample
        {
            "authority_eng": {
                "quad": [1170, 632, 116, 29],
                "script": "latin",
                "value": "KIRKUK"
            }
        }
    ]
}
```

Fig. 5. Per ID ground truth in JSON format

Fig. 6. Example of Arabic field ground truth

different text editors (Fig. 6). It is common knowledge that the Perso-Arabic script is written from right to left and is cursive. In our ground truth files, the text is provided as it would be written by a person, i.e., the first letter in the Arabic text is the rightmost one. But some of the text editors do not display the text correctly, i.e., they display it as left-to-right text with letters in their isolated forms.

In total, MIDV-LAIT contains 180 unique documents and 3600 images (twenty images per document). It contains identity cards, passports, and driving licenses from 14 countries. The dataset statistics per writing system are presented in Table 2. It should be mentioned that we separate the documents with a Perso-Arabic writing system into two parts by the script (Naskh or Nastaliq). Nastaliq is used only in the identity cards of Pakistan. Also, although the main focus of our dataset is the non-Latin writing systems, we provide the statistics for Latin fields as well, as the majority of the IDs contain them. In MIDV-LAIT, we did not provide all the possible syllables as we presented only ten documents per writing system with syllables, but we did not filter the texts in any way. In contrast with many previously published papers, we did not avoid any rare forms or spellings when choosing textual information.

Table 2 shows that the majority of writing systems are presented with 10 unique IDs, but Latin and Perso-Arabic are presented with 165 and 70 items correspondingly. In general, we tried to make 10 samples per ID type if the writing system is unique, e.g. only Thai identity cards include Thai script, and 5 samples if the script

Table 2. List of writing systems and languages in MIDV-LAIT

Writing system	Number of IDs	Number of fields
thai	10	1200
perso-arabic Naskh	70	9700
perso-arabic Nastaliq	10	400
gurmukhi	10	600
tamil	10	600
telugu	10	560
bengali	10	600
gujarati	10	600
odia (oriya)	10	600
kannada	10	640
malayalam	10	600
devanagari	10	900
latin	165	23,000

Fig. 7. IDs that contain fields in both Latin and Perso-Arabic scripts

is used in several IDs. The reason for such a large number of IDs with Latin text lines is simple: the majority of documents contain at least one field with Latin script and some IDs duplicate the information in both Perso-Arabic and Latin scripts (Fig. 7). As for the Perso-Arabic writing system, not only do several IDs contain fields printed with it, but the fonts vary greatly even in the Naskh script. For example, Naskh fonts can still contain some mainly vertical ligatures that are typical for Nastaliq. Figure 8 shows the text line images with the name "Muhammad" from Iranian identity card and Iranian passport. In the second case, the first two letters form a vertical ligature. Moreover, four languages employ Perso-Arabic script in MIDV-LAIT (Arabic, Persian, Urdu, and Pushto). These languages have different alphabets and, also, Arabic and Persian languages use different numerals.

Speaking of digits and dates, MIDV-LAIT presents fields in four types of digits (Western Arabic, Eastern Arabic, Persian, and Devanagari digits) as shown in Fig. 9. Moreover, different IDs in MIDV-LAIT contain dates in five different calendars: Islamic calendar (Lunar Hijri), Iranian calendar (Solar Hijri), Thai solar calendar, Bikram Sambat (official calendar of Nepal), and Gregorian calendar.

Fig. 8. Different writing of name "Muhammad" in Iranian IDs

Fig. 9. Different types of digits in MIDV-LAIT

3.1   MIDV-LAIT Target Problems

When creating MIDV-LAIT, the central problem we wanted to solve was the lack of ID datasets for the recognition of non-Latin and non-Cyrillic based languages. But, in fact, we see a whole list of tasks that appear in ID recognition and demand new methods that can be tested only with our dataset:

- Text line images recognition of the languages from Table 1 in camera-captured images of IDs;
- Writing system identification for further recognition. MIDV-LAIT includes Indian Aadhaar Cards that can contain information in ten writing systems, eleven if counting the Latin script. Thus, to recognize the textual fields, the writing system should be identified beforehand;
- Language identification for further recognition. For example, in Indian Aadhaar cards both Hindi and Marathi employ Devanagari characters;
- Cross-checking of dates in different calendars in one ID for fraud detection and authentication tasks.

Also, MIDV-LAIT can be used for tasks that appear in ID recognition, but are not specific for the presented dataset:

- Document detection and localization in the image;
- Document type identification;
- Document layout analysis;
- Face detection;
- Integration of the recognition results [11];
- Image quality assessment and best frame choice [33];
- Super-resolution and image construction as in some frames the IDs are partially hidden or too distorted;
- Fraud detection. But MIDV-LAIT can be used only as a negative sample as the IDs do not contain protective measures, and, also, we deliberately added incorrect check digits into the machine-readable zones in passports.

4   Baseline

4.1   Tesseract OCR 4.1.0

To provide the baseline for all the writing systems and documents in MIDV-LAIT, we decided to use Tesseract OCR 4.1.0 [5]. We did so as, firstly, modern Tesseract OCR (since version 4.0.0) employs long short-term memory (LSTM) artificial neural networks (ANNs) that are an efficient modern approach to text recognition. Secondly, Tesseract OCR is open-source, so all the ANNs can be found on GitHub and retrained if the need arises. Thirdly, previously published papers [10,14,17,19,20,27,40] have shown that Tesseract OCR is popular within the community as a baseline method and demonstrate sufficient results for a system that was not specifically fine-tuned for a specific task. Moreover, in [28] Tesseract OCR 4.0.0 demonstrates the accuracy similar to the system specifically created for Kannada script recognition. Last but not least, Tesseract OCR 4.1.0 enables the recognition of all the languages and writing systems presented in our dataset. For example, the other popular recognition engine used to provide baselines - Abbyy FineReader - does not support any Indian language.

4.2   Experimental Results

To evaluate the acquired results, we calculate three measures: the per-character recognition rate PCR (Eq. 1), the per-string recognition rate (Eq. 2), and TextEval [6] character accuracy. TextEval is an freely available tool for text recognition accuracy evaluation developed by PRImA Research Lab.

$$PCR = 1 - \frac{\sum_{i=1}^{L_{total}} \min(lev(l_{i_{ideal}}, l_{i_{recog}}), len(l_{i_{ideal}}))}{\sum_{i=1}^{L_{total}} len(l_{i_{ideal}})} \tag{1}$$

where L_{total} denotes the total number of lines, $len(l_{i_{ideal}})$ is the length of the i-th line and $lev(l_{i_{ideal}}, l_{i_{recog}})$ stands for the Levenshtein distance between the recognized text and the ground truth.

$$PSR = \frac{L_{correct}}{L_{total}} \tag{2}$$

where $L_{correct}$ stands for the number of correctly recognized fields, and L_{total} is the total number of fields.

Table 3 presents the recognition results per language. It should be mentioned that in Table 3 we divide the results for Urdu into two parts: Urdu printed with Naskh script and Urdu printed with Nastaliq one as these two parts are quite different in appearance.

Table 3. Recognition results for per language

Writing system	Language	PCR (%)	PSR (%)	TextEval (%)
thai	Thai	34.09	0.0	38.23
perso-arabic	Arabic	49.27	13.20	55.26
perso-arabic	Persian	34.44	11.65	41.97
perso-arabic	Urdu	82.69	33.67	85.42
perso-arabic	Urdu (Nastaliq)	18.95	0.0	27.59
perso-arabic	Pushto	46.50	11.90	52.13
gurmukhi	Punjabi	51.81	16.33	57.15
tamil	Tamil	49.62	2.17	54.71
telugu	Telugu	50.71	23.57	56.19
bengali	Bengali	54.87	22.17	61.38
gujarati	Gujarati	66.90	32.33	71.75
odia (oriya)	Odia (Oriya)	52.45	12.83	58.74
kannada	Kannada	65.43	13.59	70.18
malayalam	Malayalam	52.22	4.50	57.14
devanagari	Hindi	76.50	23.33	82.70
devanagari	Marathi	74.29	26.67	82.70
devanagari	Nepali	41.88	13.83	46.79
latin	English	68.20	39.12	–

According to the results in Table 3, the Thai language demonstrates the lowest recognition results, except for Urdu printed with the Nastaliq script. The analysis of the recognized text has shown that the recognition engine tends to add spaces inside the text printed in Thai script. Thus, we present Table 4 that provides the results if all the spaces from the recognition results and the ground truth are deleted. As can be seen in Table 4 deletion of spaces crucially affect the resulting accuracy.

The results in Table 3 and Table 4 demonstrate that the presented dataset is challenging as even the English PSR is lower than 70%. The other important fact that these tables show is that people can speak about different measures when speaking about character accuracy. In our experiments, we employ two measures based on Levenshtein distance and they demonstrate quite different results. For example, for Marathi in Table 3 the difference is 8.41% which is tremendous as nowadays many methods differ from the state-of-the-art ones by tenths of a percent.

Table 4. Recognition results for per language

Writing system	Language	PCR (%)	PSR (%)	TextEval (%)
thai	Thai	76.11	16.58	77.57
perso-arabic	Arabic	47.74	13.74	54.55
perso-arabic	Persian	35.28	12.22	43.05
perso-arabic	Urdu	82.89	33.67	85.20
perso-arabic	Urdu (Nastaliq)	17.04	0.0	26.18
perso-arabic	Pushto	46.60	12.90	50.49
gurmukhi	Punjabi	52.81	16.33	58.30
tamil	Tamil	49.65	2.17	54.62
telugu	Telugu	50.93	23.57	55.99
bengali	Bengali	55.86	22.17	61.42
gujarati	Gujarati	67.03	32.83	71.26
odia (oriya)	Odia (Oriya)	52.64	12.83	58.15
kannada	Kannada	65.99	13.59	69.92
malayalam	Malayalam	52.00	4.59	56.70
devanagari	Hindi	79.18	23.33	83.31
devanagari	Marathi	76.44	26.67	81.83
devanagari	Nepali	41.39	15.83	46.35
latin	English	68.40	39.53	–

5 Conclusion and Future Work

As open datasets are a common way of comparing and evaluating new methods and approaches, the absence of any dataset for recognition of IDs with Perso-Arabic, Thai, and official Indian scripts creates difficulties for researchers not letting them compare the algorithms. We want to fill this gap and, therefore, propose a new dataset called MIDV-LAIT with images of IDs of India, Thailand, and several countries employing Perso-Arabic scripts. MIDV-LAIT contains 180 unique ID documents of 17 types and covers 13 scripts, 14 if dividing Perso-Arabic Naskh and Nastaliq scripts into two separate groups. The proposed dataset is of use for a variety of tasks including text line image recognition and script identification. Also, methods for problems typical for any ID processing, e.g., ID detection and identification, face detection, and image quality assessment, can employ MIDV-LAIT. To provide a baseline for MIDV-LAIT, we recognized all the textual fields with Tesseract OCR 4.1.0 and calculated three accuracy measures. The results show that the dataset is challenging as the per-string recognition rate is 39.12% for the Latin fields and is 0.0% for the Urdu Nastaliq script.

In the future, we plan to create an extension of MIDV-LAIT that would contain more scripts, for example, Sinhalese and Myanmar, more ID types for both new and already covered scripts, and more IDs per type. Besides, we consider the creation of a separate dataset for Chinese, Japanese, and Korean IDs.

References

1. Benchmark for enhancing the standard for Thai language processing. https://thailang.nectec.or.th/best. Accessed 19 Feb 2021
2. Federal law of Russian Federation "On personal data" 27.07.2006 N152-FL. http://www.kremlin.ru/acts/bank/24154. Accessed 19 Feb 2021
3. Generated Photos. https://generated.photos. Accessed 19 Feb 2021
4. Regulation (EU) 2016/679 of the European Parliament and of the Council of 27 April 2016 on the protection of natural persons with regard to the processing of personal data and on the free movement of such data, and repealing Directive 95/46/EC (General Data Protection Regulation). https://eur-lex.europa.eu/eli/reg/2016/679/oj. Accessed 19 Feb 2021
5. Tesseract OCR. https://github.com/tesseract-ocr/tesseract. Accessed 19 Feb 2021
6. Text Evaluation. http://www.primaresearch.org/tools/PerformanceEvaluation. Accessed 19 Feb 2021
7. Ahmed, S.B., Naz, S., Razzak, M.I., Yusof, R.B.: A novel dataset for English-Arabic scene text recognition (EASTR)-42K and its evaluation using invariant feature extraction on detected extremal regions. IEEE Access **7**, 19801–19820 (2019). https://doi.org/10.1109/ACCESS.2019.2895876
8. Ali, A., Pickering, M.: Urdu-text: a dataset and benchmark for Urdu text detection and recognition in natural scenes. In: 2019 International Conference on Document Analysis and Recognition (ICDAR), pp. 323–328 (2019). https://doi.org/10.1109/ICDAR.2019.00059
9. Arlazarov, V.V., Bulatov, K., Chernov, T., Arlazarov, V.L.: MIDV-500: a dataset for identity document analysis and recognition on mobile devices in video stream. Comput. Opt. **43**(5), 818–824 (2019). https://doi.org/10.18287/2412-6179-2019-43-5-818-824
10. Asad, F., Ul-Hasan, A., Shafait, F., Dengel, A.: High performance OCR for camera-captured blurred documents with LSTM networks. In: 2016 12th IAPR Workshop on Document Analysis Systems (DAS), pp. 7–12 (2016). https://doi.org/10.1109/DAS.2016.69
11. Bulatov, K.B.: A method to reduce errors of string recognition based on combination of several recognition results with per-character alternatives. Bulletin of the South Ural State University, Series: Mathematical Modelling, Programming and Computer Software **12**(3), 74–88 (2019). https://doi.org/10.14529/mmp190307
12. Bulatov, K., Arlazarov, V.V., Chernov, T., Slavin, O., Nikolaev, D.: Smart IDReader: document recognition in video stream. In: ICDAR 2017, vol. 6, pp. 39–44. IEEE Computer Society (2017). https://doi.org/10.1109/ICDAR.2017.347
13. Bulatov, K., Matalov, D., Arlazarov, V.V.: MIDV-2019: Challenges of the modern mobile-based document OCR. In: Osten, W., Nikolaev, D. (ed.) ICMV 2019, vol. 11433, pp. 1–6. SPIE, January 2020. https://doi.org/10.1117/12.2558438
14. Burie, J., et al.: ICDAR2015 competition on smartphone document capture and OCR (SmartDoc). In: 2015 13th International Conference on Document Analysis and Recognition (ICDAR), pp. 1161–1165 (2015)

15. Chabchoub, F., Kessentini, Y., Kanoun, S., Eglin, V., Lebourgeois, F.: SmartATID: a mobile captured Arabic text images dataset for multi-purpose recognition tasks. In: 2016 15th International Conference on Frontiers in Handwriting Recognition (ICFHR), pp. 120–125 (2016). https://doi.org/10.1109/ICFHR.2016.0034

16. Chandrakala, H.T., Thippeswamy, G.: Deep convolutional neural networks for recognition of historical handwritten Kannada characters. In: Satapathy, S.C., Bhateja, V., Nguyen, B.L., Nguyen, N.G., Le, D.-N. (eds.) Frontiers in Intelligent Computing: Theory and Applications. AISC, vol. 1014, pp. 69–77. Springer, Singapore (2020). https://doi.org/10.1007/978-981-13-9920-6_7

17. Chernyshova, Y.S., Sheshkus, A.V., Arlazarov, V.V.: Two-step CNN framework for text line recognition in camera-captured images. IEEE Access **8**, 32587–32600 (2020). https://doi.org/10.1109/ACCESS.2020.2974051

18. Chtourou, I., Rouhou, A.C., Jaiem, F.K., Kanoun, S.: ALTID: Arabic/Latin text images database for recognition research. In: 2015 13th International Conference on Document Analysis and Recognition (ICDAR), pp. 836–840 (2015). https://doi.org/10.1109/ICDAR.2015.7333879

19. Clausner, C., Antonacopoulos, A., Derrick, T., Pletschacher, S.: ICDAR2017 competition on recognition of early Indian printed documents - REID2017. In: 2017 14th IAPR International Conference on Document Analysis and Recognition (ICDAR), vol. 1, pp. 1411–1416 (2017). https://doi.org/10.1109/ICDAR.2017.230

20. Clausner, C., Antonacopoulos, A., Pletschacher, S.: Efficient and effective OCR engine training. Int. J. Doc. Anal. Recognit. (IJDAR) **23**, 77–83 (2020)

21. das Neves, R.B., Felipe Verçosa, L., Macêdo, D., Dantas Bezerra, B.L., Zanchettin, C.: A fast fully octave convolutional neural network for document image segmentation. In: 2020 International Joint Conference on Neural Networks (IJCNN), pp. 1–6 (2020). https://doi.org/10.1109/IJCNN48605.2020.9206711

22. Doush, I.A., AlKhateeb, F., Gharibeh, A.H.: Yarmouk Arabic OCR dataset. In: 2018 8th International Conference on Computer Science and Information Technology (CSIT), pp. 150–154 (2018). https://doi.org/10.1109/CSIT.2018.8486162

23. Dwivedi, A., Saluja, R., Sarvadevabhatla, R.K.: An OCR for classical Indic documents containing arbitrarily long words. In: 2020 IEEE/CVF Conference on Computer Vision and Pattern Recognition Workshops (CVPRW), pp. 2386–2393 (2020). https://doi.org/10.1109/CVPRW50498.2020.00288

24. Garai, A., Biswas, S., Mandal, S., Chaudhuri, B.B.: Automatic dewarping of camera captured born-digital Bangla document images. In: 2017 Ninth International Conference on Advances in Pattern Recognition (ICAPR), pp. 1–6 (2017). https://doi.org/10.1109/ICAPR.2017.8593157

25. Jaiem, F.K., Kanoun, S., Khemakhem, M., El Abed, H., Kardoun, J.: Database for Arabic printed text recognition research. In: Petrosino, A. (ed.) ICIAP 2013. LNCS, vol. 8156, pp. 251–259. Springer, Heidelberg (2013). https://doi.org/10.1007/978-3-642-41181-6_26

26. Kathiriya, K.B., Goswami, M.M.: Gujarati text recognition: a review. In: 2019 Innovations in Power and Advanced Computing Technologies (i-PACT), vol. 1, pp. 1–5 (2019). https://doi.org/10.1109/i-PACT44901.2019.8960022

27. Kišš, M., Hradiš, M., Kodym, O.: Brno mobile OCR dataset. In: 2019 International Conference on Document Analysis and Recognition (ICDAR), pp. 1352–1357 (2019). https://doi.org/10.1109/ICDAR.2019.00218

28. Kumar, H.R.S., Ramakrishnan, A.G.: Lipi Gnani: a versatile OCR for documents in any language printed in Kannada script. ACM Trans. Asian Low-Resour. Lang. Inf. Process. **19**(4) (2020). https://doi.org/10.1145/3387632

29. Madhuri, G., Kashyap, M.N.L., Negi, A.: Telugu OCR using dictionary learning and multi-layer perceptrons. In: 2019 International Conference on Computing, Power and Communication Technologies (GUCON), pp. 904–909 (2019)
30. Mathew, M., Singh, A.K., Jawahar, C.V.: Multilingual OCR for Indic scripts. In: 2016 12th IAPR Workshop on Document Analysis Systems (DAS), pp. 186–191 (2016). https://doi.org/10.1109/DAS.2016.68
31. Obaidullah, S.M., Halder, C., Santosh, K.C., Das, N., Roy, K.: PHDIndic_11: page-level handwritten document image dataset of 11 official Indic scripts for script identification. Multimedia Tools Appl. **77**(2), 1643–1678 (2017). https://doi.org/10.1007/s11042-017-4373-y
32. Ôn Vũ Ngọc, M., Fabrizio, J., Géraud, T.: Saliency-based detection of identity documents captured by smartphones. In: Proceedings of the IAPR International Workshop on Document Analysis Systems (DAS), Vienna, Austria, pp. 387–392, April 2018. https://doi.org/10.1109/DAS.2018.17
33. Polevoy, D.V., Aliev, M.A., Nikolaev, D.P.: Choosing the best image of the document owner's photograph in the video stream on the mobile device. In: ICMV 2020, vol. 11605, pp. 1–9. SPIE, January 2021. https://doi.org/10.1117/12.2586939
34. Pourreza, M., Derakhshan, R., Fayyazi, H., Sabokrou, M.: Sub-word based Persian OCR using auto-encoder features and cascade classifier. In: 2018 9th International Symposium on Telecommunications (IST), pp. 481–485 (2018). https://doi.org/10.1109/ISTEL.2018.8661146
35. Rehman, A.U., Hussain, S.U.: Large Scale Font Independent Urdu Text Recognition System (2020)
36. Sabbour, N., Shafait, F.: A segmentation-free approach to Arabic and Urdu OCR. In: Zanibbi, R., Coüasnon, B. (eds.) Document Recognition and Retrieval XX, vol. 8658, pp. 215–226. International Society for Optics and Photonics, SPIE (2013). https://doi.org/10.1117/12.2003731
37. Slimane, F., Ingold, R., Kanoun, S., Alimi, A.M., Hennebert, J.: A new Arabic printed text image database and evaluation protocols. In: 2009 10th International Conference on Document Analysis and Recognition, pp. 946–950 (2009). https://doi.org/10.1109/ICDAR.2009.155
38. Soares, A., Junior, R., Bezerra, B.: BID dataset: a challenge dataset for document processing tasks. In: Anais Estendidos do XXXIII Conference on Graphics, Patterns and Images (2020). https://doi.org/10.5753/sibgrapi.est.2020.12997
39. Suwanwiwat, H., Das, A., Saqib, M., Pal, U.: Benchmarked multi-script Thai scene text dataset and its multi-class detection solution. Multimedia Tools Appl. **80**(8), 11843–11863 (2021). https://doi.org/10.1007/s11042-020-10143-w
40. Tafti, A.P., Baghaie, A., Assefi, M., Arabnia, H.R., Yu, Z., Peissig, P.: OCR as a service: an experimental evaluation of google docs OCR, tesseract, ABBYY FineReader, and Transym. In: Bebis, G., et al. (eds.) ISVC 2016. LNCS, vol. 10072, pp. 735–746. Springer, Cham (2016). https://doi.org/10.1007/978-3-319-50835-1_66
41. Yaseen, N.O., Ganim Saeed Al-Ali, S., Sengur, A.: Development of new Anpr dataset for automatic number plate detection and recognition in North of Iraq. In: 2019 1st International Informatics and Software Engineering Conference (UBMYK), pp. 1–6 (2019). https://doi.org/10.1109/UBMYK48245.2019.8965512

Determining Optimal Frame Processing Strategies for Real-Time Document Recognition Systems

Konstantin Bulatov[1,2]([✉]) [iD] and Vladimir V. Arlazarov[1,2] [iD]

[1] Federal Research Center "Computer Science and Control" of the Russian Academy of Sciences, Moscow, Russia
[2] Smart Engines Service LLC, Moscow, Russia
{kbulatov,vva}@smartengines.com

Abstract. Mobile document analysis technologies became widespread and important, and growing reliance on the performance of critical processes, such as identity document data extraction and verification, lead to increasing speed and accuracy requirements. Camera-based documents recognition on mobile devices using video stream allows to achieve higher accuracy, however in real time systems the actual time of individual image processing needs to be taken into account, which it rarely is in the works on this subject. In this paper, a model of real-time document recognition system is described, and three frame processing strategies are evaluated, each consisting of a per-frame recognition results combination method and a dynamic stopping rule. The experimental evaluation shows that while full combination of all input results is preferable if the frame recognition time is comparable with frame acquisition time, the selection of one best frame based on an input quality predictor, or a combination of several best frames, with the corresponding stopping rule, allows to achieve higher mean recognition results accuracy if the cost of recognizing a frame is significantly higher than skipping it.

Keywords: Mobile recognition · Real-time recognition · Video stream · Anytime algorithms · Document recognition

1 Introduction

Document processing and especially documents recognition play important role in modern developed society. Technologies which allow scanning and entering document data using mobile devices, such as smart phones, continue to evolve and are deployed in many different areas of human interaction, including education [20, 28], healthcare [26], navigation [23], assistance for visually impaired [10, 14] and much more.

This work was partially financially supported by Russian Foundation for Basic Research, project 19-29-09055.

J. Lladós et al. (Eds.): ICDAR 2021, LNCS 12822, pp. 273–288, 2021.
https://doi.org/10.1007/978-3-030-86331-9_18

A particular branch of research and development of document analysis systems is directed towards automatic identity documents data entry [2,3,11] and identity documents authentication [6], which becomes imperative for such processes as mobile payments [19], customer verification within KYC (Know Your Customer) systems [7], as well as simplification of filling personal data forms during airport or hotel check-in [15].

The usage of multiple frames for extracting identity document data on a mobile device in real-time was shown to be advantageous [3,17] as it allows to achieve lower levels of recognition error by combining information from different representations of the same target object. In [5] it was shown that usage of stopping rules, which determine at run time whether a new frame should be processed, allows to reduce the mean number of required frames necessary to achieve the same mean level of accuracy. However, in these works the time required to fully process a frame was not taken into account. When the system is running on a mobile device, with camera-based input, the increased time for processing a single camera frame influences the time when the next frame from the camera will be available for processing. It is unclear whether stopping rules and frame processing strategies described in the literature still achieve the described results in such cases. Some video processing strategies, such as selection of a single best frame [17] allow to skip some of the input frame if it is known in advance that the extracted result will bear no influence to the final result, which means that selection of a method of input frames combination also influences the number and sequence of inputs processed by a system, and thus such strategies can not simply be compared by their achieved recognition result quality given the same sequence of input frames. Thus, in relation to a real-time processing systems the questions of selecting an optimal frame processing strategy, as well as selecting a stopping criterion (see Fig. 1), still need to be answered.

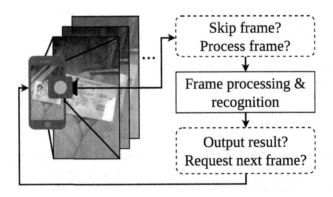

Fig. 1. Scheme of a real-time document recognition process.

In this paper, we will consider the model of a real-time video frames recognition process, the place of per-frame results combination and stopping rules within such process, and discuss the differences between frame processing strategies, which make decisions of skipping versus processing an input frame, as well as continuing versus stopping the capturing process. In Sect. 2 a basic framework for such real-time recognition process model is described, two different frame processing strategies are defined with their corresponding stopping rules, following the basic model adopted from [5]. In Sect. 3 the model and the discussed frame processing strategies are experimentally compared with different sets of model parameters.

2   Framework

2.1   Real-Time Video Frames Recognition Model

Let us consider identity document recognition process which happens on a mobile device given a sequence of input frames generated using the device's camera. The system observes a sequence of images $I_1, I_2, \ldots, I_k, \ldots$, where each image $I_k \in \mathbb{I}$. The images are generated in regular intervals, corresponding to a fixed frame rate of the camera, and the time between registrations of the images I_k and I_{k+1} is constant for all k and will be denoted as t_0.

The identity document recognition system S can process a sequence of frames in order to extract the document data. The frame processing pipeline of the document recognition system includes many different steps, including document detection, precise document location, rectification, fields segmentation and recognition [3], as well as the algorithms for per-frame combination (or accumulation) of the recognition results which yields improved results as more frames are processed. For the purposes of this simplified framework we will assume that the time t_1 required for the system S to process a single frame I_k is constant for all frames. The time required to combine the per-frame information is thus disregarded – if S processes n frames, the total run time would be $n \cdot t_1$. In the same spirit of simplification we will assume that the generation and acquisition of input frames is performed independently from the system S, and thus the camera frame I_k is instantly available as soon as its generated, but only one instance of the system S and any logical subsystem which governs the frame acquisition and processing strategy can be running at any given time.

To evaluate the quality of the accumulated result of the system S we will assume that the final target is to recognize a single text field of a document. Let us denote as $\mathbb{X}$ the set of all possible text string recognition results, and assume that the target text field of a document has a correct value $x^* \in \mathbb{X}$. An output of the system S given a single image I_k is a field recognition result $S^{(1)}(I_k) \in \mathbb{X}$, and given a sequence of images $I_1, I_2, \ldots, I_k$ it produces an accumulated result $S^{(k)}(I_1, I_2, \ldots, I_k) \in \mathbb{X}$ by means of per-frame combination using some internal procedure. A metric function $\rho : \mathbb{X} \times \mathbb{X} \to \mathbb{R}$ is defined on the set of all possible text field recognition results, which is used to determine how close the system's recognition result is to the ground truth x^*.

We will assume that the constructed process needs to exhibit the properties of anytime algorithms [27], mainly – to be *interruptible*. At any moment the "user" of the system might request immediate halt of the video stream recognition process, and after such request the result has to be returned as soon as possible. That means that if after the acquisition of a frame the decision is made to perform its full processing, such processing can not be delayed.

At any point during such recognition process the decision could be made to stop the acquisition of frames and output the currently accumulated text recognition result. If the decision to stop the process is made at time t and by this time the system S has processed the sequence of k frames $I_{j_1}, I_{j_2}, \ldots, I_{j_k}$ (a subsequence of frames acquired from the camera during the observation time t), the loss can be expressed as follows:

$$L(t) = \rho\left(x^*, S(t)\right) + c \cdot t, \tag{1}$$

where $c > 0$ is a cost of a unit of time in relation to the cost of the recognition error, and $S(t) = S^{(k)}(I_{j_1}, I_{j_2}, \ldots, I_{j_k})$ is the recognition result accumulated during time t.

The goal is to construct a stopping decision, corresponding to the model parameters t_0, t_1, c, to the strategy of feeding the frames into the system S and to the internal per-frame results accumulation method of the system S, such as to minimize the expected value of the loss (1) at stopping time.

Under the general assumption that the recognition results on average improve after the accumulation of multiple frames, and that the rate of such improvement decreases over time, in [5] it was proposed to use an approximation of myopic stopping rule with an expected distance to the next accumulated result calculated using modelling of the potential next frame result. The stopping problem model described in [5] considered each frame of the input sequence as a potential stopping point, without considering the time required to process each frame. With the loss function expressed in form (1) under the real-time process model the approximation of a myopic stopping time takes the following form:

$$
\begin{aligned}
T_\Delta &= \min\left\{t \geq t_0 + t_1 : L(t) \leq \mathrm{E}_t(L(t + \Delta t))\right\} \\
&= \min\left\{t \geq t_0 + t_1 : \rho\left(x^*, S(t)\right) - \mathrm{E}_t(\rho\left(x^*, S(t + \Delta t)\right)) \leq c \cdot \mathrm{E}_t(\Delta t)\right\} \\
&\leq \min\left\{t \geq t_0 + t_1 : \mathrm{E}_t(\rho(S(t), S(t + \Delta t))) \leq c \cdot \mathrm{E}_t(\Delta t)\right\}, \quad (2)
\end{aligned}
$$

where the condition $t \geq t_0 + t_1$ means that the stopping may only be performed after at least one frame is acquired and processed by the system S; $\mathrm{E}_t(\cdot)$ denotes conditional expectation after time t has passed; and Δt denotes a time period which will pass between the current moment for making the stopping decision and the next such moment. For the sake of generality this time period is not considered constant, thus its expected value is used in the right-hand side of the stopping criterion.

Using the approximation (2) the algorithm for making the stopping decision at time t consists of the following steps:

1. Estimate the time period Δt (or its expectation) after which another stopping decision may be made;
2. Estimate the expected distance between the current accumulated result $S(t)$ and the next one $S(t + \Delta t)$;

3. Stop the process if the estimated expected distance is less that or equal to the estimated time to next decision multiplied by the unit time cost c.

2.2   Frame Processing Strategies

The text recognition process with stopping decision was described in [5] and assumed that each frame of the input sequence is recognized and accumulated. From the point of view of the model described in Subsect. 2.1 such process implies that each new input frame I_k is passed into the system S for processing, which takes time t_1, and until the processing is done the intermediate frames have to be dropped. Let us assume for simplicity that after the system S finishes processing an image the next frame in a sequence is immediately available, i.e. the ratio t_1/t_0 has a strictly integer value. In this case, by the time t the system S will have processed and accumulated results from a total of $\lfloor (t - t_0)/t_1 \rfloor$ frames:

$$S(t) = S^{(\lfloor (t-t_0)/t_1 \rfloor)} \left(I_1, I_{1+t_1/t_0}, \dots, I_{1+(t_1/t_0)(\lfloor (t-t_0)/t_1 \rfloor - 1)} \right). \qquad (3)$$

With such frame processing strategy the stopping decision may be made after each new frame is processed and accumulated, thus the time period between decisions always equals t_1. The stopping rule (2) takes the following form:

$$T_\Delta = \min \left\{ t \geq t_0 + t_1 : E_t(\rho(S(t), S(t + t_1))) \leq c \cdot t_1 \right\}. \qquad (4)$$

Such frame processing strategy and associated stopping rule make sense if the internal per-frame recognition results accumulation algorithm employed in the system S require each input frame to be fully processed for the information to be usable. For such per-frame recognition results accumulation algorithms such as ROVER [12] or weighted ROVER modifications [17] this is true, as these algorithms require all combined recognition results to be fully available. However, this requirement is less strict for such combination methods as a selection of a single best result based on a criterion which is computable before full frame processing, or even a ROVER-like combination of a subset of best observed results, which can be regarded as a particular method of their weighting [17].

Let us consider that some quality predictor $f : \mathbb{I} \to \mathbb{R}$ defined on a set of images, and which can be used for quality estimation of an input frame before its processing by the system S. For the sake of simplicity let us disregard the time required to calculate the value of this predictor, and thus consider that the predictor value $f(I_k)$ is observed at the same time as the frame I_k itself. Let us now consider the system S which performs internal results combination by selecting a single result from a frame which had the maximal predictor value:

$$S^{(k)}(I_1, I_2, \dots, I_k) = S^{(1)} \left(\arg\max_{j=1}^{k} f(I_j) \right). \qquad (5)$$

It was shown in [24] that the same approximation of the myopic stopping rule can be applied for such results accumulation method, by additional estimation of the probability that the predictor value of the next frame will update the current maximum. But the important different when using such per-frame results combination strategy is that if the predictor value of the current frame is less than the current maximum, then there it is unnecessary to process the current frame by the system S (the accumulated result will simply not change, nor will such processing influence the further process). Thus, if such strategy is adopted, after the first frame has been processed by the system (i.e. at some time $t \geq t_0 + t_1$) if the predictor value of a frame does not exceed the currently observed maximum, then the frame can be skipped, and the next frame would be considered at time $t + t_0$. If the predictor value of the frame at time t is greater than the currently observed maximum, then this frame is processed by the system S, the maximum value is updated, and the next frame is considered at time $t + t_1$.

Such frame processing strategy changes how the approximation of the myopic stopping rule (2) should be applied. Let us denote as $P(t)$ the probability at time $t \geq t_0 + t_1$ that the next considered frame $I(t + \Delta t)$ will have a predictor value higher than the currently observed maximum. The approximation of the myopic stopping rule will take the following form:

$$T_\Delta = \min\{t \geq t_0 + t_1 : P(t) \cdot \mathrm{E}_t(\rho(S(t), S^{(1)}(I(t + \Delta t))))$$
$$\leq c \cdot (t_1 \cdot P(t) + t_0 \cdot (1 - P(t)))\}, \qquad (6)$$

where $\mathrm{E}_t(\rho(S(t), S^{(1)}(I(t + \Delta t))))$ denotes a conditional expectation of the distance between the current best result and the result which will be obtained at time $t + \Delta t$, where the maximum will be updated, thus it is guaranteed that the best frame will change. The estimation of $P(t)$ can be performed by approximating the observed values of f for the previously acquired frames with some distribution (in [24] a normal distribution was used) and then using its cumulative distribution function to calculate the probability of a new observation being higher than the current maximum. The same approach will be used in this paper for the experimental evaluation.

The selection of a single best frame, as well as combination of several best frames, were evaluated before against combination of all frame recognition results in [17], however this comparison was performed with a uniform frame processing scale. For the reasons described above, since the combination methods involving selection of one or several best samples based on predictor values allow to skip frames which will not contribute to the result, in conjunction with stopping rules they might show a significant advantage – by allowing to effectively process more frames within the same period of time. A full and fair comparison of such combination methods could be performed if the different times of frame acquisition t_0 and frame processing t_1 are taken into account, and if the stopping rule performance profiles generated by varying the unit time cost c are placed within the same axes. In the experimental section we will perform such evaluation, for different values of the t_1/t_0 ratio.

3   Experimental Evaluation

3.1   Dataset

In order to perform the experimental evaluation and comparison of the frame processing strategies and stopping methods described in Sect. 2 we have performed a set of modelling experiments in the same experimental settings as other works on the methods of stopping the text recognition process in a video, namely [5] and [24]. The evaluation was performed on an identity document video dataset MIDV-500 [1], which consists of video clips of 50 identity documents with various capturing conditions.

We analyzed the four types of text fields present in the dataset, namely numeric dates, document numbers, machine-readable zone lines, and names, for which the text recognition results of an identity document recognition system Smart IDReader are available [4]. The text recognition for each field was performed after its perspective restoration from the original frame, using a two-pass CNN framework [9].

The video clips in the MIDV-500 dataset were originally split into frames with a 10 frames per second rate, and the recognition of text fields were only performed in frames where the identity document was fully visible. Within the experimental scope of this paper, in order not to deal with incomplete clips (i.e. the clips where some frames were lost due to the document falling out of frame) we only considered text line sequences with all 30 frame results present. Since, as was discussed in Sect. 2, for simplicity we considered that the target of the system S is to recognize a single field, some of the video clips from MIDV-500 were used multiple times, once for each considered text field. Thus, a total of 691 clips were evaluated: 264 clips with a date field, 133 with a document number field, 226 with a name written in Latin alphabet, and 68 clips where a single machine-readable line was considered.

Given 10 frames per second sampling rate, when using this dataset to model the real-time document recognition process described in Sect. 2 the time t_0 between the acquisition of next frame has to be set to 0.1 s.

3.2   Frame Processing Time

The time t_1 required for the system S to fully process the frame, i.e. find or segment the document in the frame, determine its type, segment the text fields, and perform their recognition, will inevitably depend on a large number of factors, including the resolution of the input frame, the complexity of the system S and the complexity of the target document, as well as the performance parameters of the computing device. In order for the modelling experiment to comply as best as possible to a real-life scenario of identity document recognition in a video stream using a mobile device, some time characteristics of a per-frame performance of a full identity document recognition system (or at least its major component) on a mobile device needs to be obtained. Due to a high variability of mobile computing hardware such data is scarcely available. In a peer-reviewed publications it was stated that average identity document location and classification time can take 0.35 s, as determined using MIDV-500 dataset and measured using Apple iPhone 6 (released in 2014) [21]. Marketing materials of some software companies quote varying measurements of 0.1 s (bank cards recognition, [13]), 0.15 s or 0.25 s (German ID cards and USA driver's license respectively, [22]), 0.4 s (ID card, passport, travel visa or a work permit, [16], time to result, number of processed frames is unclear), however the concrete computing device is typically not stated in such materials and it is hard to evaluate their reliability even to determine an approximate range of values for t_1 in a realistic scenario.

In order to obtain a quantifiable figures to base the experiments from, we have conducted a per-frame recognition speed measurement of Smart IDReader system [3] on several different mobile devices using the printed images of sample identity documents from MIDV-500 dataset. The results of such evaluation are presented in Table 1.

Table 1. Average per-frame identity document recognition time.

Tested mobile device	Serbian passport	Azerbaijan passport	Russian internal passport	Japanese driver's license	German driver's license
Huawei Honor 8 (2016)	0.74 s	0.76 s	0.21 s	0.92 s	0.52 s
Xiaomi Pocophone F1 (2018)	0.49 s	0.42 s	0.12 s	0.61 s	0.28 s
Samsung Galaxy S10 (2019)	0.30 s	0.31 s	0.10 s	0.42 s	0.21 s
Apple iPhone XS (2018)	0.31 s	0.37 s	0.08 s	0.39 s	0.22 s
Apple iPhone SE 2 (2020)	0.20 s	0.34 s	0.07 s	0.41 s	0.18 s

It can be seen from Table 1 that the per-frame processing time varies drastically depending on the hardware used and on the type of the recognized document. Based on the measured data in the performed experiment we evaluated the real-time frame processing strategies described in Sect. 2 using the processing time value range $t_1 \in [0.1 \text{ s}, 1.0 \text{ s}]$.

3.3   Comparing Combination Strategies

To implement a full per-frame text recognition results combination strategy we used a published implementation of an extended ROVER algorithm [4] with the corresponding implementation of the stopping rule (4). The per-frame combination strategy by selecting a single best frame or several best frames as performed in [17] used focus estimation (or, rather, a joint quality estimation based on focus, contrast, and motion blur) on a level of a segmented text field image, and a confidence score obtained after recognition. The latter parameter does not allow to construct a video stream processing strategy with skipping frames, as the predictor value needs to be obtained before processing the frame by a system S. The image quality score obtained from a rectified text field image could also be problematic within the described framework, as in order to obtain such image the frame has to be specifically processed by the system S (the document needs to be detected and localized, fields need to be identified and segmented). Thus, in this paper, we used the same image focus estimation, but applied for the whole frame. A normalized generalized Levenshtein distance [25] was used as a text recognition results distance metric ρ, in the same way as in [4].

Figure 2 illustrates the profiles of accumulated recognition result quality (in terms of a distance to ground truth) at the time of the output, for three modelling experiments with different frame processing time $t_1 \in \{0.1 \text{ s}, 0.3 \text{ s}, 0.5 \text{ s}\}$ and for the three frame processing strategies. The method "Full combination", represented with a solid line in Fig. 2, corresponds to the first frame processing strategy discussed in Subsect. 2.2, where each new frame is processed by the system S and combined with the accumulated result with an extended ROVER algorithm. The method "Selecting 1 best", represented with a dotted line, correspond to the second discussed strategy, where the accumulated result in fact represents a result from a single frame with the highest value of the frame focus score estimation, and thus only when a new frame exceeds the current maximum, it gets processed by the system S to update the result. Finally, an additional method "Combining 3 best" was evaluated and represented as a dashed line, where the accumulated result of the system S is obtained by using extended ROVER to combine no more than 3 per-frame results with the highest values of the frame focus score estimation (a similar method was evaluated for the case of per-field image quality estimation in [17]). With the latter method the newly acquired frame was processed only if its score exceeds the minimal value for the current 3 best frames, and for such strategy a stopping method similar to the one for the single best frame selection (6) can be applied.

It can be seen from Fig. 2 that with the used experimental setting the differences between the profiles for the three evaluated frame processing strategies do not significantly differ. The profiles for $t_1 = 0.1$ s, drawn in green color, show that the full combination strategy generally achieves better results, especially on longer clips (i.e. if more frames are processed). Selection of the single best result in this case achieve the least accurate results, and the combination of the 3 best frames show intermediate accuracy. With frame processing time $t_1 = 0.3$ s (blue color in Fig. 2) the mutual positions of single best frame selection and full combination stays roughly the same, however the strategy of combining the 3 best frame results starts to achieve the same accuracy levels as full combination, and surpasses it in the further stages of the process. With a higher frame processing time $t_1 = 0.5$ s (red color in Fig. 2) it is harder to reliably state which strategy performs better, as their mutual position changes during the clips modelling, however the strategy of combining the best 3 frames still achieves better results at a later time.

3.4   Performance Profiles with Stopping Rules

To compare the frame processing strategies with the corresponding stopping rules their expected performance profiles [27] can be compared. The difference between the expected performance profiles and the profiles of plain combination

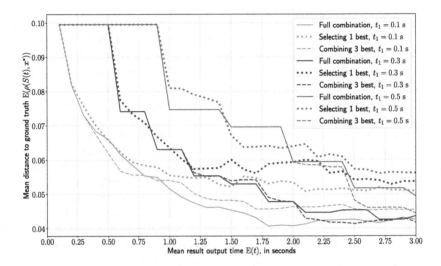

Fig. 2. Comparison of combined video stream text recognition result with different frame processing strategies and different frame processing time t_1 (stated in the legend). Time between frame acquisitions is constant for all experiments: $t_0 = 0.1$ s. (Color figure online)

(as in Fig. 2) is that each point of the profile represent a measurement of mean distance of the accumulated result to the ground truth and mean stopping time which achieves such distance, for some fixed value of the unit time cost c. By varying the unit cost time c, modelling the frame processing strategy and applying its stopping rule, we can determine which mean error levels at stopping time could be achieved by each given strategy for some mean value of time from the start of the clip to the output of the result.

Figure 3 shows the comparison of expected performance profiles for each of the three compared frame processing strategies (with their corresponding stopping rules, as defined in Subsect. 2.2) for different values of frame processing time t_1. The profiles for the lower frame processing time $t_1 = 0.1$ s (see Fig. 3a) show that with the application of an appropriate stopping rule the strategy of combining per-frame recognition results from all frames achieves lower mean error levels for all mean values of stopping time, which corresponds to this being a preferred strategy even when the stopping is not taken into account (see Fig. 2). However, if the frame processing time is increased, this strategy quickly becomes suboptimal. With increasing frame processing time t_1 the strategy of selecting the single best frame result in accordance with frame focus score becomes more preferable if the target mean stopping time is small. This is not surprising, as such strategy effectively minimizes the number of frames which have to be processed by the system S, while the selection of an input frame based on a predictor value which is in some degree correlated with the recognition quality still contributes to improvement of the result over time, whereas for the strategies which have to recognize each new frame have to delay the output of the result. It is interesting that with increasing frame processing time t_1 the strategy of combining 3 best frames yields consistently better results than the full combination, and becomes the best strategy if the target mean frame processing time is larger $(E(T) > 1.5 \text{ s})$.

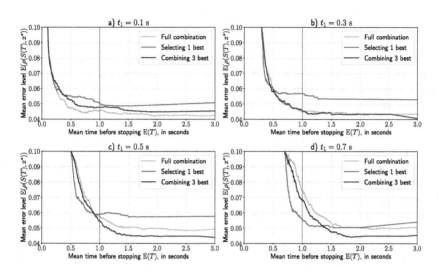

Fig. 3. Expected performance profiles for the different frame processing strategies with their corresponding stopping rules. Time between frame acquisitions is constant for all experiments: $t_0 = 0.1$ s, time of frame processing differs for subplots: a) 0.1 s, b) 0.3 s, c) 0.5 s, d) 0.7 s. (Color figure online)

In Fig. 4 the same performance profiles are presented, but for larger frame processing time $t_1 = 0.8$ s and $t_1 = 0.9$ s. With such values of t_1 the advantage of selecting a single best frame with lower target values for mean stopping time is still present, however the mutual positions of the profiles for full combination and the combination of 3 best frames change, to the point when these strategies become almost identical. This behaviour of the performance profile of the "Combining 3 best" strategy is understandable given the fact that each clip in the analyzed dataset consisted of 30 frames. Thus, the combination of 3 best frames becomes indistinguishable from full combination, given that the processing of each incurred skipping of 7 or 8 consequent frames.

Table 2 lists the achieved error values, represented as mean distance from the accumulated recognition result to the ground truth, for the three evaluated frame processing strategies and their corresponding stopping rules, for different values of frame processing time t_1. The values compared in Table 2 were achieved using the stopping rules configured such that the mean stopping time $E(T)$ for the clips in the dataset amounted to 1.0 s (this target mean stopping time is represented as a blue vertical line in Figs. 3 and 4). From this listing it is clearly seen how the optimal frame processing strategy changes with the increase of

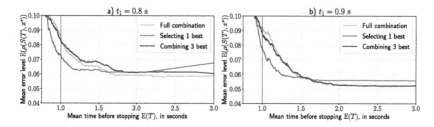

Fig. 4. Expected performance profiles for the different frame processing strategies with their corresponding stopping rules, for higher frame processing time. Time between frame acquisitions is constant for all experiments: $t_0 = 0.1$ s, time of frame processing differs for subplots: a) 0.8 s, b) 0.9 s. (Color figure online)

frame processing time – from full combination if the system S has high per-frame performance, to combining several best frames, to the selection of a single best frame. Such effect is not clearly evident if the stopping rules are not taken into account (see e.g. plots in Fig. 2 for $t_1 = 0.3$ s and $t_1 = 0.5$ s), as for each individual clip the result obtained using these two strategies at time $t = 1.0$ s might by almost the same. The evaluation of stopping rules show the difference more clearly, as they take into account the mean recognition result accuracy obtained on the dataset if the *mean* stopping time equals to 1.0 s.

Table 2. Achieved mean distance from the combined result to the ground truth for the stopping rules configured to yield mean stopping time $E(T) = 1.0$ s.

t_1	Full combination	Selecting 1 best	Combining 3 best
0.1 s	**0.046**	0.051	0.048
0.2 s	0.044	0.052	**0.041**
0.3 s	0.046	0.057	**0.045**
0.4 s	0.056	0.054	**0.053**
0.5 s	0.057	0.059	**0.054**
0.6 s	0.056	0.056	**0.054**
0.7 s	0.075	**0.056**	0.069
0.8 s	0.085	**0.073**	0.083
0.9 s	0.092	**0.077**	0.088
1.0 s	0.100	0.100	0.100

The code and data required to reproduce the experiments are available via the following link: https://github.com/SmartEngines/realtime_stoppers_modelling.

4   Discussion

The framework of a real-time document recognition process described in Sect. 2 is applicable for different combination methods, and for different compositions of the document processing systems (denoted in the framework as S). While the evaluation was performed using a multi-stage identity document analysis system with text recognition performed as a final stage after individual field segmentation, as a future work this model can be applied for other text recognition problems not necessarily related to document analysis, such as to end-to-end text spotting methods [18] or to other general text recognition methods [8].

Depending on the system composition and the properties of the text recognition method, as well as on the specific properties of the input video stream, the optimal input frame quality predictor (denoted in the framework as f) might be different as well. The usage of focus score evaluation in this work was predicated on its simplicity and effectiveness with respect to the text recognition method used. For different methods such predictor might not correlate well with the output accuracy, and thus other predictors may have to be explored.

It is also worth mentioning that to apply the frame processing strategies of the single best frame selection, or the combination of three best frames, with the stopping rules described in Sect. 2 the exact knowledge of the frame processing time t_1 is not needed beforehand. Since the stopping rule (6) is applied only after at least one frame has been acquired and processed ($t \geq t_0 + t_1$), the model parameters t_0 and t_1 can me measured at run time of the system and the stopping decision can be "calibrated" accordingly. Such application can allow to evaluate the frame processing strategies for situations where different types of documents are recognized with different speed. It might even be feasible to construct and evaluate a composite decision system, which would change the frame processing strategy depending on the measured parameters, in order to reach a common goal of maximizing the output quality for a fixed mean running time, as the performed experiments show (see Table 2) that the selection of the strategy itself depends greatly on the frame processing time t_1.

Conclusion

In this paper, we described and evaluated a real-time video recognition system for document data extraction. Such problems as per-frame recognition results combination and deciding when the frame acquisition process should be stopped, have already been discussed in the literature, however, without taking into account the time required to process each frame and how it affects the corresponding algorithms and the input sequence.

The main contribution of this work consists of the formulation of a simple real-time text recognition process model; the extension of the stopping method described in [5] which accounts for a variable time to the next decision; the formulation of multiple frame processing strategies and the construction of the stopping rules for them; and the experimental evaluation of the frame processing strategies using a publicly available dataset, for different values of frame processing time. It was shown that the selection of a per-frame results combination method is greatly affected by the parameters of the frame recognition process, and with different ratios between the time to obtain a new frame and the time to produce the frame result, the selection of frame processing strategy (which includes a combination method and a stopping rule) is also affected. If such ratio is close to 1.0 the full processing and combination of all input frames is preferrable, and if the processing time is significantly higher (and thus leads to a loss of input frames during current frame analysis) the selection of a single best frame according to a robust predictor, or a combination of several best frames, allows to achieve better mean error levels for the same mean processing time.

References

1. Arlazarov, V.V., Bulatov, K., Chernov, T., Arlazarov, V.L.: MIDV-500: a dataset for identity document analysis and recognition on mobile devices in video stream. Comput. Optics **43**, 818–824 (2019). https://doi.org/10.18287/2412-6179-2019-43-5-818-824
2. Attivissimo, F., Giaquinto, N., Scarpetta, M., Spadavecchia, M.: An automatic reader of identity documents. In: 2019 IEEE International Conference on Systems, Man and Cybernetics (SMC), pp. 3525–3530 (2019). https://doi.org/10.1109/SMC.2019.8914438
3. Bulatov, K., Arlazarov, V.V., Chernov, T., Slavin, O., Nikolaev, D.: Smart IDReader: Document recognition in video stream. In: 14th International Conference on Document Analysis and Recognition (ICDAR), vol. 6, pp. 39–44. IEEE (2017). https://doi.org/10.1109/ICDAR.2017.347
4. Bulatov, K., Fedotova, N., Arlazarov, V.V.: Fast approximate modelling of the next combination result for stopping the text recognition in a video. arXiv. 2008.02566 (2020)
5. Bulatov, K., Razumnyi, N., Arlazarov, V.V.: On optimal stopping strategies for text recognition in a video stream as an application of a monotone sequential decision model. Int. J. Doc. Anal. Recogn. (IJDAR) **22**(3), 303–314 (2019). https://doi.org/10.1007/s10032-019-00333-0
6. Castelblanco, A., Solano, J., Lopez, C., Rivera, E., Tengana, L., Ochoa, M.: Machine learning techniques for identity document verification in uncontrolled environments: a case study. In: Figueroa Mora, K.M., Anzurez Marín, J., Cerda, J., Carrasco-Ochoa, J.A., Martínez-Trinidad, J.F., Olvera-López, J.A. (eds.) MCPR 2020. LNCS, vol. 12088, pp. 271–281. Springer, Cham (2020). https://doi.org/10.1007/978-3-030-49076-8_26
7. Chen, T.H.: Do you know your customer? bank risk assessment based on machine learning. Appl. Soft Comput. **86**, 105779 (2020). https://doi.org/10.1016/j.asoc.2019.105779
8. Chen, X., Jin, L., Zhu, Y., Luo, C., Wang, T.: Text recognition in the wild: a survey. arXiv. 2008.02566 (2020)
9. Chernyshova, Y.S., Sheshkus, A.V., Arlazarov, V.V.: Two-step CNN framework for text line recognition in camera-captured images. IEEE Access **8**, 32587–32600 (2020). https://doi.org/10.1109/ACCESS.2020.2974051
10. Cutter, M., Manduchi, R.: Towards mobile OCR: how to take a good picture of a document without sight. In: Proceedings of the ACM Symposium on Document Engineering, pp. 75–84 (2015). https://doi.org/10.1145/2682571.2797066

11. Fang, X., Fu, X., Xu, X.: ID card identification system based on image recognition. In: 2017 12th IEEE Conference on Industrial Electronics and Applications (ICIEA), pp. 1488–1492 (2017). https://doi.org/10.1109/ICIEA.2017.8283074
12. Fiscus, J.G.: A post-processing system to yield reduced word error rates: Recognizer Output Voting Error Reduction (ROVER). In: 1997 IEEE Workshop on Automatic Speech Recognition and Understanding Proceedings, pp. 347–354 (1997). https://doi.org/10.1109/ASRU.1997.659110
13. IntSig: OCR image recognition technology. https://en.intsig.com/ocr/bankcard.shtml. Accessed 20 Feb 2021
14. Jabnoun, H., Benzarti, F., Amiri, H.: A new method for text detection and recognition in indoor scene for assisting blind people. In: Proceedings of SPIE (ICMV 2016), vol. 10341 (2017). https://doi.org/10.1117/12.2268399
15. Martín-Rodríguez, F.: Automatic optical reading of passport information. In: 2014 International Carnahan Conference on Security Technology (ICCST), pp. 1–4 (2014). https://doi.org/10.1109/CCST.2014.6987041
16. Microblink: BlinkID description. https://microblink.com/products/blinkid. Accessed 20 Feb 2021
17. Petrova, O., Bulatov, K., Arlazarov, V.L.: Methods of weighted combination for text field recognition in a video stream. In: Proceedings of SPIE (ICMV 2019), vol. 11433, pp. 704–709 (2020). https://doi.org/10.1117/12.2559378
18. Qin, S., Bissaco, A., Raptis, M., Fujii, Y., Xiao, Y.: Towards unconstrained end-to-end text spotting. In: 2019 IEEE/CVF International Conference on Computer Vision (ICCV), pp. 4703–4713 (2019). https://doi.org/10.1109/ICCV.2019.00480
19. Ravneet, K.: Text recognition applications for mobile devices. J. Global Res. Comput. Sci. 9(4), 20–24 (2018)
20. Samonte, M.J.C., Bahia, R.J.D., Forlaje, S.B.A., Del Monte, J.G.J., Gonzales, J.A.J., Sultan, M.V.: Assistive mobile app for children with hearing & speech impairment using character and speech recognition. In: Proceedings of the 4th International Conference on Industrial and Business Engineering, pp. 265–270 (2018). https://doi.org/10.1145/3288155.3288182
21. Skoryukina, N., Arlazarov, V., Nikolaev, D.: Fast method of ID documents location and type identification for mobile and server application. In: 2019 International Conference on Document Analysis and Recognition (ICDAR), pp. 850–857 (2019). https://doi.org/10.1109/ICDAR.2019.00141
22. Smart Engines: Document recognition system news. https://smartengines.com/news-events/. Accessed 20 Feb 2021
23. Snoussi, S.: Smartphone Arabic signboards images reading. In: 2018 IEEE 2nd International Workshop on Arabic and Derived Script Analysis and Recognition (ASAR), pp. 52–56 (2018). https://doi.org/10.1109/ASAR.2018.8480171
24. Tolstov, I., Martynov, S., Farsobina, V., Bulatov, K.: A modification of a stopping method for text recognition in a video stream with best frame selection. In: Proceedings of SPIE (ICMV 2020), vol. 11605, pp. 464–471 (2021), https://doi.org/10.1117/12.2586928
25. Yujian, L., Bo, L.: A normalized Levenshtein distance metric. IEEE Trans. Pattern Anal. Mach. Intell. 29(6), 1091–1095 (2007). https://doi.org/10.1109/TPAMI.2007.1078
26. Zhao, L., Jia, K.: Application of CRNN based OCR in health records system. In: Proceedings of the 3rd International Conference on Multimedia Systems and Signal Processing, pp. 46–50, ICMSSP 2018 (2018). https://doi.org/10.1145/3220162.3220172

27. Zilberstein, S.: Using anytime algorithms in intelligent systems. AI Mag. **17**(3), 73–83 (1996). https://doi.org/10.1609/aimag.v17i3.1232
28. Zin, T.T., Maw, S.Z., Tin, P.: OCR perspectives in mobile teaching and learning for early school years in basic education. In: 2019 IEEE 1st Global Conference on Life Sciences and Technologies (LifeTech), pp. 173–174 (2019). https://doi.org/10.1109/LifeTech.2019.8883978

Document Analysis for Social Good

Embedded Attributes for Cuneiform Sign Spotting

Eugen Rusakov[1]([envelope]) [ORCID], Turna Somel[2] [ORCID], Gerfrid G. W. Müller[2] [ORCID],
and Gernot A. Fink[1] [ORCID]

[1] Department of Computer Science, TU Dortmund University,
44227 Dortmund, Germany
{eugen.rusakov, gernot.fink}@tu-dortmund.de
[2] Hittitology Archive, Academy of Science and Literature Mainz,
55131 Mainz, Germany
{turna.somel, gerfrid.mueller}@adwmainz.de

Abstract. In the document analysis community, intermediate representations based on binary attributes are used to perform retrieval tasks or recognize unseen categories. These visual attributes representing high-level semantics continually achieve state-of-the-art results, especially for the task of word spotting. While spotting tasks are mainly performed on Latin or Arabic scripts, the cuneiform writing system is still a less well-known domain for the document analysis community. In contrast to the Latin alphabet, the cuneiform writing system consists of many different signs written by pressing a wedge stylus into moist clay tablets. Cuneiform signs are defined by different constellations and relative positions of wedge impressions, which can be exploited to define sign representations based on visual attributes. A promising approach of representing cuneiform sign using visual attributes is based on the so-called Gottstein-System. Here, cuneiform signs are described by counting the wedge types from a holistic perspective without any spatial information for wedge positions within a sign. We extend this holistic representation by a spatial pyramid approach with a more fine-grained description of cuneiform signs. In this way, the proposed representation is capable of describing a single sign in a more detailed way and represent a more extensive set of sign categories.

Keywords: Cuneiform script · Retrieval · Visual attributes

1 Introduction

Besides Egyptian hieroglyphs, the cuneiform script is one of the earliest attested writing systems in history. Developed in Mesopotamia at the end of the 4th millennium BCE, it was used until the 1st century CE across the Near East.

© Springer Nature Switzerland AG 2021
J. Lladós et al. (Eds.): ICDAR 2021, LNCS 12822, pp. 291–305, 2021.
https://doi.org/10.1007/978-3-030-86331-9_19

Fig. 1. The left side shows a whole tablet reassembled from several fragments. The right side shows an enlarged section from the tablet. This photograph has been provided [11].

Cuneiform is a 3D script written primarily on clay tablets. It was written by pressing a stylus into moist clay surfaces in order to create signs composed of wedge-shaped impressions, henceforth to be called wedges. Figure 1 shows an example of a cuneiform tablet. The image on the left-hand side shows a typical tablet that has been reassembled by philologists by joining fragments in an attempt to restore heavy damage through the millennia. Displayed on the right side of Fig. 1 is a section of the left image where one can see the wedge impressions on the clay surface. A sign can be identified by the reader according to wedge direction, number, and positioning within the sign. As cuneiform signs may have syllabic or logographic values, the writing system contains a significantly greater number of signs than in an alphabet, with an approximate maximum of 900 signs. As this writing system was used over three millennia, it was distributed across a wide geographical range, adapted to over 10 languages. Hence, the inventory of regular signs and the variations observed within a single sign class, henceforth to be called sign variants, used in different local areas are heterogeneous. A single variant may differ from other variants in a variety of sign characteristics, namely the direction, number, and positioning of wedges. As presented in [16], a system proposed by Gottstein [5] for representing cuneiform signs based on four different wedge types and their corresponding counts fit well for a novel retrieval scenario called *Query-by-eXpression (QbX)*. However, the binary attribute vectors derived from the Gottstein-System have a limited representational power in terms of inventory size. In this work, we derive a representation based on the Gottstein-System enriched with spatial information, which encodes the presence of wedge types and counts and their approximate position within a cuneiform sign. In this way, the signs are projected into a subspace capable of representing a more extensive inventory size. We evaluate the sign representations using annotated texts written in Hittite cuneiform script in a segmentation-based *Query-by-Example (QbE)* and *Query-by-eXpression (QbX)* scenario and show the superiority of a Gottstein representation enriched with spatial information. This work is organized as follows. In the next section, related work is reviewed, followed by a brief introduction to the Gottstein-System in Sect. 3.1. The resulting attribute representation based on a pyramidal decomposition of cuneiform signs is explained in Sect. 3.2. The experiments are discussed in Sect. 4. Finally, a conclusion is drawn.

2   Related Work

In this section, related work in terms of cuneiform analysis and word embeddings is discussed. At first, we review the current publications in *cuneiform character recognition and retrieval*. Afterward, we discuss recently proposed *word embeddings* for word spotting.

2.1   Cuneiform Character Retrieval

Although the cuneiform writing system is still a less known domain in the document analysis community, several approaches based on graph representations and bag-of-features for cuneiform sign retrieval have been presented. In 2012, a method [12] for extracting 2D vector drawings from 3D cuneiform scans was proposed. Afterward, the resulting spline graphs were converted in order to compare the similarity between two cuneiform characters based on graph representations [3] or retrieve them using pattern matching on structural features [2]. With a similar goal, Massa et al. [13] described an approach to extract wedge-shaped impressions from hand-drawn cuneiform tablets. In [4] the *inkball model* [9] was adapted by Bogacz et al. to treat different wedge features as individual parts arranged in a tree structure in order to perform segmentation-free cuneiform character spotting. Next to the graph-based feature representations, Rothacker et al. [14] proposed an approach based on *Bag-of-Features* in combination with *Hidden Markov Models* integrated into a patch-based framework for segmentation-free cuneiform sign spotting.

2.2   Word Embeddings

All the approaches mentioned above require a visual example of a query and can only perform in a retrieval scenario called *Query-by-Example (QbE)*. Sudholt et al. [19] noted that this approach poses certain limitations in practical applications. A user has to identify a visual example of a query from a whole document collection which can be very tedious, especially for infrequent queries. The field of word spotting is dominated by mainly two types of spotting scenarios, namely *Query-by-Example (QbE)* and *Query-by-String (QbS)*. The *QbS* scenario is defined by queries given as word strings. Here, the word image representations and the textual representations are projected into a common subspace. The major advantage of *QbS* is that the user is not required to search for a visual example of a word in the document collection to define a query.

On the other hand, a word spotting system must learn a mapping from a textual to a visual representation first. Therefore, annotated training samples are required. Almazan et al. [1] proposed an approach for a word string embedding called *Pyramidal Histogram of Characters (PHOC)*. The authors presented a method to map a word string to a binary attribute vector representing the presence or absence of characters. Furthermore, information about character positions within a word is included by splitting the word string in a pyramidal way. Inspired by the PHOC representation, Sudholt et al. [19] proposed a CNN model based on the *VGGNet* [17] named *PHOCNet*. This model is capable of estimating a certain PHOC representation in an end-to-end fashion using *binary logistic regression* (also called *sigmoid logistic regression*) in combination with a *binary cross-entropy (BCE)* loss.

3 Cuneiform Sign Expressions

This section describes our approach of extending the representation of cuneiform signs following the Gottstein-System by including spatial information. After a brief introduction of the Gottstein-System in Sect. 3.1, the structure of spatial information encoded in our pyramidal representation is introduced in Sect. 3.2. Afterward, we explain how the sign expressions enriched with spatial information are modeled as binary attribute vectors.

3.1 Gottstein System

The Gottstein-System [5] was introduced following the idea of a unified cuneiform sign description based on commonly occur wedge types. This representation should be usable in databases without existing knowledge of cuneiform sign categories, especially in a retrieval scenario. In order to simplify the search process, Gottstein [5] proposed a system where an alphanumeric encoding represents a cuneiform sign. The Gottstein-System decompose a cuneiform sign into four different wedge types $gs = \{a, b, c, d\}$, where gs denotes the set of Gottstein wedge types. Figure 2 shows four examples of hand-drawn wedge types from the Gottstein-System. For every item from gs the count of each wedge type is given. The type a represents a vertical wedge tending from top to bottom and vice-versa. b represents a horizontal wedge type, tending from left to right. Here, the

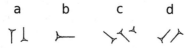

Fig. 2. Four hand drawn examples of wedge types from the Gottstein-System and their corresponding alphanumeric encodings $gs = \{a, b, c, d\}$

variant right to left does not exist. Type c represents three different variants of a wedge type, the so-called *Winkelhaken* and an oblique wedge tending diagonally from the upper left to the lower right as well as from lower right to the upper left. Finally, d represents a perpendicular wedge type to c tending from the lower left to the upper right direction and vice-versa. Fig. 3 shows three cuneiform signs and their corresponding Gottstein representations. On the far left side, a simple sign consisting of 2 wedge types is shown. The sign in the middle shows three different wedge types and $2 + 2 + 1 = 5$ wedges in total. The far-right side shows a more complex sign with 3 different wedge types and $10 + 2 + 4 = 16$ wedges in total.

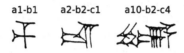

Fig. 3. Three examples of hand drawn cuneiform signs and their corresponding Gottstein representations, showing a simple sign on the left side and a more complex sign on the right side.

3.2   Gottstein with Spatial Information

Although the Gottstein system provides a holistic representation to express a particular cuneiform sign, this simplified description exhibits some shortcomings. For example, distinguishing between different sign categories consisting of the same wedge types, additional information for the relative position of a wedge is missing. Figure 4 shows the shortcoming of a holistic Gottstein representation. Here, three different cuneiform signs are described by the same Gottstein representation. The three signs, from three different categories having different visual appearances, are mapped to the same alphanumeric encoding. In order to overcome this ambiguous mapping, we decompose the signs in a pyramidal fashion. This way, a higher diversity in the mapping between categories and encodings is achieved.

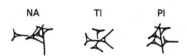

Fig. 4. Three examples of different cuneiform signs consist of the same wedge types described by the same Gottstein representation a1-b1-c2-d0.

Pyramidal Partitioning. Intuitively, writing systems are partitioned according to their reading directions like left to right in Latin scripts. The common reading direction for cuneiform texts tends from left to right and less common from top to bottom. Due to the physical process of writing cuneiform signs by pressing a wedge stylus into moist clay, a typical sequential arrangement of wedge types and positions, similar to Latin scripts, is missing. Therefore, we split the signs based on their main visual variant in a pyramidal fashion to get a Gottstein representation containing spatial information that represents cuneiform signs in a more detailed way. As the reading direction tends from left to right, horizontal partitioning is performed. We apply the splitting at level 2 and 3 which results in 2 left-right and 3 left-middle-right splits, respectively. Additional to the horizontal direction, vertical partitioning on levels 2 and 3 are applied. Figure 5 shows the proposed way of splitting the cuneiform sign *NA* in a pyramidal fashion. Assigning annotations to splits based on the visual appearance entails an issue. While most of the wedges can be assigned distinctively to a certain split, some wedge constellations lie between partitions without a visually clear assignment to a certain split. We relax the assignment between wedges and splits by defining the head of a wedge as the main orientation point. Here, a wedge-head intersecting the boundary between two splits is assigned to both splits.

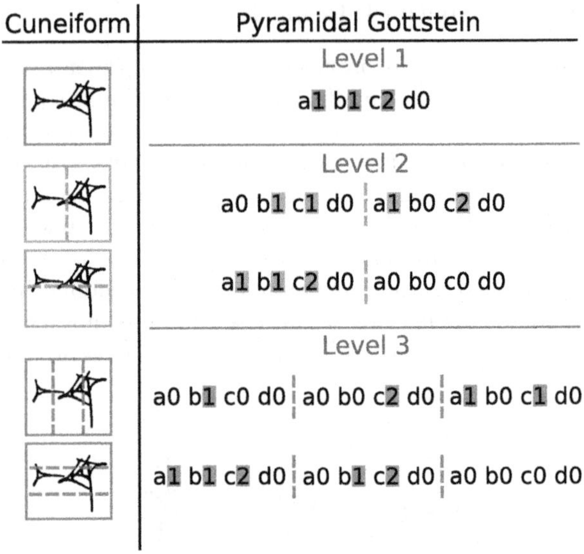

Fig. 5. Visualization of splitting the sign NA in the horizontal and the vertical direction for levels 2 and 3, respectively.

Combining Horizontal and Vertical Splits. Next to the pyramidal partitioning discussed before, a more fine-grained splitting can be achieved by combining the horizontal and vertical partitioning. Figure 6 shows three examples of possible combinations. For level 2 the horizontal and vertical partitioning can be combined into $2 \times 2 = 4$ splits. While combining the horizontal and vertical parts of level 3 results in $3 \times 3 = 9$ splits. Next to the combinations within a pyramid level, it is also possible to combine between different pyramid levels as shown in Fig. 6. In contrast to a manual partitioning process, as described in Sect. 3.2, where wedge types and counts are assigned to certain splits by a human expert, we achieve the combinations by automatically compare annotations from two overlapping splits and take the minimum count for corresponding wedge types.

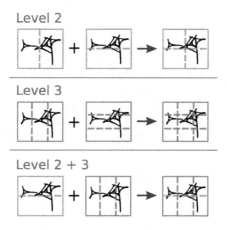

Fig. 6. Three examples of possible combinations between horizontal and vertical partitions, for level 2, 3, and a combination of 2 and 3.

Modeling Binary Attributes. In order to transfer the Gottstein-System into a binary attribute vector, we first count the minimum and the maximum number of wedges for each type. For the database (see Sect. 4.1) we use in this work, the minimum count of each wedge type denoted by a, b, c, and d is 0. For the wedge type a and b the maximum count is 10. The wedge type c has a maximum count of 12 and d have a maximum count of 2. Except for the count of 0 all counts are modeled as single attributes, respectively. The absence of a wedge type is modeled as a zero-vector. In this way, the representation results in a $10 + 10 + 12 + 2 = 34$ dimensional binary attribute vector. In the end, all attribute vectors from each split are concatenated to form the final Gottstein representation as a $34 \times N_{splits}$ dimensional vector, where N_{splits} denotes the total number of splits. For example, the partitioning from Fig. 5 will be concatenated to a $34 \times (1 + 2 + 2 + 3 + 3) = 374$ dimensional vector.

4   Evaluation

In this section, the evaluation of our approach is described. At first, our cuneiform database and the data partitioning into training and test sets are shown. Afterward, the model architecture and our decisions for different modifications are explained. In the Sects. 4.3, 4.6, and 4.5 the training setup for the model, the similarity metric, the evaluation protocol, and the evaluation metric are described. Finally, the results are presented in Sect. 4.7.

4.1   Database

In this work, we use a set of annotated cuneiform tablets collected from a total of 232 photographs, with annotated 95 040 samples of cuneiform signs, provided by [11]. These samples represent an inventory of 300 different cuneiform signs effectively. The sign images have considerable variability in size, ranging from a minimum of 11 pixel height to a maximum of 520 pixel in width, as shown in Table 1. Figure 7 shows examples of cuneiform signs. On the far left, the sign class, and next to it, the Gottstein encoding is given. In the middle of Fig. 7 hand-drawn examples and corresponding sign crops from the tablet photographs are shown. As one can see, the visual variability and quality in terms of wedge impression, image resolution, and surface damage can be tremendous.

Table 1. Statistics about image sizes given in pixel.

	Max	Min	Mean	Std
Height	319	11	85.55	±31.50
Width	520	16	106.09	±46.31

Data Partitions. As the data presented above does not have an official partitioning into training and test sets, we split the data into four equal parts. This way, we can perform four-fold cross-validation similar to [16]. First, all samples are sorted according to their category. Afterward, every fourth sample from a category is assigned to one validation set in an interleaving fashion. This assignment results in four validation sets containing approximately the same number of samples with the same distribution over sign categories. Table 2 shows the number of samples for each of the four cross-validation splits.

4.2   Model Architecture

Similar to [16] we used the successful ResNet [8] architecture. From the different versions proposed for the ResNet model, we adapted the one with 34 layers and version B, as the preliminary experiments revealed the ResNet-34B to have the best trade-off between retrieval performance, number of parameters, and

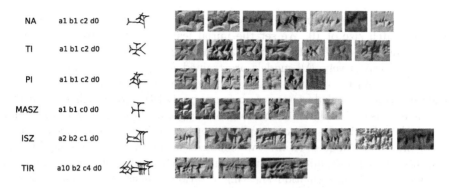

Fig. 7. Examples of six different cuneiform signs are shown. On the far right side, the sign names and the Gottstein representations are given. Next to them, the corresponding hand-drawn examples are shown. And on the right side, cropped snippets of cuneiform signs from tablet photographs are visualized.

Table 2. Number of samples for each cross-validation split.

	$Split_1$	$Split_2$	$Split_3$	$Split_4$	$Total$
# Samples	23 878	23 799	23 717	23 646	95 040

computation time. We apply some modifications to our model in order to achieve higher retrieval performance. While the original ResNet architectures perform five times downsampling, we discard the last three downsamplings and only kept the first two. One in the first convolutional layer and the second one in the first Pooling layer. In this way, all Residual-Blocks are computing on equally sized feature maps, while spatial information is preserved. After the last convolutional layer, a Temporal Pyramid Pooling (TPP) [18], in analogy to the pyramid levels presented in Sect. 3.2, with the levels $(1, 2, 3)$ is applied. We also experimented with the original Global Average Pooling (GAP) and the Spatial Average Pooling (SPP) introduced in [6], but the TPP achieved the best results in preliminary experiments. The output of the TPP layer effectively results in a $512 \times (1+2+3) = 3072$ dimensional feature vector, which is mapped to the attribute representation using a fully connected layer. As in [16], we do not rescale the images to uniform image sizes.

4.3   Training Setup

The model is trained using the *binary cross-entropy (BCE)* loss and the *Adaptive Momentum Estimation (Adam)* [10] optimizer. For the momentum mean and variance values β_1 and β_2 are set to 0.9 and 0.999 respectively, while the variance flooring value is set to 10^{-8} as recommended in [10]. The batch size is technically set to 1 while accumulating the gradients over 10 iterations, which results in an effective batch size of 10. We set the initial learning rate to 10^{-4} and divide the learning rate by 10 after 100 000 and another division by 10 after 150 000 iterations, respectively. All models are trained for 200 000 iterations in every training run. As parameter initialization, we use the strategy proposed in [7]. The weights are sampled from a zero-mean normal distribution with a variance of $\frac{2}{n_l}$, where n_l is the total number of trainable weights in layer l. In order to extend the number of training samples sampled from the database, we use some common augmentation techniques. At first, we balance the training data by sampling the sign categories from a uniform distribution, followed by augmentation techniques like perspective transformation, shearing, rotation, translation, scaling, lightness changing, and image noise generation.

4.4 Similarity Metric

For the segmentation-based scenario, retrieval is performed by computing a certain similarity between a given query and all test samples, excluding the query itself. In this work, we make use of a similarity measurement called *Probabilistic Retrieval Model* proposed in [15]. Here, the assumption is made that the binary attribute representation is a collection of d independent Bernoulli distributed variables, each having their probabilities p, which evaluates to 1. Equation 1 shows the definition of the probabilistic retrieval model.

$$p_{prm}(\mathbf{q}|\mathbf{a}) = \prod_{i=1}^{D} a_i^{q_i} \cdot (1 - a_i)^{1-q_i} \tag{1}$$

Here, $\mathbf{q}$ denotes the query vector, and $\mathbf{a}$ represents the estimated attribute representation. As the evaluation of the PRM from Eq. 1 requires to compute a product of many probabilities, we evaluate the PRM scores in the logarithmic domain in order to improve the numerical stability:

$$\log p_{prm}(\mathbf{q}|\mathbf{a}) = \sum_{i=1}^{D} q_i \log a_i + (1 - q_i) \log(1 - a_i) \tag{2}$$

4.5 Evaluation Protocol

We follow the evaluation protocol from [16] and evaluate our approach in a segmentation-based QbE and QbX scenario on the cuneiform database shown above (Sect. 4.1). As the given data is partitioned into four separate validation folds, three folds are used to train a single model, and the remaining fold is considered the test partition to perform the retrieval evaluation. For the QbE scenario, all images of cuneiform signs from a test fold, which occur at least twice, are considered as queries. Retrieval is performed by estimating the attribute representations for all sign images from a test fold given the estimated attributes from a certain image query. Afterward, the retrieval lists are obtained by evaluating the PRM scores between the list items and a given query estimate and sort them in ascending order with respect to the highest probabilities. For QbX, retrieval is performed equivalently to QbE except that a sign category is only considered once as a query. It is important to note that the relevance category determines the relevance of an item within the retrieval list. We perform retrieval based on models either trained in a one-hot encoding fashion using the sign categories as targets or in a multi-class encoding using the Gottstein attribute representations as targets.

4.6   Evaluation Metrics

To evaluate the performance of our approach, we decided to use the interpolated Average Precision (AP) as this metric is commonly used in the spotting literature [18]. The AP is defined as follows:

$$AP = \frac{\sum_{n=1}^{N} P(n) \cdot R(n)}{t} \tag{3}$$

$P(n)$ denotes the precision if we cut off the retrieval list after the i-th element. $R(n)$ is the indicator function, which evaluates to 1 if the i-th position of the retrieval list is relevant with respect to the query and 0 otherwise. N represents the length of the retrieval list, and t is the total amount of relevant elements. Similar to segmentation-based word spotting tasks, the retrieval list contains all cuneiform sign images from the respective test set. Afterward, the overall performance is obtained by computing the mean Average Precision (mAP) overall queries, where Q denotes the total amount of queries:

$$mAP_{qry} = \frac{1}{Q} \sum_{q=1}^{Q} AP(q) \tag{4}$$

We denote the standard mean Average Precision with the subscript qry, as the mean is computed across all average precision values obtained from all queries. As proposed in [16], we additionally evaluate a separate mean Average Precision for every sign category by computing the mean of all mAP_{qry} values with respect to the category:

$$mAP_{cat} = \frac{1}{C} \sum_{c=1}^{C} mAP_{qry}(c) \tag{5}$$

Where C denotes the set of sign categories and $mAP_{qry}(c)$ evaluates to the mAP value of the sign category c. In this way, the mAP_{cat} ensures a more balanced performance evaluation across all sign categories contained in our database.

4.7   Results and Discussion

Table 3, 4 and 5 list the results for the QbX and QbE experiments, given in mAP percentage. All results are averaged over four cross-validations folds. The first

Table 3. QbX and QbE experiments using a one-hot encoding as sign category representation in combination with the cross-entropy (CE) and binary cross-entropy (BCE) loss function, respectively. Sign categories are used as relevance categories. Results are shown in mAP [%].

Loss	N_{reps}	QbX	QbE	
		mAP_{cat}	mAP_{qry}	mAP_{cat}
CE	300	90.19	86.46	79.68
BCE	300	84.10	75.34	68.50

results are shown in Table 3, evaluated for a sign encoding in a one-hot fashion. Here, all sign categories are represented by a 300 dimensional output vector where every element of this vector represents a certain category. The column N_{reps} denotes the number of unique representations if all 300 sign categories are mapped to a certain representation. Here, $N_{reps} = 300$ shows that 300 sign categories are mapped to 300 unique representations (i.e. one-hot encodings). We evaluate two different loss functions, namely cross-entropy (CE) and binary cross-entropy (BCE). For the CE configuration, the Softmax-Function is applied to the output of the model. For a model trained with the BCE loss, we apply a Sigmoid-Function after the model output. Table 3 shows distinct differences in retrieval performance depending on the loss and output function used. For QbX, the CE configuration outperforms the BCE by about 6%, while for QbE the performance difference is about 11%. Table 4 shows the results for a Gottstein representation combined with different partitioning strategies. The column sign representation denotes the configuration of an attribute representation. Here, GR means Gottstein-Representation, and GR+LE means Gottstein-Representation combined with the Last Element as shown in [16]. The configuration for pyramidal splits is described by H_l for horizontal and V_l for vertical splitting, while l denotes the pyramid level in the respective direction. For example, the notation GR+H_2+V_2 means a Gottstein Representation combined with a horizontal and vertical splitting on level 2. Retrieval is performed by evaluating the retrieved items with respect to the sign representation itself. The representation estimated by the model defines the relevance of retrieved items. As in Table 3, the number of unique representations is denoted by N_{reps}. For example, GR maps all 300 sign categories to 177 unique attribute representations, which means 1.69 signs are mapped to the same representation, on average. Table 4 shows that the model can learn an arbitrary representation based on binary attributes. All representations show approximately the same performance. In Table 5, the relevance of the items in a retrieval list is defined by the sign categories. Here, the retrieval performance is correlating with the number of unique representations. The mAP values increase for higher N_{reps} numbers, which is not surprising. On the one hand, a Gottstein Representation can distinguish between 177 out of

Table 4. QbX and QbE experiments using different attribute encodings as sign category representation with the binary cross-entropy (BCE) loss function. The relevance category is defined by the sign representation itself, respectively. Results are shown in mAP [%].

Representation	N_{reps}	QbX		QbE	
		mAP_{qry}	mAP_{cat}	mAP_{qry}	mAP_{cat}
GR	177	**90.32**	88.78	80.28	75.52
GR+LE	267	89.27	88.72	**81.66**	76.93
GR+H_2	263	90.02	**89.51**	81.51	77.27
GR+H_2+V_2	287	89.12	89.01	80.77	**77.30**
GR+H_2+H_3	291	88.94	88.90	80.84	76.78
GR+H_3+V_3	297	89.85	89.79	80.89	76.78
GR+H_2+V_2+H_3+V_3	300	88.76	88.76	80.62	76.85
GR+$H_2 \times V_2$	287	88.49	88.37	78.81	74.71
GR+$H_3 \times V_3$	297	86.95	86.89	77.25	72.30
GR+$H_2 \times V_2$+$H_3 \times V_3$	300	87.40	87.40	77.02	73.02

Table 5. QbX and QbE experiments using different attribute encodings as sign category representation with the binary cross-entropy (BCE) loss function. The sign categories are used as the relevance categories. Results are shown in mAP [%].

Representation	N_{reps}	QbX		QbE	
		mAP_{qry}	mAP_{cat}	mAP_{qry}	mAP_{cat}
GR	177	55.87	55.87	60.15	60.15
GR-LE	267	80.54	80.54	77.15	71.20
GR-H_2	263	79.19	79.19	74.03	71.05
GR-H_2-V_2	287	85.93	85.93	78.34	75.00
GR-H_2-H_3	291	86.69	86.69	78.72	74.84
GR-H_3-V_3	297	**88.82**	**88.82**	80.31	75.91
GR-H_2-V_2-H_3-V_3	300	88.76	88.76	**80.62**	**76.85**
GR-$H_2 \times V_2$	287	85.31	85.31	76.51	72.53
GR-$H_3 \times V_3$	297	85.93	85.93	76.75	71.50
GR+$H_2 \times V_2$+$H_3 \times V_3$	300	87.40	87.40	77.72	73.02

300 different categories. Thus, appropriate retrieval performance is not possible. On the other hand, exploiting the position of wedges within a sign and defining a Gottstein Representation based on a spatial pyramid splitting increases the performance by 32.95% for QbX (comparing GR and GR-H_3-V_3) and 20.47% respectively 16.70% for QbE (comparing GR and GR-H_2-V_2-H_3-V_3).

5   Conclusion

In this work, we enhanced the Gottstein Representation by applying a pyramidal partitioning of cuneiform signs. This way, the signs can be expressed by wedge types and their respective counts, and the approximate positions of their wedges. We evaluate our approach on a cuneiform database consisting of 232 photographs in two segmentation-based scenarios: Query-by-Example and Query-by-eXpression. The experiments show that a Gottstein Representation enhanced with spatial information achieves better results than the Gottstein Representation proposed in [16] using item relevancy based on sign categories. Although a representation enhanced with spatial information can be superior compared to one without, this spatial information entails the problem of ambiguity, as shown in Sect. 3.2. Hence, the users' representation should be intuitively interpretable and robust against "incomplete" or noisy query definitions. Therefore, further investigations need to be made to define a cuneiform sign representation with fewer (or even without) ambiguities and an intuitive interpretation for human experts. Nevertheless, we would like to derive the conclusion that an appropriate representation based on binary attributes which encode spatial information fits well a cuneiform sign description in order to define retrieval queries.

Acknowledgements. This work is supported by the German Research Foundation (DFG) within the scope of the project Computer-unterstützte Keilschriftanalyse (CuKa).

References

1. Almazán, J., Gordo, A., Fornés, A., Valveny, E.: Word spotting and recognition with embedded attributes. TPAMI **36**(12), 2552–2566 (2014)
2. Bogacz, B., Gertz, M., Mara, H.: Character retrieval of vectorized cuneiform script. In: Proceedings of the 2015 13th International Conference on Document Analysis and Recognition (ICDAR), ICDAR 2015, pp. 326–330. IEEE Computer Society, USA (2015). https://doi.org/10.1109/ICDAR.2015.7333777
3. Bogacz, B., Gertz, M., Mara, H.: Cuneiform character similarity using graph representations (2015)
4. Bogacz, B., Howe, N., Mara, H.: Segmentation free spotting of cuneiform using part structured models. In: Proceedings of the ICFHR, pp. 301–306 (2016)
5. Gottstein, N.: Ein stringentes identifikations- und suchsystem für keilschriftzeichen (2012)
6. He, K., Zhang, X., Ren, S., Sun, J.: Spatial pyramid pooling in deep convolutional networks for visual recognition. In: Fleet, D., Pajdla, T., Schiele, B., Tuytelaars, T. (eds.) ECCV 2014. LNCS, vol. 8691, pp. 346–361. Springer, Cham (2014). https://doi.org/10.1007/978-3-319-10578-9_23
7. He, K., Zhang, X., Ren, S., Sun, J.: Delving deep into rectifiers: surpassing human-level performance on ImageNet classification. In: Proceedings of the International Conference on Computer Vision, pp. 1026–1034 (2015)
8. He, K., Zhang, X., Ren, S., Sun, J.: Deep residual learning for image recognition. In: Proceedings of the IEEE Conference on Computer Vision and Pattern Recognition, pp. 770–778 (2016)

9. Howe, N.R.: Part-structured inkball models for one-shot handwritten word spotting. In: Proceedings of the ICDAR, pp. 582–586, August 2013. https://doi.org/10.1109/ICDAR.2013.121

10. Kingma, D.P., Ba, J.L.: Adam: a method for stochastic optimization. In: Proceedings of the International Conference on Learning Representations (2015)

11. Akademie der Wissenschaften und Literatur, M., Julius-Maximilians-Universitat, W.: Hethitologie Portal Mainz (2000). http://www.hethiter.net/

12. Mara, H.: Multi-scale integral invariants for robust character extraction from irregular polygon mesh data. Ph.D. thesis (2012)

13. Massa, J., Bogacz, B., Krömker, S., Mara, H.: Cuneiform detection in vectorized raster images. In: Computer Vision Winter Workshop (CVWW) (2016)

14. Rothacker, L., Fink, G.A.: Segmentation-free query-by-string word spotting with bag-of-features HMMs. In: Proceedings of the ICDAR, pp. 661–665, August 2015. https://doi.org/10.1109/ICDAR.2015.7333844

15. Rusakov, E., Rothacker, L., Mo, H., Fink, G.A.: A probabilistic retrieval model for word spotting based on direct attribute prediction. In: Proceedings of International Conference on Frontiers in Handwriting Recognition, Niagara Falls, USA (2018). Winner of the IAPR Best Student Paper Award

16. Rusakov, E., Somel, T., Fink, G.A., Müller, G.G.W.: Towards query-by-eXpression retrieval of cuneiform signs. In: Proceedings of International Conference on Frontiers in Handwriting Recognition, Dortmund, NRW, Germany (2020)

17. Simonyan, K., Zisserman, A.: Very deep convolutional networks for large-scale image recognition. In: Proceedings of the International Conference on Learning Representations (2015)

18. Sudholt, S., Fink, G.: Evaluating word string embeddings and loss functions for CNN-based word spotting. In: Proceedings of the ICDAR, Kyoto, Japan, pp. 493–498 (2017)

19. Sudholt, S., Fink, G.A.: PHOCNet: a deep convolutional neural network for word spotting in handwritten documents. In: Proceedings of the ICFHR, Shenzhen, China, pp. 277–282 (2016). https://doi.org/10.1109/ICFHR.2016.55

Date Estimation in the Wild of Scanned Historical Photos: An Image Retrieval Approach

Adrià Molina(✉) [ID], Pau Riba [ID], Lluis Gomez [ID], Oriol Ramos-Terrades [ID], and Josep Lladós [ID]

Computer Vision Center and Computer Science Department, Universitat Autònoma de Barcelona, Bellaterra, Catalunya, Spain
adria.molinar@e-campus.uab.cat, {priba,lgomez,oriolrt,josep}@cvc.uab.cat

Abstract. This paper presents a novel method for date estimation of historical photographs from archival sources. The main contribution is to formulate the date estimation as a retrieval task, where given a query, the retrieved images are ranked in terms of the estimated date similarity. The closer are their embedded representations the closer are their dates. Contrary to the traditional models that design a neural network that learns a classifier or a regressor, we propose a learning objective based on the nDCG ranking metric. We have experimentally evaluated the performance of the method in two different tasks: date estimation and date-sensitive image retrieval, using the DEW public database, overcoming the baseline methods.

Keywords: Date estimation · Historical photographs · Image retrieval · Ranking loss · Smooth-nDCG

1 Introduction

Historical archives and libraries contain a large variability of document sources that reflect the memory of the past. The recognition of the scanned images of these documents allows to reconstruct the history. A particular type of archival data are historical photographs which are full of evidence that tells us the story of that snapshot in time. One just needs to pay attention to the subtle cues that are found in different objects that appear in the scene: the clothes that people wear, their haircut styles, the overall environment, the tools and machinery, the natural landscape, etc. All of these visual features are important cues for estimating its creation date. Apart from that, texture and color features might also be of great help to accurately estimate of image creation date since photographic techniques have evolved throughout history and have imprinted a specific date fingerprint on them.

Date estimation of cultural heritage photographic assets is a complex task that is usually performed by experts (e.g. archivists or genealogists) that exploit their expert knowledge about all the features mentioned above to provide precise date estimations for undated photographs. But their manual labor is costly

© Springer Nature Switzerland AG 2021
J. Lladós et al. (Eds.): ICDAR 2021, LNCS 12822, pp. 306–320, 2021.
https://doi.org/10.1007/978-3-030-86331-9_20

and time consuming, and automatic image date estimation models are of great interest for dealing with large scale archive processing with minimal human intervention.

Most approaches in date estimation for historical images try to directly compute the estimation through classification or regression [5,9,12]. As alternative of these classical approaches, in this paper we present a method for date estimation of historical photographs in a retrieval scenario. Thus, the date estimation of photographs is incorporated in the ranked results for a given query image. This allows to predict the date of an image contextualized regarding the other photographs of the collection. In the worst case, when the exact date is not exactly estimated, the user can obtain a relative ordering (one photograph is older than another one), which is useful in archival tasks of annotating document sources. The proposed model for historical photograph retrieval is based in a novel ranking loss function *smooth-nDCG* based on the Normalized Discounted Cumulative Gain ranking metric, which is able to train our system according to a known relevance feedback; in our case the distance in years between images.

The main idea in our approach relies on optimizing rankings such that the closer is image's date to the query's date for a certain photograph the higher will be ranked. When receiving an unknown image as query the method computes the distances towards a support dataset, consisting of a collection of images with known dates. The date is estimated assuming that the highest ranked images are images from the same date.

In contrast to the literature reviewed in Sect. 2, our method allows not only to predict but to rank images given a query. This means that considering an image from a certain year our system is capable of retrieving a list of images ordered by time proximity. This may be useful for many applications and systems that rely on retrieving information from a large amount of data.

The rest of the paper is organized as follows: In Sect. 2 we present the most relevant state of the art related to our work. The key components of the proposed model are described in Sect. 3, where we describe the learning objectives considered in the training algorithms, and in Sect. 4, where we outline the architecture and training process of the model. Section 5 provides the experimental evaluation and discussion. Finally, Sect. 6 draws the conclusions.

2   Related Work

The problem of automatic estimation of the creation date of historical photographs is receiving increased attention by the computer vision and digital humanities research communities. The first works go back to Schindler *et al.* [13,14], where the objective was to automatically construct time-varying 3D models of a city from a large collection of historical images of the same scene over time. The process begins by performing feature detection and matching on a set of input photographs, followed by structure from motion (SFM) to recover 3D points and camera poses. The temporal ordering task was formulated as a constraint satisfaction problem (CSP) based on the visibility of structural elements (buildings) in each image.

More related with the task addressed in our work, Palermo *et al.* [10] explored automatic date estimation of historical color images based on the evolution of color imaging processes over time. Their work was formulated as a five-way decade classification problem, training a set of one-vs-one linear support vector machines (SVMs) and using hand-crafted color based features (e.g. color co-occurrence histograms, conditional probability of saturation given hue, hue and color histograms, etc.). Their models were validated on a small-scale dataset consisting of 225 and 50 images per decade for training and testing, respectively.

In a similar work, Fernando *et al.* [3] proposed two new hand-crafted color-based features that were designed to leverage the discriminative properties of specific photo acquisition devices: color derivatives and color angles. Using linear SVMs they outperformed the results of [10] as well as a baseline of deep convolutional activation features [2]. On the other hand, martin *et al.* [8] also showed that the results of [10] could be slightly improved by replacing the one-vs-one classification strategy by an ordinal classification framework [4] in which relative errors between classes (decades) are adequately taken into account.

More recently, Müller *et al.* [9] have brought up to date the task of image date estimation in the wild by contributing a large-scale, publicly available dataset and providing baseline results with a state-of-the-art deep learning architecture for visual recognition [15]. Their dataset, Date Estimation in the Wild (DEW), contains more than one million Flickr[1] images captured in the period from 1930 to 1999, and covering a broad range of domains, e.g., city scenes, family photos, nature, and historical events.

Beyond the problem of unconstrained date estimation of photographs, which is the focus of this paper, there are other related works that have explored the automatic prediction of creation dates of certain objects [7,16], or of some specific types of photographs such as yearbook portraits [5,12].

In this work, contrary to all previously published methods, we approach the problem of date estimation from an image retrieval perspective. We follow the work of Brown *et al.* [1] towards differentiable loss functions for information retrieval. More precisely, our model learns to estimate the date of an image by minimizing the the Normalized Discounted Cumulative Gain.

In all our experiments we use Müller *et al.*'s Date Estimation in the Wild [9] dataset. We also share with [9] the use of state of the art convolutional neural networks in contrast to classic machine learning and computer vision approaches, such as [3].

3   Learning Objectives

As mentioned in Sect. 1 our retrieval system relies on a neural network [6] trained to minimize a differentiable ranking function [1]. Traditionally, information retrieval evaluation has been dominated by the mean Average Precision

---

[1] https://www.flickr.com/.

(mAP) metric. In contrast, we will be using the Normalized Discounted Cumulative Gain (nDCG). The problem with mAP is that it fails in measuring ranking success when the ground truth relevance of the ranked items is not binary.

In the task of date estimation the ground truth data is numerical and ordinal (e.g. a set of years in the range 1930 to 1999), thus given a query the metric should not punish equally for ranking as top result a 1-year difference image than a 20-years difference one. This problem is solved in the nDCG metric by using a relevance score that measures how relevant is a certain sample to our query. This allows us to not only to deal with date estimation retrieval but to explicitly declare what we consider a good ranking.

In this section we derive the formulation of the smooth-nDCG loss function that we will use to optimize our date estimation model.

Mean Average Precision: First, let us define *Average Precision* (AP) for a given query q as

$$\mathrm{AP}_q = \frac{1}{|\mathcal{P}_q|} \sum_{n=1}^{|\Omega_q|} P@n \times r(n), \tag{1}$$

where $P@n$ is the precision at n and $r(n)$ is a binary function on the relevance of the n-th item in the returned ranked list, $\mathcal{P}_q$ is the set of all relevant objects with regard the query q and Ω_q is the set of retrieved elements from the dataset. Then, the mAP is defined as:

$$\mathrm{mAP} = \frac{1}{Q} \sum_{q=1}^{Q} \mathrm{AP}_q, \tag{2}$$

where Q is the number of queries.

Normalized Discounted Cumulative Gain: In information retrieval, the *normalized Discounted Cumulative Gain* (nDCG) is used to measure the performance on such scenarios where instead of a binary relevance function, we have a graded relevance scale. The main idea is that highly relevant elements appearing lower in the retrieval list should be penalized. In the opposite way to mAP, elements can be relevant despite not being categorically correct with respect to the query. The *Discounted Cumulative Gain* (DCG) for a query q is defined as

$$\mathrm{DCG}_q = \sum_{n=1}^{|\Omega_q|} \frac{r(n)}{\log_2(n+1)}, \tag{3}$$

where $r(n)$ is a graded function on the relevance of the n-th item in the returned ranked list and Ω_q is the set of retrieved elements as defined above. In order to allow a fair comparison among different queries that may have a different sum of relevance scores, a normalized version was proposed and defined as

$$\mathrm{nDCG}_q = \frac{\mathrm{DCG}_q}{\mathrm{IDCG}_q}, \tag{4}$$

where IDCG_q is the ideal discounted cumulative gain, *i.e.* assuming a perfect ranking according to the relevance function. It is formally defined as

$$\mathrm{IDCG}_q = \sum_{n=1}^{|A_q|} \frac{r(n)}{\log_2(n+1)} \tag{5}$$

where A_q is the ordered set according the relevance function.

As we have exposed, we want to optimize our model at the retrieval list level, but the two classical retrieval evaluation metrics defined above are not differentiable. In the following we define two 'smooth' functions for ranking optimization that are inspired in the mAP and nDCG metrics respectively.

Ranking Function (R). Following the formulation introduced in [11], these two information retrieval metrics can be reformulated by means of the following ranking function,

$$\mathcal{R}(i,\mathcal{C}) = 1 + \sum_{j=1}^{|\mathcal{C}|} \mathbb{1}\{(s_i - s_j) < 0\}, \tag{6}$$

where $\mathcal{C}$ is any set (such as Ω_q or $\mathcal{P}_q$), $\mathbb{1}\{\cdot\}$ is the Indicator function, and s_i is the similarity between the i-th element and the query according to S. In this work we use the cosine similarity as S but other similarities such as inverse euclidean distance should work as well. Let us then define cosine similarity as:

$$S(v_q, v_i) = \frac{v_q \cdot v_i}{\|v_q\|\|v_i\|}. \tag{7}$$

Still, because of the need of a indicator function $\mathbb{1}\{\cdot\}$, with this formulation we are not able to optimize following the gradient based optimization methods and we require of a differentiable indicator function. Even though several approximations exists, in this work we followed the one proposed by Quin *et al.* [11]. Thus, we make use of the sigmoid function

$$\mathcal{G}(x; \tau) = \frac{1}{1 + e^{\frac{-x}{\tau}}}. \tag{8}$$

where τ is the temperature of the sigmoid function. As illustrated in Fig. 1 the smaller is τ the less smooth will be the function. For small values the gradient is worse defined but the approximation is better since the sigmoid becomes closer to the binary step function.

Smooth-AP: Following [1] and replacing the term $P@n \times r(n)$ in Eq. 1 by the Ranking Function introduced in Eq. 6, the AP equation becomes

$$\mathrm{AP}_q = \frac{1}{|\mathcal{P}_q|} \sum_{i \in \mathcal{P}_q} \frac{1 + \sum_{j \in \mathcal{P}_q, j \neq i} \mathbb{1}\{D_{ij} < 0\}}{1 + \sum_{j \in \Omega_q, j \neq i} \mathbb{1}\{D_{ij} < 0\}}. \tag{9}$$

where we use $D_{ij} = s_i - s_j$ for a more compact notation. Finally, replacing the indicator function by the sigmoid function (Eq. 8) we obtain a smooth approximation of AP

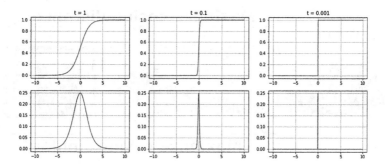

Fig. 1. Shape of the smooth indicator function (top) and its derivative (bottom) for different values of the temperature: 1 (left), 0.1 (middle), 0.001 (right).

$$\text{AP}_q \approx \frac{1}{|\mathcal{P}_q|} \sum_{i \in \mathcal{P}_q} \frac{1 + \sum_{j \in \mathcal{P}_q, j \neq i} \mathcal{G}(D_{ij}; \tau)}{1 + \sum_{j \in \Omega_q, j \neq i} \mathcal{G}(D_{ij}; \tau)}. \tag{10}$$

Averaging this approximation for all the queries in a given batch, we can define our loss as

$$\mathcal{L}_{AP} = 1 - \frac{1}{Q} \sum_{i=1}^{Q} \text{AP}_q, \tag{11}$$

where Q is the number of queries.

Smooth-nDCG: Following the same idea as above, we replace the n-th position in Eq. 3 by the ranking function, since it defines the position that the i-th element of the retrieved set, and the DCG metric is expressed as

$$\text{DCG}_q = \sum_{i \in \Omega_q} \frac{r(i)}{\log_2 \left(2 + \sum_{j \in \Omega_q, j \neq i} \mathbb{1}\{D_{ij} < 0\}\right)}, \tag{12}$$

where $r(i)$ is the same graded function used in Eq. 3 but evaluated at element i. Therefore, the corresponding smooth approximation is

$$\text{DCG}_q \approx \sum_{i \in \Omega_q} \frac{r(i)}{\log_2 \left(2 + \sum_{j \in \Omega_q, j \neq i} \mathcal{G}(D_{ij}; \tau)\right)} \tag{13}$$

when replacing the indicator function by the sigmoid one.

The smooth-nDCG is then defined by replacing the original DCG_q by its smooth approximation in Eq. 4 and the loss $\mathcal{L}_{nDCG}$ is defined as

$$\mathcal{L}_{nDCG} = 1 - \frac{1}{Q} \sum_{i=1}^{Q} nDCG_q, \tag{14}$$

where Q is the number of queries.

Fig. 2. Objective function example, images with closer ground truth should maximize the similarity in the output space.

4 Training Process

The main contribution of this paper relies on using the smooth-nDCG loss defined in the previous section for optimizing a model that learns to rank elements with non-binary relevance score. Specifically, our model learns to project images into an embedding space by minimizing the smooth-nDCG loss function. In all our experiments we use a Resnet-101 [6] convolutional neural network pretrained on the ImageNet dataset. As illustrated in Fig. 2 the idea is that in the learned embedding space the cosine similarity between image projections is proportional to the actual distance in years. We will discuss in Sect. 5 how teaching the system to rank images according to the ground truth criteria leads the embedding space to follow a certain organization (illustrated in Fig. 7).

In order to obtain such embedded representation of the photographs, we present the training algorithm (see Algorithm 1) that mainly relies on using the equations presented in the previous Sect. 3 in the output space for optimizing the CNN so it minimizes the smooth-nDCG loss function (see Eq. 14) and, consequently, maximizes the nDCG metric in our rankings, meaning that images will be closer in the output space as they are closer in the ground truth space.

Algorithm 1. Training algorithm for the proposed model.

    Input: Input data $\{\mathcal{X}, \mathcal{Y}\}$; CNN f; distance function D; max training iterations T
    Output: Network parameters w

1: **repeat**
2:      Process images to output embedding $h \leftarrow f_w(\{x_i\}_{i=1}^{N_{batch}})$
3:      Get Distance matrix from embeddings, all vs all rankings $M \leftarrow D(h)$
4:      Calculate relevance from training set Y Eq. 15
5:      Using the relevance score, $\mathcal{L} \leftarrow$ Eq. 14
6:      $w \leftarrow w - \Gamma(\nabla_w \mathcal{L})$
7: **until** Max training iterations T

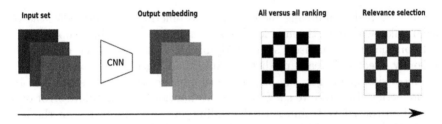

Fig. 3. Proposed baseline. Once ranked the images in the nDCG space we'll be computing smooth-nDCG in order to back-propagate the loss through the CNN.

As shown in Fig. 3, once we process the images through the CNN, we compute the rankings for each one of the batch images. We use cosine similarity as distance function, however it could be replaced by any distance function model over the vector space. Since nDCG requires a relevance score for each sample, the design of the relevance function is an important part of the proposed method. A possible relevance function could be just the inverse distance in years as defined in Eq. 16.

Additionally, we decided to consider only the closest decade in the relevance function (illustrated in Fig. 4), so every image further than 10 years is considered as relevant as any other. Equation 15 is the relevance function used in our model (see Eq. 14) but other relevance settings could be, like Eq. 17 in order to exponentially punish more the images that should not be as far as they are.

$$r(n; \gamma) = \max(0, \gamma - |y_q - y_n|) \tag{15}$$

$$r(n) \quad = \frac{1}{1 + |y_q - y_n|} \tag{16}$$

$$r(n) \quad = e^{\frac{1}{1 + |y_q - y_n|}} \tag{17}$$

where y_q and $y_n \in \mathcal{Y}$ are the dates of the query and the n-th image in the training set, respectively; and γ, in Eq. 15, is an hyper-parameter that has experimentally set to 10.

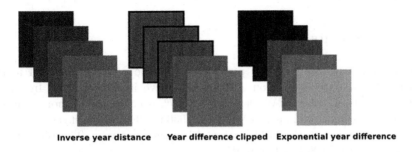

Inverse year distance   Year difference clipped   Exponential year difference

Fig. 4. Different examples of relevance functions for ground truth data.

Once trained, minimizing the retrieval loss function, we can predict the year for an unknown image through k-Nearest Neighbors (k-NN) to a support set.

In one hand, we can use a set of N train images as support set; each batch is randomly selected for each prediction. Note that, since prediction relies on the k-th most similar images, the bigger the support set the better should perform the prediction. In Fig. 5 we show the date estimation mean absolute error of our method on the DEW training set depending on the k parameter for the k-NN search. Hence, we would be using 10-Nearest Neighbors in all our experiments for the prediction task.

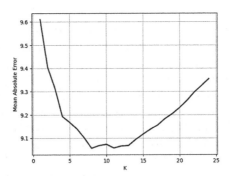

Fig. 5. Mean Absolute Error (MAE) on prediction by k-Nearest Neighbors with respect to each k using a randomly selected support set of 128 images

Additionally, we can compute the retrievals within the entire training set with Approximate Nearest Neighbours[2], which allow us to compute the retrievals efficiently. As it's shown in Fig. 6, the possibility of using the whole train set makes the k parameter more likely to be huge. We don't observe a notable divergence in the results until using a an enormous k parameter, which may indicate quite good result in terms of retrieval since worst responses are at the bottom of the dataset.

5   Experiments

In this section we evaluate the performance of the proposed method in two different tasks: date estimation and date-sensitive image retrieval. In all our experiments we use the DEW [9] dataset, which is composed of more than one million images captured in the period from 1930 to 1999. The test set is balanced across years and contains 1120 images, 16 images per year. The temperature parameter, τ, in Eq. 8, is set to 0.01.

---

[2] Using public Python library ANNOY for Approximate Nearest Neighbours https://github.com/spotify/annoy.

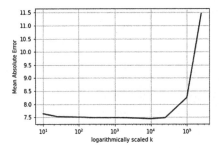

Fig. 6. Mean Absolute Error with different k for k-NN using the entire train set as support set. Note that the horizontal axis grows logarithmically.

For the task of date estimation we use the Mean Absolute Error (MAE) as the standard evaluation metric. Given a set of N images and their corresponding ground truth date annotations (years) $y = \{y_1, y_2, \ldots, y_N\}$, the MAE for a set of predictions $y' = \{y'_1, y'_2, \ldots, y'_N\}$ is calculated as follows:

$$\text{MAE} = \frac{1}{N} \sum_{i}^{N} |y_i - y'_i| \tag{18}$$

For the image retrieval task we use the mAP and nDCG metrics defined in Sect. 3. In Table 1 we present a comparison of our methods with the state of the art in the task of date estimation and with a visual similarity baseline, that ranks images using the euclidean distance between features extracted from a ResNet CNN [6] pre-trained on ImageNet. As we will discuss in the conclusions section, our approach is not directly comparable to Müller et al.'s [9]. On one hand, we are not trying to directly improve the prediction task, but a ranking one; this means that our model has to learn semantics from the images such that it learns a continuous ordination. On the other hand, the way we are estimating image's date is by using a ground truth dataset or support dataset. Nevertheless, we observe that our model isn't too far with respect to the state of the art prediction model. We consider this a good result since estimating the exact year of a image is not the main purpose of the smooth-nDCG loss function. Note that in Table 1 we mention a "cleaned test set". This is because we observed the presence of certain placeholder images indicating they are no longer available in Flickr. We computed the MAE for the weights given by Müller et al. [9][3] for a proper comparison. The presence of this kind of images may have a low impact while evaluating a classification or regression, but they can have a higher impact on predicting with k-NN, since all the retrieved images will be placeholders as well independently of their associated years.

---

[3] This is an extended version of Müller et al. [9] made publicly available by the same authors at https://github.com/TIB-Visual-Analytics/DEW-Model.

Table 1. Mean absolute error comparison of our baseline model on the test set of the Date Estimation in the Wild (DEW) dataset.

Baseline	MAE	mAP	nDCG
Müller et al. [9] GoogLeNet (regression)	7.5	–	–
Müller et al. [9] GoogLeNet (classification)	7.3	–	–
Müller et al. [9] ResNet50 (classification) (cleaned test set)	7.12	–	–
Visual similarity baseline	12.4	–	0.69
Smooth-nDCG 256 images support set	8.44	0.12	0.72
Smooth-nDCG 1M images support set	7.48	–	0.75
Smooth-nDCG 1M images support set (weighted kNN)	7.52	–	0.75

We present as well a result of predicting the year by weighted k-NN where the predicted year will be the sum of the k neighbours multiplied by their similarities to the query.

Since our model is trained for a retrieval task, we provide further qualitative examples of good and bad rankings. This exemplifies how our loss function is not mean to predict years, but finding good rankings according to query's year. Note we computed nDCG using as relevance the closeness in years between images' ground truth clipped at 10 years (see Eq. 15). Despite it may seem nDCG is so high and mAP so low; there are many subtle bias in this appearance. Since the ground truth can be interpreted as a continuous value (such as time) there are not many matches in any retrieval of N random images for a certain query regarding to query's year compared with how many negative samples are there. Then, mAP hardly punishes results in this situation. In the opposite way, the same thing happens to nDCG. Since nDCG has this 'continuous' or 'scale' sense, any retrieval will have some good samples that satisfies the relevance function (see Eq. 15), being the results way less punished if there aren't many exact responses in the retrieval. Briefly, we could say nDCG brings us a more balanced positive/negative ratio than mAP for randomly selected sets to rank (which is absolutely more realistic than forcing it to be balanced). Additionally, nDCG allows us to approach retrievals that are not based on obtaining a category but a wide number of them, or even retrieving non-categorical information such as date estimation.

Additionally, as we observe in Fig. 7, we compute the average cosine similarity between clusters from different groups of years. This shows us how our output space is distributed with respect to the ground truth data. As we expected; clusters for closer years are the most similar between themselves. This is because ranking is nothing but taking the closer points to our query; by forcing our system to take those points that satisfies the relevance function (see Eq. 15), points in the embedding space will have to be organized such that distances are proportional to ground truth's distances.

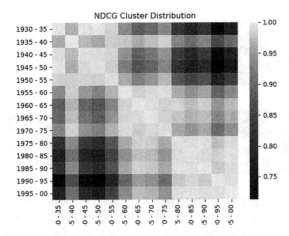

Fig. 7. Clusters average similarity for each 5 year interval. We observe a certain visible pattern in the diagonal; the closer the years the closer the clusters.

Fig. 8. Qualitative results for our date estimation model. We show a set of images from the DEW test set and indicate their ground truth year annotation and distance to the query.

Finally, in Fig. 9, we compare the top-5 retrieved images for a given query with our model and with the visual similarity baseline model, that ranks images using the distance between features extracted from a ResNet CNN [6] pre-trained on ImageNet. We appreciate that the visual similarity model retrieves images that are visually similar to the query but belong to a totally different date. On the contrary, our model performs better for the same query in terms of retrieving similar dates although for some of the top ranked images the objects layout is quite different to the query. Additionally, we present the results with a mixed model that re-ranks images with the mean ranking between both methods for each image in the test set; in this way we obtain a ranking with visually similar images from the same date as the query.

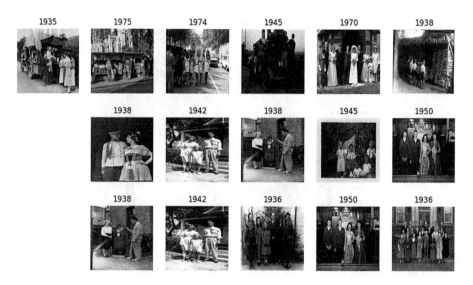

Fig. 9. Qualitative results for a retrieval system that combines our date estimation model and a visual similarity model. The top-left image is the query, in each row we show the top-5 retrieved images using a visual similarity model (top), our model (middle), and the combination of the two rankings (bottom).

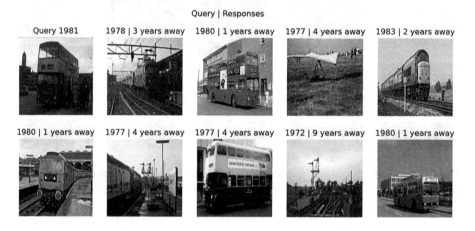

Fig. 10. Possible biased retrieval due lack of balance in the training set.

6   Conclusions

In this paper we have proposed a method for estimating the date of historic photographs. The main novelty of the presented work is that we do not formulate the solution to the problem as a prediction task, but as an information retrieval one. Hence metrics such as mean absolute error do not have to be considered as important as retrieval metrics. Even so, we managed to build an application-level model that can estimate the date from the output embedding space with a decent error according to previous baselines [9]. Since our prediction method is essentially a clustering or k-Nearest Neighbors method we are using a known support set, so predictions relies on already labeled data.

However, considering the output embedding space Fig. 7 we conclude that smooth-nDCG function works pretty well for ranking large-scale image data with deeper ground truth structures; unlike smooth-AP [1], smooth-nDCG cares about the whole sorted batch retrieved, not only how many good images are, but how are the bad ones distributed along the ranking. This allowed us to retrieval ordered information (such as years), or tree-structured information where different categories can be closer or further depending on how the tree is like.

In the case of date estimation in the wild [9] dataset, we found out some problematic patterns that can lead the retrieval task to certain bias. As there is not a clear balance between categories in the dataset, many classes such as *trains* or *vehicles* may be clustered together regardless the actual dates of the images Fig. 10. However, it is not easy to say when a certain category is most likely to be linked to a certain period of time. For example, selfies are a category way more common nowadays than 20 years ago. Something similar may happen with certain categories that could be biasing the retrieval to a pure object classification task; however we found out many good clues, Figs. 7, 8, that indicate our embedding for ranking images is working properly.

Despite further research is needed; smooth-nDCG usage for large-scale image retrieval is a promising novel approach with many practical applications. Since Brown *et al.* [1] smoothed us the path to minimize the average precision, we propose smooth-nDCG to model more complex retrieval problems where different labels should not be punished equally according to a certain criteria. As we commented in smooth-nDCG brings a new way of ranking images beyond categorical labels from a neural network approach; then we would like to consider this approach as an application example for larger information retrieval problems with a more complex structure.

Acknowledgment. This work has been partially supported by the Spanish projects RTI2018-095645-B-C21, and FCT-19-15244, and the Catalan projects 2017-SGR-1783, the Culture Department of the Generalitat de Catalunya, and the CERCA Program/Generalitat de Catalunya.

References

1. Brown, A., Xie, W., Kalogeiton, V., Zisserman, A.: Smooth-AP: smoothing the path towards large-scale image retrieval. In: Vedaldi, A., Bischof, H., Brox, T., Frahm, J.-M. (eds.) ECCV 2020. LNCS, vol. 12354, pp. 677–694. Springer, Cham (2020). https://doi.org/10.1007/978-3-030-58545-7_39
2. Donahue, J., et al.: DeCAF: a deep convolutional activation feature for generic visual recognition. In: Proceedings of the International Conference on Machine Learning, pp. 647–655 (2014)

3. Fernando, B., Muselet, D., Khan, R., Tuytelaars, T.: Color features for dating historical color images. In: Proceedings of the International Conference on Image Processing, pp. 2589–2593 (2014)

4. Frank, E., Hall, M.: A simple approach to ordinal classification. In: De Raedt, L., Flach, P. (eds.) ECML 2001. LNCS (LNAI), vol. 2167, pp. 145–156. Springer, Heidelberg (2001). https://doi.org/10.1007/3-540-44795-4_13

5. Ginosar, S., Rakelly, K., Sachs, S., Yin, B., Efros, A.A.: A century of portraits: a visual historical record of American high school yearbooks. In: Proceedings of the IEEE Conference on Computer Vision and Pattern Recognition Workshops, pp. 1–7 (2015)

6. He, K., Zhang, X., Ren, S., Sun, J.: Deep residual learning for image recognition. In: Proceedings of the IEEE Conference on Computer Vision and Pattern Recognition, pp. 770–778 (2016)

7. Lee, S., Maisonneuve, N., Crandall, D., Efros, A.A., Sivic, J.: Linking past to present: discovering style in two centuries of architecture. In: Proceedings of the International Conference on Computational Photography (2015)

8. Martin, P., Doucet, A., Jurie, F.: Dating color images with ordinal classification. In: Proceedings of International Conference on Multimedia Retrieval, p. 447 (2014)

9. Müller, E., Springstein, M., Ewerth, R.: "When was this picture taken?" – image date estimation in the wild. In: Jose, J.M., et al. (eds.) ECIR 2017. LNCS, vol. 10193, pp. 619–625. Springer, Cham (2017). https://doi.org/10.1007/978-3-319-56608-5_57

10. Palermo, F., Hays, J., Efros, A.A.: Dating historical color images. In: Fitzgibbon, A., Lazebnik, S., Perona, P., Sato, Y., Schmid, C. (eds.) ECCV 2012. LNCS, vol. 7577, pp. 499–512. Springer, Heidelberg (2012). https://doi.org/10.1007/978-3-642-33783-3_36

11. Qin, T., Liu, T.Y., Li, H.: A general approximation framework for direct optimization of information retrieval measures. Inf. Retrieval **13**(4), 375–397 (2010)

12. Salem, T., Workman, S., Zhai, M., Jacobs, N.: Analyzing human appearance as a cue for dating images. In: Proceedings of the IEEE Winter Conference on Applications of Computer Vision, pp. 1–8 (2016)

13. Schindler, G., Dellaert, F.: Probabilistic temporal inference on reconstructed 3D scenes. In: Proceedings of the IEEE Conference on Computer Vision and Pattern Recognition, pp. 1410–1417 (2010)

14. Schindler, G., Dellaert, F., Kang, S.B.: Inferring temporal order of images from 3D structure. In: Proceedings of the IEEE Conference on Computer Vision and Pattern Recognition, pp. 1–7 (2007)

15. Szegedy, C., et al.: Going deeper with convolutions. In: Proceedings of the IEEE Conference on Computer Vision and Pattern Recognition (2015)

16. Vittayakorn, S., Berg, A.C., Berg, T.L.: When was that made? In: Proceedings of the IEEE Winter Conference on Applications of Computer Vision, pp. 715–724 (2017)

Two-Step Fine-Tuned Convolutional Neural Networks for Multi-label Classification of Children's Drawings

Muhammad Osama Zeeshan, Imran Siddiqi, and Momina Moetesum[✉]

Bahria University, Islamabad, Pakistan
{imran.siddiqi,momina.buic}@bahria.edu.pk

Abstract. Developmental psychologists employ several drawing-based tasks to measure the cognitive maturity of a child. Manual scoring of such tests is time-consuming and prone to scorer bias. A computerized analysis of digitized samples can provide efficiency and standardization. However, the inherent variability of hand-drawn traces and lack of sufficient training samples make it challenging for both feature engineering and feature learning. In this paper, we present a two-step fine-tuning based method to train a multi-label Convolutional Neural Network (CNN) architecture, for the scoring of a popular drawing-based test 'Draw-A-Person' (DAP). Our proposed two-step fine-tuned CNN architecture outperforms conventional pre-trained CNNs by achieving an accuracy of 81.1% in scoring of Gross Details, 99.2% in scoring of Attachments, and 79.3% in scoring of Head Details categories of DAP samples.

Keywords: Draw-a-person test · Multi-label classification · Two-step fine-tuning · Small sample domains

1 Introduction

Drawings are one of the earliest known modes of human communication. To date drawings are preferred form of expression in a variety of situations like procedural flowcharts, engineering or architectural plans, electronic circuit diagrams and free-hand sketches. Drawings are also well tolerated among individuals from all age groups and are not limited by linguistic barriers or educational background. Due to the inherent properties of a skilled graphomotor task like drawing, it can be employed to assess consequential changes resulting from various brain dysfunctions [37]. In clinical psychology, this information is employed in a number of neuropsychological screening tests to detect early signs of various cognitive, perceptual and motor disorders both in children and adults [20,31,34]. For instance, spirals and loops are mostly employed to assess motor dysfunctions in neurodegenerative diseases like Parkinson's disease (PD) [31]. Degradation of memory and organizational abilities in Alzheimer's disease (AD) is commonly assessed by cube drawings [18] and 'Clock Draw Test' (CDT) [35]. Similarly, to analyze

J. Lladós et al. (Eds.): ICDAR 2021, LNCS 12822, pp. 321–334, 2021.
https://doi.org/10.1007/978-3-030-86331-9_21

visual-perceptual disorders, a complex figure drawing test like 'Rey-Osterrieth Complex Figure' (ROCF) [34] is used.

Drawing-based tests are especially popular among children, since they feel more comfortable in expressing via drawings than in a verbal assessment. Studies in developmental psychology [11], reveal that the periodic progression of drawing ability in children can give useful insight regarding their cognitive and emotional development. The performance of a child in a drawing task can also be linked with academic capabilities and social skills, that are common characteristics of developmental age [30]. Based on these assumptions, several drawing based tasks have been designed that include 'Bender Gestalt Test' (BGT) [1], 'House-Tree-Person' (HTP) test [2] and 'Draw-A-Person' (DAP) test [9]. These tests can be integrated into mainstream academic systems for regular assessment of children's mental and emotional health [4]. Nevertheless, such integration requires trained professionals and lengthy one-to-one sessions with each and every child. This protocol is both cost and time inefficient.

Computerized assessment of children's drawings can increase the efficiency of intervention programs by reducing load of clinical practitioners. Non-experts can be trained to use such systems, who can then compile results for the clinician to interpret. In this way, experts can focus more on the intervention instead of scoring and report formulation. Furthermore, an automated system can mitigate the adverse impact of extrinsic factors like scorer bias and inter-scorer variability. Automation can also preserve the usability of these tests and enable the practitioners to improve them. Most importantly, it can provide the collection of normative data that can later be compared with existing normative databases. However, despite the apparent advantages, several open issues need to be addressed while designing an effective system. Prominent ones include high degree of deviations, small training samples and extensive scoring standards.

In an attempt to provide an automated tool for the analysis of children's drawings (and handwriting), several techniques [3, 10, 17, 21–23, 32, 38, 39, 41, 42] have been proposed in the literature. Until recently, researchers employed heuristic-based approaches to translate domain knowledge into computational feature space. Nonetheless, it soon proved insufficient to represent all possible deviations [23]. As a result, machine learning-based solutions gained popularity in this domain despite the interpretability concerns. By employing various machine learning-based techniques, meaningful representations are learned from samples. However, like any other health-related domain, there is a lack of sufficient training data due to privacy issues. This is a major limiting factor for a data-driven approach. Nonetheless, recent advances in machine learning and more specifically deep learning, have proposed several solutions to overcome this limitation. Transfer learning [28] is one such alternative, which allows a neural network architecture to transfer learned weights across different source and target datasets. Recent studies like [22], have employed various Convolutional Neural Network (CNN) architectures pre-trained on ImageNet [19] to score children's drawings with considerable success. However, the semantic gap between the source and target datasets produced results that require further investigations.

In this paper, we propose a two-step fine tuning approach, where a CNN architecture pre-trained on a relatively large but semantically different dataset (i.e. ImageNet) is first fine-tuned on a semantically similar dataset (i.e. Sketch-Net [8]). In the second phase, the same architecture is again fine-tuned on a much smaller target dataset. The impact of a mediator dataset proved beneficial in overcoming the semantic gap between the source and the target. To assess the performance of our proposed methodology, we employ DAP drawings of 135 children between the ages of 7 to 12. DAP responses are usually scored using a 51 point extensive manual proposed by Goodenough [13]. The manual comprises ten scoring sections, out of which we will be considering 'Gross details', 'Head details' and 'Attachments', due to their complexity and diversity. It is important to mention that the purpose of this study is to ascertain the effectiveness of the proposed method and not to automate a particular manual. By changing the target dataset, our baseline architecture can be translated to another test.

The rest of the paper is organized as follows. Section 2 discusses the relevant literature. Section 3 describes our proposed methodology in detail. Section 4 states the experimental protocol and results of empirical analysis. Finally, Sect. 5 concludes the paper.

2   Related Work

Until recently, deep learning-based solutions were limited to large sample domains like image classification [19], sketch recognition [7], object detection [33] and text recognition [14]. Nonetheless, techniques such as transfer learning [28] have enabled the applicability of deep learning algorithms in small sample domains like handwriting and drawing analysis for disease detection as well. CNNs were first employed by Pereira et al. [29] to develop a tool for the detection of Parkinson's disease from spiral drawings of 308 subjects. Handwriting signals captured by a specialized Smart pen were transformed into images and then fed to a 3-layered CNN architecture. The authors reported an accuracy of 87.14% in discriminating between drawings of healthy controls and PD patients. Transfer learning was first exploited by authors in [25] on the same problem (i.e. classification of drawing and handwriting samples of PD patients). A pre-trained AlexNet [19] was employed as a feature extractor on handwriting and drawing samples of 36 PD patients and 36 healthy controls available in a benchmark database PaHaW [6]. Authors plotted the on-surface pen-positions and then produced multiple representations of raw data by employing median filter residual and edge information. The extracted features from the multiple representations were then fused to train a binary support vector classifier. The technique was employed on eight different graphomotor tasks produced by a subject and the decisions of all task-wise classifiers were combined using an ensemble approach to achieve 83% accuracy. The same technique was later adopted and extended by authors in [5,26]. Both studies advocate the effectiveness of fine-tuning over feature-extraction. Nonetheless, both highlighted the issue of overfitting due to data paucity.

Localization of scoring regions or recognition of a particular component or shape in a drawing sample is an essential pre-processing step in most visual analysis based systems. In an earlier attempt [15], to score the CDT drawing samples of 165 subjects (65 healthy and 100 patients suffering from Mild Cognitive Impairments (MCI) and Dementia), authors employed CNNs to recognize clock digits. The clock circumference was segmented into eight regions and drawn digits in each segment were then recognized. Error was reported in case of missing, multiple or wrong digits in each segment. An extensive ontology-based heuristic approach was employed to score the complete drawing. In another study [27], authors employed pre-trained CNNs to classify the nine BGT shapes. Scoring of BGT specifies a different criteria to measure an error across different shapes. Furthermore, some errors apply only on specific shapes and not all nine of them. This makes it essential to identify the template prior to scoring. Authors evaluated the performance of different pre-trained CNN architectures as feature extractors. Different classifiers were trained using the features extracted by the pre-trained CNN models and best results (98.33% accuracy) were achieved by employing an AlexNet-LDA (Linear Discriminant Analysis) combination.

Recently, in a series of related studies [22,24], authors employed pre-trained CNNs to score offline BGT test samples of 106 children. Eleven indicators of visual-perceptual maturity are targeted in both studies. To avoid extensive heuristics, authors trained deformation (error)-specific networks. In [24] fine-tuning of last few layers is applied whereas the weights of the convolutional base were kept frozen. The reported accuracies ranged between 66.5% to 87.7% for the eleven deformations. The CNN architecture employed was VGG16 [36]. Later in [22], authors evaluated various combinations of pre-trained CNN architectures as feature extractors and machine learning-based classifiers to model the standard BGT errors. The impact of depth of the CNN architecture was also evaluated. Highest accuracies across all errors were achieved by employing ResNet101 [16] as feature extractor and LDA as classifier. In both these studies, error-specific binary classifiers were employed that determined the presence/absence of a particular error in a shape. All nine BGT shapes were independently fed to each classifier and final scores were accumulated. This approach requires multiple classifiers for multiple errors that can lead to scalability issues in extensive scoring-based tests.

In all of these studies, CNNs models pre-trained on ImageNet were either employed as feature extractors or last few layers were fine-tuned on the smaller target dataset. Until recently, the idea of multiple fine-tuning was not introduced in this domain [12]. Although not a novel solution per se, nonetheless, its feasibility in small sample domains like the one under consideration is yet to be fully explored.

3   Methodology

This section presents our proposed methodology and details of the target dataset. As mentioned earlier, we are targeting small sample domain problems and therefore, a customized dataset is compiled to evaluate the effectiveness of our approach. For this purpose, we selected offline samples of Draw-A-Person (DAP) test [9]. DAP is a popular projective test for the assessment of the intellectual and emotional health of children. Unlike copying or recall tests like BGT, Spiral, and ROCF, projective tests are more challenging to analyze due to lack of a standard template. Subjects are asked to draw a person without showing them a sample or providing any specifications. This makes learning exceptionally challenging for a machine. Conventionally, Goodenough-Harris scoring system [13] assesses a DAP response against an extensive 51 point criteria. There are ten scoring categories, each focusing on different details of the drawn figure. For this study, we have considered three scoring categories, details of which are presented in the subsequent sections.

3.1 Data Preparation

To the best of our knowledge, there is no existing dataset of DAP responses. Consequently, we collected drawings samples from a cohort of children of varying ages ranging between 7 to 12 years. The drawings ranged from stick figures to head-only, revealing the diverse nature of a projective response without a visual stimulus. Figure 1-a displays some of the samples collected.

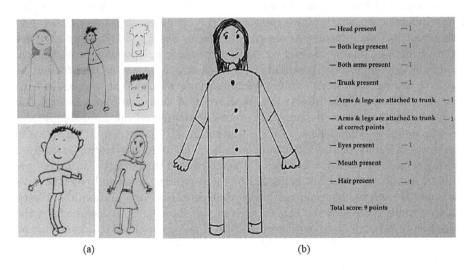

(a)   (b)

Fig. 1. (a) Examples of DAP responses collected (b) Scored response

Manual scoring of the drawings was conducted by domain experts. Three trained psychologists scored each sample. In case of disagreement, majority voting was employed. As mentioned earlier, being a pilot study, we focused on three categories of scores i.e. 'Gross Details', 'Attachments' and 'Head Details'. Details of these are outlined in Table 1. Figure 1-b, illustrates scoring of a sample against these three categories containing 9 scoring points, where each correctly drawn component is given 1 point resulting in a cumulative score of 9.

Table 1. Scoring details considered in the study

Category	Scoring Description
Gross Details	1. Head Present
	2. Legs Present
	3. Arms Present
	4. Trunk Present
Attachments	1. Both legs and arms are attached to the trunk
	2. Arms and legs are attached to the trunk at the correct Place
Head Details	1. Either or both eyes Present
	2. Mouth Present
	3. Hair Shown

For the ease of training, we design a multi-label network for each of the three scoring categories. Gross details category contains four component (head, arms, legs, and trunk) and any other component if present is ignored. For the Attachments category, we have two classes (i.e. *arms and legs are attached* and *arms and legs are attached at the correct place*). Finally, for the last category of Head details, images of head containing the three components (eyes, mouth, and hair) are employed. To extend the dataset size and to avoid over-fitting, we applied various data augmentation techniques on the original samples in each category. These include translation, application of Gaussian blur, and addition of Gaussian noise to each sample. It eventually increased the total sample size to 540 samples. Nonetheless, caution was taken during data splitting to avoid inclusion of copies of the same sample in both training and testing in a given fold (i.e. samples used for training are not repeated in testing). All images are resized to 400 × 400 before being fed to the networks. A ground truth file was created for each of the main categories that contained information about each sample (unique ID and the classes present per image). '1' is assigned for the presence of a particular component and '0' if the component is missing in the image.

3.2 Two-Step Fine Tuning

As mentioned in the introductory section, training of a deep CNN architecture requires a substantial amount of training data. This makes it quite challenging to utilize a deep ConvNet model to a small sample domain task. However, recent advent of transfer learning-based methods has enabled the adaptation of deep learning in such domains. Transfer learning allows training of a deep ConvNet on a large but not necessarily identical dataset. The weights of the network are then fine-tuned on the smaller target dataset. Although the choice of the source task can significantly impact network performance yet studies have shown that low-level features are usually common across most datasets [43]. Nonetheless, by reducing the semantic difference between the source and target dataset, we can further improve performance. In our case, we are employing CNN architectures pre-trained on ImageNet, while our target dataset comprises hand-drawn traces. To reduce the semantic gap between our source and target datasets, we introduce a mediator dataset SketchNet [7]. SketchNet database comprises 20,000 drawn sketches divided into 250 categories with 80 images in each class (sample images are presented in Fig. 2).

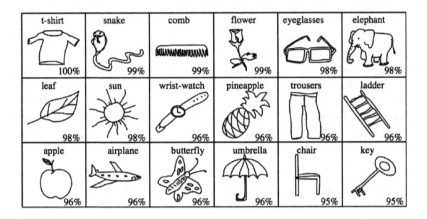

Fig. 2. Sketch database (Source: [7])

The CNN architecture pre-trained on ImageNet is first fine-tuned on Sketch-Net. Since initial layers extract low-level features that are common across most datasets, we kept these frozen. Instead, we fine-tuned the last layers (depending on the depth of the ConvNet) using SketchNet. In the next step, we fine-tuned the final few layers of the network by employing our target dataset.

3.3 Multi-label Classification

Several design considerations could have been implemented to classify the DAP responses using our fine-tuned CNNs. A straight-forward approach was to train a single multi-label network for all nine scoring points discussed earlier. However, by doing so, model training would have become too complicated and would have adversely affected the overall classification performance. It is important to mention that this is an intra-shape class variation identification problem and not an inter-shape class one. Thus, it is more challenging for the model to learn semantically similar differences.

The second option would have been to train an independent model for each scoring criteria as was done in [22,24]. This approach is only feasible for a smaller scoring criteria and cannot be scaled to an extensive one. Although in this pilot study, we are only modeling nine points of Goodenough scoring manual but our ultimate aim is to design a complete end-to-end scoring system for all the 51 points. Due to this reason, we have trained three multi-label networks for each of the three main categories (Gross Details, Attachments, and Head Details) described in Table 1. The system consists of three networks and each is responsible to perform classification on a different set of classes belonging to Gross Details (4 classes), Attachments (2 classes), and Head Details (3 classes), as shown in Fig. 3. Each network consists of multiple fine-tuned CNN layers and the last layer has been replaced with the sigmoid activation function.

Images are fed to each fine-tuned ConvNet base which then generates a feature vector. The vector is flattened and input to the multi-label dense layer where the sigmoid activation function is applied. Each model predicts the multi-label classification results for its respective classes. The output is a value ranging between 0 to 1, for every label in the image. In case the predicted score of a class is less then 0.5, it is considered to be missing from the image and is labeled as 0. On the other hand, if the value is greater than 0.5, then the class is present and is labeled as 1. At last, all the scores are accumulated and presented to the end-user with their corresponding labels and values.

4 Experimental Protocol, Results and Analysis

In this section, we present the details of the experimental protocol employed to determine the effectiveness of our proposed methodology. Since our target dataset is small, we have adopted the 10-fold cross validation technique to evaluate the performance of our models. The target dataset is divided into 10 folds, and every time one of the 10 subsets is used as the test set and rest are used for training the model. Every component present in an image is represented by 1 and those which are not present are represented by 0. We designed our experimental protocol to evaluate the following scenarios.

- Impact of two-step fine-tuning on independent multi-label networks for Gross Details, Attachments and Head Details
- Impact of two-step fine-tuning on a single multi-label network for all nine scoring points.

We first conduct experiments on CNN models pre-trained only on ImageNet to create a baseline. For this purpose, we selected VGG16 [36], ResNet101 [16], InceptionV3 [40], and InceptionResNetV2 models pre-trained on ImageNet source dataset. Last layers of the models are fine-tuned on the target DAP datasets (for Gross Details, Attachments and Head Details).Each network is

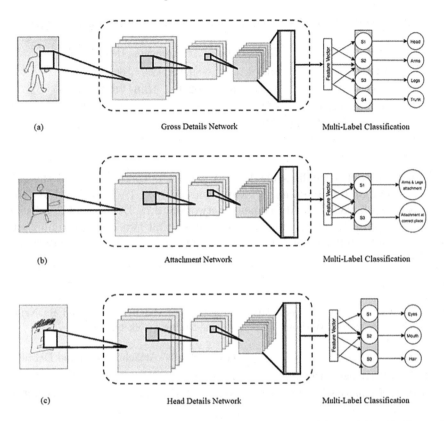

Fig. 3. (a) Gross Details Network (b) Attachments Network (c) Head Details Network

tuned using a batch size of 32, for 200 epochs, whereas the Adam optimizer (0.001) is employed in learning. The results for each category using different ConvNet models are outlined in Table 2.

It is observed that ResNet101 outperforms the rest in all three categories by achieving accuracies of 74.6% in Gross Details, 92.1% in Attachments and 73.7% in Head Details, respectively. Due to its better performance, we will consider ResNet101 to assess the effectiveness of our two-step fine-tuning method. For this purpose, we freeze the initial 20 layers of ResNet101 model and fine-tune the last 80 layers using SketchNet database. The network was trained for 200 epochs using a batch size of 32.

In the second phase, all the layers are kept frozen and two trainable classification layers are added to the network for the three DAP categories independently. Same hyper-parameters were employed as in the previous set of experiments. Each network is then evaluated using 10-fold cross validation. The summary of the results of the three networks is presented in Table 3. It can be observed that

Table 2. Classification results achieved by different onvNets Pre-Trained on ImageNet

Model	Gross Details	Attachments	Head Details
VGG16	60.3 ± 7.5	89.4 ± 12.9	66.6 ± 5.9
ResNet101	**74.6 ± 6.8**	**92.1 ± 10.6**	**73.7 ± 12.5**
InceptionV3	65.9 ± 5.3	83.1 ± 10.5	49.2 ± 11.6
InceptionResNetV2	67.2 ± 6.5	89.1 ± 11.7	73.3 ± 11.8

ResNet101 model when fine-tuned on SketchNet outperforms in all the categories as compared to baseline. A mean accuracy of 81.1% is achieved by Gross Details network, 98.2% by Attachments network and 78.3% by Head Details network. These results validate our first hypothesis that two-step fine tuning with a semantically similar mediator dataset will enhance classification performance in all scenarios.

Table 3. Comparison of Classification Results Achieved by Pre-trained ResNet101 Architecture With and Without Two-Step Fine-tuning on SketchNet

Model	Gross Details	Attachments	Head Details
ResNet101 (ImageNet Only)	74.6 ± 6.8	92.1 ± 10.6	73.7 ± 12.5
ResNet101 (ImageNet + SketchNet)	**81.1 ± 5.3**	**98.2 ± 2.0**	**78.3 ± 9.3**

Our second set of experiments was performed to ascertain that a single network may not be able to perform well as compared to multiple networks. For this purpose, we employed the ResNet101 model fine-tuned on SketchNet. However, instead of dividing classes into three main scoring categories, we trained the network on all 9 classes. Figure 4 indicates the performance of the two approaches i.e. single network for all classes and three networks for three main categories of classes. As hypothesized, it is clearly evident that the independent network approach outperforms the single network approach. To show the distribution of 10-fold cross validation accuracies of the proposed model, we present the data in the form of a box and whisker plot as illustrated in Fig. 5. It can be observed that the accuracy distribution in all of the three categories is close to their mean value.

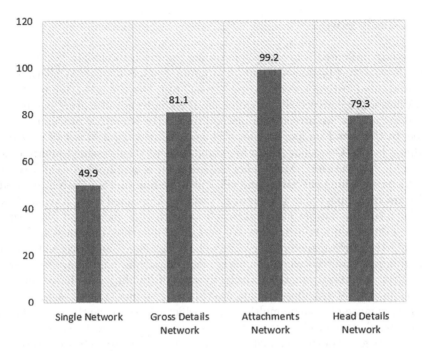

Fig. 4. Performance comparison of Single network and Multiple network approach

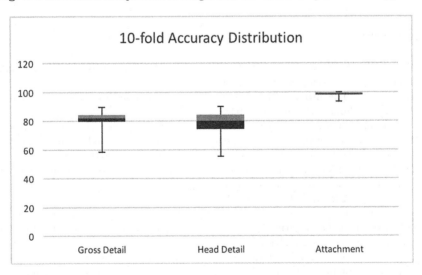

Fig. 5. Box and whisker plot representation of the results of our proposed approach

5 Conclusion

This study presented a two-step fine-tuning approach to classify DAP drawing samples according to clinical standard proposed by Goodenough. While the existing literature mostly focused on copy or recall test, this is a first attempt towards analyzing a projective task. To overcome the challenges of feature engineering and data paucity, we presented a method to effectively employ pre-trained ConvNets to this small sample domain problem. We selected a subset of categories outlined by the Goodenough scoring standard and then designed multiple networks for each category. Multi-label classification is employed in each category. To enhance learning effectively and efficiently, we fine-tuned a ResNet101 model pre-trained on ImageNet. The network was first fine-tuned using a semantically similar mediator dataset called SketchNet and then in the second phase tuned on our customized target dataset. We performed extensive empirical analysis to evaluate the performance of our approach. A significant improvement in classification was observed when compared with employing only ImageNet trained ConvNets. We also evaluated the effectiveness of dividing sub-classes into categories instead of training a single multi-label classifier. It was observed that the performance of a single multi-label classifier was significantly lower (49.9%) as compared to multiple networks. In the future, we intend to include all ten scoring categories of Goodenough standard. We also intend to test our proposed approach on other drawing-based tests like BGT.

References

1. Bender, L.: A visual motor gestalt test and its clinical use. Research Monographs, American Orthopsychiatric Association (1938)
2. Buck, J.N.: The htp technique; a qualitative and quantitative scoring manual. Journal of Clinical Psychology (1948)
3. Chindaro, S., Guest, R., Fairhurst, M., Potter, J.: Assessing visuo-spatial neglect through feature selection from shape drawing performance and sequence analysis. Int. J. Pattern Recogn. Artif. Intell. 18(07), 1253–1266 (2004)
4. De Waal, E., Pienaar, A.E., Coetzee, D.: Influence of different visual perceptual constructs on academic achievement among learners in the nw-child study. Percept. Motor Skills 125(5), 966–988 (2018)
5. Diaz, M., Ferrer, M.A., Impedovo, D., Pirlo, G., Vessio, G.: Dynamically enhanced static handwriting representation for parkinson's disease detection. Pattern Recogn. Lett. 128, 204–210 (2019)
6. Drotár, P., Mekyska, J., Rektorová, I., Masarová, L., Smékal, Z., Faundez-Zanuy, M.: Evaluation of handwriting kinematics and pressure for differential diagnosis of parkinson's disease. Artif. Intell. Med. 67, 39–46 (2016)
7. Eitz, M., Hays, J., Alexa, M.: How do humans sketch objects? ACM Trans. Graph. (TOG) 31(4), 1–10 (2012)
8. Eitz, M., Hildebrand, K., Boubekeur, T., Alexa, M.: Sketch-based image retrieval: benchmark and bag-of-features descriptors. IEEE Trans. Visual. Comput. Graph. 17(11), 1624–1636 (2011)

9. El Shafie, A.M., El Lahony, D.M., Omar, Z.A., El Sayed, S.B., et al.: Screening the intelligence of primary school children using 'draw a person' test. Menoufia Med. J. **31**(3), 994 (2018)
10. Fairhurst, M.C., Linnell, T., Glenat, S., Guest, R., Heutte, L., Paquet, T.: Developing a generic approach to online automated analysis of writing and drawing tests in clinical patient profiling. Behav. Res. Methods **40**(1), 290–303 (2008)
11. Farokhi, M., Hashemi, M.: The analysis of children's drawings: social, emotional, physical, and psychological aspects. Procedia-Soc. Behav. Sci. **30**, 2219–2224 (2011)
12. Gazda, M., Hireš, M., Drotár, P.: Multiple-fine-tuned convolutional neural networks for parkinson's disease diagnosis from offline handwriting. IEEE Transactions on Systems, Man, and Cybernetics: Systems (2021)
13. Goodenough, F.L.: Measurement of intelligence by drawings (1926)
14. Graves, A., Liwicki, M., Fernández, S., Bertolami, R., Bunke, H., Schmidhuber, J.: A novel connectionist system for unconstrained handwriting recognition. IEEE Trans. Pattern Anal. Mach. Intell. **31**(5), 855–868 (2008)
15. Harbi, Z., Hicks, Y., Setchi, R.: Clock drawing test interpretation system. Procedia Comput. Sci. **112**, 1641–1650 (2017)
16. He, K., Zhang, X., Ren, S., Sun, J.: Deep residual learning for image recognition. In: Proceedings of the IEEE Conference on Computer Vision and Pattern Recognition, pp. 770–778 (2016)
17. Khalid, P.I., Yunus, J., Adnan, R., Harun, M., Sudirman, R., Mahmood, N.H.: The use of graphic rules in grade one to help identify children at risk of handwriting difficulties. Res. Dev. Disabil. **31**(6), 1685–1693 (2010)
18. Kornmeier, J., Bach, M.: The necker cube–an ambiguous figure disambiguated in early visual processing. Vis. Res. **45**(8), 955–960 (2005)
19. Krizhevsky, A., Sutskever, I., Hinton, G.E.: Imagenet classification with deep convolutional neural networks. Commun. ACM **60**(6), 84–90 (2017)
20. Larner, A..J.. (ed.): Cognitive Screening Instruments. Springer, Cham (2017). https://doi.org/10.1007/978-3-319-44775-9
21. Moetesum, M., Aslam, T., Saeed, H., Siddiqi, I., Masroor, U.: Sketch-based facial expression recognition for human figure drawing psychological test. In: 2017 International Conference on Frontiers of Information Technology (FIT), pp. 258–263. IEEE (2017)
22. Moetesum, M., Siddiqi, I., Ehsan, S., Vincent, N.: Deformation modeling and classification using deep convolutional neural networks for computerized analysis of neuropsychological drawings. Neural Comput. Appl. **32**(16), 12909–12933 (2020). https://doi.org/10.1007/s00521-020-04735-8
23. Moetesum, M., Siddiqi, I., Masroor, U., Djeddi, C.: Automated scoring of bender gestalt test using image analysis techniques. In: 2015 13th International Conference on Document Analysis and Recognition (ICDAR), pp. 666–670. IEEE (2015)
24. Moetesum, M., Siddiqi, I., Vincent, N.: Deformation classification of drawings for assessment of visual-motor perceptual maturity. In: 2019 International Conference on Document Analysis and Recognition (ICDAR), pp. 941–946. IEEE (2019)
25. Moetesum, M., Siddiqi, I., Vincent, N., Cloppet, F.: Assessing visual attributes of handwriting for prediction of neurological disorders–a case study on parkinson's disease. Pattern Recogn. Lett. **121**, 19–27 (2019)
26. Naseer, A., Rani, M., Naz, S., Razzak, M.I., Imran, M., Xu, G.: Refining parkinson's neurological disorder identification through deep transfer learning. Neural Comput. Appl. **32**(3), 839–854 (2020)

27. Nazar, H.B., et al.: Classification of graphomotor impressions using convolutional neural networks: an application to automated neuro-psychological screening tests. In: 2017 14th IAPR International Conference on Document Analysis and Recognition (ICDAR), vol. 1, pp. 432–437. IEEE (2017)
28. Oquab, M., Bottou, L., Laptev, I., Sivic, J.: Learning and transferring mid-level image representations using convolutional neural networks. In: Proceedings of the IEEE Conference on Computer Vision and Pattern Recognition, pp. 1717–1724 (2014)
29. Pereira, C.R., Weber, S.A., Hook, C., Rosa, G.H., Papa, J.P.: Deep learning-aided parkinson's disease diagnosis from handwritten dynamics. In: 2016 29th SIBGRAPI Conference on Graphics, Patterns and Images (SIBGRAPI), pp. 340–346. IEEE (2016)
30. Pratt, H.D., Greydanus, D.E.: Intellectual disability (mental retardation) in children and adolescents. Primary Care Clin. Office Pract. 34(2), 375–386 (2007)
31. Pullman, S.L.: Spiral analysis: a new technique for measuring tremor with a digitizing tablet. Mov. Disord. 13(S3), 85–89 (1998)
32. Rémi, C., Frélicot, C., Courtellemont, P.: Automatic analysis of the structuring of children's drawings and writing. Pattern Recogn. 35(5), 1059–1069 (2002)
33. Ren, S., He, K., Girshick, R., Sun, J.: Faster R-CNN: Towards real-time object detection with region proposal networks. arXiv preprint arXiv:1506.01497 (2015)
34. Shin, M.S., Park, S.Y., Park, S.R., Seol, S.H., Kwon, J.S.: Clinical and empirical applications of the rey-osterrieth complex figure test. Nat. Protoc. 1(2), 892 (2006)
35. Shulman, K.I., Shedletsky, R., Silver, I.L.: The challenge of time: clock-drawing and cognitive function in the elderly. Int. J. Geriatr. Psychiatry 1(2), 135–140 (1986)
36. Simonyan, K., Zisserman, A.: Very deep convolutional networks for large-scale image recognition. arXiv preprint arXiv:1409.1556 (2014)
37. Smith, A.D.: On the use of drawing tasks in neuropsychological assessment. Neuropsychology 23(2), 231 (2009)
38. Smith, S.L., Hiller, D.L.: Image analysis of neuropsychological test responses. In: Medical Imaging 1996: Image Processing, vol. 2710, pp. 904–915. International Society for Optics and Photonics (1996)
39. Smith, S.L., Lones, M.A.: Implicit context representation cartesian genetic programming for the assessment of visuo-spatial ability. In: 2009 IEEE Congress on Evolutionary Computation, pp. 1072–1078. IEEE (2009)
40. Szegedy, C., Vanhoucke, V., Ioffe, S., Shlens, J., Wojna, Z.: Rethinking the inception architecture for computer vision. In: Proceedings of the IEEE Conference on Computer Vision and Pattern Recognition, pp. 2818–2826 (2016)
41. Tabatabaey-Mashadi, N., Sudirman, R., Guest, R.M., Khalid, P.I.: Analyses of pupils' polygonal shape drawing strategy with respect to handwriting performance. Pattern Anal. Appl. 18(3), 571–586 (2015)
42. Tabatabaey, N., Sudirman, R., Khalid, P.I., et al.: An evaluation of children's structural drawing strategies. Jurnal Teknologi, vol. 61, no. 2 (2013)
43. Yosinski, J., Clune, J., Bengio, Y., Lipson, H.: How transferable are features in deep neural networks? arXiv preprint arXiv:1411.1792 (2014)

DCINN: Deformable Convolution and Inception Based Neural Network for Tattoo Text Detection Through Skin Region

Tamal Chowdhury[1], Palaiahnakote Shivakumara[2], Umapada Pal[1], Tong Lu[3], Ramachandra Raghavendra[4(✉)], and Sukalpa Chanda[5]

[1] Computer Vision and Pattern Recognition Unit, Indian Statistical Institute, Kolkata, India
umapada@isical.ac.in

[2] Faculty of Computer Science and Information Technology, University of Malaya, Kuala Lumpur, Malaysia
shiva@um.edu.my

[3] National Key Lab for Novel Software Technology, Nanjing University, Nanjing, China
lutong@nju.edu.cn

[4] Faculty of Information Technology and Electrical Engineering, IIK, NTNU, Trondheim, Norway
raghavendra.ramachandra@ntnu.no

[5] Department of Information Technology, Østfold University College, Halden, Norway
sukalpa@ieee.org

Abstract. Identifying Tattoo is an integral part of forensic investigation and crime identification. Tattoo text detection is challenging because of its freestyle handwriting over the skin region with a variety of decorations. This paper introduces Deformable Convolution and Inception based Neural Network (DCINN) for detecting tattoo text. Before tattoo text detection, the proposed approach detects skin regions in the tattoo images based on color models. This results in skin regions containing Tattoo text, which reduces the background complexity of the tattoo text detection problem. For detecting tattoo text in the skin regions, we explore a DCINN, which generates binary maps from the final feature maps using differential binarization technique. Finally, polygonal bounding boxes are generated from the binary map for any orientation of text. Experiments on our Tattoo-Text dataset and two standard datasets of natural scene text images, namely, Total-Text, CTW1500 show that the proposed method is effective in detecting Tattoo text as well as natural scene text in the images. Furthermore, the proposed method outperforms the existing text detection methods in several criteria.

Keywords: Deep learning model · Deformable convolutional neural networks · Text detection · Crime identification · Tattoo identification · Tattoo text detection

1 Introduction

Text detection and recognition in natural scene images and video is gaining a huge popularity because of several real world application, such as surveillance, monitoring

© Springer Nature Switzerland AG 2021
J. Lladós et al. (Eds.): ICDAR 2021, LNCS 12822, pp. 335–350, 2021.
https://doi.org/10.1007/978-3-030-86331-9_22

and forensic applications [1–4]. In the same way, tattoo detection and identification is also play a vital role in person identification [5], criminal/terrorist identification [6], studying personality traits [7] and psychology of person [8]. This is because Tattoo is considered as soft biometric features for person identification. It is noted that 36 percent of Americans of 18 to 29 years old have at least once tattoo [5]. Therefore, there is a demand for tattoo detection and recognition for person identification. This is because it provides discriminative information that is complimentary to other biometric features, such as finger print, face. It can also be used for association, group affiliation, memberships, gangs, criminals and victims. However, the methods developed so far for tattoo detection and recognition focus on tattoo images and ignored the importance of text information in the tattoo images. This is because most of the methods target person identification but not personality traits, behavior and psychology of the person. We believe that tattoo is kind of expression, message and it reflects mind of the persons [7]. Therefore, in order to understand the meaning of tattoo, it is necessary to detect and recognize tattoo text. This observation motivated us to propose a method for tattoo text detection in this work.

Tattoo text image   Natural scene images

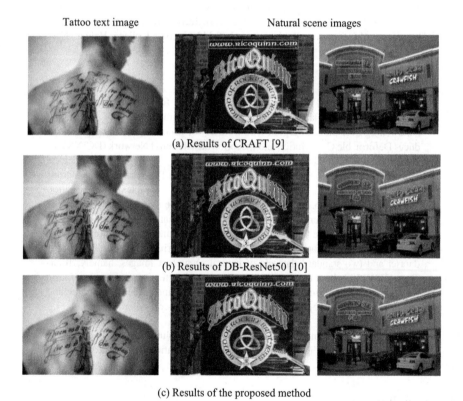

(a) Results of CRAFT [9]

(b) Results of DB-ResNet50 [10]

(c) Results of the proposed method

Fig. 1. Examples of tattoo text and natural scene text detection of the proposed and state-of-the-art methods.

When we look at sample images show in Fig. 1, tattoo text detection is complex compared to natural scene text detection because of free style handwriting on deformable skin region. Due to deformable skin, tattoo text quality degrades. In addition, dense tattoo text with decoration and arbitrary orientation make the tattoo detection complex and interesting. Therefore, the existing natural scene text detection methods [9, 10], which use deep learning model may not work well for such images as shown in Fig. 1(a) and 1(b). The main reason is that tattoo text does not exhibit regular pattern, such as uniform color pixels, uniform spacing between character and words, finite shape of the characters and predictable background. On the other hand, the proposed method detects tattoo text as well as natural scene text properly compared to the existing methods. This is because the proposed method reduces the complexity of the problem by detecting deformable skin region which contains tattoo text. For the next step of tattoo text detection, the proposed method uses skin region as input but not the whole image. For tattoo text detection, we explore deep learning model as it has ability to addresses challenges of tattoo text detection. The main contribution of the proposed work is to explore the combination of skin region detection and deformable convolutional and inception based neural network for tattoo text detection. To our knowledge this is the first work of its kind.

The structure of the paper is follows. We present review of text detection in natural scene images in Sect. 2. The proposed skin region detection using color based model and tattoo text detection using DCINN are presented in Sect. 3. In Sect. 4, experimental results of the proposed and existing methods, and discussion on different datasets are presented. Conclusion and future work is given in Sect. 5.

2 Related Work

Since tattoo text detection is closely related to text detection in natural scene images and there is no work on tattoo text detection, the state-of-the-art methods on text detection in natural scene images are reviewed here.

Wang et al. [11] proposed a Progressive Scale Expansion Network (PSENet) for text detection in natural scene images. The approach involves in segmentation-based detectors for predicting multiple text instances. Ma et al. [12] proposed a rotation region proposal network for text detection in natural scene images. It considers rotation of region interest as pooling layers for the classifier. Angular or directional information is good for text of many character, else direction may yield incorrect results. Long et al. [4] proposed a flexible architecture for text detection in natural scene images. It considers text instances as a sequence of ordered, and finds symmetric axes with radius and orientation information. When the image contains dense text with non-uniforms spacing between the text lines, the method may perform well.

Feng et al. [13] proposed a method for arbitrarily-oriented text spotting in natural scene images. The approach introduces a RoISlide operator, which connects series of quadrangular texts. The method follows the idea of Long et al.'s method [5] for extracting instances of quadrangles. Since it depends on directions of texts, it may not work well for tattoo text in the images. Baek et al. [9] proposed a character awareness-based method for text detection in natural scene images. It finds the relationship between characters for detecting texts in images. Liao et al. [14] proposed a single shot oriented scene text detector for scene text detection in natural scene images. The method focusses on single network for achieving accuracy as well as efficiency. Raghunandan et al. [15] proposed a method for text detection in natural scene, video and Born digital images. The method uses bit plane and convex deficiency concepts for detecting text candidates. Xu et al. [16] proposed a method for irregular scene text detection in natural scene images. It finds the relationship between current text and its neighbor boundary to fix the bounding box of any orientation. However, the method is not tested on the image which contains dense text lines and text with decorative symbols. Cai et al. [17] proposed an Inside-to-Outside Supervision Network (IOS-Net) for text detection in natural scene images. It designs a hierarchical supervision module to capture texts of different aspect ratios, and then multiple scale supervision for a stacked hierarchical supervision module. The Mask-R-CNN has been studied for text detection in natural scene images as it is popular for improving instance segmentation by predicting accurate masks [18, 19]. Lyu et al. [20] proposed an end-to-end trainable neural network for text spotting in natural scene images. This method generates shape masks of objects and detects text by segmenting instance regions. This approach may not be effective for tattoo text images.

Roy et al. [3] proposed a method for text detection from multi-view images of natural scenes. The method uses Delaunay triangulation for extracting features from estimating similarity and dissimilarity between components in different views. Wang et al. [21] proposed a quadrilateral scene text detector for text detection in natural scene images. The approach uses two stages for achieving better results. However, for arbitrary oriented and irregular texts, the quadrilateral proposal network may not be effective. Liu et al. [22] proposed a method for text spotting in natural scene images based on an adaptive Bezier curve network. Wang et al. [23] proposed text detection from multi-view of natural scene images. It finds correspondence between multi-views for achieving results. To find correspondence, the method uses similarity and dissimilarity estimation between text components in multi-views. The method requires multiple views.

In the above discussions on the methods for text detection in natural scene images, it is noted that most methods use direction and aspect ratio of character information for designing deep learning architectures to address the problem of arbitrary orientation. It is observed that none of the methods considered tattoo images for detection. In case of deformable region contained text, one can expect words of short length compared to text in natural scene images, partial occlusion due to deformation effect, body movements and free style writing due to handwriting on skin region. As a result, characters may not preserve actual structures. Therefore, natural scene text detection methods may not be good enough to addresses the challenges of tattoo text images.

To reduce the complexity of the problems in sports images, some methods are proposed to use multimodal concepts like face, skin, torso, and human body parts information for achieving better results for sports images. Ami et al. [24] proposed a method for text detection in marathon images using face information. The approach first detects face and then torso which contains bib numbers for detection. As long as face is visible in the images, the method works well. Otherwise, it does not. To overcome this limitation, Shivakumara et al. [25] proposed torso segmentation without face information for bib number detection in marathon images. However, the above methods only detect texts in torso regions but not from the other parts of human bodies. To improve bib number and text detection performances for sports images, Nag et al. [26] proposed a method to detect human body parts rather than relining only on torso regions. However, the performance of the method depends on the success of human body parts detection. Similarly, Kamlesh et al. [27] used text information in marathon images for person re-identification. The method recognizes text in the marathon images for person re-identification. It is observed from experimental results that the method works well for high quality images but not for poor-quality ones. In summary, the methods that use multi-modal concept addressed a few issues of text detection in sports images and the performances of the methods depend on pre-processing steps like face, torso and human body parts detection. The methods ignore deformable information for detecting text in the images. None of the methods use tattoo images for text detection.

In this work, we propose to explore deformable convolution neural network for tattoo text detection through skin region. In order reduce the complexity of the tattoo text detection problem and we use color based model for skin region detection and it is considered as context information for tattoo text detection. This makes sense because tattoo text usually written on only skin and it is common for different type tattoo images. Inspired by the success of deep learning model for natural scene text detection, we explore the combination of deformable convolution neural network and differential binarization network [10] for tattoo text detection by considering skin region as input. The way the proposed method combines color based model of skin region detection, deformable convolutional neural network for generating binary map using differential binarization network is the main contribution for tattoo text detection compared to the state-of-the-art methods.

3   Proposed Approach

The proposed method consists of two stages, namely, skin region detection step and deep learning model for text detection from the extracted skin region. Since the proposed work aims to detect Tattoo text in the images, one can expect tattoo text embedded over skin region, therefore we propose to explore skin region separation for detecting tattoo text region from the input image. It is noted from the method [28] that the pixels of skin region share similar color values or properties. This intuition motivated us to propose color based heuristic model, which uses YCbCr and HSV color spaces of the input image for skin region separation from the images. The skin region detection helps to reduce complexity of problem of tattoo text detection because most of the background information is removed from the input image and hence it helps the deep learning-based feature extractor to detect tattoo text accurately regardless of the challenges posed by the tattoo images mentioned above. This is an advantage of skin region detection step.

Inspired by the promising results by the deep learning models for text detection in natural scene images in the literature, we propose deformable convolution and inception module based deep learning model for feature extraction from the skin region, which generates probability map for each input image. As noted from the tattoo image shown in Fig. 1, fixing exact bounding box is not so easy due to cursive nature of the text containing decorative characters and dense text line. To overcome this problem, we have explored the Differential Binarization technique as proposed by Liao et al. [10] that not only enables adaptive thresholding but also helps differentiating text instances from complex background. The overall block diagram of the proposed method can be seen in Fig. 2, where path-1 follows the steps of tattoo text detection while the path-2 follows scene text detection from natural scene images. Since natural scene text images do not necessarily contain skin region and in order to show the proposed method can be used for natural scene text detection also, one can also follow path-2 of this work.

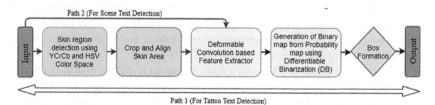

Fig. 2. Overall block diagram of the proposed method. Path 1 is followed for tattoo text detection whereas Path 2 is followed for text detection from natural scene image.

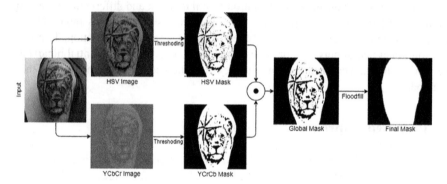

Fig. 3. Steps of the proposed skin region detection using HSV and YCbCr color spaces

3.1 Skin Region Detection Using YCrCb and HSV Color Space

YCbCr and HSV color spaces are used for skin region detection as shown in Fig. 3, where one can notice that each color space represents skin region in a different way. More interesting is that from each color space, we can see clear separation from skin region and non-skin region in the image. This observation motivated us to use color values for skin pixels' classification. The method uses standard intensity ranges as defined in Eqs. (1) and (2) for classifying skin pixels from non-pixels. The binary masks obtained from these two different color spaces are combined using elementwise dot product for a more accurate mask. However, due to the presence of tattoo text in the skin region, the skin region detection step may miss pixels in tattoo text region, which results skin region with the hole as shown by the global mask in Fig. 3. To fill the hole in the region, we use simple flood fill algorithm to fill the region, where it is noted that skin region is considered as one connected component or region of interest. The skin regions are then cropped out from the original image which is further used for feature extraction using the deep learning network. The cropped areas are aligned and resized to a fixed dimension before sending it to the feature extractor network.

$$0/15/0 \leq H/S/V \leq 17/170/255 \tag{1}$$

$$0/85/135 \leq Y/Cb/Cr \leq 255/135/180 \tag{2}$$

3.2 Deep Model for Tattoo Text Detection from Skin Region

In this section, the proposed Deformable Convolution and Inception Module-based feature Extractor Network (DCINN) for scene/tattoo text is described. The architecture can be seen in Fig. 4. Further, inspired by the work of Liao et al. [10] differentiable binarization technique is used for generating the approximate segmentation map. The description and more details about the architecture are presented below.

Downsampling Path: The proposed deep learning model is a symmetrical Encoder-Decoder (E-D) network which is being used for feature extraction. The E-D network consists of one downsampling path followed by an upsampling path. The downsampling path contains 4 feature extractor block each of which is further a combination of two inception modules [29] followed by a Max-pooling layer. The inception modules consist of four parallel paths of convolutional layers each with different field of view (kernel size) as shown in Fig. 5. Inception modules of the first two block utilizes traditional convolutional layers whereas the inception modules of the latter two blocks utilizes deformable convolutional layers [30]. 32, 64, 128 and 256 number of feature maps are generated in each successive block where each path of the inception module generates exactly $1/4th$ number of the total output feature maps. The Max-pooling layer in each block reduces the dimension of the input by half. Hence, the output of the downsampling path is ($256 \times 8 \times 8$). We call this as the feature vector. This path encodes the semantic information of the image. **Upsampling Path:** The feature vector obtained from the downsampling path is passed to the upsampling path or the decoder that captures the spatial information and also increases the spatial resolution. Similar to the downsampling path the upsampling path also consists of 4 upsampling blocks each of which is a combination of two inception modules followed by a transpose convolution layer to increase the feature dimension. Here also, the inception modules of the initial two blocks are made of simple convolutional layers whereas for the latter two blocks we have used deformable convolutional layers. At each level of the upsampling path the output is further concatenated with the feature maps of the corresponding downsampling path by means of a transition layer for maintaining the coherence between these two paths and minimizing the information loss as shown in the figure. The transition layer is a simple convolution layer with specified no of feature maps. The resolution of the input gets doubled after each upsampling block giving a final feature map of dimensions ($64 \times 64 \times 32$).

The outputs the last two blocks of the decoder are upsampled to output feature map shape and passed through (1×1) convolution layers individually to obtain feature maps of shape $(128 \times 128 \times 1)$. They are concatenated with the output of the last layer as shown and passed through another (1×1) convolution layer to get the final feature map. Inspired by the work [8], this final feature map is used to generate a segmentation probability map which is further used for label generation. More details about these individual modules can be found from [10].

Loss Function and Optimizer: The overall loss of the network L is defined as the weighted sum of two losses L_p and L_b as:

$$L = \alpha L_p + \beta L_b \tag{3}$$

Where L_p and L_b are losses for probability map and binary map respectively. Both of them are formulated as focal loss which is defined as:

$$FL(p) = -(1 - p)^\gamma \log(p) \tag{4}$$

From the scales of these losses and their weightage both α and β are chosen to be 0.5 and 0.5 respectively.

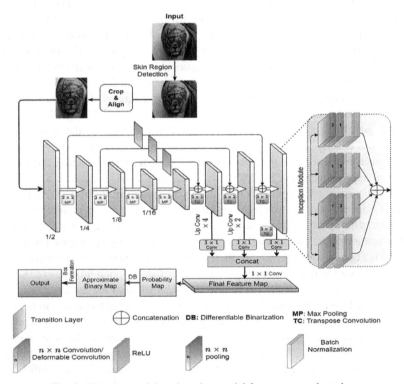

Fig. 4. The proposed deep learning model for tattoo text detection

4 Experimental Results

As per our knowledge, this is the first work on tattoo text detection. Therefore, there is no standard dataset available in the literature for evaluation. We create our own dataset for experimentation in this work. **Tattoo-Text**: The dataset consists of 500 RGB images of Tattoos text embedded over skin region in the images. All the images are collected from internet and other resources. The created dataset is more challenging due to several adverse factors, such as freehand writing style, unconventional orientation, dense text lines with decorative characters and symbols and complex background as shown sample images in Fig. 5. In Fig. 5, we can see the complexity of tattoo text detection and the images are affected by several adverse factors. Thus, the created dataset is fair enough for evaluating the proposed method for tattoo text detection. In the same way, to show objectiveness of the proposed method, we also use standard datasets of natural scene text detection for testing the proposed method in this work. **CTW1500 Dataset** [31]: It is a popular dataset for scene text detection dataset that contains a total of 1500 images of curved text. The dataset is divided into training and test sets that consist of 1000 and 500 images respectively where all the text instances are annotated at a text line level. In particular, it provides 10,751 cropped text instances, with 3,530 being curved texts. **Total-Text** [32]: Total-Text is another popular benchmark dataset for scene text detection with 1255 training images and 300 testing images. The images contain English language text instances of different orientations, such as horizontal, multi-oriented and curved. The annotations are provided at a word level.

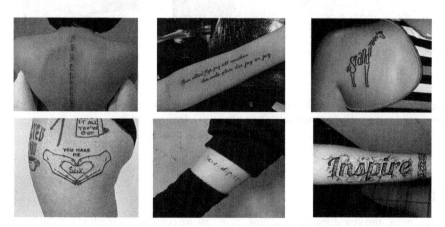

Fig. 5. Sample tattoo images of our tattoo-text dataset

For evaluating effectiveness of the proposed method, we run the following existing natural scene text detection methods for comparing with the proposed method. The CRAFT [9] approach, which is Character Region Awareness for Text Detection (CRAFT), DB-Net [10], which is Differential Binarization Network. The reason for using the above methods for comparative study is that the methods are state-of-the-art methods and addresses several challenges of natural scene text detection, which are similar to tattoo text detection. For evaluating the performance of the methods, we use the standard measures, namely, Recall (R), Precision (P) and F-measures (F) for all the experiments.

Implementation Details: The proposed method is optimized using Adam algorithm. We trained our model on both CTW1500 and Total-Text Dataset and evaluated the performance on their respective test sets. The skin region separation module is omitted here and the images are directly passed to the E-D network. Initially the model is trained on the SynthText dataset for 50 k epochs with a constant learning rate of 10^{-3}. This dataset is used only for pretraining our model to learn the global feature of texts. Images are resized to (128×128) and a batch size of 64 is used during this pre-training phase. The trained model weights are further used for fine tuning on the CTW1500 and Total-Text Dataset. 10% of the total training data is used as validation set with dynamic learning rate and early stopping criterion for better regularization.

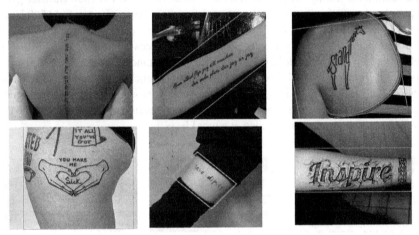

Fig. 6. Skin region detection of the proposed method for images shown in Fig. 5. The skin regions are marked using green rectangular bounding boxes (Color figure online)

4.1 Ablation Study

In this work, the key steps for achieving the result are color based method for skin region detection and Deformable Convolution and Inception based Neural Network (DCINN) for text detection. To show the effectiveness of the skin region detection and DCINN, we conduct experiments for calculating measures by passing the input images directly to DCINN omitting the skin region separation step. The results are reported in Table 1, where it is noted that the results are promising but not higher than the proposed method (with skin detection module). This shows that the skin region detection is effective and it plays an important role for addressing the challenges of tattoo text detection in this work. Since this step helps in reducing background complexity by removing almost all non-text information, the precision of the proposed method improves for tattoo text detection.

Sample qualitative results of skin region detection are shown in Fig. 6, where one can see for all the images with different complexities, the proposed skin region detection works well. The results of the proposed method with skin region reported in Table 1 show that the DCINN is effective in achieving better results for tattoo text detection. Therefore, we can conclude that the combination of skin region detection and DCINN is the best for tattoo text detection.

Table 1. Assessing the effectiveness of skin region detection step for tattoo text detection on our tattoo dataset

Steps	P	R	F
Proposed DCINN without skin region detection	80.3	81.8	80.1
Proposed DCINN with skin region detection (Proposed method)	82.7	82.2	82.3

Fig. 7. Sample results of our proposed method on tattoo image dataset

4.2 Experiments for Tattoo Text Detection

Qualitative results of the proposed method for tattoo text detection are shown in Fig. 7, where we can see the proposed method detects tattoo text accurately for the images of different levels of complexities. This shows that the proposed method is capable of tackling the challenges of tattoo text detection. Quantitative results of the proposed and existing methods are reported in Table 2, where it is noted that the performance of the proposed method is better than existing methods in terms of recall, precision and F-measures. It is also observed that precision is higher than recall and F-measure. Therefore, the proposed method is reliable compared to the existing methods. The reason for the relatively lower results of the existing methods is that the methods are developed and tuned for text detection in natural scene images but not for tattoo images. On the other hand, since the proposed method involves skin region detection to reduce background complexity of the tattoo images, DCINN for feature extraction and fixing proper bounding boxes for tattoo text lines, the proposed method is has ability to handle the challenges of tattoo text detection.

Table 2. Performance of the proposed and existing methods for tattoo text detection using our dataset

Methods	P	R	F
CRAFT [9]	67.8	69.4	68.6
DB-ResNet-50 [10]	79.8	81.1	80.7
Proposed DCINN (with skin detector)	**82.7**	**82.2**	**82..3**

(a) Total-Text dataset

(b) CTW1500 dataset

Fig. 8. Example for text detection of the proposed method on different benchmark dataset natural scene images

4.3   Experiments on Natural Scene Text Dataset

To show the robustness of the proposed method for text detection in natural scene images, two benchmark datasets, namely, CTW1500 and Total-Text, which provide curved text line images fort text detection. Note: for text detection in natural scene images, the input images are fed to DCINN network directly without skin region detection step.

This is because the natural scene images do not contain skin region. Therefore, rather than supplying the results of skin region detection, we feed input images directly to DCINN for text detection in this work. Qualitative results of the proposed method for the above two datasets are shown in Fig. 8, where we can see the proposed method detect text accurately for different cases and complexities. This shows that the proposed method is robust and has generalization ability for detecting text in images of different applications in spite of the proposed method is developed for tattoo text detection. This is the advantage of the proposed method and it makes difference compared to existing work. Quantitative results of the proposed and existing methods for the same two datasets are reported in Tables 3 and 4. It is observed from Table 3 that the proposed method achieves the best recall and F-measure compared to the existing methods. However, the CRAFT is better than the proposed and other existing methods in terms of recall. This method does not misses text instances for text detection as it works based on character region awareness based network. But the precision and F-measures are lower than the proposed method.

In the same way, the results of the proposed and existing methods reported in Table 4 show that the proposed method outperforms the existing methods in terms of recall, precision and F-measure. The main reason for poor results of the existing methods compared to the proposed method is that confined scope, constraints on learning, parameter tuning and the number of samples chosen for learning. On the other hand, the proposed combination of skin region detection and DCINN make difference compared to the existing methods to achieve better results.

Table 3. Performance of the proposed and existing method for text detection in natural scene image on CTW1500 dataset

Method	P	R	F
TextDragon [13]	84.5	82.8	83.6
TextField [16]	83.0	79.8	81.4
PSENet-1s [11]	84.8	79.7	82.2
CRAFT [9]	**86.0**	81.1	83.5
DB-ResNet-50 [10]	85.9	83.0	84.4
Proposed	85.8	**83.4**	**84.1**

As shown in Fig. 9, in few occasions the proposed method fails to detect tattoo text. This happened due to the failure of the skin region detection step to tackle low contrast images.. When skin region detection step does not work well, it affects for tattoo text detection in the images. Therefore, there is a scope for the improvement to extend the proposed method further.

Table 4. Performance of the proposed and existing method for text detection in natural scene image on Total-Text dataset

Method	P	R	F
TextSnake [4]	82.7	74.5	78.4
TextDragon [13]	85.6	75.7	80.3
TextField [16]	81.2	79.9	80.6
PSENet-1s [11]	84.0	77.9	80.8
CRAFT [9]	87.6	79.9	83.6
DB-ResNet-50 [10]	86.5	84.9	85.7
Proposed	**87.9**	**85.2**	**86.6**

Fig. 9. Examples of some failure cases of the propose model on our tattoo dataset

5   Conclusion and Future Work

We have presented a novel idea for detecting tattoo text in the images by exploring the combination of skin region detection and Deformable Convolutional and Inception based Neural Network (DCINN) for tattoo text detection in the images. The main challenge of tattoo text detection is that the tattoo text is cursive in nature containing decorative characters and symbols. In addition, dense text lines with non-uniform spacing makes the problem more challenging. To reduce the background complexity of the problem, we have proposed color based model for skin region detection, which results in skin region containing tattoo text. Further, the proposed method explores DCINN for tattoo text detection from skin region. Experimental results on our dataset of tattoo text and two benchmark datasets of natural scene text show that the proposed method outperforms the existing methods. The results on both tattoo text dataset and natural scene text dataset reveal that the proposed method is robust to tattoo text and natural scene text detection. As discussed in the experimental section, when the tattoo text has text as background with different fonts and size, the performance of the proposed method degrades. Therefore, our next target is to extend the proposed idea for addressing this issue to enhance the tattoo text detection performance.

References

1. Nandanwar, L., et al.: Forged text detection in video, scene, and document images. IET Image Process. **14**(17), 4744–4755 (2021)

2. Nag, S., et al.: A new unified method for detecting text from marathon runners and sports players in video (PR-D-19-01078R2). Pattern Recogn. **107**, 107476 (2020)
3. Roy, S., Shivakumara, P., Pal, U., Lu, T., Kumar, G.H.: Delaunay triangulation based text detection from multi-view images of natural scene. Pattern Recogn. Lett. **129**, 92–100 (2020)
4. Long, S., Ruan, J., Zhang, W., He, X., Wu, W., Yao, C.: TextSnake: a flexible representation for detecting text of arbitrary shapes. In: Ferrari, V., Hebert, M., Sminchisescu, C., Weiss, Y. (eds.) ECCV 2018. LNCS, vol. 11206, pp. 19–35. Springer, Cham (2018). https://doi.org/10. 1007/978-3-030-01216-8_2
5. Sun, Z.H., Baumes, J., Tunison, P., Turek, M., Hoogs, A.: Tattoo detection and localization using region based deep learning. In: Proceedings of the ICPR, pp. 3055–3060 (2016)
6. Han, H., Li, J., Jain, A.K., Shan, S., Chen, X.: Tattoo image search at scale: joint detection and compact representation learning. IEEE Trans. Pattern Anal. Mach. Intell. **41**(10), 2333–2348 (2019)
7. Molloy, K., Wagstaff, D.: Effects of gender, self-rated attractiveness, and mate value on perceptions tattoos. Personality Individ. Differ **168**, 110382 (2021)
8. Di, X., Patel, V.M.: Deep tattoo recognition. In: Proceedings of the ICVPRW, pp. 119–126 (2016)
9. Baek, Y., Lee, B., Han, D., Yun, S., Lee, H.: Character region awareness for text detection. In: Proceedings of the CVPR, pp. 9365–9374 (2019)
10. Liao, M., Wan, Z., Yao, C., Chen, K., Bai, X.: Real-time scene text detection with differentiable binarization. In: Proceedings of the AAAI (2020)
11. Wang, W., et al.: Shape robust text detection with progressive scale expansion network. In: Proceedings of the CVPR, pp. 9328–9337 (2019)
12. Ma, J., et al.: Arbitrary-oriented scene text detection via rotation proposals. IEEE Trans. Multimedia **20**, 3111–3122 (2018)
13. Feng, W., He, W., Yin, F., Zhang, X.Y., Liu, C.L.: TextDragon: an end-to-end framework for arbitrary shaped text spotting. In: Proceedings of the ICCV, pp. 9076–9084 (2019)
14. Liao, M., Shi, B., Bai, X.: TextBoxes++: a single-shot oriented scene text detector. IEEE Trans. Image Process. **27**(8), 3676–3690 (2018)
15. Raghunandan, K.S., Shivakumara, P., Roy, S., Kumar, G.H., Pal, U., Lu, T.: Multi-script-oriented text detection and recognition in video/scene/born digital images. IEEE Trans. Circuits Syst. Video Technol. **29**, 1145–1162 (2019)
16. Xu, Y., Wang, Y., Zhou, W., Wang, Y., Yang, Z., Bai, X.: TextField: learning a deep direction field for irregular scene text detection. IEEE Trans. Image Process. **28**, 5566–5579 (2019)
17. Cai, Y., Wang, W., Chen, Y., Ye, Q.: IOS-Net: an inside-to-outside supervision network for scale robust text detection in the wild. Pattern Recogn. **103**, 107304 (2020)
18. He, K., Gkioxari, G., Dollár, P., Girshick, R.: Mask R-CNN. In: ICCV, pp. 2980–2988 (2017)
19. Huang, Z., Zhong, Z., Sun, L., Huo, Q.: Mask R-CNN with pyramid attention network for scene text detection. In: Proceedings of the ICCV, pp. 764–772 (2019)
20. Lyu, P., Liao, M., Yao, C., Wu, W., Bai, X.: Mask TextSpotter: an end-to-end trainable neural network for spotting text with arbitrary shapes. In: Proceedings of the ECCV, pp. 71–78 (2018)
21. Wang, S., Liu, Y., He, Z., Wang, Y., Tang, Z.: A quadrilateral scene text detector with two-stage network architecture. Pattern Recogn. **102**, 107230 (2020)
22. Liu, Y., Chen, H., Shen, C., He, T., Jin, L., Wang, L.: ABCNet: real-time scene text spotting with adaptive Bezier curve network. In: Proceedings of the CVPR (2020)
23. Wang, C., Fu, H., Yang, L., Cao, X.: Text co-detection in multi-view scene. IEEE Trans. Image Process. **29**, 4627–4642 (2020)
24. Ami, I.B., Basha, T., Avidan, S.: Racing bib number recognition. In: Proceedings of the BMCV, pp. 1–12 (2012)

25. Shivakumara, P., Raghavendra, R., Qin, L., Raja, K.B., Lu, T., Pal, U.: A new multi-modal approach to bib number/text detection and recognition in Marathon images. Pattern Recogn. **61**, 479–491 (2017)
26. Nag, S., Ramachandra, R., Shivakumara, P., Pal, U., Lu, T., Kankanhalli, M.: CRNN based jersey number/text recognition in sports and marathon images. ICDAR, pp. 1149–1156 (2019)
27. Kamlesh, Xu, P., Yang, Y., Xu, Y.: Person re-identification with end-to-end scene text recognition. In: Proceedings of the CCCV, pp. 363–374 (2017)
28. Paracchini, M., Marcon, M., Villa, F., Tubaro, S.: Deep skin detection on low resolution grayscale images. Pattern Recogn. Lett. **131**, 322–328 (2020)
29. Szegedy, C., Liu, W., Jia, Y., Sermanet, P., Reed, S.: Going deeper with convolutions. In: Proceedings of the CVPR, pp. 1–9 (2015)
30. Zhu, X., Hu, H., Lin, S., Dai, J.:Deformable convnets v2: more deformable, better results. In: Proceedings of the CVPR 2019, pp. 9300–9308 (2019)
31. Yuliang, L., Lianwen, J., Shuaitao, Z., Sheng, Z.: Detecting curve text in the wild: new dataset and new solution. arXiv:1712.02170 (2017)
32. Chng, C.K., Chan, C.S.: Total-text: a comprehensive dataset for scene text detection and recognition. In: Proceedings of the ICDAR, pp. 935–942 (2017)

Sparse Document Analysis Using Beta-Liouville Naive Bayes with Vocabulary Knowledge

Fatma Najar$^{(\boxtimes)}$ and Nizar Bouguila

Concordia Institute for Information and Systems Engineering (CIISE),
Concordia University, Montreal, QC, Canada
{fatma.najar,nizar.bouguila}@concordia.ca

Abstract. Smoothing the parameters of multinomial distributions is an important concern in statistical inference tasks. In this paper, we present a new smoothing prior for the Multinomial Naive Bayes classifier. Our approach takes advantage of the Beta-Liouville distribution for the estimation of the multinomial parameters. Dealing with sparse documents, we exploit vocabulary knowledge to define two distinct priors over the "observed" and the "unseen" words. We analyze the problem of large-scale and sparse data by enhancing Multinomial Naive Bayes classifier through smoothing the estimation of words with a Beta-scale. Our approach is evaluated on two different challenging applications with sparse and large-scale documents namely: emotion intensity analysis and hate speech detection. Experiments on real-world datasets show the effectiveness of our proposed classifier compared to the related-work methods.

Keywords: Multinomial Naive Bayes · Beta-Liouville distribution · Sparse data · Document analysis · Text recognition

1 Introduction

Multinomial Naive Bayes (MNB), mostly used for document classification, is known to be an effective and successful solution [5,13,18]. The MNB algorithm is based on Bayes' theorem where the probability of document is generated by multinomial distribution. The strength of this approach lies in the inference speed and the principle of classifying new training data incrementally using prior belief. The MNB was largely applied to sentiment analysis [1], spam detection [16], short text representation [27], and mode detection [24]. However, assuming the features to be independent still presents a limitation for its deployment in real-life applications. Given this central problem in text retrieval, different solutions have been proposed to improve the performance of MNB classifier. Some of these solutions are based on weighting heuristics as in [21], words are associated with term frequencies where common words are given less weight and rare words are accorded to an increased term. Other transformations are based on word normalization such as document length normalization to reduce the influence of

© Springer Nature Switzerland AG 2021
J. Lladós et al. (Eds.): ICDAR 2021, LNCS 12822, pp. 351–363, 2021.
https://doi.org/10.1007/978-3-030-86331-9_23

common words [2]. Indeed, the most popular heuristic transformation applied to multinomial model is the term-frequency inverse-document-frequency (TF-IDF) associated with log-normalization for all the documents which makes them have the same length and consequently the same influence on the model parameters. Additional alternatives have been applied to enhance the multinomial such as the complement class modeling [21] which solves the problem of unseen words from becoming zero through smoothing the model parameters within a class. Other different smoothing techniques have been associated to MNB such as Laplace smoothing, Jelinek-Mercer (JM), Absolute Discounting (DC), Two-stage (TS) smoothing, and Bayesian smoothing using Dirichlet priors [3,27,29].

Among the above approaches, the Dirichlet smoothing gained a lot of attention in information retrieval and different tasks such as novelty detection [30], text categorization [17], texture modeling [10], emotion recognition [20], and cloud computing [26]. With the explosion of online communications, text classification poses huge challenges related to high-dimensional and sparseness problems due to the new variety of available texts including tweets, short messages, customer feedback, and blog spots. Considering large-scale data collections with short and sparse text, text representation based on Multinomial Naive Bayes classifier and the Dirichlet smoothing approaches are not anymore robust enough to deal with these issues. In this context, other priors have been proposed for smoothing the parameters of multinomial such as the generalized Dirichlet distribution [4,9], the Beta-Liouville [6,8], and the scaled Dirichlet [28]. By presenting a prior over the training words in a text representation, we ignore the fact that words could be unseen in the vocabulary which correspond to 0 values in the features domain. This fact leads to the sparsity in text categorization. One possible way to cope with this challenge is to exploit a hierarchical prior over all the possible sets of words. In this regards, a hierarchical Dirichlet prior over a subset of feasible outcomes drawn from the multinomial distribution was proposed in [22]. Such knowledge about the uncertainty over the vocabulary of words improves estimates in sparse documents. Yet, still the Dirichlet prior suffers form limitations concerning the strictly negative covariance structure where applying this prior in case of positively correlated data, the modeling will be inappropriate. In this matter, motivated by the potential structure of hierarchical priors [7,12], we propose a novel hierarchical prior over the multinomial estimates of the Naive Bayes classifier. We introduce the Beta-Liouville prior over all the possible subsets of vocabularies where we consider two assumptions for the "observed" and "unseen" words. We examine the merits of the proposed approach on two challenging applications involving emotion intensity analysis and hate speech detection which are characterized by sparse documents due to the nature of the short text and the large-scale documents.

The structure of the rest of this paper is as follows. In Sect. 2, we present some preliminaries about the Dirichlet smoothing and the Multinomial Naive Bayes that we will consider in this work. We propose, in Sect. 3, the Beta-Liouville Naive Bayes with vocabulary knowledge approach. Section 4 demonstrates the effectiveness and the capability of our proposed approach in categorizing sparse documents. Finally, we give the summary in Sect. 5.

2   The Dirichlet Smoothing for Multinomial Bayesian Inference

In a Bayesian classification problem, the categorization of an observed data $\mathcal{X}$ with N instances $\vec{X}_1, \ldots, \vec{X}_N$ is done by calculating the conditional probability $p(c_j|\mathcal{X})$ for all class c_j. Then, selecting the class that maximizes the following posterior:

$$p(c_j|\mathcal{X}) \propto p(c_j)p(\vec{X}_1, \ldots, \vec{X}_N|c_j) \tag{1}$$

where $p(c_j)$ is the proportion of each class c_j and $p(\vec{X}_1, \ldots, \vec{X}_N|c_j)$ is the probability that represents the instances within the class. If we consider that the instances are independents within the class c_j, this probability will be given as: $p(\vec{X}_1, \ldots, \vec{X}_N|c_j) = \prod_{i=1}^{N} p(\vec{X}_i|c_j)$. Let $\mathcal{X} = (\vec{X}_1, \ldots, \vec{X}_N)$ be a set of N independent draws of $\vec{X} = (x_1, \ldots, x_{D+1})$ from an unknown multinomial distribution with parameters $(P_1, \ldots, P_D)$ defined as follows:

$$p(\vec{X}_i|c_j, \vec{P}_j) = \frac{(\sum_{d=1}^{D+1} x_{id})!}{\prod_{d=1}^{D+1} x_{id}!} \prod_{d=1}^{D+1} P_{dj}^{x_{id}} \tag{2}$$

where $P_{D+1j} = 1 - \sum_{d=1}^{D} P_{dj}$, $\sum_{d=1}^{D} P_{dj} < 1$.

The objective of Bayesian parameter estimation for multinomial distributions is to find an approximation to the parameters $(P_{1j}, \ldots, P_{Dj})$ for each class j which can be interpreted as calculating the probability of a possible outcome $\vec{X}_{N+1}$ where:

$$p(\vec{X}_{N+1}|\vec{X}_1, \ldots, \vec{X}_N, \zeta) = \int p(\vec{X}_{N+1}|\vec{P}, \zeta)p(\vec{P}|\mathcal{X}, \zeta)d\vec{P} \tag{3}$$

where ζ is the context variables and $p(\vec{P}|\mathcal{X}, \zeta)$ is the posterior probability of $\vec{P}$. By Bayes' theorem, this conditional probability is given by:

$$p(\vec{P}_j|\mathcal{X}, \zeta) = p(\vec{P}_j|\zeta) \prod_{i=1}^{N} p(\vec{X}_i|\vec{P}_j, \zeta) \tag{4}$$

$$\propto p(\vec{P}_j|\zeta) \prod_{d=1}^{D+1} P_{dj}^{M_d} \tag{5}$$

where $M_d = \sum_{i=1}^{N} x_{id}$.

Using the Dirichlet properties, the conjugate prior distribution $p(\vec{P}_j|\zeta)$ is specified by hyperparameters $(\alpha_1, \ldots, \alpha_{D+1})$ as follows:

$$p(\vec{P}_j|\vec{\alpha}) = \frac{\Gamma(\sum_{d=1}^{D+1} \alpha_d)}{\prod_{d=1}^{D+1} \Gamma(\alpha_d)} \prod_{d=1}^{D+1} P_{dj}^{\alpha_d-1} \tag{6}$$

Using the fact that when $\vec{P}$ follows a Dirichlet $D(\alpha_1, \ldots, \alpha_D; \alpha_{D+1})$ and $p(\vec{X}|\vec{P})$ follows a multinomial distribution then the posterior density $p(\vec{P}|\vec{X})$ is also a Dirichlet distribution with parameters $(\alpha'_1, \ldots, \alpha'_D; \alpha'_{D+1})$ where $\alpha'_d = \alpha_d + M_d$ for $d = 1, \ldots, D+1$, the estimate of P^*_{dj} is then given by:

$$P^*_{dj} = \frac{\alpha_d + M_{dj}}{\sum_{d=1}^{D+1} \alpha_d + \sum_{d=1}^{D+1} M_{dj}} \tag{7}$$

where M_{dj} is the number of occurrence of x_{id} in document of class j.

3 The Liouville Assumption Using Vocabulary Knowledge

The Liouville family of distributions is an extension of Dirichlet distribution and is proven to be a conjugate prior to the multinomial [14,15]. The Liouville distribution characterized by positive parameters $(a_1, \ldots, a_D)$ and a generating density function $f(.)$ is defined by:

$$p(\vec{P}_j|\vec{a}) = \frac{\Gamma(a^*)}{\prod_{d=1}^{D} \Gamma(a_d)} \frac{\prod_{d=1}^{D} P_{dj}^{a_d-1}}{(\sum_{d=1}^{D} P_{dj})^{a^*-1}} f\left(\sum_{d=1}^{D} P_{dj}\right) \tag{8}$$

where $a^* = \sum_{d=1}^{D} a_d$. The probability density function is defined in the simplex $\{(P_{1j}, \ldots, P_{jD}); \sum_{d=1}^{D} P_{dj} \leq u\}$ if and only if the generating density $f(.)$ is defined in $(0, u)$. A convenient choice for compositional data is the Beta distribution where the resulted distribution is commonly known as Beta-Liouville distribution that presents several well-known properties [6,23,25]. To name a few, in contrast to the Dirichlet smoothing, the Beta-Liouville has a general covariance structure which can be positive or negative. Further, with more flexibility the Beta-Liouville distribution is defined with $(a_1, \ldots, a_D)$ and two positive parameters α and β:

$$p(\vec{P}_j|\vec{a}, \alpha, \beta) = \frac{\Gamma(a^*)\Gamma(\alpha+\beta)}{\Gamma(\alpha)\Gamma(\beta)} \left(\sum_{d=1}^{D} P_{dj}\right)^{\alpha-a^*} \left(1 - \sum_{d=1}^{D} P_{dj}\right)^{\beta-1} \prod_{d=1}^{D} \frac{P_{dj}^{a_d-1}}{\Gamma(a_d)} \tag{9}$$

The Beta-Liouville distribution with $\alpha = \sum_{d=1}^{D} a_d$ and $\beta = a_{D+1}$ is reduced to a Dirichlet $D(a_1, \ldots, a_D; a_{D+1})$.

When $\vec{P} = (P_1, \ldots, P_D)$ follows $BL(a_1, \ldots, a_D, \alpha, \beta)$ and $p(\vec{X}|\vec{P})$ follows a multinomial distribution then the posterior density $p(\vec{P}|\vec{X})$ is a Beta-Liouville distribution with hyperparameters: $a'_d = a_d + M_d$, for $d = 1, \ldots, D$, $\alpha' = \alpha + \sum_{d=1}^{D} M_d$, and $\beta' = \beta + M_{D+1}$. According to this property, the Beta-Liouville can be also a conjugate prior to the multinomial distribution where we obtain as a result the estimate of P^*_{dj}:

$$P^*_{dj} = \frac{\alpha + \sum_{d=1}^{D} M_{dj}}{\alpha + \sum_{d=1}^{D} M_{dj} + \beta + M_{D+1}} \frac{a_d + M_{dj}}{\sum_{d=1}^{D}(a_d + M_{dj})} \tag{10}$$

when $\alpha = \sum_{d=1}^{D} a_d$ and $\beta = a_{D+1}$, the estimate according to the Beta-Liouville prior is reduced to the Eq. 7.

In natural language context, we consider W a corpus of all the used words in $\mathcal{X}$ and the vocabulary $\mathcal{D}$ is a subset of W containing d words. We denote $k = |\mathcal{D}|$ the size of the vocabulary. Using this vocabulary, each text-document is described by an occurrence vector $\vec{X}_i = (x_{i1}, \ldots, x_{iD+1})$, where x_{id} denotes the number of times a word d appears in the document i. Considering the text classification problem, we denote W_j is the set of all documents of topic j. For simplicity purposes, we use the same Beta-Liouville parameters for all the words in W_j which gives:

$$P_{dj}^* = \frac{\alpha + M}{\alpha + M + \beta + M_{D+1}} \frac{a + M_{dj}}{ka + M} \tag{11}$$

where $M = \sum_{d=1}^{D} M_{dj}$.

Using vocabulary structure, unseen words are usually presented as zero values which influences the problem of sparsity. In this regard, to give the estimates of multinomial parameters, we use different vocabularies assumptions that solve the problem of unseen words from becoming zero. Assuming W_j^0 is the set of observed words in documents of topic j: $W_j^0 = \{M_{dj} > 0, \forall d\}$ and $k_j^0 = |W_j^0|$. We propose to have two different priors over the values of $\mathcal{D}$. We start with the prior over the sets containing W_j^0 where we assign a Beta-scale that assumes to see all the feasible words under a Beta-Liouville prior and a probability mass function assigned to the unseen words. Thus, by introduction of multinomial estimates, we have:

$$P_{dj}^* = p(X_{N+1} = d|W_j, \mathcal{D}) \tag{12}$$

where $\mathcal{D} \subseteq W_j$, allowing to have the following properties:

$$p(X_{N+1} = d|W_j) = \sum_{\mathcal{D}} p(X_{N+1} = d|W_j, \mathcal{D})p(\mathcal{D}|W_j) \tag{13}$$

$$= \sum_{\mathcal{D}, k=|\mathcal{D}|} p(X_{N+1} = d|W_j, \mathcal{D})p(k|W_j)$$

and using Bayes'rule, we define the following prior over the sets with size k:

$$p(k|W_j) \propto p(W_j|k)p(k) \tag{14}$$

where:

$$p(W_j|k) = \sum_{W_j^0 \subseteq \mathcal{D}, |\mathcal{D}|=k} p(W_j|\mathcal{D})p(\mathcal{D}|k) \tag{15}$$

We introduce a hierarchical prior over the sets of $\mathcal{D}$ with size $k = 1, \ldots, D$:

$$p(\mathcal{D}|k) = \binom{D}{k}^{-1} \tag{16}$$

and using the Beta-Liouville prior properties, we define the probability $p(W_j|\mathcal{D})$ as follows:

$$p(W_j|\mathcal{D}) = \frac{\Gamma(ka)}{\Gamma(ka+M)} \frac{\Gamma(\alpha+\beta)\Gamma(\alpha+M)\Gamma(\beta+M_{D+1})}{\Gamma(\alpha+\beta+M')\Gamma(\alpha)\Gamma(\beta)} \prod_{d\in W_j^0} \frac{\Gamma(a+M_{dj})}{\Gamma(a)}$$

(17)

Thus summing up over the sets of observed words when $d \in W_j^0$, we have the following probability:

$$p(W_j|k) = \binom{D-k^0}{k-k^0}\binom{D}{k}^{-1} \frac{\Gamma(ka)}{\Gamma(ka+M)}\Gamma(a)^{-k^0} \prod_{d\in W_j^0} \Gamma(a+M_{dj})$$

(18)

$$\left[\frac{\Gamma(\alpha+\beta)\Gamma(\alpha+M)\Gamma(\beta+M_{D+1})}{\Gamma(\alpha+\beta+M')\Gamma(\alpha)\Gamma(\beta)}\right]$$

$$= \left[\frac{(D-k^0)!}{D!}\Gamma(a)^{-k^0}\prod_{d\in W_j^0}\Gamma(a+M_{dj})\frac{\Gamma(\alpha+\beta)\Gamma(\alpha+M)\Gamma(\beta+M_{D+1})}{\Gamma(\alpha+\beta+M')\Gamma(\alpha)\Gamma(\beta)}\right]$$

$$\frac{k!}{(k-k^0)!}\frac{\Gamma(ka)}{\Gamma(ka+M)}$$

It is noteworthy to mention that the parameters inside the brackets have no influence on the choice of k. For that, we assume the probability is given by:

$$p(k|W_j) \propto \frac{k!}{(k-k^0)!}\frac{\Gamma(ka)}{\Gamma(ka+M)}p(k)$$

(19)

As a result, we have the estimates of Beta-Liouville multinomial parameters for $d \in W_j^0$ defined by:

$$p(X_{N+1}=d|W_j) = \sum_{k=k^0}^{D}\frac{\alpha+M}{\alpha+M+\beta+M_{D+1}}\frac{a+M_{dj}}{ka+M}\frac{k!}{(k-k^0)!}\frac{\Gamma(ka)}{\Gamma(ka+M)}p(k)$$

(20)

Thus, we define the estimates of Beta-Liouville multinomial parameters using two hypothesis over the "observed" and the "unseen" words as the following:

$$P_{dj}^* = \begin{cases} \frac{\alpha+M}{\alpha+\beta+M'}\frac{a+M_{dj}}{k^0a+M}\mathcal{B}_l(k,\mathcal{D}) & \text{if } d \in W_j^0 \\ \frac{1}{D-k^0}(1-\mathcal{B}_l(k,\mathcal{D})) & \text{if } d \notin W_j^0 \end{cases}$$

(21)

where $M' = M + M_{D+1}$ and $\mathcal{B}_l(k,\mathcal{D})$ is the Beta-scale:

$$\mathcal{B}_l(k,\mathcal{D}) = \sum_{k=k^0}^{D}\frac{k^0a+M}{ka+M}p(k|W_j)$$

(22)

4   Experimental Results

In order to evaluate the proposed approach, we consider two applications that are marked with sparseness of its data namely emotion intensity analysis and hate speech detection; both of them in tweets. We compare the novel Beta-Liouville Naive Bayes with vocabulary knowledge so-called (BLNB-VK) with the related-works: Multinomial Naive Bayes (MNB), the Dirichlet smoothing (DS), the Dirichlet smoothing with vocabulary knowledge (DS-VK), and the other models applied on the considered datasets. In our experiments, we assign an exponential prior for the word size $p(k) \propto \gamma^k$, where we set $\gamma = 0.9$. The results obtained in these two applications are achieved with the following hyper-parameters: $a = 1$, $\alpha = 0.2$, $\beta = 0.2$. Details on the implementation code are available on https://github.com/Fatma-Nj.

4.1   Emotion Intensity Analysis

With the explosion of the new communications means, speaking out our feeling takes new formats than verbal and facial expressions. Social media allows people to share their emotions, opinions, and even their attitudes. These online expressions give the opportunity for entrepreneur, producer, and politician to predict the preference and the tendency of their community. In our work, we consider a dataset that considers not only emotions in tweets but also the intensity dataset [19]. The tweet emotions intensity (EmoInt) dataset contains four focus emotions: *anger, joy, fear*, and *sadness* where tweets are associated with different intensities of each emotion. For example, for the subset *anger*, 50 to 100 terms are associated with that focus emotion as *mad, frustrated, furry, peeved*, etc. The EmoInt dataset [1] contains 7,097 of tweets split into training, development, and testing sets. We consider in our experiments the testing set for comparison issues and we randomly split the data into 70/30 to generate the vocabulary. In Fig. 1, we show a vocabulary of 352,000 words with 600 features size.

We explore the influence of the features dimensions and the number of words on the performance of the proposed approach. Accordingly, we measure the accuracy of classifiers. Figure 2 shows the out performance of the BLNB-VK when the feature of texts are less than 200 and when we have also documents with more than 1000 size of vocabulary. This proves the ability of our algorithm to recognize short texts as well as high-dimensional documents. We mention that number of words affect also the classification of the tweets emotions where we have a vocabulary with less words is not able to recognize properly the emotions intensity. Yes, with an adequate number of words as 150,000, the tweets are accurately detected by the mentioned classifiers: BLNB-VK, DS-VK, DS, and MNB. In Table 1, the performance of different classifiers is compared according to Pearson correlation and accuracy percentage where the best overall results are again achieved by BLNB-VK.

---

[1] http://saifmohammad.com/WebPages/EmotionIntensity-SharedTask.html.

Fig. 1. Visualization of the vocabulary of words for EmoInt dataset

Table 1. Pearson (Pr) correlations and Accuracy percentages of emotion intensity predictions using different classifier methods

Model	Pr correlation	Accuracy
Word embedding (WE) [19]	0.55	–
Lexicons (L) [19]	0.63	–
WE + L [19]	0.66	–
MNB	0.59	72.12
DS	0.61	71.51
DS-VK	0.64	73.63
BLNB-VK	0.68	76.36

4.2   Hate Speech Detection

Recently, social media becomes a platform of expressing racism, religious beliefs, sexual orientation, and violence. Violence and related crimes are on rise due to the spreading of online hate speech. Researchers and social media companies as Facebook and Tweeter conduct efforts to detect and delete offensive materials. We are interested in this type of data where haters express their beliefs in short tweets or messages. In this regard, we study the performance of our proposed approach on Tweeter hate speech dataset [11]. The dataset contains 24, 802 tweets categorized into hate speech, offensive but non-hate, and neither.

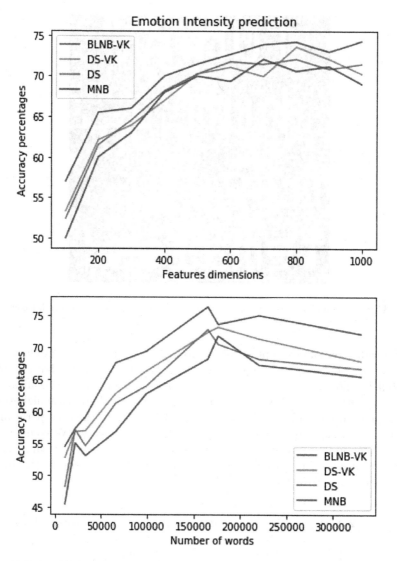

Fig. 2. Influence of feature dimension and the number of words on the performance of the proposed approach BLNB-VK and comparison with Naive-based related works on EmoInt dataset

Each tweet was encoded by three experts based on Hate-base lexicon where only 5% were encoded as hate speech, 76% are considered to be offensive, and the remainder are considered neither offensive nor hate-speech. We split the dataset into training and testing sets to form the vocabulary.

Fig. 3. Visualization of the vocabulary of words for Hate speech dataset

Figure 3 shows an example of set of words observed in the vocabulary of the dataset. By comparing performance results in Fig. 4, we can see that size of the vocabulary and the overall number of words have influenced the classification of hate speech texts. Indeed, we mention that BLNB-VK achieves the highest classification accuracy while Multinomial Naive Bayes attains the least performance in both cases: short texts and high-dimensional documents. We compare also the classification results against previously reported results on the same dataset where our proposed approach outperforms the Logistic regression in [11] by 10% with respect to accuracy metric (Table 2).

Table 2. Classification results of different methods on hate speech dataset

Model	Accuracy
Logistic regression [11]	82.33
MNB	89
DS	90.5
DS-VK	90
BLNB-VK	92.85

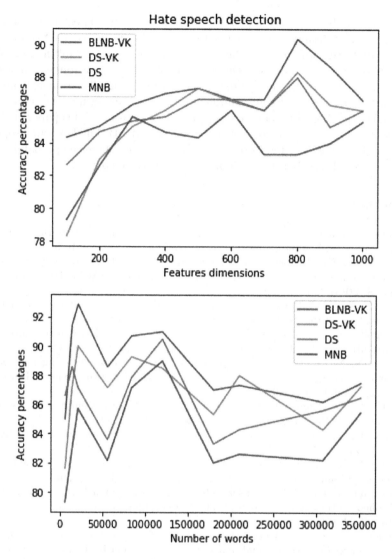

Fig. 4. Influence of feature dimension and the number of words on the performance of the proposed approach BLNB-VK and comparison with Naive-based related works on hate speech dataset

5   Conclusion

In this work, we investigated the problem of sparse document analysis using Bayesian classifier. First, we presented an alternative Liouville prior for the Multinomial Naive Bayes: the Beta-Liouville distribution. Then, we incorporated the vocabulary knowledge to BLNB for the purpose of taking into account the unseen words in the vocabulary. We introduced hierarchical priors over all the possible sets of the vocabulary. Evaluation results illustrate how our proposed approach was able to analyze emotion intensities and to detect hate speech in Tweeter. Better results have been obtained by the proposed BLNB-VK with comparison to the other related Bayes classifier, for instance, Multinomial Naive Bayes, Dirichlet smoothing, and Dirichlet smoothing with vocabulary knowledge in both applications. In our approach, we assume that all parameters are the same for all words for inference simplicity. However, fixing model's parameters could affect the flexibility. For that, a promising extension of this paper could be to investigate the inference of parameters of the BLNB. Further, our proposed model could be combined with reinforcement learning where the probabilities of each state could use Beta-Liouville multinomial parameters.

References

1. Abbas, M., Memon, K.A., Jamali, A.A., Memon, S., Ahmed, A.: Multinomial Naive Bayes classification model for sentiment analysis. IJCSNS **19**(3), 62 (2019)
2. Amati, G., Van Rijsbergen, C.J.: Probabilistic models of information retrieval based on measuring the divergence from randomness. ACM Trans. Inf. Syst. (TOIS) **20**(4), 357–389 (2002)
3. Bai, J., Nie, J.Y., Paradis, F.: Using language models for text classification. In: Proceedings of the Asia Information Retrieval Symposium, Beijing, China (2004)
4. Bouguila, N.: Clustering of count data using generalized Dirichlet multinomial distributions. IEEE Trans. Knowl. Data Eng. **20**(4), 462–474 (2008)
5. Bouguila, N.: A model-based approach for discrete data clustering and feature weighting using MAP and stochastic complexity. IEEE Trans. Knowl. Data Eng. **21**(12), 1649–1664 (2009)
6. Bouguila, N.: Count data modeling and classification using finite mixtures of distributions. IEEE Trans. Neural Netw. **22**(2), 186–198 (2010)
7. Bouguila, N.: Infinite Liouville mixture models with application to text and texture categorization. Pattern Recognit. Lett. **33**(2), 103–110 (2012)
8. Bouguila, N.: On the smoothing of multinomial estimates using Liouville mixture models and applications. Pattern Anal. Appl. **16**(3), 349–363 (2013)
9. Bouguila, N., Ghimire, M.N.: Discrete visual features modeling via leave-one-out likelihood estimation and applications. J. Vis. Commun. Image Represent. **21**(7), 613–626 (2010)
10. Bouguila, N., Ziou, D.: Unsupervised learning of a finite discrete mixture: applications to texture modeling and image databases summarization. J. Vis. Commun. Image Represent. **18**(4), 295–309 (2007)
11. Davidson, T., Warmsley, D., Macy, M., Weber, I.: Automated hate speech detection and the problem of offensive language. In: Proceedings of the International AAAI Conference on Web and Social Media, vol. 11 (2017)
12. Epaillard, E., Bouguila, N.: Proportional data modeling with hidden Markov models based on generalized Dirichlet and Beta-Liouville mixtures applied to anomaly detection in public areas. Pattern Recognit. **55**, 125–136 (2016)
13. Eyheramendy, S., Lewis, D.D., Madigan, D.: On the Naive Bayes model for text categorization (2003)
14. Fan, W., Bouguila, N.: Learning finite Beta-Liouville mixture models via variational Bayes for proportional data clustering. In: Rossi, F. (ed.) IJCAI 2013, Proceedings of the 23rd International Joint Conference on Artificial Intelligence, Beijing, China, 3–9 August 2013, pp. 1323–1329. IJCAI/AAAI (2013)

15. Fan, W., Bouguila, N.: Online learning of a Dirichlet process mixture of Beta-Liouville distributions via variational inference. IEEE Trans. Neural Networks Learn. Syst. **24**(11), 1850–1862 (2013)

16. Kadam, S., Gala, A., Gehlot, P., Kurup, A., Ghag, K.: Word embedding based multinomial Naive Bayes algorithm for spam filtering. In: 2018 Fourth International Conference on Computing Communication Control and Automation (ICCUBEA), pp. 1–5. IEEE (2018)

17. Madsen, R.E., Kauchak, D., Elkan, C.: Modeling word burstiness using the Dirichlet distribution. In: Proceedings of the 22nd International Conference on Machine Learning, pp. 545–552 (2005)

18. McCallum, A., Nigam, K., et al.: A comparison of event models for Naive Bayes text classification. In: AAAI-98 Workshop on Learning for Text Categorization, vol. 752, pp. 41–48. Citeseer (1998)

19. Mohammad, S., Bravo-Marquez, F.: Emotion intensities in tweets. In: Proceedings of the 6th Joint Conference on Lexical and Computational Semantics (*SEM 2017), pp. 65–77. Association for Computational Linguistics, Vancouver, Canada, August 2017

20. Najar, F., Bouguila, N.: Happiness analysis with fisher information of Dirichlet-multinomial mixture model. In: Goutte, C., Zhu, X. (eds.) Canadian AI 2020. LNCS (LNAI), vol. 12109, pp. 438–444. Springer, Cham (2020). https://doi.org/10.1007/978-3-030-47358-7_45

21. Rennie, J.D., Shih, L., Teevan, J., Karger, D.R.: Tackling the poor assumptions of Naive Bayes text classifiers. In: Proceedings of the 20th International Conference on Machine Learning (ICML 2003), pp. 616–623 (2003)

22. Singer, N.F.Y.: Efficient Bayesian parameter estimation in large discrete domains. Adv. Neural. Inf. Process. Syst. **11**, 417 (1999)

23. Sivazlian, B.: On a multivariate extension of the gamma and beta distributions. SIAM J. Appl. Math. **41**(2), 205–209 (1981)

24. Willems, D., Vuurpijl, L.: A Bayesian network approach to mode detection for interactive maps. In: Ninth International Conference on Document Analysis and Recognition (ICDAR 2007), vol. 2, pp. 869–873. IEEE (2007)

25. Wong, T.T.: Alternative prior assumptions for improving the performance of Naïve Bayesian classifiers. Data Min. Knowl. Disc. **18**(2), 183–213 (2009)

26. Xiao, Y., Lin, C., Jiang, Y., Chu, X., Shen, X.: Reputation-based QoS provisioning in cloud computing via Dirichlet multinomial model. In: 2010 IEEE International Conference on Communications, pp. 1–5. IEEE (2010)

27. Yuan, Q., Cong, G., Thalmann, N.M.: Enhancing Naive Bayes with various smoothing methods for short text classification. In: Proceedings of the 21st International Conference on World Wide Web, pp. 645–646 (2012)

28. Zamzami, N., Bouguila, N.: A novel scaled Dirichlet-based statistical framework for count data modeling: unsupervised learning and exponential approximation. Pattern Recogn. **95**, 36–47 (2019)

29. Zhai, C., Lafferty, J.: A study of smoothing methods for language models applied to information retrieval. ACM Trans. Inf. Syst. (TOIS) **22**(2), 179–214 (2004)

30. Zhang, J., Ghahramani, Z., Yang, Y.: A probabilistic model for online document clustering with application to novelty detection. Adv. Neural. Inf. Process. Syst. **17**, 1617–1624 (2004)

Automatic Signature-Based Writer Identification in Mixed-Script Scenarios

Sk Md Obaidullah[1]([✉]), Mridul Ghosh[2], Himadri Mukherjee[3], Kaushik Roy[3], and Umapada Pal[4]

[1] Aliah University, Kolkata, India
sk.obaidullah@aliah.ac.in
[2] Shyampur Sidheswari Mahavidyalaya, Howrah, India
[3] West Bengal State University, Barasat, India
[4] Indian Statistical Institute, Kolkata, India
umapada@isical.ac.in

Abstract. Automated approach for human identification based on biometric traits has become popular research topic among the scientists since last few decades. Among the several biometric modalities, handwritten signature is one of the very common and most prevalent approaches. In the past, researchers have proposed different handcrafted feature-based techniques for automatic writer identification from offline signatures. Currently huge interests towards deep learning-based solutions for several real-life pattern recognition problems have been found which revealed promising results. In this paper, we propose a light-weight CNN architecture to identify writers from offline signatures written by two popular scripts namely Devanagari and Roman. Experiments were conducted using two different frameworks which are as follows: (i) In first case, signature script separation has been carried out followed by script-wise writer identification, (ii) Secondly, signature of two scripts was mixed together with various ratios and writer identification has been performed in a script independent manner. Outcome of both the frameworks have been analyzed to get the comparative idea. Furthermore, comparative analysis was done with recognized CNN architectures as well as handcrafted feature-based approaches and the proposed method shows better outcome. The dataset used in this paper can be freely downloaded from the link: https://ieee-dataport.org/open-access/multi-script-handwritten-signature-roman-devanagari for research purpose.

Keywords: Writer identification · Signature identification · Mixed script · Deep learning

1 Introduction

Signature is one of the most popular and widely used behavioral biometric traits used for authentication of a person. Since decades researchers are paying attention to develop various automatic signature identification systems to identify individuals. It is expected that signature of two persons will differ significantly

© Springer Nature Switzerland AG 2021
J. Lladós et al. (Eds.): ICDAR 2021, LNCS 12822, pp. 364–377, 2021.
https://doi.org/10.1007/978-3-030-86331-9_24

and two writers can be distinguished by their signatures. In earlier days, human interventions were employed to do this identification tasks by analyzing their writing patterns. Later computational approaches were developed where hand-crafted feature based technique was used mostly to automate signature based writer identification. The main goal of automating these systems is to reduce the uncertainty of manual endorsement process as well as reducing false-positive cases as much as possible by means of dual checking by both experts and the automatic systems [1].

Depending upon the various input methods signature based writer identification tasks can be classified into two broad ways: (a) static or offline writer identification and (b) dynamic or online writer identification. In static mode of identification, users write their signature on plain papers, and then digitize the same using a scanner or camera, and then it is recognized by its overall shape and structure mostly holistically. This approach is also known as offline because the identification does not happen in real time. On contrary, in case of dynamic writer identification, the information processing is done in real time. Digital devices like tablets or pads are used to capture the signature which is nothing but the dynamic information like: spatial coordinates in x and y axis, pressure, inclination, stroke angle to name a few. This approach is known as online because all information collected and processed are done in real time. To compare the performances of both offline and online system, it is found that, in general online system are more accurate because they work on the real time dynamic data. But availability of online data capturing devices are not very common in most of the real life scenarios which leads to the development of offline system a pressing need. In offline system all the dynamic information are lost and it only works on the two dimensional image data. Holistic or component level approaches are followed to find the writer patterns from offline signature data. That's why offline writer identification from signature is still very challenging compared to online one. In this paper, we worked on offline Devanagari and Roman script signature based writer identification system.

As per 8th schedule of Indian constitution there are 13 official scripts including Roman which are used to write 23 different languages [2]. Devanagari and Roman are the two most popular and widely used scripts in India. Devanagari, which is a part of the Brahmic family of scripts, is used by several languages like Hindi, Marathi, Nepali, and Sanskrit for writing. The script composed of 47 primary characters out of which 14 are vowels and 33 consonants. Among the structural properties, presence of 'shirorekha' is one of the most prominent properties which join several components/characters while writing to form a larger connected component [3]. Roman is another popular script which is used to write English language. Roman is a true alphabetic script which consists of 26 primary characters where 5 are vowels and rests 21 are consonants. Research development on writer identification on regional Indic scripts is still very much limited which motivates us to work on these two popular scripts of India for the said problem.

During writer identification, signature of a writer is provided to the system, and then the system compares the input signature with all the available signatures in the dataset and measures a similarity score. The best score refers to the identified writer. Writer identification tasks are broadly classified into two types: (a) script specific writer identification and (b) script independent writer identification. In script specific writer identification script separation is done at first, followed by script-wise writer identification. In later scenario, i.e. script independent writer identification, signature of two scripts was mixed together with various ratios and writer identification has been performed in a script independent manner. There is another aspect of script independent writer identification where training has been done with one script signature data and testing is carried out with another script signature data of the same writer. But the main constraint for such experiment is we need a dataset where same writer contributes their signature on different scripts. Due to unavailability of such signature data we limited our experiment with former two cases of writer identification problem.

During our survey of the state-of-the-arts we found majority of the available works were based on hand crafted feature computation. Several works have been reported on Indic and non-Indic scripts. These works are either holistic or component level approach. Well known global features like signature window size, projection profile, signature contour, energy [4], centroid and geometric information [5], statistical texture information [6], quad tree based structural information [7], SURF [8], SIFT [9] like global descriptor, wavelet based features [10], grid and geometry based information [11], fuzzy modeling based feature fusion [12], HOG [13] were used for various signature recognition task on non-Indic scripts. Among the popular works on Indic scripts gradient and Zernike moments [14], contour information based on chain code [15], LBP [16] were considered. Among the deep learning based approaches convolutional siamese network [17], deep convolutional generative adversarial networks (DCGANs) [18], writer independent CNN [19] are some popular CNN based works available in literature on Indic and non-Indic scripts.

This paper presents a deep neural network based approach for automatic writer identification from two popular and most widely used scripts namely Devanagari and Roman. We designed lightweight CNN architecture for the present problem. The increased computational cost of CNN is because it contains several intermediate layers with higher dimensions like convolution, pooling, fully connected layer, softmax/logistics. To reduce the computational cost, in our model we used three lower dimensional convolution layers (5×5, 4×4 and 3×3, respectively) and a pooling layer in between the 2nd and 3rd convolutional layer. We used a dataset of 3929 signatures from 80 writers in Roman script and 1504 signatures from 46 writers in Devanagari scripts. Altogether, from both the scripts 5433 signatures of 126 writers are considered for experiments. Two different frameworks namely: script specific and script independent mixed type are used. Furthermore performance of the proposed network was compared with state-of-the-art handcrafted features as well as some well-known CNN architectures.

Rests of the papers are organized as follows: In Sect. 2, we describe the design of the proposed CNN architecture. In Sect. 3, we discuss about the dataset considered, experimental setup, experimental outcome, error analysis and comparative study. In Sect. 4, we conclude the paper mentioning some future plans.

2   Proposed Method

In the present work we have addressed the problem of writer identification from signatures in a holistic approach. Our primary concern was to study different writer identification frameworks for Indic scripts, i.e. script dependent and script independent manner. We have employed a six-layered deep learning based architecture which computes the feature vector automatically and classify accordingly. In CNN, three different types of layers are present namely: convolution, pooling and dense layer. Features are computed from the input images by the convolution layers by applying some predefined filters. The generated feature vector is passed through a pooling layer to reduce the spatial size of feature map. This in turn reduces the number of parameters and computational cost of the network. Dense layer sends all out-puts from the previous layer to all its neurons and each neuron further produces one output to the next layer. Use of CNN for writer identification from various Indic scripts is still very much limited [1]. The major issue is the computational cost and system overhead while designing any CNN architecture. We designed the proposed model a light-weight architecture which not only computationally efficient but also produces comparable results, thus making a trade-off between computational cost and experimental outcome. In the following section we elaborate the design of proposed architecture

2.1   Proposed Architecture

Proposed multi script writer identification network is a six-layered CNN architecture. Experimentally we set the dimensions of first three convolution layers as 32, 16 and 8, respectively, to keep the network lightweight. The filter sizes are considered as 5×5, 4×4 and 3×3, respectively. In between the 2nd and 3rd convolution layers a max pooling layer having a window size of 2×2 is employed. Lower dimensional convolution layers are used in order to ensure lower computational cost. After the last convolution layer there are two dense layers having dimensions of 256 and nb_classes, respectively, where nb_classes denotes number of classes for the particular experiment (in our experiments number of classes are the number of scripts/writers). For example, for our script identification problem the value of nb_classes was set to 2 as Devanagari and Roman scripts are consider. Similarly, for Devanagari and Roman writer identification problem the value of nb_classes was set to 46 and 80, respectively, as number of writers in both the scripts were 46 and 80, respectively. ReLU activation function (as presented by Eq. 1) is used in all the convolution layers along with the first dense layer. In final layer a softmax function is used which is presented by Eq. 2.

$$f(x) = max(0; x) \tag{1}$$

where x is the input to a neuron.

$$\sigma(z)_j = \frac{e^{z_j}}{\sum_{k=1}^{K} e_k^z} \tag{2}$$

where z is the jth input vector of length K.

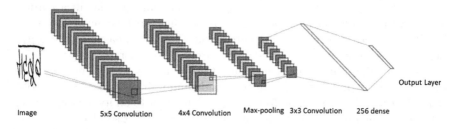

Fig. 1. Block diagram of the proposed CNN architecture with six layers (3 convolution, 1 pooling and 2 dense).

Figure 1 shows the schematic diagram of the proposed network. We started our experiment with an initial value of 100 batch size (bs) and 100 epoch size (es) keeping other parameters like image dimension, pooling window size and dropout rate as default. To obtain the best hyper parameters, we varied the values of bs and es up to 500 with an interval of 100. The system was evaluated with n-fold cross validation where we choose the value of n as 5 experimentally. The hyper parameters were further tuned to obtained performance gains which are presented hereafter:

Table 1. Number of parameters used in different layers for the proposed architecture

Layer	#Parameters
Convolution 1	2432
Convolution 2	8208
Convolution 3	1160
Dense 1	39,65184
Dense 2	19,789
Total	**39,96,773**

Table 1 shows the parameters used by the proposed network. Number of parameters by three convolution layers is 2432, 8208 and 1160, respectively. Then first dense layer (256 dimensional) performs highest computation with 39,65,184 parameters followed by 19,789 parameters considered by the final dense layer. Total number of parameters considered is 39,96,773. Finally, we called the proposed architecture as lightweight because

- Only three convolution layers were used with a pooling layer in between 2nd and 3rd convolution layers.
- The dimensions of the convolution layers used are 32, 16 and 8 only.
- During the parameters fine tuning fewer number of parameters were produced by the proposed architecture compared to other standard models (experimentally found).

3 Experiments

We start our discussion describing the dataset development, and then we will discuss the experimental setup used, followed by discussion on finding best results with ablations studies on parameters. Comparative results with state-of-the-art handcrafted features as well as CNN architectures are also discussed here.

3.1 Dataset

Writer identification dataset availability on Indic scripts is a major issue to carry forward research in this domain. Devanagari and Roman are two most popular and widely used scripts of India. We are in process of building a multi-script offline signature dataset and the present dataset is a subset of that. For present work a total of 5433 signatures of 126 writers are considered for experiments, out of which 3929 signatures from 80 writers in Roman script and 1504 signatures from 46 writers in Devanagari scripts. Script-wise per writer 49 signatures from Roman and 32 signatures from Devanagari are considered making an average of 43 signatures per writer on whole dataset. Table 2 shows a glimpse of the dataset used.

Table 2. Dataset details of Roman and Devanagari scripts signature

Script name	Total signatures	Total writers	Average (signatures/writer)
Roman	3929	80	49
Devanagari	1504	46	32
Roman + Devanagari	5433	126	43

(a)   (b)   (c)   (d)

Fig. 2. Sample Roman (a & b) and Devanagari (c & d) signatures of four different writers from our dataset which was built in-house

The data were collected from persons with varying sex, educational background, and age. In addition, to ensure the natural variability in the handwriting, the samples were collected in diverse time intervals. No constrained on ink type, pen type was imposed on the writers. Few sample signatures from our dataset are shown in Fig. 2. The entire dataset used in this paper will be available to the researchers in given link mentioned earlier.

To evaluate the performance of our system we have used n-fold cross validation approach where experimentally the value of n was chosen as 5. The performance was measured by means of overall recognition accuracy which is calculated by the following equation:

$$accuracy = \frac{(tp + tn)}{(tp + fp + fn + tn)} \tag{3}$$

Where tp, tn, fp, and fn are the number of true positive, true negative, false positive, and false negative samples, respectively. All the experiments were carried out using a system with configurations: Intel Core i5-10400H processor with 2.60 GHz clock speed, 32 GB RAM and NVIDIA RTX 5000 GPU with 16 GB memory. In software part Python 3.7.0 was used to write the code. The network parameters were set based on experimental trials which are discussed below.

3.2 Experimental Framework and Results with Ablations Studies on Parameters

As mentioned earlier, for the present work we have considered two different experimental frameworks. Firstly, script separation has been carried out followed by script-wise writer identification. Secondly, signature of two scripts was mixed together with different ratios and writer identification has been performed in a script independent manner. Both the frameworks are discussed below:

Script Dependent Framework (SDF): Script dependent framework consists of two steps: (i) script separation and (ii) writer identification. During script separation experiment we find the results with ablations studies on parameters namely batch size and epoch. After finding the best values of batch size and epoch, rests of the experiments was carried out with these batch size and epoch values. Table 3 and 4 shows the outcome of varying batch sizes and epoch, respectively.

Script Separation: From the Table 3 and 4 it is found that, the best script separation results generated at a batch size of 200 and epoch of 200. At batch

Table 3. Script identification performance of different batch sizes of 100, 200, 300, 400 and 500. The best results obtained at batch size 200.

Batch sizes	100	200	300	400	500
Accuracy (%)	98.18	**98.91**	98.14	96.69	98.03

size 200 we found 98.91% script separation accuracy, whereas at epoch value of 200 we found 98.86% script separation accuracy. In next experiment, we carried out the script separation using these parameter values (batch size 200 and epoch 200) and the maximum script separation accuracy of 98.95% was obtained. The confusion matrix for the said experiment is shown in Table 5.

Table 4. Script identification performance of different training iterations (epoch) of 100, 200, 300, 400 and 500. The best results obtained at epoch size 200.

Training epoch	100	200	300	400	500
Accuracy (%)	98.18	**98.86**	98.49	98.18	98.79

Table 5. Confusion matrix of script separation between Roman and Devanagari scripts. This experiment was carried out using batch size 200 and epoch 200 which produced best results in Table 3 and 4. We obtained highest script separation accuracy of 98.95%.

Script name	English	Hindi
Roman	3883	46
Devanagari	11	1493
Accuracy	98.95%	

Writer Identification: Our next experiment under SDF was script-wise writer identification separately for Roman and Devanagari. The best hyper parameters obtained earlier were used for these experiments too. For Roman script, we found an overall writer identification results of 98.50% and for Devanagari we found 97.81% identification results. On an average the writer identification accuracy of both the scripts is 98.15%. These outcomes are represented in Table 6.

Table 6. Script wise (separately Roman and Devanagari) writer identification results with batch size 200 and epoch 200.

Script name	Batch size	Epoch	Writer identification accuracy (%)
Roman	200	200	98.50
Devanagari	200	200	97.81
Roman + Devanagari			**98.15**

Script Independent Framework (SIF): In SIF, signature of two scripts was mixed together with different ratios and writer identification has been performed in a script independent manner. In our dataset we have heterogeneous number of writers for both the scripts. We conducted two types of experiments: (i) equal numbers of writers are taken from both the Devanagari and Roman scripts, i.e. 46 writers from Devanagari and 46 from Roman. (ii) all the writers from both the scripts are taken together and writer identification has been done. In later case we have taken 126 (46 from Devanagari and 80 from Roman) number of writers. We received an average writer identification accuracy of 98.49% and 97.94%, respectively, for both the above scenarios (i) and (ii), respectively. It is to be noted that all the experiments were carried out with the hyper parameters as we considered earlier, i.e. batch size 200 and epoch 200. All the results are summarized in Table 7.

Table 7. Comparison of script dependent and script independent frameworks.

Experimental framework		#Writers	Accuracy (%)
Script dependent	Roman	80	**98.50**
	Devanagari	46	97.81
Script independent	Equal writers	92 (46+46)	**98.49**
	All writers	126 (46+80)	97.94

Observations: Upon comparing the outcome of two frameworks (SDF & SIF) what we found is though the highest writer identification accuracy was obtained during SDF but that is marginally ahead in comparison to SIF. In addition, for SDF there is an extra layer of script identification module which increases overall computational overhead to the system. In SIF, the best performance obtained while considered number of writers for both the scripts are balanced, i.e. the first scenario in our experiment.

3.3 Comparative Study with State-of-the-Art and Analysis

Comparison with Handcrafted Features. We compare the performance of the proposed architecture with some well-known state-of-the-art texture based handcrafted features namely: Gray level co-occurrence matrix (GLCM) [20], Zernike moments [21], Histogram of oriented gradient (HOG) [13], Local binary pattern (LBP) [20], Weber local descriptor (WLD) [22] and Gabor Wavelet Transform (GWT) [30]. For GLCM, 22 dimensional feature vectors were generated applying different statistical properties and it produced a writer identification accuracy of 62.10% for Roman and 72.73% for Devanagari scripts, respectively. A 20 dimensional feature set were extracted using Zernike moments which produced an overall accuracy of 78% and 79.58%, respectively, for Roman and Devanagari scripts. Initially a feature vector of size 4356 was generated by HOG and after applying PCA [23] we reduced the feature dimension to 200. Further applying this 200 dimensional feature set we received a very good writer identification accuracy of 94.93% and 96.94%, respectively, for Roman and Devanagari scripts. LBP is one of the most widely used and popular texture based technique used for many applications in document analysis. Considering four parameters namely radius, neighbors, x and y axis information we have extracted 59 dimensional feature set which produced an accuracy of 66.37% and 75.86%, respectively, for Roman and Devanagari scripts. The application of WLD as a texture based feature gained considerable attention among the researchers since its inception from 2009. We have tested the performance of WLD in our applications too. Varying two parameters namely radius and neighbors we extracted a 960 dimensional feature vector initially and applying PCA we retained 200 features for final classification. During our experiment WLD produced an accuracy of 58.31% and 74.26%, respectively, for Roman and Devanagari scripts. We have used Gabor wavelet transform (GWT) based method with 3 scales, 2 orientations and filter size 39×39 to extract 3750 features. After applying PCA, 100 features were retained for classification. Using GWT and PCA we found an accuracy of 95.36% for Roman and 95.81% for Devanagari. Random forest (RF) classifiers are used for all the above experiments as it performs well (fast training and good accuracy) during some of our earlier experiments [24]. Table 8 shows the summarized comparative results as discussed.

Comparison Other CNN Architectures. In the previous subsection we discussed about the comparison of proposed method with well-known handcrafted features and we found the proposed method outperforms. In addition, we compared the performance with other CNN architectures too. As one of our primary objectives was to design a lightweight CNN architecture which can be deployed in low resource environment so we have chosen two existing state-of-the-art models which are logically comparable. In this regard, we have tested MovileNetV2 [25], EfficientNet-B0 and VGG16 [27] with our bi-script writer identification dataset. MobileNetV2 is well known to be architecture of next generation mobile vision applications [26]. The EfficientNet-B0 network is based on the inverted bottleneck residual blocks of MobileNetV2 architecture. VGG16 uses only 3×3

Table 8. Comparison of the proposed system with some of the state-of-the-art texture based handcrafted features based approach. The feature dimensions of each handcrafted features also mentioned.

Script & techniques	Handcrafted features & (dimensions)						Proposed system
	GLCM [20] (22)	Zernike [21] (20)	HOG [13] (200)	LBP [20] (59)	WLD [22] (200)	GWT [30] (100)	
Roman	62.10	78.00	94.93	66.37	58.31	95.36	**98.50**
Devanagari	72.73	79.58	96.94	75.86	74.26	95.81	**97.81**

convolutional layers stacked on top of each other by depth-wise increasing. There are two fully connected layers, having 4096 nodes in each of them. The output is sent to a softmax classifier in order to normalize the classification vector.

Table 9. Comparison of the proposed network with two existing models namely: MobileNetV2 and EfficientNet-B0.

Model	#Parameters	Accuracy (%)
MobileNetV2 [25]	80,11,996	92.91
EfficientNet-B0 [26]	41,67,416	96.60
VGG16 [27]	147,61,884	98.23
Proposed method	39,96,773	**98.49**

During this experiment, to understand the mixed script type scenarios we have taken both Roman and Devanagari signatures on equal basis, i.e. 46 writers from Roman and 46 writers from Devanagari are considered. Table 9 shows the comparative analysis outcome. We found that, the proposed architecture is 5.58% and 1.89% more efficient compared to MobileNetV2 and EfficientNet-B0 architectures, respectively. While comparing the numbers of hyper parameters generated, the proposed method is more than 100% faster than MobileNetV2 and almost 42% faster than EfficientNetB0 architecture.

3.4   Error Analysis

We have proposed two different frameworks for writer identification namely: script dependent framework and script independent framework. In the first framework, we have performed script separation followed by writer identification. In this subsection we have shown few of the misclassification instances occurred during the script separation process. In Fig. 3, we have shown some of the misclassified samples from our experiments. Figure 3(a) and 3(b) are the samples from Devanagari signatures which are misclassified with Roman. Similarly,

Fig. 3(c) and 3(d) are two samples of Roman signatures which are misclassified as Devanagari. We have observed that, Roman script components/characters have dominance of lines, whereas Devanagari script components/characters have a prevalence of curve-like shapes. In the samples shown below we found presence of lines like structure in the Devanagari scripts which caused the misclassification. On the other hand, due to the presence of curvature or loop-like structure few Roman signature samples are misclassified as Devanagari. We will investigate these issues in more detail in future by applying some script dependent core features like presence or absence of 'shirorekha' in the scripts.

| (a) | (b) | (c) | (d) |

Fig. 3. Sample misclassified instance from our experiments: (a) & (b) Devanagari signatures identified as Roman, (c) & (d) Roman signatures identified as Devanagari

4   Conclusions and Future Directions

The problem of writer identification on multi-script scenario is addressed in this paper. Two well-known and popular Indic scripts namely Devanagari and Roman were considered for experiment. We explored the possibility of two different frameworks for the said problem. Firstly, a script dependent framework where script separation is done a prior, followed by script-wise writer identification. During script separation, by ablations studies of parameters, we have chosen best performer batch sizes and epochs. Using those batch sizes and epoch values we found the best script separation accuracy of 98.95%. Applying the same setup we found the writer identification accuracy of 98.50% and 97.81% for Roman and Devanagari script, respectively. Furthermore we continued our investigation using a script independent manner where two scripts are mixed with different ratios. Using equal number of writers (46 signatures both in Devanagari and Roman) from both the scripts we found a writer identification accuracy of 98.49%. While using all the writers (46 Devanagari and 80 Roman) from both the scripts the accuracy drops slight marginally and we found 97.94% accuracy. In addition, we compared the performance of our method with handcrafted features and some well-known CNN architectures and the proposed method outperforms.

Considering the future scopes of the present work there are few directions: Firstly, there is a scarcity of multi-script writer identification dataset for all official scripts of India which need to be addressed. Secondly, script dependent writer identification features need to be found out to improve the writer identification performance at optimum level. For example, writing style of 'shirorekha' can be a distinguishing factor while considering Devanagari data because presence of 'shirorekha' is a script dependent feature. Last but the not least, nowadays research emphasis is going for low resource computer vision application development where light weight design is a primary concern with good recognition accuracy. We need to address this optimization issue also as much as possible [28]. Like any other works, present work has also few limitations: firstly, in addition to writer identification, we need to address the problem of writer verification with forge signature data also [29]. Another issue is: we have not considered the signature data of same writer for different scripts due to inadequacy of such dataset. This task will also be addressed.

Acknowledgement. The first author of this paper is thankful to Science and Engineering Research Board (SERB), DST, Govt. of India for funding this research through Teachers Associateship for Research Excellence (TARE) grant TAR/2019/000273.

References

1. Diaz, M., Ferrer, M.A., Impedovo, D., Malik, M.I., Pirlo, G., Plamondon, R.: A perspective analysis of handwritten signature technology. ACM Comput. Surv. **51**(6) (2019). https://doi.org/10.1145/3274658. Article No. 117
2. https://rajbhasha.gov.in/en/languages-included-eighth-schedule-indian-constitution . Accessed 20 Feb 2021
3. Obaidullah, S.M., Goswami, C., Santosh, K.C., Halder, C., Das, N., Roy, K.: Separating Indic scripts with 'matra' as a precursor for effective handwritten script identification in multi-script documents. Int. J. Pattern Recogn. Artif. Intell. (IJPRAI) **31**(4), 1753003 (17 pages) (2017). World Scientific
4. Nguyen, V., Blumenstein, M., Leedham, G.: Global features for the off-line signature verification problem. In: 10th International Conference on Document Analysis and Recognition (ICDAR 2009), pp. 1300–1304 (2009)
5. Schafer, B., Viriri, S.: An off-line signature verification system. In: International Conference on Signal and Image Processing Applications (ICSIPA 2009), 95–100 (2009)
6. Vargas, J.F., Ferrer, M.A., Travieso, C.M., Alonso, J.B.: Off-line signature verification based on grey level information using texture features. Pattern Recogn. **44**(2), 375–385 (2011)
7. Serdouk, Y., Nemmour, H., Chibani, Y.: Handwritten signature verification using the quad-tree histogram of templates and a support vector-based artificial immune classification. Image Vis. Comput. **66**(2017), 26–35 (2017)
8. Malik, M.I., Liwicki, M., Dengel, A., Uchida, S., Frinken, V.: Automatic signature stability analysis and verification using local features. In 14th International Conference on Frontiers in Handwriting Recognition (ICFHR 2014), pp. 621–626. IEEE (2014)
9. Deng, H.-R., Wang, Y.-H.: On-line signature verification based on correlation image. In: International Conference on Machine Learning and Cybernetics, vol. 3, pp. 1788–1792 (2009)
10. Falahati, D., Helforoush, M.S., Danyali, H., Rashidpour, M.: Static signature verification for Farsi and Arabic signatures using dynamic time warping. In: 19th Iranian Conference on Electrical Engineering (ICEE 2011), pp. 1–6 (2011)
11. Mamoun, S.: Off-line Arabic signature verification using geometrical features. In: National Workshop on Information Assurance Research, pp. 1–6 (2016)

12. Darwish, S., El-Nour, A.: Automated offline Arabic signature verification system using multiple features fusion for forensic applications. Arab J. Forensic Sci. Forensic Med. **1**(4), 424–437 (2016)
13. Dutta, A., Pal, U., Lladós, J.: Compact correlated features for writer independent signature verification. In: 23rd International Conference on Pattern Recognition (ICPR), Cancún Center, Cancún, México, 4–8 December 2016 (2016)
14. Pal, S., Pal, U., Blumenstein, M.: Off-line verification technique for Hindi signatures. IET Biometr. **2**(4), 182–190 (2013)
15. Pal, S., Reza, A., Pal, U., Blumenstein, M.: SVM and NN based offline signature verification. Int. J. Comput. Intell. Appl. **12**(4), 1340004 (2013)
16. Pal, S., Alaei, A., Pal, U., Blumenstein, M.: Performance of an off-line signature verification method based on texture features on a large Indic-script signature dataset. In: 12th IAPR Workshop on Document Analysis Systems (DAS 2016), pp. 72–77. IEEE (2016)
17. Dey, S., Dutta, A., Toledo, J.I., Ghosh, S.K., Lladós, J., Pal, U.: SigNet: convolutional siamese network for writer independent offline signature verification. Corr (2017). arXiv:1707.02131
18. Soleimani, A., Araabi, B.N., Fouladi, K.: Deep multitask metric learning for offline signature verification. Pattern Recogn. Lett. **80**(2016), 84–90 (2016)
19. Zhang, Z., Liu, X., Cui, Y.: Multi-phase offline signature verification system using deep convolutional generative adversarial networks. In: 9th International Symposium on Computational Intelligence and Design (ISCID 2016), vol. 2, pp. 103–107 (2016)
20. Wang, G.D., Zhang, P.L., Ren, G.Q., Kou, X.: Texture feature extraction method fused with LBP and GLCM. Comput. Eng. **38**, 199–201 (2012)
21. Sharma, N., Chanda, S., Pal, U., Blumenstein, M.: Word-wise script identification from video frames. ICDAR **2013**, 867–871 (2013)
22. Chen, J., Shan, S., He, C., et al.: WLD: a robust local image descriptor. IEEE Trans. Pattern Anal. Mach. Intell. **32**, 1705–1720 (2010)
23. Jolliffe I., Principal component analysis. In: Lovric, M. (eds.) International Encyclopedia of Statistical Science. Springer, Heidelberg (2011). https://doi.org/10.1007/978-3-642-04898-2_455
24. Obaidullah, S.M., Santosh, K.C., Halder, C., Das, N., Roy, K.: Automatic Indic script identification from handwritten documents: page, block, line and word-level approach. Int. J. Mach. Learn. Cybern. **10**(1), 87–106 (2019)
25. Sandler, M., Howard, A., Zhu, M., Zhmoginov, A., Chen, L.C.: MobileNetV2: inverted residuals and linear bottlenecks. arXiv preprint. arXiv:1801.04381 (2018)
26. Tan, M., Le, Q.V.: EfficientNet: rethinking model scaling for convolutional neural networks (2019). https://arxiv.org/pdf/1905.11946.pdf
27. Simonyan, K., Zisserman, A.: Very deep convolutional networks for large-scale image recognition. arXiv:1409.1556 (2015)
28. Jain, A., Singh, S.K., Singh, K.P.: Handwritten signature verification using shallow convolutional neural network. Multimed. Tools Appl. **79**, 19993–20018 (2020)
29. Gideona, S.J., Kandulna, A., Abhishek, A., Diana, K.A., Raimond, K.: Handwritten signature forgery detection using convolutional neural networks. Procedia Comput. Sci. **143**(2018), 978–987 (2018)
30. Gabor Wavelet Transform. https://bit.ly/3bl03Sx. Accessed 10 May 2021

Indexing and Retrieval of Documents

Learning to Rank Words: Optimizing Ranking Metrics for Word Spotting

Pau Riba$^{(\boxtimes)}$ [ID], Adrià Molina [ID], Lluis Gomez [ID], Oriol Ramos-Terrades [ID], and Josep Lladós [ID]

Computer Vision Center and Computer Science Department,
Universitat Autònoma de Barcelona, Catalunya, Spain
{priba,lgomez,oriolrt,josep}@cvc.uab.cat, adria.molinar@e-campus.uab.cat

Abstract. In this paper, we explore and evaluate the use of ranking-based objective functions for learning simultaneously a word string and a word image encoder. We consider retrieval frameworks in which the user expects a retrieval list ranked according to a defined relevance score. In the context of a word spotting problem, the relevance score has been set according to the string edit distance from the query string. We experimentally demonstrate the competitive performance of the proposed model on query-by-string word spotting for both, handwritten and real scene word images. We also provide the results for query-by-example word spotting, although it is not the main focus of this work.

Keywords: Word spotting · Smooth-nDCG · Smooth-AP · Ranking loss

1 Introduction

Word spotting, also known as keyword spotting, was introduced in the late 90's in the seminal papers of Manmatha *et al.* [19,20]. It emerged quickly as a highly effective alternative to text recognition techniques in those scenarios with scarce data availability or huge style variability, where a strategy based on full transcription is still far from being feasible and its objective is to obtain a ranked list of word images that are relevant to a user's query. Word spotting has been typically classified in two particular settings according to the target database gallery. On the one hand, there are the segmentation-based methods, where text images are segmented at word image level [8,28]; and, on the other hand, the segmentation-free methods, where words must be spotted from cropped text-lines, or full documents [2,26]. Moreover, according to the query modality, these methods can be classified either query-by-example (QbE) [25] or query-by-string (QbS) [3,13,28,33], being the second one, the more appealing from the user perspective.

The current trend in word spotting methods is based on learning a mapping function from the word images to a known word embedding spaces that can be handcrafted [28,33] or learned by another network [8]. These family of

J. Lladós et al. (Eds.): ICDAR 2021, LNCS 12822, pp. 381–395, 2021.
https://doi.org/10.1007/978-3-030-86331-9_25

approaches have demonstrated a very good performance in the QbS task. However, this methods focus on optimizing this mapping rather than making use of a ranking-based learning objective which is the final goal of a word spotting system.

In computer vision, siamese or triplet networks [31] trained with margin loss have been widely used for the image retrieval task. Despite their success in a large variety of tasks, instead of considering a real retrieval, these architectures act locally in pairs or triplets. To overcome this drawback, retrieval-based loss functions have emerged. These novel objective functions aim to optimize the obtained ranking. For instance, some works have been proposed to optimize the average precision (AP) metric [4,24]. Other works, such as Li *et al.* [18] suggested to only consider negative instances before positive ones to learn ranking-based models. Traditionally, word spotting systems are evaluated with the mean average precision (mAP) metric. However, this metric only takes into account a binary relevance function. Besides mAP, in these cases where a graded relevance score is required, information retrieval research has proposed the normalized Discounted Cumulative Gain (nDCG) metric. This is specially interesting for image retrieval and, in particular, word spotting because from the user perspective, the expected ranking list should follow such an order that is relevant to the user even though not being a perfect match. This metric has also been previously explored as a learning objective [30].

Taking into account the previous comments, the main motivation of this work is to analyze and evaluate two retrieval-based loss functions, namely Smooth-AP [4] and Smooth-nDCG, to train our QbS word spotting system based on the mAP and nDCG metrics respectively. We hypothesize that, with different purposes from the application perspective, training a word spotting system according to its retrieval list should lead to more robust embeddings. Figure 1 provides an overview of the expected behavior for each loss. On the one hand, the Smooth-AP loss expects to find the relevant images at the top ranking position *i.e.* these words with the same transcription as the query. On the other hand, Smooth-nDCG aims at obtaining a more interpretable list from the user's perspective. For instance, the graded relevance function can consider the string edit distance or any other semantic similarity among strings depending on the final application.

To summarize, the main contributions of this work are:

- We propose a word spotting system which is trained solely with retrieval-based objective functions.
- We introduce a novel ranking loss function, namely, *Smooth-nDCG*, which is able to train our system according to a known relevance feedback. Therefore, we can present the results in a more pleasant way.
- We conduct extensive evaluation to demonstrate, on the one hand, the advantages and disadvantages of the proposed learning objectives and, on the other hand, the boost in word spotting performance for the QbS settings.

The rest of this paper is organized as follows. Section 2 reviews the previous works that are relevant to the current task. Afterwards, Sect. 3 presents the

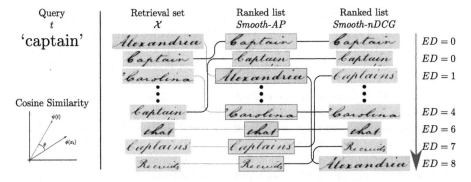

Fig. 1. Overview of the behavior of the proposed model. Given a query t and the retrieval set $\mathcal{X}$, the proposed word spotting system uses the cosine similarity to rank the retrieval set accordingly. Observe that Smooth-AP considers a binary relevance score whereas in the case of Smooth-nDCG, the ranked list is graded according a non-binary relevance score such as the string edit distance.

embedding models for both, images and strings. Section 4 conducts an extensive evaluation on the proposed word spotting system. Finally, Sect. 5 draws the conclusions and the future work.

2   Related Work

2.1   Word Spotting

In this section, we introduce word spotting approaches that are relevant to the current work. Word spotting has been a hot research topic from its origin. The first successful attempts on using neural networks for word spotting were done by Frinken *et al.* [7]. They adapted the CTC Token passing algorithm to be applied to word spotting. In that work, the authors apply a BLSTM network to segmented text line images and then the CTC Token passing algorithm. It's main limitation is that it can only perform QbS tasks but not QbE.

Later, the research challenge evolved to integrate visual and textual information to perform the retrieval in a QbS setting. In this regard, one of the first works proposing this integration was Aldavert *et al.* [1]. In their work, the textual description is built based on n-gram histograms while the visual description is built by a pyramidal representation of bag of words [16]. Both representations are then projected to a common space by means of Latent Semantic Analysis (LSA) [6].

Another relevant work, introduced by Almazan *et al.* [3], propose a pyramidal histogram of characters, which they called PHOC, to represent in the same embedding space segmented word images and their corresponding text transcription. This shared representation enables exchange the input and output modalities in QbS and QbE scenarios. This seminal work inspired the current state-of-the-art methods [13,28,33] on using the PHOC representation. In [28],

Sudholt *et al.*propose the PHOCNet architecture to learn an image embedding representation of segmented words. This network architecture adapts the previous PHOC representation to be learned using a deep architecture. To deal with changing image sizes, the authors make us of a spatial pyramid pooling layer [9]. Similarly, Wilkinson *et al.* [33] train a triplet CNN followed a fully connected 2-layer network to learn an image embedding representation in the space of word embeddings. They evaluate two hand-crafted embeddings, the PHOC and the discrete cosine transform. To learn a shared embedding space, they used the cosine loss to train the network.

Krishnan *et al.* [13] also proposed a deep neuronal architecture to learn a common embedding space for visual and textual information. The authors use a pre-trained CNN on synthetic handwritten data for segmented word images [15] and the PHOC as attribute-based text representation. Later, they presented an improved version of this work in [14]. On the one hand, the architecture is improved to the ResNet-34. They also exploit the use of synthetic word images at test time.

However, all these methods focus on training architectures to rank first relevant images but without ranking them into a retrieval list.

More recently, Gomez *et al.* [8] faced the problem of obtaining a more appealing retrieval list. First, they proposed a siamese network architecture that is able to learn a string embedding that correlates with the Levenshtein edit distance [17]. Second, they train a CNN model inspired from [11] to train an image embedding close to the corresponding string embedding. Thus, even though they share our motivation, they do not exploit this ranking during training.

2.2 Ranking Metrics

In word spotting and, more generally in information retrieval, several metrics have been carefully designed to evaluate the obtained rankings. We briefly define the two main ones, that are considered in this work.

Mean Average Precision: Word spotting performance has been traditionally measured by the *mean Average Precision* (mAP), a classic information retrieval metric [27]. The *Average Precision* given a query q (AP_q) is formally defined as the mean

$$AP_q = \frac{1}{|\mathcal{P}_q|} \sum_{n=1}^{|\Omega_q|} P@n \times r(n), \tag{1}$$

where $P@n$ is the precision at n, $r(n)$ is a binary function on the relevance of the n-th item in the returned ranked list, $\mathcal{P}_q$ is the set of all relevant objects with regard the query q and Ω_q is the set of retrieved elements from the dataset. Then, the mAP is defined as:

$$mAP = \frac{1}{Q} \sum_{q=1}^{Q} AP_q, \tag{2}$$

where Q is the number of queries.

Normalized Discounted Cumulative Gain: In information retrieval, the *normalized Discounted Cumulative Gain* (nDCG) is used to measure the performance on such scenarios where instead of a binary relevance function, we have a graded relevance scale. The main idea is that highly relevant elements appearing lower in the retrieval list should be penalized. The *Discounted Cumulative Gain* (DCG) for a query q is defined as

$$\text{DCG}_q = \sum_{n=1}^{|\Omega_q|} \frac{r(n)}{\log_2(n+1)}, \tag{3}$$

where $r(n)$ is a graded function on the relevance of the n-th item in the returned ranked list and Ω_q is the set of retrieved elements as defined above. However, to allow a fair comparison among different queries, a normalized version was proposed and defined as

$$\text{nDCG}_q = \frac{\text{DCG}_q}{\text{IDCG}_q}, \tag{4}$$

where IDCG_q is the ideal discounted cumulative gain, *i.e.* assuming a perfect ranking according to the relevance function. It is formally defined as

$$\text{IDCG}_q = \sum_{n=1}^{|\Lambda_q|} \frac{r(n)}{\log_2(n+1)} \tag{5}$$

where Λ_q is the ordered set according the relevance function.

3 Architecture

3.1 Problem Formulation

Let $\{\mathcal{X}, \mathcal{Y}\}$ be a word image dataset, containing word images $\mathcal{X} = \{x_i\}_{i=0}^N$, and their corresponding transcription strings $\mathcal{Y}$. Let $\mathcal{A}$ be the alphabet containing the allowed characters. Given the word images $\mathcal{X}$ as a retrieval set or gallery for searching purposes on the one hand, and a given text string t such that its characters $t_i \in \mathcal{A}$ on the other hand; the proposed embedding functions $\phi(\cdot)$ and $\psi(\cdot)$ for word images and text strings respectively, have the objective to map its input in such a space that a given similarity function $S(\cdot, \cdot)$ is able to retrieve a ranked list with the relevant elements. Traditionally, the evaluation of word spotting divides this list in two partitions, namely the positive or relevant word images *i.e.* their transcription matches with the query t, and the negative or non-relevant word images. Thus, the aim of a word spotting system is to rank the positive elements at the beginning of the retrieval list. However, from the user perspective, the non-relevant elements might be informative enough to require them to follow some particular order. Therefore, we formulate the word spotting problem in terms of a relevance score $R(\cdot, \cdot)$ for each element in the list. Finally, given a query t we can formally define this objective as

$$S(\phi(x_i), \psi(t)) > S(\phi(x_j), \psi(t)) \iff R(y_i, t) > R(y_j, t), \tag{6}$$

where $x_i, x_j \in \mathcal{X}$ are two elements in the retrieved list and $y_i, y_j \in \mathcal{Y}$ their corresponding transcriptions.

3.2   Encoder Networks

As already explained, the proposed word spotting system consists of a word image encoding $\phi(\cdot)$ and a textual encoding $\psi(\cdot)$. The outputs of these encoding functions are expected to be in the same space.

Word Image Encoding: Given a word image $x_i \in \mathcal{X}$, it is firstly resized into a fixed height, but we keep the original aspect ratio. As the backbone of our image encoding, we use a ImageNet pre-trained ResNet-34 [10] network. As a result of the average pooling layer, our model is able to process images of different widths. Finally, we L_2-normalize the final 64D embedding.

String Encoder: The string encoder embeds a given query string t in a 64D space. Firstly, we define a character-wise embedding function $e\colon \mathcal{A} \to \mathbb{R}^m$. Thus, the first step of our encoder is to embed each character $c \in t$ into the character latent space. Afterwards, each character is feed into a Bidirectional Gated Recurrent Unit (GRU) [5] with two layers. In addition, a linear layer takes the last hidden state of each direction to generate our final word embedding. Finally, we also L_2-normalize the final 64D embedding.

Although both embeddings can be compared by means of any arbitrary similarity measure $S(\cdot, \cdot)$, in this work we decided to use the cosine similarity between two embeddings v_q and v_i, which is defined as:

$$S(v_q, v_i) = \frac{v_q \cdot v_i}{\|v_q\|\|v_i\|}. \tag{7}$$

From now on and for the sake of simplicity, given a query q its corresponding similarity against the i-th element of the retrieval set is denoted as $s_i = S(v_q, v_i)$.

3.3   Learning Objectives

As already stated in the introduction, we analyze two learning objectives which provide supervision at the retrieval list level rather than at sample, pair or triplet level. The proposed losses are inspired by the classical retrieval evaluation metrics introduced in Sect. 2.

Let us first define some notations and concepts.

Ranking Function ($\mathcal{R}$). Following the notation introduced in [22], these information retrieval metrics can be reformulated by means of the following ranking function,

$$\mathcal{R}(i, \mathcal{C}) = 1 + \sum_{j=1}^{|\mathcal{C}|} \mathbb{1}\{D_{ij} < 0\}, \tag{8}$$

where $\mathcal{C}$ is any set (such as Ω_q or $\mathcal{P}_q$), $D_{ij} = s_i - s_j$ and $\mathbb{1}\{\cdot\}$ is an Indicator function.

However, with this formulation, we are not able to optimize following the gradient based optimization methods and an smooth version of the Indicator is required. Even though several approximations exist, in this work we followed the one proposed by Quin *et al.* [22], which make use of the sigmoid function

$$\mathcal{G}(x;\tau) = \frac{1}{1 + e^{\frac{-x}{\tau}}}. \tag{9}$$

Smooth-AP: The smoothed approximation of AP, namely Smooth-AP and proposed by Brown *et al.* [4], has shown a huge success on image retrieval. There, the authors replace the $P@n \times r(n)$ term in Eq. 1 by the ranking function and the exact AP equation becomes

$$AP_q = \frac{1}{|\mathcal{P}_q|} \sum_{i \in \mathcal{P}_q} \frac{1 + \sum_{j \in \mathcal{P}_q, j \neq i} \mathbb{1}\{D_{ij} < 0\}}{1 + \sum_{j \in \Omega_q, j \neq i} \mathbb{1}\{D_{ij} < 0\}}. \tag{10}$$

Therefore, with this notation we can directly obtain an smooth approximation making use of Eq. 9 as

$$AP_q \approx \frac{1}{|\mathcal{P}_q|} \sum_{i \in \mathcal{P}_q} \frac{1 + \sum_{j \in \mathcal{P}_q, j \neq i} \mathcal{G}(D_{ij};\tau)}{1 + \sum_{j \in \Omega_q, j \neq i} \mathcal{G}(D_{ij};\tau)}. \tag{11}$$

Averaging this approximation for all the queries in the batch, we can define our loss as

$$\mathcal{L}_{AP} = 1 - \frac{1}{Q} \sum_{i=1}^{Q} AP_q, \tag{12}$$

where Q is the number of queries.

Smooth-nDCG: Following the same idea as above, we replace the n-th position in Eq. 3 by the ranking function, since it defines the position that the i-th element of the retrieved set, and the DCG metric is expressed as

$$DCG_q = \sum_{i \in \Omega_q} \frac{r(i)}{\log_2 \left(2 + \sum_{j \in \Omega_q, j \neq i} \mathbb{1}\{D_{ij} < 0\} \right)}, \tag{13}$$

where r is the same graded function used in Eq. 3 but evaluated at element i. Therefore, the corresponding smooth approximation is

$$DCG_q \approx \sum_{i \in \Omega_q} \frac{r(i)}{\log_2 \left(2 + \sum_{j \in \Omega_q, j \neq i} \mathcal{G}(D_{ij};\tau) \right)} \tag{14}$$

when replacing the indicator function by the sigmoid one.

The smooth-nDCG is then defined by replacing the original DCG_q by its smooth approximation in Eq. 4 and the loss $\mathcal{L}_{nDCG}$ is defined as

$$\mathcal{L}_{nDCG} = 1 - \frac{1}{Q} \sum_{i=1}^{Q} nDCG_q, \tag{15}$$

where Q is the number of queries.

Algorithm 1. Training algorithm for the proposed model.

    Input: Input data $\{\mathcal{X}, \mathcal{Y}\}$; alphabet $\mathcal{A}$; max training iterations T
    Output: Networks parameters $\Theta = \{\Theta_\phi, \Theta_\psi\}$.

1: **repeat**
2:     Get word images $X = \{x_i\}_{i=1}^{N_B}$ and its corresponding transcription $Y = \{y_i\}_{i=1}^{N_B}$
3:     $\mathcal{L} \leftarrow \mathcal{L}_{img} + \mathcal{L}_{str} + \mathcal{L}_{cross} + \alpha \frac{1}{N_B} \sum_{i=1}^{N_B} \mathcal{L}_{L_1}$
4:     $\Theta \leftarrow \Theta - \Gamma(\nabla_\Theta \mathcal{L})$
5: **until** Max training iterations T

3.4 End-to-End Training

Considering a batch of size N_B of word images $X = \{x_i\}_{i=1}^{N_B}$ and their corresponding transcriptions $Y = \{y_i\}_{i=1}^{N_B}$, the proposed word image and string encoder is trained considering all the elements in both, the query and the retrieval set. Bearing in mind that Smooth-AP cannot be properly used to train the string encoder alone, we combine the Smooth-AP and Smooth-nDCG loss functions into the following three loss functions:

$$\mathcal{L}_{\text{img}} = \mathcal{L}_{AP}(X) + \mathcal{L}_{nDCG}(X) \tag{16}$$
$$\mathcal{L}_{\text{str}} = \mathcal{L}_{nDCG}(Y) \tag{17}$$
$$\mathcal{L}_{\text{cross}} = \mathcal{L}_{AP}(Y, X) + \mathcal{L}_{nDCG}(Y, X) \tag{18}$$

Moreover, the L_1-loss $\mathcal{L}_{L_1}$ between each image embedding and its corresponding word embedding is used to force both models to lay in the same embedding space. This loss is multiplied by a parameter α that we set experimentally to 0.5. Note that the gradients of this loss will only update the weights of the word image encoder. Algorithm 1 depicts the explained training algorithm in the situation in which both losses are considered. $\Gamma(\cdot)$ denotes the optimizer function.

4 Experimental Evaluation

In this section, we present an exhaustive evaluation of the proposed model in the problem of word spotting. All the code necessary to reproduce the experiments is available at github.com/priba/ndcg_wordspotting.pytorch using the PyTorch framework.

4.1 Implementation Details

The proposed model is trained for 50 epochs where in each epoch 15,000 samples are drawn by means of a random weighted sampler. This setting ensures that the data is balanced among the different classes during the training of the model. In addition, data augmentation is applied in the form of a random affine transformation. The possible transformations have been kept small to ensure that the text is not altered. Thus, we only allow rotations and shear up to 5° and a scale factor in the range of 0.9 and 1.1.

For all the experiments the Adam optimizer [12] has been employed. The learning rate, starting from 1e−4 has been decreased by a factor of 0.25 at epochs 25 and 40.

For evaluation purposes, we used the classic ranking metrics introduced in Sect. 2, *i.e.* mAP and nDCG considering the full retrieval set. In particular, in our setup we define the following relevance function $r(n)$ to the nDCG metric, these words with edit distances 0, 1, 2, 3 and 4 receive a score of 20, 15, 10, 5 and 3. In addition, the smooth-nDCG loss is trained with the following relevance function,

$$r(n; \gamma) = \max(0, \gamma - \mathrm{Lev}(q, y_n)), \tag{19}$$

where q and $y_n \in \mathcal{Y}$ are the transcriptions of the query at the n-th ranking of the retrieval list and $\mathrm{Lev}(\cdot, \cdot)$ corresponds to the Levenshtein distance [17] between two strings. Moreover, γ is an hyperparameter that has been set experimentally to 4 in our case of study.

4.2   Experimental Discussion

The proposed model has been evaluated in two different datasets. First, the GW dataset, which is composed of handwriting word images and, second, the IIIT5K dataset, which is composed of words cropped from real scenes.

George Washington (GW) [23]. This database is based on handwritten letters written in English by George Washington and his associates during the American Revolutionary War in 1755[1]. It consists of 20 pages with a total of 4, 894 handwritten words. Even though several writers were involved, it presents small variations in style and only minor signs of degradation. There is not official partition for the GW dataset, therefore, we follow the evaluation adopted by Almazan *et al.* [3]. Thus, the dataset is splitted in two sets at word level containing 70% of the words for training purpose and the remaining 25% for test. This setting is repeated four times following a cross validation setting. Finally, we report the average result of these four runs.

The IIIT 5K-Word Dataset (IIIT5K) Dataset [21]. This dataset provides 5,000 cropped word images from scene texts and born digital images obtained from Google Image engine search. The authors of this dataset provide an official partition containing two subsets of 2,000 and 3,000 images for training and testing purposes. In addition, the each word is associated with two lexicon subsets of 50 and 1,000 words, respectively. Moreover, a global lexicon [32] of more than half a million words can be used for word recognition. In this work, none of the provided lexicons have been used.

---

[1] George Washington Papers at the Library of Congress from 1741–1799, Series 2, Letterbook 1, pages 270–279 and 300–309, https://www.loc.gov/collections/george-washington-papers/about-this-collection/.

For experimental purposes, we have evaluated three different combinations of the introduced learning objectives, namely, Smooth-AP, Smooth-nDCG and Join, *i.e.* a combination of the previous losses.

Table 1. Mean of ranking metrics, mAP and nDCG, for state-of-the-art query-by-string (QbS) and query-by-example (QbE) methods in the GW and IIIT5K datasets.

Method	GW				IIIT5K			
	QbS		QbE		QbS		QbE	
	mAP	nDCG	mAP	nDCG	mAP	nDCG	mAP	nDCG
Aldavert *et al.* [1]	56.54	–	–	–	–	–	–	–
Frinken *et al.* [7]	84.00	–	–	–	–	–	–	–
Almazán *et al.* [3]	91.29	–	93.04	–	66.24	–	63.42	–
Goméz *et al.* [8]	91.31	–	–	–	–	–	–	–
Sudholt *et al.* [28]	92.64	–	96.71	–	–	–	–	–
Krishnan *et al.* [13]	92.84	–	94.41	–	–	–	–	–
Wilkinson *et al.* [33]	93.69	–	**97.98**	–	–	–	–	–
Sudholt *et al.* [29]	98.02	–	97.78	–	–	–	–	–
Krishnan *et al.* [14]	**98.86**	–	98.01	–	–	–	–	–
Smooth-AP *et al.* [4]	96.79	88.99	97.17	87.64	81.25	80.86	80.60	76.13
Smooth-nDCG	98.25	**96.41**	97.94	**94.27**	88.99	90.63	87.53	85.11
Join	98.38	96.40	**98.09**	**94.27**	**89.14**	**90.64**	**87.60**	**85.16**

Table 1 provides a comparison with the state-of-the-art techniques for word spotting. Most of these techniques have been only evaluated in the handwritten case corresponding to the GW dataset. The best performing model for QbS in the GW dataset is the method proposed by Krishnan *et al.* [14] that obtains a mAP slightly better than our Join setting. Even though, our proposed model has been pre-trained with ImageNet, Krishnan *et al.* makes use of a huge dataset of 9M synthetic word images. Besides the good performance for QbS task, our model is able to perform slightly better in the QbE task. From the same table, we can observe that our model is able to generalize to real scene word images. In such dataset, we outperform the work of Almazán *et al.* by more than 20 points for both tasks, QbS and QbE.

Observe that exploiting the Smooth-nDCG loss, our model is able to enhance the performance of the architecture trained with the Smooth-AP objective function alone. This remarks the importance of not only considering those images whose transcription matches with the query word. In addition, in terms of the nDCG, the performance is also boosted by this loss. Thus, the results are more appealing to the final user following the pre-defined relevance function, in this example, the string edit distance.

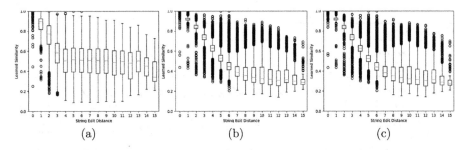

Fig. 2. Box plots for (a) smooth-AP; (b) smooth-nDCG; (c) Join, losses.

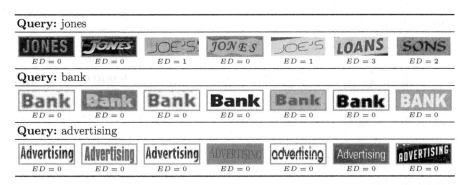

Fig. 3. Qualitative retrieval results for the IIIT5K dataset using the model trained with the Join loss. In green, the exact matches. (Color figure online)

The achieved performance can be explained by the box plots depicted in Fig. 2. This figure shows the correlation between the real string edit distance and the learned similarity of two words. On the one hand, note that the model trained by the smooth-AP loss is able to specialize on detecting which is the matching image given a query. On the other hand, both nDCG or Join losses are able to describe a better ranking when increasing the real string edit distance.

Figures 3 and 4 demonstrates qualitatively the performance of the proposed setting. Observe that the proposed model is able to rank in the first positions the exact match given the query word. Moreover, introducing the Smooth-nDCG we observe that the edit distance is in general smaller than in the Smooth-AP case.

Finally, Fig. 5 provides an overview of the average edit distance of the top-n results given a query. For instance, the Ideal case, tells us that considering the ground-truth, in average, all the retrieved images in the top-50 have an edit distance smaller than 3.5. Here, we can clearly identify that both learning objectives Smooth-nDCG and Join are able to close the gap between the Ideal case and the Smooth-AP loss.

Fig. 4. Qualitative results for the GW dataset. In green, the exact matches. (Color figure online)

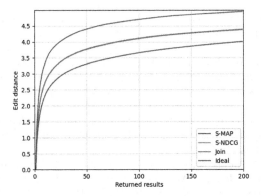

Fig. 5. Average edit distance among the top-n returned results. Note that Smooth-nDCG and Join overlaps in the plot being Smooth-nDCG slightly closer to the Ideal case.

5 Conclusions

In this work, we have presented a word spotting framework completely based on ranking-based losses. The proposed approach learns directly from the retrieval list rather than pairs or triplets as most of the state-of-the-art methodologies. In addition, we do not require any prelearned word embedding. From the application point of view, we have shown the competitive performance of our model against the state-of-the-art methods for word spotting in handwritten and real scene text images. Overall, we have demonstrated the importance of considering not only the corresponding image/transcription pair but also, they relation between the different elements in the batch thanks to a graded relevance score.

As future work, we plan to perform an exhaustive evaluation on the different hyperparameters of the proposed smooth-nDCG objective such as, the relevance and the indicator functions. Moreover, the final loss can be weighted between the smooth-AP and the smooth-nDCG losses. Finally, we would like to explore how this framework extends to other multi-modal retrieval tasks.

Acknowledgment. This work has been partially supported by the Spanish projects RTI2018-095645-B-C21, and FCT-19-15244, and the Catalan projects 2017-SGR-1783, the Culture Department of the Generalitat de Catalunya, and the CERCA Program/Generalitat de Catalunya.

References

1. Aldavert, D., Rusiñol, M., Toledo, R., Lladós, J.: Integrating visual and textual cues for query-by-string word spotting. In: Proceedings of the International Conference on Document Analysis and Recognition, pp. 511–515 (2013)
2. Almazán, J., Gordo, A., Fornés, A., Valveny, E.: Segmentation-free word spotting with exemplar SVMs. Pattern Recogn. **47**(12), 3967–3978 (2014)

3. Almazán, J., Gordo, A., Fornés, A., Valveny, E.: Word spotting and recognition with embedded attributes. IEEE Trans. Pattern Anal. Mach. Intell. **36**(12), 2552–2566 (2014)
4. Brown, A., Xie, W., Kalogeiton, V., Zisserman, A.: Smooth-AP: smoothing the path towards large-scale image retrieval. In: Vedaldi, A., Bischof, H., Brox, T., Frahm, J.-M. (eds.) ECCV 2020. LNCS, vol. 12354, pp. 677–694. Springer, Cham (2020). https://doi.org/10.1007/978-3-030-58545-7_39
5. Chung, J., Gulcehre, C., Cho, K., Bengio, Y.: Empirical evaluation of gated recurrent neural networks on sequence modeling. In: Proceedings of the NeurIPS Workshop on Deep Learning (2014)
6. Deerwester, S., Dumais, S., Furnas, G., Landauer, T., Harshman, R.: Indexing by latent semantic analysis. J. Am. Soc. Inf. Sci. **41**, 391–407 (1990)
7. Frinken, V., Fischer, A., Manmatha, R., Bunke, H.: A novel word spotting method based on recurrent neural networks. IEEE Trans. Pattern Anal. Mach. Intell. **34**(2), 211–224 (2011)
8. Gómez, L., Rusinol, M., Karatzas, D.: LSDE: Levenshtein space deep embedding for query-by-string word spotting. In: Proceedings of the International Conference on Document Analysis and Recognition, vol. 1, pp. 499–504 (2017)
9. He, K., Zhang, X., Ren, S., Sun, J.: Spatial pyramid pooling in deep convolutional networks for visual recognition. IEEE Trans. Pattern Anal. Mach. Intell. **37**(9), 1904–1916 (2015)
10. He, K., Zhang, X., Ren, S., Sun, J.: Deep residual learning for image recognition. In: Proceedings of the IEEE Conference on Computer Vision and Pattern Recognition, pp. 770–778 (2016)
11. Jaderberg, M., Simonyan, K., Vedaldi, A., Zisserman, A.: Synthetic data and artificial neural networks for natural scene text recognition. arXiv preprint 1406.2227 (2014)
12. Kingma, D.P., Ba, J.: Adam: a method for stochastic optimization. arXiv preprint arXiv:1412.6980 (2014)
13. Krishnan, P., Dutta, K., Jawahar, C.: Deep feature embedding for accurate recognition and retrieval of handwritten text. In: Proceedings of the International Conference on Frontiers in Handwriting Recognition, pp. 289–294 (2016)
14. Krishnan, P., Dutta, K., Jawahar, C.: Word spotting and recognition using deep embedding. In: Proceedings of the International Workshop on Document Analysis Systems, pp. 1–6 (2018)
15. Krishnan, P., Jawahar, C.V.: Matching handwritten document images. In: Leibe, B., Matas, J., Sebe, N., Welling, M. (eds.) ECCV 2016. LNCS, vol. 9905, pp. 766–782. Springer, Cham (2016). https://doi.org/10.1007/978-3-319-46448-0_46
16. Lazebnik, S., Schmid, C., Ponce, J.: Beyond bags of features: spatial pyramid matching for recognizing natural scene categories. In: Proceedings of IEEE Conference on Computing Vision Pattern Recognition, vol. 2, pp. 2169–2178 (2006)
17. Levenshtein, V.I.: Binary codes capable of correcting deletions, insertions, and reversals. Soviet Physics Doklady **10**(8), 707–710 (1966)
18. Li, Z., Min, W., Song, J., Zhu, Y., Jiang, S.: Rethinking ranking-based loss functions: only penalizing negative instances before positive ones is enough. arXiv preprint (2021)
19. Manmatha, R., Han, C., Riseman, E.M.: Word spotting: a new approach to indexing handwriting. In: Proceedings of the IEEE Conference on Computer Vision and Pattern Recognition, pp. 631–637 (1996)

20. Manmatha, R., Han, C., Riseman, E.M., Croft, W.B.: Indexing handwriting using word matching. In: Proceedings of the ACM International Conference on Digital Libraries, pp. 151–159 (1996)
21. Mishra, A., Alahari, K., Jawahar, C.V.: Scene text recognition using higher order language priors. In: Proceedings of the British Machine Vision Conference (2012)
22. Qin, T., Liu, T.Y., Li, H.: A general approximation framework for direct optimization of information retrieval measures. Inf. Retr. **13**(4), 375–397 (2010)
23. Rath, T.M., Manmatha, R.: Word spotting for historical documents. Int. J. Doc. Anal. Recogn. **9**(2–4), 139–152 (2007)
24. Revaud, J., Almazán, J., Rezende, R.S., Souza, C.R.D.: Learning with average precision: training image retrieval with a listwise loss. In: Proceedings of the IEEE International Conference on Computer Vision, pp. 5107–5116 (2019)
25. Rusinol, M., Aldavert, D., Toledo, R., Lladós, J.: Browsing heterogeneous document collections by a segmentation-free word spotting method. In: Proceedings of the International Conference on Document Analysis and Recognition, pp. 63–67 (2011)
26. Rusiñol, M., Aldavert, D., Toledo, R., Lladós, J.: Efficient segmentation-free keyword spotting in historical document collections. Pattern Recogn. **48**(2), 545–555 (2015)
27. Rusiñol, M., Lladós, J.: A performance evaluation protocol for symbol spotting systems in terms of recognition and location indices. Int. J. Doc. Anal. Recogn. **12**(2), 83–96 (2009)
28. Sudholt, S., Fink, G.A.: PHOCNet: a deep convolutional neural network for word spotting in handwritten documents. In: Proceedings of the International Conference on Frontiers in Handwriting Recognition, pp. 277–282 (2016)
29. Sudholt, S., Fink, G.A.: Evaluating word string embeddings and loss functions for CNN-based word spotting. In: Proceedings of the International Conference on Document Analysis and Recognition, vol. 1, pp. 493–498 (2017)
30. Valizadegan, H., Jin, R., Zhang, R., Mao, J.: Learning to rank by optimizing NDCG measure. Adv. Neural. Inf. Process. Syst. **22**, 1883–1891 (2009)
31. Weinberger, K.Q., Saul, L.K.: Distance metric learning for large margin nearest neighbor classification. J. Mach. Learn. Res. **10**, 207–244 (2009)
32. Weinman, J.J., Learned-Miller, E., Hanson, A.: Scene text recognition using similarity and a lexicon with sparse belief propagation. IEEE Trans. Pattern Anal. Mach. Intell. **31**(10), 1733–1746 (2009)
33. Wilkinson, T., Brun, A.: Semantic and verbatim word spotting using deep neural networks. In: Proceedings of the International Conference on Frontiers in Handwriting Recognition, pp. 307–312 (2016)

A-VLAD: An End-to-End Attention-Based Neural Network for Writer Identification in Historical Documents

Trung Tan Ngo⬤, Hung Tuan Nguyen$^{(\boxtimes)}$⬤, and Masaki Nakagawa⬤

Tokyo University of Agriculture and Technology, Tokyo, Japan
ntuanhung@gmail.com, nakagawa@cc.tuat.ac.jp

Abstract. This paper presents an end-to-end attention-based neural network for identifying writers in historical documents. The proposed network does not require any preprocessing stages such as binarization or segmentation. It has three main parts: a feature extractor using Convolutional Neural Network (CNN) to extract features from an input image; an attention filter to select key points; and a generalized deep neural VLAD model to form a representative vector by aggregating the extracted key points. The whole network is trained end-to-end by a combination of cross-entropy and triplet losses. In the experiments, we evaluate the performance of our model on the HisFragIR20 dataset that consists of about 120,000 historical fragments from many writers. The experiments demonstrate better mean average precision and accuracy at top-1 in comparison with the state-of-the-art results on the HisFragIR20 dataset. This model is rather new for dealing with various sizes of historical document fragments in the writer identification and image retrieval.

Keywords: Writer identification · Image retrieval · Historical documents · Convolutional neural network

1 Introduction

For many decades of collection, the number of excavated historical documents becomes countless. They require a huge effort of experts to read, analyze, and process. Therefore, the requirement of automatic artificial systems is indisputable. Writer identification between documents is a valuable piece of information. Due to aging, however, these documents are damaged, stained, degraded, faded, and even split into fragments, which adds serious difficulty.

Writer identification has been theoretically and practically a critical topic in document analysis. There are two kinds of writer identification: text-dependent and text-independent. The text-dependent methods compare samples with the same text contents. Meanwhile, the text-independent methods compare samples without any assumption on input contents. Generally, the latter is more challenging.

Through years of research, writer identification observes continuous development. The recent techniques have shown promising results on benchmark databases, such as IAM [1], Firemaker [2], ICDAR2013 WI [3]. However, handwritten patterns in these

© Springer Nature Switzerland AG 2021
J. Lladós et al. (Eds.): ICDAR 2021, LNCS 12822, pp. 396–409, 2021.
https://doi.org/10.1007/978-3-030-86331-9_26

databases are collected under controlled conditions by a small number of writers, which produces clean output images.

Nevertheless, many state-of-the-art (SOTA) techniques on these databases do not have equivalent results in historical ones. For example, their results are worse in datasets, such as ICDAR2017-Historical-WI [4], ICDAR2019 HDRC-IR [5], and HisFragIR20 [6], all of which are historical documents for writer identification.

Due to the achievements of deep learning techniques, many deep learning models are applied for writer identification tasks in recent years. Although these models accomplish high performance in the benchmark databases, they often perform worse than handcrafted feature extraction techniques [5, 6]. In the HisFragIR20 competition, most of the participated methods are based on convolution neural networks (CNN), but the best result is marginally better than the baseline method using handcrafted features. It achieves below 35% of mean average precision (mAP). There is still room for improvement of the deep learning approach in the writer identification, especially historical ones.

In this paper, we focus on the large-scale historical document fragment retrieval regarding writer identification. In handcrafted feature-based methods, the extracted features from inputs are aggregated by encoding techniques, such as bag-of-word (BOW), Fisher kernel [7], and VLAD [8]. The encoding vectors are then used to represent the handwriting style on those documents. Meanwhile, the deep learning-based techniques extract features from a full image or predefined image segments. They exploit the high effectiveness of CNNs to extract discriminative features in handwriting.

Our main contribution is to propose an attention-based neural network, namely A-VLAD, to extract handwriting style from noisy and variously sized samples. We also introduce a technique for fusing attention filters into the NetVLAD model. With the advantage of the VLAD algorithm, the attention mechanism, and the CNN architecture, our model achieves the SOTA performance on the HisFragIR20 database. Furthermore, we also apply the online triplet techniques in the training process.

This paper is organized as follows. Section 2 describes the related work of writer identification in terms of handcraft feature-based and deep neural network-based (DNN-based) methods. Section 3 presents an overview of the A-VLAD model. Section 4 describes the experiments on the HisFragIR20 database. Section 5 reports our experimental results. Finally, Sect. 6 draws conclusions.

2 Related Work

Many methods have been proposed for writer identification. In this paper, we summarize those related to text-independent identification. Generally, the writer identification methods follow the pipeline: preprocessing, feature extraction, and classification/sample matching. We divide the next section based on feature extraction: handcraft-based and DNN-based.

2.1 Handcrafted Feature-Based Methods

Researchers have been inventing writer-specific features in handwritten text. These features can be categorized into global and local features. Handwriting may have texture properties so that frequency transformation methods can extract the global features of a handwritten pattern. One sample of this approach is Oriented Basic Image Feature (oBIF) [9], which was applied to ICDAR2017-Historical-WI and HisFragIR20 competitions and got mAP at 56.82% and 24.1%, respectively. Other methods that follow this approach are the Gabor domain [10] and Local binary patterns (LBP) [11]. Extracting features from a whole image helps avoid text segmentation from the document background, thus preventing the accumulated error from this step.

Other methods focus on local features, such as edges, key points, text lines. The extracted local features are generally aggregated to form a representative vector for a whole image by codebook algorithms, such as Bag-of-Word (BoW), Fisher vectors, and VLAD [8]. Scale-invariant feature transform (SIFT) [12–14] is a well-known local feature descriptor, which has been widely used not only in writer identification topics but also in other image retrieval tasks. Especially in the ICDAR2019 HDRCIR competition, the Pathlet+SIFT+bVLAD [12] method gained the SOTA mAP at 92.5%.

Handcrafted feature-based methods achieve higher performance and comparable with the SOTA results [11, 12]. The advantage that needs less or neither ground truth nor data augmentation for training helps these methods be applied to a wide range of datasets. However, they need careful preprocessing, like slant or orientation correction and binarization. The processing pipeline also contains many sub-tasks, which can lead to a local minimum.

2.2 Deep Neural Network-Based Methods

In recent years, CNN models have proven their effectiveness in extracting complex features from images, and they are being applied for writer identification tasks. For example, the CNN models extract high-level features from words [15], patches [16, 17], and pages [18]. With the lack of sufficient training data, the CNN model [19] is trained using pseudo-classes, which are obtained by clustering extracted SIFT features. With the high discriminative power of the CNN model, FragNet [15] showed high accuracy. It achieved top-1 accuracy at 72.2% with only images of a few words in the IAM dataset.

As mentioned, however, the CNNs perform the impaired results in historical writer identification [5, 6]. In the HisFragIR20 competition [6], FragNet and the modified model of [19], namely ResNet20$_{ssl}$ got only 6.4% and 33.7% in mAP, respectively. Another approach in this competition, TwoPath, used both writer labels and page labels to train the ResNet50 model. It just got 33.5% in mAP. The high adaptation of deep neural networks may lead to overfitting, thereby tending to extract the background information, such as paper materials, pen types, and text contexts. Therefore, we propose the use of an attention mechanism as a filter for eliminating the background information while training the deep neural networks.

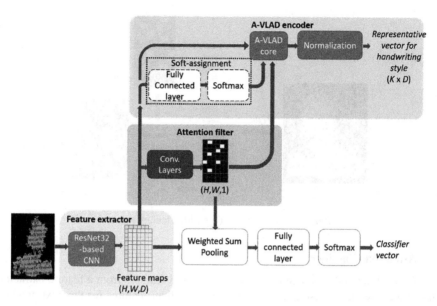

Fig. 1. Network architecture of A-VLAD.

3 Proposed Method

We seek another solution to overcome the issues mentioned in Sect. 2 by replacing the SIFT+VLAD pipeline in handcrafted-based methods with the deep neural network. The A-VLAD model has three main parts: a feature extractor, an attention classifier, and an A-VLAD core, as shown in Fig. 1. At first, the CNN module extracts local features from an image. The attention filter helps to find the key points so that useful local features are obtained. The A-VLAD network aggregates the selected local features to form a global representation.

In order to efficiently train the A-VLAD model, we apply both online triplet loss with an online mining strategy and softmax classification. The prediction output of classification supports the training process but not the inference process. We describe all of them in the next sections.

3.1 CNN-Based Feature Extraction

As mentioned above, CNNs effectively extract features from handwritten text [15, 16]. In this work, we use a CNN module as the local feature extractor. The CNN module is constructed by stacking convolutional, batch norm, and max pool layers. This part is designed to extract features in small local areas, preventing the CNN module from overfitting to background information. All local features of a whole image are fed to the classification module and the A-VLAD core. The output F is a matrix of the size (H, W, D), where H, W, D are, in turn, height, width, and feature dimension.

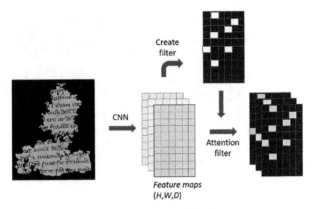

Fig. 2. The attention filter with a feature map.

3.2 Attention Classifier

In traditional methods, we need to select useful local features from a feature extractor. This process reduces noise and chooses only the salient features for the next process. With SIFT key points, they can be selected by the intensity of contrast with a threshold or top-k points. The key points are also extracted in predefined text areas [12]. However, the heuristic methods may lead to a local minimum. These methods are also hard to apply for end-to-end deep neural networks. We apply spatial attention to the extracted output F to overcome this issue.

Figure 2 shows the filtering process where the feature map determines the attention value at each point of the filter grid. We build up the attention filter by stacking up convolutions with the filter size 1×1 and ReLU functions. Features at points with zero attention values will be cut off in the next step. This process simulates the selection in traditional methods.

This attention filter will be used in the aggregation part in A-VLAD. Furthermore, we train it with another classification output to help the training process converge smoothly. Note that this classifier is not used in the inference step. Let A denote the attention matrix, which has the size $(H, W, 1)$. We have Eq. (1):

$$hidden_{ij} = f_{ij} * a_{ij}, f_{ij} \in F \tag{1}$$

$$sumpooling^d = \sum\nolimits_{i,j} hidden_{ij}^d \tag{2}$$

where $hidden_{ij}^d$ is the output features after filtering at the position i, j, and d^{th} feature. The term a_{ij} is the attention value at the position i, j and the grid $hidden_{ij}$ is then applied to the global sum pooling as Eq. (2). Next, the *sumpooling* vector with the length D is fed through a few fully connected layers. The final output vector has the same length as the number of writers in the training set. Finally, we apply the Softmax function and train it with the Negative log-likelihood Loss. Furthermore, the final loss function of this part is as follows:

$$L_{classifier} = CrossEntropy(Softmax(sumpooling)) \tag{3}$$

3.3 A-VLAD Core Architecture

The VLAD [8] encoding methods face the local-minimum problem when chaining separated modules in a system. NetVLAD [20], a differentiable VLAD layer, is proposed to connect with any CNN structures to deal with this problem. It is differentiable and can be trained in end-to-end systems. The pipeline of the VLAD algorithm has two main steps. Firstly, N feature vectors with the size of D are clustered to create a K-word codebook. The residuals between image descriptors and their corresponding cluster center are summed regarding each cluster. Then, this result is used as the final representative vector V with the size $(K \times D)$.

To simulate the VLAD algorithm, the NetVLAD model creates K D-dimension vectors used as the cluster centers. The input for this model is a set of N local image descriptors X. The sum residual between these descriptors and each cluster center is calculated as Eq. (4).

$$V_k^d = \sum_{i=0}^{N} s_k(x_i) * \left(x_i^d - c_k^d \right), x_i \in X \tag{4}$$

where x_i^d, c_k^d indicate, in turn, the values at d^{th} dimension of an encoded vector x_i and a cluster center c_k, the function $s_k()$ describes the probability that the vector x_i belongs to the cluster c_k.

In VLAD, the residual of each descriptor vector is calculated with only its closest cluster center. However, NetVLAD uses the soft-cluster assignment instead of the hard-cluster assignment of VLAD as Eq. (5) to make this model differentiable:

$$s_k(x_i) = \frac{e^{w_k^T x_i + b_k}}{\sum_{k'}^{K} e^{w_{k'}^T x_i + b_{k'}}} \tag{5}$$

where w_k and b_k are, in turn, a weight and a bias of the fully connected layer. Furthermore, Eq. (5) is the SoftMax function for the K cluster centers.

Both VLAD and NetVLAD equally use all the input descriptors, which leads to ineffectiveness when dealing with noisy data. Therefore, we apply the attention filter to focus on the salient descriptors. We modify Eq. (4) into Eq. (6) as follows:

$$V_k^d = \sum_{i=0}^{H} \sum_{j=0}^{W} s_k(f_{ij}) * \left(f_{ij}^d - c_k^d \right) * a_{ij} \tag{6}$$

The residual matrix is then normalized as the NetVLAD model. The attention filter reduces the effect of noise points, which contain less handwriting style information in the final representative vector. Then, we use the online triplet training method in the training process. In the testing process, we also apply the PCA-whitening to reduce the size of embedding vectors. It helps to decrease the noisy features and avoid the curse of dimensionality.

3.4 Online Triplet Training

The triplet loss is commonly used to train image retrieval models [21]. It keeps intra-class distances being smaller than inter-class distances. Let q be a query sample; p and n be reference samples in the training dataset. A triple of $\{q, p, n\}$ is valid if p has a similar class with q while n has a different class with q. Equation (7) describes the triplet loss for a valid triplet.

$$L(q, p, n) = \max\{dis(q, p) - dis(q, n) + margin, 0\} \tag{7}$$

where $dis()$ is a distance function; *margin* is a predefined minimum gap between the two kinds: a positive pair (q, p) and a negative pair (q, n). The distance function plays an important role in training A-VLAD. Generally, there are some typical distances for writer identification, such as Euclidean, Chi-Square, Mahalanobis, and Manhattan. Because the Euclidean distance provides a lower contrast between the farthest and nearest neighbors while the Manhattan distance gives a higher contrast with increasing dimensionality [22]. Thus, the Manhattan distance is preferable for high-dimensional data mining applications. Because the final encoded vector V of A-VLAD ($K \times D$) is high dimensional, we use the Manhattan distance as the $dis()$ in Eq. (7).

The optimizing of the triplet loss depends on the triplet selection. There are two ways to prepare triplets [23]. The first one is offline mining, in which the triplet sets are predefined before the training process. For effectiveness, it requires manual selection rather than random. The manual selection finds the triplets, in which negative pairs are visually similar while positive pairs are visually different. Therefore, they force neural networks to learn salient features without overfitting to wrong ones.

The other one is online mining, which selects the triplets during the training process. It computes a batch of encoded vectors and then selects all possible valid triplets from this batch. In this research, let T be all triplet losses built-up from the batch, and $\overline{T}$ be hard triplet losses, which are greater than 0. We describe the triplet loss function as Eq. (8) and the final loss for the whole A-VLAD model as Eq. (9), where α is a weight for balancing the two losses.

$$L_{triplet} = \frac{\sum \overline{T}}{|\overline{T}|} \tag{8}$$

$$Loss = \alpha * L_{classifier} + (1 - \alpha) * L_{triplet} \tag{9}$$

4 Experiments

To evaluate the proposed A-VLAD model, we conducted experiments on the HisFra-gIR20 competition dataset [6]. The dataset details, evaluation methods, and implementation details are described in the following subsections.

4.1 Datasets

The HisFrag20 dataset consists of both training and test sets. The training set consists of fragments extracted from the images belonging to the public Historical-IR19 test set. These images come from 3 sources: manuscripts, charters, and letters. They are split into smaller fragments by random and rectangular shapes. The fragments without any text pixels are filtered out. The test set consists of different manuscript and letter images, which have not been appeared in the training set.

The training set comprises 101,706 fragments of 8,717 writers from 17,222 document images. The HisFrag20 competition allows using additional data from other historical documents datasets for the training process. However, we only use the provided training set for our experiments. We also split the original training dataset into sub-training and validation sets. The sub-training set has 90,639 fragments by 7,845 writers, and the validation subset has 11,067 fragments by 872 writers. We evaluate the A-VLAD model on the HisFrag20 test set, which consists of 20,019 fragments from 2,732 images written by 1,152 writers. The sub-training, the validation, and the test sets are disjoint.

During the training process, we apply augmentation every iteration, such as random affine transforms and Gaussian noises, to increase the amount of data. The input batches are standardized by randomly cropping 512×512 patches from the origin fragments. Meanwhile, we use full fragments in the testing process.

4.2 Evaluation Methods

Each test fragment is used as a query sample, while the remaining ones in the test set are considered reference samples. We use the Manhattan distance in Eq. (7) to calculate similarities between the query and reference samples and create an ascending ranked list of similarities. The lists of all test fragments are then combined into a $20,019 \times 20,019$ matrix. We apply the same evaluation in the HisFrag-IR20 competition to compare with other methods.

The average precision for a query q is computed as Eqs. (10) and (11).

$$Pr_q(r) = \frac{R_q(r)}{r} \tag{10}$$

$$AP_q = \frac{\sum_{r=1}^{S} Pr_q(r) \cdot rel_q(r)}{\sum_{r'=1}^{S} rel_q(r')} \tag{11}$$

where q is the index of the query; $Pr_q(r)$ is the precision at rank r; $R_q(r)$ is the number of relevant samples for a query q in top-r samples; $rel_q(r)$ indicates the ground truth relation between a reference sample at rank r and the query q, one if relevant and zero otherwise; and S is the size of the ranked list.

Then, the mean Average Precision (mAP) is defined as Eq. (12).

$$mAP = \frac{1}{N} \sum_{q=0}^{N} AP_q \tag{12}$$

where N is the number of query samples.

The accuracy at top 1 is calculated. The mean precision at top-10 (Pr@10) and that at top-100 (Pr@100) are also computed as Eq. (13).

$$Pr@K = \frac{1}{N} \sum_{q=1}^{N} \frac{R_q(K)}{\min\big(R_q(S), K\big)} \tag{13}$$

4.3 A-VLAD Configuration

We derive the pre-trained ResNet32 for the CNN-based feature extractor, which allows our model to have good initialization for training. In detail, we use only the first 13 layers of ResNet32 instead of the full 32 layers. We also integrate a convolution layer to control the dimension of the feature descriptors. We use two convolutional layers with Rectified Linear Unit (ReLU) activation functions for the attention filter. Tables 1 and 2 describe the detailed configurations for the CNN-based feature extractor and the attention filter.

Table 1. Configuration of our CNN model in the feature extractor.

Type	Configurations
Input	$H \times W \times 3$
Convs_ResNet32	First 13 layers of ResNet32
Conv14	Number of kernels: d (depend on the setting), Kernel size: 1×1, strike: 1, padding size: 1

Table 2. Configuration of our attention filter.

Type	Configurations
Input	$H \times W \times d$ (depend on the setting) feature matrix
Conv1 – ReLU	Number of kernels: 512, kernel size: 1×1, strike: 1, padding size: 0
Conv2 – ReLU	Number of kernels: 1, kernel size: 1×1, strike: 1, padding size: 0

For training the A-VLAD model, each training batch has 18 samples with only 12 distinct classes, and the validated batch's size is 12 with only 8 distinct classes because we apply the online triplet mining method. In Eq. (8), α is set at 0.15. For updating the A-VLAD parameters, we use an Adam optimizer with a learning rate of 5e−5. The model is trained with 100 epochs. In the testing process, we apply PCA-whitening to reduce the dimension of the encoded vectors to 500.

5   Results and Discussion

We compare different hyperparameter configurations of local descriptors and clusters in Sect. 5.1. We demonstrate our model results and the attention filter on HisFragIR20 in Sect. 5.2.

5.1 Hyperparameter Finetuning

To evaluate the effect of the number of channels in descriptors and the number of clusters, we conducted experiments with seven different settings. The best model of each hyperparameter configuration is chosen using the accuracy on the validation set. The results of all settings on the test set are declared in Table 3.

Table 3. Performance (%) using the different hyperparameter settings.

Model index	#channels	#clusters	mAP	top-1	Pr@10	Pr@100
S1	32	64	37.31	78.21	57.71	55.17
S2	32	128	39.10	78.86	59.29	57.03
S3	64	32	41.44	82.04	61.78	58.61
S4	64	64	43.25	82.19	63.12	60.71
S5	64	128	40.57	81.21	61.15	57.73
S6	128	64	36.75	79.17	57.64	53.78
S7	**128**	**128**	**46.61**	**85.21**	**65.76**	**63.25**

The highest mAP of 46.61% is from the S7 setting with 128 channels and 128 clusters. With the same number of 128 channels, the S6 setting achieves 36.75% mAP. It implies that the number of clusters should be as large as the number of channels. Meanwhile, with the same number of 128 clusters, the results based on 32 and 64 channels show that it lacks features to identify handwriting style.

The settings from S1 to S5 show the effect of the smaller numbers of channels. The settings with 32 channels perform much lower than the settings with 64 channels. Among the settings with 64 channels, the S4 setting achieves the highest mAP of 43.25%. It again demonstrates that the number of clusters should be as large as the number of channels.

For top-1 accuracy, Pr@10 and Pr@100 precisions, their highest values are consistent with the mAP results.

5.2 Performance on HisFragIR20 Dataset

Table 4 shows the A-VLAD model with the S7 setting compared to other methods in the HisFragIR20 competition. The results imply that A-VLAD outperforms the state-of-the-art method from the previous work [6] in mAP, top-1 accuracy, and Pr@10, Pr@100. Among the traditional methods, the SRS-LBP model achieves the highest performance in the HisFragIR20 competition. The A-VLAD model gains 13.2% points higher than SRS-LBP in mAP. To compare with the other deep neural networks, the A-VLAD model performs around 13% points higher than the ResNet20$_{ssl}$ model.

Table 4. Performance (%) on the HisFragIR20 test set.

Method	mAP	top-1	Pr@10	Pr@100
SRS-LBP [11]	33.4	60.0	46.8	45.9
FragNet [15]	6.4	32.5	16.8	14.5
TwoPath$_{writer}$ [6]	33.5	77.1	53.1	50.4
TwoPath$_{page}$ [6]	25.2	61.1	41.2	44.1
ResNet20$_{ssl}$ [6, 19]	33.7	68.9	52.5	46.5
oBIF [9]	24.1	55.4	39.2	37.9
Our model (S7)	**46.61**	**85.21**	**65.76**	**63.25**

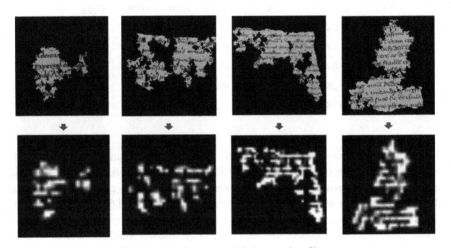

Fig. 3. Input images and their attention filter.

Furthermore, we visualize the attention filters, as shown in Fig. 3. The first row is composed of the fragment images, while the second row is composed of the attention filters. These attention filters seem to focus on the text areas and ignore the backgrounds. It demonstrates the effectiveness of the attention filters in the A-VLAD model.

In Fig. 4, we show the top-3 results of the best and worst queries in the HisFragIR20 test set. Fragment images from the same writer with the query are bounded in green. Meanwhile, those of different writers are bounded in red. We can see that all three worst samples lack text.

Query image   Top-3 result

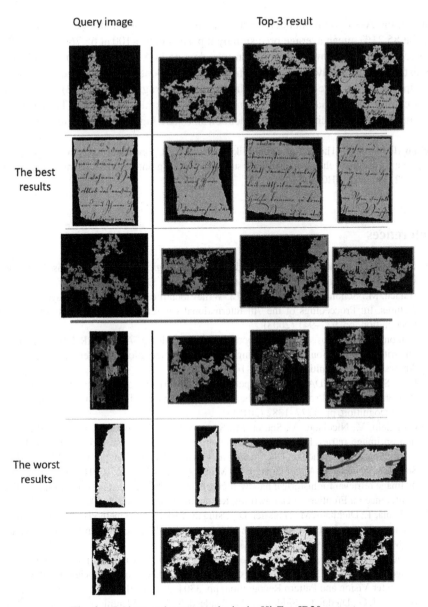

The best
results

The worst
results

Fig. 4. The best and worst results in the HisFragIR20 test set.

6   Conclusion

This paper presented an attention-based VLAD, A-VLAD in short, for identifying writers in historical document fragments. Following the experiments on the HisFragIR20 dataset, the A-VLAD model achieved mean average precision at 46.61%, top-1 accuracy at 85.21%, mean average precision by top-10, and top-100 at 65.76% and 63.25%, respectively. The following conclusions are drawn: 1) A-VLAD is trained end-to-end for extracting the handwriting style of historical documents in various size fragments. 2) It outperforms the state-of-the-art methods in the HisFragIR20 dataset. 3) The attention filter can find appropriate locations on fragments to extract features. 4) The relation of the number of feature channels and codebook size has a high effect on the retrieval result.

Acknowledgement. The authors would like to thank Dr. Cuong Tuan Nguyen for his valuable comments. This research is being partially supported by the grant-in-aid for scientific research (S) 18H05221 and (A) 18H03597.

References

1. Marti, U.V., Bunke, H.: The IAM-database: an English sentence database for offline handwriting recognition. Int. J. Doc. Anal. Recognit. **5**, 39–46 (2003)
2. Bulacu, M., Schomaker, L., Vuurpijl, L.: Writer identification using edge-based directional features. In: Proceedings of the 7th International Conference on Document Analysis and Recognition, pp. 937–941 (2003)
3. Louloudis, G., Gatos, B., Stamatopoulos, N., Papandreou, A.: ICDAR 2013 competition on writer identification. In: Proceedings of the 12th International Conference on Document Analysis and Recognition, pp. 1397–1401 (2013)
4. Fiel, S., et al.: ICDAR2017 competition on historical document writer identification (Historical-WI). In: Proceedings of the 14th International Conference on Document Analysis and Recognition, pp. 1377–1382 (2018)
5. Christlein, V., Nicolaou, A., Seuret, M., Stutzmann, D., Maier, A.: ICDAR 2019 competition on image retrieval for historical handwritten documents. In: Proceedings of the 15th International Conference on Document Analysis and Recognition, pp. 1505–1509 (2019)
6. Seuret, M., Nicolaou, A., Stutzmann, D., Maier, A., Christlein, V.: ICFHR 2020 competition on image retrieval for historical handwritten fragments. In: Proceedings of 17th International Conference on Frontiers in Handwriting Recognition, pp. 216–221 (2020)
7. Jaakkola, T., Diekhans, M., Haussler, D.: Using the Fisher kernel method to detect remote protein homologies. In: Proceedings of the 7th International Conference on Intelligent Systems for Molecular Biology, pp. 149–158 (1999)
8. Jégou, H., Douze, M., Schmid, C., Pérez, P.: Aggregating local descriptors into a compact image representation. In: Proceedings of the 23th IEEE Computer Society Conference on Computer Vision and Pattern Recognition, pp. 3304–3311 (2010)
9. Abdeljalil, G., Djeddi, C., Siddiqi, I., Al-Maadeed, S.: Writer identification on historical documents using oriented basic image features. In: Proceedings of 16th International Conference on Frontiers in Handwriting Recognition, pp. 369–373 (2018)
10. He, Z., You, X., Zhou, L., Cheung, Y., Du, J.: Writer identification using fractal dimension of wavelet subbands in gabor domain. Integr. Comput. Aided Eng. **17**, 157–165 (2010)
11. Nicolaou, A., Bagdanov, A.D., Liwicki, M., Karatzas, D.: Sparse radial sampling LBP for writer identification. In: Proceedings of the 13th International Conference on Document Analysis and Recognition, pp. 716–720 (2015)
12. Lai, S., Zhu, Y., Jin, L.: Encoding pathlet and SIFT features with bagged VLAD for historical writer identification. IEEE Trans. Inf. Forensics Secur. **15**, 3553–3566 (2020)

13. Fiel, S., Sablatnig, R.: Writer identification and writer retrieval using the fisher vector on visual vocabularies. In: Proceedings of the 12th International Conference on Document Analysis and Recognition, pp. 545–549 (2013)
14. Khan, F.A., Khelifi, F., Tahir, M.A., Bouridane, A.: Dissimilarity Gaussian mixture models for efficient offline handwritten text-independent identification using SIFT and RootSIFT descriptors. IEEE Trans. Inf. Forensics Secur. **14**, 289–303 (2018)
15. He, S., Schomaker, L.: FragNet: writer identification using deep fragment networks. IEEE Trans. Inf. Forensics Secur. **15**, 3013–3022 (2020)
16. Xing, L., Qiao, Y.: DeepWriter: a multi-stream deep CNN for text-independent writer identification. In: Proceedings of the 15th International Conference on Frontiers in Handwriting Recognition, pp. 584–589 (2016)
17. Nguyen, H.T., Nguyen, C.T., Ino, T., Indurkhya, B., Nakagawa, M.: Text-independent writer identification using convolutional neural network. Pattern Recognit. Lett. **121**, 104–112 (2019)
18. Tang, Y., Wu, X.: Text-independent writer identification via CNN features and joint Bayesian. In: Proceedings of the 15th International Conference on Frontiers in Handwriting Recognition, pp. 566–571 (2016)
19. Christlein, V., Gropp, M., Fiel, S., Maier, A.: Unsupervised feature learning for writer identification and writer retrieval. In: Proceedings of the 14th International Conference on Document Analysis and Recognition, pp. 991–997 (2017)
20. Arandjelovic, R., Gronat, P., Torii, A., Pajdla, T., Sivic, J.: NetVLAD: CNN architecture for weakly supervised place recognition. IEEE Trans. Pattern Anal. Mach. Intell. **40**, 1437–1451 (2018)
21. Bui, T., Ribeiro, L., Ponti, M., Collomosse, J.: Compact descriptors for sketch-based image retrieval using a triplet loss convolutional neural network. Comput. Vis. Image Underst. **164**, 27–37 (2017)
22. Aggarwal, C.C., Hinneburg, A., Keim, D.A.: On the surprising behavior of distance metrics in high dimensional space. In: Van den Bussche, J., Vianu, V. (eds.) ICDT 2001. LNCS, vol. 1973, pp. 420–434. Springer, Heidelberg (2001). https://doi.org/10.1007/3-540-44503-X_27
23. Schroff, F., Kalenichenko, D., Philbin, J.: FaceNet: a unified embedding for face recognition and clustering. In: Proceedings of the 28th IEEE Computer Society Conference on Computer Vision and Pattern Recognition, pp. 815–823 (2015)

Manga-MMTL: Multimodal Multitask Transfer Learning for Manga Character Analysis

Nhu-Van Nguyen$^{(\boxtimes)}$, Christophe Rigaud , Arnaud Revel ,
and Jean-Christophe Burie

Laboratory L3i, SAIL Joint Lab, La Rochelle University,
17042 La Rochelle, CEDEX 1, France
nhu-van.nguyen@univ-lr.fr

Abstract. In this paper, we introduce a new pipeline to learn manga character features with visual information and verbal information in manga image content. Combining these set of information is crucial to go further into comic book image understanding. However, learning feature representations from multiple modalities is not straightforward. We propose a multitask multimodal approach for effectively learning the feature of joint multimodal signals. To better leverage the verbal information, our method learn to memorize the content of manga albums by additionally using the album classification task. The experiments are carried out on Manga109 public dataset which contains the annotations for characters, text blocks, frame and album metadata. We show that manga character features learnt by the proposed method is better than all existing single-modal methods for two manga character analysis tasks.

Keywords: Manga image analysis · Multimodal learning · Auxiliary task learning · Transfer learning

1 Introduction

Digital comics and manga (Japanese comics) wide spread culture, education and recreation all over the world. They were originally all printed but nowadays, their digital version is easier to transport and allows on-screen reading with computers and mobile devices. To deliver digital comics content with an accurate and user-friendly experience on all mediums, it might be necessary to adapt or augment the content [2]. This processing will help to create new digital services in order to retrieve very precise information in a corpus of images. For instance, comic character retrieval and identification would be useful as copyright owners need efficient and low-cost verification systems to assert their copyrights and detect possible plagiarisms [24]. In our paper, we use the term "character" as

This work is supported by the Research National Agency (ANR) in the framework of the 2017 Lab-Com program (ANR 17-LCV2-0006-01).

© Springer Nature Switzerland AG 2021
J. Lladós et al. (Eds.): ICDAR 2021, LNCS 12822, pp. 410–425, 2021.
https://doi.org/10.1007/978-3-030-86331-9_27

the person in comic books. From this point on, the term "character" in character retrieval, character classification, character clustering should be understood as "comic/manga character".

Current work [11,30] on comic/manga character analysis often rely on transfer learning [19]. When it is hard to perform a target task directly on comic character images due to the missing of labels, one can learn the character visual features basing on related tasks then use the learnt features to perform the target task. In these methods, a visual feature extractor (uni-modal) is learnt using the character classification task and then it is used to extract features to perform character retrieval, identification or clustering tasks.

However, single modality (e.g. graphics) is limited and can not profit from all the information we can extract or infer from static images. In comic/manga image analysis domain, recent works state that learning using only a single modality, as very often the image, can not cope with feature changes in some images and suggest to take advantage of another modality (text in their study) simultaneously with the first one [30]. Existing works have often ignored the implicit verbal information in the comic book images which are presented in form of speech text or narrative text. This source of verbal information is important, especially since the text recognition technology is strong nowadays.

Theoretically, multimodal analysis is able to retrieve more information and of an higher level than the uni-modal scenario, with an overall better performance. However, training from multiple modalities is actually hard and often results with lower performance than the best uni-modal model. The first reason is that it is easy to overfit: the learnt patterns from a train set that do not generalize to the target distribution [28]. Another reason is that one of the modality might interacts with the target task in a more complex way than the other [22]. For example, in our case, the image modality is relevant for characters retrieval/clustering because we can identify the character name and the character emotion by looking at the image. In contrast, the text modality is more adapted for album retrieval/clustering or character relationship analysis than the character retrieval/clustering. This is due to the text in each album that does not describe a specific character but instead tells a specific story involving many characters. While both modalities seem likely to be complementary, there are essential for other elements such as emotion analysis that can be jointly extracted from character face expression, speech balloon shape and speech text processing.

Multitasks analysis [18,22] is a powerful approach to train a deep model, especially when the target task, in a single task context, is difficult to train to differentiate between relevant and irrelevant features. In our work, we propose to use an auxiliary task to facilitate the multimodal training. The multitask multimodal methods we propose for learning the joint feature of multimodal signals (text and image) can be used subsequently in different analysis tasks for comic and manga image analysis such as character clustering, character retrieval and emotion recognition.

The contributions of this paper are summarized as follows:

- Instead of using only visual information in comic/manga images, we propose a new pipeline which uses the implicit verbal information in the images to learn multimodal manga character features.
- Different methods for learning the comic/character feature are analysed. To achieve the best performance, we propose a self-supervised strategy with the album classification task. This strategy forces the model to remember the manga content which can improve the learnt multimodal features.
- The effectiveness of the learnt multimodal feature is demonstrated by transferring it to character retrieval and clustering tasks.

2   Related Work

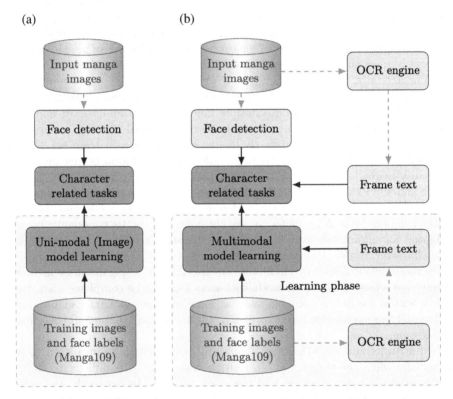

a. Current methods: Uni-modal manga character learning

b. Our pipeline: Multimodal manga character learning

Fig. 1. Current methods (a) compare with the proposed pipeline (b): Blue boxes and arrows show our contributions, other parts are not in the scope of this paper. (Color figure online)

Our work focuses on comic character feature extraction to perform related tasks such as character retrieval, clustering and emotion recognition. Most of the previous works have focused on these tasks independently and using only image-based features (uni-modal).

Regarding the character retrieval task, large handcrafted methods have started to be proposed in 2017 [12]. Later and similarly, convolution neural networks (CNN) and fine-tuning of two CNN on manga face images have been applied [15]. They used with-and-without screen-tone images to perform accurate query- and sketch- based image retrieval. Recently, the challenge of long-tail distribution data like manga character (face) images, inappropriate for using the usual cross-entropy softmax loss function, have been tackled [11]. To do so, the authors proposed a dual loss from dual ring and dual adaptive re-weighting losses which improved the mean average precision score by 3.16%.

Few methods have been proposed for character (face) clustering. Tsubota et al. [25] proposed to fine tune a CNN model using deep metric learning (DML) in order to get help from domain knowledge such as page and frame information. They include pseudo positive and negative pairs into the training process and use k-means clustering given the number of characters for the final clustering. The method applies its learning stage for each specific album of comic to learn the character features before realizing the character clustering for the album.

Yanagisawa et al. [30] analysed different clustering algorithms on the learnt features after training a CNN model on the character classification task. The HDBSCAN clustering algorithm [3] was selected because of its better performance and no requirement of the number of cluster priors (unlike k-means). The authors mention that "learning using only images cannot cope with feature changes in some character face images" and suggest to "utilize word information contained in manga simultaneously with images".

This suggestion makes the link with text analysis and natural language processing techniques that can be combined for retrieving character names, relationships, emotions etc. In 2018, a survey mentions that "research has been done on emotion detection based on facial image and physiological signals such as electroencephalogram (EEG) while watching videos [...]. However, such research has not been conducted while reading comics." [2].

From our knowledge, there is only one study about recognizing emotion from text and facial expression in manga images. This work is part of an end-to-end comic book analysis and understanding method able to text-to-speech a comic book image with respect of character identification, gender, age and emotion of each speech balloon [29]. It combines several modalities using computer vision, natural language processing and speech processing techniques.

Multimodal approaches are usually developed for image/video classification and retrieval where multiple modalities are available explicitly and strong for the task [16]. From our knowledge, there is no multimodal work on comic/manga book images where the verbal information is included in the image modality. Moreover, the verbal information in comic/manga images is very weak for the character classification task so it is not trivial to apply multimodal learning methods to comic book images. To overcome this issue, we may think about finding another suitable task to learn the features from both modalities, but it is not a good choice. Firstly, because it is hard to find a suitable task with necessary labels for learning. Secondly, the character classification task is naturally a relevant task to learn the comic character features. Another solution is to add a useful task which will guide the model to learn relevant features to boost the performance [26].

3   Proposed Method

3.1   Overview

The digital comic book datasets are often composed by images partially annotated [1,7,10,17]. This is the reason why most of researches in comic book images focus on exploiting the visual information to extract the features and apply to other tasks in this domain. In reality, the comic book may come with the album metadata which includes some basic information such as the book name, the author, published date, a short summary of the story etc. Thus, the information from metadata is limited. Some datasets provide also text transcriptions based on an Optical Character Recognition system (OCR) [10,17].

Concerning the feature extraction in comic book images, researchers often ignore the text in the comic book images, including speech text (the text that a character speaks) and narrative text (the text describing the flow of the situation in the comic story). However, text and images can be combined to form singular units of expressions in comics as mentioned by Cohn et al. [4], so it is very important to take into account the verbal information while performing comic analysis tasks, including character retrieval and recognition.

The high quality of text recognition technology nowadays gives us a big opportunity to take advantages of digitized text as a source of verbal information [14]. In this work, we consider the comic character face recognition and the comic text recognition to be performed previously. Hence, we have these two sources of information to study the related tasks. To simplify the experiments, and allow others to reproduce them, we directly use the ground-truth information of the face and the text from Manga109 dataset to learn the character feature and analyze our multimodal approach. To associate the face image with text and form a multimodal data sample, we associate a face image with all the text in the same comic frame. We assume that the text in the frame is related directly to the character whose the face is presented in the frame (speaking the text or receiving the speech). Our proposed pipeline is illustrated in Fig. 1.

Assuming that we have a set of comic character samples where each sample contains two modalities $text$ and $image$ defined as T and V, respectively. The existing works [11,18,30] use V as the only information to learn or manually create the feature extractor which is then used to extract the features and realize the character related task such as multimodal and/or cross-modal character retrieval, character clustering, character recognition, character emotion recognition, etc. All of the learning approaches use character classification task to train the feature extractor which aims at classifying all samples (face image and associated text) in the database of multiple albums and authors.

In our work, the objective is to learn the character extractor from both T and V which are then used to perform the character related tasks. In single-modality setting, we learn the two modalities separately: the verbal feature extractor $E_T : T \rightarrow F_T$ and the visual feature extractor $E_V : V \rightarrow F_V$. In multimodality setting, we can combine the two sources of information to learn a combined feature extractor $E_{mix} : (T, V) \rightarrow F_{mix}$.

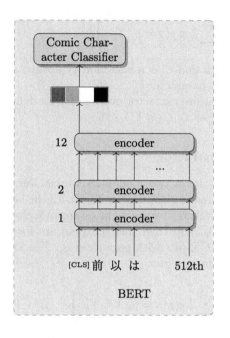
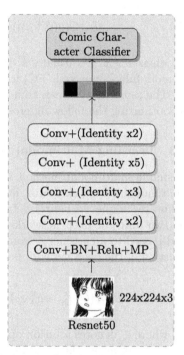

a. Text classification by BERT [5]   b. Image classification by Resnet

Fig. 2. Learning two feature extractors separately

In the next sections, we analyze different strategies corresponding to several settings to learn from the two modalities of comic book images. We will focus on the late fusion strategy because in this case the early fusion strategy does not work well, which will be explained later. We will explain and prove by empirical experimentation that the multimodal training using the character classification does not work. Then we will propose a new way to learn the multimodal feature extractor.

3.2   Learning Uni-modal Feature Extractor

In the single modality setting, we learn the feature representations separately (Fig. 2). Given a training set of a modality, for example the visual modal V: $C_v = \{X_1^v, ..., X_n^v, y_1, ..., y_n\}$, where X_i^v is the visual information of i-th training example and y_i is its true label (the character identity), we can learn the single modality feature extractor by training a neural network with respect to the classification task. The loss function can then be the cross entropy loss, an ubiquitous loss in modern deep neural networks.

$$L_v(C(\theta(X^v), y)) = -\frac{1}{N}\left(\sum_{i=1}^{N} \log p(y = y_i | X_i^v)\right) \tag{1}$$

where $\theta^v(X^v)$ is usually a deep network and C is a classifier, typically one or more fully-connected (FC) layers with a parameter θ_c^v.

The same way, we can train another feature extractor on the text modality by optimizing the following cross entropy loss $L_t(C(\theta(X^t), y))$.

3.3   Learning Multimodal Feature Extractor

A simple multimodal method is to jointly learn two modalities by training a multimodal feature extractor, based on character classification task with the late fusion technique. In neural network architecture, two modalities are processed by two different deep networks θ_X^t and θ_X^v, and then their outputs are fused and passed to a classifier C composed of FC layers with parameter θ_c. The loss function of the deep network is the cross entropy loss:

$$L_{mix}(C(\theta_X^t \oplus \theta_X^v, y)) = -\frac{1}{N}\left(\sum_{i=1}^{N} \log p(y = y_i | X_i^v, X_i^t)\right) \tag{2}$$

where $\oplus$ denotes a fusion operator (e.g. max, concatenation, addition, etc.).

One can argue that this multimodal feature extractor must be better than or equal to the best single-modal feature extractor because in the worst case the model will learn to mute all the parameters in one modality (θ_X^t or θ_X^v) and then becomes a single-modal model.

In our case, the multimodal approach is also worst than the single-modal based on the visual information (see Sect. 4). This bad performance is due to two problems. According to Wang et al. [28], the main problem is overfitting. When training any multimodal model, the deep network has nearly twice as many parameters as a single-modal model, and one may suspect that the overfitting is caused by the increased number of parameters.

We understand that another important problem comes from the target classification task used to train the multimodal deep network. All mentioned approaches do not work if the signals in the two modalities are very different in term of the target classification task: one modality is dominant for the task while another one is weak or complex. For comic character analysis tasks, the signal in face image are very strong, we can identify the character name and the character emotion by looking at the image. The signal in the text is however more complex. To decide if the two texts come from the same character (in the context of a multiple-albums dataset), one needs to integrate a lot of knowledge to reason, including for each album, the story, the relationship between characters, names, places, events etc. So when we train the multimodal deep network with the two modalities on the character classification task, the model parameter θ_X^t often stucks at a bad local maximum because of the weak and complex signal in the text, which leads to the sub-optimum of the parameter θ_c. One can check the signal of a modality by training a single-modal model on that modality to see the performance. In our case, the performance of text-modal model on the character classification task is bad (see Sect. 4).

To overcome this difficulty and benefit from the advantages of the text modality, we propose to learn the text features by a self-supervised learning method. As mentioned before, we have to read all the manga albums to identify the character from a piece of text, so we are going to train the model to remember the contents of all the albums in the training dataset. In this work, we propose to use an already-available information to learn the strong signal in the text: *the album name*. The text signal is strong to identify the albums as the text in each book tells a different story. Training the model with the album classification task is an effective way to learn relevant information from manga albums. It is also helpful to differentiate the characters identity because the characters from different comic stories are different. We know the album names of each image from the comic book images dataset [6].

In the next section, we propose a multimodal multi-task learning (M-MTL) architecture which can learn a good feature extractor for character related tasks from the two modalities with two tasks: character classification and album classification.

3.4   Manga M-MTL Modal for Comic Character Analysis Tasks

We propose a new multimodal architecture with three outputs, two for the main task character classification and one for the auxiliary album classification task. We use popular techniques to reduce overfitting such as Dropout [23], EarlyStopping, and Transfer learning as suggested in [28].

The Manga M-MTL architecture is shown in Fig. 3. It consists in two different feature components: the image network for learning visual features and the text network for learning the textual features. Three classifiers are added into our architecture. Each classifier, composed of FC layers, is used to perform the character classification task and the album classification task based on the visual feature, the textual feature, and the combined (concatenated or averaged) features. The character classification loss for the combined features is the main loss of the model, while the two other losses are the auxiliary losses which can help the former to learn better combined features.

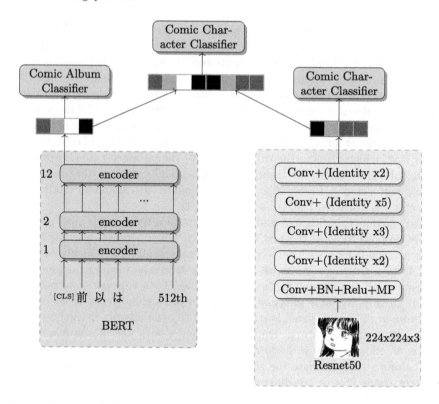

Fig. 3. Manga M-MTL architecture: multimodal learning with the auxiliary task.

Text Network: The transfer learning is well studied in natural language processing (NLP) domain. Strong pre-trained models such as BERT [5], GPT2 [20] are successfully used to solve many sub-tasks. We use the popular architecture Transformer and pre-trained weights BERT for the text network. Transformer is a strong architecture which can learn the feature representations of text based on its famous attention mechanism. The architecture we used is the BERT base architecture which has 12 attention layers, 768-dim hidden units, 12 heads.

Image Network: As for the text network, we adopt a common CNN architecture in the vision domain for our image network. Resnet [8] is widely used in the vision domain because of the capacity of avoiding the gradient vanishing and the fast computation capacity. We use the 50 layers Resnet architecture.

3.5   Implementation Details

Training Feature Extractor Models. We use WordPiece tokenizer [5] to process the text input before feeding them into the transformer network. All face images are resized to (224×224). We use image augmentation during training, including random crop (192×192) and rotate. The pre-trained weights using Japanese Wikipedia (performed by the Inui Laboratory, Tohoku University) are used for the BERT network. The ImageNet pre-trained Resnet weights are used for the CNN network. Dropout is applied to the BERT output layer and the Resnet output layer with a probability of 0.2. Early stopping is considered within 20 epochs. We train our models using SGD optimizer with a momentum of 0.9, a learning rate of 0.0001 for 100 epochs. For the multimodal models, we use the pre-trained weights of uni-modal models and the concatenation as the fusion operator, which is the best choice according to our experiments.

Extracting Features to Use in Manga Character Related Tasks. We extract the second-last FC layer in the two classifiers: album classifier and the final character classifier. Then we apply L2 regularization before concatenate these two features for character clustering and character retrieval tasks. The character clustering task is applied for each album as described by Yanagisawa et al. [30]. It is a measure of the intra-album character identification capacity of the learnt features. The character retrieval task is applied for all character in the test albums which gives a measure of the inter-album character identification capacity of the learnt feature, as presented by Li et al. [11].

For realizing the character clustering, we need to reduce the dimensions of the extracted multimodal features of manga characters to reduce the computation cost in the clustering process. We follow the same setting and the parameters configuration as described by Yanagisawa et al. [30] where the dimension reduction algorithm is UMAP [13] and the clustering algorithm is HDBSCAN [3].

To do the character retrieval, we rank the retrieval results by the cosine similarity of the extracted multimodal features, same as described by Li et al. [11].

4   Results

4.1   Experiment Protocol

Dataset: We use the large-scale Manga109 dataset which is the largest comic/manga dataset with ground truth information for characters [1]. This dataset consists of 109 manga albums, 118,715 faces, 147,918 texts and 103,900 frames. Following the work of Yanagisawa et al. [30], we make the training dataset of face images and text of characters appearing in 83 manga each drawn by different authors and remaining 26 albums are used for testing. The characters who appear less than 10 times in each book are ignored. The total manga character in this training dataset is 76,653. It is then divided into two sets for training and validation. The 26 test albums consist of 26,653 face images with or without associated texts (see Table 1). This multimodal character samples are used for testing the two tasks: characters clustering and character retrieval. For character retrieval, we use 2000 samples as queries and the 24,653 samples as retrieval dataset. For character clustering, we use the same 11 test albums as in the reference work [30] (a subset of the 26 test albums). We will open all the information for the community[1].

Table 1. Statistics of the training and test sets in our setting

#train album	#test album	#train face	#test face	#train character	#test character
83	26	76,653	26,653	1,114	319

Metric for Evaluation: We evaluate the clustering performance by the V-measure, ARI, and AMI metrics [9, 21, 27]. These values all range from 0.0 to 1.0, and the better clustering result, the higher the value is. For each metric, we average the values over 11 selected test albums as used in the work of Yanagisawa et al. [30]. We evaluate the character retrieval performance by rank-1, rank-5 precision, and mean Average Precision (mAP) as in the work of Li et al. [11].

Different Training Models: We have trained 5 models with different configuration as shown in Table 2. Basing on these trained models, in the next sections, we will 1) compare the performance of multimodal models and uni-modal models on two manga character tasks: *character clustering* and *character retrieval* (including other feature extraction models of existing work [11, 30]). 2) show that the manually concatenation of different learnt features (multimodal or uni-modal) is a simple yet effective method to improve the subsequent tasks in character comic analysis.

Table 2. Our different manga character feature models.

Model name	Image	Text	Training objectives (tasks)
M1 (Uni-modal Image)	✓		Manga character classification
M2 (Uni-modal Text)		✓	Manga character classification
M3 (Uni-modal Text)		✓	Manga album classification
M4 (Multimodal)	✓	✓	Manga character classification
M5 (Multimodal with auxiliary task)	✓	✓	Manga character classification and album classification

4.2   Multimodal vs. Uni-modal Models

Character Clusteringm. We compare six different learnt features, using our five trained models and the uni-modal model reported in the work of Yanagisawa et al. [30], where the authors have used only image modality to train a feature

---

[1] https://gitlab.univ-lr.fr/nnguye02/paper-icdar2021-mmtl.

extractor. In our experiments, we re-implement the method proposed in the work [30] and train the model M1 in the same way as described.

In Table 3, the multimodal multitask model (M5) outperforms all other models with big improvements (3.8% of ARI, 4.61% of AMI and 3.84% of V-measure). The M2 model using text alone performs poorly so it is not surprise that concatenating these noisy features to the features in the image domain will degrade the performance of the system (M4) compared to the image-only model M1. These results also show that our proposed auxiliary task is important, compared to the base task: it helps M5 to improve by 12.9% of ARI, 11.01% of AMI and 10.08% of V-measure, compared to M4.

The text-modal models (M2, M3) are the worst models but we can see that training text with album classification task (M3) can learn better features than training it with character classification task (M2). Intuitively, we can find that it is hard to distinguish manga characters inter-albums using only text. While text is good at distinguishing albums, we understand that the specific text features of each albums can be used to filter manga character inter-albums and intra-album.

Table 3. Character clustering results

Method	ARI	AMI	V-measure
Method in [30] (Image only)	0.5063	0.6160	0.6381
Image only (M1)	0.5104	0.5998	0.6225
Text only (M2)	0.0202	0.0478	0.0639
Text only (M3)	0.0678	0.2288	0.2919
Multimodal features (M4)	0.4071	0.5114	0.5383
Multimodal features (M5)	**0.5443**	**0.6621**	**0.6765**

Character Retrieval. We compare the results of different feature learning models for the character retrieval task. Li et al. [11] have applied their learnt features to the character retrieval task but they have not provided their list of 80 training albums yet. Therefore, we use the same five models presented previously (from M1 to M5). These models are trained using the list of 83 albums from the reference work [30] and the setting presented in Sect. 4.1. Although it is not the same as the training set of Li et al. [11] but as mentioned in the work of Yanagisawa et al. [30], these 83 albums come from 83 authors who are different from testing albums so it is safe to add the result of Li et al. [11] into our comparison table for reference.

In Table 4, we can see that the multimodal multitask model outperforms other uni-modal and multimodal models, thanks to the text modality. The multimodal learning without the auxiliary task is worse than the best uni-modal model (M1). This result confirms again that the base task character classification is not suitable to learn the verbal feature and our proposed auxiliary task can greatly improve the performance of the multimodal feature. Compare to the

Table 4. Character retrieval results. (*) Results in [11] is tested in unknown subset of Manga109 and a different setting compared to our experiments.

Method	Rank-1 (%)	Rank-5 (%)	mAP (%)
[11] (*)	70.55	84.30	38.88
Image only (M1)	71.70	87.65	39.15
Text only (M2)	0.30	2.05	0.79
Text only (M3)	68.85	83.25	24.62
Multimodal features (M4)	63.40	82.10	28.13
Multimodal features (M5)	**85.78**	**94.65**	**43.17**

best uni-modal model M1, our multimodal model M5 gives big improvements (15.23% of rank@1, 10.35% of rank@5 and 4.29% of mAP).

The signal in verbal information is weak to train a character classifier but strong to train an album classifier. It is confirmed by the performance of text-modal models M2 where we learnt almost zero knowledge from text using the character classification task (0.79% mAP). We understand that the verbal information is possibly good to distinguish character intra-album but it is weak to distinguish manga character inter-albums because the number of characters in an album is small while it is big in a comic/manga dataset. Using the auxiliary album classification task forces the model to remember the contents of manga albums so it can learn to distinguish the albums which can be used to distinguish characters inter-albums. It is to worth noting that, in the multitask multimodal model, the text modal is used to optimize the features by training both album and character classification tasks so it can learn to distinguish both characters and albums.

4.3   Boosting with Simple Learnt Features Combination

A part from the features learnt from end-to-end trainings, one can think about manually combine any different learnt features then apply to the related tasks. We have experimented different combinations of the learnt features to further analyse the importance and the relation of learnt features. We use the concatenation operator as the baseline combination method for different feature vectors. The setting for this evaluation keeps intact as in the previous evaluations.

We can see in Table 5 the results for character clustering task and character retrieval tasks follow the same pattern. The best complementary pair of learnt features is M3 and M5 for both tasks. For example, this combination gives improvements in all three metrics for character retrieval task, compared to the best single learnt feature (M5): 4.27% of rank-1, 2.2% of rank-5, and 5.16% of mAP. The combination of any two uni-modal features is worse than the best multimodal feature of M5. It is easy to see that M2 feature is very weak so it will pull down the performance of other features. Uni-modal text feature M3 gives improvements for any other learnt features, even the learnt multimodal

Table 5. Character retrieval and character clustering results: concatenations of different learnt features.

Combination	Character clustering			Character retrieval		
	ARI	AMI	V-measure	Rank-1 (%)	Rank-5 (%)	mAP (%)
Best single (M5)	**0.5443**	**0.6621**	**0.6765**	**85.78**	**94.65**	**43.17**
M1 + M2	0.4277	0.5278	0.5543	71.00	87.25	26.55
M1 + M3	0.5439	0.6540	0.6742	83.95	94.25	40.51
M1 + M4	0.5239	0.6158	0.6394	76.25	89.90	38.75
M1 + M5	0.5074	0.6188	0.6369	77.90	91.00	40.30
M3 + M4	0.4817	0.5868	0.6098	83.40	94.45	41.83
M3 + M5	**0.5853**	**0.6791**	**0.6965**	**90.05**	**96.85**	**49.33**
M1+ M3 + M5	0.4999	0.6197	0.6419	85.90	95.20	44.05

feature ones (see M1 + M3, M3 + M4 or M3 + M5). Our multimodal multitask feature (M5) is the best compared to single models but it also gives the best result while combining with the uni-modal text feature (M3). Combining more features does not help, as illustrated in the last row: the combination of 3 best features M1, M3 and M5 is worse than the combination of M3 and M5.

5   Conclusion

We have proposed a pipeline to leverage the implicit verbal information in the manga images. Our analysis shows that the multimodal manga character feature is better than the best uni-modal feature (visual feature). The results show that the self-supervised learning strategy with the album classification task is important. Without learning to memorize the content of the manga albums, the multimodal model gives lower performance than the best uni-modal feature which is trained using image model only.

One effective method to leverage both visual and verbal information is the simple manual combination of features from multiple single models. We have shown that this technique gives some improvements of the performance in character clustering/retrieval tasks. Therefore, the multimodal learning in this domain still need further studies. And more character related tasks need to be analysed in order to get more insights of multimodal features in this comic/manga domain, such as the emotion recognition tasks.

References

1. Aizawa, K., et al.: Building a manga dataset "Manga109" with annotations for multimedia applications. IEEE Multimedia **27**(2), 8–18 (2020)
2. Augereau, O., Iwata, M., Kise, K.: A survey of comics research in computer science. J. Imaging **4**(7), 87 (2018)

3. Campello, R.J.G.B., Moulavi, D., Sander, J.: Density-based clustering based on hierarchical density estimates. In: Pei, J., Tseng, V.S., Cao, L., Motoda, H., Xu, G. (eds.) PAKDD 2013. LNCS (LNAI), vol. 7819, pp. 160–172. Springer, Heidelberg (2013). https://doi.org/10.1007/978-3-642-37456-2_14

4. Cohn, N.: Comics, linguistics, and visual language: the past and future of a field. In: Bramlett, F. (ed.) Linguistics and the Study of Comics, pp. 92–118. Springer, London (2012). https://doi.org/10.1057/9781137004109_5

5. Devlin, J., Chang, M., Lee, K., Toutanova, K.: BERT: pre-training of deep bidirectional transformers for language understanding. In: Burstein, J., Doran, C., Solorio, T. (eds.) Proceedings of the 2019 Conference of the North American Chapter of the Association for Computational Linguistics: Human Language Technologies, NAACL-HLT 2019, Minneapolis, MN, USA, 2–7 June 2019, vol. 1 (Long and Short Papers), pp. 4171–4186. Association for Computational Linguistics (2019)

6. Fujimoto, A., Ogawa, T., Yamamoto, K., Matsui, Y., Yamasaki, T., Aizawa, K.: Manga109 dataset and creation of metadata. In: Proceedings of the 1st International Workshop on Comics Analysis, Processing and Understanding, pp. 1–5 (2016)

7. Guérin, C., et al.: eBDtheque: a representative database of comics. In: Proceedings of the 12th International Conference on Document Analysis and Recognition (ICDAR), pp. 1145–1149 (2013)

8. He, K., Zhang, X., Ren, S., Sun, J.: Deep residual learning for image recognition. In: 2016 IEEE Conference on Computer Vision and Pattern Recognition (CVPR), pp. 770–778 (2016)

9. Hubert, L., Arabie, P.: Comparing partitions. J. Classif. **2**, 193–218 (1985)

10. Iyyer, M., et al.: The amazing mysteries of the gutter: drawing inferences between panels in comic book narratives. In: 2017 IEEE Conference on Computer Vision and Pattern Recognition, CVPR 2017, Honolulu, HI, USA, 21–26 July 2017, pp. 6478–6487. IEEE Computer Society (2017)

11. Li, Y., Wang, Y., Qin, X., Tang, Z.: Dual loss for manga character recognition with imbalanced training data. In: 2020 25th ICPR International Conference on Pattern Recognition (ICPR), pp. 2166–2171 (2020)

12. Matsui, Y., et al.: Sketch-based manga retrieval using Manga109 dataset. Multimed. Tools Appl. **76**(20), 21811–21838 (2017)

13. McInnes, L., Healy, J.: UMAP: uniform manifold approximation and projection for dimension reduction. ArXiv abs/1802.03426 (2018)

14. Memon, J., Sami, M., Khan, R.A.: Handwritten optical character recognition: a comprehensive systematic literature review. IEEE Access **8**, 142642–142668 (2020)

15. Narita, R., Tsubota, K., Yamasaki, T., Aizawa, K.: Sketch-based manga retrieval using deep features. In: 2017 14th IAPR International Conference on Document Analysis and Recognition (ICDAR), vol. 3, pp. 49–53. IEEE (2017)

16. Ngiam, J., Khosla, A., Kim, M., Nam, J., Lee, H., Ng, A.Y.: Multimodal deep learning. In: Proceedings of the 28th Int. Conf. on Machine Learning, ICML 2011, Madison, WI, USA, pp. 689–696. Omnipress (2011)

17. Nguyen, N., Rigaud, C., Burie, J.: Digital comics image indexing based on deep learning. J. Imaging **4**(7), 89 (2018)

18. Nguyen, N.-V., Rigaud, C., Burie, J.-C.: Comic MTL: optimized multi-task learning for comic book image analysis. Int. J. Doc. Anal. Recogn. (IJDAR) **22**(3), 265–284 (2019). https://doi.org/10.1007/s10032-019-00330-3

19. Pan, S.J., Yang, Q.: A survey on transfer learning. IEEE Trans. Knowl. Data Eng. **22**(10), 1345–1359 (2010)

20. Radford, A., Wu, J., Child, R., Luan, D., Amodei, D., Sutskever, I.: Language models are unsupervised multitask learners (2019)
21. Rosenberg, A., Hirschberg, J.: V-measure: a conditional entropy-based external cluster evaluation measure. In: EMNLP-CoNLL, Prague, Czech Republic, pp. 410–420, June 2007
22. Ruder, S.: An overview of multi-task learning in deep neural networks. ArXiv abs/1706.05098 (2017)
23. Srivastava, N., Hinton, G., Krizhevsky, A., Sutskever, I., Salakhutdinov, R.: Dropout: a simple way to prevent neural networks from overfitting. J. Mach. Learn. Res. 15(56), 1929–1958 (2014)
24. Sun, W., Kise, K.: Detection of exact and similar partial copies for copyright protection of manga. Int. J. Doc. Anal. Recogn. (IJDAR) 16(4), 331–349 (2013). https://doi.org/10.1007/s10032-013-0199-y
25. Tsubota, K., Ogawa, T., Yamasaki, T., Aizawa, K.: Adaptation of manga face representation for accurate clustering. In: SIGGRAPH Asia Posters, pp. 1–2 (2018)
26. Vafaeikia, P., Namdar, K., Khalvati, F.: A brief review of deep multi-task learning and auxiliary task learning. ArXiv abs/2007.01126 (2020)
27. Vinh, N.X., Epps, J., Bailey, J.: Information theoretic measures for clusterings comparison: variants, properties, normalization and correction for chance. J. Mach. Learn. Res. 11, 2837–2854 (2010)
28. Wang, W., Tran, D., Feiszli, M.: What makes training multi-modal classification networks hard? In: IEEE Conference on Computing Vision and Pattern Recognition (CVPR), Los Alamitos, CA, USA, pp. 12692–12702. IEEE Computer Society, June 2020
29. Wang, Y., Wang, W., Liang, W., Yu, L.F.: Comic-guided speech synthesis. ACM Trans. Graph. (TOG) 38(6), 1–14 (2019)
30. Yanagisawa, H., Kyogoku, K., Ravi, J., Watanabe, H.: Automatic classification of manga characters using density-based clustering. In: International Workshop on Advanced Imaging Technology(IWAIT) 2020, vol. 11515, p. 115150F. Int. Society for Optics and Photonics (2020)

Probabilistic Indexing and Search for Hyphenated Words

Enrique Vidal[1] and Alejandro H. Toselli[2]([✉])

[1] PRHLT Research Center, Universitat Politècnica de València,
Valencia, Spain
evidal@prhlt.upv.es
[2] College of Computer and Information Science, Northeastern University,
Boston, MA, U.S.A.
a.toselli@northeastern.edu

Abstract. Hyphenated words are very frequent in historical manuscripts. Reliable recognition of (the prefix and suffix fragments of) these words is problematic and has not been sufficiently studied so far. If the aim is to transcribe text images, a sufficiently accurate character-level recognition of the fragments might be an admissible transcription result. However, if the goal is to allow searching for words or "keyword spotting", this is not acceptable at all because users need to query entire words, rather than possible fragments of these words. The situation becomes even worse if the aim is to index images for lexicon-free searching for any arbitrary text. To start with, this makes it necessary to know whether the concatenation of two-word fragments may constitute a regular word, or each fragment is instead a word by itself. We propose a probabilistic model to deal with these complications and present a first development of this model, based only on lexicon-free probabilistic indexing of the text images. Albeit preliminary, it already allows to very accurately find both entire and hyphenated forms of arbitrary query words by using just the entire forms of the words. Experiments carried out on a representative part of a huge historical collection of the National Archives of Finland, confirm the usefulness of the proposed methods.

Keywords: Handwritten text recognition · Probabilistic indexing · Hyphenated words

1 Introduction

We consider the problem of searching for hyphenated words in handwritten text images of historical documents, by means of entire word textual queries. We do not know, or care, how each query word may have been split into two fragments in the different handwritten instances of this word in the document.

In the handwritten text recognition (HTR) literature, this problem is generally ignored [13,19], or sidestepped [2,11,12,14,15]. In the first case, hyphenated word fragments are considered "tokens", akin to regular words, which obviously lead to a significant undesired vocabulary growth. Moreover, by ignoring the real words formed by pairs of hyphenated fragments, the resulting transcripts

© Springer Nature Switzerland AG 2021
J. Lladós et al. (Eds.): ICDAR 2021, LNCS 12822, pp. 426–442, 2021.
https://doi.org/10.1007/978-3-030-86331-9_28

are inadequate for many applications where the real, entire words are needed. In the second case, all the works relay on some form of hyphenation dictionary, such as those provided by some software hyphenation tools[1], to explicitly enlarge the lexicon and enrich the language model used in the HTR decoding process. A main drawback of this idea is that hyphenation dictionaries or software tools only exist for some modern languages and, moreover, as discussed in Sect. 2, hyphenation rules are rarely used or consistently followed in most historical manuscripts. Finally, there are other works like [24], where spotting of hyphenated words is addressed by trying to detect their hyphenation symbols. However, the common cases of hyphenation without explicit hyphens, or when hyphens are just too small and/or difficult to be reliably detected, remain unconsidered.

The hyphenation problem becomes more acute if we consider searching for, rather than transcribing handwritten words. In contrast with automatic transcription, where users may accept results that just contain transcripts of hyphenated word fragments, in a word searching framework users need to query entire words, not possible fragments of these words. For instance, consider the word "Isabella", which can appear hyphenated or not in some document collection. It would clearly be inadequate to ask users to formulate a query such as:

$$\text{Isabella} \vee (\text{I} \wedge \text{sabella}) \vee (\text{Isa} \wedge \text{bella}) \vee (\text{Isabel} \wedge \text{la})$$

Moreover, such a query would probably miss images where this word has been hyphenated as Is-abella or Isabe-lla. And still worse, this query would probably result in many false alarm images which contain some of these pairs of fragments, but not actually constituting a hyphenated word.

In this work we provide adequate solutions to the hyphenated word searching problem within the Probabilistic Indexing (PrIx) framework [10,21,23]. PrIx draws from ideas and concepts previously developed for keyword spotting (KWS). However, rather than caring for "key" words, any element in an image which is likely enough to be interpreted as a word is detected and stored, along with its relevance probability and its location in the image. PrIx easily allows complex queries that go far beyond the typical single keyword queries of traditional KWS. In particular, full support for standard multi-word boolean and word-sequence queries has been developed [20] and is being used in many real PrIx search applications for very large collections of handwritten text documents[2].

To deal with the hyphenation problem in this framework, we propose a probabilistic hyphenation model and present a first development of this model which does support very accurate search for both entire and hyphenated forms of arbitrary query words by using just the entire forms of the words. Experiments on a representative part of a huge historical collection of the National Archives of Finland are presented which confirm the usefulness of the proposed methods.

---

[1] Like "PyHyphen" (https://pypi.org/project/PyHyphen/), which determines hyphenation points for several (mostly Roman) languages.

[2] See http://prhlt-carabela.prhlt.upv.es/PrIxDemos for a list of PrIx live demonstrators.

2   Hyphenation in Historical Manuscripts

A *hyphenated word (HyW)* is a pair of text *fragments (HwF)* which result from splitting an *entire word* into a *prefix* and a *suffix* for line breaking. Figure 1 shows examples of entire words and prefix and suffix HwFs of HyWs.

Fig. 1. All the prefix and suffix hyphenated word fragments in this image region are marked with color bounding boxes. The remaining text tokens (without bounding boxes) are entire words.

Hyphenation appears even in the earliest known handwritten documents. HyWs are often marked by tagging the prefix with a special mark (such as "-", called *hyphen*), but many other hyphenation mark-up styles abound in historical manuscripts: from tag-free, to tagging both the prefix and the suffix with different special marks. Here are a few fairly common examples:

Ma / ria, Ma- / ria, Ma / -ria, Ma- / -ria, Ma_ / ria, Ma: / ria,
Ma_/ :ria, Ma: / _ria, Ma= / ria, Ma= / ria, Ma~ / ria, etc. ...

In general, hyphenation styles are inconsistently adopted across documents, often within a given document, or even within a single page!

Hyphenation rules may specify the admissible ways of word splitting. However these rules vary widely across languages and time periods: from free splitting with no rules at all, to deterministic, syllable-based word splitting, as adopted in many modern languages.

Also, hyphenation rates vary broadly across documents and languages: from no hyphenation, to using hyphenation every other text line. For example, in the Finish Court Records collection considered in the experiments of this paper, 42% of the lines contain at least one HwF, 9% of running tokens are HwFs, and 20% of the different tokens (lexicon) are HwFs.

All these facts raise a series of challenges for the indexing and search of HyWs in handwritten document images. First, users wish to query entire words, *not* possible HwFs of these words. And, of course, in a lexicon-free framework, we do not know in advance which are the words the users may want to query. Note that this requirement is much milder in HTR, where users are generally content with results that just contain sufficiently accurate transcripts of HwFs. Second, while automatic hyphenation tools exist (see Footnote 1), they are different for every language. Even worst, they cannot deal with the variable, inconsistent, or unexisting rules used in historic or legacy handwritten text. And last, but not least, using hyphenation tools at query time can be very problematic for the following reasons: a) this would increases software complexity of the query user interface; b) it would also increase, perhaps prohibitively, the search computation demands and/or query response time; c) text fragments that are not prefix/suffix pairs of a HyW should be discarded, which entails complex probabilistic geometric reasoning that would further complicate the user interfaces and increase computation at query time.

The approach proposed in this paper adequately deals with all these challenges and provides useful results.

3   Probabilistic Indexing of Handwritten Text Images

The Probabilistic Indexing (PrIx) framework was proposed to deal with the intrinsic word-level uncertainty generally exhibited by handwritten text in images and, in particular, images of historical manuscripts. It draws from ideas and concepts previously developed for keyword spotting (KWS), both in speech signals and text images. However, rather than caring for "key" words, any element in an image which is likely enough to be interpreted as a word is detected and stored, along with its *relevance probability* (RP) and its location in the image. These text elements are referred to as *"pseudo-word spots"*.

KWS can be seen as the binary classification problem of deciding whether a particular image region x is *relevant* for a given query word v, i.e. try to answer the following question: "Is v actually written in x?". As in [10,21], we denote this image-region word RP as $P(R=1 \mid X=x, V=v)$, but for the sake of conciseness, we will omit the random variable names, and for $R = 1$, we will simply write R. As discussed in [23], this RP can be simply approximated as:

$$P(R \mid x, v) = \sum_{b \sqsubseteq x} P(R, b \mid x, v) \approx \max_{b \sqsubseteq x} P(v \mid x, b) \tag{1}$$

where b is any small, word-sized image sub-region or bounding box (BB), and with $b \sqsubseteq x$ we mean the set of all BBs contained in x. $P(v \mid x, b)$ is just the posterior probability needed to "recognize" the BB image (x, b). Therefore, assuming the computational complexity entailed by the maximization in (1) is algorithmically managed, any sufficiently accurate isolated word classifier can be used to obtain $P(R \mid x, v)$.

Image region word RPs do not explicitly take into account where the considered words may appear in the region x, but the precise positions of the words within x are easily obtained as a by-product. According to Eq. (1), the best BB for v in the image region x can be obtained as:

$$\hat{b}_v = \arg\max_{b \sqsubseteq x} P(v \mid x, b) \tag{2}$$

If image regions are small (for example text-line regions), it is unlikely that interesting words appear more than once in the region. Therefore the BB obtained in Eq. (2) generally provides good information about the location and size of v within x. This can be straightforwardly extended to cases where the image regions are larger (for example page images) and/or if multiple instances of the same interesting word are expected to appear in x. To do this, rather than finding the (single) best BB in Eq. (2), the n-best BBs and the corresponding relevance probabilities can be obtained as:

$$\hat{b}_{v1},\dots,\hat{b}_{vn} = \underset{b \sqsubseteq x}{n\text{–best}}\, P(v\,|\,x,b)\,, \quad P(R,\hat{b}_{vi}\,|\,x,v) = P(v\,|\,x,\hat{b}_{vi})\,, \; 1\leq i\leq n \quad (3)$$

By setting n large enough[3], all the sufficiently relevant location and size hypotheses for a word v in x are obtained. As a result, the PrIx of a image x consists of a list of *"spots"* of the form:

$$[x,v,P_{vi},\hat{b}_i]\,, \quad P_{vi} \overset{\text{def}}{=} P(v\,|\,x,\hat{b}_{vi})\,, \quad v \in V, \; 1 \leq i \leq n \quad (4)$$

where V is a set of relevant words or "pseudo-words" (see below).

An alternative to Eq. (1) to compute $P(R\,|\,x,v)$ is to use a suitable segmentation-free *word-sequence* recognizer [10,21,23]:

$$P(R\,|\,x,v) = \sum_w P(R,w\,|\,x,v) = \sum_{w:v\in w} P(w\,|\,x) \quad (5)$$

where w is the sequence of words of any possible transcript of x and with $v \in w$ we mean that v is one of the words of w. So the RP can be computed using state-of-the-art optical and language models and processing steps similar to those employed in handwritten text recognition (cf. Sect. 5.4), even though no actual text transcripts are explicitly produced in PrIx. Note that Eq. (5) can be extended, as in Eq. (2)–(3), to obtain also the best and n-best BBs for v.

In the PrIx approach character-level optical and language models are generally adopted, but good word-level performance is achieved by determining RPs for *"pseudo-words"*, which are arbitrary character sequences that are likely-enough to correspond to actual words. Pseudo-words are automatically "discovered" in the very test images being indexed, as discussed in [10,17].

The PrIx framework is not limited to just single-word queries; by adequately combining relevance probabilities, it also accommodates Boolean combinations of words and sequences of words or characters [10,20].

This word-level indexing approach has proved to be very robust and flexible. So far it has been used to very successfully index several large iconic manuscript collections, such as the medieval French CHANCERY collection [1], the BENTHAM PAPERS [18], the Spanish CARABELA collection [22], and the massive collection of FINNISH COURT RECORDS, among others (see Footnote 2).

---

[3] In practice, n is not fixed but rather adaptatively chosen so that $P(v\,|\,x,\hat{b}_{vi}) > p_0$, $1\leq i\leq n$, where $p_0 < 1$ is determined empirically.

4   Proposed Probabilistic Hyphenation Model

To deal with the challenges discussed in Sect. 2, we capitalize on the following general ideas:

- Keep off hyphenation software or rules; instead, try to discover likely prefix/suffix HwFs in the images themselves
- Compute relevance probabilities of entire words composed of pairs of pseudo-words which are likely prefix/suffix pairs of HyWs

To achieve these goals, a probabilistic hyphenation model is proposed which relies on spots of HwFs indexed in the PrIx's. Assume a word v has been hyphenated in the text of an image x and we want to compute the relevance probability $P(R \mid x, v)$. Let r and s be a possible prefix and suffix of v (i.e., $v = rs$) and let b_r, b_s possible BBs of r and s, respectively. As in Eq. 1, $P(R \mid x, v)$ can be approximated as:

$$
\begin{aligned}
P(R \mid x, v) &= \sum_{b_r, b_s \sqsubseteq x} P(R, b_r, b_s \mid x, v) \\
&\approx \max_{b_r, b_s \sqsubseteq x} P(b_r \mid x)\, P(b_s \mid x, b_r)\, P(v \mid x, b_r, b_s)
\end{aligned}
\tag{6}
$$

where the posterior probability $P(v \mid, x, b_r, b_s)$ can in turn be approximated as:

$$
\begin{aligned}
P(v \mid x, b_r, b_s) &= \sum_{r,s} P(v, r, s \mid x, b_r, b_s) \approx \max_{\substack{r,s: \\ rs=v}} P(r \mid x, b_r)\, P(s \mid x, b_s) \\
&\approx \max_{\substack{r,s: \\ rs=v}} \min(P(r \mid x, b_r), P(s \mid x, b_s))
\end{aligned}
\tag{7}
$$

The last approximation may compensate to some extent inaccuracies due to assuming s is independent of r given b_s. It also supports the intuitive idea that a query for the word v can be seen as a Boolean AND query for the HwFs r and s and, according to [20], AND relevance probability can be better approximated using the minimum rather than the product. Finally, from Eq. (6) and (7):

$$
P(R \mid x, v) \approx \max_{\substack{r, b_r, s, b_s: \\ rs=v,\, b_r, b_s \sqsubseteq x}} P(b_r \mid x)\, P(b_s \mid x, b_r)\, \min(P(r \mid x, b_r), P(s \mid x, b_s))
\tag{8}
$$

The first factor, $P(b_r \mid x)$, is a *prior probability* for the position of the prefix b_r in x and the second, $P(b_s \mid x, b_r)$, is the *conditional probability* of the position of the suffix b_s, given the position of the corresponding prefix, b_r. The last two factors, $P(r \mid x, b_r)$ and $P(s \mid x, b_s)$ are the relevance probabilities of the pseudo-words r and s, which are directly provided by the PrIx of the image x.

4.1   Simplifications

The computation of Eq. (8) entails several complications. First, it requires to consider all pairs of BBs which likely contain HwFs (r, s) of the word v. This amounts to consider all pairs of spots $([x, r, P_{ri}, b_{ri}], [x, s, P_{sj}, b_{rj}])$ $1 \le i, j \le n$, in the PrIx of x, such that $rs = v$ and b_r, b_s lay in likely positions in x, according to $P(b_r \mid x)$ and $P(b_s \mid x, b_r)$.

Moreover, since we want to avoid this computation at query time, the query word v is not known, and all likely pairs of pseudo-words in the PrIx of x would need to be considered as candidates to be r and s. So, in fact, rather than actually performing the maximization of Eq. (8), all the terms in this maximization are used to derive new entire-word spots to be added to the PrIx, as discussed below (Subsect. 4.4). In order to avoid adding prohibitive amounts of (low probability and maybe unwished) spots, we rely on the two following ideas:

1. Accurately predict prefix/suffix HwFs. Rather than considering all spots of x, we try to consider only those which likely correspond to prefix and suffix HwFs. To this end we adopt a special form of Optical and Language Modeling to allow distinguishing entire words from HwFs. As a result, PrIx's will contain special spots for predicted HwFs, as discussed below (Subsect. 4.2).

2. Apply BB (probabilistic) geometric reasoning. Again, rather than considering all pairs of spots (or only all pairs likely to correspond to HwFs), we consider only those for which the geometric probability of being paired is high enough. For this purpose, the prior and conditional probabilities of HwF positions, $P(b_r \mid x)$ and $P(b_s \mid x, b_r)$ can be used to predict which PrIx BBs are likely enough to be considered in the computation of Eq. (8), as discussed below (Subsect. 4.3).

4.2   HwF Prediction: Optical and Language Modeling

We train special optical and language models to predict whether a token (sequence of characters) is likely to be a prefix or a suffix of a HwF, or rather a normal entire word. To this end, training transcripts are tagged with the special symbol ">" *after* each prefix HyF and *before* the corresponding suffix HwF. The usual ground-truth annotation of HyWs is by writing a dash "-" after the prefix HwF (both if there is a hyphenation mark in the image or not), but nothing is annotated in the corresponding suffix HwF. We rely on the correct reading order of the line transcripts (which is annotated by the transcribers) in order to adequately tag both fragments with the ">" tag, as illustrated in Fig. 2.

Then, for *Optical Modeling*, we just train a standard character optical model with the HyW-tagged training set. For *Language Modeling*, we use a special character N-gram which deals both with the probabilistic modeling of character and word concatenation regularities, and with enforcing two deterministic hyphenation-derived constraints; namely: a prefix HwF can only appear at the end of a detected line, while a suffix HwF can only appear at the beginning of a line. No HwF-tags are allowed in the middle of a line. These constraints can be easily enforced by training a conventional N-gram character language model and manually editing the trained back-off parameters.

It is expected that the (often blank) right and left context of HyW prefixes and suffixes help the Optical and Language models learn to distinguish HwFs from normal text.

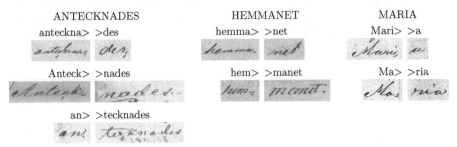

ANTECKNADES   HEMMANET   MARIA

anteckna> >des   hemma> >net   Mari> >a

Anteck> >nades   hem> >manet   Ma> >ria

an> >tecknades

Fig. 2. Examples of hyphenated handwritten words. The symbols ">" is used to tag the training transcripts of the corresponding prefix and suffix HwFs.

4.3 Geometric Reasoning

Prior and conditional probabilities of HwF BBs, $P(b_r \mid x)$ and $P(b_s \mid x, b_r)$, could be estimated from training examples of images with annotated HwFs. However, in the present work we have just considered these probabilities to be *uniform* for all BBs which fulfill the following *knowledge-based* geometry constraints (and *zero* otherwise):

- b_s should be *a little below* b_r in x (say, no more than 200 pixels below)
- b_s should be *on the left* of b_r in x (typically far to the left)

4.4 Unite Prefix/Suffix HwF Pairs

Let $[x, r, P_r, b_r]$, $[x, s, P_s, b_s]$ be a pair of spots such that r and s are HwF prefix and suffix predictions, respectively, and b_r, b_s meet the above geometric constraints. Then a new "entire-word spot" $[x, v, P_v, b_{rs}]$ is produced, where v is the concatenation of r and s (after removing their ">" HwF tags) and, according to Eq. (7) or (8), $P_v = \min(P_r, P_s)$.

The special "entire-word BB" b_{rs} should be understood as the union of two typically disjoint and distant BBs, b_r and b_s. So, in practice, rather than adding this special-form spot to the PrIx of x, two regular spots are added; namely: $[x, v, P_v, b_r]$ and $[x, v, P_v, b_s]$.

5   Dataset, Assessment and Empirical Settings

This section presents the dataset[4], evaluation measures, query sets, and experimental setup adopted to assess empirically the performance of the proposed approach for hyphenated word PrIx and search.

---

[4] Available from: https://doi.org/10.5281/zenodo.4767732.

5.1   Dataset

The Finnish Court Records (FCR) collection encompasses 785 manuscripts from the "Renovated District Court Records" held by the National Archives of Finland. Many of these manuscripts were scanned into double-page images, amounting to 630, 388 images of about one million pages in total. The manuscripts date from the 18th century and consist of records of deeds, mortgages, traditional life-annuity, among others. They were written by many hands mostly in Swedish. Figure 3 show examples of FCR images, where HwFs are highlighted.

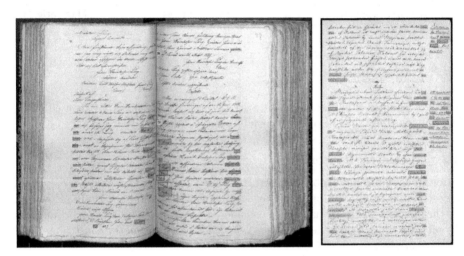

Fig. 3. Two FCR image examples showing bounding boxes of HwFs. Colors correspond to relevance probabilities of PrIx spots for these HwFs (red: low; green: high). (Color figure online)

In addition to the many hands and writing styles involved, one of the main challenges of this collection stems from the heavy warping of many double-page images, as in Fig. 3-left. Despite these challenges, the complete collection has recently been made fully free-text searchable using PrIx technology.[5] Another important challenge is, in point of fact, the very large proportion of words that are hyphenated. Indexing and search for these HyWs was unsolved so far, which significantly hindered the search for important facts in this collection.

The dataset used in the experiments of this work is a (mostly random) selection of 600 FCR images, which is referred to as FCR-HYP. All these images were accurately transcribed, including careful annotation of hyphenated words, as discussed in Sect. 4.2 (see Fig. 2 and also Fig. 3, above).

Text line baselines were automatically detected, but no other layout analysis information (marginalia, double column, double-page, or block detection) was used to obtain the PrIx's of these images. As illustrated by Fig. 3-left, this is highly challenging for the proposed HwFs detection and pairing techniques described in

---

[5] http://prhlt-kws.prhlt.upv.es/fcr.

Sect. 4. Line detection was carried out with the *P2Pala* tool[6] and (few) baselines detection errors were manually fixed using the *Transkribus* platform.[7]

Following a common practice in search engines and KWS systems alike, our system retrieves all instances of a given query word without matter of unessential spelling nuances. To this end, ground-truth transcripts were transliterated. As in [5], case and diacritic folding was applied, and some symbols were mapped onto ASCII equivalences to facilitate their typing on standard keyboards.

Table 1 shows basic statistics of the FCR-HYP dataset, including training-test partition and the corresponding proportion of HwFs.

Table 1. Basic statistics of the FCR-HYP experimental dataset and their HwFs. All the text has been transliterated and the punctuation marks ignored.

Dataset partition	Dataset		HwFs		
	Train-Val	Test	Train-Val	Test	Overall %
Images	400	200	–	–	–
Lines	25 989	13 341	10 973	5 609	42%
Running words	147 118	73 849	13 081	6 589	9%
Lexicon size	20 710	13 955	4 091	2 677	20%
ALLWORDS query set	–	10 416	–	–	–
MAYBEHYPH query set	–	1 972	–	–	–

5.2 Evaluation Measures

Since HyWs are indexed to allow them to be searched by using entire-word queries, performance evaluation can be carried out in a uniform way, without needing to distinguish entire and hyphenated words.

The standard *recall* and *interpolated precision* [6] measures are used to evaluate search performance. Results are reported in terms of both global and mean *average precision* (AP and mAP, respectively) [7]. In addition, we report word error rate (WER) to indirectly evaluate the quality of optical and language modeling through simple handwritten text recognition experiments using these models (see Sect. 6.1).

5.3 Query Selection

Following a common practice in KWS (see e.g. [18]), most of the test-set (entire) words are used as keywords, which allows mAP to be properly computed. In addition to HwFs, all the test-set words with length shorter than 1 character are excluded, and the remaining words are transliterated as discussed in Sect. 5.1.

To allow distinguishing the relative impact of entire and hyphenated word instances on the results, two query sets were defined: one named MAYBEHYPH,

---

[6] https://github.com/lquirosd/P2PaLA.
[7] https://readcoop.eu/transkribus.

with 1 972 keywords for which at least one instance in the test set is hyphenated; and other, ALLWORDS, with 10 416 keywords, which also includes the MAYBE-HYPH set (see Table 1).

5.4   Experimental Settings

To obtain PrIx relevance probabilities according to Eq. (5), the posterior $P(w \mid x)$ is computed using character-level optical and language models. Then, rather than plain decoding as for HTR, a *character lattice* (CL) is obtained for each line-shaped image region x. Finally, $P(R \mid v, x)$ is computed following Eq. (5), using the techniques introduced in [21, 23] and [10].

Optical modeling was based on Convolutional-Recurrent Neural Networks (CRNN), trained using the PyLaia Toolkit [9][8] The CRNN architecture comprises four convolutional layers followed by three recurrent bidirectional long-short-term memory (BLSTM) layers. Layer parameters (convolution kernel size, max-pooling, activation function, etc.) were setup according to Table 2. The final output layer with softmax activation is in charge to compute the probabilities of each of the 59 characters in the training alphabet plus a non-character symbol. The RMSProp training method [16] was used with a mini-batch of 24 examples, using a preset learning rate of 0.0003 to minimize the so called CTC loss [3]. The optimization procedure was stopped when the error on the development set did not decrease for 20 epochs. Moreover, data augmentation was used, applying on the fly random affine transformations on each training line image. This allows the optical model to generalize better on unseen data.

Table 2. CRNN setup for FCR-HYP optical modeling.

Input	# Channels: 1,		Line image height: 115px		
	Convolution kernels:	3×3	3×3	3×3	3×3
	Feature maps:	12	24	48	48
4 Convol. Layers	Max-pooling kernels:	2×2	2×2	0×0	2×2
	Batch normalization:	True	True	True	True
	Activation functions:	L-Relu	L-Relu	L-Relu	L-Relu
3 BLSTM Layers	# Units per layer: 256, 256, 256,		RNN dropout prob.: 0.5		
Output layer		Type:	Full connected (FC)		
	FC dropout prob.:		0.5		
	Output Activation:		Softmax		
	# Output Units (+ non-char symbol):		59 + 1		

On the other hand, a 8-gram character language model(with Kneser-Ney back-off smoothing) [4]) was also estimated from the transliterated transcripts.

As commented in Sect. 4.2, both optical and language models were trained on transliterated text with the prefix/suffix HwFs tagged with the ">" symbol. The trained CRNN and character language model were subsequently used by the Kaldi decoder [8], with a search beam of 15, to produce CLs for all the test-set lines images.

---

[8] https://github.com/jpuigcerver/PyLaia.

Finally, following the character lattice based indexing methods developed in [10], a large set of most probable word-like subpaths are extracted from the CL of each line image. Such subpaths define the character sequences referred to as "pseudo-words" which, along with their geometric locations and the corresponding relevance probabilities, constitute the entries of the resulting PrIx.

In addition, the first-best character sequence on each CL was obtained, which is equivalent to the standard (1-best) decoding used in conventional HTR. The resulting transcripts were used in the initial evaluation tests described in Sect. 6.1.

6 Experiments and Results

Two kind of experiments were carried out. First the quality of the trained optical and language models was indirectly assessed though conventional HTR experiments, where HwFs are considered individual tokens rather than parts of HyWs. Then, standard search and retrieval performance for entire-word queries was evaluated through experiments with different amounts of HyWs.

6.1 Model Assessment Through HTR Performance

For an initial assessment of the quality of the trained optical and language models, the test-set images were automatically transcribed as discussed in Sect. 6. Table 3 shows the transcription WER achieved using only the optical model and using also an 8-gram character language model. Both the global WER (all test-set tokens, including all entire words and HwFs) and the WER considering only HwFs, are reported. All the results are reasonably good, and clearly show a significant positive impact of explicitly using a good language model.

Table 3. FCR-HYP test-set transcription WER (in %) with/without using a character 8-gram language model.

Language model	no LM	char 8-gr
All tokens (including HwFs)	31.1	23.0
Only HwFs	44.0	39.3

6.2 Hyphenated Word Probabilistic Indexing and Search Results

We evaluate the *HyW PrIx* approach proposed in Sect. 4, which is able to spot both entire and hyphenated words, using only entire-word queries. Results are compared with those of the baseline, *Plain PrIx* method outlined in Sect. 3, which can only find entire words.

Figure 4 shows the recall-precision curves for these two methods and for both query sets, ALLWORDS and MAYBEHYPH described in Sect. 5.3, along with the corresponding AP and mAP scalar results.

The *HyW PrIx* approach clearly outperforms *Plain PrIx* for both query sets, both in terms of AP and mAP. As expected, the AP/mAP (and overall recall) improvements achieved by *HyW PrIx* are more remarkable if we focus on searching only for potentially hyphenated words. Note that, as discussed in Sect. 5.1, the query set MAYBEHYPH has an hyphenation rate of 13%, while it is only 4.5% for ALLWORDS. Even though isolated HwFs are hard to predict, geometric constraints discussed in Sect. 4.3 do help to correctly pair prefix/suffix fragments, leading to good search performance for HyWs.

These results suggest that a good practical search and retrieval experience is expected when spotting hyphenated instances of words using the methods proposed in this paper.

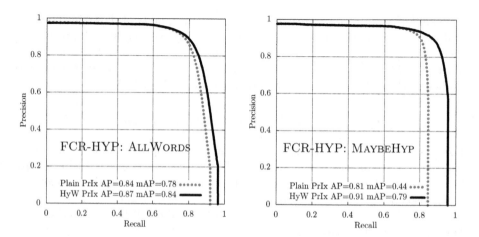

Fig. 4. *Plain PrIx* and *HyW PrIx* keyword search recall-precision curves for the query sets ALLWORDS and MAYBEHYPH on the FCR-HYP test set.

6.3   Live Demonstrator and Search Examples

A demonstrator of proposed HyW probabilistic indexing and search approach is available for public testing at: http://prhlt-kws.prhlt.upv.es/fcr-hyp .

Figure 5 shows an example of search results using this demonstrator for the *AND* query *"ohansson ombudet bonden klanderfria sjundedels tillost"*. With a confidence level of 10%, the only page image found to be relevant to this query is the one shown in the figure. There are no misses in this page and there is only a false alarm for the entire word *"den"* which is mistakenly spotted as an (hyphenated) instance of the query word *"bonden"*.

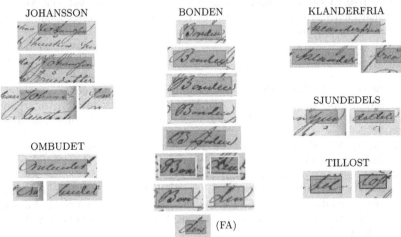

Fig. 5. FCR-HYP image found with the *AND* query *"Johansson ombudet bonden klanderfria sjundedels tillost"* with confidence level set to 10%. All the spotted words are shown in detail below. Bounding box colors represent relevance probabilities (low: red, high: green). There are no misses and all the spots are correct, except for a false alarm (FA) of the word *"bonden"* on the lower part of the left page, caused by wrongly pairing the entire word *"den"* with the prefix HwF *"bon"* on the right page marginalia. (Color figure online)

7 Conclusion

Based on lexicon-free probabilistic indexing of text images, a model for indexing and search hyphenated words has been proposed. The proposed approach avoids using hyphenation software or rules by training the optical model to predict whether a text token or "pseudo-word" is likely to be an entire word or a prefix or suffix fragment of a hyphenated word. Using a simplified, knowledge-based version of the geometry probability distributions of the model, it has been tested on a relatively small dataset from the massive historical manuscript collection known as *The Finnish Court Records*. This dataset is characterized by having many hyphenated word instances, which represent a very significant proportion compared with the total number of running words.

Empirical results show that words with many hyphenated instances can be very reliably spotted by issuing queries that just need the entire-word versions of these words. This is consistent with the good user searching experience that can be observed using a real, publicly available demonstrator.

In future works, we aim to refrain from applying knowledge-based hard geometric constraints, and use instead training data to estimate the explicit geometry distributions of the proposed probabilistic hyphenation model.

Acknowledgments. Work partially supported by the BBVA Foundation through the 2018–2019 Digital Humanities research grant "HistWeather – Dos Siglos de Datos Climáticos", by the Spanish MCIU project DocTIUM (Ref. RTI2018-095645-B-C22), and by Generalitat Valenciana under project DeepPattern (PROMETEO/2019/121). Computing resources were provided by the EU-FEDER Comunitat Valenciana 2014–2020 grant IDIFEDER/2018/025. Special thanks are due to the National Archives of Finland that kindly provided us with the FCR images and the corresponding ground-truth transcripts with hyphenation markup.

References

1. Bluche, T., et al.: Preparatory KWS experiments for large-scale indexing of a vast medieval manuscript collection in the HIMANIS project. In: 14th International Conference on Document Analysis and Recognition (ICDAR), vol. 01, pp. 311–316 (2017)
2. Bluche, T., Ney, H., Kermorvant, C.: The LIMSI handwriting recognition system for the HTRtS 2014 contest. In: 13th International Conference on Document Analysis and Recognition (ICDAR), pp. 86–90 (2015)
3. Graves, A., Fernández, S., Gomez, F., Schmidhuber, J.: Connectionist temporal classification: labelling unsegmented sequence data with recurrent neural networks. In: 23rd International Conference on Machine Learning, ICML 2006, NY, USA, pp. 369–376. ACM (2006)
4. Kneser, R., Ney, H.: Improved backing-off for N-gram language modeling. In: International Conference on Acoustics, Speech and Signal Processing, vol. 1, pp. 181–184 (1995)
5. Lang, E., Puigcerver, J., Toselli, A.H., Vidal, E.: Probabilistic indexing and search for information extraction on handwritten German parish records. In: 16th International Conference on Frontiers in Handwriting Recognition (ICFHR), pp. 44–49, August 2018

6. Manning, C.D., Raghavan, P., Schtze, H.: Introduction to Information Retrieval. Cambridge University Press, New York (2008)
7. Perronnin, F., Liu, Y., Renders, J.M.: A family of contextual measures of similarity between distributions with application to image retrieval. In: Conference on Computer Vision and Pattern Recognition, pp. 2358–2365, June 2009
8. Povey, D., et al.: The Kaldi speech recognition toolkit. In: International Workshop on Automatic Speech Recognition and Understanding. IEEE Signal Processing Society (2011)
9. Puigcerver, J.: Are multidimensional recurrent layers really necessary for handwritten text recognition? In: 14th International Conference on Document Analysis and Recognition (ICDAR), vol. 01, pp. 67–72 (2017)
10. Puigcerver, J.: A probabilistic formulation of keyword spotting. Ph.D. thesis, Univ. Politècnica de València (2018)
11. Sánchez, J.A., Romero, V., Toselli, A.H., Vidal, E.: ICFHR 2014 competition on handwritten text recognition on transcriptorium datasets (HTRtS). In: 14th International Conference on Frontiers in Handwriting Recognition, pp. 785–790 (2014)
12. Sanchez, J.A., Romero, V., Toselli, A.H., Vidal, E.: ICFHR 2016 competition on handwritten text recognition on the READ dataset. In: 15th International Conference on Frontiers in Handwriting Recognition (ICFHR), pp. 630–635 (2016)
13. Sánchez, J.A., Romero, V., Toselli, A.H., Villegas, M., Vidal, E.: A set of benchmarks for handwritten text recognition on historical documents. Pattern Recogn. **94**, 122–134 (2019)
14. Sanchez, J.A., Toselli, A.H., Romero, V., Vidal, E.: ICDAR 2015 competition HTRtS: handwritten text recognition on the tranScriptorium dataset. In: 13th International Conference on Document Analysis and Recognition (ICDAR), pp. 1166–1170 (2015)
15. Swaileh, W., Lerouge, J., Paquet, T.: A unified French/English syllabic model for handwriting recognition. In: 15th International Conference on Frontiers in Handwriting Recognition (ICFHR), pp. 536–541 (2016)
16. Tieleman, T., Hinton, G.: Lecture 6.5-rmsprop: divide the gradient by a running average of its recent magnitude. COURSERA Neural Netw. Mach. Learn. **4**(2), 26–31 (2012)
17. Toselli, A.H., Puigcerver, J., Vidal, E.: Two methods to improve confidence scores for Lexicon-free word spotting in handwritten text. In: 15th International Conference on Frontiers in Handwriting Recognition (ICFHR), pp. 349–354 (2016)
18. Toselli, A., Romero, V., Vidal, E., Sánchez, J.: Making two vast historical manuscript collections searchable and extracting meaningful textual features through large-scale probabilistic indexing. In: 15th International Conference on Document Analysis and Recognition (ICDAR) (2019)
19. Toselli, A.H., Vidal, E.: Handwritten text recognition results on the Bentham collection with improved classical N-Gram-HMM methods. In: 3rd International Workshop on Historical Document Imaging and Processing (HIP 2015), pp. 15–22 (2015)
20. Toselli, A.H., Vidal, E., Puigcerver, J., Noya-García, E.: Probabilistic multi-word spotting in handwritten text images. Pattern Anal. Appl. **22**(1), 23–32 (2019)
21. Toselli, A.H., Vidal, E., Romero, V., Frinken, V.: HMM word graph based keyword spotting in handwritten document images. Inf. Sci. **370–371**, 497–518 (2016)
22. Vidal, E., et al.: The Carabela project and manuscript collection: large-scale probabilistic indexing and content-based classification. In: 17th International Conference on Frontiers in Handwriting Recognition (ICFHR), pp. 85–90 (2020)

23. Vidal, E., Toselli, A.H., Puigcerver, J.: A probabilistic framework for lexicon-based keyword spotting in handwritten text images. CoRR abs/2104.04556 (2021). https://arxiv.org/abs/2104.04556
24. Villegas, M., Puigcerver, J., Toselli, A., Sánchez, J.A., Vidal, E.: Overview of the ImageCLEF 2016 handwritten scanned document retrieval task. In: CLEF (2016)

Physical and Logical Layout Analysis

SandSlide: Automatic Slideshow Normalization

Sieben Bocklandt$^{(\boxtimes)}$ ⓘ, Gust Verbruggen ⓘ, and Thomas Winters ⓘ

Department of Computer Science, Leuven.AI, KU Leuven, Leuven, Belgium
{sieben.bocklandt,gust.verbruggen,thomas.winters}@kuleuven.be

Abstract. Slideshows are a popular tool for presenting information in a structured and attractive manner. There exists a wide range of different slideshows editors, often with their own proprietary encoding that is incompatible with other editors. Merging slideshows from different editors and making the slide design consistent is a nontrivial and time-intensive task. We introduce SandSlide, the first system for automatically normalizing a deck of slides from a PDF file into an editable Power-Point file that adheres to the default slide templates, and is thus able to fully leverage the flexible layout capabilities of modern slideshow editors. SandSlide achieves this by labeling objects, using a qualitative representation to find the most similar slide layout and aligning content from the slide with this layout. To evaluate SandSlide, we collected and annotated slides from different slideshows. Our experiments show that a greedy search is able to obtain high responsiveness on supported and almost supported slides, and that a significant majority of slides fall into this category. Additionally, our annotated dataset contains fine-grained annotations on different properties of slideshows to further incentivize research on all aspects of the problem of slide normalization.

Keywords: Slideshow normalization · Document annotation

1 Introduction

Slideshows are one of the most popular methods for transferring information in a structured and attractive manner to an audience. Many different applications allow users to easily create their own slideshows, the most popular of which are *PowerPoint, KeyNote, Google Slides, Beamer* and *Prezi*. Each of these applications has the same goal of helping a user to design a beautiful deck of slides with text, images and other multimedia content. In order to do so, they provide design templates that control the look and feel of all slides. These design templates allow the user to focus on their content, rather than requiring them to spend hours overlooking the individual layout of each slide.

One large issue is that most applications use their own proprietary encoding format and compatibility between these tools is therefore extremely low.

S. Bocklandt and G. Verbruggen—Equally contributed.

J. Lladós et al. (Eds.): ICDAR 2021, LNCS 12822, pp. 445–461, 2021.
https://doi.org/10.1007/978-3-030-86331-9_29

For example, it is borderline impossible to edit a given *KeyNote* presentation in *PowerPoint* and vice versa. Merging slides from different editors into a single slideshow with a uniform design requires the user to manually fix each slide, often making it easier to create the uniform slides from scratch. Even the simple task of presenting a slideshow on a machine that does not have the correct application will often cause it not to be displayed correctly.

To address this last issue, there is one type of file that almost all editors can export slides to: the Portable Document Format (PDF). While sacrificing some of the functionality provided by each tool, this encoding allows for slideshows to be opened on platforms that do not have the slideshow editor itself. The main drawback of exporting a slideshow as a PDF is losing the ability to edit the slideshow in a slide editor. Several tools are available for converting PDF slideshows back into editable slideshows, but they tend to bluntly convert the content without any regard for the templates that slides were initially designed with. The user can then edit text or replace images, but they cannot easily change the overall design of the slideshow. For example, the font of each piece of text is attached to that specific piece of text, as opposed to having a uniform font for the whole slideshow as provided by the design template.

We propose a new normalization process for converting PDF slideshows back into an editable slideshow that respects the templates provided by slideshow editors. In the resulting slideshow, each object is assigned a specific role using placeholders and the editor decides where it places objects based on this role and the design template that the user selected. Examples of such roles could be *title* or *leftmost column of a slide with multiple columns*. Changing the template then changes the overall design of the slideshow. This allows interoperability between different applications, as it allows different slideshows to be merged seamlessly or the design of different slideshows to be made uniform with minimal effort.

As the exported PDF does not contain semantic information about the objects it contains, nor where they came from in the original editor, this task is highly nontrivial. Even worse, the words that make up a single paragraph are often split up as different objects. The normalization process thus needs to discover the high-level objects that make up a slide, assign them a semantic label and align these labeled objects with those of one of the slide layouts to discover their role.

In this paper, we make the following contributions:

- We identify and introduce the problem of normalizing slides by assigning a specific role to each object.
- We describe and implement an algorithm based on qualitative representations of slides to obtain this assignment of objects to roles.
- We create and release a benchmark dataset to evaluate the novel problem of slide normalization.

2   Background

We start by providing background on the structure of slideshows and describing the limitations of existing converters.

2.1   Slideshow Structure

A slideshow is a sequential set of slides filled with objects. All slides generally follow the same *design template* that specifies the default fonts, colors, backgrounds and style of other components for each type of slide layout. Each slide is instantiated using a particular slide layout, which defines the objects that it contains and their roles. These roles are implicit and the user interacts with them by editing *placeholders* for the objects of each role. A placeholder is thus an object that the user interacts with that has an implicit role defined by the slide layout.

Example 1. Examples of slide layouts are *Title slide*, *Section header* and *Title with double content*. The latter of these has three placeholders, one for the title, one for its left content and one for its right content.

Most design templates provided by editors support the same slide layouts. When changing to a different design template, objects in placeholders automatically move to the positions defined by their role in the corresponding layout of the new template. Objects not in a placeholder remain in the same position.

Example 2. Figure 1 shows three slide layouts in the default design template and their placeholders. The user can edit these placeholders and fill them with content. By selecting a new design template, all slides are automatically adapted to the layouts defined by this new template.

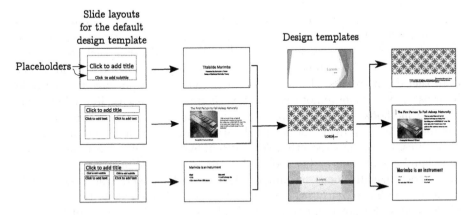

Fig. 1. Steps a user encounters when creating a new slideshow in PowerPoint. Slides are created by filling placeholders in slide templates and the layout is decided by choosing a specific slide master.

2.2   PDF to PPT Converters

Several online tools for converting PDF files into editable PowerPoint files already exist, such as `smallpdf`, `ilovepdf`, `pdf2go`, `simplypdf` and `online2pdf`.[1] For each page in the PDF, these applications first create a slide with the empty slide layout and then convert all objects of the PDF page into PowerPoint objects. Not only do they not use placeholders, some applications even convert text into images, making it even harder to change the design of the slides. The inability of dealing with a change of design template is illustrated in Fig. 2.

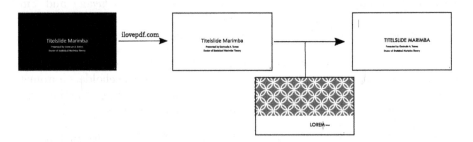

Fig. 2. A slide converted with `ilovepdf.com` does not adapt its layout when changing the new design template.

2.3   Automatic Document Annotation

Automatic document annotation enables the retrieval of structured textual content from PDF files. In this field, algorithms are crafted or trained to automatically labeling parts of the document with their function, such as *abstract*, *title* and *author*. These annotations are then used to improve the performance of further downstream tasks, such as search. While most document annotation tools use heuristics for annotating the text [5,10], some also use limited machine learning methods—such as support vector machines—to additionally annotate metadata [11]. To provide the required data to train such document annotation algorithms, researchers recently released PubLayNet, a dataset containing multiple hundreds of thousands annotated documents [14]. While document annotation tools usually focus on annotating formal documents, the task of slide segmentation has recently been gaining attention. Slide segmentation tools like SPaSe and WiSe are build to automatically segmenting pictures of slides by detecting regions using neural architectures [6,7].

---

[1] Found at their `name.com`.

3   Slideshow Normalisation

In this section, we describe the problem of slide normalization and define the scope of this paper.

3.1   Problem Statement

The goal of normalizing a slideshow is to allow users to easily change its design by editing or changing the design template. This is trivial when all content is added in a placeholder and thus has a semantic role. We therefore define a *normal slide* as a slide in which all objects are stored in placeholders. A *normal slideshow* consists of only normal slides.

Given a slide and a set of design templates, the problem of slide normalization is to select one of these templates such that each object of the slide is assigned to exactly one placeholder and each placeholder is assigned exactly one object in such a way that objects from the input slide fulfill the role in the normalized slide that the user intended them to. Evaluating whether an object is assigned *the correct* role is impossible without human intervention. It is the task of the normalization algorithm to try and maximize the probability of this happening.

Example 3. Three examples of slides and their normalization in the default design template are shown in Fig. 3. It is important that *"Lorem"* is assigned to the placeholder which fulfills the title role and not to the subtitle placeholder.

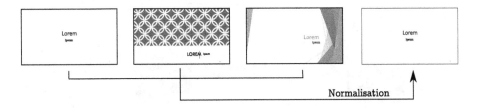

Fig. 3. Three slides that share the same normalization

3.2   Scope

In this paper, we limit the scope of object types considered in the normalization process. Our system currently deals with text objects (normal text, bullet lists, captions, footers, slide numbers) and pictures. More complicated objects—such as SmartArt, mathematical equations and charts—are thus out of scope for this paper. Some of these are easy to integrate with the correct preprocessing, such as charts, and others are interesting pointers for future work, such as arrows and equations.

4 SandSlide: Automatic Slideshow Normalization

We propose and implement a method called SandSlide (*Searching to Automatically Normalize Decks of Slides*) for automatically obtaining normal slideshows from PDF exports. SandSlide is based on three main components. First, it detects high-level objects and annotates them with a semantic label. Second, it uses spatial relations between these objects to obtain a qualitative representation of a slide, which allows for aligning the objects across two slides and quantitatively comparing them. Third, it searches for the best alignment between the objects of a slide and a layout. The obtained alignment can be used to convert the slide into its normalized equivalent.

4.1 Detecting and Labeling Objects in Slides

In the first step, SandSlide looks for and semantically labels objects from the PDF slide. This is achieved in four steps, which are illustrated in Fig. 4 and briefly covered in the following paragraphs.

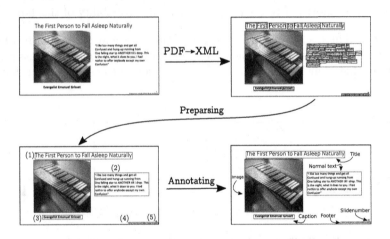

Fig. 4. Converting a page of a PDF file to annotated slide objects.

PDF to Objects. First, it converts the input PDF file into XML using pdf2txt[2]. SandSlide then removes all objects that are not text or images; either because they are artifacts from the tool, like rectangles around text boxes and single-color pictures, or because they are out of the scope of the system, like curves and SmartArt. The remaining objects are grouped into high-level objects. In Fig. 4, this corresponds to the PDF $\rightarrow$ XML and Preparsing steps.

---

[2] https://github.com/euske/pdfminer/blob/master/tools/pdf2txt.py.

Annotating Objects. Each of the remaining objects is then assigned a label representing its most likely type of content. Heuristic approaches are often used in document annotation methods and have as the main benefit that they do not require training data [5,10,11]. Each object is assigned a score for each label using a local heuristic, computed as the sum of several smaller heuristics. An overview of these heuristics for all labels is shown in Table 1. The label assigned to an object is that with the highest score. The *mean title* is the weighted mean of the bounding boxes of the title in previous slides, with the slide number as weights.

Example 4. The normalized feature vectors for the text objects of the slide in Fig. 4 are shown in Table 2. The assigned labels are printed in bold.

Table 1. Local heuristics for assigning a label to objects in a slide, and whether or not it is used in the qualitative representation later (denoted by Q). Bold words are used to refer to objects in Table 2.

Label	Heuristic	Weight	Q
Background	Area covers more than 80% of the slide	1.0	No
Listing image	Two or more identical images are placed in a vertical line	1.0	No
Normal image	None of the above	1.0	Yes
Slide **number**	Positioned on the outside 20% of the slide	0.33	No
	Follows \d+ or \d+[/\]\d+	0.66	
Title	Largest font size in slide	0.14	Yes
	Positioned in the upper 40% of the slide	0.29	
	More than 20% overlap between object and *mean title*	0.57	
	First slide and largest font size	1.0	
	Largest font size is similar to font size of title in first slide	0.57	
	Not in upper 40% of the slide	0.43	
Footer	Positioned in the outside 10% of the slide	0.33	No
	Either contains a word from {src, source,...} or is a URL	0.66	
Listing	Over 20% of the sentences start with a listing character or listing image	0.66	Yes
	Contains more than one line of text	0.33	
Caption	Image is positioned 20% of the slide height above the object with a 90% horizontal overlap	0.75	No
	Is a single line of text	0.25	
Normal text	None of the above	0.5	Yes

4.2 Qualitative Representation

Annotated slide objects are now converted into a qualitative representation. Such a representation provides a level of abstraction on top of the numerical properties of objects and is typically used in systems that need to reason about concepts in space and/or time [9]. In the context of this paper, it is hard to compare a slide

Table 2. Feature vectors for the example slide in Fig. 4

Object	Number	Title	Footer	Listing	Caption	Normal
1	0.00	**0.53**	0.00	0.00	0.16	0.31
2	0.00	0.00	0.00	0.40	0.00	**0.60**
3	0.00	0.00	0.00	0.00	**0.67**	0.33
4	0.00	0.00	**0.57**	0.00	0.14	0.29
5	**0.48**	0.00	0.16	0.00	0.12	0.24

with a template based on the exact positions of their objects. Simply knowing whether an object is placed *above* or *left of* another object then provides enough information to compare two slides and align their objects.

The Allen relations [1] describe seven configurations of two intervals, as illustrated in Fig. 5a. By projecting two objects on the x- and y- axes defined by the top and left border of a slide, their relative position can be uniquely described by exactly two of these relations, one in each dimension. We write $r_x(a, b)$ if relation r holds between objects a and b along the x-axis.

Example 5. In Fig. 4, we can see that before$_x(2,3)$ and during$_y(3,2)$ hold.

The qualitative representation of a slide is obtained by the Allen relations between all pairs of objects that are not trivially aligned. Slide numbers and footers, for example, are trivially aligned across two slides. The Q column in Table 1 indicates for every object whether it is included in the qualitative representation or not. Additionally, we extend this set of relations with the unary title(o) and background(o) predicates and whether or not the slide is the first slide of the slideshow. The title serves as an anchor for comparing two slides, as it is present in all slide layouts that contain other content. A full example of the qualitative representation of a slide is shown in Fig. 5b. We write R_a to describe the representation of a slide a.

4.3   Searching for Slide Layouts

The final and most important step is finding the most suitable slide layout for each slide. This is achieved by first creating possible reference slides for each slide layout and then (qualitatively) moving objects in the given slide until its representation is equal to that of one of the reference slides. This process is represented schematically in Fig. 6.

Generating Reference Slices. For each of the given slide layouts, SandSlide generates reference slides by filling their placeholders with content of varying size and selecting different design templates. Each reference slide is converted to a qualitative representation using the method described in the previous section. Obtaining more reference slides can be interpreted as a *backward search* that

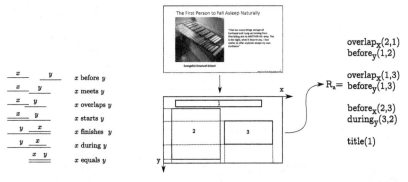

(a) Allen relations   (b) Qualitative representation using the Allen relations

Fig. 5. Allen relations are used to obtain a qualitative representation of a slide.

yields more target slides to be found at the cost of requiring more comparisons after each moved object.

Qualitatively Comparing Slides. Two slides are equal if their representation is equal, but this is not invariant to a permutation of the identifiers for objects in slides. For example, two slides with representations

$$R_1 = \{\text{before}_x(1,2), \text{during}_y(1,2)\} \quad \text{and} \quad R_2 = \{\text{before}_x(2,1), \text{during}_y(2,1)\}$$

are equal, but their representations are not. A *substitution* is a transformation of a representation that simultaneously replaces an object identifier with another identifier in all of its relations. For example, the substitution $\{1 \to 2, 2 \to 1\}$ can be applied to either R_1 or R_2 to make them equal. More generally, when comparing two slides a and b, we say that they are equal if and only if there exists a substitution θ such that $\theta(R_a) = R_b$.

Finding such a substitution is called the *set unification problem* and is shown to be NP-complete [8]. Many algorithms for set unification have been presented and exact details are considered out of the scope for this paper. Anchoring the title across two slides allows us to greatly prune the search space and we use a brute force approach.

Qualitatively Transforming Slides. Rarely will an arbitrary slide be equal to one of the reference slides. SandSlide will thus perform qualitative transformations of the representation of the given slide and match the transformed slide with references. Let R be the set of Allen relations. A *transformation* t is a function that replaces one or more relations $r_d(o_1, o_2) \in R$ with a new relation $r'_d(o_1, o_2) \in I$ along the same axis d. The length of a transformation $|t|$ is the number of replacements that it performs.

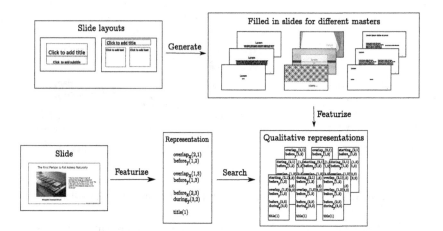

Fig. 6. Discovering the most suitable slide layout type for a slide based on its qualitative representation.

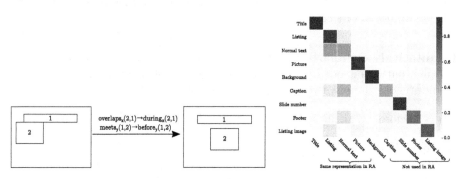

Fig. 7. Qualitative transformation of objects.

Fig. 8. Confusion matrix of annotations with local heuristics.

Example 6. An example of a transformation of length 2 is shown in Fig. 7.

Not all transformations yield a qualitative representation for which there exists an actual configuration of objects. Checking if such a configuration exists is called the *satisfaction problem for interval algebra* and it is also shown to be NP-complete [12]. We opt to simply not check whether a transformed representation is consistent as these intermediate, inconsistent representations serve as stepping stones to find an exact match.

Searching for Slides. Given a slide s and a set of slide layouts $\mathbf{L}$ with each layout $L_i \in \mathbf{L}$ described by a set of representations $l_i^j \in L_i$, we then look for the smallest transformation t such that there exists a representation $l_i^j \in L_i$ such that $t(s)$ is equal to l_i^j.

A slide with n objects is represented by $2n$ relations. Each relation can be transformed into six new relations. There are then $\binom{n(n-1)}{d} \times 6^d$ different transformations of length d. Each transformed slide has to be compared with all reference slides. Even for small d, this quickly becomes intractable to compute.[3]

We therefore use a heuristic algorithm based on the similarity between a slide and all reference slides. The Jaccard similarity $J(A, B) = \frac{|A \cap B|}{|A \cup B|}$ is a common way to compute the similarity between two sets A and B. In order to obtain a similarity between two slides a and b that respects an optimal alignment between objects, we compute

$$sim(a, b) = \max_{\theta \in \Theta} J(\theta(a), b)$$

with Θ all possible substitutions. The score for a transformation on the original slide is

$$score(t) = \max_{L \in \mathbf{L}} \max_{l \in L} sim(l, t(\text{slide})) \tag{1}$$

with $\mathbf{L}$ the set of all layouts, L a specific layout, l a possible representation of that layout and $t(\text{slide})$ the representation of the transformed slide. Note that the similarity for equal slides is 1 and computing the heuristic also informs us when an exact match is found. We do not need to find the optimal substitution twice as the heuristic is obtained when checking for a solution.

SandSlide performs a greedy search using the heuristic. At each step, the best transformation so far according to Eq. 1 is expanded. All six possible transformations of a relation are considered at once. Relations across different axes are transformed independently. Loops are prevented by first generating all transformations of length one and combining these to form larger transformations.

If a solution is not found after a few transformations, any solution is not likely to be closely related to the intended solution. Search is therefore cut short after a predefined number of comparisons, in other words, the number of times an optimal substitution has to be computed. The layout assigned to a slide is then given by

$$solution(s, \mathbf{L}) = \arg\max_{L_i \in \mathbf{L}} \max_{l \in L_i} \max_{t \in T} sim(l, t(s)) \tag{2}$$

where T is the set of all transformations evaluated during search.

5   Evaluation

We performed several experiments to answer the following research questions.

Q1 Do we require search or can we simply look for the most similar reference slide according to the heuristic?

Q2 What is the effect of performing a deeper backward search at the cost of requiring more comparisons in each iteration?

Q3 Does our algorithm guarantee that objects are assigned their intended role?

---

[3] For a slide with five objects, there are 25920 transformations of length 3. For as little as 40 reference slides, that is over a million comparisons.

5.1   Experimental Setup

Data. In order to answer these questions, we start from a set of 1000 slideshows provided by the U.S. Library of Congress [3]. We filtered out all files that were not PowerPoint files and slideshows that are completely out of scope for slideshow normalization, which leaves 640 slideshows.

From those slideshows, we manually annotated a randomly sampled set of 500 slides using VIA [4] after the PDF → XML and the preparsing steps. Two types of annotations were recorded; the slide as a whole and individual objects. For slides, we make the following annotations; (1) objects in scope and corresponds to a layout; (2) objects in scope and superset of a layout; (3) objects not in scope, but would otherwise fit 1 or 2; (4) does not match a layout but would make an interesting layout or (5) completely out of scope. Slides of type (1) and (2) are used in our evaluation. For objects, we made precise annotations of their semantic type and their intended role. We use these last annotations as a human evaluation of the algorithm, as it allows to check if the annotations assigned by the algorithm match those assigned by humans.

The role of an object denotes the positions on a slide where it is allowed to end up after transformation. For example: when there are two images besides each other, the one on the left should stay on the left or when there is a title and a subtitle, the subtitle should be used as a subtitle and not as content. This implies that the image on the left should never be the rightmost piece of content and that the subtitle should be below the title and above the content.

These annotations are more precise than required for our experiments. Our goal is to encourage research on different aspects of slideshow normalization and to work towards an established benchmark for this problem. We make the code, original slides, results after the preparsing and annotation steps and all ground truth annotations publicly available.[4,5]

Evaluation. We measure two properties of aligning a slide with a reference. The *responsiveness* is the proportion of objects that can be assigned to a placeholder. A random assignment of objects to placeholders trivially yields high responsiveness. We therefore use annotated roles to compute the *sensibility* of an alignment as the proportion of alignments that are allowed with a set of handcrafted rules. The sensibility measure acts as qualitative evaluation, as it checks if all the elements stay in allowed positions.

We do not compare SandSlide with existing online tools (2.2) using these measures as the results are trivial: they do not assign objects to placeholders and place the objects on the same position as in the PDF, which results in zero responsiveness and undefined sensibility.

Reference Slides. Most design templates support ten layouts. Using these layouts, we created three levels of reference slides. The **baseline** set considers only

---

[4] https://github.com/zevereir/SandSlide_data.

[5] https://github.com/zevereir/SandSlide.

the placeholders from the default design theme, without filling them with content, resulting in one reference for each layout. The **learned** set considers the default design theme, where placeholders are filled in with content of different lengths. This results in 137 representations. The **masters** set also considers different design themes, which results in 1117 unique representations. Limiting the search based on the number of comparisons was done to ensure a fair comparison between the sets of references.

5.2   Results

We did not perform an extensive evaluation of object annotation using local heuristics, as we found them to be sufficiently accurate for our purposes. A confusion matrix of assignments is shown in Fig. 8. Each row represents the normalized number of assignments made by the local heuristics. Listings, normal text and pictures are all content and no distinction is made between them in the qualitative representation. Some objects are not used in the representation, either because they are trivially aligned or no placeholder exists for them. Our experiments focus on evaluating the qualitative representation and search.

Responsiveness and sensibility for all sets of reference slides are plotted in function of the maximal number of comparisons in Fig. 9. Figure 10 shows the distribution of the slides over the different transformation lengths. We use these figures to answer the research questions in the following paragraphs.

Q1: Requiring search. We can see that the heuristic alone obtains respectable results, but that even a little bit of search significantly improves results, especially for the learned set. Searching for longer transformations increases the number of alignments—the responsiveness increases—but that some faulty alignments are introduced, which can be seen from the decreasing sensibility. Greedy search is able to correctly and quickly identify almost all slides with supported layouts.

Q2: Backward search. More references yields improved results, with the learned set outperforming the other sets in both methods. Adding references from different designs does not improve responsiveness, as many comparisons are wasted on the excess references. The masters set only catches up in responsiveness after a high number of comparisons, at the cost of significantly more wrong alignments. The effect of the backward search on the transformation lengths is as expected: more possible references causes less transformations needed to complete the search.

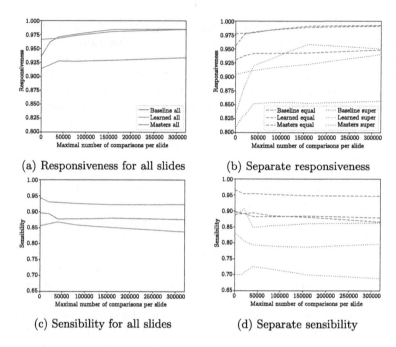

(a) Responsiveness for all slides

(b) Separate responsiveness

(c) Sensibility for all slides

(d) Separate sensibility

Fig. 9. Left: Main results for all slides (`all`) – Right: Main results for slides that exactly match a layout (`equal`) and supersets of layouts (`super`). Each data point is the average over all slides for an experiment with a specific parameter—the maximal number of allowed comparisons per slide.

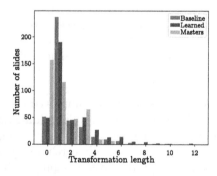

Fig. 10. Distribution of the number of replacements to find the best alignment in the experiment with the learned set and maximal 15000 comparisons per slide.

Q3: Alignments. As expected, deeper searches yield more alignments at the cost of making more mistakes. In general, our greedy method obtains very sensible alignments, even for searches with many comparisons. Adding references from different design templates has a negative effect on the sensitivity as most slides follow classic layouts.

5.3   Rebuilding Slides

From the layout and alignment, SandSlide can build a new slide. An example that obtained perfect responsiveness and sensibility is shown in the top part of Fig. 11. The logos on top are copy-pasted onto the new slide, to avoid losing any information, and the slide number is trivially aligned with a placeholder, as can be seen after applying a new template. The bottom of Fig. 11 shows an example of an imperfect normalization, as the subtitle in the original slide is converted to left content in the normalized one. The new slide has a perfect responsiveness, but only two out of three elements in the slide were assigned a sensible role.

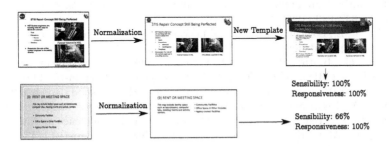

Fig. 11. (top) Perfectly normalized slide with new template. (bottom) Normalized slide with imperfect sensibility as the subtitle is assigned the role of left content.

6   Conclusion and Future Work

This paper introduces the nontrivial problem of slideshow normalization and implemented a first system for solving it, called SandSlide. This system is able to transform a given PDF file representing a slideshow into a normalized slideshow with a responsive slide layout using placeholders. The resulting slideshow can easily be edited and transformed into another design template. Our experiments show that search is a necessary component, that the heuristic works well and that it is beneficial to generate more varied representations for each slide layout, but that too many reduces the quality of the alignment.

Future Work. An immediate pointer for follow-up is to support more complex objects—like curves and word art—and objects that require anchors to existing objects—like arrows. This will make the search more complicated and using relational learners [2] would allow to easily incorporate background knowledge about these objects. Our qualitative representation provides an excellent starting point for such learners. As SandSlide allows to efficiently annotate objects and retrieve their semantic role, it can be used to improve downstream tasks involving slideshows. For example, it can be used to create datasets for approaches that learn to automatically generate slideshows [13].

Acknowledgments. This work has received funding from the European Research Council (ERC) under the European Union's Horizon 2020 research and innovation programme (grant agreement No [694980] SYNTH: Synthesising Inductive Data Models). Thomas Winters is supported by the Research Foundation-Flanders (FWO-Vlaanderen, 11C7720N). We would like to thank Luc De Raedt for his frustration with merging slideshows, which led to the master's thesis that formed the basis for this research.

References

1. Allen, J.F.: Maintaining knowledge about temporal intervals. Commun. ACM **26**(11), 832–843 (1983)
2. De Raedt, L.: Logical and Relational Learning. Springer Science & Business Media (2008). https://doi.org/10.1007/978-3-540-68856-3
3. Dooley, C.: In the library's web archives: 1,000 u.s. government powerpoint slide decks (2018). https://blogs.loc.gov/thesignal/2019/11/in-the-librarys-web-archives-1000-u-s-government-powerpoint-slide-decks/. Visited 04 Feb 2021
4. Dutta, A., Zisserman, A.: The VIA annotation software for images, audio and video. In: Proceedings of the 27th ACM International Conference on Multimedia. MM 2019, ACM, New York, NY, USA (2019). https://doi.org/10.1145/3343031.3350535
5. Ferrés, D., Saggion, H., Ronzano, F., Bravo, A.: PDFdigest: an adaptable layout-aware PDF-to-XML textual content extractor for scientific articles. In: Proceedings of the Eleventh International Conference on Language Resources and Evaluation (LREC 2018). European Language Resources Association (ELRA), Miyazaki, Japan, May 2018. https://www.aclweb.org/anthology/L18-1298
6. Haurilet, M., Al-Halah, Z., Stiefelhagen, R.: Spase-multi-label page segmentation for presentation slides. In: 2019 IEEE Winter Conference on Applications of Computer Vision (WACV), pp. 726–734. IEEE (2019)
7. Haurilet, M., Roitberg, A., Martinez, M., Stiefelhagen, R.: Wise–slide segmentation in the wild. In: 2019 International Conference on Document Analysis and Recognition (ICDAR), pp. 343–348. IEEE (2019)
8. Kapur, D., Narendran, P.: NP-completeness of the set unification and matching problems. In: Siekmann, J.H. (ed.) CADE 1986. LNCS, vol. 230, pp. 489–495. Springer, Heidelberg (1986). https://doi.org/10.1007/3-540-16780-3_113
9. Kuipers, B.: Qualitative reasoning: modeling and simulation with incomplete knowledge. Automatica **25**(4), 571–585 (1989). https://www.sciencedirect.com/science/article/pii/000510988990099X
10. Luong, M.T., Nguyen, T.D., Kan, M.Y.: Logical structure recovery in scholarly articles with rich document features. Int. J. Digit. Libr. Syst. **1**, 1–23 (2010)
11. Tkaczyk, D., Szostek, P., Fedoryszak, M., Dendek, P.J., Bolikowski, Ł: CERMINE: automatic extraction of structured metadata from scientific literature. Int. J. Doc. Anal. Recogn. (IJDAR) **18**(4), 317–335 (2015). https://doi.org/10.1007/s10032-015-0249-8
12. Vilain, M., Kautz, H., Van Beek, P.: Constraint propagation algorithms for temporal reasoning: a revised report. In: Readings in Qualitative Reasoning About Physical Systems, pp. 373–381. Elsevier (1990)

13. Winters, T., Mathewson, K.W.: Automatically generating engaging presentation slide decks. In: Ekárt, A., Liapis, A., Castro Pena, M.L. (eds.) EvoMUSART 2019. LNCS, vol. 11453, pp. 127–141. Springer, Cham (2019). https://doi.org/10.1007/978-3-030-16667-0_9
14. Zhong, X., Tang, J., Yepes, A.J.: Publaynet: largest dataset ever for document layout analysis. In: 2019 International Conference on Document Analysis and Recognition (ICDAR), pp. 1015–1022. IEEE (2019)

Digital Editions as Distant Supervision for Layout Analysis of Printed Books

Alejandro H. Toselli, Si Wu, and David A. Smith[✉]

Khoury College of Computer Sciences, Northeastern University,
Boston, MA 02115, USA
{toselli,wu.si2,davi.smith}@northeastern.edu

Abstract. Archivists, textual scholars, and historians often produce digital editions of historical documents. Using markup schemes such as those of the Text Encoding Initiative and EpiDoc, these digital editions often record documents' semantic regions (such as notes and figures) and physical features (such as page and line breaks) as well as transcribing their textual content. We describe methods for exploiting this semantic markup as distant supervision for training and evaluating layout analysis models. In experiments with several model architectures on the half-million pages of the *Deutsches Textarchiv* (DTA), we find a high correlation of these region-level evaluation methods with pixel-level and word-level metrics. We discuss the possibilities for improving accuracy with self-training and the ability of models trained on the DTA to generalize to other historical printed books.

Keywords: Layout analysis · Distant supervision · Evaluation

1 Introduction

Expanding annotated data for training and evaluation has driven progress in automatic layout analysis of page images. Most commonly, these annotated datasets are produced by manual annotation or by aligning the input documents with the typesetting information in PDF and similar formats [24].

This paper describes methods for exploiting a further source of information for training and testing layout analysis systems: **digital editions with semantic markup**. Many researchers in archival and literary studies, book history, and digital humanities have focused on digitally encoding books from the early modern period (from 1450) and the nineteenth century [7]. These editions have often employed semantic markup—now usually expressed in XML—to record logical

Authors contributed equally. We would like to thank the Deutsches Textarchiv, the Women Writers Project, and Rui Dong for collecting data and running the baseline OCR system. This work was supported in part by a National Endowment for the Humanities Digital Humanities Advancement Grant (HAA-263837-19) and the Andrew W. Mellon Foundation's Scholarly Communications and Information Technology program. Any views, findings, conclusions, or recommendations expressed do not necessarily reflect those of the NEH or Mellon.

© Springer Nature Switzerland AG 2021
J. Lladós et al. (Eds.): ICDAR 2021, LNCS 12822, pp. 462–476, 2021.
https://doi.org/10.1007/978-3-030-86331-9_30

components of a document, such as notes and figures, as well as physical features, such as page and line breaks.

Common markup schemes—as codified by the Text Encoding Initiative, EpiDoc, or others—have been mostly used for "representing those features of textual resources which need to be identified explicitly in order to facilitate processing by computer programs" [20, p. xvi]. Due to their intended uses in literary and linguistic analysis, many digital editions abstract away precise appearance information. The typefaces used to distinguish footnotes from body text, for example, and the presence of separators such as horizontal rules or whitespace, often go unrecorded in digital editions, even when the semantic separation of these two page regions is encoded.

After discussing related work on modeling layout analysis (Sect. 2), we describe the steps in our procedure for exploiting digital editions with semantic markup to produce annotated data for layout analysis.[1]

First (Sect. 3), we analyze the markup in a corpus of digital editions for those elements corresponding to page-layout features. We demonstrate this analysis on the *Deutsches Textarchiv* (DTA) in German and the *Women Writers Online* (WWO) and *Text Creation Partnership* (TCP) in English.

Then (Sect. 4), we perform forced alignment to link these digital editions to page images and to link regions to subareas on those page images. For the DTA, which forms our primary test case, open-license images are already linked to the XML at the page level; for the WWO, we demonstrate large-scale alignment techniques for finding digital page images for a subset of books in the Internet Archive. For pages with adequate baseline OCR, we also align OCR output with associated page coordinates with text in regions in the ground-truth XML. Some page regions, such as figures, are not adequately analyzed by baseline OCR, so we describe models to locate them on the page.

In experimental evaluations (Sect. 5), we compare several model architectures, pretrained, fine-tuned, and trained from scratch on these bootstrapped page annotations. We compare region-level detection metrics, which can be computed on a whole semantically annotated corpus, to pixel- and word-level metrics and find a high correlation among them.

2   Related Work

Perhaps the largest dataset proposed recently for document layout analysis is PubLayNet [24]. The dataset is obtained by matching XML representations and PDF articles of over 1 million publicly available academic papers on PubMed Central[TM]. This dataset is then used to train both Faster-RCNN and Mask-RCNN to detect text, title, list, table, and figure elements. Both models use ResNeXt-101-64 × 4d from Detectron as their backbone. Their Faster-RCNN and Mask-RCNN achieve macro mean average precision (MAP) at intersection over union (IOU) [0.50:0.95] of 0.900 and 0.907 respectively on the test set.

---

[1] For data and models, see https://github.com/NULabTMN/PrintedBookLayout.

Newspaper Navigator [11] comprises a dataset and model for detecting non-textual elements in the historic newspapers in the *Chronicling America* corpus. The model is a finetuned R50-FPN Faster-RCNN from Detectron2 and is trained to detect photographs, illustrations, maps, comics/cartoons, editorial cartoons, headlines, and advertisements. The authors report a MAP of 63.4%.

U-net was first proposed for medical image segmentation [16]. Its architecture, based on convolutional layers, consists of a down-sampling analysis path (encoder) and an up-sampling synthesis path (decoder) which, unlike regular encoder-decoders, are not decoupled. There are skip connections to transfer fine-grained information from the low-level layers of the analysis path to the high-level layers of the synthesis path as this information is required to accurately generate fine-grained reconstructions. In this work, we employ the U-net implementation *P2PaLa*[2] described in [15] for detection and semantic classification of both text regions and lines. This implementation has been trained and tested on different publicly available datasets: cBAD [5] for baseline detection, and Bozen [19] and OHG [14] for both text region classification and baseline detection. Reported mean intersection over union results are above 84% for region and baseline detection on the Bozen dataset. It is worth noting that the U-net implementation is provided with a weighted loss function mechanism [13], which can mitigate possible class imbalance problems.

Kraken, an OCR system forked from *Ocropy*, uses neural networks to perform both document layout analysis and text recognition.[3] For pixel classification in layout analysis, Kraken's network architecture was designed for fewer memory resources than U-net. Roughly, it comprises down-sampling convolutional layers with an increasing number of feature maps followed by BLSTM blocks for processing such feature maps in both horizontal and vertical directions [9]. The final convolutional layer, with sigmoid activation function, outputs probability maps of regions and text lines. Kraken's model for baseline detection has been trained and tested on the public dataset BADAM [10] and also on the same datasets as P2PaLA. For region detection, Kraken obtained mean intersection over union figures are 0.81 and 0.49 for Bozen and OHG datasets respectively.

Several evaluation metrics have been commonly employed for document layout analysis. The Jaccard Index, also known as intersection over union (iu), is one of the most popular pixel-level evaluation measures used in ICDAR's organized competitions related with document layout analysis as [4,8]. Likewise this measure has also served as a way to consider when there is a match between detected objects and their references as in [17].

3   Analyzing Ground Truth Markup

Scholars of early printed books create digital editions for several purposes, from enabling full-text search to studying language change and literary style to studying the work practices of letterpress compositors.

---

[2] https://github.com/lquirosd/P2PaLA.
[3] See http://kraken.re and https://github.com/ocropus/ocropy.

In this paper, we focus on the "top-level" units of layout, which for brevity we call **regions**. Within each region, one or more lines follow each other in a sequential reading order (e.g., top-to-bottom or right-to-left). *Among* regions on a page, no such total order constraint necessarily holds. Page numbers and running titles, for instance, whether at the top or bottom or a page, do not logically "precede" or "follow" the main body or footnote text.

We analyze the conventions of encoding these top-level regions in three broad-coverage corpora of historical printed books. The *Deutsches Textarchiv* (DTA) [3] comprises transcriptions of 1406 German books in XML following Text Encoding Initiative (TEI) [20] conventions along with over 500,000 page images. The *Women Writers Online* (WWO) [22] corpus contains, as of this writing, 336 books of womens' writing in English, transcribed in TEI XML but with no page images of the editions transcribed. In Sect. 4 below, we discuss a forced alignment process to link a subset of the WWO gold-standard XML to page images from the Internet Archive. The *Text Creation Partnership* (TCP) [1] contains TEI XML transcriptions of 32,853 books from the proprietary microfilm series *Early English Books Online, Eighteenth-Century Collections Online*, and *Evans Early American Imprints*.

Table 1 summarizes the XML encoding conventions used for eight top-level regions in these three corpora. For precision, we use XPath notation [6]. All three corpora include some top-level regions such as **body**, **figure**, and **note**. The source texts that were transcribed to compile a corpus may still of course contain regions not reflected in the XML edition: for example, **running titles** are present in books from all three corpora, but the DTA is the only corpus that transcribes then.

Table 1. Summary of page zone markup in TEI editions from the *Deutsches Textarchiv* (DTA), *Text Creation Partnership* (TCP), and *Women Writers Online* (WWO). We remove trailing `text()` functions from the XPath selectors for simplicity.

Corpus	Caption	Catchword
DTA	`//figure/*`	`//fw[@type='catch']`
TCP	`//figure/*[not(self::figDesc)]`	–
WWO	`//figure/*[not(self::figDesc)]`	`//mw[@type='catch']`
	Column head	Figure
DTA	`//cb[substring(@n,1,1)!='[']/@n`	`//figure`
TCP	–	`//figure`
WWO	–	`//figure`
	Note	Pagination
DTA	`//note`	`//pb[substring(@n,1,1)!='[']/@n`
TCP	`//note`	–
WWO	`//notes/note`	`//mw[@type='pageNum']`
	Running title	Signature
DTA	`//fw[@type='head']`	`//fw[@type='sig']`
TCP	–	–
WWO	–	`//mw[@type='sig']`

Many small elements from the skeleton of the printing forme and other marginal matter are present in early modern books [21]. Both the DTA and WWO record printed **signature** marks and **catchwords** at the bottom of the page, which aided printers in assembling printed sheets into books. The DTA and WWO both transcribe printed **page numbers**. The DTA also encodes inferred page numbers, e.g., when the pagination is not explicitly printed on the first page of a chapter, by enclosing the number in square brackets. The DTA transcribes **running titles** at the top of the page and the **column heads** that appear in some multi-column page layouts (e.g., in dictionaries). The TCP does not record any of these minor elements.

All three corpora transcribe **notes**. The DTA and TCP insert `<note>` elements near the reference mark printed in the body text for footnotes and endnotes. They also transcribe marginal notes inline. The WWO transcribes all notes in a separate `<notes>` section at the beginning of each text and links the child `<note>` elements to references in the body with XML IDREFs. The text of the notes in the WWO must therefore be associated with the appropriate page record. In all three corpora, some foot- and endnotes continue onto the next page. We therefore assign each part of the text of these run-on notes with the appropriate page.

We define the **body** text as almost all contents of the `<text>` element that are not described one of the floating or extraneous elements described above and summarized in Table 1. The few exceptions to this definition in the three corpora we examine are elements recording editorial interventions in the text: the `<corr>` element in the WWO for corrected spelling and the `<gap>` element in the TCP for recording gaps in the transcription due to unreadable microfilm images. The body text is broken into different zones by page breaks (`<pb>`) and column breaks (`<cb>`). The DTA and WWO record line breaks in the editions they transcribe with `<lb>` milestones although the TCP does not. Although these line breaks might provide some slight improvement to the forced alignment we describe below, we do not depend on them.

The three corpora we examined provide further encoding of layout information beyond the top-level regions we focus on in this paper. All three mark header lines within the running text—often distinguished by larger type and centering—with `<head>`. The DTA and WWO record changes of typeface within running text, both at the level of appearance (e.g., roman vs. italic, or Fraktur vs. Antiqua), and at the semantic level (e.g., proper names are often italicized in roman text and in expanded type in Fraktur, but in roman type when surrounded by italics). The DTA encodes the row and cell structure of some tables but not others. We do not evaluate table layout analysis in this paper due to this inconsistency in the ground truth.

Based on this analysis, we started the process of bootstrapping annotated data for layout analysis with the DTA. Besides consistently encoding all top-level regions, both its XML transcriptions and page images are available under an open-source license. We can therefore release the annotations on the DTA produced for this paper as a benchmark dataset. In addition to experiments on the DTA, we also compiled page-level alignments for a subset of the WWO to test the generalization of models trained on the DTA. Since the TCP only transcribes a few of the main page regions, we leave further analysis of that corpus for future work.

4   Annotation by Forced Alignment

To train and evaluate layout models, we must link digital editions to page images. This coarse-grained page-level alignment allows us to evaluate models' retrieval accuracy, supporting user queries for images [11] or footnotes [2]. Most models and evaluations of layout analysis, however, require a finer-grained assignment of rectangular or polygonal image zones to particular regions. For both page-level and pixel-level image annotation, we perform forced alignment between the text of digital editions and the output of a baseline OCR system.

For **page-level** annotation, the DTA already links open-source page images to each page of its 1406 XML editions. For the WWO, we aligned the ground-truth editions with a corpus of 347,428 OCR'd early modern books from the Internet Archive. We applied the `passim` [18] text-reuse analysis system to the ABBYY FineReader transcripts of pages in these books and the 336 XML editions in the WWO. Processing the pairwise alignments between pages in the IA and in the WWO produced by `passim`, we selected pairs of scanned and transcribed books such that 80% of the pages in the scanned book aligned to the XML and 80% of the pages in the XML aligned with the scanned book. Furthermore, we required that less than 10% of the pages in the scanned book align to more than one page in the XML. This last condition was necessary to exclude editions with pagination differing from that transcribed in the WWO. In the end, this process produced complete sets of page images for 23 books in the WWO.

Prior to **pixel-level** image annotation, we have the transcripts of the page regions described above (Sect. 3) but not their locations on physical page image. We run the Tesseract OCR engine[4] on all DTA page images for text line detection and recognition using its publicly available pretrained German model. The OCR output is then aligned with the ground-truth transcripts from DTA XML in two steps: first, we use `passim` to perform a line-level alignment of the OCR output with the DTA text. Next, we perform a character-level forced alignment of the remaining not-yet-aligned OCR output, as well the already aligned text, with the ground-truth text to correct possible line segmentation issues. In this way, we align regions with one short line—such as page or column number, signature, catchword, and short headings and figure captions—for which `passim` failed due to limited textual context. This cleanup pass corrected, for example, alignments between a main body region and a note region placed on the left or right.

Once ground-truth transcripts for each text region had been aligned with the OCR output, region boundaries can be inferred from bounding boxes of the OCR'd text lines. Assuming that ground-truth transcripts of a region are

---

[4] https://github.com/tesseract-ocr/tesseract.

in reading order, we combined in this order all the bounding boxes and the boundary of the resulting combination is taken as that of the region.

In the digital editions we have examined, figures are not annotated with their exact coordinates or sizes. Pretrained models such as PubLayNet and Newspaper Navigator can extract figures from page images; however, since they are trained, respectively, on scientific papers and newspapers, which have different layouts from books, the figure detected sometimes also includes parts of other elements such as caption or body near the figure. To bootstrap annotations for the DTA, we ran Newspaper Navigator on all pages images where the ground truth contained a `<figure>` element. Since Newspaper Navigator produces overlapping hypotheses for elements such as figure at decoding time, we check the true number of figures in the ground truth for the page and then greedily select them in descending order of posterior probability, ignoring any bounding boxes that overlap higher-ranked ones.

The final location accuracy of regions in a page depends on how well Tesseract detected and recognized lines in that page image, how accurate the forced alignment was on noisy OCR output, and how accurately the baseline figure-detection model works. We therefore manually checked a subset of pages in the DTA for the accuracy of the pixel-level region annotation. For efficiency, we asked annotators only for binary judgments about the correctness of all regions on a page, rather than asking them to correct bounding boxes or polygons. We then split the page images into training and test sets (Table 2). Since the DTA and Internet Archive images are released under open-source licenses, we release these annotations publicly.

Table 2. Pages and regions in the force-aligned, manually checked DTA dataset

Region Type	Train	Test
Pages	318	136
Body	340	146
Caption	33	11
Catchword	17	4
Figure	53	23
Note	318	125
PageNum	313	135
Signature	33	22
Title	279	122

5   Experiments

Having produced alignments between ground-truth editions and page images at the pixel level for the DTA and at the page level for a subset of the WWO, we train and benchmark several page-layout analysis models on different tasks. First, we consider common pixel-level evaluation metrics. Then, we evaluate the ability of layout analysis models to retrieve the positions of words in various page regions. In this way, we aim for a better proxy for end-to-end OCR performance than pixel-level metrics, which might simply capture variation in the parts of regions not overlapping any text. Then, we evaluate the ability of layout models to retrieve page elements in the full dataset, where pixel-level annotations are not available but the ground-truth provides a set of regions to be detected on each page. We find a high correlation of these region-level and word-level evaluations with the more common pixel-level metrics. We close by measuring the possibilities for improving accuracy by self-training and the generalization of models trained on the DTA to the WWO corpus.

5.1 Models

We trained four models on the training portion of the DTA annotations produced by the forced alignment in Sect. 4. The process produced polygonal boundaries for some regions. For some experiments, as noted below, we computed the rectangular bounding boxes of these polygons to train on.

Initially, we ran the pretrained PubLayNet [24] model on the DTA test set, but it failed to find any regions. We then fine-tuned the **PubLayNet** F-RCNN weights provided on the DTA training set. PubLayNet's original training set of over 1 million PDF is only annotated for body, title, list, table, and figures, so it does not produce output for the other region classes. The best model, using the COCO primary challenge metric mean average precision (mAP = 0.824), results from a learning rate of 0.001, batch size of 128, and iteration of 1800.

We trained our own Faster-RCNN (**F-RCNN**) from scratch on the DTA training set. Our F-RCNN model is based on the ResNet50-FNP-3X baseline provided by Model Zoo[5] and was trained with Detectron2 [23]. The best performing model has a learning rate of 0.00025, a batch size of 16, and was trained for 30 epochs.

We also trained two models more directly specialized for page layout analysis: **Kraken** and **U-net** (P2PaLA). We adopted both systems' default architecture definitions and training hyperparameters. Page images were binarized and scaled to a height of 1200 and 1024 pixels for Kraken and U-net, respectively. Both models were trained with binary cross-entropy loss and the Adam optimizer, with learning rate 20^{-5} for 50 epochs with Kraken, and 10^{-4} for 200 epochs with U-net. To allow the models to generalize better on unseen samples, data augmentation was used by applying on-the-fly random transformations on each training image.

5.2 Pixel-Level Evaluations

To investigate whether regions annotated with polygonal coordinates have some advantage over annotation with rectangular coordinates, we trained the Kraken

---

[5] https://github.com/facebookresearch/detectron2/blob/master/MODEL_ZOO.md.

and U-net models on both annotation types. (The F-RCNN models only infer rectangles.) These models were trained and evaluated on the data defined in Table 2. Table 3 reports figures for standard region segmentation metrics [12]: pixel accuracy (p_acc), mean pixel accuracy (m_acc), mean Jaccard Index (m_iu), and frequency-weighted Jaccard Index (f_iu), for evaluating layout models for systems trained on different annotation types.

Table 3. Evaluation on four pixel-level metrics: PubLayNet fine-tuned and F-RCNN, Kraken, and U-net trained on DTA data. The first two models require rectangular bounding boxes at training time; the latter two may use polygons or rectangles.

	PubLayNet	F-RCNN	Kraken		U-net	
	Rect	Rect	Poly	Rect	Poly	Rect
p_acc	0.966	0.975	0.909	0.938	0.960	0.960
m_acc	0.973	0.894	0.511	0.537	0.928	0.946
m_iu	0.890	0.781	0.480	0.516	0.810	0.790
f_iu	0.886	0.881	0.858	0.907	0.932	0.933

Table 4. Pixel-level evaluation by region type: PubLayNet fine-tuned and F-RCNN, Kraken, and U-net trained on DTA data. PubLayNet does not output region types not in its original training data; Kraken produces no output for the smaller regions.

Region	PubLayNet		F-RCNN		Kraken		U-net	
	p_acc	iu	p_acc	iu	p_acc	iu	p_acc	iu
Body	0.96	0.92	0.97	0.94	0.93	0.91	0.96	0.94
Caption	–	–	0.91	0.77	–	–	0.78	0.60
Catchword	–	–	0.50	0.40	–	–	0.51	0.33
Figure	0.99	0.90	0.98	0.89	–	–	0.94	0.74
Note	–	–	0.98	0.92	0.82	0.77	0.93	0.88
PageNum	–	–	1.00	0.86	–	–	0.94	0.68
Signature	–	–	0.82	0.60	–	–	0.37	0.26
Title	0.97	0.84	0.99	0.87	–	–	0.96	0.74

From the results of Table 3, we can see there is not a significant difference between using rectangular or polygonal annotation for regions, but there is a substantial difference between the performance of the systems. Not shown in the table is the out-of-the-box PubLayNet, which is not able to detect any content in the dataset, but its performance improved dramatically after fine-tuning. Our own F-RCNN provides comparable results for the regions detectable in the fine-tuned PubLayNet, while it also detects 5 other regions. The differences among systems are more evident in Table 4, where Kraken's predictions detected only "body" and "note" and failed for the remaining (small) regions and the fine-tuned PubLayNet likewise predicted only a subset of the page regions. For this reason, we consider only the F-RCNN and U-net models in later experiments.

Table 5. Using F-RCNN on our annotated test set. AP @ iu [0.5:0.95].

Region type	F-RCNN AP
Body	0.888
Caption	0.638
Catchword	0.523
Figure	0.788
Note	0.868
PageNum	0.829
Signature	0.454
Title	0.792
OVERALL(mAP)	0.723

5.3   Word-Level Evaluations

While pixel-level evaluations focus on the layout analysis task, it is also worthwhile to measure a proxy for end-to-end OCR performance.

Using the positions of word tokens in the DTA test set as detected by Tesseract, we evaluate the performance of regions predicted by the U-net model considering how many words of the reference region fall inside or outside the boundary of the predicted region. Table 6 shows word-level retrieval results in terms of recall (Rc), precision (Pr) and F-measure (F1) metrics for each region type.

Table 6. Word-level retrieval results for the different region types predicted by the U-net model.

Region Type	Rc	Pr	F1
Body	0.91	0.98	0.94
Caption	0.63	0.73	0.66
Catchword	0.33	0.33	0.33
Note	0.85	0.98	0.91
PageNum	0.81	0.81	0.81
Signature	0.26	0.44	0.31
Title	0.92	0.97	0.94

5.4   Region-Level Evaluations

This is a simpler evaluation since it does not require word-position coordinates as the word-level case, considering only for each page whether its predicted region types are or not in the page ground-truth. Therefore, we can use the already trained layout models for inferring the regions on the entire DTA collection (composed of 500K page images) and also on the out-of-sample WWO dataset containing more than 5,000 pages with region types analogous to DTA.

Since PubLayNet and Kraken do not detect all the categories we want to evaluate, we perform this region-level evaluation using only the U-net and F-RCNN models, which were already trained on the 318 annotated pages of the DTA collection. To evaluate the performance over the entire DTA dataset and on WWO data, we use region-level precision, recall, and F1 metrics. Table 7 reports these evaluation metrics for the regions detected by these two models on the entire DTA and WWO datasets.

Table 7. Region-level retrieval results (Pr, Rc and F1) on the entire DTA collection and WWO data using U-net and F-RCNN. IoU threshold detection meta-parameter of F-RCNN model was set up to 0.9 and 0.5 for DTA and WWO respectively. The WWO does not annotate running titles or column heads, and the WWO test books contain figures but no captions.

Reg. Type	U-net						F-RCNN					
	DTA			WWO			DTA			WWO		
	Rc	Pr	F1	Rc	Pr	F1	Rc	Pr	F1	Rc	Pr	F1
Body	0.89	0.89	0.89	0.95	0.95	0.95	0.92	1.00	0.96	0.99	1.00	0.99
Caption	0.85	0.37	0.52	–	–	–	0.06	0.88	0.80	–	–	–
Catchword	0.89	0.63	0.74	0.77	0.52	0.62	0.26	0.96	0.41	0.54	0.48	0.52
ColNum	0.20	0.25	0.22	–	–	–	0.81	0.84	0.82	–	–	–
Figure	0.77	0.81	0.79	0.53	0.43	0.48	0.59	0.32	0.41	0.50	0.00	0.00
Note	0.97	0.97	0.97	0.84	0.84	0.84	0.61	0.64	0.62	0.42	0.18	0.26
PageNum	0.96	0.68	0.80	0.98	0.45	0.61	0.81	0.79	0.80	0.61	0.23	0.34
Signature	0.65	0.48	0.56	0.32	0.25	0.28	0.12	0.85	0.21	0.13	0.14	0.13
Title	0.97	0.69	0.81	–	–	–	0.52	0.78	0.54	–	–	–

The F-RCNN model can find all the graphic figures in the ground truth; however, since it also has a high false positive value, the precision for `figure` is 0 at confidence threshold of 0.5. In general, as can be observed in Table 7, F-RCNN seems to generalize less well than U-net on several region types in both the DTA and WWO.

5.5   Improving Accuracy with Self-training

While the amount of data we can manually label at the pixel level is small, the availability of page-level information on regions in the whole corpus allows us to improve these models by self-training. Instead of simply adding in potentially noisy automatically labeled images to the training set, we can restrict the new training examples to those pages where all regions have been successfully detected. In analyzing one iteration of this procedure, we find that overall pixel-level metrics improve slightly, but improve substantially for particular regions (Table 8).

Table 8. Pixel-level evaluation results for U-net with two self-training rounds on the 136 annotated pages of the DTA test set: globally (left) and by region (right).

Metric	Round 1	Round 2
p_acc	0.960	0.964
m_acc	0.928	0.934
m_iu	0.810	0.845
f_iu	0.932	0.937

Region type	Round 1		Round 2	
	p_acc	iu	p_acc	iu
Body	0.960	0.940	0.968	0.952
Caption	0.783	0.596	0.704	0.548
Catchword	0.513	0.334	0.657	0.447
Figure	0.937	0.735	0.966	0.701
Note	0.928	0.880	0.942	0.902
PageNum	0.937	0.683	0.960	0.740
Signature	0.369	0.262	0.481	0.410
Title	0.961	0.735	0.920	0.827

Likewise, we see similar improvements in many region types in the word- and region-level evaluations (Tables 9 and 10). Notably, the accuracy of detecting figures declines with self-training, which we find is due to only two images with figures appearing in the set of pages with all regions correctly detected. Balancing different page layouts during self-training might mitigate this problem.

Table 9. Word-level retrieval results for U-net with two self-training rounds on the 136 annotated pages of the DTA test set.

Region Type	Round 1			Round 2		
	Rc	Pr	F1	Rc	Pr	F1
Body	0.91	0.98	0.94	0.94	1.00	0.97
Catchword	0.33	0.33	0.33	0.83	1.00	0.89
Figure	0.63	0.73	0.66	0.52	0.53	0.50
Note	0.85	0.98	0.91	0.89	0.95	0.92
PageNum	0.81	0.81	0.81	0.87	0.87	0.87
Signature	0.26	0.44	0.31	0.41	0.56	0.44
Title	0.92	0.97	0.94	0.88	0.98	0.91

Table 10. Region-level retrieval results for U-net with self-training on the 136 annotated pages of DTA's test set.

Region Type	Round 1			Round 2		
	Rc	Pr	F1	Rc	Pr	F1
Body	0.99	0.99	0.99	1.00	1.00	1.00
Caption	1.00	0.82	0.90	1.00	0.62	0.77
Catchword	1.00	1.00	1.00	1.00	1.00	1.00
Figure	0.86	0.86	0.86	0.82	0.82	0.82
Note	1.00	1.00	1.00	0.96	0.96	0.96
PageNum	1.00	0.92	0.96	0.99	0.95	0.97
Signature	0.91	0.57	0.70	0.95	0.72	0.82
Title	1.00	0.92	0.96	1.00	0.96	0.98

5.6   Correlation of Word/Region-Level and Pixel-Level Metrics

To analyze the correlation of word/region- and pixel-level metrics, the Pearson correlation coefficient has been computed for each region between word/region-level F-score (F1) and pixel Jaccard index (iu) values obtained on the test set of 136 annotated pages (see Table 2). The reported linear correlation coefficients in Table 11 in general show significant correlations (with p-values lower than 10^{-10}) between these different metrics.

Table 11. Pearson correlation coefficients between the Jaccard-Index pixel-level metric (iu), and word- & region-level F-score (F1) metrics. This study was conducted on DTA test set with the U-net model.

Region Type	Word Level	Region Level
Body	0.980	0.906
Note	0.970	0.903
Title	0.812	0.861
PageNum	0.316	0.569
Signature	0.524	0.702
Figure	0.962	0.966
Catchword	0.792	0.916

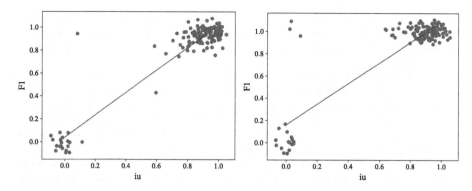

Fig. 1. Scatter plots with linear correlation of Jaccard Index (iu) versus word-level (left) and region-level F-score (right) for region type "note". These were produced using U-net on DTA's test set data.

6   Conclusions

We found that several broad-coverage collections of digital editions can be aligned to page images in order to construct large testbeds for document layout analysis. We manually checked a sample of regions annotated at the pixel level by forced alignment. We benchmarked several state-of-the-art methods and showed a high correlation of standard pixel-level evaluations with word- and region-level evaluations applicable to the full corpus of a half million images from the DTA. We publicly released the annotations on these open-source images at https://github.com/NULabTMN/PrintedBookLayout. Future work on these corpora could focus on standardizing table layout annotations; on annotating sub-regions, such as section headers, poetry, quotations, and contrasting type-faces; and on developing improved layout analysis models for early modern print.

References

1. Text Creation Partnership (1999–2020). https://textcreationpartnership.org/
2. Abuelwafa, S., et al.: Detecting footnotes in 32 million pages of ECCO. Journal of Cultural Analytics (2018). https://doi.org/10.31235/osf.io/7m8ue
3. Berlin-Brandenburgische Akademie der Wissenschaften: Deutsches Textarchiv. Grundlage für ein Referenzkorpus der neuhochdeutschen Sprache (2021). https://www.deutschestextarchiv.de/
4. Burie, J., et al.: ICDAR2015 competition on smartphone document capture and OCR (SmartDoc). In: International Conference on Document Analysis and Recognition (ICDAR), pp. 1161–1165 (2015). https://doi.org/10.1109/ICDAR.2015.7333943
5. Diem, M., Kleber, F., Sablatnig, R., Gatos, B.: cBAD: ICDAR2019 competition on baseline detection. In: International Conference on Document Analysis and Recognition (ICDAR), pp. 1494–1498 (2019). https://doi.org/10.1109/ICDAR.2019.00240

6. Dyck, M., Robie, J., Spiegel, J.: XML path language (XPath) 3.1. W3C recommendation, W3C (2017). https://www.w3.org/TR/2017/REC-xpath-31-20170321/
7. Franzini, G.: A catalogue of digital editions (2012). https://github.com/gfranzini/digEds_cat
8. Gao, L., Yi, X., Jiang, Z., Hao, L., Tang, Z.: ICDAR2017 competition on page object detection. In: International Conference on Document Analysis and Recognition (ICDAR), pp. 1417–1422 (2017). https://doi.org/10.1109/ICDAR.2017.231
9. Kiessling, B.: A modular region and text line layout analysis system. In: International Conference on Frontiers in Handwriting Recognition (ICFHR), pp. 313–318 (2020). https://doi.org/10.1109/ICFHR2020.2020.00064
10. Kiessling, B., Ezra, D.S.B., Miller, M.T.: BADAM: a public dataset for baseline detection in Arabic-script manuscripts. In: Proceedings of the 5th International Workshop on Historical Document Imaging and Processing (HIP), pp. 13–18 (2019). https://doi.org/10.1145/3352631.3352648
11. Lee, B.C.G., et al.: The Newspaper Navigator dataset: extracting and analyzing visual content from 16 million historic newspaper pages in Chronicling America. In: KDD (2020)
12. Long, J., Shelhamer, E., Darrell, T.: Fully convolutional networks for semantic segmentation. In: CVPR, pp. 3431–3440 (2015)
13. Paszke, A., Chaurasia, A., Kim, S., Culurciello, E.: ENet: a deep neural network architecture for real-time semantic segmentation. ArXiv abs/1606.02147 (2016)
14. Quirós, L., Bosch, V., Serrano, L., Toselli, A.H., Vidal, E.: From HMMs to RNNs: computer-assisted transcription of a handwritten notarial records collection. In: International Conference on Frontiers in Handwriting Recognition (ICFHR), pp. 116–121 (2018). https://doi.org/10.1109/ICFHR-2018.2018.00029
15. Quirós, L.: Multi-task handwritten document layout analysis (2018)
16. Ronneberger, O., Fischer, P., Brox, T.: U-net: convolutional networks for biomedical image segmentation. In: Medical Image Computing and Computer-Assisted Intervention (MICCAI), pp. 234–241 (2015)
17. Shi, B., et al.: ICDAR2017 competition on reading Chinese text in the wild (RCTW). In: International Conference on Document Analysis and Recognition (ICDAR), pp. 1429–1434 (2017). https://doi.org/10.1109/ICDAR.2017.233
18. Smith, D.A., Cordel, R., Dillon, E.M., Stramp, N., Wilkerson, J.: Detecting and modeling local text reuse. In: IEEE/ACM Joint Conference on Digital Libraries, pp. 183–192 (2014). https://doi.org/10.1109/JCDL.2014.6970166
19. Sánchez, J.A., Romero, V., Toselli, A.H., Vidal, E.: ICFHR2016 competition on handwritten text recognition on the READ dataset. In: International Conference on Frontiers in Handwriting Recognition (ICFHR), pp. 630–635 (2016). https://doi.org/10.1109/ICFHR.2016.0120
20. TEI Consortium (ed.): TEI P5: Guidelines for Electronic Text Encoding and Interchange. 4.1.0. TEI Consortium, August 2020. http://www.tei-c.org/Guidelines/P5/ (2021-02-22)
21. Werner, S.: Studying Early Printed Books 1450–1800: A Practical Guide. Wiley Blackwell, New York (2019)
22. Women Writers Project: Women Writers Online (1999–2016). https://www.wwp.northeastern.edu
23. Wu, Y., Kirillov, A., Massa, F., Lo, W.Y., Girshick, R.: Detectron2 (2019). https://github.com/facebookresearch/detectron2
24. Zhong, X., Tang, J., Jimeno Yepes, A.: PubLayNet: largest database ever for document layout analysis. In: ICDAR (2019)

Palmira: A Deep Deformable Network for Instance Segmentation of Dense and Uneven Layouts in Handwritten Manuscripts

S. P. Sharan⊙, Sowmya Aitha⊙, Amandeep Kumar⊙, Abhishek Trivedi⊙, Aaron Augustine⊙, and Ravi Kiran Sarvadevabhatla$^{(\boxtimes)}$⊙

Centre for Visual Information Technology, International Institute of Information Technology, Hyderabad 500032, India
ravi.kiran@iiit.ac.in
https://ihdia.iiit.ac.in/Palmira

Abstract. Handwritten documents are often characterized by dense and uneven layout. Despite advances, standard deep network based approaches for semantic layout segmentation are not robust to complex deformations seen across semantic regions. This phenomenon is especially pronounced for the low-resource Indic palm-leaf manuscript domain. To address the issue, we first introduce Indiscapes2, a new large-scale diverse dataset of Indic manuscripts with semantic layout annotations. Indiscapes2 contains documents from four different historical collections and is 150% larger than its predecessor, Indiscapes. We also propose a novel deep network PALMIRA for robust, deformation-aware instance segmentation of regions in handwritten manuscripts. We also report Hausdorff distance and its variants as a boundary-aware performance measure. Our experiments demonstrate that PALMIRA provides robust layouts, outperforms strong baseline approaches and ablative variants. We also include qualitative results on Arabic, South-East Asian and Hebrew historical manuscripts to showcase the generalization capability of PALMIRA.

Keywords: Instance segmentation · Deformable convolutional network · Historical document analysis · Document image segmentation · Dataset

1 Introduction

Across cultures of the world, ancient handwritten manuscripts are a precious form of heritage and often serve as definitive sources of provenance for a wide variety of historical events and cultural markers. Consequently, a number of research efforts have been initiated worldwide [6,11,20,23] to parse image-based versions of such manuscripts in terms of their structural (layout) and linguistic (text) aspects. In many cases, accurate layout prediction greatly facilitates downstream processes such as OCR [18]. Therefore, we focus on the problem of obtaining high-quality layout predictions in handwritten manuscripts.

© Springer Nature Switzerland AG 2021
J. Lladós et al. (Eds.): ICDAR 2021, LNCS 12822, pp. 477–491, 2021.
https://doi.org/10.1007/978-3-030-86331-9_31

Among the varieties of historical manuscripts, many from the Indian subcontinent and South-east Asia are written on palm-leaves. These manuscripts pose significant and unique challenges for the problem of layout prediction. The digital versions often reflect multiple degradations of the original. Also, a large variety exists in terms of script language, aspect ratios and density of text and non-text region categories. The Indiscapes Indic manuscript dataset and the deep-learning based layout parsing model by Prusty et al. [23] represent a significant first step towards addressing the concerns mentioned above in a scalable manner. Although Indiscapes is the largest available annotated dataset of its kind, it contains a rather small set of documents sourced from two collections. The deficiency is also reflected in the layout prediction quality of the associated deep learning model.

To address these shortcomings, we introduce Indiscapes2 dataset as an expanded version of Indiscapes (Sect. 3). Indiscapes2 is **150%** larger compared to its predecessor and contains two additional annotated collections which greatly increase qualitative diversity (see Fig. 1, Table 1a). In addition, we introduce a novel deep learning based layout parsing architecture called Palm leaf Manuscript Region Annotator or PALMIRA in short (Sect. 4). Through our experiments, we show that PALMIRA outperforms the previous approach and strong baselines, qualitatively and quantitatively (Sect. 6). Additionally, we demonstrate the general nature of our approach on out-of-dataset historical manuscripts. Intersection-over-Union (IoU) and mean Average Precision (AP) are popular measures for scoring the quality of layout predictions [10]. Complementing these area-centric measures, we report the boundary-centric Hausdorff distance and its variants as part of our evaluation approach (Sect. 6).

The source code, pretrained models and associated material are available at this link: https://ihdia.iiit.ac.in/Palmira.

2   Related Work

Layout analysis is an actively studied problem in the document image analysis community [4,6,14]. For an overview of approaches employed for historical and modern document layout analysis, refer to the work of Prusty et al. [23] and Liang et al. [16]. In recent times, large-scale datasets such as PubLayNet [30] and DocBank [15] have been introduced for document image layout analysis. These datasets focus on layout segmentation of modern language printed magazines and scientific documents.

Among recent approaches for historical documents, Ma et al. [18] introduce a unified deep learning approach for layout parsing and recognition of Chinese characters in a historical document collection. Alaasam et al. [2] use a Siamese Network to segment challenging historical Arabic manuscripts into main text, side text and background. Alberti et al. [3] use a multi-stage hybrid approach for segmenting text lines in medieval manuscripts. Monnier et al. [20] introduce docExtractor, an off-the-shelf pipeline for historical document element extraction from 9 different kinds of documents utilizing a modified U-Net [25]. dhSegment [21] is a similar work utilizing a modified U-Net for document segmentation

Table 1. Document collection level, region level statistics of Indiscapes2 dataset.

(a) Collection level stats.

	Train	Validation	Test	Total	Indiscapes (old)
PIH	285	70	94	449	193
Bhoomi	408	72	96	576	315
ASR	36	11	14	61	–
Jain	95	40	54	189	–
Total	824	193	258	1275	508

(b) Region count statistics.

	Character Line Segment (CLS)	Character Component (CC)	Hole (Virtual) (Hv)	Hole (Physical) (Hp)	Page Boundary (PB)	Library Marker (LM)	Decorator/ Picture (D/P)	Physical Degradation (PD)	Boundary Line (BL)
PIH	5105	1079	–	9	610	52	153	90	724
Bhoomi	5359	524	8	737	547	254	8	2535	80
ASR	673	59	–	–	52	41	–	81	83
Jain	1857	313	93	38	166	7	–	166	292
Combined	12994	1975	101	784	1375	354	161	2872	1179

of medieval era documents. Unlike our instance segmentation formulation (i.e. a pixel can simultaneously have two distinct region labels), existing works (except dhSegment) adopt the classical segmentation formulation (i.e. each pixel has a single region label). Also, our end-to-end approach produces page and region boundaries in a single stage end-to-end manner without any postprocessing.

Approaches for palm-leaf manuscript analysis have been mostly confined to South-East Asian scripts [22,29] and tend to focus on the problem of segmented character recognition [11,19,24]. The first large-scale dataset for palm leaf manuscripts was introduced by Prusty et al. [23], which we build upon to create an even larger and more diverse dataset.

Among deep-learning based works in document understanding, using deformable convolutions [7] to enable better processing of distorted layouts is a popular choice. However, existing works have focused only on tabular regions [1,26]. We adopt deformable convolutions, but for the more general problem of multi-category region segmentation.

3   Indiscapes2

We first provide a brief overview of Indiscapes dataset introduced by Prusty et al. [23]. This dataset contains 508 layout annotated manuscripts across two collections - Penn-in-Hand (from University of Pennsylvania's Rare Book Library) and Bhoomi (from libraries and Oriental Research Institutes across India). The images span multiple scripts, exhibit diversity in language and text line density, contain multiple manuscripts stacked in a single image and often contain non textual elements (pictures and binding holes).

Although Indiscapes was the first large-scale Indic manuscript dataset, it is rather small by typical dataset standards. To address this limitation and enable advanced layout segmentation deep networks, we build upon Indiscapes to create Indiscapes2. For annotation, we deployed an instance of HInDoLA [28] - a multi-feature annotation and analytics platform for historical manuscript layout processing. The fully automatic layout segmentation approach from Prusty et al. [23] is available as an annotation feature in HInDoLA. The annotators utilize the same to obtain an initial estimate and edit the resulting output, thus minimizing the large quantum of labour involved in pure manual annotation. HInDoLA also provides a visualization interface for examining the accuracy of annotations and tagging documents for correction.

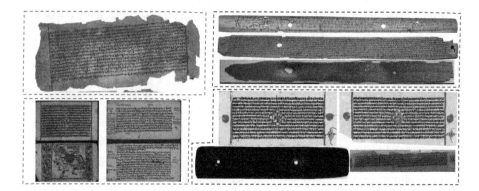

Fig. 1. Representative manuscript images from Indiscapes2 - from newly added ASR collection (top left, pink dotted line), Penn-in-Hand (bottom left, blue dotted line), Bhoomi (green dotted line), newly added Jain (brown dotted line). Note the diversity across collections in terms of document quality, region density, aspect ratio and non-textual elements (pictures). (Color figure online)

In the new dataset, we introduce additional annotated documents from the Penn-in-Hand and Bhoomi book collections mentioned previously. Above this, we also add annotated manuscripts from two new collections - ASR and Jain. The ASR documents are from a private collection and contain 61 manuscripts written in Telugu language. They contain 18–20 densely spaced text lines per document (see Fig. 1). The Jain collection contains 189 images. These documents contain 16–17 lines per page and include early paper-based documents in addition to palm-leaf manuscripts.

Altogether, Indiscapes2 comprises of 1275 documents - a 150% increase over the earlier Indiscapes dataset. Refer to Tables 1a,1b for additional statistics related to the datasets and Fig. 1 for representative images. Overall, Indiscapes2 enables a greater coverage across the spectrum of historical manuscripts - qualitatively and quantitatively.

4   Our Layout Parsing Network (Palmira)

In their work, Prusty et al. [23] utilize a modified Mask-RCNN [10] framework for the problem of localizing document region instances. Although the introduced framework is reasonably effective, the fundamental convolution operation throughout the Mask-RCNN deep network pipeline operates on a fixed, rigid spatial grid. This rigidity of receptive fields tends to act as a bottleneck in obtaining precise boundary estimates of manuscript images containing highly deformed regions. To address this shortcoming, we modify two crucial stages of the Mask R-CNN pipeline in a novel fashion to obtain our proposed architecture (see Fig. 1). To begin with, we briefly summarize the Mask R-CNN approach adopted by Prusty et al. We shall refer to this model as the Vanilla Mask-RCNN model. Subsequently, we shall describe our novel modifications to the pipeline.

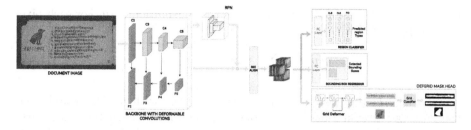

Fig. 2. A diagram illustrating PALMIRA's architecture (Sect. 4). The orange blocks in the backbone are deformable convolutions (Sect. 4.2). Refer to Fig. 3b, Sect. 4.3 for additional details on Deformable Grid Mask Head which outputs region instance masks. (Color figure online)

4.1 Vanilla Mask-RCNN

Mask R-CNN [10] is a three stage deep network for object instance segmentation. The three stages are often referred to as Backbone, Region Proposal Network (RPN) and Multi-task Branch Networks. One of the Branch Networks, referred to as the Mask Head, outputs individual object instances. The pipeline components of Mask-RCNN are modified to better suit the manuscript image domain by Prusty et al. [23]. Specifically, the ResNet-50 used in Backbone is initialized from a Mask R-CNN trained on the MS-COCO dataset. Within the RPN module, the anchor aspect ratios of 1:1,1:3,1:10 were chosen keeping the peculiar aspect ratios of manuscript images in mind and the number of proposals from RPN were reduced to 512. The various thresholds involved in other stages (objectness, NMS) were also modified suitably. Some unique modifications were included as well – the weightage for loss associated with the Mask head was set to twice of that for the other losses and focal-loss [17] was used for robust labelling.

We use the modified pipeline described above as the starting point and incorporate two novel modifications to Mask-RCNN. We describe these modifications in the sections that follow.

4.2 Modification-1: Deformable Convolutions in Backbone

Before examining the more general setting, let us temporarily consider 2D input feature maps x. Denote the 2D filter operating on this feature map as w and the convolution grid operating on the feature map as $\mathcal{R}$. As an example, for a 3×3 filter, we have:

$$\mathcal{R} = \left\{ \begin{matrix} (-1,-1) & (-1,0) & (-1,1) \\ (0,-1) & (0,0) & (0,1) \\ (1,-1) & (1,0) & (1,1) \end{matrix} \right\} \tag{1}$$

Let the output feature map resulting from the convolution be y. For each pixel location p_0, we have:

$$y(p_0) = \sum_{p_n \in \mathcal{R}} w(p_n) \cdot x(p_0 + p_n) \tag{2}$$

where n indexes the spatial grid locations associated with $\mathcal{R}$. The default convolution operation in Mask R-CNN operates via a fixed 2D spatial integer grid as described above. However, this setup does not enable the grid to deform based on the input feature map, reducing the ability to better model the high inter/intra-region deformations and the features they induce.

As an alternative, Deformable Convolutions [7] provide a way to determine suitable local 2D offsets for the default spatial sampling locations (see Fig. 3a). Importantly, these offsets $\{\Delta p_n; n = 1, 2 \ldots\}$ are adaptively computed as a function of the input features for each reference location p_0. Equation 2 becomes:

$$y(p_0) = \sum_{p_n \in \mathcal{R}} w(p_n) \cdot x(p_0 + p_n + \Delta p_n) \tag{3}$$

Since the offsets Δp_n may be fractional, the sampled values for these locations are generated using bilinear interpolation. This also preserves the differentiability of the filters because the offset gradients are learnt via backpropagation through the bilinear transform. Due to the adaptive sampling of spatial locations, broader and generalized receptive fields are induced in the network. Note that the overall optimization involves jointly learning both the regular filter weights and weights for a small additional set of filters which operate on input to generate the offsets for input feature locations (Fig. 3a).

4.3   Modification-2: Deforming the Spatial Grid in Mask Head

The 'Mask Head' in Vanilla Mask-RCNN takes aligned feature maps for each plausible region instance as input and outputs a binary mask corresponding to the predicted document region. In this process, the output is obtained relative to a 28×28 regular spatial grid representing the entire document image. The output is upsampled to the original document image dimensions to obtain the final region mask. As with the convolution operation discussed in the previous section, performing upsampling relative to a uniform (integer) grid leads to poorly estimated spatial boundaries for document regions, especially for our challenging manuscript scenario.

Similar in spirit to deformable convolutions, we adopt an approach wherein the output region boundary is obtained relative to a deformed grid [8] (see Fig. 3b). Let $F \in \mathbb{R}^{256 \times 14 \times 14}$ be feature map being fed as input to the Mask Head. Denote each of the integer grid vertices that tile the 14×14 spatial dimension as $v_i = [x_i, y_i]^T$. Each grid vertex is maximally connected to its 8-nearest neighbors to obtain a grid with triangle edges (see 'Feature Map from ROI Align' in Fig. 3b). The Deformable Grid Mask Head network is optimized to predict the offsets of the grid vertices such that a subset of edges incident

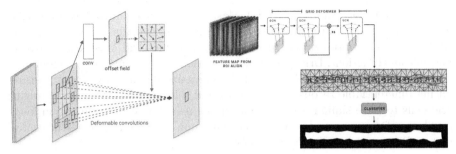

(a) Deformable convolution(Sec. 4.2).   (b) Deformable Grid Mask Head (Sec. 4.3).

Fig. 3. Our novel modifications to the Vanilla Mask-RCNN framework (Sect. 4).

on the vertices form a closed contour which aligns with the region boundary. To effectively model the chain-graph structure of the region boundary, the Mask Head utilizes six cascaded Residual Graph Convolutional Network blocks for prediction of offsets. The final layer predicts binary labels relative to the deformed grid structure formed by the offset vertices (i.e. $v_i + [\Delta x_i, \Delta y_i]$). The resulting deformed polygon mask is upsampled to input image dimensions via bilinear interpolation to finally obtain the output region mask.

4.4 Implementation Details

Architecture: The Backbone in PALMIRA consists of a ResNet-50 initialized from a Mask R-CNN network trained on the MS-COCO dataset. Deformable convolutions (Sect. 4.2) are introduced as a drop-in replacement for the deeper layers C3-C5 of the Feature Pyramid Network present in the Backbone (see Fig. 3a). Empirically, we found this choice to provide better results compared to using deformable layers throughout the Backbone. We use $0.5, 1, 2$ as aspect ratios with anchor sizes of $32, 64, 128, 256, 512$ within the Region Proposal Network. While the Region Classifier and Bounding Box heads are the same as one in Vanilla Mask-RCNN (Sect. 4), the conventional Mask Head is replaced with the Deformable Grid Mask Head as described in Sect. 4.3.

Optimization: All input images are resized such that the smallest side is 800 pixels. The mini-batch size is 4. During training, a horizontal flip augmentation is randomly performed for images in the mini-batch. To address the imbalance in the distribution of region categories (Table 1a), we use repeat factor sampling [9] and oversample images containing tail categories. We perform data-parallel optimization distributed across 4 GeForce RTX 2080 Ti GPUs for a total of 15000 iterations. A multi-step learning scheduler with warmup phase is used to reach an initial learning rate of 0.02 after a linear warm-up over 1000 iterations. The learning rate is decayed by a factor of 10 at 8000 and 12000 iterations. The optimizer used is stochastic gradient descent with gamma 0.1 and momentum 0.9.

Except for the Deformable Grid Mask Head, other output heads (Classifier, Bounding Box) are optimized based on choices made for Vanilla Mask-RCNN [23]. The optimization within the Deformable Grid Mask Head involves multiple loss functions. Briefly, these loss functions are formulated to (i) minimize the variance of features per grid cell (ii) minimize distortion of input features during differentiable reconstruction (iii) avoid self-intersections by encouraging grid cells to have similar area (iv) encourage neighbor vertices in region localized by a reference central vertex to move in same spatial direction as the central vertex. Please refer to Gao et al. [8] for details.

5   Experimental Setup

5.1   Baselines

Towards fair evaluation, we consider three strong baseline approaches.

Boundary Preserving Mask-RCNN [5], proposed as an improvement over Mask-RCNN, focuses on improving the mask boundary along with the task of pixel wise segmentation. To this end, it contains a boundary mask head wherein the mask and boundary are mutually learned by employing feature fusion blocks.

CondInst [27] is a simple and effective instance segmentation framework which eliminates the need for resizing and RoI-based feature alignment operation present in Mask RCNN. Also, the filters in CondInst Mask Head are dynamically produced and conditioned on the region instances which enables efficient inference.

In recent years, a number of instance segmentation methods have been proposed as an alternative to Mask-RCNN's proposal-based approach. As a representative example, we use PointRend [12] - a proposal-free approach. PointRend considers image segmentation as a rendering problem. Instead of predicting labels for each image pixel, PointRend identifies a subset of salient points and extracts features corresponding to these points. It maps these salient point features to the final segmentation label map.

5.2   Evaluation Setup

We partition Indiscapes2 dataset into training, validation and test sets (see Table 1a) for training and evaluation of all models, including PALMIRA. Following standard protocols, we utilize the validation set to determine the best model hyperparameters. For the final evaluation, we merge training and validation set and re-train following the validation-based hyperparameters. A one-time evaluation of the model is performed on the test set.

Table 2. Document-level scores for various performance measures. The baseline models are above the upper separator line while ablative variants are below the line. PALMIRA's results are at the table bottom.

Model	Add-On	HD ↓	HD_{95} ↓	Avg. HD ↓	IoU ↑	AP ↑	AP_{50} ↑	AP_{75} ↑
PointRend [12]	–	252.16	211.10	56.51	69.63	41.51	66.49	43.49
CondInst [27]	–	267.73	215.33	54.92	69.49	42.39	62.18	43.03
Boundary Preserving MaskRCNN [5]	–	261.54	218.42	54.77	69.99	42.65	68.23	**44.92**
Vanilla MaskRCNN [23]	–	270.52	228.19	56.11	68.97	41.46	68.63	34.75
Vanilla MaskRCNN	Deformable Convolutions	229.50	202.37	51.04	65.61	41.65	65.97	44.90
Vanilla MaskRCNN	Deformable Grid Mask Head	**179.84**	153.77	45.09	71.65	42.35	69.49	43.16
Palmira : Vanilla MaskRCNN	Deformable Conv., Deformable Grid Mask Head	184.50	**145.27**	**38.24**	**73.67**	**42.44**	**69.57**	42.93

5.3 Evaluation Measures

Intersection-over-Union (IoU) and Average Precision (AP) are two commonly used evaluation measures for instance segmentation. IoU and AP are area-centric measures which depend on intersection area between ground-truth and predicted masks. To complement these metrics, we also compute boundary-centric measures. Specifically, we use Hausdorff distance (HD) [13] as a measure of boundary precision. For a given region, let us denote the ground-truth annotation polygon by a 2D point set $\mathcal{X}$. Let the prediction counterpart be $\mathcal{Y}$. The Hausdorff Distance between these point sets is given by:

$$\text{HD} = d_H(\mathcal{X}, \mathcal{Y}) = max \left\{ \max_{x \in \mathcal{X}} \min_{y \in \mathcal{Y}} d(x, y), \max_{y \in \mathcal{Y}} \min_{x \in \mathcal{X}} d(x, y) \right\} \quad (4)$$

where $d(x, y)$ denotes the Euclidean distance between points $x \in \mathcal{X}$, $y \in \mathcal{Y}$. The Hausdorff Distance is sensitive to outliers. To mitigate this effect, the Average Hausdorff Distance is used which measures deviation in terms of a symmetric average across point-pair distances:

$$\text{Avg. HD} = d_{AH}(\mathcal{X}, \mathcal{Y}) = \left(\frac{1}{|\mathcal{X}|} \sum_{x \in \mathcal{X}} \min_{y \in \mathcal{Y}} d(x, y) + \frac{1}{|\mathcal{Y}|} \sum_{y \in \mathcal{Y}} \min_{x \in \mathcal{X}} d(x, y) \right) / 2 \quad (5)$$

Note that the two sets may contain unequal number of points ($|\mathcal{X}|, |\mathcal{Y}|$). Additionally, we also compute the 95th percentile of Hausdorff Distance (HD_{95}) to suppress the effect of outlier distances.

For each region in the test set documents, we compute HD, HD_{95}, IoU, AP at IoU thresholds of 50 (AP_{50}) and 75 (AP_{75}). We also compute overall AP by averaging the AP values at various threshold values ranging from 0.5 to 0.95 in steps of 0.05. We evaluate performance at two levels - document-level and region-level. For

Table 3. Document-level scores summarized at collection level for various performance measures.

Collection name	# of test images	HD ↓	HD_{95} ↓	Avg. HD ↓	IoU ↑	AP ↑	AP_{50} ↑	AP_{75} ↑
PIH	94	66.23	46.51	11.16	76.78	37.57	59.68	37.63
BHOOMI	96	220.38	175.52	46.75	69.83	30.40	50.53	29.03
ASR	14	629.30	562.19	169.03	67.80	51.02	73.09	64.27
JAIN	54	215.14	159.88	38.91	76.59	48.25	70.15	50.34
OVERALL	258	184.50	145.27	38.24	73.67	42.44	69.57	42.93

Table 4. PALMIRA's overall and region-wise scores for various performance measures. The HD-based measures (smaller the better) are separated from the usual measures (IoU, AP etc.) by a separator line.

Metric	Overall	Character Line Segment (CLS)	Character Component (CC)	Hole (Virtual) (Hv)	Hole (Physical) (Hp)	Page Boundary (PB)	Library Marker (LM)	Decorator/ Picture (D/P)	Physical Degradation (PD)	Boundary Line (BL)
HD_{95}	171.44	34.03	347.94	70.79	88.33	52.01	289.81	593.99	851.02	111.97
AVG HD	45.88	8.43	103.98	18.80	16.82	13.19	73.23	135.46	255.86	17.95
IoU (%)	72.21	78.01	54.95	74.85	77.21	92.97	67.24	50.57	27.68	61.54
AP	42.44	58.64	28.76	45.57	56.13	90.08	27.75	32.20	03.09	39.72
AP_{50}	69.57	92.73	64.55	81.20	90.53	93.99	55.18	54.23	12.47	81.24
AP_{75}	42.93	92.74	64.55	81.20	90.52	93.99	55.18	54.24	12.47	81.24

a reference measure (e.g. HD), we average its values across all regions of a document. The resulting numbers are averaged across all the test documents to obtain document-level score. To obtain region-level scores, the measure values are averaged across all instances which share the same region label. We use document-level scores to compare the overall performance of models. To examine the performance of our model for various region categories, we use region-level scores.

6   Results

The performance scores for our approach (PALMIRA) and baseline models can be viewed in Table 2. Our approach clearly outperforms the baselines across the reported measures. Note that the improvement is especially apparent for the boundary-centric measures (HD, HD_{95}, Avg. HD). As an ablation study, we also evaluated variants of PALMIRA wherein the introduced modifications were removed separately. The corresponding results in Table 2 demonstrate the collective importance of our novel modifications over the Vanilla Mask-RCNN model.

To understand the results at collection level, we summarize the document-level scores of PALMIRA in Table 3. While the results across collections are mostly consistent with overall average, the scores for ASR are suboptimal. This is due to the unusually closely spaced lines and the level of degradation encountered for these documents. It is easy to see that reporting scores in this manner is useful for identifying collections to focus on, for improvement in future.

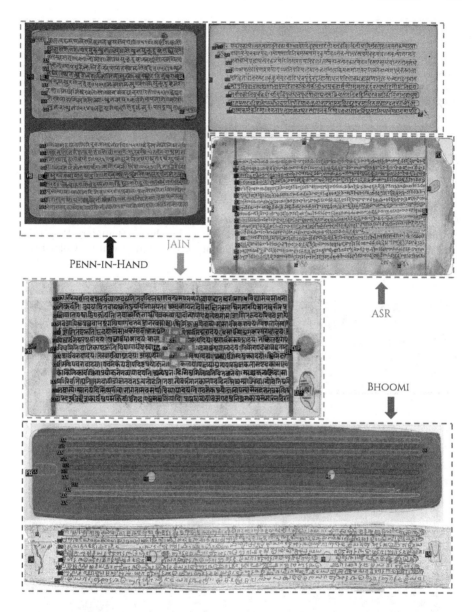

Fig. 4. Layout predictions by PALMIRA on representative test set documents from Indiscapes2 dataset. Note that the colors are used to distinguish region instances. The region category abbreviations are present at corners of the regions. (Color figure online)

We also report the performance measures for PALMIRA, but now at a per-region level, in Table 4. In terms of the boundary-centric measures (HD_{95}, Avg. HD), the best performance is seen for the most important and dominant region category - Character Line Segment. The seemingly large scores for some categories ('Picture/Decorator', 'Physical Degradation') are due to the drastically

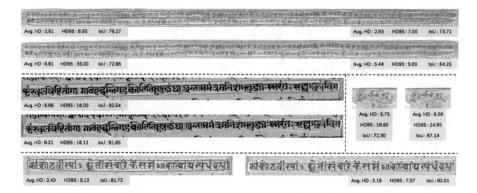

Fig. 5. A comparative illustration of region-level performance. PALMIRA's predictions are in red. Predictions from the best model among baselines (Boundary-Preserving Mask-RCNN) are in green. Ground-truth boundary is depicted in white. (Color figure online)

Fig. 6. Layout predictions by PALMIRA on out-of-dataset handwritten manuscripts.

small number of region instances for these categories. Note that the scores for other categories are reasonably good in terms of boundary-centric measures as well as the regular ones (IoU, AP).

A qualitative perspective on the results can be obtained from Fig. 4. Despite the challenges in the dataset, the results show that PALMIRA outputs good quality region predictions across a variety of document types. A comparative illustration of region-level performance can be viewed in Fig. 5. In general, it can be seen that PALMIRA's predictions are closer to ground-truth. Figure 6 shows PALMIRA's output for sample South-East Asian, Arabic and Hebrew historical manuscripts. It is important to note that the languages and aspect ratio (portrait) of these documents is starkly different from the typical landscape-like aspect ratio of manuscripts used for training our model. Clearly, the results demonstrate that PALMIRA readily generalizes to out of dataset manuscripts without requiring additional training.

7 Conclusion

There are three major contributions from our work presented in this paper. The *first* contribution is the creation of Indiscapes2, a new diverse and challenging dataset for handwritten manuscript document images which is 150% larger than its predecessor, Indiscapes. The *second* contribution is PALMIRA, a novel deep network architecture for fully automatic region-level instance segmentation of handwritten documents containing dense and uneven layouts. The *third* contribution is to propose Hausdorff Distance and its variants as a boundary-aware measure for characterizing the performance of document region boundary prediction approaches. Our experiments demonstrate that PALMIRA generates accurate layouts, outperforms strong baselines and ablative variants. We also demonstrate PALMIRA's out-of-dataset generalization ability via predictions on South-East Asian, Arabic and Hebrew manuscripts. Going ahead, we plan to incorporate downstream processing modules (e.g. OCR) for an end-to-end optimization. We also hope our contributions assist in advancing robust layout estimation for handwritten documents from other domains and settings.

Acknowledgment. We wish to acknowledge the efforts of all annotators who contributed to the creation of Indiscapes2.

References

1. Agarwal, M., Mondal, A., Jawahar, C.: Cdec-net: Composite deformable cascade network for table detection in document images. In: ICPR (2020)
2. Alaasam, R., Kurar, B., El-Sana, J.: Layout analysis on challenging historical arabic manuscripts using siamese network. In: 2019 International Conference on Document Analysis and Recognition (ICDAR), pp. 738–742. IEEE (2019)
3. Alberti, M., Vögtlin, L., Pondenkandath, V., Seuret, M., Ingold, R., Liwicki, M.: Labeling, cutting, grouping: an efficient text line segmentation method for medieval manuscripts. In: 2019 International Conference on Document Analysis and Recognition (ICDAR), pp. 1200–1206. IEEE (2019)

4. Barman, R., Ehrmann, M., Clematide, S., Oliveira, S.A., Kaplan, F.: Combining visual and textual features for semantic segmentation of historical newspapers. arXiv preprint arXiv:2002.06144 (2020)

5. Cheng, T., Wang, X., Huang, L., Liu, W.: Boundary-preserving mask R-CNN. In: Vedaldi, A., Bischof, H., Brox, T., Frahm, J.-M. (eds.) ECCV 2020. LNCS, vol. 12359, pp. 660–676. Springer, Cham (2020). https://doi.org/10.1007/978-3-030-58568-6_39

6. Clausner, C., Antonacopoulos, A., Pletschacher, S.: Icdar 2019 competition on recognition of documents with complex layouts-rdcl2019. In: 2019 International Conference on Document Analysis and Recognition (ICDAR), pp. 1521–1526. IEEE (2019)

7. Dai, J., et al.: Deformable convolutional networks. In: Proceedings of the IEEE International Conference on Computer Vision, pp. 764–773 (2017)

8. Gao, J., Wang, Z., Xuan, J., Fidler, S.: Beyond fixed grid: learning geometric image representation with a deformable grid. In: Vedaldi, A., Bischof, H., Brox, T., Frahm, J.-M. (eds.) ECCV 2020. LNCS, vol. 12354, pp. 108–125. Springer, Cham (2020). https://doi.org/10.1007/978-3-030-58545-7_7

9. Gupta, A., Dollár, P., Girshick, R.B.: LVIS: A dataset for large vocabulary instance segmentation. CoRR abs/1908.03195 (2019). http://arxiv.org/abs/1908.03195

10. He, K., Gkioxari, G., Dollár, P., Girshick, R.B.: Mask R-CNN. In: ICCV, pp. 2980–2988 (2017)

11. Kesiman, M.W.A., Pradnyana, G.A., Maysanjaya, I.M.D.: Balinese glyph recognition with gabor filters. J. Phys. Conf. Ser. **1516**, 012029 (2020). https://doi.org/10.1088/1742-6596/1516/1/012029

12. Kirillov, A., Wu, Y., He, K., Girshick, R.: Pointrend: image segmentation as rendering. In: Proceedings of the IEEE/CVF Conference on Computer Vision and Pattern Recognition, pp. 9799–9808 (2020)

13. Klette, R., Rosenfeld, A. (eds.) Digital Geometry. The Morgan Kaufmann Series in Computer Graphics, Morgan Kaufmann, San Francisco (2004)

14. Lee, J., Hayashi, H., Ohyama, W., Uchida, S.: Page segmentation using a convolutional neural network with trainable co-occurrence features. In: 2019 International Conference on Document Analysis and Recognition (ICDAR), pp. 1023–1028. IEEE (2019)

15. Li, M., et al.: Docbank: a benchmark dataset for document layout analysis. In: Proceedings of the 28th International Conference on Computational Linguistics, pp. 949–960 (2020)

16. Liang, J., Hu, Q., Zhu, P., Wang, W.: Efficient multi-modal geometric mean metric learning. Pattern Recogn. **75**, 188–198 (2018)

17. Lin, T.Y., Goyal, P., Girshick, R., He, K., Dollár, P.: Focal loss for dense object detection. In: ICCV, pp. 2980–2988 (2017)

18. Ma, W., Zhang, H., Jin, L., Wu, S., Wang, J., Wang, Y.: Joint layout analysis, character detection and recognition for historical document digitization. In: 2020 17th International Conference on Frontiers in Handwriting Recognition (ICFHR), pp. 31–36. IEEE (2020)

19. Made Sri Arsa, D., Agung Ayu Putri, G., Zen, R., Bressan, S.: Isolated handwritten balinese character recognition from palm leaf manuscripts with residual convolutional neural networks. In: 2020 12th International Conference on Knowledge and Systems Engineering (KSE), pp. 224–229 (2020). https://doi.org/10.1109/KSE50997.2020.9287584

20. Monnier, T., Aubry, M.: docExtractor: An off-the-shelf historical document element extraction. In: ICFHR (2020)

21. Oliveira, S.A., Seguin, B., Kaplan, F.: dhsegment: a generic deep-learning approach for document segmentation. In: 2018 16th International Conference on Frontiers in Handwriting Recognition (ICFHR), pp. 7–12. IEEE (2018)

22. Paulus, E., Suryani, M., Hadi, S.: Improved line segmentation framework for sundanese old manuscripts. J. Phys. Conf. Ser. **978**, 012001. IOP Publishing (2018)

23. Prusty, A., Aitha, S., Trivedi, A., Sarvadevabhatla, R.K.: Indiscapes: instance segmentation networks for layout parsing of historical indic manuscripts. In: 2019 International Conference on Document Analysis and Recognition (ICDAR), pp. 999–1006. IEEE (2019)

24. Puarungroj, W., Boonsirisumpun, N., Kulna, P., Soontarawirat, T., Puarungroj, N.: Using deep learning to recognize handwritten thai noi characters in ancient palm leaf manuscripts. In: Ishita, E., Pang, N.L.S., Zhou, L. (eds.) ICADL 2020. LNCS, vol. 12504, pp. 232–239. Springer, Cham (2020). https://doi.org/10.1007/978-3-030-64452-9_20

25. Ronneberger, O., Fischer, P., Brox, T.: U-Net: convolutional networks for biomedical image segmentation. In: Navab, N., Hornegger, J., Wells, W.M., Frangi, A.F. (eds.) MICCAI 2015. LNCS, vol. 9351, pp. 234–241. Springer, Cham (2015). https://doi.org/10.1007/978-3-319-24574-4_28

26. Siddiqui, S., Malik, M., Agne, S., Dengel, A., Ahmed, S.: DECNT: deep deformable CNN for table detection. IEEE Access **6**, 74151–74161 (2018)

27. Tian, Z., Shen, C., Chen, H.: Conditional convolutions for instance segmentation. In: Vedaldi, A., Bischof, H., Brox, T., Frahm, J.-M. (eds.) ECCV 2020. LNCS, vol. 12346, pp. 282–298. Springer, Cham (2020). https://doi.org/10.1007/978-3-030-58452-8_17

28. Trivedi, A., Sarvadevabhatla, R.K.: Hindola: a unified cloud-based platform for annotation, visualization and machine learning-based layout analysis of historical manuscripts. In: 2019 International Conference on Document Analysis and Recognition Workshops (ICDARW), vol. 2, pp. 31–35. IEEE (2019)

29. Valy, D., Verleysen, M., Chhun, S., Burie, J.C.: Character and text recognition of khmer historical palm leaf manuscripts. In: 2018 16th International Conference on Frontiers in Handwriting Recognition (ICFHR), pp. 13–18. IEEE (2018)

30. Zhong, X., Tang, J., Yepes, A.J.: Publaynet: largest dataset ever for document layout analysis. In: 2019 International Conference on Document Analysis and Recognition (ICDAR), pp. 1015–1022. IEEE (2019). https://doi.org/10.1109/ICDAR.2019.00166

Page Layout Analysis System for Unconstrained Historic Documents

Oldřich Kodym$^{(\boxtimes)}$ and Michal Hradiš

Brno University of Technology, Brno, Czech Republic
ikodym@fit.vutbr.cz

Abstract. Extraction of text regions and individual text lines from historic documents is necessary for automatic transcription. We propose extending a CNN-based text baseline detection system by adding line height and text block boundary predictions to the model output, allowing the system to extract more comprehensive layout information. We also show that pixel-wise text orientation prediction can be used for processing documents with multiple text orientations. We demonstrate that the proposed method performs well on the cBAD baseline detection dataset. Additionally, we benchmark the method on newly introduced PERO layout dataset which we also make public.

Keywords: Layout analysis · Historic documents analysis · Text line extraction

1 Introduction

Preserving historic documents in a digitized form has become an important part of digital humanities. Although (semi-)automatic OCR can be used to obtain document transcriptions with high accuracy, they require extraction of individual text layout elements as the first step. A lot of attention has been given to the task of text baseline detection in recent years [6] but the detected baselines do not provide additional layout information such as font heights and text line grouping into text blocks. This layout information can be extracted using document-specific rules from simple printed documents but the task becomes more challenging in case of handwritten documents or documents with irregular layout.

In this work, we propose a CNN model for joint detection of text baselines, text line polygons and text blocks in a broad range of printed and handwritten documents. We create the text lines by locally estimating font height together with baselines and then we cluster them into text blocks based on local text block boundary estimation. The system processes documents with arbitrary text directions by combining layout detections in multiple orientations using a dedicated dense text orientation estimation model.

Most of the current text baseline detection methods are based on binary segmentation using CNN [5,6]. Currently best performing approach, ARU-net, proposed by Gruening et al. [8], relies on a multi-scale CNN architecture. The CNN

© Springer Nature Switzerland AG 2021
J. Lladós et al. (Eds.): ICDAR 2021, LNCS 12822, pp. 492–506, 2021.
https://doi.org/10.1007/978-3-030-86331-9_32

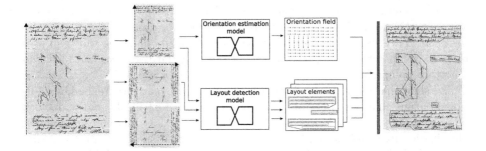

Fig. 1. The proposed system jointly extracts text baselines, text line polygons and text blocks from unconstrained documents. Local text orientation is estimated by a dedicated model.

Fig. 2. Typographic parts of textual glyphs used in text detection.

outputs binary segmentation maps of baselines and baseline endpoints which are then converted to individual baselines using superpixel clustering. Although it reaches high baseline detection accuracy even on challenging documents, it does not allow direct extraction of text lines and text blocks.

To generate text line bounding polygons, most of the existing works rely on text binarization as the first step. Ahn et al. [1] use simple grouping of connected components followed by skew estimation and re-grouping to produce text lines. A seam-carving algorithm has been proposed by Alberti et al. [2] to correctly resolve vertical overlaps of binarized text. These approaches, however, make assumptions about the document, such as constant text orientation, and rely on only processing a single text column at time. Other authors rely on segmenting parts of the text body, such as baseline or x-height area (see Fig. 1), using CNNs first [12,13]. To convert the detected text x-height areas to text line polygons, Pastor-Pellicer et al. [17] use local extreme points of binarized text to assign ascender and descender lines to each detected text line. Vo et al. [20] use x-height areas and binarized text to group the text pixels belonging to a text line using line adjacency graphs. In these cases, document-specific rules are also required during post-processing, limiting the generalization of these approaches.

Extraction of text blocks is significant for structuring the detected text into correct reading order. Approaches based on image processing methods, such as connected component analysis, texture analysis or run length smearing can be used for text block detection, but only in case of simple printed documents [3]. Deep learning-based approaches, on the other hand, offer more robustness to the artifacts often encountered in historical documents, although they may suffer from inadequate training datasets [4]. Most of the recent works formulate this task as semantic segmentation followed by a global post-processing step [18,23], although this method is susceptible to merging near regions of the same type. Ma et al. [11] instead only detects straight text block boundaries and then uses them to group detected text characters into rectangular text lines and blocks in a bottom-up manner.

We show that an ARU-net-like CNN architecture [8] can be extended by additional output channels containing information about text height and text block boundary. The proposed text line extraction method is therefore completely binarization-free and does not require any document-specific post-processing steps. Similarly to Ma et al. [11], we also construct text blocks by grouping the detected text lines, guided by the additional CNN output. However, we do not limit the layout to be strictly grid-based, which makes the proposed approach suitable for virtually any type of document. Additionally, we also propose a dense text orientation prediction model which allows the proposed method to process documents with multiple text orientations.

2   Joint Text Line and Block Detection

The proposed text line and block detection method comprises of a fully convolutional neural network (ParseNet) and a set of post-processing steps that estimates the individual text baselines and text line bounding polygons directly from the network outputs. Text blocks are then constructed in bottom-up manner by clustering text lines while taking into account estimated local block boundary probability.

The proposed ParseNet is inspired by the ARU-net [8] which detects baselines using two binary segmentation output channels: the *baseline* channel and the *baseline endpoint* channel. We extend the outputs of ARU-net by three additional channels. The *ascender height* channel and the *descender height* channel provide information about the spatial extent of a text line above and below the baseline. The *text block boundary* channel provides information about neighbouring text lines adjacency likelihood.

2.1   ParseNet Architecture and Training

ParseNet architecture is illustrated in Fig. 3. It processes an input image in three scales by a detection U-net which extracts the five layout output channels. The three detection U-net instances share the same weights. The two downsampled outputs are upscaled to the original resolution by nearest neighbour upsampling and a single convolutional layer. These final convolutions learn to locally interpolate the output maps and they also learn to scale the ascender and descender height channels accordingly. A smaller U-net extracts pixel-wise weights which are used to combine the scale-wise layout detection channels by a pixel-wise weighted average.

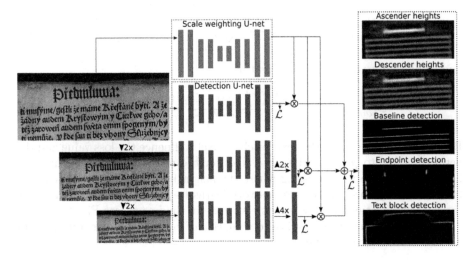

Fig. 3. Architecture of the proposed model is mainly based on the ARU-net model [8] with several modification, such as producing the final output directly using weighted averaging and computing loss at each scale output. The model outputs two text height channels, baseline and endpoint detection channels and a text block boundary detection channel.

Our approach differs from ARU-net [8] in several aspects. ARU-net uses residual connections in the convolutional blocks of the detection U-net and processes images in 6 scales. Perhaps most importantly, ARU-net upsamples and fuses features and computes segmentation outputs on the fused feature maps and the scale weights are extracted separately from each scale. Fusing the output layout detection channels on each scale in case of ParseNet allows us to add auxiliary loss at each scale as well as at the final output. This forces the detection U-net to learn a wider range of text sizes while still focusing on the optimal scale, which improves convergence and which additionally regularizes the network.

The losses are computed using five text element ground truth maps (C_{GT}). For the baseline (C^{base}), endpoint (C^{end}) and text block boundary (C^{block}) detection channels, the ground truth maps contain binary segmentation maps with the respective objects as seen in Fig. 3. The ascender (C^{asc}) and descender (C^{des}) height ground truth maps are created by multiplying baseline map foreground pixels by the heights of the corresponding text lines. The height units are pixels in the network input and output resolution. We refer to the model output maps as C_{pred}.

We train the model using a weighted sum of masked MSE losses for the two text line height channels and Dice losses [14] for the three binary segmentation channels. The text line ascender height loss for one image sample with N pixels is defined as

$$\mathcal{L}_{MaskedMSE}^{asc} = \frac{\sum_{i=1}^{N}[C_{pred}^{asc}(i) - C_{GT}^{asc}(i)]^2 \cdot C_{GT}^{base}(i)}{\sum_{i=1}^{N} C_{GT}^{base}(i)} \tag{1}$$

and similarly for the descender height as

$$\mathcal{L}_{MaskedMSE}^{des} = \frac{\sum_{i=1}^{N}[C_{pred}^{des}(i) - C_{GT}^{des}(i)]^2 \cdot C_{GT}^{base}(i)}{\sum_{i=1}^{N} C_{GT}^{base}(i)}. \tag{2}$$

Due to the masking, only the text baseline pixels influence the loss value and the network is free to predict arbitrary values for the rest of the image. This is in line with output post-processing where the height information is collected only from baseline pixels. In fact, the network learns to predict correct heights near the baseline location and low values elsewhere in the image (see Fig. 3).

The Dice segmentation loss is a combination of individual segmentation channel losses with equal weights

$$\mathcal{L}_{Dice} = Dice(C_{pred}^{base}, C_{GT}^{base}) + Dice(C_{pred}^{end}, C_{GT}^{end}) + Dice(C_{pred}^{block}, C_{GT}^{block}). \tag{3}$$

We believe that weighting of the individual channels is not necessary as Dice loss compensates for class imbalance.

The final loss is computed as weighted sum of the two components:

$$\mathcal{L} = \lambda(\mathcal{L}_{MaskedMSE}^{asc} + \mathcal{L}_{MaskedMSE}^{des}) + \mathcal{L}_{Dice} \tag{4}$$

where λ is empirically set to 0.01 to compensate for generally higher magnitudes of the MSE gradients. The loss is computed for the output channels of each scale as well as for the final weighted average with equal weights.

The network architectures follow U-net with 32 initial features and 3 max-pooling steps in case of the detection network and 16 initial features with 3 max-pooling steps in case of the scale weighting network. We use Adam optimizer with learning rate of 0.0001 to train the model for 300 000 steps on batches of size 6. Each batch sample is a random image crop of size 512×512.

We make use of color transformations, rotation, and scaling to augment the set of training images by factor of 50 offline before the training. We also experimented with blur and noise augmentations but found their effect negligble. Scale augmentations are of big significance in our setting as the model needs to learn to robustly estimate the font ascender and descender heights. Therefore, we normalize the scale of each image so that the median text ascender height is 12 pixels. Then, for each image, we sample further random scaling factor $s = 2^x$ where x is sampled randomly from normal distribution with zero mean and unit standard deviation, resulting in 66% of images having median text ascender height between 6 and 24 pixels.

During inference, the whole input image is processed at once. In our TensorFlow implementation, a 5 Mpx image requires around 5 GB of GPU RAM. The images of documents coming from different sources can vary substantially in their DPI and font sizes. Although the model is trained to work on a broad range of font sizes due to the scale augmentations, the optimal processing resolution corresponds to median ascender height of 12 pixels. Therefore, we perform adaptive input scaling which runs inference two times. First, the image is processed in the original resolution (or with a constant preset downsample factor) and we compute median ascender height from the ascender height channel masked by the raw baseline detection channel. This estimation is used to compute an optimal scaling factor for the input image, such that the median ascender height is close to 12 pixels. The final output is obtained by processing the input image in this optimal resolution.

2.2   Text Baseline Detection

Baseline detection is the first step that follows the CNN inference. The baseline and endpoint detection channels are used in this step. First, the baseline detection channel is processed using smoothing filter with size 3 and vertical non-maxima suppression with kernel size 7 to obtain objects with single pixel thickness. Next, we simply subtract the endpoint detection channel from the baseline detection channel to better separate adjacent baselines.

We threshold the result and use connected component analysis to obtain separate baselines. The local connectivity neighborhood is a centered rectangular region 5 pixels wide and 9 pixels high. This enables the connected components to ignore small discontinuities which may result from the non-maxima suppression of slightly tilted baselines, especially in y dimension. We use threshold value of 0.3 which gives consistently good results on broad range of datasets. We filter any baseline shorter than 5 pixels. Detected baselines are represented as linear splines with up to 10 uniformly spaced control points along the baseline.

2.3   Text Line Polygon Estimation

Each detected baseline is used to create the corresponding text line polygon using the ascender and descender detection channels. We assume that each text line has a constant ascender and descender height and these values are obtained for each text line by computing 75th percentile of the ascender and descender values at the baseline pixels. We found that in practice, the 75th percentile gives better estimate of text height than average or median because the model tends to locally underestimate the heights in some parts of text line. The text line polygon is created by offsetting the baseline by ascender and descender values in direction locally perpendicular to the baseline.

2.4   Clustering Text Lines into Text Blocks

This step clusters the text lines into meaningful text blocks with sequential reading order. We use a bottom-up approach that makes use of the text line polygons that are detected in the previous steps and the text block boundary detection channel. We assume that two neighbouring text lines belong to the same text block if they are vertically adjacent and there is no strong text block boundary detected between them.

Fig. 4. Example of the neighbourhood penalization areas for two potentially neighbouring text line pairs (left). The penalty is computed by summing the text block boundary detection channel output (right) in the highlighted areas and normalizing it by their length. In this case, the upper text line pair (green areas) will be split into different blocks, while the lower pair (pink areas) will be assigned to the same block. (Color figure online)

The algorithm starts by finding all neighbouring pairs of text lines. We consider two lines to be neighbours if they overlap horizontally and their vertical distance is less than their text height. Two neighbouring lines are clustered into single text block if their adjacency penalty is lower than a specific threshold. The adjacency penalty is computed from text block boundary detection channel area between the two lines. Specifically, we use two areas which span the horizontal intersection of the two lines. These areas are placed on the descender and ascender line of the upper and lower text line, respectively (see Fig. 4) and are 3 pixels thick. The adjacency penalty is computed by summing over each of these areas in the text block boundary detection channel and normalizing by their lengths. The two lines are then considered neighbours if both of the adjacency penalties are lower than the threshold value, which we empirically set to 0.3.

The text blocks are finally created so that all pairs of neighbouring text lines are always assigned to the same text block. The text block polygons are formed as an alpha shape of all the corresponding text line polygon points. Additionally, we perform merging of lines in each text block as an additional text line post processing step. Two lines belonging to the same text block are merged if they are horizontally adjacent and have similar vertical position.

3   Multi-orientation Text Detection

The proposed system as previously described can not correctly process documents containing both vertical and horizontal text. Although the ParseNet model could possibly learn to detect vertical text lines, they would be discarded during post-processing in the vertical non-maxima suppression step. For pages with arbitrary but consistent text orientation, it is possible to estimate the dominant text orientation and compensate for it in a pre-processing step [6]. We propose a solution designed for documents which contain multiple text directions on a single page. We use separate model for local text orientation estimation which is used to merge layouts detected in multiple rotations of the input image.

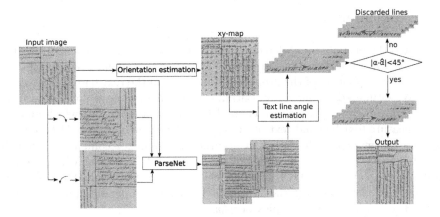

Fig. 5. Multi-orientation text extraction pipeline. The input image is processed with ParseNet in three different orientations, which results in some vertical text being detected duplicitly. Using the orientation estimation model output, an angle estimation $\hat{\alpha}$ is computed for each line. The line is discarded if the difference of the estimated angle $\hat{\alpha}$ differs from the actual angle α of detected line by more than $45°$.

The text orientation estimation model uses a single U-net network to process the image and has two output channels. The orientation of the text at each pixel is represented as x and y coordinates on a unit circle. This representation allows us to use masked MSE loss to train the model, similarly to the ParseNet text height channels. In case of the orientation estimation model, the loss is computed at all pixels inside text line polygon. The rest of the training strategy is almost identical to the ParseNet, except for stronger rotation transformations with rotation angles sampled from uniform distribution between $-110°$ and $110°$. This particular rotation range introduces vertical text into the orientation estimation training set, but does not introduce upside-down text which is uncommon in real world data.

An overview of the multi-orientation text extraction pipeline is illustrated in Fig. 5. To make use of the orientation predictions, we extend the text detection framework described in the previous section by processing the image in three orientations: $0°$, $90°$ and $270°$ and by combining the three respective layouts. Naive combination of the layouts would for example result in duplicate detections of near vertical lines since the ParseNet model usually cannot distinguish between regular and upside-down text. To solve this and similar problems, we test each line if its orientation is consistent with the processing orientation. The estimated text line orientation $\hat{\alpha}$ is computed as

$$\hat{\alpha} = atan2(\tilde{y}, \tilde{x}), \tag{5}$$

where $\tilde{x}$ and $\tilde{y}$ is median of the output x and y channels computed over the text line polygon. If the text line estimated orientation angle $\hat{\alpha}$ differs from the processing orientation angle α (i.e. $0°$, $90°$ or $270°$), by more than $45°$, the text line is discarded.

4 Experiments

We tested the proposed text detection method in ablation experiments using the cBAD 2019 baseline detection dataset [6]. This allowed us to measure the effect of individual components of the proposed framework on the text baseline detection task. We evaluated quality of all outputs of the method on a novel dataset of historic documents for which we created full layout annotations which allowed us to measure text line and text block polygon detection accuracy. Although the layout element geometric detection accuracy is useful for comparing different layout detection methods, it does not necessarily fully correlate with practical usability in an automatic OCR pipeline. Because of that we evaluated how much the automatic layout detection changes OCR output compared to manual layout annotations on a set of 9 complex newspaper pages.

4.1 CBAD 2019 Text Baseline Detection

The cBAD 2019 text baseline detection dataset [6] contains 755 training, 755 validation, and 1511 testing images with text baseline annotations. We additionally manually annotated text line and text block polygons on 255 of the training images (the annotations are included in the PERO layout dataset). During the model training, the original 755 training images were used for training of the text baseline and endpoint detection channels, while the additionally annotated 255 training images were used for training of all 5 output channels. We report the precision (P-value), recall (R-value) and F-value averaged over the 1511 testing images to adhere to the original competition format.

To set a methodological baseline for the ablations experiments we trained the ParseNet model only with the text baseline and endpoint detection channels. Because this model does not output any information about the image resolution or font size and therefore adaptive scaling during the inference cannot be exploited, a constant preset downsample factor of 5 was used for this evaluation. This ensured that the per-page average text height was also 12 pixels. This model reached F-value of 0.873, showing good detection accuracy.

Adding the text line height and region boundary channels to the output resulted in a slight drop of accuracy, showing that multi-task learning itself does not have significantly positive impact on the training. However, when using the additional text height information to compute the optimal scaling factor during model inference, the accuracy increased slightly to 0.879.

One of the most common sources of error in baseline detection is undesired tearing of the baselines. Using the ability to cluster the text lines into text blocks makes it possible to repair some of these errors by merging adjacent lines belonging to the same text region. This simple post-processing step increased the F-value to 0.886.

The ParseNet model with multi-orientation processing and orientation estimation described in Sect. 3, which is able to process vertical text lines, further improved the accuracy, resulting in F-value of 0.902. Table 1 shows the sum-

Table 1. ParseNet text baseline detection performance on cBAD 2019 dataset [6]. Effect of additional output channels, adaptive scaling (AS), line merging (LM) and multi-orientation processing (MO).

Method	P-value	R-value	F-value
ParseNet − regions, heights	0.900	0.847	0.873
ParseNet − regions	0.893	0.852	0.872
ParseNet	0.893	0.850	0.871
ParseNet + AS	0.899	0.860	0.879
ParseNet + AS, LM	**0.914**	0.859	0.886
ParseNet + AS, LM, MO	0.906	**0.897**	**0.902**
TJNU [6]	0.852	0.885	0.868
UPVLC [6]	0.911	0.902	0.907
DMRZ [6]	0.925	0.905	0.915
DocExtractor [15][a]	0.920	**0.931**	0.925
Planet [6]	**0.937**	0.926	**0.931**

[a] Trained on additional data

mary of the cBAD dataset experiments in context of the current state-of-the-art methods.

4.2 PERO Dataset Layout Analysis

Several public datasets of historic manuscripts, such as Diva-HisDB [19], Perzival [21] or Saint Gall [7], include manual annotations of text lines and text blocks. However, these datasets focus on single document type, preventing them from being used for development of general purpose layout detection algorithms. Recently, Monnier and Aubry [15] introduced a diverse dataset of synthetically generated documents with layout annotations for training, but the evaluation dataset is limited to a few types of manuscripts. Furthermore, the text line polygon annotations in existing datasets are often wrapped tightly around segmented text. We believe that this is an unnecessary complication as the subsequent OCR systems usually only require a rectangular image crop containing the text line [10].

We compiled a new dataset (the PERO layout dataset[1]) that contains 683 images from various sources and historical periods with complete manual text block, text line polygon and baseline annotations. The included documents range from handwritten letters to historic printed books and newspapers and contain various languages including Arabic and Russian. Part of the PERO dataset was collected from existing datasets and extended with additional layout annotations (cBAD [6], IMPACT [16] and BADAM [9]). The dataset is split into 456 training and 227 testing images.

---

[1] https://www.fit.vutbr.cz/~ikodym/pero_layout.zip.

Fig. 6. Representative examples of the text detection results on Perzival [21] (left) and Saint Gall [19] (right) dataset samples. The used model was only trained on the PERO layout dataset and no further pre-processing steps were performed.

Table 2. Detection performance of ParseNet in text baseline, text line polygon and text block detection. Polygon detection threshold is set to IoU = 0.7.

Text Element	P-value	R-value	F-value
Text baseline	0.912	0.942	0.927
Text line polygon	0.811	0.813	0.804
Text block polygon	0.625	0.686	0.633

We trained a full ParseNet model with the 5 layout detection channels using only the PERO layout training set and evaluated the performance on the test set. The layout detection was performed using adaptive scaling, line merging and multi-orientation processing. In addition to the text baselines detection metrics, we also report precision, recall and F-value for text line and text block polygons. We consider a polygon as correctly detected if the predicted and the ground-truth polygon overlap with intersection over union (IoU) > 0.7. This overlap threshold has been previously used for text line detection evaluation [11,22] and facilitates usability for subsequent OCR applications in most cases while allowing some tolerance towards exact definition of the bounding polygon. In all cases, an average value over all testing pages is computed. The results of detection performance of all text elements are presented in Table 2.

We also tested the models trained on the PERO layout dataset on several samples of Perzival and Saint Gall datasets. Qualitative results in Fig. 6 show good performance, demonstrating the ability of models trained on the PERO dataset to generalize.

4.3   Czech Newspaper OCR Evaluation

Performance of layout elements detection is usually measured using geometric overlaps, treating the document layout extraction as instance segmentation.

These metrics are, however, affected by types of errors easily recognizable by subsequent text transcription engines. For example, small false detections will often be transcribed as empty strings, introducing only small error to output transcription. Therefore, the geometric metrics do not necessarily correlate fully with the quality of final document transcription which is the only relevant metric for many use cases.

To demonstrate performance of the proposed method in practical use cases, we studied effect of the proposed automatic layout detection on quality of automatic OCR output. We used a subset of the PERO layout test set that includes 9 czech newspaper pages and an in-house transcription engine trained specifically for this type of documents. The transcription engine is based on a CNN with recurrent layers and it was trained using CTC loss [10]. We compare the output transcription computed on the detected layout elements against the output transcription on manually annotated layout elements. We report the error rates of text line detection and the introduced OCR character error. A ground truth line is considered correctly recognized if all of the following conditions are met:

- It has non-zero geometric overlap with a detected line.
- The difference of the OCR outputs of the ground truth and the corresponding detected line is less then 15%.
- Both lines have the same position in the reading order (i.e. the following line in the corresponding ground-truth text block and the following detected line in the corresponding detected text block match, or both the ground-truth and the detected lines are the last in their respective text blocks).

The results for the 9 testing newspaper pages are shown in Fig. 7 and two representative examples of detected layouts are shown in Fig. 8. The success rates of text line detection show that for 8 out of the 9 pages, the layout detection extracted almost all text lines and text blocks correctly. The decrease of the success rate in the last case was caused by undesired merging of two adjacent text blocks and subsequent merging of all the corresponding lines.

5   Discussion

We show that the proposed method is competitive with the current state-of-the-art baseline detection methods in Table 1. However, the method is not optimized for this specific dataset and it does not reach the performance of the top method, despite having similar architecture. One possible factor in this may be the strong scale augmentation during training, which is aimed at providing better robustness towards different image resolution and font sizes. While this is desirable for use of the framework on completely unconstrained data, it may be disadvantageous on the less variable cBAD test set because it prevents the model from learning to associate certain fonts with certain resolution. Furthermore, we do not make use of advanced baseline post-processing steps (such as superpixel clustering in ARU-net [8]).

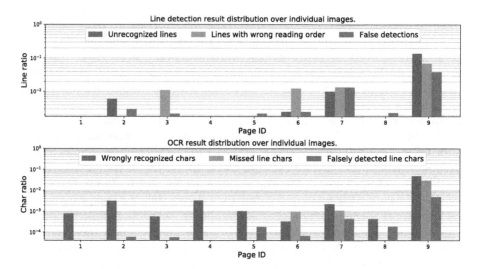

Fig. 7. Line detection results and introduced OCR errors for 9 testing czech newspapers pages. Except for one case, line detection errors are rare and introduce minimum of character errors into the OCR pipeline. Note that y-axis is logarithmic.

Fig. 8. Two examples of detected layout in czech newspaper pages.

This work puts main emphasis on practical applicability of the proposed approach as a first step before automatic text transcription. The individual components presented here can easily be applied to any other baseline detection model to extend its functionality without introducing additional constraints and assumptions about document type. We showed that adding the text height and text block boundary channels does not degrade the model performance and thus it should be possible to similarly extend other baseline detection models. We make the PERO layout dataset public to make training of the additional output channels more accessible for the whole community. Making use of the dense orientation estimation and multi-orientation text detection pipeline, models trained on mostly horizontal text can be used to extract vertical lines as well.

The results in Table 2 show that the PERO layout dataset is quite challenging and that there is still a room for improvement. However, it is worth noting that not all detection errors have detrimental effect on resulting document transcription quality. Many smaller false text line detections will be transcribed as empty strings or only introduce several erroneous characters to the complete page transcription, as demonstrated in the Fig. 7.

6   Conclusion

This work introduced a text extraction method capable of producing text line polygons grouped into text blocks which can be used directly as an input for subsequent OCR systems. We demonstrated the performance of the proposed approach on the widely used cBAD 2019 dataset as well as a newly introduced PERO layout dataset. In both cases, the proposed framework was shown to perform well and that it can be used as a robust text extraction method for wide range of documents. To our best knowledge, this is the first method that is able to extract this kind of layout information from unconstrained handwritten manuscripts as well as printed documents.

In future work, we aim to improve the performance of the proposed method by improving the backbone CNN architecture. Specialized post-processing steps can be explored for extremely challenging document layouts such as official records with tabular structure or even maps and music sheets. Another possible research direction is redefining the main text detection task from baseline detection to text x-height detection, as this has been shown to improve text detection performance [15].

Acknowledgment. This work has been supported by the Ministry of Culture Czech Republic in NAKI II project PERO (DG18P02OVV055). We also gratefully acknowledge the support of the NVIDIA Corporation with the donation of one NVIDIA TITAN Xp GPU for this research.

References

1. Ahn, B., Ryu, J., Koo, H.I., Cho, N.I.: Textline detection in degraded historical document images. EURASIP J. Image Video Process. **2017**(1), 1–13 (2017). https://doi.org/10.1186/s13640-017-0229-7
2. Alberti, M., Vogtlin, L., Pondenkandath, V., Seuret, M., Ingold, R., Liwicki, M.: Labeling, cutting, grouping: an efficient text line segmentation method for medieval manuscripts. In: ICDAR (2019)
3. Binmakhashen, G.M., Mahmoud, S.A.: Document layout analysis. ACM Computing Surveys (2020)
4. Clausner, C., Antonacopoulos, A., Pletschacher, S.: ICDAR2019 competition on recognition of documents with complex layouts - RDCL2019. In: ICDAR (2019)

5. Diem, M., Kleber, F., Fiel, S., Gruning, T., Gatos, B.: cBAD: ICDAR2017 competition on baseline detection. In: ICDAR - IAPR (2017)
6. Diem, M., Kleber, F., Sablatnig, R., Gatos, B.: cBAD: ICDAR2019 competition on baseline detection. In: ICDAR (2019)
7. Fischer, A., Frinken, V., Fornés, A., Bunke, H.: Transcription alignment of latin manuscripts using hidden markov models. In: Proceedings of the 2011 Workshop on Historical Document Imaging and Processing - HIP 2011 (2011)
8. Grüning, T., Leifert, G., Strauß, T., Michael, J., Labahn, R.: A two-stage method for text line detection in historical documents. IJDAR **22**(3), 285–302 (2019)
9. Kiessling, B., Ezra, D.S.B., Miller, M.T.: BADAM. In: Proceedings of the 5th International Workshop on Historical Document Imaging and Processing - HIP 2019 (2019)
10. Kiss, M., Hradis, M., Kodym, O.: Brno mobile OCR dataset. In: ICDAR (2019)
11. Ma, W., Zhang, H., Jin, L., Wu, S., Wang, J., Wang, Y.: Joint layout analysis, character detection and recognition for historical document digitization. In: ICFHR (2020)
12. Mechi, O., Mehri, M., Ingold, R., Amara, N.E.B.: Text line segmentation in historical document images using an adaptive u-net architecture. In: ICDAR (2019)
13. Melnikov, A., Zagaynov, I.: Fast and lightweight text line detection on historical documents. In: DAS (2020)
14. Milletari, F., Navab, N., Ahmadi, S.A.: V-net: Fully convolutional neural networks for volumetric medical image segmentation. In: 2016 Fourth International Conference on 3D Vision (3DV) (2016)
15. Monnier, T., Aubry, M.: docExtractor: An off-the-shelf historical document element extraction. In: ICFHR (2020)
16. Papadopoulos, C., Pletschacher, S., Clausner, C., Antonacopoulos, A.: The IMPACT dataset of historical document images. In: Proceedings of the 2nd International Workshop on Historical Document Imaging and Processing - HIP 2013 (2013)
17. Pastor-Pellicer, J., Afzal, M.Z., Liwicki, M., Castro-Bleda, M.J.: Complete system for text line extraction using convolutional neural networks and watershed transform. In: IAPR - DAS (2016)
18. Quirós, L.: Multi-task handwritten document layout analysis (2018)
19. Simistira, F., Seuret, M., Eichenberger, N., Garz, A., Liwicki, M., Ingold, R.: DIVA-HisDB: a precisely annotated large dataset of challenging medieval manuscripts. In: ICFHR (2016)
20. Vo, Q.N., Lee, G.: Dense prediction for text line segmentation in handwritten document images. In: ICIP (2016)
21. Wüthrich, M., et al.: Language model integration for the recognition of handwritten medieval documents. In: ICDAR (2009)
22. Xie, Z., et al.: Weakly supervised precise segmentation for historical document images. Neurocomputing **350**, 271–281 (2019). https://doi.org/10.1016/j.neucom.2019.04.001
23. Yang, X., Yumer, E., Asente, P., Kraley, M., Kifer, D., Giles, C.L.: Learning to extract semantic structure from documents using multimodal fully convolutional neural networks. In: CVPR (2017)

Improved Graph Methods for Table Layout Understanding

Jose Ramón Prieto[(✉)] and Enrique Vidal

PRHLT Research Center, Universitat Politècnica de València, Valencia, Spain
{joprfon,evidal}@prhlt.upv.es

Abstract. Recently, there have been significant advances in document layout analysis and, particularly, in the recognition and understanding of tables and other structured documents in handwritten historical texts. In this work, a series of improvements over current techniques based on graph neural networks are proposed, which considerably improve state-of-the-art results. In addition, a two-pass approach is also proposed where two graph neural networks are sequentially used to provide further substantial improvements of more than 12 F-measure points in some tasks. The code developed for this work will be published to facilitate the reproduction of the results and possible improvements.

1 Introduction

With the significant advances made in the field of Deep Learning in recent years, progress has been made in the fields of Computer Vision (CV) and Natural Language Processing (NLP) and Handwritten Text Recognition (HTR). HTR techniques have been developed that have solved many of the problems of transcription and extraction of information from handwritten documents. However, many of these documents are structured documents that provide information following some structure or logical patterns, such as tables, records, etc.

In order to obtain information on these structures, a series of techniques are needed to understand these documents. These techniques fall in the field of Layout Analysis, or more specifically, in the field of Document Understanding. A large part of these structures are tables on which it is necessary to know in which column, row, and cell lay the text considered. These are the logical substructures of a table since all the information within the same substructure will be of the same type. E.g. in notarial documents of birth records, a column can indicate each person's date of birth, while the row indicates all the information noted on the same person.

Printed tables follow an easy and regular pattern to follow and rows and columns are usually aligned and with explicit delimitation lines. However, the case of handwritten tables is different since they have several intrinsic difficulties: handwritten tables do not always have a delineation for each row or column since sometimes it has been deleted or, directly, not drawn. Moreover, the handwritten

ⓒ Springer Nature Switzerland AG 2021
J. Lladós et al. (Eds.): ICDAR 2021, LNCS 12822, pp. 507–522, 2021.
https://doi.org/10.1007/978-3-030-86331-9_33

text does seldom strictly respect column, row, or cell limits. These usually old documents often suffer from bleed through, skew, and other optical difficulties. Furthermore, HTR techniques are not always reliable in these cases. We focus on these kinds of documents; that is, tables in historic handwritten documents, and aim to recognize their substructures such as rows, columns, and cells.

The paper is structured as follows. We discussed previous related works in Sect. 2. Then our proposed approach is presented in Sect. 3 and the details of the experiments and datasets are described in Sect. 4. The results are reported in Sect. 5 and discussed in Sect. 6, Finally conclusions are drown in Sect. 7.

2   Background

In recent years, great advances have been made in Document Layout Analysis. Many of these advances have been due to neural networks, which have taken a leading role.

In [3] using a pre-trained encoder, a generic deep learning architecture is trained for various layout analysis tasks, such as page extraction and text line detection. In [9] attention methods and recursive connections with a U-Net shape network called ARU-Net are proposed, improving substantially the results on text line detection. Another work from [15] achieves comparable results while performing layout analysis and text line detection at the same time, with an U-Net shape network and a second network to regularize the training of the first one. Other works have worked with the layout not only at the graphical or "visual" level but also using text hypotheses [13], improving the results with a fusion method.

The authors in [7] use a transformation of the input image. This transformation is passed to a Region Proposal Network using Faster-RCNN (faster region convolutional neural network), which returns a series of rectangles. Using another CNN, these rectangles are classified into two classes: table or not a table. In [2] a Mask R-CNN is used as a dual backbone with deformable convolutions and a ResNeXt-101 to detect tables as another object detection problem. However, these works only focus on detecting the table, not its structure.

Similarly, other works use a Fully-Convolutional Neural Network (FCN) to analyze the structure of the tables as an object detection problem [17]. The work in [11] improves it by making a segmentation, also extracting the table, not just its structure. In [18] two modules are used to detect the grid pattern of the tables. The authors try to solve the problem by separating the rows and columns by a straight line, making projections. With a different approach, a bottom-up solution in recognition of PDF documents is used in [1]. First, (printed) words are detected with OCR software, and then the relation of each word with the rest is classified using a Multilayer Feedforward Neural Network.

Other works have used Graph Neural Networks (GNN) to find the relationships between different objects on a page. In [16] the problem of table detection is formulated as a classification of the edges between detected words in the document. Using OCR techniques, words are detected and encoded using positional information and the type of content of the cell or word. The authors of [14] follow a similar procedure. They classify each edge in a graph of connected words and use a CNN to extract visual features. In this case, rows, columns and cells are detected.

However, all of these methods are applied to printed page images, and the tables are generally assumed to be regular and straight.

In the case of handwritten documents, several problems arise. In historical manuscripts, many pages appear with slant and skew. Many rows and columns do not have a clear and established limit. Many cells are usually empty. Depending on the corpus, the transcripts are based on not very reliable hypotheses. Besides that, the variability between tables is much higher. Given all these problems, the solutions for printed-text tables can not be applied to handwritten tables.

Nonetheless, advances have also been made in the field of Table Understanding for *historical handwritten* documents. In [4] two Machine Learning techniques (Conditional Random Field and Graph Convolutional Network) are compared to detect elements in a table. A competition has also been held to detect the structure of handwritten tables. In the track B of the competition [6], one of the teams obtained results using an FCN and then looking for each cell's neighbours by constructing an adjacency graph. In [5], virtual objects are generated and classified to detect the different rows in the image. Besides, they use a version for skew images and test the model on two historical handwritten image datasets. Then, the results are improved in [12] using graph-based neural network techniques based on the conjugate graph. A graph is created for each page from the textlines already detected, interconnecting each of them with its closest and direct neighbours. Once this graph is created, each edge is classified according to the detection task (rows, columns or cells). It the end, the graph is pruned, and the connected components are used to extract the detected elements. Currently, this powerful method obtains the state of the art results. However, the creation of the initial graph connecting the lines is fragile. As the authors show, although the model succeeds in classifying all the edges, it does not reach the overall optimal result. That happens because when creating the initial graph, there might be text lines that do not "see" each other directly. Clearly, these text lines will never belong to the same connected component.

Our work builds on the techniques proposed in [12]. We improve the creation of the initial graph and propose other techniques to improve also the classification of edges in various ways, including a 2-pass approach which leads to substantially better results. In addition to the work reported in this paper, we also carried out many other preliminary experiments, with several traditional and novel table analysis approaches. All the results we achieved in these previous tests were considerably worse than those obtained with the proposed GNN methods described in the next section. Methods without creating the conjugate graph and/or without using GNNs, such as SEAL [20], have been also tried for edge classification, but all the results were much worse than those reported in [12], as well as those obtained in the present work.

3   Proposed Approach

We consider three types of Table Understanding tasks: the detection of rows, columns and cells, within a document that contain a (single or a previously detected) table. As in [12], the method used is based on two distinct phases. In the first one, an initial graph is created from previously detected text lines. In the second phase, the substructures underlying this graph are determined using GNNs, followed by connected component analysis. We improve each of these two phases separately, as discussed in the following subsections.

3.1   Building the Initial Graph

Due to the lack of strict formatting rules in handwritten tables, several situations may arise which make an initial graph created as proposed in [12] fails to contain enough connections to represent the structures we seek for.

Further, note that increasing the number of connections increases the size of the graph. If all the nodes were fully connected, the number of edges would grow quadratically, and the amount of computation required to further process the graph could become prohibitively large. Besides, increasing the number of edges can lead to model failure because, as noted in [12], false-positive edges significantly harm the results since different structures would be joined. Therefore increasing text line connections is not always beneficial. Taking these difficulties into account, we propose to improve the initial graph by wisefully increasing the number of text line connections. Two kinds of new connections are added based on the insights discussed below.

Each of the n text lines of an image is referred to as an integer, k, $1 \leq k \leq n$. The initial graph, G_0, is built by considering each text line as a *node* and by connecting two nodes with an *edge iff* the corresponding text lines are horizontal or vertical "neighbours". As in [12], we consider two text lines k' and k to be *horizontal neighbours* if the vertical spans of their bounding boxes overlap and both bounding boxes are in *"line of sight"*; i.e., there is no other bounding box horizontally in between k and k'. A similar criterion is assumed to consider two text lines to be *vertical neighbours*. Figure 1 illustrates these concepts.

A frequent error of this basic idea appears when lines fail to be in line of sight due to sloped columns and empty cells, or when there are small lines in adjacent cells which fail to fulfil the above overlapping criteria because they are not aligned horizontally or vertically. To fix this issue, we propose to relax the neighbourhood criteria by also allowing two text lines to be neighbours just if they are geometrically close enough; i.e., if the distance between the centres of their bounding boxes is smaller than a distance threshold, d_0. This is done only if the number of neighbours is smaller than a threshold κ. Let $\mathbf{a}, \mathbf{b} \in \mathbb{R}^2$, be the centers of two text lines k', k. A diagonal Mahalanobis distance is used to gauge nearness anisotropically in the horizontal and vertical directions:

$$d(\mathbf{a}, \mathbf{b}) = \left(\left(\frac{1}{\sigma_1}(a_1 - b_1) \right)^2 + \left(\frac{1}{\sigma_2}(a_2 - b_2) \right)^2 \right)^{1/2} \tag{1}$$

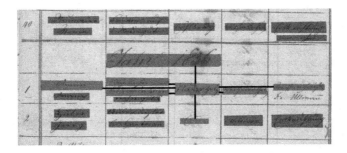

Fig. 1. Illustration of horizontally and vertically neighbouring text lines. The target textline can be seen in a red rectangle. We represent the neighbours of the target textline in blue and connected with a black edge. The rest of the textlines are in grey. (Color figure online)

where the weights $\frac{1}{\sigma_1}$ and $\frac{1}{\sigma_2}$ are determined empirically using validation data. The idea is illustrated in Fig. 2.

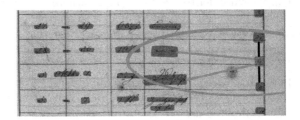

Fig. 2. The small text line in red color does not have any horizontal neighbour; therefore it would become disconnected from other textlines of the same table row. Using the distance criterion, the two text lines connected with a green edge also become neighbours, because they are within the yellow ellipse representing the points which are within d_0 Mahalanobis distance of the target text line. (Color figure online)

Another frequent error is caused by long lines which expand out of their own table cells, overlapping other rows and/or columns. Due to these overlaps, some lines which should be connected are not in the line of sight and are therefore missed. We propose to fix this kind of failures by allowing lines to connect with each other not only if they are direct neighbours but by letting them hop to the second, third, or more neighbours. Figure 3 illustrates this idea. The number of hops allowed will depend on whether the hops are horizontal or vertical since, for each task, it will be interesting to create more connections in one direction or another. It is worth pointing out that n-hops are essential to create potentially necessary edges in the initial graph, since a graph model will never be able to classify edges that did not exisit in the initial graph.

We introduce two parameters, s_h and s_v, to indicate the number of horizontal and vertical hops allowed, respectively. These parameters are empirically optimized using validation data. Note that if $s_h = s_v = 1$ and $d_0 = 0$ the initial graph will be exactly the same as the one used in [12].

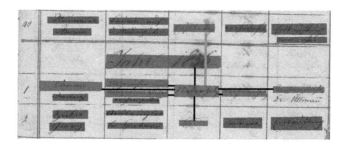

Fig. 3. This figure shows how the target textline can be connected thanks to a vertical hop. The newly added edge is displayed in green. (Color figure online)

3.2 Graph Neural Networks

Let $G_0 = (V_0, E_0)$ be the initial graph obtained as described above, where V_0 is the set of text lines, $E_0 \subseteq V_0 \times V_0$ and $m \stackrel{\text{def}}{=} |E_0|$. Following [12] and others, for structural classification, we adopt a Graph Neural Network (GNN) using the *conjugate graph* of G_0. We will refer to it as $G = (V, E)$, where $V = E_0 \stackrel{\text{def}}{=} \{1, \ldots, m\}$ and E is obtained by connecting two nodes of V with an edge $(i, j) \in V^2$ *iff* the corresponding edges of G_0 have a vertex in common. Note that this implies that each edge of G corresponds to a unique node of V_0, but two or more edges of G may correspond to the same node of V_0.

Our GNN approach is based on the module dubbed EdgeConv in [19], which differs from the module also called EdgeConv in [12]. The output of this module is a vector $\mathbf{x}'_i \in \mathbb{R}^{\phi'}$ computed as:

$$\mathbf{x}'_i = \max_{j \in \aleph(i)} \mathbf{Q}(\mathbf{x}_i, \mathbf{x}_j - \mathbf{x}_i) \tag{2}$$

where $i, j \in V$, are two nodes of the conjugate graph G, $\mathbf{x}_i, \mathbf{x}_j \in \mathbb{R}^\phi$ are feature vectors associated with (the edges of G_0 corresponding to) these nodes, and $\aleph(i)$ is the set of nodes adjacent to i in G. The vectorial function $\mathbf{Q} : \mathbb{R}^\phi \times \mathbb{R}^\phi \to \mathbb{R}^{\phi'}$ is implemented as a shared Multilayer Perceptron (MLP) and max is used as the aggregation operator for features of adjacent nodes.

Since this model does not directly take advantage of characteristics of the nodes of the V_0 (i.e., single textlines), we propose to use a module called Edge-FeaturesConv as the first layer of the GNN. The output of this module is a vector $\mathbf{x}'_i \in \mathbb{R}^{\phi'}$ computed as:

$$\mathbf{x}'_i = \max_{j \in \aleph(i)} \mathbf{W}(\mathbf{x}_i, \mathbf{x}_j - \mathbf{x}_i, \mathbf{f}_{ij}) \tag{3}$$

where $\mathbf{f}_{ij} \in \mathbb{R}^\gamma$ is a feature vector associated to the (node of G_0 corresponding to the) edge (i, j) of G and $\mathbf{W} : \mathbb{R}^\phi \times \mathbb{R}^\phi \times \mathbb{R}^\gamma \to \mathbb{R}^{\phi'}$ is implemented as an MLP in a similar way as with $\mathbf{Q}$.

Upon convergence of the GNN, a score in $[0, 1]$ is obtained for each node of G, which can be interpreted as an estimate of the probability that the corresponding edge in G_0 is a proper edge of the true graph describing the text line interconnections for the structure of the task we are interested in. Therefore, from the initial graph, G_0, we can prune out all the edges with probability lower than a threshold, p_0, which is determined empirically using validation data. This provides the resulting graph, hereafter referred to as G_1.

In addition to the improvements discussed above, here we introduce another technique, called "*2-pass GNN*", which has also proved very useful to improve detection results. In the second pass of this technique, we take the conjugate of the pruned graph G_1 as input and apply the very same GNN processing of the first pass described above, leading to an additionally "trimmed" graph, which we refer to as G_2. Of course, this additional trimming might remove potentially useful edges, which would lead to worse results. However, as shown in the experiments reported in Sect. 5, the 2-pass GNN learns to correct errors from the previous pass, generally resulting in much smaller and better graphs.

The connected components of G_2 (or G_1) directly provide the row, column, or cell text line labelling required for the considered task. Figure 4, illustrates this 2-pass GNN approach.

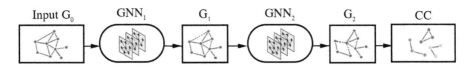

Fig. 4. The proposed two-pass approach. The graph G_0 is processed by GNN_1, yielding G_1, which is an edge-pruned version of G_0. G_1 in turn, is processsed by GNN_2, which applies a further edge prunning to yeld G_2. Finally, the connected components (CC) are extracted to obtain the required substructures.

3.3 Node and Edge Features

For each text line k, i.e., a node of G_0 corresponding to one (or more) edge(s) (i, j) of G, a total of $\gamma = 8$ features are extracted, all based on the geometric information of k and its neighbours. These features, which constitute the vector $\mathbf{f}_{ij} \in \mathbb{R}^\gamma$ of Eq. (3), are: the position (i.e., the bounding box centre), height and width of the text line, the number of horizontal and vertical neighbours, and the percentage of overlap of the text line bounding box with these neighbours on each axis.

On the other hand, for each pair of connected lines, k', k, i.e., for each edge (k', k) of G_0 which corresponds to a node i (or j) of G, $\phi = 9$ features are extracted, which constitute the vectors $\mathbf{x}_i, \mathbf{x}_j \in \mathbb{R}^\phi$ of Eqs. (2) and (3); namely, the axis through which the edge goes (horizontal or vertical), the corresponding percentage of the overlap between k' and k, the length of the edge (i.e., the distance between the centres of the bounding boxes of k' and k), the centres of the bounding boxes of k' and k, and the average of these two centres (which provides direct information about the average position of the two lines in the image).

4   Experimental Settings and Datasets

We will describe the datasets used and the improvements done in the graphs for each corpus. In order to make the experiments fully reproducible, all the code and folds are available online.[1]

4.1   Datasets

The datasets considered in this work are the same used in [12] for table understanding. These are two from Archiv des Bistums (ABP) called ABP_small and ABP_large and one from National Archive Finland (NAF). ABP_small encompasses 180 images of tables about death records. ABP_large is an extension that also contains birth and marriage records, with a total of 1098 table images. These tables usually have large cells, and there is usually not much slant or empty cells. On the other hand, the NAF dataset contains 488 table images. These tables are more heterogeneous compared to ABP tables and contain much smaller cells, with sloping rows and columns and many empty cells.

The dataset partition specifications reported in [12] are not sufficiently clear or detailed, and the corresponding data divisions do not exactly match those found in their repository. Therefore, we have created our own partitions, trying to be as close as possible to the original ones. Four folds are created for each dataset. In each experiment, the first fold is used for test and the rest for train. These folds are available online. (see Footnote 1)

4.2   Improved Graphs Parameters

For each dataset and task, the graphs have been improved using the techniques explained in Sect. 3. All the relevant parameters have been tuned using the "Oracle" method and the training set. In the Table 1, the cells tasks have been omitted in all datasets since just the baseline graphs have been used. This is equivalent to setting $s_h = s_v = 1$ and $d_0 = 0$.

Table 1. Parameters used in the graph improvements. The values of d_0 are relative percentages of page width in row's task and height in column's task.

		s_h	s_v	d_0	κ	σ_1	σ_2
ABP_small	Rows	1	1	10	5	10	1
	Columns	1	1	0	–	–	–
ABP_large	Rows	1	1	0	–	–	–
	Colums	1	2	0	–	–	–
NAF	Rows	4	1	0	–	–	–
	Colums	1	5	2	5	1	10

---

[1] https://github.com/JoseRPrietoF/TableUnderstanding_ICDAR21

Table 1 shows all the parameters tuned for each task. The neighbouring criterion based on Mahalanobis distance is only used in two cases (those in which $d_0 > 0$), namely ABP_small rows and NAF column. The hop relaxation, on the other hand, is only used in three cases (those in which $s_h > 1$ and $s_v > 1$); namely ABP_large columns and NAF both columns and rows.

For ABP_small columns, $s_h = s_v = 1$ and $d_0 = 0$ are set because no more connections are needed to achieve good results, as we will see in Sect. 5. These parameters mean that the baseline graph [12] is used, communicating each baseline only with its direct neighbours.

For the ABP_large dataset, the Mahalanobis distance neighbouring criterion for columns is not used, but text lines will communicate with up to two hops vertically. In the case of rows the baseline graph was used.

In the NAF dataset, a relatively large amount of hops have been allowed in the horizontal or vertical direction for rows and columns, respectively. On the other hand, a value of $d_0 > 0$ has only been used for the columns tasks.

To simplify hyperparameter settings, the same weights σ_1, σ_2 have been established for all the datasets for which $d_0 > 0$: $\sigma_1 = 1, \sigma_2 = 10$ for columns and the reverse for rows. The minimum number of edges, κ, is also used only with $d_0 > 0$ and is always set to 5. In all the tasks, p_0 has been set to 0.5. However, this probability and some other parameters could be further tuned, although it was not our goal in the work here presented.

4.3   GNN Parameters

As already demonstrated in [12], a deep network achieves better results than a network with few layers. We also performed preliminary experiments varing the number of layers and the results of these tests fully agreed with those of [12]. Therefore, we will show here the best results with many layers. We have opted, in all cases, for a 10-layer network with the following number of filters: $64, 64, 64, 64, 32, 32, 32, 16, 16, 2$. It starts with an EdgeFeaturesConv layer with 64 input channels and the rest of the layers are EdgeConv. Between each layer Batch Normalization is used. Each convolution is implemented as an MLP, according to Eq. (2) for EdgeConv and (3) for EdgeFeatureConv. Each MLP includes two linear layers with a ReLU activation function between them.

It should be noted that the values of the parameters ϕ and γ, introduced in Sect. 3.2, are those described in Sect. 3.3 only in the first layer of each GNN. From the second layer onward, only ϕ is used since each of these layers is of type EdgeConv (Eq. (2)) and the value of ϕ is obviously set to the number of output filters ϕ' of the previous layer.

Since the number of layers or filters used in GNN_2 was not observed to change the results significantly, the same 10-layers architecture of GNN_1 was also adopted for GNN_2.

The parameters of each layer were initialized following [8] and trained for 6000 epochs and a patience of 1000 epochs, according to the cross-entropy loss. The ADAM optimizer [10] was used with a learning rate of 0.001 and a batch size of 8.

4.4 Assessment Measures

Although an exact comparison is not possible because, as explained above, the test partitions are not identical, we try to provide results which are as comparable as possible. To this end, the same metric used in [12] has been adopted for assessing sub-structure detection. This metric uses Intersection over Union (IoU) and a threshold to indicate whether the partitions are equal or not. In this work, we will only consider the threshold of 100%, which is more strict because it implies a perfect sub-structure match. Using the IoU calculated in this way, precision and recall are computed and their harmonic mean, F1@100, is finally computed and presented as the only assessment figure (always in %).

We will also show "Oracle results". These are results using the precise table description ground truth to obtain an Oracle Initial Graph, which contains all and only the edges strictly needed to connect the text lines according to the task (rows, columns, or cells). In this way, the maximum F1 achievable by the graph is obtained subject to classifying all the edges correctly in each graph. The number of edges is also shown since the improved initial graphs might be considerably larger than the initial graphs obtained as in [12].

5 Results

This section will follow the Oracle method to compare the results of the improved initial graphs discussed in Sect. 4.2 with those of the baseline graphs. Afterwards, results using the trained models with the baseline graphs, the improved graphs when applicable and the 2-pass approach presented in this work will be reported.

5.1 Oracle Results

Oracle results are reported in Table 2.

Table 2. Results (in %) of the Oracle method in all datasets. In two tasks no Improved Graph was found to overcome the Baseline Graph according to F1@100 metric (in %).

	Graph:	Baseline		Improved	
		F1@100	#Edges	F1@100	#Edges
ABP_small	Rows	93.2	35 833	99.7	45 780
	Colums	99.0	35 833	–	–
ABP_large	Rows	97.5	219 992	–	–
	Colums	91.1	219 992	98.8	291 665
NAF	Rows	84.9	126 860	91.9	216 427
	Colums	87.6	126 860	97.9	330 445

As mentioned earlier, in the case of ABP_small only the graph for rows has been improved. This is because the baseline graph for columns already has excellent results. Table 2 shows the detection performance result of the baseline graph and its number of edges. The graph improvements for rows lead to a F_1@100 value higher than 99%, improving more than 6 points the result of the baseline graph. On the other hand, the number of edges increases by around 27%, which is computationally acceptable.

In the ABP_large dataset, no improved initial graph has overcome the baseline graph for rows, but for columns the improved graph achieves more than 7 points over the baseline graph, reaching a F_1@100 value of 98.8%. In this case, the number of edges increases by around 30%.

Finally, in the NAF dataset, the improved initial graphs for both rows and columns overcome the baseline results. F1@100 improves more than 7 points for rows, with an increment in the number of edges on of 70%. In this case, there is still considerable room for improvement, but the increase in the number of edges would probably become prohibitive. The improvement for columns is higher than 10 points, reaching an F1@100 value of 97.9, with an increment in the number of edges of 173% over the baseline graph. It is worth mentioning that the NAF dataset contains much larger tables with many small text lines in their cells. These tables usually have also many more rows and columns, increasing the number of edges connecting text lines and growing much faster than in the ABP datasets. Even thus, the sizes of the improved initial graphs still make computationally feasible their use in further processing steps.

In cell detection, the optimal result is reached in all datasets with the baseline graph. Therefore, the result is reported in the tables, and the baseline graphs will be used in the next experiments. This is because text lines that are from the same cell tend to be very close to each other and the graph size does not need to be increased, which makes the problem easier to solve.

5.2 GNN Results

Now we report the results of training the models presented in Sect. 3. To keep results as comparable with those in [12] as possible, we also provide results using only the first part of the model (GNN$_1$), trained using the baseline graphs. The results are not far apart from the results reported in [12], even though the dataset partitions used here are not exactly the same.

Table 3 shows the results for all the datasets. The first column shows the results of training GNN$_1$ using the baseline graphs. The G_1 column shows the results of training GNN$_1$ using the improved graph. And the G_2 column shows the results of applying the 2-pass approach, depicted in Fig. 4. We can differentiate the improvements achieved, on the one hand, by improving the initial graph, and on the other, by using the 2-pass approach presented in this work.

For the ABP_small dataset, the initial graph has been improved only for the rows task, where the baseline results are overcome by more than 3 points of F1@100. The 2-pass approach provides better results in all tasks. Substantial improvements by almost 7 points for the rows G_2 and more than 10 points for

Table 3. Results of training the GNN models for all the datasets. The baseline graph has been used where no result is shown in the G_1 column. The G_2 column corresponds to the 2-pass output all the results correspond to the F1@100 metric in %.

	Graph	Baseline	G_1	G_2
ABP_small	Rows	83.6	87.2	94.0
	Colums	94.2	–	95.6
	Cells	97.3	–	97.7
ABP_large	Rows	83.2	–	90.3
	Colums	84.3	93.5	95.2
	Cells	96.6	–	97.1
NAF	Rows	71.9	80.0	84.5
	Colums	79.3	89.1	91.5
	Cells	98.4	–	98.6

G_1 are achieved. When the baseline or G_1 results are already high, the margin for better results is smaller. The F1@100 results for the columns task are better by more than 1 point, while for the cells task, the improvement is negligible.

For the ABP_large dataset, only the initial graph for columns has been improved. The F1@100 results for the columns task go from 84.3 to 93.5, improving by more than 9 points in total. The F1@100 of the 2-pass approach reaches 95.2, improving by 1.7 additional points. In the rest of the cases, the baseline graph is used. The result for rows task is 7.1 points better, while for the cell task, only a marginal improvement of 0.5 points is achieved.

Finally, regarding the NAF dataset, substantial improvements over the baseline are achieved, 8.1 F1@100 points for rows and 9.8 for columns. For the row task, F1@100 results go from 71.9 to 80, while the column task, they fo from 79.3 to 89.1. Using the 2-pass approach, up to 4.5 additional points are achieved for the row task, reaching 84.5, while for the columns task, 91.5 F1@100 is reached. The baseline graph is used for the cell task, and only marginally better results are obtained using the 2-pass approach.

The sizes of the baseline graphs used to obtain the row and column results in Table 3 are those shown in Table 2. G_1 sizes are about 30% to 50% smaller than those of the corresponding baseline or improved graphs, and the sizes of G_2 are about 1% to 4% smaller than those of the corresponding G_1 graphs.

Figures 5 and 6 show how using the improved initial graphs the rows and columns have been well segmented in two difficult examples discussed in Sect. 3.1.

Fig. 5. One case in which the columns could not be well classified without applying the hops criterion. The baseline results can be seen on the left, and the results with the improved graph on the right side. The segmentation problem illustrated in Fig. 1, due to horizontal overlap, is corrected in these results

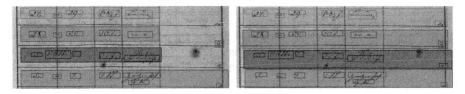

Fig. 6. One case in which without applying the distance criterion to add edges, the rows could not be well classified. The baseline results can be seen on the left, and the results with the improved graph on the right side. The segmentation problem illustrated in Fig. 2, due to close text lines not being in line of sight, is corrected in these results.

6   Discusion

In both ABP datasets, the graphs have been improved for two tasks. However, in the NAF dataset, the row and column tasks are substantially more difficult, due to the differently slanted rows and columns, with many empty cells and large numbers of rows and columns per table. Better results have been achieved by creating improved initial graphs for each task.

Producing larger initial graphs is not always beneficial. Adding more ways to connect text lines can make graphs to grow very fast. For instance, if in the NAF table images we connect each text line to all the other text lines within a large (Mahalanobis) distance threshold, without putting any limitation, the resulting graphs would not be computationally feasible. It should be noted that connecting nodes within the same substructure is not problematic. False-negatives can not affect the result as long as part of a substructure is not completely isolated. However, false-positives have a very negative effect when searching for substructures since for a single false-positive, two complete substructures are joining. Also, the more edges are added to the graph, the more difficult this is to happen. Adding edges makes the model more sensitive to false-positives but more robust to false-negatives.

In any case, the results show that using the proposed 2-pass approach tends to improve the results significantly. While GNN_1 seeks to classify the edges of G_0 so as to minimize the number of false-positives in the resulting G_1 graph, GNN_2 tries to correct the remaining false-positives of G_1 to obtain G_2. This

2-pass strategy has demonstrated that it can deal with amounts of added edges without threatening the final results because of possible false-positives.

Note that false-positives produced by the GNN_1 network are not as damaging as false-negatives. False-negative edges in G_1 may separate sub-structures that GNN_2 could never put together. However, GNN_2 can re-classify false-positives. What is sought by having two different networks is that each one classifies different types of edges. While the first network does a great deal of pruning, the second network can classify those edges that the first has failed as false-positives.

7   Conclusions

We have proposed several techniques which provide substantially better Table Understanding results with respect to the results reported in [12]. These include better techniques to build the initial graph and a 2-pass approach which finally detects remarkably improved substructures by adding another GNN that learns to remove possible false-positives.

Significantly better detection results have been achieved in all tasks where the results were considerably lower compared to the other tasks, such as the results of rows detection. The greater the difficulty of the task, the more significant impact our methods have had.

It is important to notice that for the practical and final use of these techniques, such as information extraction, considerably higher accuracy is needed. Nevertheless, we believe that the present results may be sufficient to reliably extract information from well-structured documents, which will be a next step.

As future work, word-level methods will be tested using probabilistic indices, trying to avoid possible errors of line segmentation. Likewise, solutions will be sought to solve the spans in the rows and columns, which has been ignored until now.

Acknowledgments. Work partially supported by the Universitat Politècnica de València under grant FPI-I/SP20190010 (Spain). Work partially supported by the BBVA Foundation through the 2018–2019 Digital Humanities research grant "Hist-Weather – Dos Siglos de Datos Climáticos" and also supported by the Generalitat Valenciana under the EU- FEDER Comunitat Valenciana 2014–2020 grant "Sistemas de fabricación inteligente para la industria 4.0", by Ministerio de Ciencia, Innovación y Universidades project DocTIUM (Ref. RTI2018-095645-B-C22) and by Generalitat Valenciana under project DeepPattern (PROMETEO/2019/121)

Computing resources were provided by the EU-FEDER Comunitat Valenciana 2014–2020 grant IDIFEDER/2018/025.

References

1. Adiga, D., Bhat, S., Shah, M., Vyeth, V.: Table structure recognition based on cell relationship, a bottom-up approach. In: International Conference Recent Advances in Natural Language Processing, (RANLP 2019), pp. 1–8 (2019)

2. Agarwal, M., Mondal, A., Jawahar, C.V.: CDeC-net: composite deformable cascade network for table detection in document images. arXiv, pp. 9491–9498 (2020)
3. Ares Oliveira, S., Seguin, B., Kaplan, F.: DhSegment: a generic deep-learning approach for document segmentation. In: Proceedings of International Conference on Frontiers in Handwriting Recognition, ICFHR, pp. 7–12 (2018)
4. Clinchant, S., Dejean, H., Meunier, J.L., Lang, E.M., Kleber, F.: Comparing machine learning approaches for table recognition in historical register books. In: Proceedings - 13th IAPR International Workshop on Document Analysis Systems, DAS 2018, pp. 133–138 (2018)
5. Dejean, H., Meunier, J.L.: Table rows segmentation. In: Proceedings of the International Conference on Document Analysis and Recognition, ICDAR, pp. 461–466 (2019)
6. Gao, L., et al.: ICDAR 2019 competition on table detection and recognition (cTDaR). In: Proceedings of the International Conference on Document Analysis and Recognition, ICDAR, pp. 1510–1515 (2019)
7. Gilani, A., Qasim, S.R., Malik, I., Shafait, F.: Table detection using deep learning. In: Proceedings of the International Conference on Document Analysis and Recognition, ICDAR 1, pp. 771–776 (2017)
8. Glorot, X., Bengio, Y.: Understanding the difficulty of training deep feedforward neural networks. J. Mach. Learn. Res. **9**, 249–256 (2010)
9. Grüning, T., Leifert, G., Strauß, T., Michael, J., Labahn, R.: A two-stage method for text line detection in historical documents. Int. J. Doc. Anal. Recogn. **22**(3), 285–302 (2019)
10. Kingma, D.P., Ba, J.L.: Adam: a method for stochastic optimization. In: 3rd International Conference on Learning Representations, ICLR 2015 - Conference Track Proceedings (2015)
11. Paliwal, S.S., Vishwanath, D., Rahul, R., Sharma, M., Vig, L.: TableNet: deep learning model for end-to-end table detection and tabular data extraction from scanned document images. In: Proceedings of the International Conference on Document Analysis and Recognition, ICDAR, pp. 128–133 (2019)
12. Prasad, A., Dejean, H., Meunier, J.L.: Versatile layout understanding via conjugate graph. In: Proceedings of the International Conference on Document Analysis and Recognition, ICDAR, pp. 287–294 (2019)
13. Prieto, J.R., Bosch, V., Vidal, E., Stutzmann, D., Hamel, S.: Text content based layout analysis. In: Proceedings of International Conference on Frontiers in Handwriting Recognition, ICFHR, pp. 258–263 (2020)
14. Qasim, S.R., Mahmood, H., Shafait, F.: Rethinking table recognition using graph neural networks. In: Proceedings of the International Conference on Document Analysis and Recognition, ICDAR, pp. 142–147 (2019)
15. Quirós, L.: Multi-task handwritten document layout analysis, pp. 1–23 (2018). http://arxiv.org/abs/1806.08852
16. Riba, P., Dutta, A., Goldmann, L., Fornes, A., Ramos, O., Llados, J.: Table detection in invoice documents by graph neural networks. In: Proceedings of the International Conference on Document Analysis and Recognition, ICDAR, pp. 122–127 (2019)
17. Siddiqui, S.A., Fateh, I.A., Rizvi, S.T.R., Dengel, A., Ahmed, S.: DeepTabStR: deep learning based table structure recognition. In: Proceedings of the International Conference on Document Analysis and Recognition, ICDAR, pp. 1403–1409 (2019)
18. Tensmeyer, C., Morariu, V.I., Price, B., Cohen, S., Martinez, T.: Deep splitting and merging for table structure decomposition. In: Proceedings of the International Conference on Document Analysis and Recognition, ICDAR, pp. 114–121 (2019)

19. Wang, Y., Sun, Y., Liu, Z., Sarma, S.E., Bronstein, M.M., Solomon, J.M.: Dynamic graph CNN for learning on point clouds. ACM Trans. Graph. **38**(5) (2019). Article 146. https://doi.org/10.1145/3326362
20. Zhang, M., Chen, Y.: Link prediction based on graph neural networks. In: Proceedings of the 32nd International Conference on Neural Information Processing Systems, NeurIPS, pp. 5171–5181. Curran Associates Inc., Montreal (2018). https://doi.org/10.5555/3327345.3327423

Unsupervised Learning of Text Line Segmentation by Differentiating Coarse Patterns

Berat Kurar Barakat[2(✉)], Ahmad Droby[2], Raid Saabni[1], and Jihad El-Sana[2]

[1] Academic College of Telaviv Yafo, Tel-Aviv, Israel
raidsa@mta.ac.il
[2] Ben-Gurion University of the Negev, Beer-Sheva, Israel
{berat,drobya,el-sana}@post.bgu.ac.il

Abstract. Despite recent advances in the field of supervised deep learning for text line segmentation, unsupervised deep learning solutions are beginning to gain popularity. In this paper, we present an unsupervised deep learning method that embeds document image patches to a compact Euclidean space where distances correspond to a coarse text line pattern similarity. Once this space has been produced, text line segmentation can be easily implemented using standard techniques with the embedded feature vectors. To train the model, we extract random pairs of document image patches with the assumption that neighbour patches contain a similar coarse trend of text lines, whereas if one of them is rotated, they contain different coarse trends of text lines. Doing well on this task requires the model to learn to recognize the text lines and their salient parts. The benefit of our approach is zero manual labelling effort. We evaluate the method qualitatively and quantitatively on several variants of text line segmentation datasets to demonstrate its effectivity.

Keywords: Text line segmentation · Text line extraction · Text line detection · Unsupervised deep learning

1 Introduction

Text line segmentation is a central task in document image analysis. Basically text line segmentation can be represented as text line detection and text line extraction. Text line detection is a coarse representation of text lines in terms of baselines or blob lines. Text line extraction is a fine grained representation of text lines in terms of pixel labels or bounding polygons. Once the text line detection is achieved, text line extraction is trivial using standard tools. However, text line detection is challenging due to the prevalence of irregular texture regions in handwriting.

Given a document image patch, it contains a coarse trend of text lines. Human visual system can easily track these trend lines (Fig. 2), but a computer algorithm cannot track them due to the textured structure of each text line at fine details. Inspired by this fact, we hypothesize that a convolutional network

© Springer Nature Switzerland AG 2021
J. Lladós et al. (Eds.): ICDAR 2021, LNCS 12822, pp. 523–537, 2021.
https://doi.org/10.1007/978-3-030-86331-9_34

Fig. 1. The proposed method learns an embedding space in an unsupervised manner such that the distances between the embedded image patches correspond to the similarity of the coarse text line pattern they include.

can be trained in an unsupervised manner to map document image patches to some vector space such that the patches with the same coarse text line pattern are proximate and the patches with different coarse text line pattern are distant. We can assume that two neighbouring patches contain the same coarse text line pattern and contain different coarse text line pattern if one of them is rotated 90°. Doing well on this task requires the model to learn to recognize the text lines and their salient parts. Hence the embedded features of document patches can also be used to discriminate the differences in the horizontal text line patterns that they contain. Clustering the patches of a document page by projecting their vectors onto three principle directions yields a pseudo-rgb image where coarse text line patterns correspond to similar colours (Fig. 1). The pseudo-rgb image can then be thresholded into blob lines that strikethrough the text lines and guide an energy minimization function for extracting the text lines.

The proposed method has been evaluated on two publicly available handwritten documents dataset. The results demonstrate that this unsupervised learning method provides interesting text line segmentation results on handwritten document images.

Age	5	7	10	11
Perceived text lines				

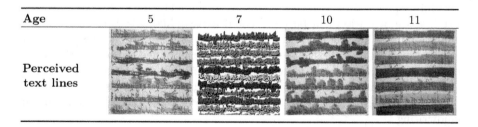

Fig. 2. Human visual system can easily perceive the coarse trend of handwritten text lines. Children can segment the text lines although written in a language they are not its reader.

2   Related Work

The recent trend in solving the handwritten text line segmentation problem is to employ deep networks that learn the representation directly from the pixels of the image rather than using engineered features [7]. These methods use a large dataset of labelled text lines to acquire the texture variances due to different font types, font sizes, and orientations.

Early attempts formulate text line segmentation as a supervised binary dense prediction problem. Given a document image, a Fully Convolutional Network (FCN) [17] is trained to densely predict whether a pixel is a text line pixel or not. However, the question that arises here is: Which pixels belong to a text line? Foreground pixels definitely can not discriminate a text line from the others because FCN output is a semantic segmentation where multiple instances of the same object are not separated. Very recently, text line segmentation has been formulated as an instance segmentation problem using Mask-RCNN [10], and its results are available in [14]. However, when using FCN, each text line is represented as a single connected component. This component can be either a blob line [14,16,18,20,22] strikes through the main body area of the characters that belong to a text line or a baseline [9] passes through the bottom part of the main body of the characters that belong to a text line. FCNs are very successful at detecting handwritten text lines [7]. However, scarcity of labelled data causes rarely occurring curved text lines to be poorly detected. This problem has been handled via augmentation [9] or learning-free detection [13].

Both the text line representations, blob line and baseline, are coarse grained representations and do not fully label all the pixels of a text line but only detect the spatial location of a text line. There are metrics that can evaluate the detected baselines [7,18,20] or blob lines [16]. Alternatively, detected spatial location of text lines are utilized to further extract the pixels of text lines. Some of these extraction methods assume horizontal text lines [22] whereas some can extract text lines at any orientation, with any font type and font size [14]. Text line extraction is evaluated by classical image segmentation metrics [7].

Deep networks have apparently increased handwritten text line segmentation performance by their ability to learn comprehensive visual features. However, they need to leverage large labelled datasets, which in turn brings costly human annotation effort. Learning-free algorithms would be a natural solution but still they do not achieve state of the art [12] except used in hybrid with deep networks [1]. Another solution would be unsupervised learning methods. However, the main concern is to find an objective function that will use a representation to capture text lines, although they are not labelled. Kurar *et al.* [15] formulated this concern as the answer to the question of whether a given document patch contains a text line or space line. The answer is based on a human adjusted score. In this paper, we propose an unsupervised text line segmentation method that trains a deep network to answer whether two document image patches contain the same coarse text line pattern or different coarse text line pattern. The network is urged to learn the salient features of text lines in order to answer this question.

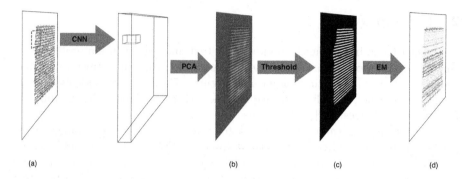

Fig. 3. Given a handwritten document image (a), first stage extracts feature vectors of image patches such that the patches with similar text line trends are close in the space. Second stage clusters the patches of a document image according to the first three principal components of their feature vectors. This stage outputs the a pseudo-rgb image (b) which is then thresholded onto blob lines (c) that strike through text lines. Energy minimization with the assistance of detected blob lines extracts the pixel labels of text lines (d).

3   Method

Unsupervised learning of text line segmentation is a three stage method (Fig. 3). The first stage relies on a deep convolutional network that can predict a relative similarity for a pair of patches and embed the patches into feature vectors. The similarity of two patches in document images correlates with their text line orientation assuming that the neighbouring patches contain the same orientation. The second stage generates a pseudo-rgb image using the three principals of the feature vectors obtained from the first stage. The pseudo-rgb image is further thresholded to detect the blob lines that strike through the text lines. Final stage performs pixel labelling for text lines using an energy minimization function that is assisted by the detected blob lines.

3.1   Deep Convolutional Network

Convolutional networks are well known to learn complex image representations from raw pixels. We aim the convolutional network to learn the coarse trend of text lines. We train it to predict the similarity for a pair of patches in terms of text line orientation. In a given document image neighbouring patches would contain the same coarse trend of text lines. Therefore, the network is expected to learn a feature embedding such that the patches that contain the same text line pattern would be close in the space.

To achieve this we use a pair of convolutional networks with shared weights such that the same embedding function is computed for both patches. Each convolutional branch processes only one of the patches hence the network performs most of the semantic reasoning for each patch separately. Consequently, the

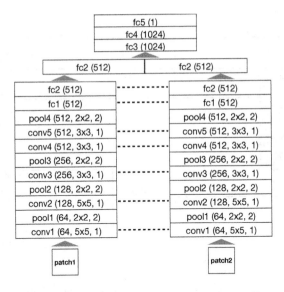

Fig. 4. Convolutional network architecture for pair similarity. Dotted lines stand for identical weights, conv stands for convolutional layer, fc stands for fully connected layer and pool is a max pooling layer.

feature representations are concatenated and fed to fully connected layers in order to predict whether the two image patches are similar or different.

The architecture of the branches is based on AlexNet [11] and through experiments we tune the hyperparameters to fit our task. Each of the branches has five convolutional layers as presented in Fig. 4. Dotted lines indicate identical weights, and the numbers in parentheses are the number of filters, filter size and stride. All convolutional and fully connected layers are followed by ReLU activation functions, except fc5, which feeds into a sigmoid binary classifier.

Pair Generation. Given a document image, we sample the first patch uniformly from regions containing foreground pixels. Given the position of the first patch we sample the second patch randomly from the eight possible neighbouring locations. We include a gap and jitter between patches in order to prevent cues like boundary patterns or lines continuing between patches. Neighbouring patches in a document image can be assumed to contain the same text line orientation and are labeled as similar pairs. Different pairs are generated by rotating the second patch 90°. Additionally for both, the similar pairs and the different pairs, the second patches are randomly rotated 0° or rotated 180° or flipped. Pair generation is demonstrated in Fig. 5. In case of fluctuating or skewed text lines, the similarity does not correlate with the proximity. However in a document image with almost all horizontal text lines these dissimilar and close patches are rare.

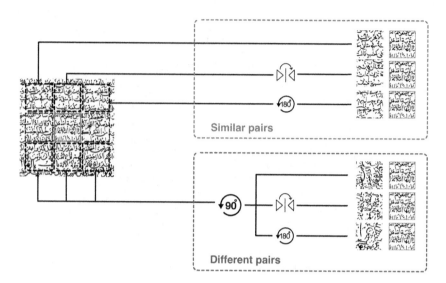

Fig. 5. The pairs are generated with the assumption that neighbouring patches contain similar text line trends. Different pairs are generated by rotating one of the patches 90°. Both, the similar and different, pairs are augmented by randomly rotating one of the patches 0° or 180° or flipping.

Training. For each dataset we train the model from scratch using n_p pairs:

$$n_p = \frac{h_a \times w_a}{p \times p} \times n_d \tag{1}$$

where h_a and w_a are the average document image height and width in the dataset, p is the patch size, and n_d is the number of document images in the set. The learning rate is 0.00001, the batch size is 8 and the optimizing algorithm is Adam. We continue training until there is no improvement on the validation accuracy with a patience of 7 epochs and save the model with the best validation accuracy for the next stage.

3.2 Pseudo-RGB Image

The convolutional network performs most of the semantic reasoning for each patch separately because only three layers receive input from both patches. Hence we can use a single branch to extract the significant features of patches. This embeds every patch into a feature vector of 512 dimensions. To visualize the features of a complete document image, a sliding window of the size $p \times p$ is used, but only the inner window of the size $w \times w$ is considered to increase the representation resolution. We also pad the document image with background pixels at its right and bottom sides if its size is not an integer multiple of the sliding window size. An additional padding is added at four sides of the document image for considering only the central part of the sliding window. Resultantly, a document image with the size $h_d \times w_d$ is mapped to a representation matrix of the size $\frac{h_d}{w} \times \frac{w_d}{w} \times 512$. We project $512D$ vectors into their three principle components and use these components to construct pseudo-rgb image in which similar patches are assigned the similar colors (Fig. 3(b)). Binary blob lines image is an outcome of thresholded pseudo-rgb image (Fig. 3(c)).

3.3 Energy Minimization

We adopt the energy minimization framework [4] that uses graph cuts to approximate the minimal of an arbitrary function. We adapt the energy function to be used with connected components for extracting the text lines. Minimum of the adapted function correspond to a good extraction which urges to assign components to the label of the closest blob line while straining to assign closer components to the same label (Fig. 3(d)). A touching component c among different blob lines is split by assigning each pixel in c to the label of the closest blob line.

Let $\mathcal{L}$ be the set of binary blob lines, and $\mathcal{C}$ be the set of components in the binary document image. Energy minimization finds a labeling f that assigns each component $c \in \mathcal{C}$ to a label $l_c \in \mathcal{L}$, where energy function $\mathbf{E}(f)$ has the minimum.

$$\mathbf{E}(f) = \sum_{c \in \mathcal{C}} D(c, \ell_c) + \sum_{\{c,c'\} \in \mathcal{N}} d(c, c') \cdot \delta(\ell_c \neq \ell_{c'}) \qquad (2)$$

The term D is the data cost, d is the smoothness cost, and δ is an indicator function. Data cost is the cost of assigning component c to label l_c. $D(c, \ell_c)$ is defined to be the Euclidean distance between the centroid of the component c and the nearest neighbour pixel in blob line l_c for the centroid of the component c. Smoothness cost is the cost of assigning neighbouring elements to different labels. Let $\mathcal{N}$ be the set of nearest component pairs. Then $\forall \{c, c'\} \in \mathcal{N}$

$$d(c, c') = \exp(-\beta \cdot d_c(c, c')) \qquad (3)$$

where $d_c(c, c')$ is the Euclidean distance between the centroids of the components c and c', and β is defined as

$$\beta = (2 \langle d_c(c, c') \rangle)^{-1} \qquad (4)$$

$\langle \cdot \rangle$ denotes expectation over all pairs of neighbouring components [5] in a document page image. $\delta(\ell_c \neq \ell_{c'})$ is equal to 1 if the condition inside the parentheses holds and 0 otherwise.

4 Experiments

In this section we first introduce the datasets used in the experiments. We define the parameters of the baseline experiment, and investigate the influence of patch size and central window size on the results. Then we visualize patch saliency for understanding the unsupervised learning of text line segmentation. Finally we discuss the limitations of the method.

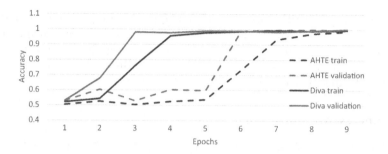

Fig. 6. Train and validation logs on the VML-AHTE and ICDAR2017 datasets.

4.1 Data

The experiments cover five datasets that are different in terms of the challenges they pose. The VML-AHTE dataset [14] consists of Arabic handwritten documents with crowded diacritics and cramped text lines. The Pinkas dataset [3] contains slightly edge rounded and noisy images of Hebrew handwritten documents. Their ground truth is provided in PAGE xml format [6,19]. The Printed dataset is our private and synthetic dataset that is created using various font types and sizes. The ICFHR2010 [8] is a dataset of modern handwriting that is heterogeneous by document resolutions, text line heights and skews. The ICDAR2017 dataset [21] includes three books, CB55, CSG18, and CSG863. In this dataset we run our algorithm on presegmented main text regions by the given ground truth. The VML-MOC dataset [2] is characterized by multiply oriented and curved handwritten text lines.

4.2 Baseline Experiment

We choose to experiment on five datasets with different challenges in order to verify that the method generalizes. Therefore, we define a baseline experiment that set the parameter values. There is no best set of parameters that fit all challenges and one can always boost the performance on a particular dataset by ad-hoc adjusting. However we wish to propose a baseline experiment that can fit all challenges as much as possible. The baseline experiment sets the input patch size $p = 350$, and the sliding central window size $w = 20$. The results are shown in Fig. 7. The convolutional network easily learns the embedding function. The validation accuracy almost always reaches over 99% (Fig. 6). We have preliminary experiment which suggest that increasing the number of layers until VGG-16 and then until VGG-19 leads to successful blob detection as well as AlexNet do. However, a deeper network such as ResNet does not detect blobs, probably because the reception field of its last convolutional layer is larger.

Pinkas	ICFHR	AHTE	MOC

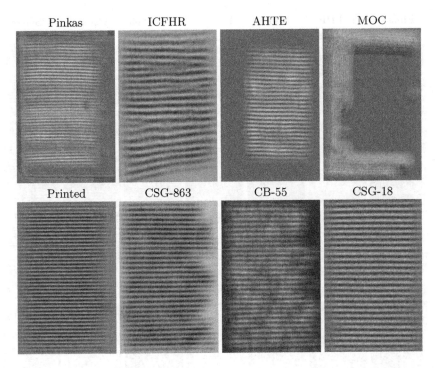

Printed	CSG-863	CB-55	CSG-18

Fig. 7. The results of baseline experiment are shown overlapped with the input images. The result on the VML-MOC dataset is a mess because the method assumes almost horizontal text lines when labeling the similar and different pairs.

4.3 Effect of Patch Size (p)

We have experimented with different patch sizes and found 350×350 performs well while keeping memory overhead manageable. Figure 8 shows results using patches of variable sizes. One can see that larger patch sizes lead to compact and well separated clusters of blob lines. Obviously at some point the performance is expected to decrease, if the patch size is increased further, because the assumption that the neighbouring patches are similar will gradually decrease. On the other hand the small patches do not contain a coarse trend of text line patterns therefore the blob lines fade out.

4.4 Effect of Central Window Size (w)

Consider that the input document that is downsampled by a factor of central window size should still be containing the text lines in an apartable form. Input document image size is downsampled by a factor of the central window size of the sliding window. Therefore this factor is effective on the representability of text lines in the pseudo-rgb image. This factor has to be small enough so the text lines in the downsampled images will not be scrambled. Otherwise it is impossible to

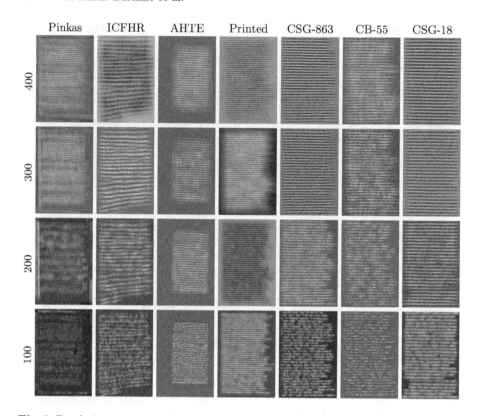

Fig. 8. Patch size comparison by qualitative results. Each row shows an example output from different datasets using a patch size. A patch size larger than 400 pixels could not be experimented due to memory overhead. Vertical observation illustrates that the method is insensitive to small variations in the patch size. Very small patches lead blob lines to fade out because they don't contain a coarse trend of text line patterns.

represent the detected blob lines that strike through the scrambled text lines (Fig. 9). On the other hand, the computation time is inversely proportional to the central window size. We have experimented with central window sizes and found $w = 20$ is efficient and effective well enough.

4.5   Patch Saliency Visualization

We visualize the features from last convolutional layer of a single branch to gain insight into the regions that the network looks at the decision of the classifier. The output from the last convolutional layer is a matrix of the size $m \times m \times 512$ where m is determined by the number of pooling layers and the input patch size p. We consider this matrix as $n = m \times m$ vectors each with 512 dimensions. Then, we get the first three components of these multidimensional vectors and visualize them as a pseudo-rgb image. No matter the transformation on the patch, the network recognizes the similar salient features on every patch (Fig. 10). As a

20   10   5

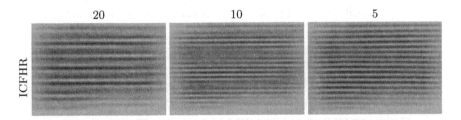

ICFHR

Fig. 9. Visualization of the effect of central window size. From left to right shows the results with a decreasing central window size. Central window has to be small enough so the text lines in the downsampled images will not be scrambled. Otherwise blob lines that strike through the text lines will be scrambled.

result of this, it can segment the text lines in a document image that is entirely transformed (Fig. 11).

4.6 Limitations

Extracting the features of a document image at patch level is a computationally intensive task and time consuming. Especially the consumed time is inversely proportional to the central window size which has to be small enough to represent the well separated blob lines. Severely skewed or curved text lines do not comply with the assumption that neighbouring patches contain similar coarse trends of text lines. Therefore the method cannot segment a multiply oriented and curved dataset such as the VML-AHTE.

5 Results

This section provides quantitative results on the VML-AHTE dataset and the ICDAR 2017 dataset. The results are compared with some other supervised and unsupervised methods. Note that the proposed method uses the same parameters of the baseline experiment on all the datasets. The performance is measured using the text line segmentation evaluation metrics, LIU and PIU, of the ICDAR2017 competition on layout analysis [21].

5.1 Results on the VML-AHTE Dataset

We compare our results with those of supervised learning methods, Mask-RCNN [14] and FCN+EM [14], and an unsupervised deep learning method, UTLS [15]. Mask-RCNN is an instance segmentation algorithm which is fully supervised using the pixel labels of the text lines. FCN+EM method [14] is fully supervised by human annotated blob lines. It uses energy minimization to extract the pixel labels of text lines. The comparison in terms of LIU and PIU are reported in Table 1. On the VML-AHTE dataset, the proposed method outperforms the compared methods in terms of LIU metric, and is competitive in terms of the

| | Normal | Flipped | Rotated 180 | Rotated 90 |

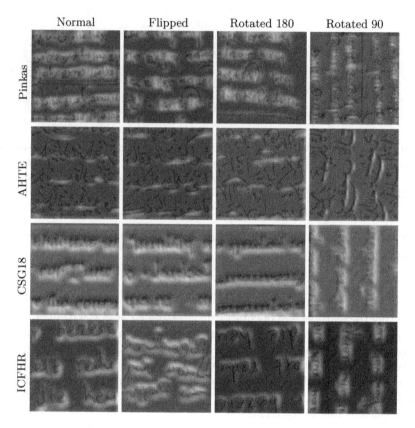

Fig. 10. Visualization of the features from the last convolutional layer. No matter the transformation on the patch, the network recognizes the similar salient features on every patch.

PIU metric. The error cases arise from few number of touching blob lines. Such errors can easily be eliminated but this is out of the focus of this paper. The advantage of the proposed method on the supervised methods is zero labelling effort. Also UTLS [15] has zero labelling effort, however it requires to adjust a heuristic formula. The proposed method eliminates this formula by assuming the neighbouring patches contain the same text line patterns.

5.2 Results on the ICDAR2017 Dataset

The second evaluation is carried out on the ICDAR2017 dataset [21]. We run our algorithm on presegmented text block areas by the given ground truth. Hence, we can compare our results with unsupervised System 8 and System 9 which are based on a layout analysis prior to text line segmentation. The comparison in terms of LIU and PIU are reported in Table 2. The main challenge in this dataset for the proposed method is the text line parts that are single handed and not

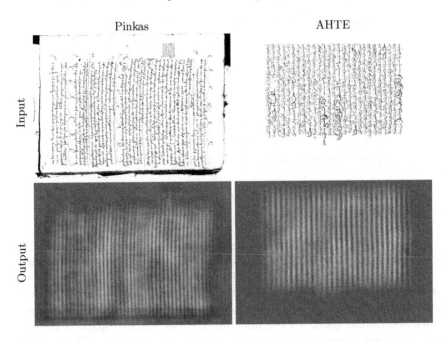

Fig. 11. The trained machine can segment an input document image that is entirely rotated by 90°.

accompanied by other text lines in their above and below. Since this is a rare case, the learning system recognizes as an insignificant noise. The performance of the proposed method on the ICDAR dataset is on par with the performances of two unsupervised methods, but these methods probably will need to be readjusted for each new dataset. However, the proposed method has been tested using the same parameters on all the considered datasets.

Table 1. LIU and PIU values on the VML-AHTE dataset.

	LIU	PIU
Unsupervised		
UTLS [15]	**98.55**	88.95
Proposed method	90.94	83.40
Supervised		
Mask-RCNN [14]	93.08	86.97
FCN+EM [14]	94.52	**90.01**

Table 2. LIU and PIU values on the ICDAR2017 dataset

	CB55		CSG18		CSG863	
	LIU	PIU	LIU	PIU	LIU	PIU
Unsupervised						
UTLS [15]	80.35	77.30	94.30	95.50	90.58	89.40
System-8	**99.33**	93.75	94.90	94.47	96.75	90.81
System-9+4.1	98.04	**96.67**	96.91	**96.93**	**98.62**	**97.54**
Proposed method	93.45	90.90	**97.25**	96.90	92.61	91.50

6   Conclusion

We presented a novel method for unsupervised deep learning of handwritten text line segmentation. It is based on the assumption that in a document image of almost horizontal text lines, the neighbouring patches contain similar coarse pattern of text lines. Hence if one of the neighbouring patches is rotated by 90°, they contain different coarse pattern of text lines. A network that is trained to embed the similar patches close and the different patches apart in the space, can extract interpretable features for text line segmentation. The method is insensitive to small variations in the input patch size but requires a careful selection of the central window size. We also demonstrated that entirely rotated document images can also be segmented with the same model. The method is effective at detecting cramped, crowded and touching text lines and can surpass the supervised learning methods whereas it has comparable results in terms of text line extraction.

Acknowledgment. The authors would like to thank Gunes Cevik and Hamza Barakat for helping in data preparation. This research was partially supported by The Frankel Center for Computer Science at Ben-Gurion University.

References

1. Alberti, M., Vögtlin, L., Pondenkandath, V., Seuret, M., Ingold, R., Liwicki, M.: Labeling, cutting, grouping: an efficient text line segmentation method for medieval manuscripts. In: ICDAR, pp. 1200–1206. IEEE (2019)
2. Barakat, B.K., Cohen, R., Rabaev, I., El-Sana, J.: VML-MOC: segmenting a multiply oriented and curved handwritten text line dataset. In: ICDARW, vol. 6, pp. 13–18. IEEE (2019)
3. Barakat, B.K., El-Sana, J., Rabaev, I.: The Pinkas dataset. In: ICDAR, pp. 732–737. IEEE (2019)
4. Boykov, Y., Veksler, O., Zabih, R.: Fast approximate energy minimization via graph cuts. IEEE Trans. Pattern Anal. Mach. Intell. **23**(11), 1222–1239 (2001)
5. Boykov, Y.Y., Jolly, M.P.: Interactive graph cuts for optimal boundary & region segmentation of objects in nd images. In: ICCV, vol. 1, pp. 105–112. IEEE (2001)

6. Clausner, C., Pletschacher, S., Antonacopoulos, A.: Aletheia-an advanced document layout and text ground-truthing system for production environments. In: ICDAR, pp. 48–52. IEEE (2011)
7. Diem, M., Kleber, F., Fiel, S., Grüning, T., Gatos, B.: cBAD: ICDAR 2017 competition on baseline detection. In: ICDAR, vol. 1, pp. 1355–1360. IEEE (2017)
8. Gatos, B., Stamatopoulos, N., Louloudis, G.: ICFHR 2010 handwriting segmentation contest. In: ICFHR, pp. 737–742. IEEE (2010)
9. Grüning, T., Leifert, G., Strauß, T., Michael, J., Labahn, R.: A two-stage method for text line detection in historical documents. IJDAR **22**(3), 285–302 (2019)
10. He, K., Gkioxari, G., Dollár, P., Girshick, R.: Mask R-CNN. In: ICCV, pp. 2961–2969 (2017)
11. Krizhevsky, A., Sutskever, I., Hinton, G.E.: ImageNet classification with deep convolutional neural networks. In: Advances in Neural Information Processing Systems, pp. 1097–1105 (2012)
12. Kurar Barakat, B., Cohen, R., Droby, A., Rabaev, I., El-Sana, J.: Learning-free text line segmentation for historical handwritten documents. Appl. Sci. **10**(22), 8276 (2020)
13. Kurar Barakat, B., Cohen, R., El-Sana, J.: VML-MOC: segmenting a multiply oriented and curved handwritten text line dataset. In: ICDARW, vol. 6, pp. 13–18. IEEE (2019)
14. Kurar Barakat, B., Droby, A., Alaasam, R., Madi, B., Rabaev, I., El-Sana, J.: Text line extraction using fully convolutional network and energy minimization. In: PatReCH, pp. 3651–3656. IEEE (2020)
15. Kurar Barakat, B., et al.: Unsupervised deep learning for text line segmentation. In: ICPR, pp. 3651–3656. IEEE (2020)
16. Kurar Barakat, B., Droby, A., Kassis, M., El-Sana, J.: Text line segmentation for challenging handwritten document images using fully convolutional network. In: ICFHR, pp. 374–379. IEEE (2018)
17. Long, J., Shelhamer, E., Darrell, T.: Fully convolutional networks for semantic segmentation. In: CVPR, pp. 3431–3440 (2015)
18. Mechi, O., Mehri, M., Ingold, R., Amara, N.E.B.: Text line segmentation in historical document images using an adaptive u-net architecture. In: ICDAR, pp. 369–374. IEEE (2019)
19. Pletschacher, S., Antonacopoulos, A.: The page (page analysis and ground-truth elements) format framework. In: ICPR, pp. 257–260. IEEE (2010)
20. Renton, G., Soullard, Y., Chatelain, C., Adam, S., Kermorvant, C., Paquet, T.: Fully convolutional network with dilated convolutions for handwritten text line segmentation. IJDAR **21**(3), 177–186 (2018)
21. Simistira, F., et al.: ICDAR 2017 competition on layout analysis for challenging medieval manuscripts. In: ICDAR, vol. 1, pp. 1361–1370. IEEE (2017)
22. Vo, Q.N., Kim, S.H., Yang, H.J., Lee, G.S.: Text line segmentation using a fully convolutional network in handwritten document images. IET Image Proc. **12**(3), 438–446 (2017)

Recognition of Tables and Formulas

Rethinking Table Structure Recognition Using Sequence Labeling Methods

Yibo Li[1,2], Yilun Huang[1], Ziyi Zhu[1], Lemeng Pan[3], Yongshuai Huang[3],
Lin Du[3], Zhi Tang[1], and Liangcai Gao[1(✉)]

[1] Wangxuan Institute of Computer Technology, Peking University, Beijing, China
{yiboli,huangyilun,1800012988,tangzhi,glc}@pku.edu.cn
[2] Center for Data Science, Peking University, Beijing, China
[3] Huawei AI Application Research Center, Huawei, China
{panlemeng,huangyongshuai,dulin09}@huawei.com

Abstract. Table structure recognition is an important task in document analysis and attracts the attention of many researchers. However, due to the diversity of table types and the complexity of table structure, the performances of table structure recognition methods are still not well enough in practice. Row and column separators play a significant role in the two-stage table structure recognition and a better row and column separator segmentation result can improve the final recognition results. Therefore, in this paper, we present a novel deep learning model to detect row and column separators. This model contains a convolution encoder and two parallel row and column decoders. The encoder can extract the visual features by using convolution blocks; the decoder formulates the feature map as a sequence and uses a sequence labeling model, bidirectional long short-term memory networks (BiLSTM) to detect row and column separators. Experiments have been conducted on PubTabNet and the model is benchmarked on several available datasets, including PubTabNet, UNLV ICDAR13, ICDAR19. The results show that our model has a state-of-the-art performance than other strong models. In addition, our model shows a better generalization ability. The code is available on this site (www.github.com/L597383845/row-col-table-recognition).

Keywords: Table structure recognition · Encoder-decoder · Row and column separators segmentation · Sequence labeling model

1 Introduction

Table structure recognition is an important task in document analysis. Tables are commonly found in research papers, books, invoices, and financial documents. A Table contains structured information with the arrangement of rows and columns. The manual extraction of structured information is often a tedious and time-consuming process. Therefore, extracting table structure automatically has attracted many researchers' attention in recent years.

© Springer Nature Switzerland AG 2021
J. Lladós et al. (Eds.): ICDAR 2021, LNCS 12822, pp. 541–553, 2021.
https://doi.org/10.1007/978-3-030-86331-9_35

However, due to the diversity of table types and the complexity of table structure, table structure recognition is not a easy problem to solve. Different background colors of the table, the absence of some ruling lines, and the existence of row-span or column-span cells are all challenges to the table structure recognition. Besides, text content of the cell may be centered, left-aligned, or right-aligned, which influences some methods' performance.

In the past few decades, many table recognition methods have been proposed. These proposed methods can be divided into two categories, end-to-end methods and two-stage methods. Chris et al. [18] and Saqib et al. [10] proposed that table structure recognition could be divided into splitting stage and merging stage. The splitting stage predicts the row and column separators and the merging stage predicts which grid elements should be merged to recover cells that span multiple rows or columns. Chris et al. [18] presented Split, which uses a convolution network with novel projection pooling to get the rows and columns segmentation. But this model is not stable in some datasets. Saqib et al. presented BiGRU, which uses image pre-processing and two BiGRU networks to get row and column separators segmentation. However, the performance of BiGRU is limited to the result of the pre-processing stage.

In this paper, we propose a novel encoder-decoder model to recognize row and column separators. The encoder uses a convolution block that contains several dilated convolutions to extract the visual feature. A $H \times W$ image will be converted to a $H \times W$ feature map after the encoder. The feature map is formulated as a sequence of length H or a sequence of length W, so the row and column separators segmentation task can be solved by sequence labeling methods. The encoder is followed by two parallel BiLSTM branches for 1) Segmentation of the row separators and 2) Segmentation of the column separators. Besides, the results of different sequence labeling methods have been compared in the experiments.

We have trained our model on the PubTabNet-Train [25] dataset and evaluated its performance on several available datasets, including PubTabNet-Val, UNLV [16], ICDAR13 [5] and ICDAR19 [4], demonstrating that our approach outperforms other methods in row and column separators segmentation marginally. The code will be publicly released to GitHub.

In summary, the primary contributions of this paper are as follows:

1) The segmentation of row and column separators is firstly formulated as a sequence labeling problem.
2) A unified encoder-decoder architecture for the segmentation of both row and column separators simultaneously using convolutional and recursive neural networks is proposed in this paper and achieved a state-of-the-art performance on the publicly available PubTabNet, UNLV, ICDAR13, and ICDAR19 table structure recognition datasets.

The rest of the paper is organized as follows: Sect. 2 provides the overall of the related work on table structure segmentation. Section 3 describes the details of our method. Section 4 outlines the details about four datasets, how to use the annotation, and the evaluation metric. Section 5 will outline experiment details and results. Finally, the conclusion and future work are presented in Sect. 6.

2   Related Work

2.1   Table Structure Recognition

In the beginning, some researchers used heuristics-based methods to solve this task. T-Recs system [3,11,12], proposed by Kieninger et al. is one of the earliest works to extract tabular information. This system takes words bounding boxes as input. It groups words into columns by their horizontal ruling lines and divides them into cells based on column margins. Then, Wang et al. [20] proposed a system that relied on probability optimization to tackle the table structure understanding problem. Jianying et al. [8] used hierarchical clustering for column detection. Their system uses lexical and spatial criteria to classify headers of tables. Chen et al. [1] used Min-Cut/Max-Flow algorithm to decompose tabular structures into 2-D grids of table cells.

With the development of deep learning, neural networks have been successfully applied in various tasks [22–24]. Thus, some works try to utilize neural networks to solve the table structure recognition. Currently, most of them are two-stage methods and can be divided into cell detection and extraction of cell relationships. Shah et al. [14] used OCR technology to get the bounding box of cells and used CNN to extract the feature of every cell. Then a graph neural network was applied to classify the relationship between two cells. Devashish et al. [13] proposed an automatic table recognition method for interpretation of tabular data in document images, CascadeTabNet. It is a Cascade mask Region-based CNN High-Resolution Network (Cascade mask R-CNN HRNet) based model that detects the regions of tables and recognizes the structural body cells from the detected tables at the same time. Then it uses a rule-based method to recover the table structure. TabStruct-Net, proposed by Sachin et al. [15], is an approach for table structure recognition that combines cell detection and interaction modules to localize the cells and predicts their row and column associations with other detected cells. Structural constraints are incorporated as additional differential components to the loss function for cell detection. Shoaib et al. [17] proposed DeepTabStR, which uses a deformable convolution to detect rows, columns and cells at the same time. But it doesn't perform well on tables that span rows or columns.

Chris et al. [10] and Saqib et al. [18] proposed that table structure recognition could be divided into splitting stage and merging stage. The splitting stage predicts the row and column separators and the merging stage predicts which grid elements should be merged to recover cells that span multiple rows or columns. The former uses a convolution network with novel projection pooling to get the rows and columns segmentation, and the latter uses two BiGRU networks to solve it.

2.2   Sequence Labeling

Sequence labeling is a classical problem in the field of natural language processing (NLP), various neural models have been introduced to solve it. The proposed neural models usually contain three components: word embedding layer, context encoder layer, and decoder layer. In this paper, the convolutional block will be used to extract the feature map which can be regarded as the word embedding; the linear layer followed by the softmax layer is the decoder layer. Therefore, the detailed infomation will be introduced in the following paragraphs.

Recurrent Neural Networks (RNNs) are widely employed in NLP tasks due to their sequential characteristic, which is aligned well with the language. Specifically, bidirectional long short-term memory network (BiLSTM) [7] is one of the most widely used RNN structures. BiLSTM and Conditional Random Fields (CRF) were applied to sequence labeling tasks in [9]. Owing to BiLSTM's high power to learn the contextual representation of sequence, it has been adopted by the majority of sequence labeling models as the encoder. Cho et al. [2] proposed Gated Recurrent Unit (GRU) to solve Machine Translation problem. As a variant of LSTM, GRU combines forget gate and input gate into a single update gate. The GRU model is simpler than the standard LSTM model and is a very popular RNN structure.

Recently, Transformer began to prevail in various NLP tasks, like machine translation, language modeling and language pretraining models [19]. The Transformer encoder adopts a fully-connected self-attention structure to model the long-range context, which is the weakness of RNNs. Moreover, Transformer has better parallelism ability than RNNs. Some researchers began to use Transformer to solve sequence labeling tasks. Guo et al. [6] tested the performance of transformer on sequence labeling firstly. Then, Yan et al. [21] proved that Transformer lost the directivity of the sequence and proposed an Adapting Transformer Encoder to improve the performance of Transformer on solving sequence labeling problems.

3   Method

We formulate the problem of table row and column separators segmentation as a sequence labeling problem. An $H \times W$ image can be regarded as a sequence of length H with W features when segmenting the row separators or as a sequence of length W with H features when segmenting the column-separators. The proposed model, illustrated in Fig. 1 which contains an encoder and two paralleled decoders. This model takes a table image as input and outputs the basic row separators and column separators segmentation.

3.1   Encoder

The encoder is responsible for extracting the visual features of the image. There is a common 3×3 convolution at the beginning. Then we use several 3×3

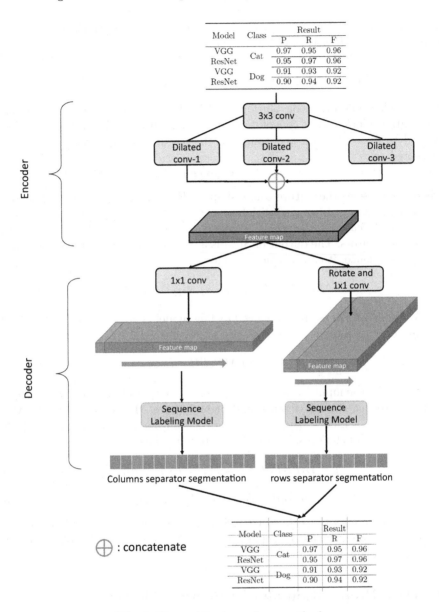

Fig. 1. The architecture of our method

dilation convolution operators with $[1, 2, 5]$ dilated rates and $[1, 2, 5]$ padding size to enlarge the receptive field of each pixel. After this, a concatenation operator is applied to fusion features. Using such a model, an image can be converted to a feature map. The encoder can be represented as:

$$F = \varphi(x) \tag{1}$$

where x represents the input whose size is $H \times W$, φ denotes the encoder and F represents the $H \times W \times C$ feature map.

3.2   Decoder

The decoder contains a convolution block and a sequence labeling model. The convolution block uses a convolution operator with a 1×1 kernel. This block is used to fusion feature and compress the feature to one channel. It can be presented as:

$$S = Conv(F). \tag{2}$$

where S represents the feature whose shape is $H \times W$. Then S can be regarded as a sequence $[s_0, s_1, s_2, \cdots, s_{H-1}]$ with feature of length W or a sequence $[s_0, s_1, s_2, \cdots, s_{W-1}]$ with feature of length H. A sequence labeling model (SLM) is followed to label separators. The output of SLM is a sequence which has the same length as input sequence. It can be calculated as:

$$Result = SLM([s_0, s_1, s_2, \cdots, s_{H-1}]) \tag{3}$$

We use BiLSTM as the sequence labeling model. The output of a LSTM cell h_t can be represented as:

$$h_t = LSTMCell(h_{t-1}, s_t) \tag{4}$$

Therefore, for the input $[s_0, s_1, s_2, \cdots, s_{H-1}]$, the output of LSTM is $[h_0, h_1, h_2, \cdots, h_{H-1}]$. To get the left feature and right feature, the bi-direction LSTM is applied. Then a Linear Layer and a information operator are used to get the final sequence labeling result. It can be represented as:

$$y = BiLSTM(S) = [h_0, h_1, h_2, \cdots, h_{H-1}] \tag{5}$$

$$Result = Softmax(Linear(y)) \tag{6}$$

Where y represents the result of BiLSTM whose size is $H \times 2 * hidden\_size$. Because there are two labels, "Separator" and "No-Separator", $Result$ represents the sequence labeling result with the size of $H \times 2$.

4   Data Preparation and Evaluation Metric

We conduct experiments on PubTabNet [25] and evaluate our model on ICDAR13 [5], ICDAR19 [4], and UNLV [16]. However, the annotation of these datasets is not suitable for the method. Therefore, we convert the annotations to the appropriate format. And we use the evaluation metric represent in [10]. The details are as follows:

4.1 Data Preparation

Four datasets are used in the experiments, including PubTabNet, UNLV, ICDAR13 and ICDAR19. The summary of these datasets is on Table 1.

Table 1. Summary of datasets used in our experiments

DataSet	Usage	Numbers
PubTabNet (training set)	Train	500,777
PubTabNet (validation set)	Test	9,115
UNLV	Test	557
ICDAR13	Test	225
ICDAR19	Test	145

Images in PubTabNet are extracted from the scientific publications included in the PubMed Central Open Access Subset (commercial use collection). It contains heterogeneous tables in both image and HTML format. The HTML representation encodes both the structure of the tables and the content in each table cell. Position (bounding box) of table cells is also provided to support more diverse model designs. There are $500,777$ images in the training set and $9,115$ images in the validation set. We decode the cell information including bounding box, row and column position with row and column spans from the HTML and JSON annotation. Then we compute the index of row-separators and column-separators from the cell information. When obtaining the row separator of the table, for an image of size $[H, W]$, we create a matrix M of the same size and set all the values in it to 1. Then, set the corresponding position of the cells that does not span rows to 0. When obtaining the column-separators of the table, we will set the corresponding position of the cell that does not span columns to 0. The index of all 1 rows and columns in the M matrix is the index of the row and column separator. A visualized result is shown in Fig. 2. However, considering that the bounding box of the cell marked by PubTabNet has some overlaps (the lower bound of the upper cell is greater than the previous one of the lower cell in the adjacent two rows of cells), so when we calculate the row separator, we will reduce the lower bound of every cell.

The ICDAR13 and ICDAR19 datasets present the table position, cell bounding boxes and their ["start – row", "start – col", "end – col", "end – row"]. The ICDAR13 dataset includes a further collection of EU and US documents. The modern dataset in ICDAR19 comes from different kinds of PDF documents such as scientific journals, forms, financial statements, etc. They are both documents images. In order to adapt to this method, we extract the image of a single table from the document images. We compute the row-span and column-span of every cell and get the ground truth by using the same algorithm as the PubTabNet. There are 225 table images in the ICDAR13 dataset and 145 table images in the ICDAR19 dataset.

images

Variable	Male		Female	
	%	95% CI	%	95% CI
Sensitivity	39.13	31.55 to 47.12	37.50	30.49 to 44.92
Specificity	93.92	92.68 to 95.00	92.42	91.18 to 93.54
Positive Likelihood Ratio	6.43	4.92 to 8.41	4.95	3.89 to 6.29
Negative Likelihood Ratio	0.65	0.57 to 0.73	0.68	0.60 to 0.76
Disease prevalence	8.61	7.37 to 9.97	8.36	7.23 to 9.59
Positive Predictive Value	37.72	30.35 to 45.54	31.08	25.04 to 37.63
Negative Predictive Value	94.25	93.04 to 95.31	94.19	93.07 to 95.18

Normal
mask

Row-cell
mask

Col-cell
mask

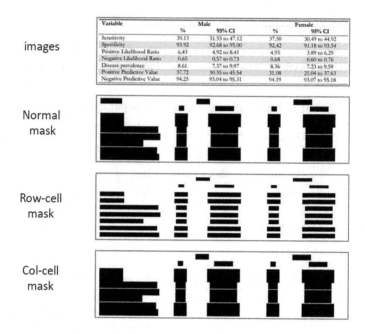

Fig. 2. The mask of cells for row and column separators

The UNLV dataset consists of a variety of documents, including technical reports, business letters, newspapers, and magazines. The dataset contains 2,889 scanned documents, 403 of which contain tables. It presents the coordinate of the table and its basic row and column separators. Thus, we can extract the table in the images directly by using the annotation. There are 557 table images in total.

4.2   Evaluation Metric

Various researchers have used different evaluation metrics ranging from simple precision and recall to more complex evaluation algorithms. In this paper, we use the methodology proposed by Shahab et al. [16]. Saqib et al. [10] detailed the methodology when benchmarking the row or column segmentation. This metric contains six values, including Correct Detections, Partial Detections, Missed Detections, Over Segmented Detections, Under Segmented Detections, and False Position Detections. The Correct Detections shows the total number of ground truth segments that have a large intersection with a detected segment and the detected segment does not have a significant overlap with any other ground truth segment. It is the most important one among the six measures.

5   Experiments and Result

Table 2. The row and column segmentation results on PubTabNet dataset by using different sequence labeling models.

	SLM	Accuracy %					
		Correct	Partial	Missed	Over-Seg	Under-Seg	False-Pos
Row	BiLSTM (ours)	**95.23**	0.63	0.03	1.93	2.49	2.25
	BiGRU	93.26	0.73	0.05	3.14	2.77	3.17
	Transformer [19]	44.79	8.15	0.33	29.10	20.48	2.22
	Ada transformer [21]	89.39	0.49	0.01	5.12	2.75	4.61
Column	BiLSTM (ours)	**97.39**	0.37	0.01	1.32	1.23	0.87
	BiGRU	95.26	0.85	0.01	2.61	1.93	1.10
	Transformer	33.11	8.46	0.12	17.73	17.21	6.24
	Ada transformer	91.21	1.12	0.00	5.84	2.47	0.93

Table 3. The results of evaluating our method and other models on four datasets. The following benchmasrk is for row segmentation.

Test dataset	Model	Accuracy %					
		Correct	Partial	Missed	Over-seg	Under-seg	False-pos
PubTabNet (Val Set)	BiGRU [10]	56.93	3.51	0.03	17.38	21.95	7.23
	Split [18]	84.07	2.56	0.02	7.01	4.95	4.68
	Ours	**95.23**	0.63	0.03	1.93	2.49	2.25
UNLV	BiGRU	26.02	3.77	0.23	28.18	22.13	5.60
	Split	48.47	3.26	0.23	26.19	17.06	5.07
	Ours	**54.80**	2.28	0.22	14.24	16.42	5.29
ICDAR13	BiGRU	25.56	2.34	0.05	23.71	30.53	5.86
	Split	55.68	5.68	0.00	23.78	23.42	10.36
	Ours	**57.78**	3.60	0.00	27.55	27.98	7.46
ICDAR19	BiGRU	29.59	4.43	0.62	32.55	32.07	5.86
	Split	8.74	4.76	2.83	54.69	48.29	2.76
	Ours	**70.43**	3.85	0.52	14.06	9.31	4.32

The sequence labeling model is used in our method to get the row and column separators segmentation. We compare the performance of different sequence labeling models on PubTabNet, including BiLSTM, BiGRU, Transformer Encoder, Adaptive Transformer Encoder. The results of these sequence labeling models are on Table 2. Comparing with other SLMs, the BiLSTM model achieves the best results in correct row and column detections. On the other hand, the value of other measures is the minimum among these SLMs.

To verify the effectiveness of this method, the PubtabNet training set is used to train our model and two baseline models (BiGRU [10], Split [18]). And all table images have been resized to 600 × 600. The comparison between different models on four table recognition datasets are shown in Table 3 and Table 4.

Table 4. The results of evaluating our method and other models on four datasets. The following benchmasrk is for column segmentation.

Test dataset	Model	Accuracy %					
		Correct	Partial	Missed	Over-seg	Under-seg	False-pos
PubTabNet (Val Set)	BiGRU [10]	60.98	6.98	0.28	25.08	12.08	1.56
	Split [18]	88.35	1.36	0.00	8.61	2.91	0.96
	Ours	**97.39**	0.37	0.01	1.32	1.23	0.87
UNLV	BiGRU	34.24	18.96	4.11	33.80	25.99	1.66
	Split	60.53	9.65	0.05	20.86	16.77	2.99
	Ours	**71.85**	5.69	0.00	13.39	12.28	3.62
ICDAR13	BiGRU	32.46	7.30	1.09	36.16	26.16	5.78
	Split	71.65	10.24	0.00	16.67	17.34	0.40
	Ours	**90.63**	2.56	0.00	4.93	4.60	0.81
ICDAR19	BiGRU	38.21	14.33	2.95	37.19	26.09	0.00
	Split	12.22	7.81	1.44	48.41	42.94	6.16
	Ours	**81.09**	3.10	0.00	10.74	7.03	2.57

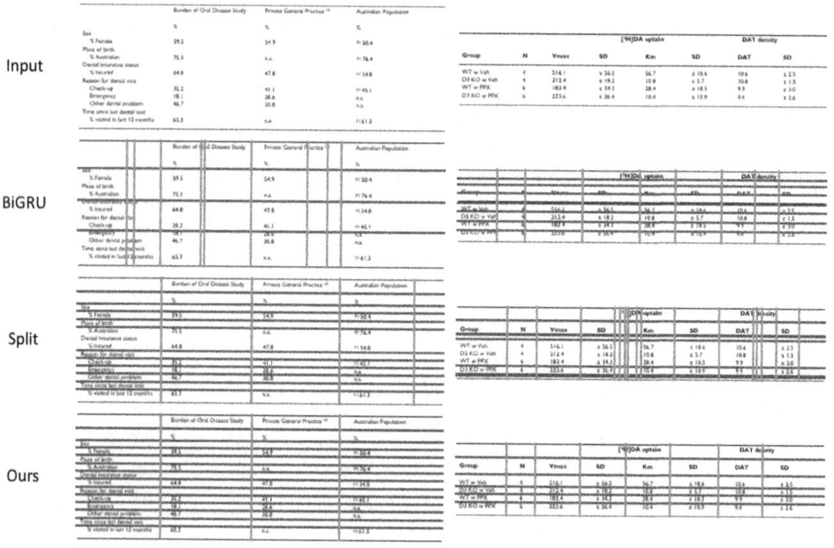

Fig. 3. Two samples of the results. The green lines represent the column separators and the red lines represent the row separators. (Color figure online)

Through the results, we can find the following things. When using BiGRU, it is obvious that there is a large quantity of Over Segmentation Detections and Under Segmentation Detections contributing to the lowest correct detections. And when evaluating the performance on UNLV, ICDAR13, and ICDAR19 datasets, the Correct Detections will be much lower. The Split has excellent performance on PubTabNet, UNLV, and ICDAR13 datasets. However, this model fails when testing on the ICDAR19 dataset. Our model achieves state-of-the-art performance than other models.

No matter on which data set, the Correct Detections of our model is much higher than BiGRU and Split. Two samples are represented on Fig. 3. It is obvious that the Over Segmentation Detection will occur when the blank space between two rows or two columns is large using BiGRU and Split. And when two rows or two columns are too close, BiGRU and Split will have higher Under Segmentation detections with lower Correct Detections. Our model solves these problems and achieves state-of-the-art performance in four datasets.

6 Conclusion

This paper proposes a novel encoder-decoder architecture for row and column separators segmentation. It uses a convolutional block as an encoder to extract feature. The decoder formulates the problem of labeling row and column separators as a sequence labeling problem and uses two parallel SLMs. In this paper, we make a comparison with the performances of different sequence labeling models. Besides, our model is evaluated on several available table recognition datasets. Our method achieves state-of-the-art performance on row and column segmentation and has a better generalization ability than other proposed methods.

In the future, a graph neural network will be applied to merge the basic cells to recognize the accurate table structure.

Acknowledgement. This work is supported by the projects of National Key R&D Program of China (2019YFB1406303) and National Natural Science Foundation of China (No. 61876003), which is also a research achievement of Key Laboratory of Science, Technology and Standard in Press Industry (Key Laboratory of Intelligent Press Media Technology).

References

1. Chen, J., Lopresti, D.P.: Model-based tabular structure detection and recognition in noisy handwritten documents. In: 2012 International Conference on Frontiers in Handwriting Recognition, ICFHR 2012, pp. 75–80 (2012)
2. Cho, K., et al.: Learning phrase representations using RNN encoder-decoder for statistical machine translation. In: Proceedings of the 2014 Conference on Empirical Methods in Natural Language Processing, EMNLP 2014, pp. 1724–1734 (2014)
3. Dengel, A., Kieninger, T.: A paper-to-HTML table converting system. In: Proceedings of Document Analysis Systems, pp. 356–365 (1998)
4. Gao, L., et al.: ICDAR 2019 competition on table detection and recognition (CTDAR). In: 2019 International Conference on Document Analysis and Recognition, ICDAR 2019, pp. 1510–1515 (2019)
5. Göbel, M.C., Hassan, T., Oro, E., Orsi, G.: ICDAR 2013 table competition. In: 12th International Conference on Document Analysis and Recognition, ICDAR 2013, pp. 1449–1453 (2013)
6. Guo, Q., Qiu, X., Liu, P., Shao, Y., Xue, X., Zhang, Z.: Star-transformer. In: Proceedings of the 2019 Conference of the North American Chapter of the Association for Computational Linguistics: Human Language Technologies, NAACL-HLT 2019, pp. 1315–1325 (2019)

7. Hochreiter, S., Schmidhuber, J.: Long short-term memory. Neural Comput. **9**, 1735–1780 (1997)
8. Hu, J., Kashi, R.S., Lopresti, D.P., Wilfong, G.T.: Table structure recognition and its evaluation. In: Document Recognition and Retrieval VIII, 2001. SPIE Proceedings, vol. 4307, pp. 44–55 (2001)
9. Huang, Z., Xu, W., Yu, K.: Bidirectional LSTM-CRF models for sequence tagging. CoRR abs/1508.01991 (2015)
10. Khan, S.A., Khalid, S.M.D., Shahzad, M.A., Shafait, F.: Table structure extraction with bi-directional gated recurrent unit networks. In: 2019 International Conference on Document Analysis and Recognition, ICDAR 2019, pp. 1366–1371 (2019)
11. Kieninger, T., Dengel, A.: The T-Recs table recognition and analysis system. In: Lee, S., Nakano, Y. (eds.) Document Analysis Systems: Theory and Practice, Third IAPR Workshop, DAS 1998. vol. 1655, pp. 255–269 (1998)
12. Kieninger, T., Dengel, A.: Table recognition and labeling using intrinsic layout features. In: International Conference on Advances in Pattern Recognition, pp. 307–316 (1999)
13. Prasad, D., Gadpal, A., Kapadni, K., Visave, M., Sultanpure, K.: CascadeTabNet: an approach for end to end table detection and structure recognition from image-based documents. In: 2020 IEEE/CVF Conference on Computer Vision and Pattern Recognition, CVPR Workshops 2020, pp. 2439–2447 (2020)
14. Qasim, S.R., Mahmood, H., Shafait, F.: Rethinking table recognition using graph neural networks. In: 2019 International Conference on Document Analysis and Recognition, ICDAR 2019, pp. 142–147 (2019)
15. Raja, S., Mondal, A., Jawahar, C.V.: Table structure recognition using top-down and bottom-up cues. In: Vedaldi, A., Bischof, H., Brox, T., Frahm, J.-M. (eds.) ECCV 2020. LNCS, vol. 12373, pp. 70–86. Springer, Cham (2020). https://doi.org/10.1007/978-3-030-58604-1_5
16. Shahab, A., Shafait, F., Kieninger, T., Dengel, A.: An open approach towards the benchmarking of table structure recognition systems. In: The Ninth IAPR International Workshop on Document Analysis Systems, DAS 2010. pp. 113–120 (2010)
17. Siddiqui, S.A., Fateh, I.A., Rizvi, S.T.R., Dengel, A., Ahmed, S.: DeepTabStR: deep learning based table structure recognition. In: 2019 International Conference on Document Analysis and Recognition, ICDAR 2019, pp. 1403–1409 (2019)
18. Tensmeyer, C., Morariu, V.I., Price, B.L., Cohen, S., Martinez, T.R.: Deep splitting and merging for table structure decomposition. In: 2019 International Conference on Document Analysis and Recognition, ICDAR 2019, pp. 114–121 (2019)
19. Vaswani, A., et al.: Attention is all you need. In: Advances in Neural Information Processing Systems (NIPS) 2017, pp. 5998–6008 (2017)
20. Wang, Y., Phillips, I.T., Haralick, R.M.: Table structure understanding and its performance evaluation. Pattern Recognit. **37**(7), 1479–1497 (2004)
21. Yan, H., Deng, B., Li, X., Qiu, X.: TENER: adapting transformer encoder for named entity recognition. CoRR abs/1911.04474 (2019)
22. Yan, Z., Ma, T., Gao, L., Tang, Z., Chen, C.: Persistence homology for link prediction: an interactive view. arXiv preprint arXiv:2102.10255 (2021)
23. Yuan, K., He, D., Jiang, Z., Gao, L., Tang, Z., Giles, C.L.: Automatic generation of headlines for online math questions. In: Proceedings of the AAAI Conference on Artificial Intelligence, vol. 34, pp. 9490–9497 (2020)

24. Yuan, K., He, D., Yang, X., Tang, Z., Kifer, D., Giles, C.L.: Follow the curve: arbitrarily oriented scene text detection using key points spotting and curve prediction. In: 2020 IEEE International Conference on Multimedia and Expo (ICME), pp. 1–6. IEEE (2020)
25. Zhong, X., ShafieiBavani, E., Jimeno Yepes, A.: Image-based table recognition: data, model, and evaluation. In: Vedaldi, A., Bischof, H., Brox, T., Frahm, J.-M. (eds.) ECCV 2020. LNCS, vol. 12366, pp. 564–580. Springer, Cham (2020). https://doi.org/10.1007/978-3-030-58589-1_34

TABLEX: A Benchmark Dataset for Structure and Content Information Extraction from Scientific Tables

Harsh Desai$^{(\boxtimes)}$, Pratik Kayal, and Mayank Singh

Indian Institute of Technology, Gandhinagar, India
{pratik.kayal,singh.mayank}@iitgn.ac.in

Abstract. Information Extraction (IE) from the tables present in scientific articles is challenging due to complicated tabular representations and complex embedded text. This paper presents TABLEX, a large-scale benchmark dataset comprising table images generated from scientific articles. TABLEX consists of two subsets, one for table structure extraction and the other for table content extraction. Each table image is accompanied by its corresponding LaTeX source code. To facilitate the development of robust table IE tools, TABLEX contains images in different aspect ratios and in a variety of fonts. Our analysis sheds light on the shortcomings of current state-of-the-art table extraction models and shows that they fail on even simple table images. Towards the end, we experiment with a transformer-based existing baseline to report performance scores. In contrast to the static benchmarks, we plan to augment this dataset with more complex and diverse tables at regular intervals.

Keywords: Information Extraction · LaTeX · Scientific articles

1 Introduction

Tables are compact and convenient means of representing relational information present in diverse documents such as scientific papers, newspapers, invoices, product descriptions, and financial statements [12]. Tables embedded in the scientific articles provide a natural way to present data in a structured manner [5]. They occur in numerous variations, especially visually, such as with or without horizontal and vertical lines, spanning multiple columns or rows, non-standard spacing, alignment, and text formatting [6]. Besides, we also witness diverse semantic structures and presentation formats, dense embedded text, and formatting complexity of the typesetting tools [23]. These complex representations and formatting options lead to numerous challenges in automatic tabular information extraction (hereafter, *'TIE'*).

Table Detection vs. Extraction: In contrast to table detection task (hereafter, *'TD'*) that refers to identifying tabular region (e.g., finding a bounding box that encloses the table) in an document, TIE refers to two post identification

© Springer Nature Switzerland AG 2021
J. Lladós et al. (Eds.): ICDAR 2021, LNCS 12822, pp. 554–569, 2021.
https://doi.org/10.1007/978-3-030-86331-9_36

tasks: (i) table structure recognition (hereafter, *'TSR'*) and (ii) table content recognition (hereafter, *'TCR'*). TSR refers to the extraction of structural information like rows and columns from the table, and TCR refers to content extraction that is embedded inside the tables. Figure 1 shows an example table image with its corresponding structure and content information in the TEX language. Note that, in this paper, we only focus on the two TIE tasks.

Limitations in Existing State-of-the-Art TIE Systems: There are several state-of-the-art tools (Camelot[1], Tabula[2], PDFPlumber[3], and Adobe Acrobat SDK[4]) for text-based TIE. On the contrary, Tesseract-based OCR [24] is commercially available tool which can be used for image-based TIE. However, these tools perform poorly on the tables embedded in the scientific papers due to the complexity of tables in terms of spanning cells and presence of mathematical content. The recent advancements in deep learning architectures (Graph Neural Networks [31] and Transformers [27]) have played a pivotal role in developing the majority of the image-based TIE tools. We attribute the limitations in the current image-based TIE primarily due to the training datasets' insufficiency. Some of the critical issues with the training datasets can be (i) the size, (ii) diversity in the fonts, (iii) image resolution (in dpi), (iv) aspect ratios, and (v) image quality parameters (blur and contrast).

Our Proposed Benchmark Dataset: In this paper, we introduce TABLEX, a benchmark dataset for information extraction from tables embedded inside scientific documents compiled using LATEX-based typesetting tools. TABLEX is composed of two subsets—a table structure subset and a table content subset—to extract the structure and content information. The table structure subset contains more than three million images, whereas the table content subset contains over one million images. Each tabular image accompanies its corresponding ground-truth program in TEX macro language. In contrast to the existing datasets [3,29,30], TABLEX comprises images with 12 different fonts and multiple aspect ratios.

Main Contributions: The main contributions of the paper are:

1. Robust preprocessing pipeline to process the scientific documents (created in TEX language) and extract the tabular spans.
2. A large-scale benchmark dataset, TABLEX, comprising more than three million images for structure recognition task and over one million images for the content recognition task.
3. Inclusion of twelve font types and multiple aspect ratios during the dataset generation process.
4. Evaluation of state-of-the-art computer vision based baseline [7] on TABLEX.

---

[1] https://github.com/camelot-dev/camelot.

[2] https://github.com/chezou/tabula-py.

[3] https://github.com/jsvine/pdfplumber.

[4] https://www.adobe.com/devnet/acrobat/overview.html.

The Paper Outline: We organize the paper into several sections. Section 2 reviews existing datasets and corresponding extraction methodologies. Section 3 describes the preprocessing pipeline and presents TABLEX dataset statistics. Section 4 details three evaluation metrics. Section 5 presents the baseline and discusses the experimental results and insights. Finally, we conclude and identify the future scope in Sect. 6.

Nuclide	NPA Dimension	FCI Dimension
^{52}Fe	350	1.1×10^8
^{53}Fe	6106	2.2×10^8
^{54}Fe	706	3.5×10^8
^{56}Fe	1276	5.0×10^8

(a) Table Image

{ | c | c | c | }\hline CELL & CELL & CELL \\ \hline CELL & CELL & CELL \\ \hline CELL & CELL & CELL \\ \hline CELL & CELL & CELL \\ \hline CELL & CELL & CELL \\ \hline

(b) Structure Information

Nuclide¦&NPA¦Dimension¦&FCI¦Dimension ¦\\ $^{52¦}$Fe¦&350¦&$1.1¦\times10¦^8¦$\\ $^{53¦}$Fe¦&6106¦&$2.2¦\times10¦^8¦$\\ $^ {54¦}$Fe¦&706¦&$3.5¦\times10¦^8¦$\\ $^{56¦ }$Fe¦&1276¦&$5.0¦\times10¦^8¦$\\

(c) Content Information

Fig. 1. Example of table image along with its structure and content information from the dataset. Tokens in content information are character-based, and the "¦" token acts as a delimiter to identify words out of a continuous stream of characters.

2   The Current State of the Research

In recent years, we witness a surge in the digital documents available online due to the high availability of communication facilities and large-scale internet infrastructure. In particular, the availability of scientific research papers that contain complex tabular structures has grown exponentially. However, we witness fewer research efforts to extract embedded tabular information automatically. Table 1 lists some of the popular datasets for TD and TIE tasks. Among the two tasks, there are very few datasets available for the TIE, specifically from scientific domains containing complex mathematical formulas and symbols. As the current paper primarily focuses on the TIE from scientific tables, we discuss some of the popular datasets and their limitations in this domain.

Scientific Tabular Datasets: Table 2Latex [3] dataset contains 450k scientific table images and its corresponding ground-truth in LATEX. It is curated from

Table 1. Datasets and methods used for Table Detection (TD), Table Structure Recognition (TSR) and Table Recognition (TR). * represents scientific paper datasets.

Datasets	TD	TSR	TR	Format	# Tables	Methods
Marmot [25]*	✓	×	×	PDF	958	Pdf2Table [21], TableSeer[15],[10]
PubLayNet [30]*	✓	×	×	PDF	113k	F-RCNN [19], M-RCNN [11]
DeepFigures [22]*	✓	×	×	PDF	1.4m	Deepfigures [22]
ICDAR2013 [9]	✓	✓	✓	PDF	156	Heuristics+ML
ICDAR2019 [8]	✓	✓	×	Images	3.6k	Heuristics+ML
UNLV [20]	✓	✓	×	Images	558	T-Recs [13]
TableBank [14]*	✓	✓	×	Images	417k (TD)	F-RCNN [19]
TableBank [14]*	✓	✓	×	Images	145k (TSR)	WYGIWYS [2]
SciTSR [1]*	×	✓	✓	PDF	15k	GraphTSR [1]
Table2Latex [3]*	×	✓	✓	Images	450k	IM2Tex [4]
Synthetic data [18]	×	✓	✓	Images	Unbounded	DCGNN [18]
PubTabNet [29]*	×	✓	✓	Images	568k	EDD [29]
TABLEX (ours)*	×	✓	✓	Images	1m+	TRT [7]

the $arXiv^5$ repository. To the best of our knowledge, Table2Latex is the only dataset that contains ground truth in TEX language but is not publicly available. TableBank [14] contains 145k table images along with its corresponding ground truth in the HTML representation. TableBank [14] contains table images from both Word and LATEX documents curated from the internet and $arXiv$ (see footnote 5), respectively. However, it does not contain content information to perform the TCR task. PubTabNet [29] contains over 568k table images and corresponding ground truth in the HTML language. PubTabNet is curated from the PubMed Central repository[6]. However, it is only limited to the biomedical and life sciences domain. SciTSR [1] contains 15k tables in the PDF format with table cell content and the coordinate information in the JSON format. SciTSR has been curated from the $arXiv$(see Footnote 5) repository. However, manual analysis tells that 62 examples out of 1000 randomly sampled examples are incorrect [28].

TIE Methodologies: The majority of the TIE methods employ encoder-decoder architecture [3,14,29]. Table2Latex [3] uses IM2Tex [4] model where encoder consists of convolutional neural network (CNN) followed by bidirectional LSTM and decoder consists of standard LSTM. TableBank [14] uses WYGIWS [2] model, an encoder-decoder architecture, where encoder consists of CNN followed by a recurrent neural network (RNN) and decoder consists of standard RNN. PubTabNet [29] uses a proposed encoder-dual-decoder (EDD) [29] model which consists of CNN encoder and two RNN decoders called structure and cell decoder, respectively. In contrast to the above works, SciTSR [1]

---

[5] http://arxiv.org/.
[6] https://pubmed.ncbi.nlm.nih.gov/.

proposed a graph neural network-based extraction methodology. The network takes vertex and edge features as input and computes their representations using graph attention blocks, and performs classification over these edges.

TIE Metrics: Table 2Latex [3] used BLEU [16] score (a text-based metric) and exact match accuracy (a vision-based metric) for evaluation. TableBank [14] also conducted BLEU [16] metric for evaluation. Tesseract OCR [24] uses the Word Error Rate (WER) metric for evaluation of the output. SciTSR [1] uses micro- and macro-averaged precision, recall, and F1-score to compare the output against the ground truth, respectively. In contrast to the above standard evaluation metrics in NLP literature, PubTabNet [29] proposed a new metric called Tree-Edit-Distance-based Similarity (TEDS) for evaluation of the output HTML representation.

The Challenges: Table 1 shows that there are only two datasets for Image-based TCR from scientific tables, that is, Table 2Latex [3] and PubTabNet [29]. We address some of the challenges from previous works with TABLEX which includes (i) large-size for training (ii) Font-agnostic learning (iii) Image-based scientific TCR (iv) domain independent

3   The TABLEX Dataset

This section presents the detailed curation pipeline to create the TABLEX dataset. Next, we discuss the data acquisition strategy.

3.1   Data Acquisition

We curate data from popular preprint repository *arXiv* (see Footnote 5). We downloaded paper (uploaded between Jan 2019–Sept 2020) source code and corresponding compiled PDF. These articles belong to eight subject categories. Figure 2 illustrates category-wise paper distribution. As illustrated, the majority of the papers belong to three subject categories, physics (33.93%), computer science (25.79%), and mathematics (23.27%). Overall, we downloaded 347,502 papers and processed them using the proposed data processing pipeline (described in the next section).

3.2   Data Processing Pipeline

The following steps present a detailed description of the data processing pipeline.

LATEX Code Pre-processing Steps

1. **Table Snippets Extraction:** A table snippet is a part of LATEX code that begins with '\begin{tabular}' and ends with '\end{tabular}' command. During its extraction, we removed citation command '\cite{}', reference command '\ref{}', label command '\label{}', and graphics command

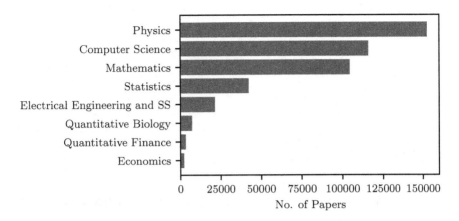

Fig. 2. Total number of papers in arXiv's subject categories. Here SS denotes Systems Science.

'\includegraphics[]{}', along with $\sim$ symbol (preceding these commands). Also, we remove command pairs (along with the content between them) like '\begin{figure}' and '\end{figure}', and '\begin{subfigure}' and '\end{subfigure}', as they cannot be predicted from the tabular images. Furthermore, we also remove the nested table environments. Figure 3a and 3b show an example table and its corresponding LATEX source code, respectively.

2. **Comments Removal:** Comments are removed by removing the characters between '%' token and newline token '\n '. This step was performed because comments do not contribute to visual information.

3. **Column Alignment and Rows Identification:** We keep all possible alignment tokens ('l' , 'r' , and 'c') specified for column styling. The token ' | ' is also kept to identify vertical lines between the columns, if present. The rows are identified by keeping the '\\' and '\tabularnewline' tokens. These tokens signify the end of each row in the table. For example, Fig. 3c shows extracted table snippet from LATEX source code (see Fig. 3b) containing the column and rows tokens with comment statements removed.

4. **Font Variation:** In this step, the extracted LATEX code is augmented with different font styles. We experiment with a total 12 different LATEX font packages[7]. We use popular font packages from PostScript family which includes 'courier', 'helvet', 'palatino', 'bookman', 'mathptmx', 'utopia' and also other font packages such as 'tgbonum', 'tgtermes', 'tgpagella', 'tgschola', 'charter' and 'tgcursor'. For each curated image, we create 12 variations, one in each of the font style.

5. **Image Rendering:** Each table variant is compiled into a PDF using a LATEX code template (described in Fig. 3d). Here, *'table snippet'* represents the extracted tabular LATEX code and *'font package'* represents the LATEX font

---

[7] https://www.overleaf.com/learn/latex/font_typefaces.

```
% Please add the following required packages to your document
\begin{table}[t!]
    \small
    \centering
    \begin{tabular}{lcc}
        \hline
        \multicolumn{1}{l}{\textbf{Technique}} &
        \multicolumn{1}{c}{\textbf{\approach{}}} &
        \multicolumn{1}{c}{\textbf{\approach{} (BERT)}}
        \\ \hline \hline
        Base model & 40.5\% & 53.9\% \\
        % \hspace{2mm}+EL & 47.1\% & 60.3\% \\
        \hspace{2mm}+SL & 48.5\% & 60.3\% \\
        \hspace{2mm}+SL + MEM & 51.3\% & 60.6\% \\
        \hspace{2mm}+SL + MEM + CF & 53.2\% & 61.9\% \\
        \hline
    \end{tabular}
    \caption{\label{tab:ablation_results} Ablation study
    results. Base model means that we does not perform schema
    linking (SL), memory augmented pointer network (MEM) and the
    coarse-to-fine framework (CF) on it.}
\end{table}
```

Technique		(BERT)
Base model	40.5%	53.9%
+SL	48.5%	60.3%
+SL + MEM	51.3%	60.6%
+SL + MEM + CF	53.2%	61.9%

(a) A real table image

(b) Corresponding LaTeX source code

```
\documentclass{standalone}
\usepackage{booktabs}
\usepackage{array}
\usepackage{multirow}
\usepackage{longtable}
\usepackage{graphicx}
\usepackage{amsmath}
\usepackage{amssymb}
\usepackage{amsbsy}
\usepackage{amsthm}
% font package

\begin{document}

% table snippet

\end{document}
```

```
\begin{tabular}{lcc}
    \hline
    \multicolumn{1}{l}{\textbf{Technique}} &
    \multicolumn{1}{c}{\textbf{\approach{}}} &
    \multicolumn{1}{c}{\textbf{\approach{} (BERT)}} \\
    \hline \hline
    Base model & 40.5\% & 53.9\% \\
    \hspace{2mm}+SL & 48.5\% & 60.3\% \\
    \hspace{2mm}+SL + MEM & 51.3\% & 60.6\% \\
    \hspace{2mm}+SL + MEM + CF & 53.2\% & 61.9\% \\
    \hline
\end{tabular}
```

(c) Extracted table snippet

(d) LaTeX code template

```
\documentclass{standalone}
\usepackage{booktabs}
\usepackage{array}
\usepackage{multirow}
\usepackage{longtable}
\usepackage{graphicx}
\usepackage{amsmath}
\usepackage{amssymb}
\usepackage{amsbsy}
\usepackage{amsthm}
\usepackage{charter}

\begin{document}
\begin{tabular}{lcc}
\hline
\multicolumn{1}{l}{\textbf{Technique}} &
\multicolumn{1}{c}{\textbf{\approach{}}} &
\multicolumn{1}{c}{\textbf{\approach{} (BERT)}} \\
\hline \hline
Base model & 40.5\% & 53.9\% \\
\hspace{2mm}+SL & 48.5\% & 60.3\% \\
\hspace{2mm}+SL + MEM & 51.3\% & 60.6\% \\
\hspace{2mm}+SL + MEM + CF & 53.2\% & 61.9\% \\
\hline
\end{tabular}
\end{document}
```

(e) Code with CHARTER font package

Technique		(BERT)
Base model	40.5%	53.9%
+SL	48.5%	60.3%
+SL + MEM	51.3%	60.6%
+SL + MEM + CF	53.2%	61.9%

(f) The final generated table image

Fig. 3. Data processing pipeline.

package used to generated the PDF. The corresponding table PDF files are then converted into JPG images using the Wand library[8] which uses

---

[8] https://github.com/emcconville/wand

ImageMagick [26] API to convert PDF into images. Figure 3e shows the embedded code and font information within the template. Figure 3f shows the final table image with CHARTER font. During conversion, we kept the image's resolution as 300 dpi, set the background color of the image to white, and removed the transparency of the alpha channel by replacing it with the background color of the image. We use two types of aspect ratio variations during the conversion, described as follows:

(a) *Conserved Aspect Ratio:* In this case, the bigger dimension (height or width) of the image is resized to 400 pixels[9]. We then resize the smaller dimension (width or height) by keeping the original aspect ratio conserved. During resizing, we use a blur factor of 0.8 to keep images sharp.

(b) *Fixed Aspect Ratio:* The images are resized to a fixed size of 400 × 400 pixels using a blur factor of 0.8. Note that this resizing scheme can lead to extreme levels of image distortions.

LaTeX Code Post-processing Steps for Ground-Truth Preparation

1. **Filtering Noisy LaTeX tokens:** We filter out LaTeX environment tokens (tokens starting with '\') having very less frequency (less than 5000) in the overall corpus compared to other LaTeX environment tokens and replaced it with '\LATEX_TOKEN' as these tokens will have a little contribution. This frequency-based filtering reduced the corpus's overall vocabulary size, which helps in training models on computation or memory-constrained environments.

2. **Structure Identification:** In this post-processing step, we create Table Structure Dataset (TSD). It consists of structural information corresponding to the table images. We keep tabular environment parameters as part of structure information and replace tokens representing table cells' content with a placeholder token 'CELL'. This post-processing step creates ground truth for table images in TSD. Specifically, the vocabulary comprises digits (0–9), '&', 'CELL', '\\', '\hline', '\multicolumn', '\multirow', '\bottomrule', '\midrule', '\toprule', '|', 'l', 'r', 'c', '{', and '}' . A sample of structure information is shown in Fig. 1b, where the 'CELL' token represents a cell structure in the table. Structural information can be used to identify the number of cells, rows, and columns in the table using 'CELL', '\\' and alignment tokens ('c', 'l', and 'r'), respectively. Based on output sequence length, we divided this dataset further into two variants TSD-250 and TSD-500, where the maximum length of the output sequence is 250 and 500 tokens, respectively.

3. **Content Identification:** Similar to TSD, we also create a Table Content Dataset (TCD). This dataset consists of content information including alphanumeric characters, mathematical symbols, and other LaTeX environment tokens. Content tokens are identified by removing tabular environment

---

[9] We use '400' pixels as an experimental number.

Table 2. TABLeX Statistics. ML denotes the maximum length of output sequence. Samples denotes the total number of table images. T/S denotes the average number of tokens per sample. VS denotes the vocabulary size. AR and AC denote the average number of rows and columns among all the samples in the four dataset variants, respectively.

Dataset	ML	Samples	Train/Val/Test	T/S	VS	AR	AC
TSD-250	250	2,938,392	2,350,713/293,839/293,840	74.09	25	6	4
TSD-500	500	3,191,891	2,553,512/319,189/319,190	95.39	25	8	5
TCD-250	250	1,105,636	884,508/110,564/110,564	131.96	718	4	4
TCD-500	500	1,937,686	1,550,148/193,769/193,769	229.06	737	5	4

parameters and keeping only the tokens that identify table content. Specifically, the vocabulary includes all the alphabets (a–z and A–Z), digits (0–9), LaTeX environment tokens ('\textbf', \hspace', etc.), brackets ('(', ')', '{', '}', etc.) and all other possible symbols ('$', '&', etc.). In this dataset, based on output sequence length, we divided it further into two variants TCD-250 and TCD-500, where the maximum length of the output sequence is 250 and 500 tokens, respectively. A sample of content information is shown in Fig. 1c.

3.3   Dataset Statistics

We further partition each of the four dataset variants TSD-250, TSD-500, TCD-25, and TCD-500, into training, validation, and test sets in a ratio of 80:10:10. Table 2 shows the summary of the dataset. Note that the number of samples present in the TSD and the corresponding TCD can differ due to variation in the output sequence's length. Also, the average number of tokens per sample in TCD is significantly higher than the TSD due to more information in the tables' content than the corresponding structure. Figure 4 demonstrates histograms representing the token distribution for the TSD and TCD. In the case of TSD, the majority of tables contain less than 25 tokens. Overall the token distribution shows a long-tail behavior. In the case of TCD, we witness a significant proportion between 100–250 tokens. The dataset is licensed under Creative Commons Attribution 4.0 International License and available for download at https://www.tinyurl.com/tablatex.

4   Evaluation Metrics

We experiment with three evaluation metrics to compare the predicted sequence of tokens (T_{PT}) against the sequence of tokens in the ground truth (T_{GT}). Even though the proposed metrics are heavily used in NLP research, few TIE works have leveraged them in the past. Next, we discuss these metrics and illustrate the implementation details using a toy example described in Fig. 5.

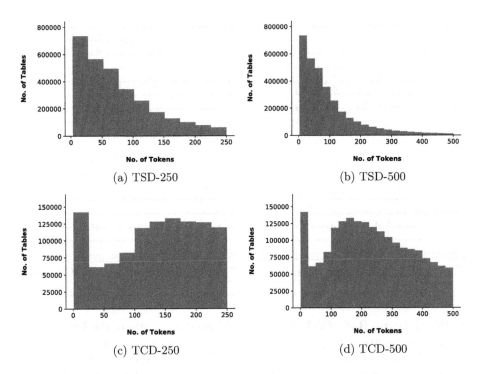

Fig. 4. Histograms representing the number of tokens distribution for the tables present in the dataset.

1. **Exact Match Accuracy (EMA):** EMA outputs the fraction of predictions with exact string match of T_{PT} against T_{GT}. A model having a higher value of EMA is considered a good model. In Fig. 5, for the TSR task, T_{PT} misses '\hline' token compared to the corresponding T_{GT}, resulting in EMA = 0. Similarly, the T_{PT} exactly matches the T_{GT} for the TCR task, resulting in EMA = 1.

2. **Bilingual Evaluation Understudy Score (BLEU):** The BLEU [16] score is primarily used in Machine Translation literature to compare the quality of the translated sentence against the expected translation. Recently, several TIE works [2,4] have adapted BLEU for evaluation. It counts the matching n-grams in T_{PT} against the n-grams in T_{GT}. The comparison is made regardless of word order. The higher the BLEU score, the better is the model. We use SacreBLEU [17] implementation for calculating the BLEU score. We report scores for the most popular variant, BLEU-4. BLEU-4 refers to the product of brevity penalty (BP) and a harmonic mean of precisions for unigrams, bigrams, 3-grams, and 4-grams (for more details see [16]). Figure 5, there is an exact match between T_{PT} and T_{GT} for TCR task yielding BLEU = 100.00. In the case of TSR, the missing '\hline' token in T_{PT} yields a lower BLEU = 89.66.

Table Content:
a ¦ & b ¦ & c ¦ & d ¦ \\10 ¦ & 20 ¦ & 30
¦ & 40 ¦ \\20 ¦ & 30 ¦ & 40 ¦ & 60 ¦

a	b	c	d
10	20	30	40
20	30	40	60

(a) A Toy Table

Table Structure:
{ | c | c | c | c | } CELL & CELL & CELL &
CELL \\ \hline CELL & CELL & CELL & CELL
\\ CELL & CELL & CELL & CELL

(b) Ground Truth

a ¦ & b ¦ & c ¦ & d ¦ \\10 ¦ & 20 ¦
& 30 ¦ & 40 ¦ \\20 ¦ & 30 ¦ & 40
¦ & 60 ¦

(c) Table Content Output

{ | c | c | c | c | } CELL & CELL & CELL
& CELL \\ CELL & CELL & CELL & CELL
\\ CELL & CELL & CELL & CELL

(d) Table Structure Output

Fig. 5. (a) A toy table, (b) corresponding ground truth token sequence for TSR and TCR tasks, (c) model output sequence for TCR task, and (d) model output sequence for TSR task.

3. **Word Error Rate (WER):** WER is defined as a ratio of the Levenshtein distance between the T_{PT} and T_{GT} to the length of T_{GT}. It is a standard evaluation metric for several OCR tasks. Since it measures the rate of error, models with lower WER are better. We use jiwer[10] Python library for WER computation. In Fig. 5, for the TCR task, the WER between T_{PT} and T_{GT} is 0. Whereas in the case of TSR, the Levenshtein distance between T_{PT} and T_{GT} is one due to the missing '\hline' token. Since the length of T_{GT} is 35, WER comes out to be 0.02.

5 Experiments

In this section, we experiment with a deep learning-based model for TSR and TCR tasks. We adapt an existing model [7] architecture proposed for the scene text recognition task and train it from scratch on the TABLEX dataset. It uses partial ResNet-101 along with a fully connected layer as a feature extractor module for generating feature embeddings with a cascaded Transformer [27] module for encoding the features and generating the output. Figure 6 describes the detailed architecture of the model. Note that in contrast to the scene image as an input in [7], we input a tabular image and predict the LaTeX token sequence. We term it as the TIE-ResNet-Transformer model (*TRT*).

5.1 Implementation Details

For both TSR and TCR tasks, we train a similar model. We use the third intermediate hidden layer of partial ResNet-101 and an FC layer as a feature

---

[10] https://github.com/jitsi/jiwer.

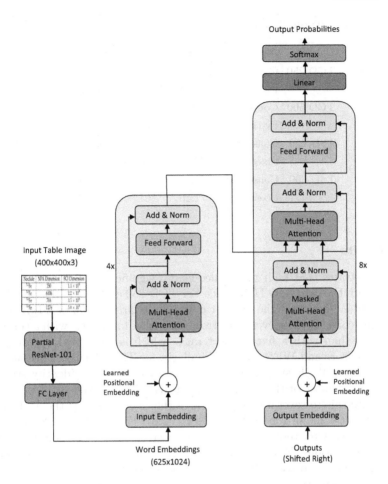

Fig. 6. TIE-ResNet-Transformer model architecture.

extractor module to generate a feature map. The generated feature map (size = 625×1024) is further embedded into 256 dimensions, resulting in reduced size of 625×256. We experiment with four encoders and eight decoders with learnable positional embeddings. The models are trained for ten epochs with a batch size of 32 using cross-entropy loss function and Adam Optimizer with an initial learning rate of 0.1 and 2000 warmup training steps, decreasing with Noam's learning rate decay scheme. A dropout rate of 0.1 and $\epsilon_{ls} = 0.1$ label smoothing parameter is used for regularization. For decoding, we use the greedy decoding technique in which the model generates an output sequence of tokens in an auto-regressive manner, consuming the previously generated tokens to generate the next token. All the experiments were performed on $4 \times$ NVIDIA V100 graphics card.

Table 3. TRT results on the TABLEX dataset.

Dataset	Aspect ratio	EMA(%)	BLEU	WER(%)
TCD 250	Conserved	21.19	95.18	15.56
	Fixed	20.46	96.75	14.05
TCD 500	Conserved	11.01	91.13	13.78
	Fixed	11.23	94.34	11.12
TSD 250	Conserved	70.54	74.75	3.81
	Fixed	74.02	70.59	4.98
TSD 500	Conserved	70.91	82.72	2.78
	Fixed	71.16	61.84	9.34

5.2   Results

Table 3 illustrates that higher sequence length significantly degrades the EMA score for both the TSR and TCR tasks. For TCD-250 and TCD-500, high BLEU scores (>90) suggest that the model can predict a large chunk of LATEX content information correctly. However, the errors are higher for images with a conserved aspect ratio than with a fixed aspect ratio, which is confirmed by comparing their BLEU score and WER metrics. Similarly, for TSD-250 and TSD-500, high EMA and lower WER suggest that the model can correctly identify structure information for most of the tables. TCD yields a lower EMA score than TSD. We attribute this to several reasons. One of the reasons is that the TCR model fails to predict some of the curly braces ('{' and '}') and dollar ('$') tokens in the predictions. After removing curly braces and dollar tokens from the ground truth and predictions, EMA scores for conserved and fixed aspect ratio images in TCD-250 increased to 68.78% and 75.33%, respectively. Similarly, for TCD-500, EMA for conserved and fixed aspect ratio images increases to 49.23% and 59.94%, respectively. In contrast, TSD do not contain dollar tokens, and curly braces are only present at the beginning of the column labels, leading to higher EMA scores. In the future, we believe the results can be improved significantly by proposing better vision-based DL architectures for TSR and TCR tasks.

6   Conclusion

This paper presents a benchmark dataset TABLEX for structure and content information extraction from scientific tables. It contains tabular images in 12 different fonts with varied visual complexities. We also proposed a novel prepro-cessing pipeline for dataset generation. We evaluate the existing state-of-the-art transformer-based model and show excellent future opportunities in developing algorithms for IE from scientific tables. In the future, we plan to continuously augment the dataset size and add more complex tabular structures for training and prediction.

Acknowledgment. This work was supported by The Science and Engineering Research Board (SERB), under sanction number ECR/2018/000087.

References

1. Chi, Z., Huang, H., Xu, H., Yu, H., Yin, W., Mao, X.: Complicated table structure recognition. CoRR abs/1908.04729 (2019). http://arxiv.org/abs/1908.04729
2. Deng, Y., Kanervisto, A., Rush, A.M.: What you get is what you see: a visual markup decompiler. ArXiv abs/1609.04938 (2016)
3. Deng, Y., Rosenberg, D., Mann, G.: Challenges in end-to-end neural scientific table recognition. In: 2019 International Conference on Document Analysis and Recognition (ICDAR), pp. 894–901 (2019). https://doi.org/10.1109/ICDAR.2019.00148
4. Deng, Y., Kanervisto, A., Ling, J., Rush, A.M.: Image-to-markup generation with coarse-to-fine attention. In: Proceedings of the 34th International Conference on Machine Learning, ICML 2017, vol. 70, pp. 980–989. JMLR.org (2017)
5. Douglas, S., Hurst, M., Quinn, D., et al.: Using natural language processing for identifying and interpreting tables in plain text. In: Proceedings of the Fourth Annual Symposium on Document Analysis and Information Retrieval, pp. 535–546 (1995)
6. Embley, D.W., Hurst, M., Lopresti, D.P., Nagy, G.: Table-processing paradigms: a research survey. Int. J. Doc. Anal. Recognit. **8**(2–3), 66–86 (2006)
7. Feng, X., Yao, H., Yi, Y., Zhang, J., Zhang, S.: Scene text recognition via transformer. arXiv preprint arXiv:2003.08077 (2020)
8. Gao, L., et al.: ICDAR 2019 competition on table detection and recognition (CTDAR). In: 2019 International Conference on Document Analysis and Recognition (ICDAR), pp. 1510–1515 (2019). https://doi.org/10.1109/ICDAR.2019.00243
9. Gbel, M., Hassan, T., Oro, E., Orsi, G.: ICDAR 2013 table competition. In: 2013 12th International Conference on Document Analysis and Recognition, pp. 1449–1453 (2013). https://doi.org/10.1109/ICDAR.2013.292
10. Hao, L., Gao, L., Yi, X., Tang, Z.: A table detection method for pdf documents based on convolutional neural networks. In: DAS, pp. 287–292 (2016)
11. He, K., Gkioxari, G., Dollár, P., Girshick, R.B.: Mask R-CNN. CoRR abs/1703.06870 (2017). http://arxiv.org/abs/1703.06870
12. Kasar, T., Bhowmik, T.K., Belad, A.: Table information extraction and structure recognition using query patterns. In: 2015 13th International Conference on Document Analysis and Recognition (ICDAR), pp. 1086–1090 (2015). https://doi.org/10.1109/ICDAR.2015.7333928
13. Kieninger, T., Dengel, A.: A paper-to-html table converting system. Proc. Doc. Anal. Syst. (DAS) **98**, 356–365 (1998)
14. Li, M., Cui, L., Huang, S., Wei, F., Zhou, M., Li, Z.: TableBank: table benchmark for image-based table detection and recognition. CoRR abs/1903.01949 (2019). http://arxiv.org/abs/1903.01949
15. Liu, Y., Bai, K., Mitra, P., Giles, C.L.: Tableseer: Automatic table metadata extraction and searching in digital libraries. In: Proceedings of the 7th ACM/IEEE-CS Joint Conference on Digital Libraries, JCDL 2007, New York, NY, USA, pp. 91–100. Association for Computing Machinery (2007). https://doi.org/10.1145/1255175.1255193

16. Papineni, K., Roukos, S., Ward, T., Zhu, W.J.: Bleu: a method for automatic evaluation of machine translation. In: Proceedings of the 40th Annual Meeting of the Association for Computational Linguistics, Philadelphia, Pennsylvania, USA, pp. 311–318. Association for Computational Linguistics, July 2002. https://doi.org/10.3115/1073083.1073135. https://www.aclweb.org/anthology/P02-1040

17. Post, M.: A call for clarity in reporting BLEU scores. In: Proceedings of the Third Conference on Machine Translation: Research Papers, Belgium, Brussels, pp. 186–191. Association for Computational Linguistics October 2018. https://www.aclweb.org/anthology/W18-6319

18. Qasim, S.R., Mahmood, H., Shafait, F.: Rethinking table parsing using graph neural networks. CoRR abs/1905.13391 (2019). http://arxiv.org/abs/1905.13391

19. Ren, S., He, K., Girshick, R., Sun, J.: Faster R-CNN: towards real-time object detection with region proposal networks. In: Cortes, C., Lawrence, N., Lee, D., Sugiyama, M., Garnett, R. (eds.) Advances in Neural Information Processing Systems, vol. 28, pp. 91–99. Curran Associates, Inc. (2015). https://proceedings.neurips.cc/paper/2015/file/14bfa6bb14875e45bba028a21ed38046-Paper.pdf

20. Shahab, A., Shafait, F., Kieninger, T., Dengel, A.: An open approach towards the benchmarking of table structure recognition systems. In: Proceedings of the 9th IAPR International Workshop on Document Analysis Systems, DAS 2010, New York, NY, USA pp. 113–120. Association for Computing Machinery (2010). https://doi.org/10.1145/1815330.1815345

21. Shigarov, A., Mikhailov, A., Altaev, A.: Configurable table structure recognition in untagged pdf documents. In: Proceedings of the 2016 ACM Symposium on Document Engineering, DocEng 2016, New York, NY, USA, pp. 119–122. Association for Computing Machinery (2016). https://doi.org/10.1145/2960811.2967152

22. Siegel, N., Lourie, N., Power, R., Ammar, W.: Extracting scientific figures with distantly supervised neural networks. In: Proceedings of the 18th ACM/IEEE on Joint Conference on Digital Libraries, JCDL 2018, New York, NY, USA, pp. 223–232. Association for Computing Machinery (2018). https://doi.org/10.1145/3197026.3197040

23. Singh, M., Sarkar, R., Vyas, A., Goyal, P., Mukherjee, A., Chakrabarti, S.: Automated early leaderboard generation from comparative tables. In: Azzopardi, L., Stein, B., Fuhr, N., Mayr, P., Hauff, C., Hiemstra, D. (eds.) ECIR 2019. LNCS, vol. 11437, pp. 244–257. Springer, Cham (2019). https://doi.org/10.1007/978-3-030-15712-8_16

24. Smith, R.: An overview of the tesseract OCR engine. In: Ninth International Conference on Document Analysis and Recognition (ICDAR 2007), vol. 2, pp. 629–633. IEEE (2007)

25. Tao, X., Liu, Y., Fang, J., Qiu, R., Tang, Z.: Dataset, ground-truth and performance metrics for table detection evaluation. In: IAPR International Workshop on Document Analysis Systems, Los Alamitos, CA, USA, pp. 445–449. IEEE Computer Society, March 2012. https://doi.org/10.1109/DAS.2012.29

26. The ImageMagick Development Team: Imagemagick. https://imagemagick.org

27. Vaswani, A., et al.: Attention is all you need. In: Advances in Neural Information Processing Systems, pp. 5998–6008 (2017)

28. Wu, G., Zhou, J., Xiong, Y., Zhou, C., Li, C.: TableRobot: an automatic annotation method for heterogeneous tables. Personal Ubiquit. Comput. 1–7 (2021). https://doi.org/10.1007/s00779-020-01485-1

29. Zhong, X., ShafieiBavani, E., Jimeno Yepes, A.: Image-based table recognition: data, model, and evaluation. In: Vedaldi, A., Bischof, H., Brox, T., Frahm, J.-M. (eds.) ECCV 2020. LNCS, vol. 12366, pp. 564–580. Springer, Cham (2020). https://doi.org/10.1007/978-3-030-58589-1_34

30. Zhong, X., Tang, J., Jimeno-Yepes, A.: PublayNet: largest dataset ever for document layout analysis. CoRR abs/1908.07836 (2019). http://arxiv.org/abs/1908.07836

31. Zhou, J., Cui, G., Zhang, Z., Yang, C., Liu, Z., Sun, M.: Graph neural networks: a review of methods and applications. CoRR abs/1812.08434 (2018). http://arxiv.org/abs/1812.08434

Handwritten Mathematical Expression Recognition with Bidirectionally Trained Transformer

Wenqi Zhao[1] , Liangcai Gao[1(✉)], Zuoyu Yan[1], Shuai Peng[1], Lin Du[2],
and Ziyin Zhang[2]

[1] Wangxuan Institute of Computer Technology, Peking University, Beijing, China
{gaoliangcai,yanzuoyu,pengshuaipku}@pku.edu.cn
[2] Huawei AI Application Research Center, Beijing, China
{dulin09,zhangziyin1}@huawei.com

Abstract. Encoder-decoder models have made great progress on handwritten mathematical expression recognition recently. However, it is still a challenge for existing methods to assign attention to image features accurately. Moreover, those encoder-decoder models usually adopt RNN-based models in their decoder part, which makes them inefficient in processing long LaTeX sequences. In this paper, a transformer-based decoder is employed to replace RNN-based ones, which makes the whole model architecture very concise. Furthermore, a novel training strategy is introduced to fully exploit the potential of the transformer in bidirectional language modeling. Compared to several methods that do not use data augmentation, experiments demonstrate that our model improves the ExpRate of current state-of-the-art methods on CROHME 2014 by 2.23%. Similarly, on CROHME 2016 and CROHME 2019, we improve the ExpRate by 1.92% and 2.28% respectively.

Keywords: Handwritten mathematical expression recognition ·
Transformer · Bidirection · Encoder-decoder model

1 Introduction

The encoder-decoder models have shown quite effective performance on various tasks such as scene text recognition [6] and image captioning [25]. Handwritten Mathematical Expression Recognition (HMER) aims to generate the math expression LaTeX sequence according to the handwritten math expression image. Since HMER is also an image to text modeling task, many encoder-decoder models [34,36] have been proposed for it in recent years.

However, existing methods suffer from the lack of coverage problem [36] to varying degrees. This problem refers to two possible manifestations: over-parsing and under-parsing. Over-parsing means that some regions of HME image are

J. Lladós et al. (Eds.): ICDAR 2021, LNCS 12822, pp. 570–584, 2021.
https://doi.org/10.1007/978-3-030-86331-9_37

redundantly translated multiple times, while under-parsing denotes that some regions remain untranslated.

Most encoder-decoder models are RNN-based models, which have difficulty modeling the relationship between two symbols that are far apart. Previous study [2] had noted this long-term dependency problem caused by gradient vanishing. This problem is exposed more obviously in HMER task. Compared to traditional natural language processing, LaTeX is a markup language designed by human, and thus has a clearer and more distinct syntactic structure, e.g., "{" and "}" are bound to appear in pairs. When dealing with long LaTeX sequences, it is difficult for RNN-based models to capture the relationship between two distant "{" and "}" symbol, resulting in lack of awareness of the LaTeX syntax specification.

Traditional autoregressive models [34,36] use left-to-right (L2R) direction to predict symbols one by one in the inference phase. Such approaches may generate unbalanced outputs [15], which prefixes are usually more accurate than suffixes. To overcome this problem, existing study [15] employ two independent decoders, trained for left-to-right and right-to-left directions, respectively. This usually leads to more parameters and longer training time. Therefore, an intuitive attempt is to adapt a single decoder for bi-directional language modeling.

In this paper, we employ the transformer [22] decoder into HMER task, alleviating the lack of coverage problem [36] by using positional encodings. Besides, a novel bidirectional training strategy is proposed to obtain a **B**idirectionally **T**rained **TR**ansformer (BTTR) model. The strategy enables single transformer decoder to perform both L2R and R2L decoding. We further show that our BTTR model outperforms RNN-based ones in terms of both training parallelization and inferencing accuracy. The main contributions of our work are summarized as follows:

- To the best of our knowledge, it is the first attempt to use end-to-end trained transformer decoder for solving HMER task.
- The combination of image and word positional encodings enable each time step to accurately assign attention to different regions of input image, alleviating the lack of coverage problem.
- A novel bidirectional training strategy is proposed to perform bidirectional language modeling in a single transformer decoder.
- Compared to several methods that do not use data augmentation, experiments demonstrate that our method obtains new SOTA performance on various dataset, including an ExpRate of 57.91%, 54.49%, and 56.88% on the CROHME 2014 [18], CROHME 2016 [19], and CROHME 2019 [17] test sets, respectively.
- We make our code available on the GitHub.[1]

---

[1] https://github.com/Green-Wood/BTTR.

2   Related Work

2.1   HMER Methods

In the last decades, many approaches [4, 13, 27–29, 32, 37] related to HMER have been proposed. These approaches can be divided into two categories: grammar-based and encoder-decoder based. In this section, we will briefly review the related work in both categories.

Grammar Based. These methods usually consist of three parts: symbol segmentation, symbol recognition, and structural analysis. Researchers have proposed a variety of predefined grammars to solve HMER task, such as stochastic context-free grammars [1], relational grammars [16], and definite clause grammars [3,5]. None of these grammar rules are data-driven, but are hand-designed, which could not benefit from large dataset.

Encoder-Decoder Based. In recent years, a series of encoder-decoder models have been widely used in various tasks [26,30,31]. In HMER tasks, Zhang et al. [36] observed the lack of coverage problem and proposed WAP model to solve the HMER task. In the subsequent studies, DenseWAP [34] replaced VGG encoder in the WAP with DenseNet [11] encoder, and improved the performance. Further, DenseWAP-TD [35] enhanced the model's ability to handle complex formulas by substituting string decoder with a tree decoder. Wu et al. [24] used stroke information and formulated the HMER as a graph-to-graph (G2G) modeling task. Such encoder-decoder based models have achieved outstanding results in several CROHME competitions [17–19].

2.2   Transformer

Transformer [22] is a neural network architecture based solely on attention mechanisms. Its internal self-attention mechanism makes transformer a breakthrough compared to RNN in two aspects. Firstly, transformer does not need to depend on the state of the previous step as RNN does. Well-designed parallelization allows transformer to save a lot of time in the training phase. Secondly, tokens in the same sequence establish direct one-to-one connections through the self-attention mechanism. Such a mechanism fundamentally solves the gradient vanishing problem of RNN [2], making transformer more suitable than RNN on long sequences. In recent years, RNN is replaced by transformer in various tasks in computer vision and natural language processing [8,20].

Recently, transformer has been used in offline handwritten text recognition. Kang et al. [14] first adopted transformer networks for the handwritten text recognition task and achieved state-of-the-art performance. For the task of mathematical expression, "Univ. Linz" method in CROHME 2019 [17] used a Faster R-CNN detector and a transformer decoder. Two subsystems were trained separately.

2.3 Right-to-Left Language Modeling

To solve the problem that traditional autoregressive models can only perform left-to-right (L2R) language modeling, many studies have attempted right-to-left (R2L) language modeling. Liu et al. [15] trained a R2L model separately. During the inference phase, hypotheses from L2R and R2L models are re-ranked to produce the best candidate. Furthermore, Zhang et al. [38] used R2L model to regularize L2R model in the training phase to obtain a better L2R model. Zhou et al. [39] proposed SB-NMT that utilized a single decoder to generate sequences bidirectionally. However, all of these methods increase model complexity. On the other hand, our approach achieves bidirectional language modeling on a single decoder while keeping the model concise.

3 Methodology

In this section, we will detailed introduce the proposed BTTR model architecture, as illustrated in Fig. 1. In Sect. 3.1, we will briefly describe the DenseNet [11] model used in the encoder part. In Sect. 3.2 and 3.3, the positional encodings and transformer model used in the encoder and decoder part will be described in detail. Finally in Sect. 3.4, we will introduce the proposed novel bidirectional training strategy, which allows to perform bidirectional language modeling in a single transformer decoder.

3.1 CNN Encoder

In the encoder part, DenseNet is used as the feature extractor for HME images. The main idea of DenseNet is to increase information flow between layers by introducing direct connections between each layer and all its subsequent layers. In this way, given output features $\mathbf{x}_0, \mathbf{x}_1, \ldots, \mathbf{x}_{l-1}$ from 0th to $(l-1)^{th}$ layer, the output feature of lth layer can be computed by:

$$\mathbf{x}_\ell = H_\ell\left([\mathbf{x}_0; \mathbf{x}_1; \ldots; \mathbf{x}_{\ell-1}]\right) \tag{1}$$

where $[\mathbf{x}_0; \mathbf{x}_1; \ldots; \mathbf{x}_{\ell-1}]$ denotes the concatenation operation of all the output features, and $H_\ell(\cdot)$ denotes a composite function of three consecutive layers: a batch normalization (BN) [12] layer, followed by a ReLU [9] layer and a 3×3 convolution (Conv) layer.

Through concatenation operation in the channel dimension, DenseNet enables better propagation of gradient. The paper [11] states that by this dense connection, DenseNet can achieve better performance with fewer parameters compared to ResNet [10].

In addition to DenseNet, we also add a 1×1 convolution layer in the encoder part to adjust the image feature dimension to the size of embedding dimension d_{model} for subsequent processing.

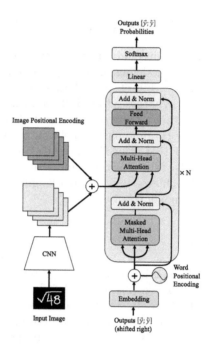

Fig. 1. The architecture of BTTR model. L2R and R2L sequences $[\overrightarrow{y}; \overleftarrow{y}]$ are concatenated through the batch dimension as the input to decoder part.

3.2 Positional Encoding

The positional information of image features and word vectors can effectively help the model to identify regions that need to attend. In the previous studies [34, 36], although the RNN-based model inherently takes the order of word vectors into account, it neglects the positional information of image features.

In this paper, since the transformer model itself doesn't have any sense of position for each input vector, we use two types of positional encodings to address this information. In detail, we use image positional encodings and word positional encodings to represent the image feature position and word vector position, respectively.

We refer to image features and word vector features as *content-based*, and the two types of positional encodings as *position-based*. As illustrated in Fig. 1, *content-based* and *position-based* features are summed up, as the input to transformer decoder.

Word Positional Encoding. Word positional encoding is basically the same as sinusoidal positional encoding proposed in the original transformer work [22]. Given position *pos* and dimension d as input, the word positional encoding vector $\mathbf{p}_{pos,d}^{W}$ is defined as:

$$\mathbf{p}_{pos,d}^{W}[2i] = \sin(pos/10000^{2i/d}) \tag{2}$$

$$\mathbf{p}_{pos,d}^{W}[2i + 1] = \cos(pos/10000^{2i/d}) \tag{3}$$

where i is the index in dimension.

Image Positional Encoding. A 2-D normalized positional encoding is used to represent the image position features. We first compute sinusoidal positional encoding $\mathbf{p}_{pos,d/2}^{W}$ in each of the two dimensions and then concatenate them together. Given a 2-D position tuple (x, y) and the same dimension d as the word positional encoding, the image positional encoding vector $\mathbf{p}_{x,y,d}^{I}$ is represented as:

$$\bar{x} = \frac{x}{H}, \quad \bar{y} = \frac{y}{W} \tag{4}$$

$$\mathbf{p}_{x,y,d}^{I} = [\mathbf{p}_{\bar{x},d/2}^{W}; \mathbf{p}_{\bar{y},d/2}^{W}] \tag{5}$$

where H and W are height and width of input images.

3.3 Transformer Decoder

For the decoder part, we use the standard transformer model [22]. Each basic transformer decoder layer module consists of four essential parts. In the following, we will describe the implementation details of these components.

Scaled Dot-Product Attention. This attention mechanism is essentially using the query to obtain the value from key-value pairs, based on the similarity between the query and key. The output matrix can be computed in parallel by the query $\mathbf{Q}$, key $\mathbf{K}$, value $\mathbf{V}$, and dimension d_k.

$$\text{Attention}(\mathbf{Q}, \mathbf{K}, \mathbf{V}) = \text{softmax}(\frac{\mathbf{Q}\mathbf{K}^{T}}{\sqrt{d_k}})\mathbf{V} \tag{6}$$

Multi-Head Attention. With multi-head mechanism, the scaled dot-product attention module can attend to feature-map from multiple representation subspaces jointly. With projection parameter matrices $\mathbf{W}_i^{Q} \in \mathbb{R}^{d_{\text{model}} \times d_k}, \mathbf{W}_i^{K} \in \mathbb{R}^{d_{\text{model}} \times d_k}, \mathbf{W}_i^{V} \in \mathbb{R}^{d_{\text{model}} \times d_v}$, we first project the query $\mathbf{Q}$, key $\mathbf{K}$, and value $\mathbf{V}$ into a subspace to compute the head $\mathbf{H}_i$.

$$\mathbf{H}_i = \text{Attention}\left(\mathbf{Q}\mathbf{W}_i^{Q}, \mathbf{K}\mathbf{W}_i^{K}, \mathbf{V}\mathbf{W}_i^{V}\right) \tag{7}$$

Then all the heads are concatenated and projected with a parameter matrix $\mathbf{W}^{O} \in \mathbb{R}^{hd_v \times d_{\text{model}}}$ and the number of heads h.

$$\text{MultiHead}(\mathbf{Q}, \mathbf{K}, \mathbf{V}) = [\mathbf{H}_1; \dots; \mathbf{H}_h]\mathbf{W}^{O} \tag{8}$$

Masked Multi-Head Attention. In the decoder part, due to the autoregressive property, the next symbol is predicted based on the input image and previously generated symbols. In the training phase, a lower triangle mask matrix is used to enable the self-attention module to restrict the attention region for each time step. Due to masked multi-head attention mechanism, the whole training process requires only one forward computation.

Position-Wise Feed-Forward Network. Feed-Forward Network (FNN) consists of three operations: a linear transformation, a ReLU activation function and another linear transformation.

$$\text{FFN}(\mathbf{x}) = \max\left(0, \mathbf{x}\mathbf{W}_1 + \mathbf{b}_1\right)\mathbf{W}_2 + \mathbf{b}_2 \tag{9}$$

After multi-head attention, the information between positions has been fully exchanged. FFN enables each position to integrate its own internal information separately.

3.4 Bidirectional Training Strategy

First, two special symbols "$\langle SOS \rangle$" and "$\langle EOS \rangle$" are introduced in the dictionary to denote the start and end of the sequence. For the target LATEX sequence $y = \{y_1, \ldots, y_T\}$, we denote the target sequence from left to right (L2R) as $\overrightarrow{y} = \{\langle SOS \rangle, y_1, \ldots, y_T, \langle EOS \rangle\}$, and the right-to-left (R2L) target sequence as $\overleftarrow{y} = \{\langle EOS \rangle, y_T, \ldots, y_1, \langle SOS \rangle\}$.

Conditioned on image x and model parameter θ, the traditional autoregressive model need to compute the probability distribution:

$$p\left(\overrightarrow{y}_j \mid \overrightarrow{y}_{<j}, x, \theta\right) \tag{10}$$

where j is the index in target sequence.

In this paper, since the transformer model itself does not actually care about the order of input symbols, we can use a single transformer decoder for bidirectional language modeling. Modeling both Eq. (10) and Eq. (11) at the same time.

$$p\left(\overleftarrow{y}_j \mid \overleftarrow{y}_{<j}, x, \theta\right) \tag{11}$$

To achieve this goal, a simple yet effective bidirectional training strategy is proposed, in which for each training sample, we generate two target sequences, L2R and R2L, from the target LATEX sequence, and compute the training loss in the same batch (details in Sect. 4.2). Compared with unidirectional language modeling, our approach trains a model to perform bidirectional language modeling without sacrificing model conciseness. Experiment results in Sect. 5.3 verify the effectiveness of our bidirectional training strategy.

4   Implementation Details

4.1   Networks

In the encoder part, to make a fair comparison with the previous state-of-the-art method, we use the same DenseNet feature extractor as the DenseWAP model [34]. Specifically, three bottleneck layers are used in the backbone network and transition layers are added in between to reduce the number of feature-maps. In each bottleneck layer, we set the growth rate to $k = 24$, the depth of each block to $D = 16$, and the compression hyperpatameter of the transition layer to $\theta = 0.5$.

In the decoder part, we use the standard transformer model. We set the embedded dimension and model dimension to $d_{\mathrm{model}} = 256$, the number of heads in the multi-head attention module to $H = 8$, the dimension of intermediate layers in the FFN to $d_{ff} = 1024$, and the number of transformer decoder layer to $N = 3$. The 0.3 dropout rate is used to prevent overfitting.

4.2   Training

Our training objective is to maximize the predicted probability of the ground truth symbols in Eq. (10) and Eq. (11), so we use the standard cross-entropy loss function to calculate the loss between the predicted probabilities w.r.t. the ground truth at each decoding position. Given the training sample $\left\{x^{(z)}, y^{(z)}\right\}_{z=1}^{Z}$, the objective function for optimization is shown as follows:

$$\overrightarrow{\mathcal{L}}_{j}^{(z)}(\theta) = -\log p(\overrightarrow{y}_{j}^{(z)} \mid \overrightarrow{y}_{<j}^{(z)}, x^{(z)}, \theta) \tag{12}$$

$$\overleftarrow{\mathcal{L}}_{j}^{(z)}(\theta) = -\log p(\overleftarrow{y}_{j}^{(z)} \mid \overleftarrow{y}_{<j}^{(z)}, x^{(z)}, \theta) \tag{13}$$

$$\mathcal{L}(\theta) = \frac{1}{2ZL} \sum_{z=1}^{Z} \sum_{j=1}^{L} \left(\overrightarrow{\mathcal{L}}_{j}^{(z)}(\theta) + \overleftarrow{\mathcal{L}}_{j}^{(z)}(\theta) \right) \tag{14}$$

The model is trained from scratch using the Adadelta algorithm [33] with a weight decay of 10^{-4}, $\rho = 0.9$, and $\epsilon = 10^{-6}$. PyTorch framework is used to implement our model. The model is trained on four NVIDIA 1080Ti GPUs with 11×4 GB memory.

4.3   Inferencing

In the inference phase, we aim to generate the most likely LATEX sequence conditioned on the input image. Which can be fomulated as follows:

$$\hat{\mathbf{y}} = \underset{\mathbf{y}}{\mathbf{argmax}} \; p\left(\mathbf{y} \mid \mathbf{x}, \theta\right) \tag{15}$$

where $\mathbf{x}$ denotes the input image and θ denotes the model parameter.

Unlike the training phase where a lower triangular mask matrix is used to generate the prediction for all time steps simultaneously. Since we have no ground truth of the previously predicted symbol, we can only predict symbols one by one until the "End" symbol appears or the predefined maximum length is reached.

Obviously, we cannot search for all possible sequences, thus a heuristic beam search is proposed to balance the computational cost with the quality of decoding. Further, taking advantage of the fact that our decoder is capable of bidirectional language modeling, approximate joint search [15] is used to improve the performance. The basic idea consists of three steps: (1) Firstly, a beam search is performed on L2R and R2L directions to obtain two k-best hypotheses. (2) Then, we reverse L2R hypotheses to R2L direction and R2L hypotheses to L2R direction and treat these hypotheses as ground truth to compute the loss values for each of them as in the training phase. (3) Finally, those loss values are added to their original hypothesis scores to obtain the final scores, which is then used to find the best candidate. In practice, we set beam size $k = 10$, the maximum length to 200, and length penalty $\alpha = 1.0$.

5   Experiments

5.1   Datasets

We use the Competition on Recognition of Online Handwritten Mathematical Expressions (CROHME) benchmark, which currently is the largest dataset for handwritten mathematical expression to validate the proposed model. We use the same training dataset but different test datasets. The training set contains 8836 handwritten mathematical expressions, while the test sets of CROHME 2014/2016/2019 contain 986/1147/1199 expressions respectively.

In the CROHME dataset, each handwritten math expression is saved in a InkML file, which contains handwritten stroke trajectory information and ground truth in both MathML and LaTeX formats. We transform the handwritten stroke trajectory information in the InkML files to offline images in bitmap format for training and testing. With official evaluation tools provided by the CROHME 2019 [17] organizers and the ground truth in symLG format, we convert the predicted LaTeX sequences to symLG format and evaluate the performance.

5.2   Compare with State-of-the-Art Results

The results of some models on the CROHME 2014/2016/2019 datasets are shown in Table 1. To ensure the fairness of performance comparison, the methods we show all use only the officially provided 8836 training samples. Neither we nor the methods we compare use data augmentation.

We first provide results of three traditional handwritten mathematical formula recognition methods based on tree grammar in CROHME 2014 as the baseline, denoted as I, VI, and VII. For CROHME 2016, we provide the best performing "TOKYO" method as the baseline, which only used official training samples. For CROHME 2019 official methods, we provide "Univ. Linz" [17] method as the baseline. For Image-to-LaTeX methods, we use the previous state-of-the-art "WYGIWYS" [7], "PAL-v2" [23], "WAP" [36], "Weakly supervised WAP" (WS WAP) [21], "DenseWAP" [34] as well as the tree decoder-based "DenseWAP-TD" [35] method. The "Ours-Uni" and "Ours-Bi" methods denote the model trained using the vanilla transformer decoder and the model trained with the bidirectional training strategy on top of it.

In Table 1, by comparing "DenseWAP" with "Ours-Uni", both are unidirectional string decoder based models, we can obtain 4.29% average performance improvement in ExpRate by simply replacing the RNN-based decoder with the vanilla transformer decoder.

Compared with other methods, our proposed BTTR model outperforms the previous state-of-the-art methods in nearly all metrics and is about 3.4% ahead of the "DenseWAP-TD" method in ExpRate, which explicitly encodes our prior knowledge about the LaTeX grammar through a tree decoder.

5.3 Ablation Study

In Table 2, the first "IPE" column denotes whether to use image positional encoding or not. Secondly, the "Bi-Trained" column shows whether the bidirectional training strategy is used. On the "AJS" column, ✓ indicates the use of approximate joint search [15], while ✗ represents L2R search. The last "Ensemble" column denotes whether the ensemble method is used.

First, we can see that whether to use image positional encoding or not makes huge difference on the CROHME 2019 test set. This shows that image positional encoding improves the generalization ability of our model in different scales.

Comparing the 2nd and the 3rd rows in each dataset, we can see that the model trained using the bidirectional training strategy still outperforms the unidirectionally trained model by about 2.52% in ExpRate, though both of them using L2R search. This shows that while training a bidirectional language model, the bidirectional training strategy also helps the whole model to extract information from the images more comprehensively.

Further, using the properties of the bidirectional language model, we evaluate the decoding results of both L2R and R2L directions using approximate joint search, resulting in an improvement of about 4.68% in ExpRate.

Finally we report the results using the ensemble method, showing that this can significantly improve the overall recognition performance by about 3.35% in ExpRate. Specifically, We train five models initialized with different random seeds and average their prediction probabilities at each decoding step.

5.4 Case Study

As can be seen in Fig. 2, we give several case studies for the "DenseWAP", "Ours-Uni" and "Ours-Bi" models. These three models use the same DenseNet encoder. The difference between these three models is that "DenseWAP" uses an RNN-based decoder, "Ours-Uni" adapts a vanilla transformer decoder, and "Ours-Bi" uses the bidirectional training strategy to train transformer decoder.

Table 1. Performance comparison of single models on the CROHME 2014/2016/2019 test sets (in %), where "ExpRate", "≤ 1 error" and "≤ 2 error" columns mean expression recognition rate when zero to two structural or symbol errors can be tolerated. "StruRate" column means structure recognition rate.

Dataset	Model	ExpRate	≤ 1 error	≤ 2 error	StruRate
CROHME14	I	37.22	44.22	47.26	–
	VI	25.66	33.16	35.90	–
	VII	26.06	33.87	38.54	–
	WYGIWYS	36.4	–	–	–
	WAP	40.4	56.1	59.9	–
	DenseWAP	43.0	57.8	61.9	63.2
	PAL-v2	48.88	64.50	69.78	–
	DenseWAP-TD	49.1	64.2	67.8	68.6
	WS WAP	53.65	–	–	–
	Ours-Uni	48.17	59.63	63.29	65.01
	Ours-Bi	**53.96**	**66.02**	**70.28**	**71.40**
CROHME16	TOKYO	43.94	50.91	53.70	61.6
	WAP	37.1	–	–	–
	DenseWAP	40.1	54.3	57.8	59.2
	PAL-v2	49.61	64.08	**70.27**	–
	DenseWAP-TD	48.5	62.3	65.3	65.9
	WS WAP	51.96	**64.34**	70.10	–
	Ours-Uni	44.55	55.88	60.59	61.55
	Ours-Bi	**52.31**	63.90	68.61	**69.40**
CROHME19	Univ. Linz	41.49	54.13	58.88	60.02
	DenseWAP	41.7	55.5	59.3	60.7
	DenseWAP-TD	51.4	**66.1**	69.1	69.8
	Ours-Uni	44.95	56.13	60.47	60.63
	Ours-Bi	**52.96**	65.97	**69.14**	**70.06**

Firstly, comparing the prediction results between "DenseWAP" and "Ours-Uni", we can see that for input images with complex structure, the "DenseWAP" model cannot predict all the symbols completely. Moreover, for input image with discontinuous structure, the right half is unnecessarily predicted twice. Problems mentioned above reflecting under-parsing and over-parsing phenomenon. However, the "Ours-Uni" and "Ours-Bi" models employed with positional encodings are able to identify all the symbols in these images accurately.

Secondly, by comparing the prediction results of "Ours-Uni" and "Ours-Bi", we find that "Ours-Bi" gives more accurate predictions. Owing to the re-score mechanism in approximate joint search procedure, "Ours-Bi" avoids the asymmetry of "{" and "}" in the prediction results given by "Ours-Uni".

Table 2. Ablation study on the CROHME 2014/2016/2019 test sets (in %)

Dataset	IPE	Bi-Trained	AJS	Ensemble	ExpRate
CROHME14	✗	✗	✗	✗	45.13
	✓	✗	✗	✗	48.17
	✓	✓	✗	✗	49.49
	✓	✓	✓	✗	53.96
	✓	✓	✓	✓	57.91
CROHME16	✗	✗	✗	✗	43.33
	✓	✗	✗	✗	44.55
	✓	✓	✗	✗	46.90
	✓	✓	✓	✗	52.31
	✓	✓	✓	✓	54.49
CROHME19	✗	✗	✗	✗	20.77
	✓	✗	✗	✗	44.95
	✓	✓	✗	✗	48.79
	✓	✓	✓	✗	52.96
	✓	✓	✓	✓	56.88

Image	DenseWAP	Ours-Uni	Ours-Bi
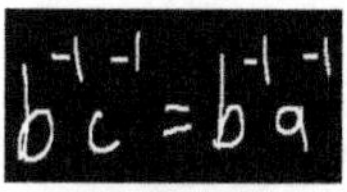	b ^ { - 1 - 1 } = b ^ { - 1 } a ^ { - 1 } <EOS>	b ^ { - 1 } c ^ { - 1 } = b ^ { - 1 } a ^ { - 1 } <EOS>	b ^ { - 1 } c ^ { - 1 } = b ^ { - 1 } a ^ { - 1 } <EOS>
	\tan \alpha _ { i } = \alpha _ { i n } <EOS>	\tan \alpha _ { i } <EOS>	\tan \alpha _ { i } <EOS>
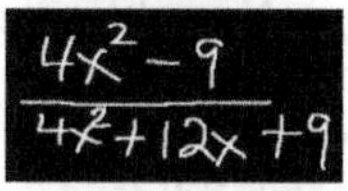	\frac { 4 x ^ { 2 } - 9 } { 4 x + 1 2 x + 9 } <EOS>	\frac { 4 x ^ { 2 } - 9 } { 4 x ^ { 2 } + 1 2 x + 9 <EOS>	\frac { 4 x ^ { 2 } - 9 } { 4 x ^ { 2 } + 1 2 x + 9 } <EOS>
	\frac { 1 } { \sqrt { 2 } } (\frac { b } { \sqrt { z } - 0 }) <EOS>	\frac { 1 } { \sqrt { 2 } } (\frac { b } { \sqrt { 2 } } } - 0) <EOS>	\frac { 1 } { \sqrt { 2 } } (\frac { b } { \sqrt { 2 } } - 0) <EOS>

Fig. 2. Case studies for the "DenseWAP" [34], "Ours-Uni" and "Ours-Bi" models. The red symbols represent incorrect predictions, while the green symbols represent correct predictions. (Color figure online)

6   Conclusion

In this paper, a novel bidirectionally trained trans former model is proposed for handwritten mathematical expression recognition task. Compared to previous approaches, our proposed BTTR model has the following three advantages: (1) Through image positional encoding, the model could capture the location information of the feature-map to guide itself to reasonably assign attention and alleviate the lack of coverage problem. (2) We take the advantage of the transformer model's permutation-invariant property to train a decoder with bidirectional language modeling capability. The bidirectional training strategy enables BTTR model to make predictions in both L2R and R2L directions while ensuring model simplicity. (3) The RNN-based decoder is replaced by the transformer decoder to improve the parallelization of training process. Experiments demonstrate the effectiveness of our proposed BTTR model. Concretely speaking, our model gets ExpRate scores of 57.91%, 54.49%, and 56.88% on the CROHME 2014, CROHME 2016, and CROHME 2019 respectively.

Acknowledgments. This work is supported by the projects of National Key R&D Program of China (2019YFB1406303) and National Natural Science Foundation of China (No. 61876003), which is also a research achievement of Key Laboratory of Science, Technology and Standard in Press Industry (Key Laboratory of Intelligent Press Media Technology).

References

1. Alvaro, F., Sánchez, J.A., Benedí, J.M.: Recognition of on-line handwritten mathematical expressions using 2D stochastic context-free grammars and hidden Markov models. Pattern Recogn. Lett. **35**, 58–67 (2014)
2. Bengio, Y., Frasconi, P., Simard, P.: The problem of learning long-term dependencies in recurrent networks. In: IEEE International Conference on Neural Networks, pp. 1183–1188. IEEE (1993)
3. Chan, K.F., Yeung, D.Y.: An efficient syntactic approach to structural analysis of on-line handwritten mathematical expressions. Pattern Recogn. **33**(3), 375–384 (2000)
4. Chan, K.F., Yeung, D.Y.: Mathematical expression recognition: a survey. Int. J. Doc. Anal. Recogn. **3**(1), 3–15 (2000)
5. Chan, K.F., Yeung, D.Y.: Error detection, error correction and performance evaluation in on-line mathematical expression recognition. Pattern Recogn. **34**(8), 1671–1684 (2001)
6. Cheng, Z., Bai, F., Xu, Y., Zheng, G., Pu, S., Zhou, S.: Focusing attention: towards accurate text recognition in natural images. In: Proceedings of the IEEE International Conference on Computer Vision, pp. 5076–5084 (2017)
7. Deng, Y., Kanervisto, A., Rush, A.M.: What you get is what you see: a visual markup decompiler. arXiv preprint arXiv:1609.04938 10, 32–37 (2016)
8. Devlin, J., Chang, M.W., Lee, K., Toutanova, K.: Bert: pre-training of deep bidirectional transformers for language understanding. arXiv preprint arXiv:1810.04805 (2018)

9. Glorot, X., Bordes, A., Bengio, Y.: Deep sparse rectifier neural networks. In: Proceedings of the Fourteenth International Conference on Artificial Intelligence and Statistics, pp. 315–323 (2011)
10. He, K., Zhang, X., Ren, S., Sun, J.: Deep residual learning for image recognition. In: Proceedings of the IEEE Conference on Computer Vision and Pattern Recognition, pp. 770–778 (2016)
11. Huang, G., Liu, Z., Van Der Maaten, L., Weinberger, K.Q.: Densely connected convolutional networks. In: Proceedings of the IEEE Conference on Computer Vision and Pattern Recognition, pp. 4700–4708 (2017)
12. Ioffe, S., Szegedy, C.: Batch normalization: accelerating deep network training by reducing internal covariate shift. arXiv preprint arXiv:1502.03167 (2015)
13. Jiang, Z., Gao, L., Yuan, K., Gao, Z., Tang, Z., Liu, X.: Mathematics content understanding for cyberlearning via formula evolution map. In: Proceedings of the 27th ACM International Conference on Information and Knowledge Management, pp. 37–46 (2018)
14. Kang, L., Riba, P., Rusiñol, M., Fornés, A., Villegas, M.: Pay attention to what you read: non-recurrent handwritten text-line recognition. arXiv preprint arXiv:2005.13044 (2020)
15. Liu, L., Utiyama, M., Finch, A., Sumita, E.: Agreement on target-bidirectional neural machine translation. In: Proceedings of the 2016 Conference of the North American Chapter of the Association for Computational Linguistics: Human Language Technologies, pp. 411–416 (2016)
16. MacLean, S., Labahn, G.: A new approach for recognizing handwritten mathematics using relational grammars and fuzzy sets. Int. J. Doc. Anal. Recogn. (IJDAR) 16(2), 139–163 (2013)
17. Mahdavi, M., Zanibbi, R., Mouchere, H., Viard-Gaudin, C., Garain, U.: ICDAR 2019 CROHME + TFD: competition on recognition of handwritten mathematical expressions and typeset formula detection. In: 2019 International Conference on Document Analysis and Recognition (ICDAR), pp. 1533–1538. IEEE (2019)
18. Mouchere, H., Viard-Gaudin, C., Zanibbi, R., Garain, U.: ICFHR 2014 competition on recognition of on-line handwritten mathematical expressions (CROHME 2014). In: 2014 14th International Conference on Frontiers in Handwriting Recognition, pp. 791–796. IEEE (2014)
19. Mouchère, H., Viard-Gaudin, C., Zanibbi, R., Garain, U.: ICFHR 2016 CROHME: competition on recognition of online handwritten mathematical expressions. In: 2016 15th International Conference on Frontiers in Handwriting Recognition (ICFHR), pp. 607–612. IEEE (2016)
20. Parmar, N., et al.: Image transformer. In: International Conference on Machine Learning, pp. 4055–4064. PMLR (2018)
21. Truong, T.N., Nguyen, C.T., Phan, K.M., Nakagawa, M.: Improvement of end-to-end offline handwritten mathematical expression recognition by weakly supervised learning. In: 2020 17th International Conference on Frontiers in Handwriting Recognition (ICFHR), pp. 181–186. IEEE (2020)
22. Vaswani, A., et al.: Attention is all you need. Adv. Neural. Inf. Process. Syst. 30, 5998–6008 (2017)
23. Wu, J.W., Yin, F., Zhang, Y.M., Zhang, X.Y., Liu, C.L.: Handwritten mathematical expression recognition via paired adversarial learning. Int. J. Comput. Vis., 1–16 (2020)
24. Wu, J.W., Yin, F., Zhang, Y., Zhang, X.Y., Liu, C.L.: Graph-to-graph: towards accurate and interpretable online handwritten mathematical expression recognition. In: AAAI 2021 (2021)

25. Xu, K., et al.: Show, attend and tell: neural image caption generation with visual attention. In: International Conference on Machine Learning, pp. 2048–2057 (2015)
26. Yan, Z., Ma, T., Gao, L., Tang, Z., Chen, C.: Persistence homology for link prediction: an interactive view. arXiv preprint arXiv:2102.10255 (2021)
27. Yan, Z., Zhang, X., Gao, L., Yuan, K., Tang, Z.: ConvMath: a convolutional sequence network for mathematical expression recognition. In: 2020 25th International Conference on Pattern Recognition (ICPR), pp. 4566–4572. IEEE (2021)
28. Yuan, K., Gao, L., Jiang, Z., Tang, Z.: Formula ranking within an article. In: Proceedings of the 18th ACM/IEEE on Joint Conference on Digital Libraries, pp. 123–126 (2018)
29. Yuan, K., Gao, L., Wang, Y., Yi, X., Tang, Z.: A mathematical information retrieval system based on RankBoost. In: Proceedings of the 16th ACM/IEEE-CS on Joint Conference on Digital Libraries, pp. 259–260 (2016)
30. Yuan, K., He, D., Jiang, Z., Gao, L., Tang, Z., Giles, C.L.: Automatic generation of headlines for online math questions. In: Proceedings of the AAAI Conference on Artificial Intelligence, vol. 34, pp. 9490–9497 (2020)
31. Yuan, K., He, D., Yang, X., Tang, Z., Kifer, D., Giles, C.L.: Follow the curve: arbitrarily oriented scene text detection using key points spotting and curve prediction. In: 2020 IEEE International Conference on Multimedia and Expo (ICME), pp. 1–6. IEEE (2020)
32. Zanibbi, R., Blostein, D.: Recognition and retrieval of mathematical expressions. Int. J. Doc. Anal. Recogn. (IJDAR) 15(4), 331–357 (2012)
33. Zeiler, M.D.: ADADELTA: an adaptive learning rate method. arXiv preprint arXiv:1212.5701 (2012)
34. Zhang, J., Du, J., Dai, L.: Multi-scale attention with dense encoder for handwritten mathematical expression recognition. In: 2018 24th International Conference on Pattern Recognition (ICPR), pp. 2245–2250. IEEE (2018)
35. Zhang, J., Du, J., Yang, Y., Song, Y.Z., Wei, S., Dai, L.: A tree-structured decoder for image-to-markup generation. In: ICML (2020, in Press)
36. Zhang, J., et al.: Watch, attend and parse: an end-to-end neural network based approach to handwritten mathematical expression recognition. Pattern Recogn. 71, 196–206 (2017)
37. Zhang, X., Gao, L., Yuan, K., Liu, R., Jiang, Z., Tang, Z.: A symbol dominance based formulae recognition approach for pdf documents. In: 2017 14th IAPR International Conference on Document Analysis and Recognition (ICDAR), vol. 1, pp. 1144–1149. IEEE (2017)
38. Zhang, Z., Wu, S., Liu, S., Li, M., Zhou, M., Xu, T.: Regularizing neural machine translation by target-bidirectional agreement. In: Proceedings of the AAAI Conference on Artificial Intelligence, vol. 33, pp. 443–450 (2019)
39. Zhou, L., Zhang, J., Zong, C.: Synchronous bidirectional neural machine translation. Trans. Assoc. Comput. Linguist. 7, 91–105 (2019)

TabAug: Data Driven Augmentation for Enhanced Table Structure Recognition

Umar Khan[1(✉)], Sohaib Zahid[1], Muhammad Asad Ali[1], Adnan Ul-Hasan[1], and Faisal Shafait[1,2]

[1] Deep Learning Laboratory, National Center of Artificial Intelligence, Islamabad, Pakistan
{umar.khan1,sohaib.zahid,muhammad.asadali,adnan.ulhassan,
faisal.shafait}@seecs.edu.pk
[2] School of Electrical Engineering and Computer Science,
National University of Sciences and Technology (NUST), Islamabad, Pakistan

Abstract. Table Structure Recognition is an essential part of end-to-end tabular data extraction in document images. The recent success of deep learning model architectures in computer vision remains to be non-reflective in table structure recognition, largely because extensive datasets for this domain are still unavailable while annotating new data is expensive and time-consuming. Traditionally, in computer vision, these challenges are addressed by standard augmentation techniques that are based on image transformations like color jittering and random cropping. As demonstrated by our experiments, these techniques are not effective for the task of table structure recognition. In this paper, we propose TabAug, a re-imagined Data Augmentation technique that produces structural changes in table images through replication and deletion of rows and columns. It also consists of a data-driven probabilistic model that allows control over the augmentation process. To demonstrate the efficacy of our approach, we perform experimentation on ICDAR 2013 dataset where our approach shows consistent improvements in all aspects of the evaluation metrics, with cell-level correct detections improving from 92.16% to 96.11% over the baseline.

Keywords: Table Structure Recognition · Table augmentation · Data augmentation · Table data extraction · Probabilistic model · Data-driven model · Table segmentation · Deep learning

1 Introduction

Document structure analysis and parsing are some of the most crucial parts of document image processing for digitization and information extraction. One

---

U. Khan and S. Zahid—These authors have contributed equally.

© Springer Nature Switzerland AG 2021
J. Lladós et al. (Eds.): ICDAR 2021, LNCS 12822, pp. 585–601, 2021.
https://doi.org/10.1007/978-3-030-86331-9_38

of the most important components of documents is tables. Tables are a structured way of representing data allowing for visual and logical grouping of data in a highly comprehensible manner. Tables in documents are frequently used to present key information such as financial records, receipts, and data forms. Extracting this information can be vital and beneficial to most businesses around the world. However, tabular data extraction can be a challenging task as more often this data is found as scanned document images that contain no structural or content metadata.

Tabular data extraction consists of three major tasks, Table Detection, Table Structure Recognition, and Semantic Understanding. Table detection is the task of locating tables in a given document, while table structure recognition aims towards the segmentation of tables into rows and columns for layout understanding. Semantic understanding involves the assignment of information such as row or column header (e.g., Unit price, Stock value, etc.) to a particular cell. Our focus in this work is the table structure recognition part, where the aim is to extract rows and columns in a given table image.

In practice, Convolutional Neural Network (CNN) is used as an effective tool for extracting meaningful features from visual information. Given that scanned document images contain only visual information, CNNs become increasingly important for Table Detection and Table Structure Recognition alike. State-of-the-art CNNs are efficient at extracting visual features from data; however, they can be highly data demanding in nature. This precondition of CNNs coupled with the unavailability of large Table Structure Recognition datasets presents a challenging problem as annotating large datasets can be both expensive and time-consuming. It is, therefore, crucial to focus on methods for improving the data-efficiency of deep learning models to achieve better results for Table Structure Recognition on smaller datasets.

Augmentation techniques are widely used to improve the data efficiency in Deep Neural networks. It is the practice of adding slight realistic changes to the original data to increase the diversity of the training data. This concept helps model regularize and avoid over-fitting in small datasets. There are several image transformation techniques such as scaling, rotating, smearing, etc. that are widely used in Computer Vision to augment data. For the purposes of comparison, we choose to apply Random cropping, (first employed in AlexNet [14]) and Color Jitter (used in re-implementation of ResNet [9] by Facebook AI Research) as the *standard* augmentation. Please refer to the bottom row of Fig. 2 for example of these transformations on ICDAR 2013 dataset. However, as shown by our experimentation in Sect. 4, they are not effective for table structure recognition as they do not alter the structure of a table but instead produce unrepresentative data, which results in a decreased performance.

To overcome these challenges, we propose a novel method of data augmentation based on two key operations of *Replication* and *Deletion* on rows and columns, illustrated in Fig. 1 and sample images of these illustrated operations can be seen in Fig. 2. We also introduce a data-driven probabilistic model for generating parameters that control the outlook of the augmented data. The aug-

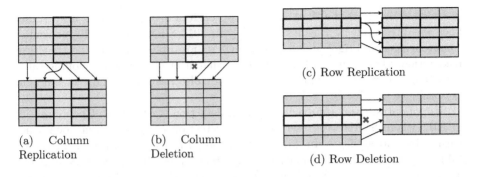

(a) Column
Replication

(b) Column
Deletion

(c) Row Replication

(d) Row Deletion

Fig. 1. Visual depiction of proposed augmentation operations

mented variations generated by this technique better represent the real-world variation in tables. Please refer to Sect. 3 for further details about our methodology.

Our experiments on ICDAR 2013 dataset demonstrate that our proposed augmentation technique yields a higher data efficiency and out-performs both the non-augmented and standard image augmentation approaches. Please refer to Sect. 4 for further details on experimental evaluation.

These results set a strong foundation for more future work in this direction. To facilitate this, we have decided to make our implementation of structural data augmentation named *TabAug* and associated experimentation code open-source[1]. We have also developed the existing code to be fairly simple to plug into various existing models with support for T-Truth annotation format [20] with no training overhead.

The rest of paper is structured into following sections. Section 2, outlines literature review of relevant works in the domain of table structure recognition and tabular data extraction in general. In Sect. 3 we present our methodology and implementation of the proposed augmentation technique. In Sect. 4 we report on our experiments and comparative results. Finally, in Sect. 5 we conclude our research with future direction for our work.

2   Related Work

Tabular data extraction is an old and recognized problem with solutions maturing from over past 20 years. Starting with one of the earliest works in table detection in text files using heuristics was done by [24] in 1996 followed by Pyreddy et al. [15] in 1997. The following years saw a novel heuristic-based approach of detecting tables in document images from Keinenger et al. [11,12] and [13] forming a combined table plotting and recognition system named TRECS. Zanibbi et al. [25] presented a comprehensive survey paper that focused on table recognition literature writing down extensive observations on various aspects of popular

---

[1] https://github.com/sohaib023/splerge-tab-aug.

methods available at the time. A novel approach of detecting tables from ruling lines was presented by Basilios et al. [6]. In 2010 Shahab et al. [20] presented comprehensive and rigorous evaluation metrics for table detection and structure recognition. Significant work was done by Shafait et al. [19] in developing table detection algorithm for multiple layouts, they also introduced more meaningful performance metrics for table detection.

Use of data-driven approaches in table detection started from Chen et al. [3] in 2011 where SVMs were used for table detection in handwritten documents with noise and artifacts. In 2013 Kasar et al. [10] also made use of SVMs for table region detection by classifying ruling lines. The following year Anukriti et al. [2] used Conditional Random Fields (CRFs) with encoded foreground block characteristics and the contextual information as features for learning layouts and labeling table and non-table regions in documents.

From 2015 onward, Deep Learning models have been used extensively for solving the detection and recognition problem. [1,8,18] and [22] made use of Faster R-CNN [17] with their own take on handcrafted features and pre-processing methodologies. These methods were successful in producing state-of-the-art results for table detection but failed to produce significant results in Structure Recognition. Schreiber et al. [18] mentions the unnatural approach of detecting rows and columns through Faster R-CNN [17] and instead proposes fine-grained image segmentation through FCN-X's architecture by Shelhamer et al. [21], however, their proposed technique heavily relies upon image stretching for expanding background pixels as a pre-processing methodology.

In more recent works from Qasim et al. [16] and Tensmeyer et al. [23] on table structure recognition, we see a more stable and natural approach towards the formulation of problem. Qasim et al. [16] made use of Graph Neural Networks for generating cell adjacency matrix for all existing OCR detected words and Tensmeyer et al. [23] formulating the problem of structure recognition as a combination of row, column splits and cell mergings for defining the structure of a table. Qasim et al. [16] mentions that lack of large datasets has been a major hindrance for Deep Learning methods in table structure recognition and makes use of synthetic data to show the effectiveness of their network. However, synthetic data generated randomly is limited in capturing the visual and general characteristics of the target dataset, due to which the model fails to perform on the target dataset despite the optimal results on the synthetic data. Similarly for the split model from Tensmeyer et al. [23], even though more data-efficient than Qasim et al. [16] GNN model, makes use of large proprietary dataset to train the network for an effective model to be tested on ICDAR 2013. Our experiments of training and testing of split model from Tensmeyer et al. [23] on ICDAR 2013 dataset reveals sub-optimal results than the network potential due to the lack of diverse and large training dataset.

With newer deep learning algorithms, we are seeing a trend of increase in performance on larger datasets; however, their results translate to sub-optimal results on smaller datasets thus limiting their potential use cases. Traditionally in computer vision, these problems are addressed with image transformation based data augmentation techniques defined as standard data augmentation for increasing data diversity in training data. Albeit successful in natural images, standard augmentation hold an insignificant and unstable impact on table images. In this paper, we propose a new re-imagined Tabular Data Augmentation inspired from [4] and [7] by replicating and removing table structure elements (rows and columns) to form augmented tables all while maintaining their visual artifacts, we have also introduced a data-driven probabilistic model (similar to [5]) for decision parameters of our method of Tabular Data Augmentation.

3 Methodology

We adopt an elementary approach of structural data augmentation for table structure recognition. We consider cells to be the building block of a table, that is, a table is defined as a combination of cells. Therefore, it should theoretically be possible to achieve a large number of combinations given a small number of cells. However, combining different cells into a table poses many limitations with regards to compatibility. First, widths and heights of cells can vary greatly, and thus combining them without due deliberation can lead to unnatural tables. Second, cells tend to inherit their formatting and styling from their adjacent cells. In a fully randomized collection of cells, one cell might have left-justified text while the next might have right-justified text, making for a confusing and unnatural table.

To overcome these limitations, we redefine a table to be a combination of rows and columns rather than cells. This helps keep intact the co-relation of cell styling within a row and column. To further simplify the problem, we limit these combinations to randomization of rows and columns within the same table. Thus, we can generate a randomized permutation of the rows and columns within a table to obtain a structurally augmented version of the table. This augmentation technique is then combined with a data-driven probabilistic sampler for maximizing its effectiveness.

3.1 Augmentation Operations

To achieve randomization of rows and columns, we define four basic operations that we can apply to each table:

- Row deletion
- Column deletion
- Row replication
- Column replication

These operations are depicted visually in Fig. 1 for better understanding. These are the core atomic operations that we utilize during our augmentation pipeline. They can be applied sequentially in random orders on a table to achieve increasingly varying versions of the same table.

(a) Original

Age	Enrollment	%
14-17	231,000	1.3
18-19	3,769,000	21.2
20-21	3,648,000	20.5
22-24	3,193,000	18.0
25-29	2,401,000	13.5
30-34	1,409,000	7.9
Over 35	3,107,000	17.5
Total	17,758,000	100

(b)

Age	Enrollment	Enrollment
25-29	2,401,000	2,401,000
14-17	231,000	231,000
18-19	3,769,000	3,769,000
20-21	3,648,000	3,648,000
22-24	3,193,000	3,193,000
25-29	2,401,000	2,401,000
30-34	1,409,000	1,409,000

(c)

Age	Enrollment	%
14-17	231,000	1.3
18-19	3,769,000	21.2
Total	17,758,000	100
22-24	3,193,000	18.0
30-34	1,409,000	7.9
Over 35	3,107,000	17.5
Total	17,758,000	100

(d)

Age	Enrollment	%
14-17	231,000	1.3
18-19	3,769,000	21.2
20-21	3,648,000	20.5
22-24	3,193,000	18.0
25-29	2,401,000	13.5
30-34	1,409,000	7.9

(e)

Age	Enrollment	%
14-17	231,000	1.3
18-19	3,769,000	21.2
20-21	3,648,000	20.5
22-24	3,193,000	18.0
25-29	2,401,000	13.5
30-34	1,409,000	7.9
ıver 35	3,107,000	17.5

Fig. 2. Sub-figure 2a contains an original table sample from ICDAR 2013 dataset. Sub-figures 2b and 2c show variations of the table generated by TabAug. Sub-figures 2d and 2e show the results of image transformation based augmentations.

3.2   Augmentation Sub-operations

Each of the proposed augmentation operations consists of several sub-operations, which are further explained below. For simplicity, we will only consider Column deletion and replication, as Row deletion and replication can be directly inferred from it.

Source Selection: Before doing any replication or removal, we need to select a column on which we may apply the operation. We randomly select an index c in the range of $\{1..C-1\}$, where C is the total number of columns in a given table. We purposefully leave out the $0-th$ index as the first column can have row headers (similarly column headers in the case of row selection), which should be retained in its location to maintain a natural table.

Furthermore, in case if spanning cells exist in the selected column, we must ensure that no partial/broken cells are copied.

$$\text{span}_{\text{min}} = \min(\{\text{cells}_{r\,c}^{\text{Start Column}} \ \forall r \in \{0..R-1\}\}) \qquad (1)$$

$$\text{span}_{\text{max}} = \max(\{\text{cells}_{r\,c}^{\text{End Column}} \ \forall r \in \{0..R-1\}\}) \qquad (2)$$

For the selected column c, the values of span_{min} and span_{max} are calculated using Eqs. 1 and 2. These values provide us with column indices for a self contained convex block (as depicted in Fig. 3a and 3b). In case if even after performing the above steps the selected block is non-convex (as depicted in Fig. 3c) we abort the operation and return the table unaltered. Otherwise our selection is defined by $c_{max} = \text{span}_{max}$ and $c_{min} = \text{span}_{min}$ which is used by next sub-operations.

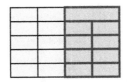
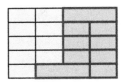

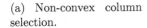

(a) Non-convex column selection.

(b) Expansion of selection to retrieve convex block.

(c) Example where expansion fails to retrieve convex block.

Fig. 3. A non-convex selection 3a cannot be replicated or deleted and hence it is expanded so as to make it convex 3b. Sub-figure 3c represents the case where expansion also results in a non-convex selection.

Target Location Selection (only for Replication): This sub-operation is similar to the previous one with few modifications. We select an index d in the range of $\{1 .. C\}$ randomly, where C is the total number of columns in a given table. The target location for column replication will be at the start of d-th column. So we purposefully leave out the $0 - th$ index as we do not want to move the first column. Furthermore, if $d = C$ then the target location is after the last column at the end of the image. We perform a check to see if the target location d is intersecting with a spanning cell. If so, placing a column at that location will cause the existing spanning cell to split apart, and hence we change d according to the equations below:

$$d = \begin{cases} \text{span}_{min}, & \text{if abs}(d - \text{span}_{min}) <= \text{abs}(d - \text{span}_{max}) \\ & \text{and span}_{min} \neq 0 \\ \text{span}_{max} + 1, & \text{otherwise} \end{cases} \qquad (3)$$

where span_{min} and span_{max} are calculated using Eqs. 1 and 2 by setting $c = d$.

Similar to the previous sub-operation, if after the correction of d it intersects with a spanning cell we abort the operation and return the table unaltered.

Execution: To execute the operation, whether deletion or replication, we need to alter both the table image and the ground-truth of the table, the specifics of which vary between the two operations. Following values are required in both operations:

$$x_{min} = \text{columns}[c_{min}].x1 \quad , \quad x_{max} = \text{columns}[c_{max}].x2 \quad , \quad w = x_{max} - x_{min}$$

where $x1$ denotes starting x-coordinate of a column and $x2$ denotes the ending x-coordinate of a column.

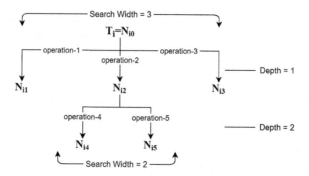

Fig. 4. An example of an augmentation search tree. Each of the nodes N_{ij} is obtained by applying an augmentation operation to its parent node, where the root of the tree is the original table T_i.

Deletion: Firstly, we remove the columns with indices in the range (c_{min}, c_{max}) from the ground-truth, including all the contained cells and span information. For the columns and cells having indices greater than c_{max}, we subtract w from their x-coordinates, to move them to the left.

For the image, we cut out the image from x_{min} to x_{max} and then move left the image pixels to the right side of x_{max} by w.

Replication: Firstly, we calculate the x-coordinate of the target location where the replicated column is to be placed.

$$x_{dst} = \text{columns}[d].x1 \tag{4}$$

Then we copy the columns with indices in the range (c_{min}, c_{max}) from the ground-truth, including all the contained cells and span information. For the copied columns and cells we offset their x-coordinates by $x_{dst} - x_{min}$, so as to move it to the target location. Further, the columns and cells having indices greater than d, we add w to their x-coordinates, to move them to the right, clearing up space for the replicated column.

For the image, we move the image pixels to the right of x_{dst} by w, so as to clear up space for crop of the replicated columns. Then we copy the image pixels from x_{min} to x_{max} and move them by $x_{dst} - x_{min}$, so as to move it to destination location x_{dst}.

3.3   Augmentation Pipeline

Having defined the augmentation operations, an effective pipeline must be established that can utilize the proposed operations for training a model, therefore we propose a pipeline in the following.

Augmentation Tree Exploration: Before training a model, we pre-compute a sample set N_i of augmented versions of a given table T_i. Each member of N_i

Number of Columns

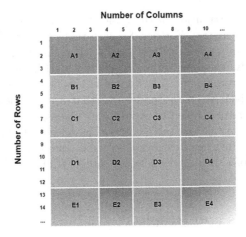

Fig. 5. A grid enumerating all of the possible table categories. Each table is assigned one of these categories based on its number of rows and columns

is considered to be a node (represented as N_{ij}), which is achievable by applying a series of augmentation operations (edges in the tree). An example of such a partial tree is shown in Fig. 4. By applying this tree search, we can extract a controlled sample set of all achievable augmented tables of a given root table T_i. To avoid inundation of samples, we apply pruning to the tree search to keep the number within a reasonable limit. First, only the nodes having tree depth greater than 5 and less than 10 are kept as a part of this sample set. Further, the maximum search width for each depth level is also restricted for pruning purposes. For depth $= 1$, maximum search width is 8, for depth $= 2$, it is 4, for depth $= \{3..5\}$ it is 2, and finally for depth $= \{6..10\}$ it is 1. Furthermore, out of the obtained sample set, any node is ignored that has a pixel height or width greater than 1.5 times the original table T_i, to ensure unnaturally large tables are not used for training.

Categorization of Tables and Their Nodes: As a pre-processing step, both the tables and their nodes are mapped into categories based on their number of rows and columns. The mapping table is of size 5×4 as depicted in Fig. 5. An example of such a mapping is that a node with 4 rows and 5 columns will be assigned to $B2$ category. Further, if the number of columns in N_{ij} are greater than 10, it is still assigned category $\{A - E\}4$. Similarly, if the number of rows in N_{ij} are greater than 14, it is assigned category $E\{1 - 4\}$.

After the completion of this step each T_i and N_{ij} will have a category associated with it which is referenced as T_i^{cat} and N_{ij}^{cat} respectively. It must be noted that the decision of category boundaries is purely empirical.

Probability Based Selection: For each table T_i we construct a probability distribution P_i using which T_i's nodes will be sampled during training. Firstly, a global frequency distribution of the tables over table categories is generated, giving us F^G a 5×4 matrix. Further, for each table T_i, a frequency distribution of all N_{ij} over the table categories is generated, giving us F_i, which is another 5×4 matrix. Lastly, we generate another 5×4 matrix $gauss_i$ which is a 2-D Gaussian centered around T_i^{cat}. The spread of this Gaussian distribution allows control over the diversity of the nodes N_{ij} that are sampled during training. These 3 matrices are multiplied to obtain $P_i = \text{Gauss}_i * F^G * F_i$.

During training, a random table category is selected using P_i as the probability distribution. Once the category is selected, one node is randomly sampled from all the nodes of T_i having the selected category. This selected node is passed for training.

4   Experiments and Results

To evaluate the efficacy of our proposed augmentation methodology, we train the Split model proposed by [23] on ICDAR 2013 dataset. We train the model using three different methodologies:

1. **Non-Augmented**: Images fed for training without any modification.
2. **Standard**: Basic image transformations, such as, hue saturation and brightness jitter in combination with image cropping.
3. **TabAug:** Our proposed augmentation methodology.

The dataset has a total of 128 pages and 156 tables, which is divided into train, test and validation splits with a proportion of $0.72 : 0.2 : 0.08$ respectively. We train the model with 4 different percentages of the training set, that is, 25%, 50%, 75% and 100%. It must be noted that this is a ratio of the total training set utilized while training and not a ratio of training samples divided by total samples. Further, we repeat each experiment 3 times to get a better estimate of the average results and their deviation.

For training, we use a batch size of 1, and a learning rate of 0.00075. After every 15 epoch, we decay the learning rate by a factor of 0.8. Lastly, we train the models for a total of $27,500$ iterations, which we empirically found to be enough for convergence of the model.

4.1   Ground Truth

For ICDAR 2013 dataset, ground-truths are provided as bounding boxes of cells, along with their starting and ending row/column indices. This ground-truth format was converted to T-Truth format [20], however, the resulting row and column separators were not aligned with the ruling lines of the table, hence, we re-annotated the dataset using T-Truth to align the separators and then cross-checked it to ensure that the annotations were coherent. The models are trained and tested on this re-annotated dataset, however, as they have been manually cross-checked, the evaluation is applicable to original annotations as well.

During the re-annotation phase, we upgraded the T-Truth annotation tool to fix several bugs and make it user-friendly. The upgraded version has been made publicly available[2].

While T-Truth annotation format is used for the augmentation process, Split model [23] requires pixel-wise annotations. To generate such ground-truth, the separators of T-Truth annotations are expanded to the nearest words (horizontally in the case of column ground-truth, and vertically in the case of row ground-truth).

4.2   Performance Measures

We chose the evaluation metrics proposed by Shahab et al. [20] primarily for two reasons. First, these metrics are comprehensive, painting a complete picture of a model's performance. Second, these are general-purpose metrics and can be applied to any type of segment such as rows, columns, and cells.

In the evaluation metrics proposed by Shahab et al. [20], first the ground truth segments and the predicted segments are numbered. A correspondence matrix of shape n x m is created where n is the number of ground truth segments while m is the number of predicted segments. Each entry $[i, j]$ in the matrix stores the number of pixels that are intersecting between the given ground-truth segment G_i and predicted segment S_j, that is $|G_i \cap S_j|$. Further, the sum of i-th row in the correspondence matrix gives the total number of pixels in ground-truth segment G_i and the sum of j-th column gives the total number of pixels in predicted segment S_j. Once this correspondence matrix is generated, we define the following evaluation metrics using the threshold value of $T = 0.1$ (consequently $1 - T = 0.9$):

1. **Correct Detections**: The total number of ground-truth segments that have a one-to-one mapping with a predicted segment and a major overlap. Concretely, a given ground-truth segment G_i is considered to be a correct detection if it has a major overlap with a predicted segment S_j and S_j does not have a significant overlap with any other ground-truth segment $(G_k; \forall k \neq i)$. That is:
$$\frac{|G_i \cap S_j|}{G_i} > 1 - T \text{ and } \frac{|G_k \cap S_j|}{G_k} < T \;\; ; \forall k \neq i$$

2. **Over-Segmentations**: The total number of ground-truth segments that have a significant overlap with more than one predicted segment. That is, a ground-truth segment G_i is over-segmented if:
$$T < \frac{|G_i \cap S_j|}{G_i} < 1 - T \text{ and } T < \frac{|G_i \cap S_k|}{G_i} < 1 - T \text{ where}; k \neq j$$

3. **Under-Segmentations**: Total number of predicted segments that have a significant overlap with more than one ground-truth segments. That is, a

---

[2] https://github.com/sohaib023/T-Truth.

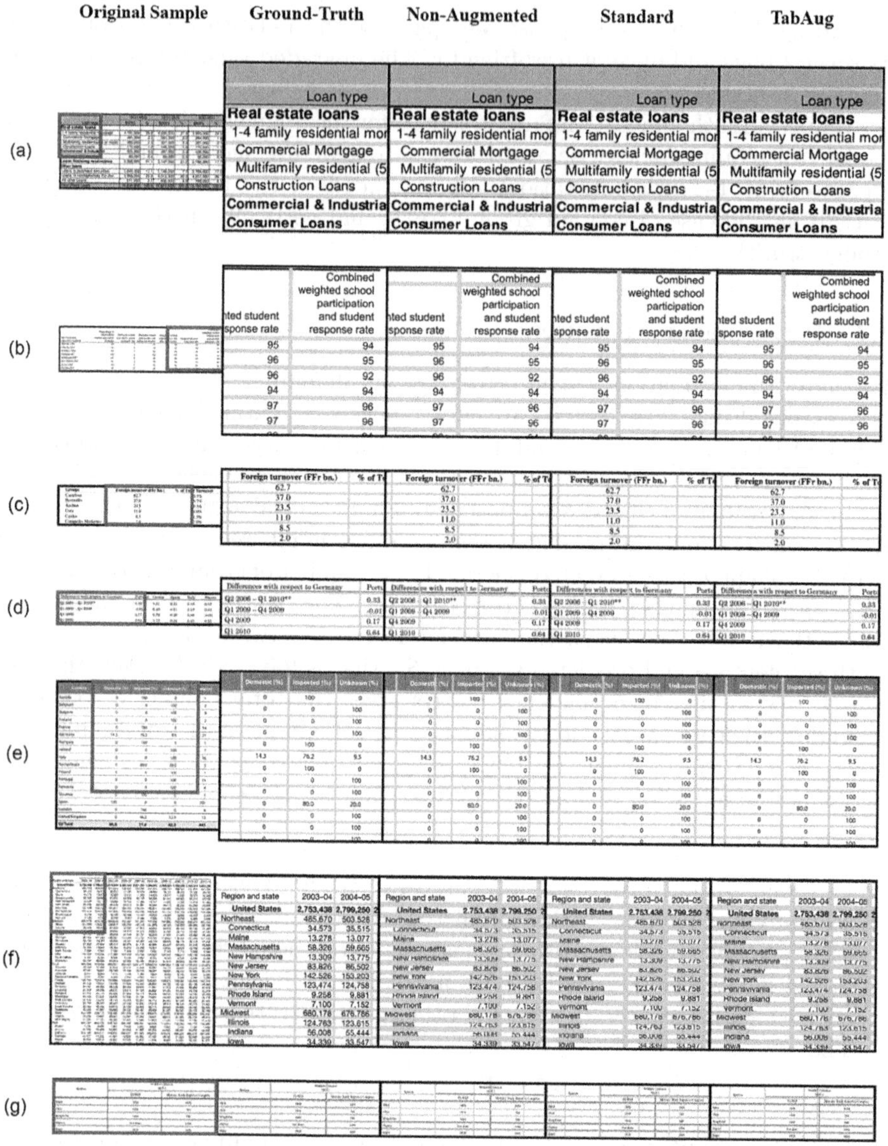

Fig. 6. Sample outputs of all three approaches for a comparative analysis. Each row provides a separate test sample. The red boxes highlight the image region that is displayed for visualization of ground-truth and the predictions. Ground-truth is displayed as blue lines while predictions are displayed as green lines. (Color figure online)

predicted segment S_j is under-segmented if:

$$T < \frac{|G_i \cap S_j|}{G_i} < 1 - T \text{ and } T < \frac{|G_k \cap S_j|}{G_k} < 1 - T \text{ where; } k \neq i$$

Table 1. Results of evaluating the trained models on 20% of ICDAR 2013 dataset reserved for testing.

		Non-augmented	Standard	TabAug
Row	Correct (%)	96.44 ± 1.13	97.86 ± 0.80	**97.86 ± 0.35**
	Over-segmented (%)	2.64 ± 0.54	1.71 ± 0.46	**1.71 ± 0.35**
	Under-segmented (%)	0.71 ± 0.56	0.21 ± 0.18	**0.21 ± 0.18**
Column	Correct (%)	92.12 ± 1.11	86.38 ± 1.54	**94.44 ± 0.25**
	Over-segmented (%)	4.84 ± 0.76	5.20 ± 1.11	**3.77 ± 0.76**
	Under-segmented (%)	1.43 ± 0.25	3.95 ± 1.01	**0.90 ± 0.25**
Cell	Correct (%)	92.16 ± 3.84	82.12 ± 6.76	**96.11 ± 1.61**
	Over-segmented (%)	1.58 ± 0.25	1.13 ± 0.28	**1.02 ± 0.37**
	Under-segmented (%)	3.37 ± 1.99	6.90 ± 2.21	**1.46 ± 0.72**

All of these evaluation metrics are normalized by division with the total number of ground-truth segments, which are presented as percentages in the results. We intentionally leave out the Partial Detections, Missed Segments, and False Positive Detections from our evaluation metrics, as due to the problem formulation of Split Model [23] they always evaluate to zero.

4.3   Results and Analysis

We evaluate all the trained models on a consistent 20% test split of ICDAR 2013 dataset. An evaluation of the performance measures described in Sect. 4.2 is provided in Table 1. Our proposed approach outperforms all other approaches in all performance measures for each of the Cell, Row, and Column recognition. Further, the models trained with the proposed approach show less standard deviation in the evaluation metrics, which is indicative of more stability in training. Moreover, as depicted in Table 2 and Fig. 7 our proposed approach consistently shows improvements across a range of training dataset sizes.

The standard augmentation approach manages to improve upon the evaluation metrics for Row, however, happens to be deficient in Cell and Column evaluations. This is explained by the fact that Row segmentation, especially in ICDAR 2013 dataset, is an easier task with less visual variability and complexity. Hence, it can be learned mostly based on white-spaces, for which image transformations provide an ample amount of augmentation. Column segmentation contains many complex scenarios and requires learning of more abstract and structural features, for which image transformations turn out to be insufficient.

We have further shown a comparison of the predictions generated by the models trained on each of the three approaches in Fig. 6. In Fig. 6a, we see the case where both standard augmentation and TabAug approaches successfully segment the rows, however, the non-augmented approach under-segments a row

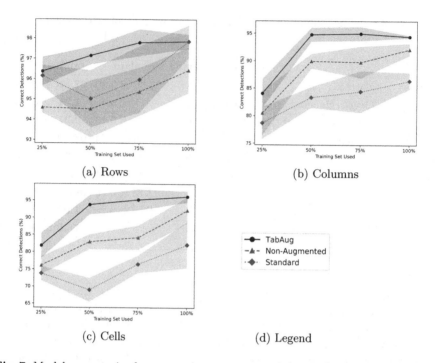

(a) Rows

(b) Columns

(c) Cells

(d) Legend

Fig. 7. Models were trained on a varying percentage of the total training dataset and then evaluated on a consistent 20% test split. The graphs show the progression of correct detection with an increase in training data for all three approaches.

in the header region. Conversely, in Fig. 6b, we see a case of over-segmentation by the non-augmented approach. In Fig. 6c, non-augmented and standard augmentation approaches fail to recognize the boundaries of the columns correctly, due to the white-spaces that exist between the protruding words of the header. However, TabAug robustly recognizes the boundaries of the column and predicts a single column separator. In Fig. 6d, we see a scenario where the TabAug approach fails, as it confuses vertically consistent breaks in the text as a column breakage. Regardless, we see that it still does a decent job at making logically sound column segmentations as compared to segmentations from the other two approaches. Figure 6e depicts another sample that is correctly segmented by TabAug, however causes significant confusion in column segmentation for the other two approaches. Finally, in Fig. 6g the first column is over-segmented by all of the approaches, as the column header has no overlap with its content below. This demonstrates a hard sample that requires a higher cognitive understanding for correct prediction.

Table 2. Correct detection percentages achieved by models trained on different percentages of the training dataset evaluated on a consistent 20% test split of ICDAR 2013 dataset.

	Training data used	Correct detections (%)		
		Non-augmented	Standard	TabAug
Row	25%	94.59 ± 0.27	96.15 ± 0.52	$\mathbf{96.37 \pm 0.70}$
	50%	94.52 ± 1.40	95.02 ± 1.41	$\mathbf{97.15 \pm 0.44}$
	75%	95.37 ± 1.19	94.73 ± 3.38	$\mathbf{97.79 \pm 0.66}$
	100%	96.44 ± 1.13	97.86 ± 0.80	$\mathbf{97.86 \pm 0.35}$
Column	25%	80.46 ± 4.08	78.67 ± 2.92	$\mathbf{84.05 \pm 2.65}$
	50%	89.96 ± 1.34	83.34 ± 1.91	$\mathbf{94.80 \pm 1.27}$
	75%	89.79 ± 2.88	84.41 ± 3.80	$\mathbf{94.98 \pm 1.26}$
	100%	92.12 ± 1.11	86.38 ± 1.54	$\mathbf{94.44 \pm 0.25}$
Cell	25%	76.07 ± 2.56	73.76 ± 2.08	$\mathbf{81.84 \pm 3.68}$
	50%	82.92 ± 2.17	68.99 ± 3.50	$\mathbf{93.73 \pm 2.81}$
	75%	84.17 ± 3.51	76.45 ± 2.55	$\mathbf{95.14 \pm 3.01}$
	100%	92.16 ± 3.84	82.12 ± 6.76	$\mathbf{96.11 \pm 1.61}$

5   Conclusion

In this paper, we presented TabAug, a novel augmentation technique capable of producing structural changes in a table through replication and deletion of rows and columns. A data-driven probabilistic model is used in conjunction with the augmentation technique to control the augmentation process. Following the promising results of Split-model [23] trained on the publicly available ICDAR 2013 dataset using TabAug, we believe our work provides a strong foundation for numerous future extensions. In future, we plan to explore ideas for cross-table augmentation through statistical layout matching.

Acknowledgement. This work has been partially funded by the Higher Education Commission of Pakistan's grant for National Center of Artificial Intelligence (NCAI).

References

1. Arif, S., Shafait, F.: Table detection in document images using foreground and background features. Digital Image Comput. Tech. Appl. **2018**, 1–8 (2018)
2. Bansal, A., Harit, G., Dutta Roy, S.: Table extraction from document images using fixed point model. In: ICVGIP 2014: Proceedings of the 2014 Indian Conference on Computer Vision Graphics and Image Processing, pp. 1–8 (2014)
3. Chen, J., Lopresti, D.: Table detection in noisy off-line handwritten documents. In: 2011 International Conference on Document Analysis and Recognition, Beijing, China, pp. 399–403 (2011)

4. Dwibedi, D., Misra, I., Hebert, M.: Cut, paste and learn: surprisingly easy synthesis for instance detection. In: 2017 IEEE International Conference on Computer Vision (ICCV), pp. 1310–1319 (2017)

5. Fang, H., Sun, J., Wang, R., Gou, M., Li, Y., Lu, C.: InstaBoost: boosting instance segmentation via probability map guided copy-pasting. In: 2019 IEEE/CVF International Conference on Computer Vision (ICCV), pp. 682–691 (2019)

6. Gatos, B., Danatsas, D., Pratikakis, I., Perantonis, S.J.: Automatic table detection in document images. In: Singh, S., Singh, M., Apte, C., Perner, P. (eds.) ICAPR 2005. LNCS, vol. 3686, pp. 609–618. Springer, Heidelberg (2005). https://doi.org/10.1007/11551188_67

7. Ghiasi, G., et al.: Simple copy-paste is a strong data augmentation method for instance segmentation. ArXiv (2020)

8. Gilani, A., Qasim, S.R., Malik, I., Shafait, F.: Table detection using deep learning. In: 14th International Conference on Document Analysis and Recognition, pp. 771–776 (2017)

9. He, K., Zhang, X., Ren, S., Sun, J.: Deep residual learning for image recognition. In: 2016 IEEE Conference on Computer Vision and Pattern Recognition (CVPR), pp. 770–778 (2016)

10. Kasar, T., Barlas, P., Adam, S., Chatelain, C., Paquet, T.: Learning to detect tables in scanned document images using line information. In: Twelfth International Conference on Document Analysis and Recognition, pp. 1185–1189 (2013)

11. Kieninger, T., Dengel, A.: A paper-to-HTML table converting system. In: Proceedings of Document Analysis Systems, pp. 356–365 (1998)

12. Kieninger, T., Dengel, A.: Table recognition and labeling using intrinsic layout features. In: International Conference on Advances in Pattern Recognition, pp. 307–316 (1999)

13. Kieninger, T., Dengel, A.: Applying the T-Recs table recognition system to the business letter domain. In: International Conference on Document Analysis and Recognition, p. 0518 (2001)

14. Krizhevsky, A., Sutskever, I., Hinton, G.E.: ImageNet classification with deep convolutional neural networks. In: Pereira, F., Burges, C.J.C., Bottou, L., Weinberger, K.Q. (eds.) Advances in Neural Information Processing Systems, vol. 25 (2012)

15. Pyreddy, P., Croft, W.B.: TINTI: a system for retrieval in text tables TITLE2: Technical report, University of Massachusetts, USA (1997)

16. Qasim, S.R., Mahmood, H., Shafait, F.: Rethinking table recognition using graph neural networks. In: 2019 International Conference on Document Analysis and Recognition (ICDAR), pp. 142–147 (2019)

17. Ren, S., He, K., Girshick, R., Sun, J.: Faster R-CNN: towards real-time object detection with region proposal networks. IEEE Trans. Pattern Anal. Mach. Intell. **39**, 1137–1149 (2015)

18. Schreiber, S., Agne, S., Wolf, I., Dengel, A., Ahmed, S.: DeepDeSRT: deep learning for detection and structure recognition of tables in document images. In: Fourteenth International Conference on Document Analysis and Recognition, vol. 1, pp. 1162–1167 (2017)

19. Shafait, F., Smith, R.: Table detection in heterogeneous documents. In: Proceedings of the 9th IAPR International Workshop on Document Analysis Systems, pp. 65–72. Document analysis systems (2010)

20. Shahab, A., Shafait, F., Kieninger, T., Dengel, A.: An open approach towards the benchmarking of table structure recognition systems. In: Document Analysis Systems, pp. 113–120 (2010)

21. Shelhamer, E., Long, J., Darrell, T.: Fully convolutional networks for semantic segmentation. IEEE Trans. Pattern Anal. Mach. Intell. **39**(4), 640–651 (2017)
22. Siddiqui, S., Malik, M., Agne, S., Dengel, A., Ahmed, S.: DeCNT: deep deformable CNN for table detection. IEEE Access **6**, 74151–74161 (2018)
23. Tensmeyer, C., Morariu, V.I., Price, B., Cohen, S., Martinez, T.: Deep splitting and merging for table structure decomposition. In: 2019 International Conference on Document Analysis and Recognition (ICDAR), pp. 114–121 (2019)
24. Tupaj, S., Shi, Z., Chang, D.H.: Extracting tabular information from text files. In: EECS Department, Tufts University (1996)
25. Zanibbi, R., Blostein, D., Cordy, J.: A survey of table recognition. IJDAR **7**, 1–16 (2004)

An Encoder-Decoder Approach to Handwritten Mathematical Expression Recognition with Multi-head Attention and Stacked Decoder

Haisong Ding🆔, Kai Chen$^{(\boxtimes)}$🆔, and Qiang Huo

Microsoft Research Asia, Beijing, China
{hadin,kaic,qianghuo}@microsoft.com

Abstract. Encoder-decoder framework with attention mechanism has become a mainstream solution to handwritten mathematical expression recognition (HMER) since "watch, attend and parse (WAP)" approach was proposed in 2017, where a convolutional neural network is used as encoder and a gated recurrent unit with attention is used in decoder. Inspired by the recent success of Transformer in many applications, in this paper, we adopt the design of multi-head attention and stacked decoder in Transformer to improve the decoder part of the WAP framework for HMER. Experimental results on CROHME tasks show that multi-head attention can boost the expression recognition rate (ExpRate) of WAP from 54.32%/58.05% to 56.76%/59.72% and stacked decoder can further improve ExpRate to 57.72%/61.38% on CROHME 2016/2019 test sets.

Keywords: Encoder-decoder · Multi-head attention · Stacked decoder · Transformer · Mathematical expression recognition

1 Introduction

Handwritten mathematical expression recognition (HMER) is an important enabling technology to empower many high-value scenarios, especially in online education, such as math question searching and solving, automatic grading, and technical writing assistant. Due to the pandemic of COVID-19 since 2020, millions of students all around the world have been learning online at home through internet. The relevant education apps and services with better HMER capability have become even more important to improve students' learning efficiency and reduce teachers' workloads.

Compared with traditional handwriting text recognition, HMER is a more challenging pattern recognition task. Besides various writing styles, it also needs to handle various two-dimensional structures and stronger contextual information dependencies [1]. Therefore, HMER can be divided into two tasks: symbol recognition and structure analysis [2,3]. Traditional methods often solve these

© Springer Nature Switzerland AG 2021
J. Lladós et al. (Eds.): ICDAR 2021, LNCS 12822, pp. 602–616, 2021.
https://doi.org/10.1007/978-3-030-86331-9_39

two problems sequentially or globally. For sequential methods [4,5], the input expression is segmented into math symbols first, then structure analysis is conducted based on symbol segmentation and recognition results. For global methods [6,7], these two tasks are solved simultaneously with global information. Both kinds of methods require lots of domain knowledge and the system design is quite complex.

Since the goal of HMER is to convert math expression images into representations of a math description language such as LaTeX understandable by a computer, HMER can also be formulated as an image-to-sequence problem and solved by an encoder-decoder with attention mechanism [8,9]. This idea was first tried in [10,11]. In [10], a so-called "watch, attend and parse (WAP)" approach was proposed, where a convolutional neural network (CNN) is used as encoder and a gated recurrent unit (GRU) with attention is used in decoder to convert math expression images into LaTeX command sequences directly. Experimental results on CROHME 2014 dataset showed that WAP outperformed traditional methods significantly.

The success of WAP proves the potential of encoder-decoder solution to HMER. Since then, a lot of research efforts are spent on this new paradigm from several perspectives including data augmentation, model architectures, training strategies, loss functions, etc. In [12], multi-scale attention was tried in WAP framework to recognize math symbols in different scales and restore fine-grained details dropped by pooling operations. Paired adversarial learning was used in [13,14] to force a model to focus on semantic-invariant features of patterns to conquer the difficulties caused by writing-style variation and small sample size, which can improve the generalization capability of attention-based encoder-decoder model. [15] introduced a transition probability matrix into decoder to capture long-range dependencies, which helps the model to learn syntactical rules. [19] replaced string decoder with a tree-structured decoder to explicitly model parent-child relationship of mathematical expression trees. In [16], scale augmentation and drop attention were verified to be effective in improving HMER performance. In [17], pattern generation strategies are proposed to improve HMER. [18] introduced a loss function of symbol recognition task to provide weak supervision signal to improve WAP performance. [20] integrated a global context block into CNN feature extractor and used a Transformer [22] encoder to encode positional information, which helps improve the recognition accuracy of math expressions. In [21], a CNN based decoder was used in an encoder-decoder based printed math expression recognition (MER) system to improve decoding efficiency.

As an emerging state-of-the-art, attention-based encoder-decoder solution to many tasks including natural language processing (e.g. [22,23]), computer vision (e.g. [24,25]), and speech recognition (e.g. [26,27]), Transformer has attracted more and more researchers' attention. In this paper, we investigate the performance of transformer-based model on HMER task and leverage its design insights to improve WAP-like attention-based encoder-decoder model. It is known that the key aspects of Transformer design include multi-head atten-

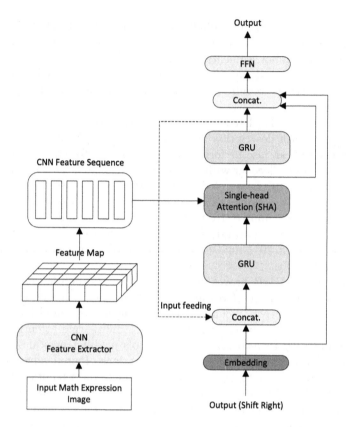

Fig. 1. Overview of baseline WAP-based HMER model with single-head attention.

tion and stacked decoder. Multi-head attention helps the decoder to make effective use of multiple encoder outputs simultaneously, and the stacked decoder can gradually refine model prediction to reduce recognition errors. Therefore, we first change the attention mechanism of WAP to multi-head version, which can boost the expression recognition rate (ExpRate) from 54.32%/58.05% to 56.76%/59.72% on CROHME 2016/2019 test sets. Then we stacked the decoder part several times, which can further improve the ExpRate to 57.72%/61.38%.

The remainder of this paper is organized as follows. In Sect. 2, we introduce our proposed approach. Experimental results are presented in Sect. 3. We conclude the paper in Sect. 4.

2   Our Approach

In a typical encoder-decoder framework with attention mechanism for HMER (e.g. [10]), given an image I, it will be first processed by an encoder to extract features. Then a decoder is used to predict the output sequence in a step-by-step manner, with the help of attention mechanisms to gather relevant features at each step. In this section, we introduce two baseline HMER models based on WAP and CNN-Transformer. Then, we adopt the design insights of Transformer decoder and propose a WAP-based model with multi-head attention (MHA) and stacked decoder for HMER.

2.1 WAP-Based HMER Model

First, we introduce our WAP-based baseline model [28]. Given an input math expression image, let $X \in \mathcal{R}^{D \times H \times W}$ denote extracted CNN feature maps, where D, H and W represent channel number, height and width, respectively. These feature maps can be viewed as a feature vector sequence $x = \{x_1, x_2, \cdots, x_L\}$, where $x_i \in \mathcal{R}^D$ and $L = H \times W$. Let $y = \{y_1, y_2, \cdots, y_T\}$ denote the target output sequence, where y_T represents a special end-of-sentence (EOS) symbol. As shown in Fig. 1, this baseline model uses two unidirectional GRU layers and a single-head attention (SHA) mechanism which works as follows:

$$h_t^{(0)} = \text{GRU}^{(0)}([y'_{t-1}; h_{t-1}^{(1)}]; h_{t-1}^{(0)}) \tag{1}$$

$$c_t = \text{Attention}(h_t^{(0)}, x) \tag{2}$$

$$h_t^{(1)} = \text{GRU}^{(1)}(c_t; h_{t-1}^{(1)}). \tag{3}$$

The first GRU layer uses previous step's y'_{t-1} and output $h_{t-1}^{(1)}$ as inputs, where $y'_{t-1} \in \mathcal{R}^{d_{\text{embed}}}$ is a trainable embedding of y_{t-1}. Using additional $h_{t-1}^{(1)}$ as input to the next step is known as "input feeding" [29]. The first GRU layer produces $h_t^{(0)} \in \mathcal{R}^{d_{\text{GRU}}}$ which works as the query of the attention mechanism. Then, the attention mechanism uses x as key and value and produces context vector $c_t \in \mathcal{R}^{d_{\text{out}}}$. Finally, the second GRU layer uses c_t as input and produces $h_t^{(1)} \in \mathcal{R}^{d_{\text{GRU}}}$.

We use a similar attention mechanism as in WAP, and the context vector c_t is calculated as follows:

$$e_{t,i} = v_a^T \tanh(W_h h_t^{(0)} + W_x x_i + f_{t,i} \cdot v_c + b_a) \tag{4}$$

$$\alpha_{t,i} = \frac{\exp(e_{t,i})}{\sum_{j=1}^{L} \exp(e_{t,j})} \tag{5}$$

$$c_t = \sum_{i=1}^{L} \alpha_{t,i} W'_x x_i, \tag{6}$$

where $v_a \in \mathcal{R}^{d_{\text{att}}}$, $W_h \in \mathcal{R}^{d_{\text{att}} \times d_{\text{GRU}}}$, $W_x \in \mathcal{R}^{d_{\text{att}} \times D}$, $W'_x \in \mathcal{R}^{d_{\text{out}} \times D}$, $v_c \in \mathcal{R}^{d_{\text{att}}}$, $b_a \in \mathcal{R}^{d_{\text{att}}}$ are trainable parameters, and f_t is the coverage vector calculated as follows:

$$f_t = \sum_{k=1}^{t-1} \alpha_k. \tag{7}$$

To calculate the output posterior probability, the previous step's symbol embedding y'_{t-1}, current step's context vector c_t and the output of the second

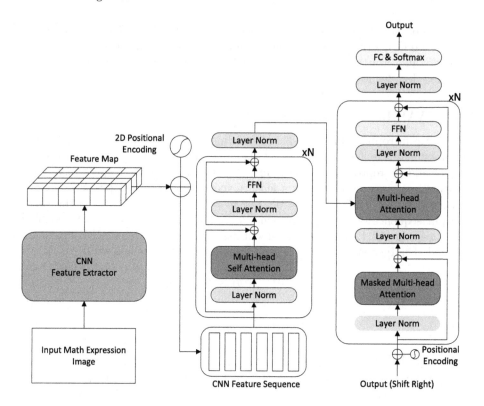

Fig. 2. Overview of baseline CNN-Transformer based HMER model.

GRU layer $h_t^{(1)}$ are used together as input to a 2-layer feed-forward network (FFN) with PReLU [30] and Softmax activations, respectively:

$$o_t = \text{PReLU}(\boldsymbol{W}_1[\boldsymbol{y}'_{t-1}; \boldsymbol{c}_t; \boldsymbol{h}_t^{(1)}] + \boldsymbol{b}_1) \tag{8}$$

$$p(y_t|y_1, \cdots, y_{t-1}; \boldsymbol{x}) = \text{Softmax}(\boldsymbol{W}_2 \boldsymbol{o}_t + \boldsymbol{b}_2), \tag{9}$$

where $\boldsymbol{W}_1 \in \mathcal{R}^{d_{\text{embed}} \times (d_{\text{embed}} + d_{\text{out}} + d_{\text{GRU}})}$, $\boldsymbol{W}_2 \in \mathcal{R}^{K \times d_{\text{embed}}}$, $\boldsymbol{b}_1 \in \mathcal{R}^{d_{\text{embed}}}$, $\boldsymbol{b}_2 \in \mathcal{R}^K$ and K is the number of total symbols.

2.2   CNN-Transformer Based HMER Model

Next, we introduce our CNN-Transformer based baseline model. As shown in Fig. 2, given the extracted CNN feature maps $\boldsymbol{X}$, it is viewed as a feature vector sequence after adding a positional encoding. Due to the nature of two-dimensional input, a two-dimensional positional encoding is used, which is calculated as

$$\text{PE}_{2\text{D}}(i, j) = [\text{PE}(i); \text{PE}(j)], \tag{10}$$

where each $\text{PE}(\cdot)$ denotes a regular sinusoidal positional encoding and (i, j) denotes the coordinate of each position. Then a regular Transformer encoder

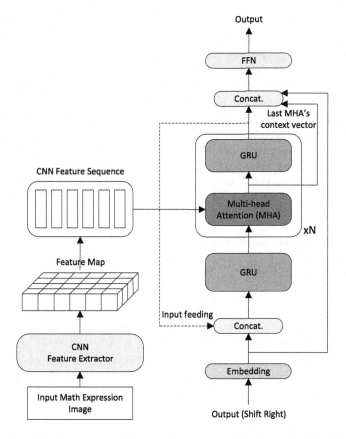

Fig. 3. Overview of our proposed HMER model with MHA and stacked decoder.

consisting of stacked Transformer encoder layers is applied to the CNN feature sequence. Each transformer encoder layer contains a multi-head self-attention module using scaled dot-product attention, an FFN, layer normalization layers, and residual connections. Finally, the Transformer encoded output is fed to a regular Transformer decoder which contains stacked Transformer decoder layers with masked multi-head self-attention and multi-head attention modules. Different from the original Transformer architecture [22], layer normalization is applied before attention and FFN modules instead of after as shown in Fig. 2, which is known as a "Pre-LN" Transformer [31].

2.3   WAP-based HMER Model with MHA and Stacked Decoder

Inspired by the success of WAP and Transformer, we adopt the design insights of Transformer to improve WAP-like encoder-decoder models for HMER. Specifically, we propose to apply multi-head attention (MHA) and stacked decoder to the WAP model as shown in Fig. 3.

Multi-head Attention. Compared with SHA, multi-head attention (MHA) uses N_h independent attention heads that run in parallel. Each attention head uses different linear projections for query, key and value. The output context vectors are concatenated together to produce a final attention output. With MHA, different attention heads have a potential to capture different information effectively. HMER models need to perform complex 2-dimensional structure pattern recognition including visible symbol recognition (e.g., letters, symbols and numbers) and relationship predictions (e.g., subscript and superscript). MHA could be useful when predicting symbols if different attention heads capture and combine different context information.

To apply MHA to WAP, we simply replace SHA in Eq. (2) with MHA as

$$c_t = \text{MHA}(h_t^{(0)}, x) \tag{11}$$

by using the following mechanism:

$$c_{t,i} = \text{Attention}(h_t^{(0)}, x; W_{h,i}, W_{x,i}, W'_{x,i}, v_{a,i}, v_{c,i}, b_{a,i}) \text{ for } 1 \leq i \leq N_h \tag{12}$$

$$c_t = [c_{t,1}; c_{t,2}; \cdots, c_{t,N_h}]. \tag{13}$$

In MHA, the i-th output head $c_{t,i}$ is calculated as in Eqs. (4, 5, 6) using its own trainable parameters $v_{a,i} \in \mathcal{R}^{d'_{\text{att}}}$, $W_{h,i} \in \mathcal{R}^{d'_{\text{att}} \times d_{\text{GRU}}}$, $W_{x,i} \in \mathcal{R}^{d'_{\text{att}} \times D}$, $W'_{x,i} \in \mathcal{R}^{d'_{\text{out}} \times D}$, $v_{c,i} \in \mathcal{R}^{d'_{\text{att}}}$, $b_{a,i} \in \mathcal{R}^{d'_{\text{att}}}$ and coverage vector, where $d'_{\text{att}} = \frac{d_{\text{att}}}{N_h}$ and $d'_{\text{out}} = \frac{d_{\text{out}}}{N_h}$. It is clear that the MHA has the same number of parameters as the single-head counterpart.

Stacked Decoder. In stacked decoder, the output of current decoder block is used as a query for the next decoder block. Stacked decoder increases the model capacity and can achieve better performance than a single-layer decoder.

For HMER, we adopt this design insight and propose a decoder using stacked decoder with MHA as shown in Fig. 3. It works as follows:

$$h_t^{(0)} = \text{GRU}^{(0)}([y'_{t-1}; h_{t-1}^{(N)}]; h_{t-1}^{(0)}) \tag{14}$$

$$\begin{cases} c_t^{(n)} = \text{MHA}(h_t^{(n-1)}, x) \\ h_t^{(n)} = \text{GRU}^{(n)}(c_t^{(n)}; h_{t-1}^{(n)}) \end{cases} \text{ for } 1 \leq n \leq N, \tag{15}$$

where N is the number of stacked decoder blocks. Finally, previous step's symbol embedding y'_{t-1}, the last decoder layer's context vector $c_t^{(N)}$ and attention output $h_t^{(N)}$ are used together to calculate $p(y_t|y_1, \cdots, y_{t-1}; x)$.

In experiments, we find that it is rather difficult to optimize this stacked decoder. In practice, to facilitate training, we calculate additional output posterior probabilities based on the n-th $(1 \leq n < N)$ decoder block's context vector and output as follows:

$$o_t^{(n)} = \text{PReLU}(W_1^{(n)}[y'_{t-1}; c_t^{(n)}; h_t^{(n)}] + b_1^{(n)}) \tag{16}$$

$$p^{(n)}(y_t|y_1,\cdots,y_{t-1};\boldsymbol{x}) = \text{Softmax}(\boldsymbol{W}_2^{(n)}\boldsymbol{o}_t^{(n)} + \boldsymbol{b}_2^{(n)}). \tag{17}$$

Then, we propose to train this stacked decoder using an auxiliary training criterion as follows:

$$\mathcal{L} = -\sum_t \log p(y_t|y_1,\cdots,y_{t-1};\boldsymbol{x}) - \lambda \sum_{n=1}^{N-1}\sum_t \log p^{(n)}(y_t|y_1,\cdots,y_{t-1};\boldsymbol{x}), \tag{18}$$

where λ is a hyper-parameter. Using the above auxiliary training criterion allows HMER model to gradually refine model prediction to reduce errors. It is noted that these additional output posterior probabilities are only used in training phase. Only the final posterior probability is used in inference.

3 Experiments

3.1 Experimental Setup

We evaluate our methods on CROHME competition datasets [32–34]. The CROHME 2014 training set, which contains 8,836 mathematical expression images, is used as training set. The CROHME 2014 test set with 986 images is chosen as validation set, since its vocabulary and data distribution match the training set better. The CROHME 2016 test set with 1,147 images and CROHME 2019 test set with 1,199 images are used as testing sets. ExpRate on the validation set is used to select trained models for evaluation. We report the ExpRate as well as recognition rates when at most one, two, or three symbol errors are tolerable, which are denoted as "$\leq$1", "$\leq$1" and "$\leq$3", respectively.

During training, models are optimized with AdaDelta [35] with decay rate $\rho = 0.9$ and weight decay of 0.0001. We augment the training set using a scale augmentation method in [16] where the scaling factor is randomly chosen within $[0.7, 1.4]$. We also randomly change the aspect ratio by scaling the image width with a scaling factor within $[0.9, 1.1]$. In experiments, the data augmentation method can greatly improve the performance of WAP, which is consistent with the results in [16]. We implement our methods with PyTorch [36] and fairseq [37] toolkits.

3.2 Comparison of WAP and CNN-Transformer Baseline Models

We first evaluate the performances of WAP and CNN-Transformer based baseline models. In both models, the same DenseNet-BC [38] architecture as in [12] is used as the CNN part, which contains 3 dense blocks. The number of channels in the extracted CNN feature map is 684. For WAP's decoder part, hyperparameters are set as $d_{\text{GRU}} = d_{\text{embed}} = 256$, $d_{\text{att}} = d_{\text{out}} = 512$. Dropout with keep ratio of 0.5 is applied to $\boldsymbol{y}_t'$, $\boldsymbol{h}_t^{(*)}$, $\boldsymbol{c}_t$ and $\boldsymbol{o}_t$. For CNN-Transformer, we try various network configurations, and the model that performs the best on validation set uses 2 transformer encoder and 2 transformer decoder blocks, with 16 attention heads. The total output dimension of attention heads are 256. The number of

Table 1. Performance comparison of greedy search and beam search algorithms for two baseline models on CROHME 2014 and CROHME 2016. #Param. is the number of parameters of each model.

Model	Decoding algorithm	τ	CROHME 2014 ExpRate (%)	CROHME 2016 ExpRate (%)	#Param.
WAP	Greedy search	/	53.85	53.27	5.45M
	Beam search	1	54.36	53.62	
		2	55.27	**53.97**	
CNN-Transformer	Greedy search	/	54.06	52.58	6.90M
	Beam search	1	54.87	53.44	
		2	**55.98**	53.79	

Table 2. Performance comparison of HMER models with SHA and MHA on CROHME 2014 and CROHME 2016.

Attention mechanism	N_h	CROHME 2014 ExpRate (%)	CROHME 2016 ExpRate (%)	≤ 1 (%)	≤ 2 (%)	≤ 3 (%)
SHA (WAP)	1	55.27	53.97	70.88	78.81	83.44
MHA	2	56.09	55.27	71.05	79.95	84.57
	4	56.90	**56.41**	70.62	79.77	83.96
	8	**57.71**	55.88	70.62	80.12	85.09
	16	56.59	55.80	71.93	80.65	85.09

neurons in FFN layers is 1024. Dropout with keep ratio of 0.6 is applied to scaled dot-product attention metrics and every linear transformation output in FFNs. Warm up strategy where learning rate gradually increases from 0.0 to 1.0 in the first 5 epochs and label smoothing trick are used to better train the CNN-Transformer baseline.

During decoding, we try greedy search and beam search algorithms. In experiments, we find that the model's posterior distribution can be over-confident as observed in [39]. Therefore, we also try to "soften" the output by dividing o_t with a temperature τ. In beam search, we use a beam size of 10 in experiments.

Table 1 lists the performance comparison of these two baseline models using greedy search and beam search algorithms. It is clear that beam search achieves better performances than simple greedy search. By softening the output distribution with $\tau = 2$, we can achieve even better performances. Between these two baseline models, CNN-Transformer achieves better performance than WAP-based baseline on CROHME 2014 test set, while WAP slightly outperforms CNN-Transformer on CROHME 2016 test set. In following experiments, we report recognition results using beam search decoding algorithm with $\tau = 2$.

3.3 Effect of MHA

Built on WAP, we then evaluate the effect of MHA. We try $N_h = \{2, 4, 8, 16\}$ and the comparison results are listed in Table 2. Obviously, all models with

Table 3. Performance comparison of HMER models with single and stacked decoder on CROHME 2014 and CROHME 2016. #Param. is the number of parameters of each model.

N	N_h	λ	CROHME 2014	CROHME 2016				#Param.
			ExpRate (%)	ExpRate (%)	≤ 1 (%)	≤ 2 (%)	≤ 3 (%)	
1	1	/	55.27	53.97	70.88	78.81	83.44	5.45M
	4	/	56.90	56.41	70.62	79.77	83.96	
	8	/	57.71	55.88	70.62	80.12	85.09	
2	4	0.0	55.78	54.66	70.79	77.86	81.78	7.16M
		0.5	58.52	**57.54**	72.19	80.82	85.53	
		1.0	58.72	56.06	71.05	79.25	84.13	
	8	0.5	58.42	57.19	72.45	81.17	85.27	

MHA outperform the SHA baseline, where MHA with 4 heads achieves the best ExpRate on the CROHME 2016 test set. Compared with SHA, MHA boosts the ExpRate on CROHME 2016 test set from 53.97% to 56.41% with a 2.44% absolute improvement.

3.4 Effect of Stacked Decoder

Finally, we evaluate the effect of stacked decoder. We start with a stacked decoder HMER model with hyperparameter settings $N_h = 4$ and $N = 2$. As shown in Table 3, if we directly train this model without using auxiliary loss function ($\lambda = 0.0$), it achieves even worse performance than the model without using stacked decoder. It shows that it is difficult to optimize this deeper model. To better train this model, we try the auxiliary loss function with $\lambda = 0.5$, 1.0. The results are also listed in Table 3. Clearly, with the help of auxiliary loss function, HMER model with stacked decoder can improve the ExpRate on CROHME 2016 test set from 56.41% to 57.54%, leading to a 1.13% absolute improvement. Beside only using the final posterior probability in inference, we also try to combine the predictions of every decoder blocks and find that it achieves worse performance. For stacked decoder, we also try to add a residual connection from $h_t^{(1)}$ to $h_t^{(2)}$ and train this model without using auxiliary loss function. We find that it only achieves similar performance (ExpRate of 55.38% on CROHME 2014 test set) compared with model without residual connection. Finally, we try the stacked decoder with 8 attention heads and find that it outperforms the single decoder counterpart by 1.31% absolutely. These results verify the effectiveness of our proposed methods.

3.5 Comparison with Other HMER Systems and Error Analysis

In Table 4 and Table 5, we compare our models with other offline HMER systems on CROHME 2016 and CROHME 2019 test sets, respectively. For a fair comparison, we convert the predicted LaTeX sequences to label graphs and evaluate

Table 4. Performance comparison of HMER systems on CROHME 2016.

System	ExpRate (%)	≤1 (%)	≤2 (%)	≤3 (%)
Wiris [33]	49.61	60.42	64.69	/
Tokyo [33]	43.94	50.91	53.70	/
Sao Paolo [33]	33.39	43.50	49.17	/
Nantes [33]	13.34	21.02	28.33	/
PGS [17]	45.60	62.25	70.44	75.76
WAP [28]	46.82	64.64	65.48	/
WS-WAP [18]	48.91	57.63	60.33	61.63
PAL-v2 [14]	49.61	64.08	70.27	73.50
Li, et al. [16]	54.58	69.31	73.76	76.02
WAP (Our implementation)	54.32	69.40	75.15	77.94
CNN-Transformer	54.66	67.13	72.19	75.24
WAP + MHA ($N_h = 4$)	56.76	68.27	74.98	77.33
WAP + MHA + Stacked decoder ($\lambda = 0.5$)	**57.72**	70.01	76.37	78.90

Table 5. Performance comparison of HMER systems on CROHME 2019.

System	ExpRate (%)	≤1 (%)	≤2 (%)	≤3 (%)
DenseWAP [19]	41.7	55.5	59.3	/
DenseWAP-TD [19]	51.4	66.1	69.1	/
WAP (Our implementation)	58.05	72.31	76.81	78.48
CNN-Transformer	59.30	71.48	75.73	78.23
WAP + MHA ($N_h = 4$)	59.72	74.73	78.65	80.23
WAP + MHA + Stacked decoder ($\lambda = 0.5$)	**61.38**	75.15	80.23	82.65

the performance using official tools provided by the CROHME 2019 organizers [34]. Here we only list results for each system obtained by using a single HMER model. We do not compare our models with the competition teams in [34], since they use additional training set or some synthetic dataset besides CROHME 2014 training set. It is noted that scale augmentation technique is also used in [16].

From Table 4, we can find that there is a large gap between ExpRate and ≤1 (%). For example, the gap is about 12.29% on our proposed model. After error analysis, we find that among those wrongly recognized samples by a single error, most of them are caused by symbol substitution error. Some typical substitution error pairs include "z" and "2", "..." and "⋯", "x" and "X", "9" and "q",

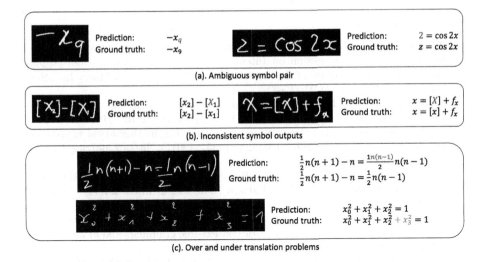

(a). Ambiguous symbol pair

(b). Inconsistent symbol outputs

(c). Over and under translation problems

Fig. 4. Some typical error cases of our proposed HMER model.

"x" and "$\times$", etc. This suggests that there is plenty of room for improvement. For example, the ExpRate could be improved further if we use discriminative features that could distinguish these symbol pairs.

In Fig. 4, we show some typical error cases. We find that (a) Some errors are caused by ambiguous symbol pairs which are difficult to distinguish; (b) The model can fail to produce consistent symbol output, for example, it produces mixed "x" and "X"; and (c) Although coverage vector is used in HMER model, it still suffers from over and under translation problems.

4 Summary

In this paper, inspired by the recent success of Transformer, we propose to improve recognition accuracies of WAP on HMER task with MHA and stacked decoder. Experimental results on CROHME datasets show that both MHA and stacked decoder can improve the recognition performances of HMER models. Our method achieves 57.72% and 61.38% expression recognition rate on CROHME 2016 and CROHME 2019 test sets, respectively. To the best of our knowledge, this is the best result on this dataset using a single model. As future work, we will continue to improve HMER performances by leveraging more advanced modeling and training methods to deal with ambiguous symbol pair, inconsistent symbol outputs, and under/over translation problems. Moreover, we only compare our proposed model and CNN-Transformer baseline on a relatively small HMER task. How they will perform on larger scale datasets need further investigations.

References

1. Anderson, R.H.: Syntax-directed recognition of hand-printed two-dimensional mathematics. In: Symposium on Interactive Systems for Experimental Applied Mathematics: Proceedings of the Association for Computing Machinery Inc., Symposium, pp. 436–459. ACM (1967)
2. Chan, K.-F., Yeung, D.-Y.: Mathematical expression recognition: a survey. Int. J. Doc. Anal. Recognit. **3**(1), 3–15 (2000). https://doi.org/10.1007/PL00013549
3. Zanibbi, R., Blostein, D.: Recognition and retrieval of mathematical expressions. Int. J. Doc. Anal. Recognit. **15**(4), 331–357 (2012). https://doi.org/10.1007/s10032-011-0174-4
4. Zanibbi, R., Blostein, D., Cordy, J.R.: Recognizing mathematical expressions using tree transformation. IEEE Trans. Pattern Anal. Mach. Intell. **24**(11), 1455–1467 (2002)
5. Álvaro, F., Sánchez, J.-A., Benedí, J.-M.: Recognition of on-line handwritten mathematical expressions using 2d stochastic context-free grammars and hidden Markov models. Pattern Recognit. Lett. **35**, 58–67 (2014)
6. Awal, A.-M., Mouchère, H., Viard-Gaudin, C.: A global learning approach for an online handwritten mathematical expression recognition system. Pattern Recognit. Lett. **35**, 68–77 (2014)
7. Álvaro, F., Sánchez, J.-A., Benedí, J.-M.: An integrated grammar-based approach for mathematical expression recognition. Pattern Recognit. **51**, 135–147 (2016)
8. Sutskever, I., Vinyals, O., Le, Q.: Sequence to sequence learning with neural networks. In: Advances in Neural Information Processing Systems (NIPS 2014)
9. Bahdanau, D., Cho, H., Bengio, Y.: Neural machine translation by jointly learning to align and translate. In: 2014 International Conference on Learning Representations (ICLR)
10. Zhang, J., Du, J., Zhang, S., et al.: Watch, attend and parse: an end-to-end neural network based approach to handwritten mathematical expression recognition. Pattern Recognit. **71**, 196–206 (2017)
11. Deng, Y., Kanervisto. A., Ling. J., et al.: Image-to-markup generation with coarse-to-fine attention. In: 2017 International Conference on Machine Learning (ICML), pp. 980–989
12. Zhang, J., Du, J., Dai, L.: Multi-scale attention with dense encoder for handwritten mathematical expression recognition. In: 2018 International Conference on Pattern Recognition (ICPR), pp. 2245–2250
13. Wu, J.-W., Yin, F., Zhang, Y.-M., Zhang, X.-Y., Liu, C.-L.: Image-to-markup generation via paired adversarial learning. In: Berlingerio, M., Bonchi, F., Gärtner, T., Hurley, N., Ifrim, G. (eds.) ECML PKDD 2018. LNCS (LNAI), vol. 11051, pp. 18–34. Springer, Cham (2019). https://doi.org/10.1007/978-3-030-10925-7_2
14. Wu, J.-W., Yin, F., Zhang, Y.-M., Zhang, X.-Y., Liu, C.-L.: Handwritten mathematical expression recognition via paired adversarial learning. Int. J. Comput. Vis. **128**, 2386–2401 (2020). https://doi.org/10.1007/s11263-020-01291-5
15. Hong, Z., You, N., Tan, J., et al.: Residual BiRNN based Seq2Seq model with transition probability matrix for online handwritten mathematical expression recognition, In: 2019 International Conference on Document Analysis and Recognition (ICDAR), pp. 635–640. IEEE
16. Li, Z., Jin, L., Lai, S., Zhu, Y.: Improving attention-based handwritten mathematical expression recognition with scale augmentation and drop attention. In: 2020 International Conference on Frontiers in Handwriting Recognition (ICFHR), pp. 175–180

17. Le, A.D., Indurkhya, B., Nakagawa, M.: Pattern generation strategies for improving recognition of handwritten mathematical expressions. Pattern Recognit. Lett. **128**, 255–262 (2019)
18. Truong, T. -N., Nguyen, C.T., Phan, K.M., et al.: Improvement of end-to-end offline handwritten mathematical expression recognition by weakly supervised learning. In: 2020 International Conference on Frontiers in Handwriting Recognition (ICFHR), pp. 181–186
19. Zhang, J., Du, J., Yang, Y., et al.: A tree-structured decoder for image-to-markup generation. In: 2020 International Conference on Machine Learning (ICML), pp 11076–11085
20. Pang, N., Yang, C., Zhu, X., et al.: Global context-based network with transformer for image2latex. In: 2020 International Conference on Pattern Recognition (ICPR)
21. Yan, Z., Zhang, X., Gao, L., et al.: ConvMath: a convolutional sequence network for mathematical expression recognition. In: 2020 International Conference on Pattern Recognition (ICPR)
22. Vaswani, A., Shazeer, N., Parmar, N., et al.: Attention is all you need. In: 2017 Advances in Neural Information Processing Systems (NeurIPS), pp. 6000–6010
23. Devlin, J., Chang, M.-W., Lee, K., et al.: BERT: Pre-training of deep bidirectional transformers for language understanding. arXiv preprint arXiv:1810.04805 (2018)
24. Dosovitskiy, A., Beyer, L., Kolesnikov, A., et al.: An image is worth 16x16 Words: transformers for image recognition at scale, arXiv preprint arXiv:2010.11929 (2020)
25. Carion, N., Massa, F., Synnaeve, G., et al.: End-to-end object detection with transformers, arXiv preprint arXiv:2005.12872 (2020)
26. Wang, Y., Mohamed, A., Le, D., et al.: Transformer-based acoustic modeling for hybrid speech recognition. In: 2020 International Conference on Acoustics, Speech and Signal Processing (ICASSP), pp. 6874–6878. IEEE
27. Gulati, A., Qin, J., Chiu, C.-C., et al.: Conformer: Convolution-augmented transformer for speech recognition, arXiv preprint arXiv:2005.08100 (2020)
28. Wang, J., Du, J., Zhang, J., et al.: Multi-modal attention network for handwritten mathematical expression recognition. In: 2019 International Conference on Document Analysis and Recognition (ICDAR), pp. 1181–1186. IEEE
29. Luong, T., Pham, H., Manning, C.D.: Effective approaches to attention-based neural machine translation. In: 2015 Empirical Methods in Natural Language Processing (EMNLP), pp. 1412–1421
30. He, K., Zhang, X., Ren, S., Sun, J.: Delving deep into rectifiers: surpassing human-level performance on ImageNet classification. In: 2015 International Conference on Computer Vision (ICCV), pp. 1026–1034. IEEE
31. Baevski, A., Auli, M.: Adaptive input representations for neural language modeling. In: 2019 International Conference on Learning Representations (ICLR)
32. Mouchère, H., Viard-Gaudin, C., Zanibbi, R., Garain, U.: ICFHR 2014 competition on recognition of on-Line handwritten mathematical expressions. In: 2014 International Conference on Frontiers in Handwriting Recognition (ICFHR), pp. 791–796
33. Mouchère, H., Viard-Gaudin, C., Zanibbi, R., Garain, U.: ICFHR2016 CROHME: competition on recognition of online handwritten mathematical expressions. In: 2016 International Conference on Frontiers in Handwriting Recognition (ICFHR), pp. 607–612
34. Mahdavi, M., Zanibbi, R., Mouchère, H., Viard-Gaudin, C., Garain, U.: ICDAR 2019 CROHME+cTFD: competition on recognition of handwritten mathematical expressions and typeset formula detection. In: 2019 International Conference on Document Analysis and Recognition (ICDAR), pp. 1533–1538

35. Zeiler, M.D.: ADADELTA: an adaptive learning rate method, arXiv preprint arXiv:1212.5701 (2012)

36. Paszke, A., Gross, S., Massa, F., et al.: Pytorch: an imperative style, high-performance deep learning library. arXiv preprint arXiv:1912.01703 (2019)

37. Ott, M., Edunov, S., Baevski, A., et al.: Fairseq: a fast, extensible toolkit for sequence modeling. In: 2019 Human Language Technology: Conference of the North American Chapter of the Association of Computational Linguistics (NAACL-HLT) Demonstrations

38. Huang, G., Liu, Z., Van Der Maaten, L., Weinberger, K.Q.: Densely connected convolutional networks. In: 2017 IEEE Conference on Computer Vision and Pattern Recognition (CVPR), pp. 4700–4708

39. Chorowski, J., Jaitly, N.: Towards better decoding and language model integration in sequence to sequence models. In: 2017 Annual Conference of the International Speech Communication Association (Interspeech), pp. 523–527

Global Context for Improving Recognition of Online Handwritten Mathematical Expressions

Cuong Tuan Nguyen[1]([envelope]) [ID], Thanh-Nghia Truong[1] [ID], Hung Tuan Nguyen[2] [ID],
and Masaki Nakagawa[1] [ID]

[1] Department of Computer and Information Sciences, Tokyo University of Agriculture and
Technology, 2-24-16 Naka-cho, Koganei-shi, Tokyo 184-8588, Japan
fx4102@go.tuat.ac.jp, nakagawa@cc.tuat.ac.jp
[2] Institute of Global Innovation Research, Tokyo University of Agriculture and Technology,
2-24-16 Naka-cho, Koganei-shi, Tokyo 184-8588, Japan

Abstract. This paper presents a temporal classification method for all three sub-tasks of symbol segmentation, symbol recognition and relation classification in online handwritten mathematical expressions (HMEs). The classification model is trained by multiple paths of symbols and spatial relations derived from the Symbol Relation Tree (SRT) representation of HMEs. The method benefits from global context of a deep bidirectional Long Short-term Memory network, which learns the temporal classification directly from online handwriting by the Connectionist Temporal Classification loss. To recognize an online HME, a symbol-level parse tree with Context-Free Grammar is constructed, where symbols and spatial relations are obtained from the temporal classification results. We show the effectiveness of the proposed method on the two latest CROHME datasets.

Keywords: Temporal classification · Online handwritten mathematical expression · Context-Free Grammar

1 Introduction

Recognizing an online handwritten mathematical expression (HME) is a task of transcribing a sequence of pen-traces as coordination points input into a LaTeX or MathML description. It provides a simple and efficient user interface for mathematical input, where users can write mathematical expressions directly on a tablet-PC or by a digital pen. The user interface provides benefits for e-learning applications. Moreover, HME recognition is also useful for automatic marking [1] and intelligent tutoring system [2].

Generally, a stroke level parse tree and Context-Free Grammar (CFG) are used to represent the two-dimensional structure of an HME [3, 4]. An HME is recognized by constructing a parse tree from the results of three subtasks: symbol segmentation, symbol recognition, and relation classification. Due to separated models for the three subtasks, however, global context is not utilized for all subtasks. Besides, these models are highly dependent on handcrafted features.

© Springer Nature Switzerland AG 2021
J. Lladós et al. (Eds.): ICDAR 2021, LNCS 12822, pp. 617–631, 2021.
https://doi.org/10.1007/978-3-030-86331-9_40

An encoder-decoder model using Deep Neural Networks is one of the solutions for HME recognition, which performs all the subtasks in a single model [5]. The model benefits from the global context by deep neural networks and avoids the dependency of handcrafted features. However, the Deep Neural Network approach currently suffers from a high cost of recognition due to its complexity. Another problem is the challenge of utilizing grammar for improving performance.

The global context could be obtained through sequential models. In handwritten text recognition, temporal recognition of characters benefits from the bidirectional context of preceding strokes and succeeding strokes by Recurrent Neural Networks (RNNs) [6]. Nguyen et al. applied a bidirectional RNN model for improving symbol segmentation and recognition of online HMEs [7]. Nonetheless, classification of spatial relations (relation classification) between symbols was separated from other subtasks, which does not benefit from the global context. Zhang et al. applied a tree-based bidirectional RNN model to learn symbol recognition and relation classification from HMEs [8]. However, the approach exhibited difficulty to learn the relation classification.

In this work, we improve the performance of the three subtasks by introducing a single recognition model for them. This model uses a bidirectional RNN, which utilizes the global context for symbol recognition and relation classification. We apply a constraint to make the model output the relation classification at a precise time step to solve both symbol segmentation and relation classification. The model is trained using raw online HMEs so that it does not depend on handcrafted features. We apply a new parsing method at the symbol level instead of the stroke level to produce the result of HME recognition. In the parsing process, missing relations are reevaluated by applying the temporal classifier.

The rest of the paper is as follows: Sect. 2 introduces the related works, Sect. 3 describes our methods, Sect. 4 shows the experiment results and discussion, Sect. 5 draws our conclusion.

2   Related Works

2.1   Tree-Based BLSTM for HME Recognition

Zhang et al. used a Stroke Label Graph (SLG) to represent an HME, in which nodes represent strokes and edges represent relations between the strokes. The graph is represented by a bidirectional Long Short-term Memory (BLSTM) or by a tree-structured BLSTM (tree-BLSTM) [8, 9]. These models took advantage of the bidirectional context by BLSTM to learn the classification directly from raw online features so that they achieved high symbol classification rates. However, the models are still troubled with learning spatial relations. Due to lack of coverage on spatial relations learned by sequential input or tree-based input, multiple BLSTM models or tree-BLSTM models were used to learn different sequences or tree derivations from the SLGs.

Tree-BLSTM firstly suffers from learning by stroke-level constraint, where the spatial relations from strokes are complex. Second, it encounters difficulty in integrating separated tree models. Consequently, there is a significant drop in the recall rate of relation classification [8, 9].

2.2 Spatial Relations Classification

For recognizing spatial relations, the common approach depends on handcrafted features such as geometrical descriptors [3], shape descriptors [10]. The handcrafted features are sensitive to ambiguous spatial relations since they are extracted from bounding boxes of symbols [3] or their centroids [4]. To deal with the problem, they modified the descriptors to add the dependency on symbol classes. The symbol classes were grouped into four groups of Ascendant, Descendant, Normal, and Big. Then, bounding boxes were modified as "body box" while vertical centers were shifted according to each group [3, 4]. These modifications also need to be carefully designed and may not be robust for various handwriting styles. Despite the modifications, the approach is still problematic with the *Subscript* relation due to its confusion to the *Right* relation.

The BLSTM and tree- BLSTM models could learn spatial relations directly from features [8, 9]. However, they perform a limited accuracy and have a problem with the *Subscript* relation.

2.3 Symbol Recognition with Bidirectional Context

Nguyen et al. applied a temporal classification method for recognizing mathematical symbols [7]. The method uses a BLSTM model to take advantage of the bi-directional context for classifying symbols. The method improves symbol classification rate and expression recognition rate of the CFG based parse tree for online HME recognition. The results showed the effectiveness of applying global context to symbol recognition. The segmentation of the online HME recognition system is improved by "junk detection," and it also benefits from the temporal classifier with the global context. For spatial relations classification, however, the method is still relying on geometrical descriptors.

3 Our Method

In this section, we propose a single model for learning all the three subtasks of online HME recognition and its integration to a parse tree for recognizing the whole HME.

3.1 Overview of the Proposed Method

Our method consists of two stages, as shown in Fig. 1. A sequence of extracted features from an online HME input is fed to a deep BLSTM symbol-relation temporal classifier to predict symbols and relations in the first stage. In the second stage, the symbols and relations from the temporal classifier are input to a symbol-level parser to build the parse tree and the recognition result is output as a Latex string.

3.2 Sequential Model for Segmentation, Recognition, and Relation Classification

The symbol relation tree (SRT) or symbol layout tree [11] is an effective way to represent the structure of an HME, where symbols are represented as nodes and spatial relations between symbols are represented as links between nodes [12]. Figure 2 shows an example

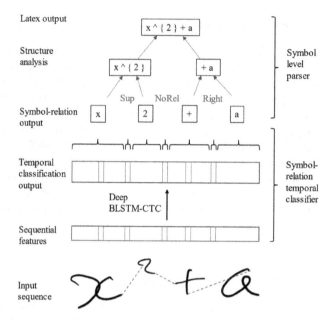

Fig. 1. Overview of the proposed method.

of an input online HME and its SRT representation in Fig. 2(a) and 2(b), respectively. Each node contains a symbol label and indexes of the strokes that belong to the symbol. A link between two nodes is a directed connection that denotes a spatial relation from the parent node to the child node. There are six types of spatial relations in SRT: *Above*, *Below*, *Sub* (subscript), *Sup* (superscript), *Right*, and *Inside* (square root).

The whole structure of an HME could be represented by derived paths of consecutive symbols and spatial relations between them from its SRT. Figure 2(c) shows derived paths by tracing from the root to all leaves of the SRT. Each derived path contains the information of the input sequence and its label sequence, which can be learned by a sequential model.

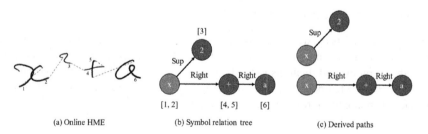

(a) Online HME   (b) Symbol relation tree   (c) Derived paths

Fig. 2. Symbol relation tree and its derived paths.

Here, we propose a sequential model for recognizing both symbols and spatial relations. The model is a temporal classifier that uses a feature sequence as input and output the sequence of symbols and spatial relations between two consecutive symbols. The temporal classifier produces the probabilities of symbols and relations for every time step of the input sequence. The input sequence consists of both strokes and off-strokes, where an off-stroke is a virtual pen trace between two consecutive strokes, connecting the end of the first stroke to the beginning of the second stroke.

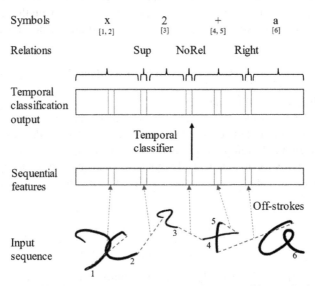

Fig. 3. Temporal classifier for symbol segmentation, symbol recognition, and relation classification.

Assuming the sequential order of input strokes and symbols are consistent (i.e., there are no delayed strokes), symbols are separated by off-strokes between them. We propose a temporal classifier that predicts symbols and spatial relations at the precise time steps of the corresponding strokes and off-strokes, as shown in Fig. 3. For off-strokes between two strokes inside a symbol, there is no output. From the output of the temporal classifier, we obtain both the segmentation positions and the spatial relations between symbols at the off-stroke locations. Then, the symbols are separated their classification are obtained. Therefore, we can solve three tasks of symbol segmentation, symbol recognition, and relation classification using the single temporal classifier.

3.3 Deep BLSTM for Segmentation, Recognition, and Relation Classification

We apply a deep BLSTM to incorporate bidirectional context for symbol and relation classification. The model is a stack of multiple BLSTM layers where each layer is a combination of two LSTM layers that process the input in forward and backward directions [13]. The forward and backward context by the two LSTM layers is combined and fed to the next BLSTM layer in the networks. The deep architecture is for extracting high-level features directly from the raw input sequence, while BLSTM benefits from the long-range context of forward and backward directions by LSTMs.

For training the deep BLSTM network, a Connectionist Temporal Classifier (CTC) [6] loss function is applied. CTC helps the network learn from the input sequence to a target sequence without explicit alignment or segmentation needed. The alignment between the input sequence and output label sequence is learned automatically with the assumption that the two sequences are in the same order.

CTC introduces a label called 'blank' that denotes no label. It defines the output y^t of RNN for each time step t with respect to an input sequence x with the length T as the probability distribution over a fixed set of classes C and the 'blank' label as shown in Eq. (1).

$$y_k^t = p(k, t|x), \forall k \in C \cup blank \tag{1}$$

where y_k^t is the output y^t for a class k.

An output label sequence l is obtained by a reduction process B over a path $\pi_{1:T} = k_1, k_2, .., k_T$ through the lattice of output labels, i.e. $k_i \in C \cup blank, i = \overline{1, T}$. The reduction process firstly removes repeated labels, then removes 'blank' labels in this path. The probability for an output label sequence l from an input sequence x is the total probability of all the paths $\pi_{1:T}$ such that each path $\pi_{1:T} \in B^{-1}(l)$ is reduced into l, as shown in Eq. (2).

$$p(l|x) = \sum_{\pi_{1:T} \in B^{-1}(l)} p(l, \pi_{1:T}|x) \tag{2}$$

where $p(l, \pi_{1:T}|x) = \prod_{t=1}^{T} p(k_t, t|x)$ is the probability of the label sequence l over the path $\pi_{1:T}$.

For a pair of an input sequence x and an output sequence l from the training dataset, the network is trained by minimizing the CTC loss obtained by Eq. (3).

$$loss_{CTC} = -log(p(l|x)) \tag{3}$$

From the sequential model, segmentation is performed by finding off-strokes with high probabilities of relations. For an online HME, let S is a sequence of n strokes of the HME as $S = (s_0, \ldots, s_{n-1})$ and O is a sequence of $(n-1)$ off-strokes as $O = (o_1, \ldots, o_{n-1})$ where o_i is an off-stroke between two strokes s_{i-1} and s_i, φ_{HME} is the BLSTM context when parsing S with the symbol-relation temporal classifier. The ith off-stroke o_i can be a relation or 'blank' character between two strokes inside a symbol. The relation between the $(i-1)^{th}$ symbol and the ith symbol is obtained from the relation probability at the ith off-stroke, which is calculated as shown in Eq. (4):

$$Rel(o_i) = \begin{cases} argmax(P_{rel}(o_i|\varphi_{HME})) if\ max(P_{rel}(o_i|\varphi_{HME})) \geq P_{blank} \\ 'blank'\ if\ max(P_{rel}(o_i|\varphi_{HME})) < P_{blank} \end{cases} \tag{4}$$

where:

- $Rel(o_i)$ is the predicted relation of ith off-stroke.
- $P_{rel}(o_i|\varphi_{HME})$ is the relation probability of the relation "rel" at o_i.
- P_{blank} is the probability of o_i being a 'blank' character.

Symbol recognition is performed by taking the maximum probability of symbols between two relation outputs. The symbol recognition for a list of t consecutive strokes $(s_i, \ldots, s_{i+t})$ is computed as shown in Eq. (5):

$$Symbol(s_{i:i+t}) = argmax\big(P_{symbol}(s_{i:i+t}|\varphi_{HME})\big) \tag{5}$$

where $P_{symbol}(s_{i:i+t}|\varphi_{HME})$ is the probability of symbol recognition from stroke s_i to stroke s_{i+t}.

3.4   Constraint for Output at Precise Time Steps

Zhang et al. used the "local CTC" learning method, which applied a constraint on the output of symbols and relations at the specific strokes and off-strokes [8]. The method, however, does not constrain the output of relations to the exact time steps so that it causes an ambiguity for learning relation classification on many possible time steps.

As BLSTMs could learn precise time steps [14], we propose a single feature point for representing each off-stroke and applying the constraint loss shown in Eq. (6). The constraint loss is in the form of a binary cross-entropy loss, which penalizes the relations output at the strokes. Thus, the model is forced to learn to output relation classification at the off-strokes.

$$loss_{CE} = -\sum_{i=0}^{n-1} log\Big(1 - \sum P_{rel}(s_i|\varphi_{HME})\Big) \tag{6}$$

The deep BLSTM model is trained using a combination of the constraint loss and CTC loss as shown in Eq. (7).

$$loss = loss_{CTC} + \lambda loss_{CE} \tag{7}$$

where λ is a weighted parameter that is determined experimentally.

3.5   Training Path Extractions

From an SRT, we extract multiple derived paths of input stroke sequences from the stroke indexes of the symbol sequence for training the temporal classifier. Each of them represents a path of strokes and off-strokes, as well as their symbol labels and the spatial relations among them.

We propose three methods of path extractions:

1. Trace all paths from the root to all leaves of an SRT
2. Trace the path by writing order. *NoRel* is added when there is no relation between two consecutive nodes.
3. Extract random paths. Algorithm 1 comprehensively presents how we randomly extract derived paths. The method randomly shuffles the order of sub-trees connected to the parent node and then combines them to simulates various writing orders.

Algorithm 1: ExtractRandomPath	
Input:	
Root node: r.	
Output:	
List of nodes as a path: p.	

1	**if** |r.childs| = 0 **then**
2	$\quad$ $p \leftarrow [r]$
3	**else if** |r.childs| = 1 **then**
4	$\quad$ $p \leftarrow [\, r, \text{ExtractRandomPath}(r.\text{childs}[0])]$
5	**else**
6	$\quad$ $p = [\quad]$
7	$\quad$ **for** $node$ in shuffle ($[r, r.\text{childs}]$) **do**
8	$\quad\quad$ **if** node $= r$ **then**
9	$\quad\quad$ $\quad$ $p \leftarrow [p, node]$
10	$\quad\quad$ **else**
11	$\quad\quad$ $\quad$ $p \leftarrow [p, \text{ExtractRandomPath}(node)]$
12	$\quad\quad$ **end if**
13	$\quad$ **end for**
14	**end if**
15	**return** p

Figures 4 and 5 show examples of how we extract derived paths from an SRT. Figure 4(a) presents the example of tracing all derived paths from the root node of the SRT. In this case, all the spatial relations can be extracted except *NoRel*. Figure 4(b) presents an example of tracing a derived path by the writing order, which may contain *NoRel*. To extract more *NoRel* relations, we propose the third path extraction methods. Figure 5 shows an example of generating derived paths by randomly shuffling child nodes of the SRT.

The path extraction methods generate multiple derived paths from a single SRT. Therefore, our proposed method can generate a large number of patterns for training the temporal classifier.

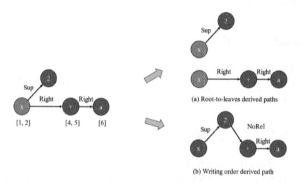

Fig. 4. Path derived from SRT.

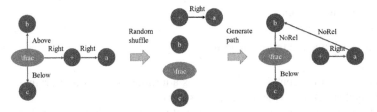

Fig. 5. Random path generation by shuffling child nodes.

3.6 Symbol Level Parse Tree

Symbol Level Syntactic Parse Tree. We use a two-dimensional Context-Free Grammar (2D-CFG) G of a 4-tuple $(\mathcal{N}, \Sigma, S, P)$, where $\mathcal{N}$ is a finite set of nonterminal symbols, Σ is a finite set of terminal symbols, $S \in \mathcal{N}$ is the start symbols of the grammar, and P is a finite set of rules in Chomsky Normal Form (CNF) as in the form of terminal rules $A \rightarrow a$ and binary rules $A \xrightarrow{r} BC$, with $A, B, C \in \mathcal{N}, a \in \sum$, and r denoting the spatial relation between B and C.

To generate the parse tree from the symbol sequence obtained from the temporal classifier, we use the bottom-up Cocke-Younger-Kasami (CYK) parsing algorithm. The probability for a terminal production $A \rightarrow a$ and that for a binary production $A \xrightarrow{r} BC$ are obtained by Eqs. (8) and (9), respectively.

$$P(A) = P(a) \tag{8}$$

$$P(A) = P(B) \times P(C) \times P(r) \tag{9}$$

where $P(B)$, $P(C)$ are obtained by the previous production recursively, $P(a)$ is the probability of terminal symbols obtained from the temporal classifier, $P(r)$ is the probability of spatial relation between two nonterminal symbols B and C that could be derived from spatial relation between two terminal symbols by the temporal classifier. If a spatial relation between two terminal symbols is not available from the temporal sequential input, we build a sequential input through the two terminal symbols and apply the temporal classification to obtain the spatial relation.

Since the temporal classifier provides many recognition candidates for symbols and relations, CYK parsing generates HME recognition candidates at the root of the parse tree with their production probabilities. We obtain the HME recognition result by selecting the candidate of the highest probability at the root of the parse tree.

Spatial Relations for Nonterminal Symbols. To obtain the probability $P(r)$ of a spatial relation between two nonterminal symbols, we need to determine the two terminal symbols representing that relation and obtain the probability from the temporal classifier. Here, we make a rule to obtain the spatial relation between two non-terminal symbols by the relation between the two terminal symbols inside them.

First, we divide the terminal symbols into two types of:

– Dominant symbols (\sqrt, \frac, \lim, \sum, \int)

– Non-dominant symbols: the remaining symbols

Then, we define the following terms:

– A component of a nonterminal symbol is a terminal symbol or a group of terminal symbols that belong to the nonterminal symbol.
– Baseline of a nonterminal symbol is a list of terminal symbols extended from the left most symbol by the *Right* relation.
– Six types of components in a nonterminal symbol: Main, Left, Right, Left-right, R-Sup, R-Sub. The Left component is the left most terminal symbol. The Right component is the right most terminal symbol in the baseline. The Left-right component is composed of Left and Right. R-Sup is the last terminal symbol extended from a Right component by the *Sup* relation. The same definition is applied for R-Sub by the *Sub* relation. The Main component is a dominant symbol in the baseline when there is no other symbol in the baseline.

For a nonterminal symbol containing the Main component, all the inbound and outbound relations are applied to the Main component. Otherwise, the inbound and outbound relations are connected by the rules presented in Fig. 6.

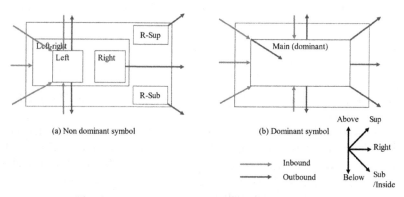

Fig. 6. Spatial relation rules for nonterminal symbols.

4   Evaluation

4.1   Dataset

We conducted the experiments on the CROHME competitions dataset [12] for both the three subtasks and HME recognition. We used only the official CROHME training set for training the temporal classifier. The number of classes for symbols is 101 and that for spatial relations is 7 composed of {*Above, Below, Sub, Sup, Right, Inside, NoRel*}.

4.2 Setup for Experiments

We use a stack of three BLSTM layers. Each BLSTM contains two LSTM layers with 128 cells. The outputs of each time step by two LSTM layers are concatenated into a feature vector of 256 dimensions before input into the next BLSTM layer. We train the networks using Stochastic Gradient Descent (SGD) with a learning rate of 0.0001 and a momentum of 0.9. The λ parameter of the combined loss function is set to 0.1.

In our experiment, an input sequence of coordinates in a stroke were sampled by the Ramer method [15]. For each sampled point, we then extracted four features: the sine and cosine of the writing directions, the normalized distance between the preceding and the succeeding points of the current point, and a binary value of pen state (pen-up/pen-down).

We measured the performance of symbol segmentation, symbol recognition, and relation classification of the model through recognizing online HMEs. The expression rate was evaluated by the provided Symbolic Label Graph (SymLGs) using the LgEval tool [16].

4.3 Experiments and Results

Subtasks Evaluation. The proposed model is firstly evaluated on the three subtasks of symbol segmentation, symbol recognition, and relation classification. For the evaluation, we obtained the prediction results from the temporal classifier as symbols and relations. Symbol segmentation and symbol recognition were evaluated on the sequential input of online HMEs in their writing order. For evaluating relation classification, we prepared 10 random sequential inputs derived from each SRT for better coverage of spatial relations.

Table 1 shows the results of the three subtasks of our method compared with the other state-of-the-art methods. As compared with the methods using separated modules of the three subtasks, our method shows the improvement in a large margin and gets a competitive result as MyScript, which used an extra dataset for training. As compared with the SRT learning approach of BLSTM or tree-BLSTM, our method shows a considerable improvement, especially on relation classification.

Table 1. Symbol level evaluation on CROHME 2016 testing set.

Type	System	Seg. (%)		Seg. + Cls. (%)		Tree rel. (%)	
		Rec.	Prec.	Rec.	Prec.	Rec.	Prec.
Separated modules	MyScript	98.89	98.95	95.47	95.53	95.11	95.11
	Wiris	96.49	97.09	90.75	91.31	90.17	90.79
	Tokyo	91.62	93.25	86.05	87.58	82.11	83.64
	Sao Paulo	92.91	95.01	86.31	88.26	81.48	84.16
	Nantes	94.45	89.29	87.19	82.42	73.20	68.72
Single module	BLSTM [9]	92.77	85.99	85.17	78.95	67.79	67.33
	Tree-BLSTM [8]	95.52	91.31	89.55	85.60	78.08	74.64
	Our system	**97.79**	**98.14**	**91.96**	**92.30**	**94.54**	**94.70**

Table 2. Confusion rate matrix for relation classification on CROHME 2016 (%).

G. Truth	Predict								
	Above	Below	Inside	Right	Sub	Sup	NoRel	NonSeg	Total
Above	**98.94**	0.00	0.07	0.00	0.00	0.00	0.59	0.40	2520
Below	0.00	**94.83**	0.00	0.52	0.00	0.00	1.47	3.18	2898
Inside	0.05	0.69	**93.07**	0.75	0.00	1.44	2.45	1.55	1569
Right	0.02	0.01	0.01	**94.55**	0.93	0.22	3.14	1.13	64733
Sub	0.00	0.28	0.02	9.00	**87.91**	0.08	1.62	1.08	5816
Sup	0.50	0.00	0.00	2.01	0.75	**92.84**	3.70	0.20	9341
NoRel	0.09	0.11	0.03	3.53	0.08	0.35	**95.47**	0.33	60493
NonSeg	0.04	0.01	0.02	0.91	0.07	0.04	0.56	**98.35**	54310

Table 2 shows the confusion matrix of relation classification on the CROHME 2016 testing set. The worst case is the *Sub* relation of 87.91% due to the confusion to the Right relation. For other relations, the classification rate is from 92.84% to 98.94%. Our model shows higher robustness with the confusion of recognizing the *Sub* relation compared with the other methods using the geometric and shape descriptors without context [10] or tree-BLSTM [8].

Expression Evaluation. Our system produced a Latex sequence for each testing sample, then we converted it into SymLGs and used the LgEval tool to obtain the expression rate and structure rate.

Table 3 shows the expression rate of the method with comparing with the other method on CROHME 2016 and CROHME 2019 testing set. We omitted the results of some teams that using offline HME recognition methods. On CROHME 2016, our system achieved 53.44%, which is more than 3 points higher compared with Wiris, and more than 9 points higher compared with BLSTM_CTC [7]. This effect is from the improvement of all three subtasks. Our method also improves the expression recognition rate in a large margin compared with tree-BLSTM and LSTM of Nantes. Our method outperforms TAP* [5] of four models ensembled with more than 3 points. The result shows the effectiveness of our model in using global context over the deep learning approach. On CROHME 2019, our method achieved 52.38%, which is a competitive result for a single model without applying a language model, ensemble models and combination of online and offline recognition models as in TAP* + WAP* + LM/USTC system. The top systems of companies: MyScript, Samsung, MathType, also benefit from using their own extra dataset for training the system. As compared with BLSTM_CTC, our single model without a language model achieves an expression rate of more than 10 points higher.

Error Analysis. We show some samples in Fig. 7, where (a), (b), and (c) are common errors in online HME recognition. Figure 7(a) presents a misrecognized symbol 'f' and a wrong relation with the following symbols of 'f', caused by a wrong writing order of

Table 3. Expression rate (%) of the state-of-the-art methods.

System	CROHME 2016		CROHME 2019	
	Correct	Structure	Correct	Structure
MyScript	67.65	88.14	79.15	90.66
Wiris/MathType	49.61	74.28	60.13	79.15
Tokyo	43.94	61.55	39.95	58.22
Sao Paulo	33.39	57.02	–	–
Nantes	13.34	21.45	–	–
Samsung R&D [17]	65.76	–	79.82	89.32
TAP* [5]	50.22	–	–	–
TAP* + WAP* + LM [5]/USTC	57.02	–	**80.73**	**91.49**
Tree-BLSTM* [8]	27.03	–	–	–
BLSTM_CTC [7]	44.81	60.94	41.28	56.80
Our system	53.44	76.02	52.38	72.64

*denotes an ensemble of models

the small bar stroke in the second symbol 'f'. Figure 7(b) shows another misrecognized symbol due to the cursive writing style. Figure 7(c) has all correctly recognized symbols, but there is a misclassified spatial relation between '\beta' and 'a' symbols. Finally, Fig. 7(d) shows a correctly recognized sample in both symbol and relation aspects, where the HME is written in a slope up direction. This shows the robustness of our method in relation classification.

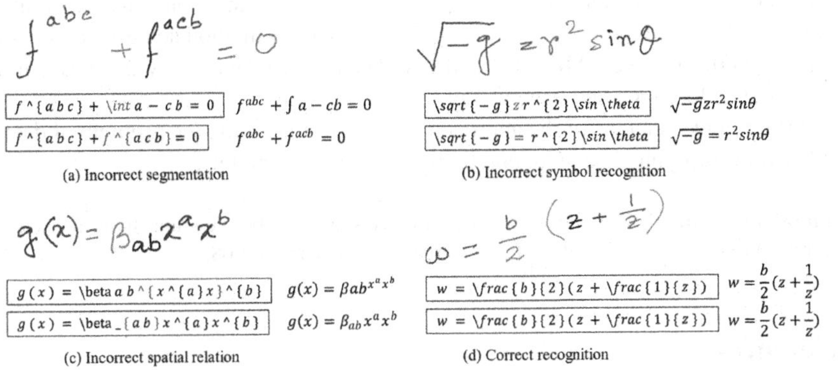

Fig. 7. Recognition results on online HME samples.

Figure 8 presents the expression recognition rates concerning the number of symbols in an HME. For the short HMEs having at most 10 symbols, the recognition rates are higher than 60%. Meanwhile, the recognition rates significantly reduce to lower than

20% for the long HMEs with more than 20 symbols. For the other HMEs consisting of between 11 to 20 symbols, the recognition rates seem adequate but might be improved in the future.

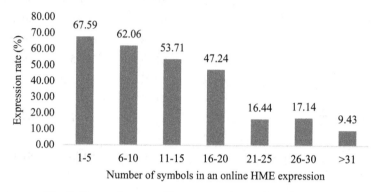

Fig. 8. Expression rate with respect to the number of symbols.

5   Conclusion

This paper proposed a temporal classification method for all three subtasks of symbol segmentation, symbol recognition, and spatial relation classification in online HME recognition. The method utilizes the global context of a bi-directional Long Short-term Memory network to improve the recognition rate of all subtasks. For recognizing a whole mathematical expression, a symbol-level parse tree built on top of the temporal classifier utilizes both the advantages of the temporal classifier and the Context-Free Grammar to improve the expression recognition rate. The proposed method achieves competitive results on both the CROHME 2016 and CROHME 2019 datasets by the single model without depending on a language model.

Further improvements of the approach are related to solving the problem of delayed handwriting, applying data augmentations and language models.

Acknowledgement. This research is being partially supported by the grant-in-aid for scientific research (A) 19H01117 and that for Early Career Research 18K18068.

References

1. Ung, H.Q., Phan, M.K., Nguyen, H.T., Nakagawa, M.: Strategy and tools for collecting and annotating handwritten descriptive answers for developing automatic and semi-automatic marking - an initial effort to math. In: 2019 International Conference on Document Analysis and Recognition Workshops (ICDARW), Sydney, NSW, Australia, pp. 13–18 (2019)
2. Anthony, L., Yang, J., Koedinger, K.R.: Toward next-generation, intelligent tutors: adding natural handwriting input. IEEE Multimedia **15**, 64–68 (2008)

3. Le, A., Nakagawa, M.: A system for recognizing online handwritten mathematical expressions by using improved structural analysis. Int. J. Doc. Anal. Recognit. **19**(4), 305–319 (2016)

4. Álvaro, F., Sánchez, J.A., Benedí, J.M.: Recognition of on-line handwritten mathematical expressions using 2D stochastic context-free grammars and hidden Markov models. Pattern Recogn. Lett. **35**, 58–67 (2014)

5. Zhang, J., Du, J., Dai, L.: Track, attend, and parse (TAP): an end-to-end framework for online handwritten mathematical expression recognition. IEEE Trans. Multimedia **21**, 221–233 (2019)

6. Graves, A., Liwicki, M., Fernández, S., Bertolami, R., Bunke, H., Schmidhuber, J.: A novel connectionist system for unconstrained handwriting recognition. IEEE Trans. Pattern Anal. Mach. Intell. **31**, 855–868 (2009)

7. Nguyen, C.T., Truong, T.N., Ung, H.Q., Nakagawa, M.: Online handwritten mathematical symbol segmentation and recognition with bidirectional context. In: Proceedings of International Conference on Frontiers in Handwriting Recognition, ICFHR, pp. 355–360. Institute of Electrical and Electronics Engineers Inc. (2020)

8. Zhang, T., Mouchère, H., Viard-Gaudin, C.: A tree-BLSTM-based recognition system for online handwritten mathematical expressions. Neural Comput. Appl. **32**(9), 4689–4708 (2018)

9. Zhang, T., Mouchere, H., Viard-Gaudin, C.: Online handwritten mathematical expressions recognition by merging multiple 1D interpretations. In: Proceedings of International Conference on Frontiers in Handwriting Recognition, ICFHR, pp. 187–192. Institute of Electrical and Electronics Engineers Inc. (2016)

10. Álvaro, F., Zanibbi, R.: A shape-based layout descriptor for classifying spatial relationships in handwritten math. In: DocEng 2013 - Proceedings of the 2013 ACM Symposium on Document Engineering, pp. 123–126. Association for Computing Machinery, New York (2013)

11. Zanibbi, R., Blostein, D.: Recognition and retrieval of mathematical expressions. Int. J. Doc. Anal. Recognit. **15**, 331–357 (2012)

12. Mouchère, H., et al.: ICFHR 2016 CROHME: competition on recognition of online handwritten mathematical expressions. In: Proceedings of the International Conference on Frontiers in Handwriting Recognition, Shenzhen, China (2016)

13. Graves, A., Schmidhuber, J.: Framewise phoneme classification with bidirectional LSTM and other neural network architectures. Neural Netw. **18**, 602–610 (2005)

14. Gers, F.A., Schmidhuber, J., Nicol, S.: Learning precise timing with LSTM recurrent networks. J. Mach. Learn. Res. **3**, 2002 (2002)

15. Ramer, U.: An iterative procedure for the polygonal approximation of plane curves. Comput. Graph. Image Process. **1**, 244–256 (1972)

16. Mahdavi, M., Zanibbi, R., Mouchère, H., Viard-Gaudin, C., Garain, U.: ICDAR 2019 CROHME + TFD: competition on recognition of handwritten mathematical expressions and typeset formula detection. In: International Conference on Document Analysis and Recognition, Sydney (2019)

17. Zhelezniakov, D., Zaytsev, V., Radyvonenko, O.: Acceleration of online recognition of 2D sequences using deep bidirectional LSTM and dynamic programming. In: Rojas, I., Joya, G., Catala, A. (eds.) IWANN 2019. LNCS, vol. 11507, pp. 438–449. Springer, Cham (2019). https://doi.org/10.1007/978-3-030-20518-8_37

Image-Based Relation Classification Approach for Table Structure Recognition

Koji Ichikawa[✉][iD]

The Japan Research Institute, Limited, Tokyo, Japan
ichikawa.koji@jri.co.jp

Abstract. In recent years, the use of tabular data has become a major area of research and development. However, the number of tables structured in a machine-readable format is still limited. A major challenge that is encountered when using tabular data is converting the table information in a free-format document into a structured format. Unlike markup languages such as HTML, XML, and JSON, free-format documents such as PDF, Word, Excel, and images generally have no tags or separators. Therefore, the table structure should be recognized from the positional information of the table elements. A major approach of table structure recognition is to classify the relationship between each pair of bounding boxes of the table elements. Recent works have achieved significant improvements by applying graph convolutional networks (GCNs) to the graph structure of the bounding boxes. However, fully recognizing a complex table structure is still a major challenge, owing to the presence of spanning cells. In this study, we propose a novel, simple image-based approach to this relation classification task. Our model efficiently exploits information such as the geometry of the table elements and ruled lines through an image cropping strategy based on the pairs of bounding boxes. We evaluate our approach on two real-world table datasets by comparing four baselines including two state-of-the-art GCN approaches. We observe that our approach significantly outperforms the baseline in the exact matching ratio for tables by up to 6.7%.

Keywords: Table structure recognition · Image recognition · Relation classification

1 Introduction

In recent years, table information retrieval has garnered substantial attention. In several cases, table data describe, explain, or complement key statements in the document; therefore, they can be utilized for various natural language processing tasks, such as question answering systems [16, 30, 34], constructing or augmenting a knowledge base [4, 22, 23], and fact-checking [1]. In particular, tables that are contained in free-format documents such as PDF, Word, Excel, and images are often critical for the above tasks, e.g., experimental data in papers; financial

© Springer Nature Switzerland AG 2021
J. Lladós et al. (Eds.): ICDAR 2021, LNCS 12822, pp. 632–647, 2021.
https://doi.org/10.1007/978-3-030-86331-9_41

performance in financial reports; and statistics in public documents, invoices, and ledgers.

However, the amount of table data available for machines is still limited; a major reason for this is that extracting the tables and modifying them into a machine-readable format is still a great challenge. This difficulty arises because free-format documents do not have tags or separators for tables similar to markup languages such as HTML, XML, and JSON; therefore, even after identifying the location of the table [6,9,24,25,29], it is necessary to structure it to a machine-readable format.

Specifically, the main issue is parsing the table elements to the machine-readable table format. Table elements can be extracted using a PDF content-stream analyzer or an optical character reader. However, these tools only provide a bounding box position for each table element. To obtain machine-readable table data, it is essential to parse these bounding boxes into a structured table format. This task is often known as table structure recognition, and is the main subject of this study.

For table structure recognition, the following difficulties prevent a simple pre-defined rule strategy: (1) the presence of spanning cells; (2) the width and height of the bounding box must vary. For instance, an intuitive approach would be to construct a parsing rule based on the relative positions of the bounding boxes; i.e., if two or more bounding boxes are aligned on a single vertical line, these boxes may belong to a single column. This rule-based approach sometimes works, especially for a simple table. In practice, however, most tables have spanning cells that belong to multiple columns or rows. Moreover, determining the box alignment is difficult because of the different widths and heights of each bounding box.

To overcome the above difficulties, recent studies have proposed deep neural network-based approaches. An earlier attempt [24] employs fully convolutional network (FCN) architecture [15] to detect the row and column regions. This approach has also been adopted in recent works [27,28], which applied the object detection framework. The advantage of this approach is that it can naturally incorporate the table structure information, such as ruled lines or margins. However, one should take care of the mechanism through which the blank cells are joined to construct the spanning cells [31,35], which is necessary for correctly recognizing the hierarchical structure of the table. In this paper, we refer to this approach as the detection-based approach.

Recently, relation classification approaches have been proposed in several studies [2,14,18,21], wherein row and column recognition is considered as a relation classification task between a pair of bounding boxes. The advantage of this method is that a joint operation is not required for constructing spanning cells. Most studies on this approach utilize the graph structure of the table elements and employ graph convolutional networks (GCNs) [13], which successfully recognize multi-rows/columns using spanning cells. However, one major disadvantage of this approach is the difficulty of feature engineering. For instance, it is difficult to fully utilize ruled line information using this approach. In this paper, we refer to this approach as the graph-based approach.

In this study, we adopt a novel and simple image-based relation classification approach for the table structure recognition task. Our idea is to employ an edge-based rectangle region formed by each pair of nodes as the input to a relation classifier. This rectangle contains essential information for the classification: the relative position, ruled lines, and the geometry of bounding boxes. Moreover, enlarging this edge-based region incorporates the global patterns of the table, which significantly improves the model accuracy. We stress that our approach has the advantages of both detection- and graph-based approaches, and succeeds in considerably reducing the complex design of pre-defined rules or feature engineering. Another advantage of our approach is that the data can be augmented through label-invariant operations. We propose novel label-invariant data augmentation techniques for the edge-based region, and demonstrate that they make significant contributions, especially when training with small amounts of data. In summary, our contributions are as follows.

- We propose a novel edge-based cropping strategy for table structure recognition.
- We introduce an edge region-based convolutional neural network (ER-CNN) that efficiently encodes the edge-based cropped images and positional information of the bounding boxes.
- We propose efficient data augmentation techniques for the edge-based cropped images.

We evaluate our approach on two real-world table datasets consisting of PDF and handwritten scanned images. We compare our approach with four baselines, including two state-of-the-art graph-based approaches. We observe that our approach significantly outperforms the baselines in the exact matching ratio for tables.

The remainder of this article is organized as follows. In Sect. 2, we briefly review related works. In Sect. 3, we define the problem that is the focus of this study. In Sect. 4, we introduce the motivation of our approach through observation. In Sect. 5, we describe our approach. We then present our experimental results in Sect. 6; finally, we provide a conclusion in Sect. 7.

2   Related Works

For table structure recognition, similar to the development of the table detection task [6, 9, 24, 25, 29], recent studies adopted a deep learning approach rather than pre-defined rules or heuristics [11, 26, 32] for structuring more complex tables. Several studies use the semantic segmentation or object detection methods to detect the columns and rows of a table [24, 27, 28]. The difference in our approach is that, while the approaches in previous studies are based on row and column detection, we adopt the relation classification approach and employ the edge-based cropping strategy for the classification.

Recently, other approaches based on relation classification have been proposed [2,14,18,21]. In this approach, the table structure is recognized via the relationship between each cell. Most works for this approach utilize the graph structure of the table elements, considering each bounding box as nodes. In [14], the graph structure is constructed using the k-nearest neighbor (k-NN) algorithm, and features for the classification are constructed via GCN [13]. In [2], a multi-head attention mechanism is incorporated. In this study, both the node and edge features are convoluted via GCN, thereby exchanging their feature propagation. [21] also convolutes the edge feature via GCN architecture. Meanwhile, [18] adopts GravNet [17] and DGCNN [33] for graph convolution. A significant difference in our approach is that, while the previous studies mainly utilize the positional information of the table element for their input feature, we incorporate information about the ruled lines and geometry of bounding boxes by adopting CNN-based architecture and an edge-based region cropping strategy.

Our approach also relates to object detection and categorization, such as R-CNN [8], Fast R-CNN [7], and Faster R-CNN [20] in that the cropped image can be considered as a proposed object, and the relation classification corresponds to the categorization. The difference is that we determine the cropping region through the combination of the nodes, and utilize both the image and position for the model input to stress the geometry of the component.

3 Problem Setting

In our problem setting, we define dataset $\mathcal{D}$ as a set of n tables: $\mathcal{D} \equiv \{T^1, \ldots, T^n\}$, where each table T^t consists of a table image I^t, set of bounding boxes $\mathcal{B}^t$ and set of relations $\mathcal{R}^t$:[1]

$$T^t \equiv \{I^t, \mathcal{B}^t, \mathcal{R}^t\}. \tag{1}$$

The table image I^t has an image with $H^t \times W^t$ pixels and C channels, i.e.,

$$I^t \in [0,1]^{H^t \times W^t \times C}. \tag{2}$$

In this study, we assume that table images are preprocessed into gray-scaled or binarized pictures with a single channel; that is, $C = 1$. Meanwhile, $\mathcal{B}^t$ is a set of bounding boxes for each table element, i.e.,

$$\mathcal{B}^t \equiv \{b_1^t, b_2^t, \ldots, b_{m^t}^t\}, \tag{3}$$

where m^t denotes the number of bounding boxes contained in table T^t. b represents a bounding box that is defined as follows:

$$b_i^t \equiv (x_{li}^t, y_{ti}^t, x_{ri}^t, y_{bi}^t). \tag{4}$$

---

[1] Note that our approach does not require additional information such as text or captions in the table.

We let each bounding box be described as a rectangle with a four-dimensional vector (x_l, y_t, x_r, y_b), where (x_l, y_t) and (x_r, y_b), respectively, represents the top-left and bottom-right position of the bounding box. Coordinate x/y increases from left/top to right/bottom; x and y satisfy $0 \leq x_l < x_r < W$ and $0 \leq y_t < y_b < H$, respectively. We also refer to (x_{ci}, y_{ci}) as the coordinate of the center of b_i. In practice, bounding boxes may be split, merged, or missing owing to incomplete identification.[2] Although such misidentification can be mitigated by improving the box identification tool, improvements of the tool are beyond the scope of this paper. Therefore, in our problem setting, we assume that the bounding boxes are ideally provided; that is, $b \in \mathcal{B}$ has a one-to-one correspondence with the table element. Finally, $\mathcal{R}^t$ represents a set of relations between pairs of boxes, which is determined by a set of triplets:

$$\mathcal{R}^t \equiv \{(b_i^t, b_j^t, y_{ij}^t) \mid b_i^t, b_j^t \in \mathcal{B}^t, y_{ij}^t \in \mathcal{L}\}, \tag{5}$$

where $\mathcal{L}$ represents a set of relation labels: $\mathcal{L} = \{\text{irrelevant, row, column}\}$. Subsequently, by analogy from the graph representation, we may refer to the bounding boxes as nodes and relations between boxes as edges. Moreover, we may omit the table index t if it is clear from the context.

The relation classification approaches for the table structure recognition are used to predict y_{ij} for b_i and b_j under a given table image I and a set of bounding boxes $\mathcal{B}$.

4   Observations

To clarify the motivation of our approach, we provide an overview of the relationship between nodes, edges, ruled lines, and other neighbor nodes, using concrete examples.

Figure 1 shows examples of the geometry of nodes in a table. The figure shows that the relationship between the two blue boxes differs depending on the geometry of the other nodes and the ruled lines, even if the relative positions of the two nodes are the same. From the upper examples in Fig. 1, most relationships can be inferred by observing the inner area of the two nodes. In (1), we can infer that a column relationship between the two blue nodes is allowed, whereas this is inappropriate in (2) because of the presence of the intermediate cell. Meanwhile, (3), (4), and (5) show the effect of the ruled lines: (3) allows for

---

[2] More specifically, the noise related to the identification of boxes can be classified into the following six types: box size, misalignment of box positions, mis-joining between boxes, unnecessary division of boxes, missing boxes, and presence of extra boxes. We expect that our data augmentation in Sect. 5.3 improves robustness against the first two cases. The rest of the cases, on the other hand, cannot be straightforwardly dealt with by the relation classification approach, and the accuracy is degraded by the noise. While we expect that the noise can be suppressed by state-of-the-art box identification tool, we also expect that it is possible to extend our approach to an end-to-end framework [19] to address them, which we see as an interesting future work.

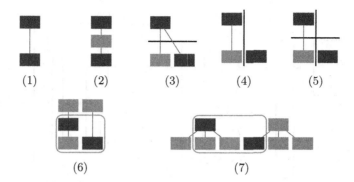

Fig. 1. Relationships between column relations and table elements. The boxes represent the bounding boxes in the table, and the two blue boxes correspond to the nodes of interest. The red lines represent the column relationships, while the thick black lines represent the table rule lines. Here, we omit the row relationships. (Color figure online)

column relationships between blue nodes, whereas (4) does not. Similarly, the combination of these ruled lines shown in (5) cuts off the column relationship between the two blue nodes.

Meanwhile, there are examples where the outer geometry influences the relationship, as shown in the lower examples in Fig. 1. In these examples, if we focus only on the inner area (orange rectangle), there could be a column relationship between the two blue nodes. However, once we increase the size of the region, such a relationship is found to be inappropriate because of the relationship with the other nodes. This observation suggests that the model should incorporate a proper range of peripheral information.

In the previous relation classification approaches, these geometrical patterns were not efficiently incorporated. This is because constructing a node or edge feature that incorporates these geometrical patterns requires hard feature engineering. Meanwhile, the image near the pair of nodes, we call it the *edge region*, naturally contains such information, which is efficiently extracted by a CNN architecture without complex feature engineering. Motivated by these observations, we propose a novel image-based relation classification approach, which is discussed in the subsequent section.[3]

5   Description of Our Approach

Our approach consists of two modules: the preprocessor, which extracts the information of the edge region, and the ER-CNN, which is employed as the classification model. An overview of our approach is shown in Fig. 2.

---

[3] We note that because our approach adopt CNN architecture, the inference speed is slower than that of the graph-based approaches. However, we believe that the table structure recognition does not necessarily require as high an inference speed as object detection tasks that are intended for applications such as automated driving.

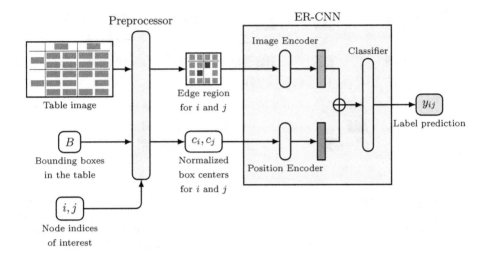

Fig. 2. Illustration of our approach.

5.1   Preprocessor

Edge-Based Cropping Region. The key idea of our approach is to employ a node pair-based image cropping strategy. Based on the observations in the previous section, we use a cropped table image with a rectangle formed by a pair of nodes as a primitive feature for the relation classification. More concretely, if node pairs have bounding boxes at $(x_{ri}, y_{ti}, x_{li}, y_{bi})$ and $(x_{rj}, y_{tj}, x_{lj}, y_{bj})$, we define the inner region of the bounding boxes as follows:

$$b^e_{ij} \equiv (\min(x_{li}, x_{lj}), \min(y_{ti}, y_{tj}), \max(x_{ri}, x_{rj}), \max(y_{bi}, y_{bj})) . \qquad (6)$$

This region encompasses information about the inner state between two nodes.

Another important aspect for relation classification is the outer status near the node pair. To incorporate this global information, we scale the width and height of the inner region b^e_{ij}. The edge-based cropping region is determined as follows:

$$b'^e_{ij} = (\min(x_{li}, x_{lj}) - rw_{ij}, \min(y_{ti}, y_{tj}) - rh_{ij},$$
$$\max(x_{ri}, x_{rj}) + rw_{ij}, \max(y_{bi}, y_{bj}) + rh_{ij}) . \qquad (7)$$

Here, w_{ij} and h_{ij} denote the width and height of b^e_{ij}, respectively, and r is a hyperparameter that defines the scale of the cropping region and we adopt $r = 1$ in this paper. If the cropping region extends outside the image, we fill in the overflow with a blank value. We cut out the rectangular region b'^e_{ij} of the original table image I, which we define as I^e_{ij}, and use for crafting an input image for ER-CNN.

Crafting the Input Image of the Model. We construct the input of ER-CNN by splitting the cropped image I_{ij}^e into three channels: the channels of bounding boxes i and j, the channel of the other bounding boxes, and the channel of the other pixels, containing the noise and ruled information of the image. In the channels for the boxes, the rectangles of the boxes are filled with a constant, and the remaining area is filled with a blank value. This channel splitting helps the model to correctly recognize the table components. Finally, we resize this channel-split image to a 64×64 square shape for the input of ER-CNN.

Box Position Extraction. In addition to the cropped image, we utilize the positional information of the bounding boxes. The sizes of the bounding boxes can vary considerably depending on the lengths and styles of the original table elements. Therefore, the pixel area occupied by a bounding box can be sometimes extremely small after the cropping and resizing procedure. In such a case, the geometry of the table element cannot be extracted properly from the image alone. To cope with this problem, we explicitly input the box position into the model. Specifically, we use the box centers of b_i and b_j normalized by the size of $b_{ij}^{'e}$: $\{c_i, c_j\}$, where

$$c_i \equiv \left(\frac{x_{ci} - x_{lij}'}{w_{ij}'}, \frac{y_{ci} - y_{tij}'}{h_{ij}'} \right) . \tag{8}$$

Here, w_{ij}' and h_{ij}' denote the width and height of $b_{ij}^{'e}$, respectively, and x_{lij}'/y_{tij}' is left/top position of $b_{ij}^{'e}$.

5.2 Model

As described in Fig. 2, the ER-CNN consists of two encoders: an image encoder and a position encoder. The outputs of these encoders are concatenated, and then, passed through a final classification layer that outputs the label probability. For the backbone architecture of the image encoder, we adopt a small pre-trained residual neural network (ResNet) model (with 18 layers) [10]. The encoded vector is obtained via the first block layer output of the ResNet (with 64 channels) and subsequent two FC layers with batch normalization and rectified linear unit (ReLU) activation. Notably, one can easily exchange this module with a larger and more complex architecture. As demonstrated in the results section, our approach performs well, even with this shallow architecture.

For the position encoder, the input is a set of box positions defined by Eq. (8). Each two-dimensional coordinate is first embedded into a d-dimensional hidden space $\mathcal{R}^d$ via transformation f: $l_i = f(c_i)$. Thereafter, the encoded position vector l_p is obtained by inputting the concatenation of l_i, l_j into function g: $l_p = h(l_i \oplus l_j)$. For the transformation functions f and g, we adopted a two-FC layer with ReLU activation. Similarly, we set up the final classification layer using two FC layers with batch normalization, ReLU activation, and a softmax function. We set the size of all hidden layers to 64 and $d = 64$.

5.3   Data Augmentation

An advantage of the image-based approach is that the amount of data can be augmented through label-invariant operations. Unlike typical image classification tasks, however, the geometry or presence of bounding boxes and ruled lines significantly affects the relation label; therefore, commonly used data augmentation, such as random crop, random erasing [36] or cutout [3], are likely to generate noisy samples.

We introduce two novel label-invariant data augmentation techniques for our approach: randomly changing the size of the bounding boxes and adding noise near the box. The former is based on the intuition that the size of the box can vary depending on the size of the characters of the table element, whereas the relationship is mostly independent of the character size. The latter incorporates noise that often occurs near the box, owing to the mismatch between the characters and bounding box. Specifically, we create the augmented images $\tilde{I}_{ij}^{e}$ table-by-table according to the process in Sect. 5.1, using the randomly rescaled bounding boxes and the table image. More precisely, we first fill the original box areas with a blank value, and then place the rescaled bounding boxes. We added one augmented data per sample for our model's training set. A new set of augmented data was generated for each training iteration

5.4   Scalability

We finally mention the scalability of our approach. If prediction is done for all combinations of boxes, $\mathcal{O}(m^2)$ computations are required in each table, which is difficult to perform for large tables. However, this computational complexity can be reduced by the fact that distant boxes are mostly irrelevant pairs. Specifically, we reduce the number of candidate pairs of boxes by using a k-NN method based on the location of the boxes [2]. This reduces the computational complexity to $\mathcal{O}(km)$, making it practically feasible. Besides, we expect that it is possible to reduce the number of actual CNN computation using techniques similar to Fast R-CNN [7], which is an interesting future work.

6   Experiments

In this section, we first review the datasets and introduce evaluation metrics. Next, we introduce the baselines and experimental details. Finally, we present the results under the three experimental settings.

6.1   Dataset

We used two real table datasets: SciTSR [2], comprising typed PDF images, and ICDAR2019 [5], comprising of handwritten scanned images.

The SciTSR dataset comprises 15,000 PDF format tables, containing bounding boxes, relationships, and table images for each table. The average number of nodes in one table is approximately 40. Because we found that some of the table relationships were missing in the bounding box in the lower-left corner, we fixed the generation code. In addition, some tables in the dataset were out of the PDF area, which were removed by imposing a simple threshold to the maximum and minimum positions of the bounding box. After filtering, 11,134 training tables and 2,801 test tables were obtained. In the test dataset, a list of complex table IDs is provided (635 tables after the preprocessing above), and we also report the performance on this list as the complex test set.

The ICDAR2019 dataset comprises 850 (600 for training and 250 for test) scanned table images with handwritten entries. The ground truth of the table area and table structure is provided in XML format. We constructed the relations by parsing these XML files. To reduce the overlap between the boxes and ruled lines, we reduced the size of the bounding box by 50%. In addition, because the images had various background colors, we gray-scaled and binarized the images using threshold values of 80 percent quantile for each table image. Sometimes one scanned image contained multiple tables; these tables were split using XML tags. In the test set, we found that images with ID numbers greater than 10,000 had significantly different properties than the other training and test data: not handwritten, captured images, approximately one-tenth the size of the images. Most models did not perform well against this test set; therefore, we decided to separately evaluate them as in-domain and out-of-domain test set. After preprocessing, we obtained 677 tables for training, 190 tables for the in-domain, and 145 for the out-of-domain test set.[4] The average number of nodes per table was approximately 300.

6.2   Evaluation Metrics

We adopted macro-averaged precision, recall, and F1 scores as metrics for our experiment. These metrics tend to achieve a high score in the relation classification of table structures. For instance, one misclassification for each table yields a F1 score of approximately 0.99. However, such misclassification, even at a rate of one per table, seriously degrades the performance of subsequent natural language processing tasks. Therefore, we employed an additional metric, the exact match. This metric yields 1 if the predicted rows or columns match the ground truth perfectly in each table. In our experiment, we measured the average ratio of the exact match for rows, columns, and tables (i.e., 1 if both rows and columns yield perfect matches).

6.3   Baselines

We compared our model performance with the following four baselines.

---

[4] We removed one XML data that contained zero-width bounding boxes.

Rule Base. A simple, but strong baseline, for constructing the table rows and columns using a pre-defined rule. Here, we adopted the following extraction algorithm based on the overlaps of the pairs of bounding boxes.

1. Select $b_i \in \mathcal{B}$, which is located at the left/top most position.
2. Select b_j, which is located at the left/top position most in $\mathcal{B}\backslash\{b_i\}$.
3. If b_i and b_j do not overlap on the vertical/horizontal-axis and overlap on the horizontal/vertical-axis above a threshold length, we set $y_{ij} = $ row/column, otherwise $y_{ij} = $ irrelevant.
4. If $y_{ij} = $ irrelevant, then remove b_j and go back to 2.
5. If b_j that satisfies $y_{ij} \neq $ irrelevant is found or all b_j is searched, then we assign $y_{ij'} = $ irrelevant to all the remaining $b_{j'}$, and restart this algorithm from 1 replacing $\mathcal{B}$ with $\mathcal{B}\backslash\{b_i\}$.

In our experiments, we set the threshold value in the step 3 as 50% of the smaller height/width of b_i and b_j. Because this rule accurately identifies the row/column relationship between two distant nodes, the algorithm achieves high prediction accuracy, although it cannot structure a spanning-cell relationship.

MLP. Multi-layer perceptron (MLP) is a class of a feed-forward network consisting of input, output, and hidden layers. In this experiment, we constructed a module with three hidden layers and ReLU activations. As the input, we fed a concatenation of the pair of node features. We will describe the node feature adopted in the experiment in Sect. 6.4.

GraphTSR. [2] incorporates node and edge features for the input of the graph neural network. The author adopts a multi-head attention layer for the graph convolution. Both node and edge features are constructed based on the positions and sizes of the nodes. We adopted the same architecture and features for our experiment.

Ties. [18] shows variations in the architecture of the graph convolution mechanism for node features. From the results of the study, we adopted the DGCNN [33] module, where the node graph structure is dynamically constructed by the hidden features of the nodes. We set the number of the vertex neighbors for the DGCNN to 10. The image information was also used for the node feature by convolving the table image and sampling the CNN feature at the node position. In addition, the authors employed an edge sampling strategy to reduce the memory complexity. In our experiment, we sampled a constant number of negative samples (i.e., irrelevant edges) for each node. We set the number of the negative samplings for each node to 10.[5]

---

[5] Ties also incorporates the textual information into the node features. In our experiment, we do not use the textual information.

Table 1. Performances on SciTSR dataset. P, R, and F_1 are precision, recall and F1 scores respectively. The numbers in parentheses represent standard deviations.

	Row [%]			Column [%]			Exact match [%]		
	$1-P$	$1-R$	$1-F_1$	$1-P$	$1-R$	$1-F_1$	Row	Column	Table
Full test set									
Rule	$0.64_{(0.00)}$	$1.17_{(0.00)}$	$1.03_{(0.00)}$	$1.40_{(0.00)}$	$3.09_{(0.00)}$	$2.40_{(0.00)}$	$86.4_{(0.0)}$	$73.4_{(0.0)}$	$70.8_{(0.0)}$
MLP	$9.13_{(0.83)}$	$12.96_{(0.51)}$	$13.27_{(0.34)}$	$1.38_{(0.06)}$	$2.74_{(0.12)}$	$2.26_{(0.11)}$	$21.2_{(1.1)}$	$68.9_{(0.6)}$	$18.9_{(1.1)}$
GraphTSR	$0.87_{(0.07)}$	$1.52_{(0.03)}$	$1.33_{(0.02)}$	$2.12_{(0.08)}$	$1.79_{(0.15)}$	$2.12_{(0.07)}$	$79.0_{(0.3)}$	$69.0_{(0.7)}$	$64.8_{(0.5)}$
Ties	$0.86_{(0.20)}$	$0.93_{(0.11)}$	$0.99_{(0.04)}$	$1.20_{(0.22)}$	$0.49_{(0.02)}$	$0.89_{(0.12)}$	$83.3_{(0.9)}$	$82.7_{(2.1)}$	$76.9_{(2.2)}$
ER-CNN	$\mathbf{0.64}_{(0.02)}$	$\mathbf{0.27}_{(0.04)}$	$\mathbf{0.49}_{(0.02)}$	$\mathbf{0.80}_{(0.03)}$	$\mathbf{0.42}_{(0.05)}$	$\mathbf{0.64}_{(0.01)}$	$\mathbf{89.2}_{(0.3)}$	$\mathbf{87.0}_{(0.3)}$	$\mathbf{83.6}_{(0.2)}$
Complex test set									
Rule	$1.06_{(0.00)}$	$5.13_{(0.00)}$	$3.54_{(0.00)}$	$3.57_{(0.00)}$	$13.07_{(0.00)}$	$8.94_{(0.00)}$	$61.7_{(0.0)}$	$11.7_{(0.0)}$	$0.5_{(0.0)}$
MLP	$12.54_{(1.17)}$	$10.20_{(0.20)}$	$12.95_{(0.65)}$	$1.81_{(0.03)}$	$7.86_{(0.24)}$	$5.26_{(0.16)}$	$9.7_{(0.9)}$	$26.4_{(1.3)}$	$4.6_{(0.5)}$
GraphTSR	$1.32_{(0.21)}$	$3.36_{(0.35)}$	$2.52_{(0.17)}$	$3.49_{(0.32)}$	$4.94_{(0.23)}$	$4.54_{(0.20)}$	$50.2_{(1.5)}$	$29.2_{(1.2)}$	$23.6_{(0.6)}$
Ties	$1.52_{(0.43)}$	$1.20_{(0.20)}$	$1.43_{(0.21)}$	$2.30_{(0.53)}$	$1.54_{(0.13)}$	$2.02_{(0.22)}$	$70.4_{(1.5)}$	$60.4_{(5.0)}$	$52.1_{(4.6)}$
ER-CNN	$\mathbf{0.95}_{(0.04)}$	$\mathbf{0.90}_{(0.13)}$	$\mathbf{0.98}_{(0.06)}$	$\mathbf{1.41}_{(0.07)}$	$\mathbf{1.28}_{(0.21)}$	$\mathbf{1.40}_{(0.07)}$	$\mathbf{79.8}_{(0.3)}$	$\mathbf{71.3}_{(1.3)}$	$\mathbf{63.1}_{(2.1)}$

Table 2. Performances on ICDAR2019 dataset.

	Row [%]			Column [%]			Exact match [%]		
	$1-P$	$1-R$	$1-F_1$	$1-P$	$1-R$	$1-F_1$	Row	Column	Table
In-domain test set									
Rule	$1.16_{(0.00)}$	$3.06_{(0.00)}$	$2.14_{(0.00)}$	$0.08_{(0.00)}$	$0.85_{(0.00)}$	$0.47_{(0.00)}$	$44.2_{(0.0)}$	$59.5_{(0.0)}$	$41.1_{(0.0)}$
MLP	$0.28_{(0.04)}$	$1.32_{(0.07)}$	$\mathbf{0.84}_{(0.04)}$	$0.10_{(0.04)}$	$0.05_{(0.00)}$	$0.08_{(0.01)}$	$50.5_{(1.1)}$	$74.0_{(2.5)}$	$49.8_{(2.2)}$
GraphTSR	$\mathbf{0.26}_{(0.13)}$	$1.85_{(0.04)}$	$1.10_{(0.07)}$	$0.08_{(0.03)}$	$\mathbf{0.04}_{(0.00)}$	$\mathbf{0.06}_{(0.01)}$	$47.9_{(1.9)}$	$\mathbf{78.2}_{(2.2)}$	$47.0_{(1.1)}$
Ties	$1.23_{(0.13)}$	$9.07_{(0.26)}$	$6.44_{(0.26)}$	$0.66_{(0.07)}$	$0.23_{(0.07)}$	$0.45_{(0.06)}$	$26.5_{(1.1)}$	$65.1_{(1.3)}$	$24.7_{(1.8)}$
ER-CNN	$0.36_{(0.11)}$	$\mathbf{1.32}_{(0.11)}$	$0.87_{(0.10)}$	$\mathbf{0.05}_{(0.04)}$	$0.20_{(0.09)}$	$0.13_{(0.03)}$	$\mathbf{56.8}_{(0.0)}$	$78.2_{(1.1)}$	$\mathbf{56.5}_{(0.6)}$
Out-of-domain test set									
Rule	$2.81_{(0.00)}$	$\mathbf{9.02}_{(0.00)}$	$\mathbf{6.40}_{(0.00)}$	$12.66_{(0.00)}$	$25.93_{(0.00)}$	$20.20_{(0.00)}$	$\mathbf{37.9}_{(0.0)}$	$8.3_{(0.0)}$	$6.2_{(0.0)}$
MLP	$43.24_{(4.50)}$	$92.16_{(1.03)}$	$87.75_{(1.60)}$	$46.83_{(4.92)}$	$91.62_{(1.36)}$	$86.77_{(1.95)}$	$1.4_{(0.0)}$	$0.0_{(0.0)}$	$0.0_{(0.0)}$
GraphTSR	$70.21_{(24.10)}$	$95.28_{(3.35)}$	$93.24_{(5.36)}$	$53.14_{(32.10)}$	$89.72_{(13.25)}$	$85.32_{(17.68)}$	$1.4_{(0.0)}$	$0.5_{(0.8)}$	$0.0_{(0.0)}$
Ties	$6.52_{(2.27)}$	$13.13_{(2.87)}$	$10.48_{(2.80)}$	$9.23_{(3.03)}$	$\mathbf{9.52}_{(2.03)}$	$9.71_{(1.55)}$	$17.7_{(6.3)}$	$23.0_{(8.2)}$	$\mathbf{10.6}_{(3.9)}$
ER-CNN	$\mathbf{2.76}_{(0.91)}$	$26.45_{(9.50)}$	$18.22_{(7.00)}$	$\mathbf{3.26}_{(1.19)}$	$11.99_{(0.16)}$	$\mathbf{8.29}_{(0.50)}$	$15.2_{(7.2)}$	$\mathbf{26.0}_{(3.3)}$	$10.6_{(3.4)}$

Table 3. Ablation study on the SciTSR dataset.

	Row [%]			Column [%]			Exact match [%]		
	$1-P$	$1-R$	$1-F_1$	$1-P$	$1-R$	$1-F_1$	Row	Column	Table
-DA	$0.68_{(0.02)}$	$0.32_{(0.04)}$	$0.54_{(0.01)}$	$0.89_{(0.09)}$	$\mathbf{0.41}_{(0.08)}$	$0.67_{(0.02)}$	$88.6_{(0.2)}$	$86.3_{(0.2)}$	$82.7_{(0.3)}$
$r=0$	$0.71_{(0.02)}$	$0.40_{(0.03)}$	$0.60_{(0.01)}$	$1.13_{(0.03)}$	$0.58_{(0.02)}$	$0.90_{(0.02)}$	$87.7_{(0.5)}$	$82.6_{(0.3)}$	$78.7_{(0.6)}$
$r=0.5$	$0.70_{(0.01)}$	$\mathbf{0.26}_{(0.00)}$	$0.52_{(0.01)}$	$0.85_{(0.04)}$	$0.47_{(0.01)}$	$0.69_{(0.02)}$	$88.9_{(0.4)}$	$85.2_{(0.3)}$	$82.0_{(0.7)}$
$r=1$	$0.64_{(0.02)}$	$0.27_{(0.04)}$	$\mathbf{0.49}_{(0.02)}$	$\mathbf{0.80}_{(0.03)}$	$0.42_{(0.05)}$	$\mathbf{0.64}_{(0.01)}$	$\mathbf{89.2}_{(0.3)}$	$\mathbf{87.0}_{(0.3)}$	$\mathbf{83.6}_{(0.2)}$
$r=2$	$\mathbf{0.59}_{(0.04)}$	$0.36_{(0.05)}$	$0.50_{(0.01)}$	$0.83_{(0.06)}$	$0.42_{(0.04)}$	$0.65_{(0.03)}$	$88.3_{(0.5)}$	$86.5_{(0.6)}$	$82.4_{(0.5)}$

6.4 Experimental Details

We split the training dataset in a ratio of 3:1; the former was used for training and the latter for validation. The training was terminated by referring to the average F1 score of the rows and columns of the validation data. We adopted the cross-entropy loss, and minimized it using the Adam optimizer [12]. For GraphTSR, and Ties, we referred to the official code, and modified or reconstructed them for our experiments, retaining their original architectures.

To construct a set of relations $\mathcal{R}$ including $y =$ irrelevant, we adopted a conventional negative sampling method: For each node i, we constructed a set of node pairs by pairing i to k-NN nodes. We assigned a row, column, or irrelevant label to each node pair by referring to the ground truth. For all baselines, we set the number of nearest neighbor nodes $k = 20$ for training and validation, and $k = 50$ for testing except for GraphTSR. For GraphTSR, we found that the same k for training and prediction yielded a higher performance, and hence we adopted $k = 20$ for prediction.

We used the following 16 node features for MLP and Ties: box positions, box centers, box width, and box height, along with these features normalized by the table size. The box center normalized by the table size was also used for the k-NN box search. For GraphTSR, the features are standardized, and the edge feature is used. We adopted the same features as the original codes.

Each approach was run three times and the average values are reported.

6.5   Performances on the Full Data

First, we tested the performance of the model when trained on all training data samples. Tables 1 and 2 summarize the results.[6] For SciTSR dataset, we observe that our method significantly outperforms the baselines for most of the metrics.[7] In ICDAR2019 dataset, the images are distorted and noisy, which is difficult for image-based approach. Nevertheless, our approach achieves a competitive performance with the baselines on the F1 scores and significantly outperforms on the exact matching ratio for rows and tables. We expect this accuracy to improve further with more sophisticated image preprocessing.

6.6   Ablation Studies

Next, we present the results of the ablation studies of our model on the SciTSR dataset. In this experiment, we tested the model performance by ablating the data augmentation (-DA) and changing scaling parameter r in the range of $(0, 0.5, 1, 2)$. The results are summarized in Table 3. Interestingly, even $r = 0$, which only contains information on the inner part, yields a competitive accuracy with the baselines, indicating the importance of the internal states of the node pairs. In contrast, the best accuracy was obtained at $r = 1$, and the accuracy decreased for larger r values. This is because, when the cropped area is excessively large, important information on the inner part is missing owing to the resizing procedure.

---

[6] We note that the benchmark micro-F1 scores [2] of ER-CNN were 0.993, 0.990, and 0.984, for SciTSR full test set, SciTSR complex test set, and ICDAR2019 full test set, respectively, although a direct comparison may be inappropriate due to the difference in preprocessing.

[7] We checked that the p-values of the F1 scores in SciTSR dataset were less than or close to 0.05. In addition, we performed two additional runs for GraphTSR, Ties, and ER-CNN and confirmed p-values < 0.05 for the F1 scores in SciTSR dataset.

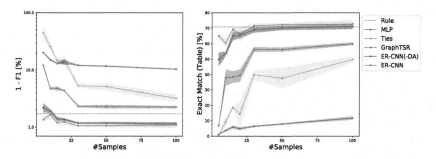

Fig. 3. Performances on the small data. The shaded regions represent the standard deviation. F1 in the left panel represents the average of the row and column F1 scores.

6.7   Performances on the Small Data

Finally, we present the model performance under a small data setting. Documents often contain sensitive information, and crowdsourcing is not always available. In such a situation, it is desirable to achieve practical accuracy with a small number of annotations. Assuming a sample size that is sufficiently small to be annotated without crowdsourcing, we sampled 5, 10, 15, 20, 30, 50 and 100 tables for training and 10 tables for validation from the SciTSR training dataset. The performance was evaluated on the same test dataset used in the full data experiment.

The results are presented in Fig. 3. In graph-based approaches (Ties and GraphTSR), the model cannot be trained well with such a small training dataset, and the performances are below that of the rule-based baseline. In contrast, our method significantly outperforms the graph-based approaches and is competitive with the rule-based baseline, even without data augmentation. With data augmentation, the performance increases further; even with five samples, the exact matching ratio is greater than 60% and with 30 samples, it is above 70%.

7   Conclusion and Discussion

In this study, we proposed a novel image-based approach for the table structure recognition task. Our model efficiently exploits information on geometry of the table elements and table ruled lines through edge-based cropped images. We evaluated our approach on two real-world table datasets, consisting of typed PDFs and handwritten scanned images, in comparison with four baselines. We have observed that our approach significantly outperforms the baselines in the exact matching ratio for tables. In addition, our experiments have confirmed that our approach works well with small amounts of data. We finally note that our approach can be easily combined with the graph convolutional architecture by exchanging the position encoder with a GCN architecture, which may help to improve robustness against noisy and distorted images.

References

1. Chen, W., et al.: Tabfact: a large-scale dataset for table-based fact verification. In: ICLR 2020
2. Chi, Z., Huang, H., Xu, H.D., Yu, H., Yin, W., Mao, X.L.: Complicated table structure recognition. arXiv preprint arXiv:1908.04729
3. Devries, T., Taylor, G.W.: Improved regularization of convolutional neural networks with cutout. arXiv preprint arXiv:1708.04552
4. Dong, X., et al.: Knowledge vault: a web-scale approach to probabilistic knowledge fusion. In: KDD 2014. https://doi.org/10.1145/2623330.2623623
5. Gao, L., et al.: ICDAR 2019 competition on table detection and recognition (CTDAR). In: ICDAR 2019. https://doi.org/10.1109/ICDAR.2019.00243
6. Gilani, A., Qasim, S.R., Malik, M.I., Shafait, F.: Table detection using deep learning. In: ICDAR 2017. https://doi.org/10.1109/ICDAR.2017.131
7. Girshick, R.B.: Fast R-CNN. In: ICCV 2015. https://doi.org/10.1109/ICCV.2015.169
8. Girshick, R.B., Donahue, J., Darrell, T., Malik, J.: Rich feature hierarchies for accurate object detection and semantic segmentation. In: CVPR 2014. https://doi.org/10.1109/CVPR.2014.81
9. Hao, L., Gao, L., Yi, X., Tang, Z.: A table detection method for PDF documents based on convolutional neural networks. In: DAS 2016. https://doi.org/10.1109/DAS.2016.23
10. He, K., Zhang, X., Ren, S., Sun, J.: Deep residual learning for image recognition. In: CVPR 2016. https://doi.org/10.1109/CVPR.2016.90
11. Kieninger, T., Dengel, A.: The T-Recs table recognition and analysis system. In: Lee, S.-W., Nakano, Y. (eds.) DAS 1998. LNCS, vol. 1655, pp. 255–270. Springer, Heidelberg (1999). https://doi.org/10.1007/3-540-48172-9_21
12. Kingma, D.P., Ba, J.: Adam: a method for stochastic optimization. In: ICLR 2015
13. Kipf, T.N., Welling, M.: Semi-supervised classification with graph convolutional networks. In: ICLR 2017
14. Li, Y., Huang, Z., Yan, J., Zhou, Y., Ye, F., Liu, X.: GFTE: graph-based financial table extraction. In: Del Bimbo, A., Cucchiara, R., Sclaroff, S., Farinella, G.M., Mei, T., Bertini, M., Escalante, H.J., Vezzani, R. (eds.) ICPR 2021. LNCS, vol. 12662, pp. 644–658. Springer, Cham (2021). https://doi.org/10.1007/978-3-030-68790-8_50
15. Long, J., Shelhamer, E., Darrell, T.: Fully convolutional networks for semantic segmentation. In: CVPR 2015. https://doi.org/10.1109/CVPR.2015.7298965
16. Pasupat, P., Liang, P.: Compositional semantic parsing on semi-structured tables. In: ACL 2015. https://doi.org/10.3115/v1/p15-1142
17. Qasim, S.R., Kieseler, J., Iiyama, Y., Pierini, M.: Learning representations of irregular particle-detector geometry with distance-weighted graph networks. Eur. Phys. J. C **79**(7), 1–11 (2019). https://doi.org/10.1140/epjc/s10052-019-7113-9
18. Qasim, S.R., Mahmood, H., Shafait, F.: Rethinking table recognition using graph neural networks. In: ICDAR 2019. https://doi.org/10.1109/ICDAR.2019.00031
19. Raja, S., Mondal, A., Jawahar, C.V.: Table structure recognition using top-down and bottom-up cues. In: Vedaldi, A., Bischof, H., Brox, T., Frahm, J.-M. (eds.) ECCV 2020, Part XXVIII. LNCS, vol. 12373, pp. 70–86. Springer, Cham (2020). https://doi.org/10.1007/978-3-030-58604-1_5
20. Ren, S., He, K., Girshick, R.B., Sun, J.: Faster R-CNN: towards real-time object detection with region proposal networks. In: NIPS 2015

21. Riba, P., Dutta, A., Goldmann, L., Fornés, A., Terrades, O.R., Lladós, J.: Table detection in invoice documents by graph neural networks. In: ICDAR 2019. https://doi.org/10.1109/ICDAR.2019.00028

22. Ritze, D., Lehmberg, O., Bizer, C.: Matching HTML tables to DBpedia. In: WIMS 2015. https://doi.org/10.1145/2797115.2797118

23. Ritze, D., Lehmberg, O., Oulabi, Y., Bizer, C.: Profiling the potential of web tables for augmenting cross-domain knowledge bases. In: WWW 2016. https://doi.org/10.1145/2872427.2883017

24. Schreiber, S., Agne, S., Wolf, I., Dengel, A., Ahmed, S.: DeepDeART: deep learning for detection and structure recognition of tables in document images. In: ICDAR 2017. https://doi.org/10.1109/ICDAR.2017.192

25. Shafait, F., Smith, R.: Table detection in heterogeneous documents. In: DAS 2010. https://doi.org/10.1145/1815330.1815339

26. Shigarov, A.O., Mikhailov, A.A., Altaev, A.: Configurable table structure recognition in untagged PDF documents. In: DocEng 2016. https://doi.org/10.1145/2960811.2967152

27. Siddiqui, S.A., Fateh, I.A., Rizvi, S.T.R., Dengel, A., Ahmed, S.: DeepTabStR: deep learning based table structure recognition. In: ICDAR 2019. https://doi.org/10.1109/ICDAR.2019.00226

28. Siddiqui, S.A., Khan, P.I., Dengel, A., Ahmed, S.: Rethinking semantic segmentation for table structure recognition in documents. In: ICDAR 2019. https://doi.org/10.1109/ICDAR.2019.00225

29. Siddiqui, S.A., Malik, M.I., Agne, S., Dengel, A., Ahmed, S.: DeCNT: deep deformable CNN for table detection. IEEE Access 6, 74151–74161 (2018). https://doi.org/10.1109/ACCESS.2018.2880211

30. Sun, H., Ma, H., He, X., Yih, W., Su, Y., Yan, X.: Table cell search for question answering. In: WWW 2016. https://doi.org/10.1145/2872427.2883080

31. Tensmeyer, C., Morariu, V.I., Price, B.L., Cohen, S., Martinez, T.R.: Deep splitting and merging for table structure decomposition. In: ICDAR 2019. https://doi.org/10.1109/ICDAR.2019.00027

32. Wang, Y., Phillips, I.T., Haralick, R.M.: Table structure understanding and its performance evaluation. Pattern Recognit. 37(7), 1479–1497 (2004). https://doi.org/10.1016/j.patcog.2004.01.012

33. Wang, Y., Sun, Y., Liu, Z., Sarma, S.E., Bronstein, M.M., Solomon, J.M.: Dynamic graph CNN for learning on point clouds. ACM Trans. Graph. 38(5), 146:1-146:12 (2019). https://doi.org/10.1145/3326362

34. Zhong, V., Xiong, C., Socher, R.: Seq2SQL: Generating structured queries from natural language using reinforcement learning. arXiv preprint arXiv:1709.00103 (2017)

35. Zhong, X., ShafieiBavani, E., Jimeno Yepes, A.: Image-based table recognition: data, model, and evaluation. In: Vedaldi, A., Bischof, H., Brox, T., Frahm, J.-M. (eds.) ECCV 2020. LNCS, vol. 12366, pp. 564–580. Springer, Cham (2020). https://doi.org/10.1007/978-3-030-58589-1_34

36. Zhong, Z., Zheng, L., Kang, G., Li, S., Yang, Y.: Random erasing data augmentation. In: AAAI (2020)

Image to LaTeX with Graph Neural Network for Mathematical Formula Recognition

Shuai Peng, Liangcai Gao, Ke Yuan, and Zhi Tang$^{(\boxtimes)}$

Wangxuan Institute of Computer Technology, Peking University, Beijing, China
{pengshuaipku,gaoliangcai,yuanke,tangzhi}@pku.edu.cn

Abstract. Mathematical formula recognition aims to automatically convert formula images into their structured description formats. Recently, some encoder-decoder models have been presented for this task, while they seldom explicitly consider spatial relationship among symbols. In this paper, we proposed a novel encoder-decoder model with Graph Neural Network (GNN) to translate mathematical formula images into LaTeX codes. In the proposed model, the symbols segmented from the raw image are used to build graphs based on their spatial connection. The encoder consists of Convolutional Neural Network (CNN) and GNN. CNN is utilized to extract the visual features from the whole formula or symbols, and GNN is used to transmit the spatial information embedded in the built graphs. The adopted decoder is a Recurrent Neural Network (RNN) model, which implements a language model to generate the output sentences based on the encoded features with attention mechanism. The experimental results on IM2LATEX-100K dataset demonstrated that the proposed model obtained a better performance than state-of-the-art approaches.

Keywords: Mathematical formula recognition · Graph neural network · Encoder-decoder architecture · image to LaTeX

1 Introduction

Mathematical formula is a common and important component of scientific or technical documents. Converting mathematical formulas into markup language description is a key issue in a wide range of applications [1–4], such as document analysis, document information retrieval and hand-written formula recognition, which brings considerable convenience to researchers and educators. Besides technical experts, the public can profit from the easy-to-use mathematical applications as well. Consequently, mathematical formula recognition has recently attracted increasing attention.

Mathematical formula recognition is a typical research topic of structured text recognition, because mathematical formula has two-dimensional layout

© Springer Nature Switzerland AG 2021
J. Lladós et al. (Eds.): ICDAR 2021, LNCS 12822, pp. 648–663, 2021.
https://doi.org/10.1007/978-3-030-86331-9_42

structure, such as nested fractions, sub and super-script notation, which empha-size the capability of capturing spatial information. Besides, different scales of mathematical symbols carry different meanings. For example, - and − are both a horizontal stroke, but the former means minus and the latter denotes fraction. In addition, sub and super-script symbols are often smaller than normal symbols, which increases the difficulty of mathematical formula recognition.

With the earliest work of mathematical formula recognition dating back to 1967 [5], vast syntactic rules have been presented for formula structure analysis. One of the most famous rule-based systems is the INFTY system, which is used to convert printed mathematical formulas into structured language like LaTeX, MathML, XHTML [6]. The recent advances in deep neural networks demonstrate the high capability of extracting feature and generating text sequence. Especially, encoder-decoder models have been applied in many image-to-text tasks such as image captioning and scene text recognition. Convolutional neural network (CNN) is often exploited to encode an image due to its good performance in capturing visual information [7–9] and recurrent neural network (RNN) with attention mechanism is utilized as language model owing to its advantage in predicting time series [10, 11]. In general, formula recognition could be considered as a special type of image caption task.

Although much improvement has been achieved by encoder-decoder mod-els in mathematical formula recognition, the spatial relationship and interaction among mathematical symbols is not fully leveraged in those models. In rule-based approaches, symbol segmentation and spatial relation analysis are neces-sary steps before formula structure generation. When it comes to deep neural network, models automatically learn patterns during the process of data-driven training. Since mathematical symbols can be easily segmented from a formula image, more specific spatial information can be utilized to enhance mathemati-cal formula recognition. Inspired by the recent successes of GNN model applied in image captioning, we propose a neural encoder-decoder model with GNN. The proposed approach is composed of five modules: (1) symbol segmentation, (2) graph construction based on spatial positions and connections of symbols, (3) encoding symbol regions and formula region into feature maps by CNN, (4) encoding feature maps with visual relationship in the built graph by GNN. (5) generating output sentence using long short-term memory (LSTM) [12] with soft attention mechanism.

The main contributions of the proposed approach include two aspects. First, by explicitly taking extra visual information of symbols into account for encod-ing, the model is more likely to focus on the relevant symbol regions. Second, we exploit GNN to enrich region representations with visual relationship in the spatial graph, which further enhances the performance of our model in the struc-ture analysis. In addition, the results on IM2LATEX-100K dataset demonstrate our model obtained a better performance than state-of-the-art approaches.

2   Related Work

A variety of methods for mathematical formula recognition have been proposed in the past 60 years [13–15]. The existing approaches can be classified into three categories: tree based, grammar based, and learning based.

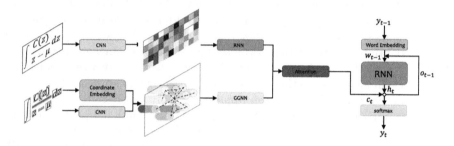

Fig. 1. Overview of our approach.

2.1   Tree Based Methods

Tree based methods extract symbol relationships as a tree to represent the formula structure. In [16], Twaakynodo et al. proposed a top-down and bottom-up combined method to construct the expression tree. Zanibbi et al. [17] designed a system using baseline structure tree to describe two-dimensional arrangement of symbols. In [6], Suzuki et al. presented an integrated OCR system to convert formula into markup formats by representing formula structure as a tree and using a minimum-cost spanning-tree algorithm for the structure analysis, which was made into the commercial software – InftyReader. The main challenge of tree-based methods is to eliminate the ambiguity in the process of building symbol trees.

2.2   Grammar Based Methods

Grammar based methods employ formal grammars and corresponding parsing rules to recognize mathematical formulas, which require researchers to predefine grammars manually. Yamamoto et al. [18] modeled handwritten formulas with a stochastic context free grammar and formulated the recognition problem as a search problem which can be solved by Cocke-Younger-Kasami (CYK) algorithm. In [19], Prusa et al. presented a method for off-line mathematical formula recognition based on structural construction paradigm and two-dimensional grammars. Alvaro et al. [20] defined the statistical framework based on two-dimensional stochastic context-free grammars which allows to jointly tackle all involved steps automatically learnt. It is difficult to manually define a universal grammar for all situations which is the main constraint of grammar-based methods.

2.3   Learning Based Methods

Learning based methods are emerging data-driven approaches with the benefit of the recent improvements in deep learning. With the inspiration of the advances made in the areas of handwriting recognition [21], OCR in natural scenes [22] and image caption generation [23], encoder-decoder model has been applied in mathematical formula recognition and achieved good results in the past few years. In [24], Zhang et al. presented a gated recurrent unit (GRU) based encoder-decoder model with attention mechanism to translate handwritten mathematical expressions to LaTeX. In [25], Deng et al. proposed a neural encoder-decoder model with coarse-to-fine attention mechanism called WYGIWYS. Leveraging the success in [25], Wang et al. [26] employed DenseNet [27] as feature extractor, and enhanced the attention mechanism with joint attention mechanism in [28]. These successes have developed handwritten mathematical expression recognition as well. Zhang et al. proposed WAP [37] and TreeDecoder [39] for offline handwritten recognition and TAP [38] for online handwritten recognition. It has been shown that structural intrinsic features captured by GNN is beneficial to downstream tasks [40]. Thus Wu et al. proposed a Graph-to-Graph approach [41] for online handwritten recognition. Almost all the learning-based methods proposed recently demonstrate the better performance than the rule-based methods.

Learning based methods usually don't require the steps like symbol segmentation, tree construction and symbol dominance analysis in the two kinds of methods mentioned above, but take raw formula image as input and produce output sentence directly, which makes model totally rely on the encoder to capture information of formula structure. Thus, we proposed an approach which employs GNN to encode features with spatial relationships among symbols to enhance the model's capability of recognizing formula structure.

3   Proposed Method

We design a novel encoder-decoder model with GNN to convert a mathematical formula image into LaTeX sequence by additionally incorporating spatial symbol relationship. The architecture of this model is illustrated in Fig. 1. Firstly, we utilize connected component analysis algorithm to segment symbols from raw formula images. Secondly, Line-Of-Sight (LOS) graph [29] is constructed using these separated symbols according to their spatial connection. Next, the symbol images and their raw source formula image are injected into feature extractor via CNN to generate feature maps. And then, GNN is employed to encode the feature maps extracted from the symbol image set with spatial graph structure, resulting in symbol representations. Finally, the encoded symbol representations combined with the feature maps of formula region are fed into the decoder based on LSTM with visual soft attention mechanism and then produce the final output LaTeX sequence.

3.1   Problem Formulation

Mathematical formula recognition is a typical sequence generation problem, which can be defined as: there is a gray-scale image $X \in \mathbb{R}^{H \times W}$ with height H and width W, the goal is to generate the corresponding LaTeX sequence $Y = [y_1, y_2, \ldots, y_T]$ consisting of T tokens that makes up the mathematical formula in the image. Hence, we can formulate this problem by minimizing the following loss function:

$$E(X, Y) = -\log P(Y|X) \tag{1}$$

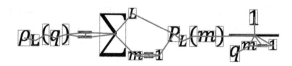

Fig. 2. Example of graph construction.

3.2   Graph Construction

In this step, symbol positional information and spatial relationships are extracted from raw formula images for the encoder. We draw inspiration from graph-based methods and construct LOS graph to retain the relationships. Specifically, we separate connected components from a formula image and record their relative coordinates as the positional information of symbol regions. And then, we generate a LOS graph over connected components. Symbols that can completely see each other share an edge in the graph. An example is shown in Fig. 2, where red boxes contain the symbol regions and blue lines represent the edges in the graph. The graph simply treats connected components as vertices and roughly represents the spatial relationships among symbols for the reason that it is constructed for GNN to refine the representation of symbols with visual relationship and doesn't need to contain accurate semantic structure.

3.3   Encoder

Convolutional Neural Network. We adopt two kinds of CNN architecture to respectively extract visual features from symbol region and formula region due to their different scales - symbol region is much smaller than the whole formula region, which requires more attention to small objects. The CNN architecture for symbol region is based on LResNet50E-IR [30], which has shown excellent performance in facial recognition tasks. Here we employ it as the feature extractor for symbol region aiming to take advantage of its high capability

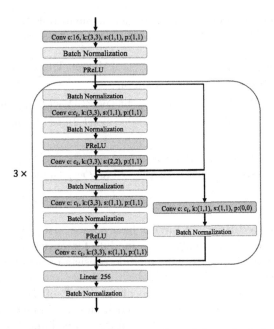

Fig. 3. The structure of LResNet50E-IR. c: number of channels, k: kernel size, s: stride size, p: padding size. $c_i = 32, 64, 128$ when $i = 1, 2, 3$

of capturing visual information from small objects. Previously, we binarize the separated symbol images and resize them into the same scale as 32×32. Feature extractor takes the processed symbol region set $\tilde{\mathbf{I}}_\mathbf{s}$ of size $L \times 1 \times 32 \times 32$ as input and produce feature map $\tilde{\mathbf{V}}_\mathbf{s}$ of size $L \times D_1$, where L represents the number of symbols in the whole formula and D_1 denotes the number of channels. The structure is illustrated in Fig. 3. As for formula region, we choose the standard CNN architecture described in [25] as the feature extractor. The CNN takes the binarized formula region $\tilde{\mathbf{I}}_\mathbf{f}$ of size $1 \times H \times W$ as input and produce feature map $\tilde{\mathbf{V}}_\mathbf{f}$ of size $D_2 \times H' \times W'$, where D_2 is set twice as much as D_1 for the convenience of the following concatenation ($D_2 = 2D_1 = 512$ in our implementation), $H \times W$ and $H' \times W'$ are the size of the original formula region and the resulted feature map, respectively. Details of its configuration are shown in Table 1.

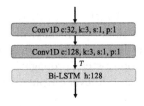

Fig. 4. The structure of coordinate embedding. c: number of channels, k: kernel size, s: stride size, p: padding size, h: hidden size

Table 1. CNN configuration for formula region. c: number of channels, k: kernel size, s: stride size, p: padding size, bn: batch normalization. The sizes are in order (H, W)

Convolution layer	Max-pooling Layer
c:512, k:(3,3), s:(1,1), p:(0,0), bn	–
c:512, k:(3,3), s:(1,1), p:(1,1), bn	po:(2,1), s:(2,1), p:(0,0)
c:256, k:(3,3), s:(1,1), p:(1,1)	po:(1,2), s:(1,2), p:(0,0)
c:256, k:(3,3), s:(1,1), p:(1,1), bn	–
c:128, k:(3,3), s:(1,1), p:(1,1)	po:(2,2), s:(2,2), p:(0,0)
c:64, k:(3,3), s:(1,1), p:(1,1)	po:(2,2), s:(2,2), p:(0,0)

Coordinate Embedding. To further incorporate spatial localization of symbols, we design coordinate embedding module for enriching global positional information of symbol regions. This module consists of one-dimensional convolution, activation and LSTM layers, which is illustrated in Fig. 4. At first two one-dimensional convolution layers, coordinate list $\tilde{C}$ of size $4 \times L$ is expanded into the matrix of size $D_1/2 \times L$. We feed it into a bidirectional LSTM layer after transposing and gain the final coordinate embedding features $\tilde{V}_c$ of size $D_1 \times L$, which are further concatenated with the symbol features $\tilde{V}_s$ from CNN. The combined features $\tilde{V}_{sc}$ are injected into the next GNN module.

Graph Neural Network. In this model, we exploit a gated graph neural network (GGNN) [31] to encode features $\tilde{V}_{sc}$ with visual relationship in the LOS graph. GGNN is a typical message passing model in spatial domain based on gated recurrent unit (GRU) and we summarize it here. Suppose we have a graph $G = (V, E)$, where V is the set of vertices and E is the set of edges. Each vertex $v_i \in V$ retains a state h_i, and each edge $e_{ij} \in E$ represents a directed edge $v_i \rightarrow v_j$. The propagation process of GGNN is:

$$m_{ij} = f^*(h_i^t, h_j^t, e_{ij}) \tag{2}$$

$$m_i = \sum_j m_{ji} \tag{3}$$

$$h_i^{t+1} = GRU(h_i^t, m_i) \tag{4}$$

where m_{ji} means the message which v_j send to v_i through edge e_{ji} and f^* is the message transmitting function. Specifically, we treat $\tilde{V}_{sc}$ as the initial state $h^0 = \{h_1^0, h_2^0, \ldots, h_L^0\}$ and utilize the LOS graph to transmit messages between vertices. In this model, we adopt 2-steps propagation which performs best in the comparative experiment. For graph-level outputs, we produce output vector h_G as

$$h_G = \tanh(W^G[h^T, h^0]) \tag{5}$$

Finally, we get the feature grid $\tilde{V}'_{sc}$ with size of $L' \times 2D_1$.

Row Encoder. We adopt the RNN architecture to localize the formula features $\tilde{\mathbf{V}}_{\mathbf{f}}$ by running it over each of the rows of feature grid, which has been shown to be effective in [25]. The CNN feature vectors $\tilde{\mathbf{V}}_{\mathbf{f}}$ are first partitioned into $\mathbf{V_1}, \mathbf{V_2}, \dots, \mathbf{V_H}$ by row. In order to capture sequential order information, we prepare an embedding layer for each row to generate positional embeddings, which is regarded as an initial hidden state for RNN. After that, we run RNN on each vector $\mathbf{V_i}$ and combine the output feature vectors together into a new feature grid $\tilde{\mathbf{V}}'_{\mathbf{f}}$ with size of $H' \times D_2$.

3.4   Decoder

Sequence Embedding. We utilize the ground truth of LaTeX sequence to guide the prediction of output sequence in the training stage. Assuming that LaTeX sequence Y is split into T tokens $y_1, y_2, \dots, y_T$, the purpose is to represent them as computable vectors for neural networks. Considering the high similarity between machine translating and this task in generating sequence, we adopt word embedding [32] layer which is widely used in natural language processing (NLP) tasks to capture the interrelationship between tokens, where a LaTeX token y_t is transformed into a high-dimensional vector $\mathbf{w_t}$.

Recurrent Neural Network. Formally, RNN is a parameterized function that recursively update hidden state with old hidden state and input vector one step at each time, which can be represented as $h_t = RNN(h_{t-1}, v_t)$. Owing to its advantage in predicting time series, RNN has been generally applied in sequence prediction tasks. Here, we choose LSTM network as our decoder for the reason that LSTM demonstrate better performance in capturing long term dependencies than the standard RNN.

We propose two layers of LSTM as the language model. Multi-layers increase the depth of the language model and thus enhance the ability of capturing complex language semantics. Normally, the output of encoder RNN is treated as the initial hidden state and cell state of the decoder. Hence, here we use the output of row encoder to initialize the LSTM network. The updating procedure is as

$$h_t = LSTM(h_{t-1}, [w_{t-1}, o_{t-1}]) \tag{6}$$

$$o_t = \tanh(W^C[h_t, c_t]) \tag{7}$$

where $\mathbf{h_t}$ is the hidden state of LSTM unit, $\mathbf{w_t}$ is the embedded vector, and $\mathbf{c_t}$ is the context vector which will be introduced in the next part.

Attention. The initial state in RNN is not sufficient to retain all the information from encoder particularly when the features contain the GNN output that is not included by the initial state. Therefore, we add attention mechanism to further capture mapping relations between LaTeX tokens and visual features.

Concretely, we take the encoder output features $\tilde{\mathbf{V}}'_{\mathbf{sc}}$ and $\tilde{\mathbf{V}}'_{\mathbf{f}}$ as the input. Note that the number of channels D_2 is twice as much as D_1, we are able to concatenate them into a new feature grid $\mathbf{V} = [\mathbf{v_1}, \mathbf{v_2}, \ldots, \mathbf{v_{L+H'}}]$ with size of $(L + H') \times D_2$. Attention mechanism attempts to focus on the semantic-related feature vectors by assigning different weights to them. To calculate the attention contribution, we define it as:

$$\alpha_{i,t} = softmax(a_{i,t}) \tag{8}$$

where $\mathbf{a_{i,t}}$ is the unnormalized attention weights. Here we adopt the formula developed by Luong [33] as:

$$a_{i,t} = \beta^T \tanh(W_1 h_t + W_2 v_i) \tag{9}$$

Based on the attention contribution, we can calculate out the context vector $\mathbf{c_t}$ by aggregating all the features weighted with attention:

$$c_t = \sum_i^{L+H'} \alpha_{i,t} v_i \tag{10}$$

The process of predicting output sequence is defined as:

$$p(y_t|y_1, \ldots, y_{t-1}, V) = softmax(W^{out} o_t) \tag{11}$$

which represents the probability of the next token at each time step.

4   Experiments

To verify the effectiveness of our approach, we conduct the experiments and evaluate our model on IM2LATAX-100K dataset [25]. In this section, we will introduce the dataset and the data preprocessing, and then the details of implementation, followed by evaluation metrics in the end.

4.1   Datasets and Preprocessing

We conducted experiments on IM2LATEX-100K dataset provided by [25], which contains a total of 103,556 formula images and the corresponding LaTeX sequences collected from the LaTeX sources in publications. The LaTeX formulas range from 38 to 997 characters, with mean 118 and median 98. We render the LaTeX codes into the PDF format and then convert them into PNG images of resolution 533×1466. The rendered dataset is randomly separated into training set (75,275 equations), validation set (17,926 equations) and test set (10,355 equations).

Before training, we need to construct a token vocabulary for word embedding and sentence prediction. Here we follow the same rules as [25] to split LaTeX sequences into minimal meaningful tokens. In addition, two special tokens $< BOS >$ and $< EOS >$ are added into the vocabulary as signs of *the beginning of the sequence* and *the end of the sequence*, respectively. Finally, we get a vocabulary with 583 tokens.

4.2 Implementation Details

The configurations of feature extractors in encoder are shown in Fig. 3 and Table 1. Note that $\frac{D_2}{D_1} = 2, \frac{H}{H'} = \frac{W}{W'} = 4$. The hidden state of the coordinate embedding RNN is of size 128, row encoder RNN is of size 256, decoder RNN is of size 512, and token embeddings is of size 80.

In the training stage, we use the mini-batch stochastic gradient descent to train the parameters in our model. The initial learning rate is set to 0.1, and decays by 0.5 every epoch. The batch size is 10 limited by GPU memory. We train the model for 18 epochs, and choose the best model according to the validation perplexity. In the inference stage, in order to achieve global optimum in sentence generation, we adopt the beam search strategy [34] and set the beam size as 5.

4.3 Evaluation Metrics

We adopt three types of evaluation metrics: BLEU-4 [35], edit distance, exact match. BLEU score is originally designed to measure the performance of a machine translation system, which is a form of precision of word n-grams. Edit distance reflects the similarity between two sentences by counting the minimum number of operations required to transform one into the other. Exact match means the ratio of the predicted images to be exactly the same as the standard ones.

5 Results

Table 2. Results on IM2LATEX-100K

	BLEU-4	Exact match	Edit distance
INFTY [6]	66.65	15.60	24.14
WYGIWYS [25]	87.73	77.46	18.36
DenseNet [26]	88.25	–	–
Double attention [36]	88.42	79.81	–
Our method	**90.19**	**81.82**	**11.85**

5.1 Performance Comparison

Our method is evaluated against a rule-based method INFTY and three deep learning methods - WYGIWYS, DenseNet, Double Attention. The experimental results are presented in Table 2. Overall, the results across three evaluation metrics demonstrate that our method achieves superior performance than the other four approaches. Concretely, all four deep learning methods make significant improvement than INFTY, which implies great advantages of encoder-decoder architecture in image-to-text task. Compared with the deep learning baseline WYISWYS, BLEU score and exact match accuracy are respectively increased by 2.8% and 5.63%, and edit distance is decreased by 35.6%. Moreover, the performances of the advanced models [26,36] are also lower than our model. It indicates the advantage of bringing in the spatial information of symbols. In addition, we utilize GNN to refine the representation of symbols by leveraging graph structure, which further boosts up the performance.

Table 3. Results on test set with noises

	BLEU-4	Exact match	Edit distance
WYGIWYS [25]	58.12	43.00	68.99
Our Method	**65.09**	**45.59**	**49.31**

5.2 Robustness Analysis

In real-world applications, the formula images to be recognized are not so clear with random noises, which requires high robustness of the approach. To verify the robustness of our model, we randomly divide the test set into three parts averagely and add salt noise, gaussian noise, gaussian blur respectively. We conduct an experiment on this new test set using the model trained on the clear train set. The results are shown in Table 3. As we can see, the performance of both methods drops much on three evaluation metrics, but still our method retain superiority compared with WYGIWYS. In addition, note that BLEU score of our method drops by 27.8% while WYGIWYS drops by 33.8%, which indicates the higher robustness of our method than WYGIWYS. We owe it to the extra symbol information added into encoder, which enhances the capability of capturing correct regions and decreases the possibility of being disturbed by noises.

Table 4. Contribution of different modules

Module	BLEU-4
0-layer GGNN	76.73
1-layer GGNN	90.32
2-layer GGNN	**90.50**
3-layer GGNN	89.29
Replaced by LSTM	86.93
Simple graph	87.49
No CoorEmbed	90.12

5.3   Contribution Analysis

To investigate the contributions of different modules for the overall performance, we conduct further comparative experiments. The results are shown in Table 4.

Contribution of GNN. As is shown in the table, BLEU score drops to 76.73 when the propagation step is set to 0, i.e., without GNN, which is even far less than the baseline model. It suggests that the ability of model learning symbol representation and formula structure cannot be improved by simply splicing the features of symbols and formula. Besides, BLEU score drops 3.57 when GNN encoder is replaced by LSTM, which implies GNN performs better than simple LSTM in incorporating 2-D spatial structure and enhances the ability of model in extracting and refining symbol representation.

Contribution of GGNN Layer. For GGNN, shallow network is not conductive to the propagation of node features, while deep network makes the model over-smoothing. After comparative experiments, we adopt 2 layers GGNN in our model.

Contribution of LOS Graph. To verify the effectiveness of LOS graph in extracting spatial relationship among symbols, we replace it with a simple graph where nodes are sequentially connected for comparison. The experimental results point to remarkable improvement by LOS graph, which means the graph constructed from spatial information has positive significance to extract formula structure.

Contribution of Coordinate Embedding. With the help of coordinate embedding, BLEU score increases by 0.38. Considering the excellent performance of the model, this improvement is considerable. Coordinate embedding module provides the model with location and scale information of symbol regions, especially when the symbol regions injected into CNN are in the same size in our model.

5.4   Case Study

Two examples of recognition results are given in Fig. 5, where errors are highlighted in red and missing parts are painted in yellow. Overall, both methods achieve quite satisfying performance but differ in detail. In the first example, owing to explicitly taking symbol regions into account, our method generates completely correct LaTeX sequence without missing any symbols, while WYGI-WYS misses $\prod$. The second example is a fairly long sequence, where WYGIWYS misses the superscript '2'. Although our method doesn't miss this small symbol, it still wrongly regards '2' as the superscript of '4' rather than 'x'. A possible solution will be to incorporate semantic information in our model in the future.

	LaTeX code	Image
Ground Truth	Z=\sum_{spins}\prod_{cubes}W(ale,f,glb,c,dlh),	$Z = \sum_{spins}\prod_{cubes} W(a|e,f,g|b,c,d|h),$
WYGIWYS	Z=\sum_{spins cubes}W(ale,f,glb,c,dlh),	$Z = \sum_{spins cubes} W(a|e,f,g|b,c,d|h),$
Our Method	Z=\sum_{spins}\prod_{cubes}W(ale,f,glb,c,dlh),	$Z = \sum_{spins\ cubes}\prod W(a|e,f,g|b,c,d|h),$
Ground Truth	\frac{3}{2}(\partial^{\mu}\partial^{\nu}h_{\mu\nu}-\partial_{\mu}\partial^{\mu}\tilde{h})-\frac{15}{2}\partial_{4}\sigma\partial_{4}\tilde{de}{h}-\frac{3}{2}\frac{\partial^{2}\tilde{h}}{\partial x^{4}}^{2})-\frac{3}{2}\partial_{\mu}\partial^{\mu}\phi+30k^{2}\phi-12k\phi\tilde{delta}=0.	$\frac{3}{2}\left(\partial^{\mu}\partial^{\nu}h_{\mu\nu}-\partial_{\mu}\partial^{\mu}\tilde{h}\right)-\frac{15}{2}\partial_{4}\sigma\partial_{4}\tilde{h}-\frac{3}{2}\frac{\partial^{2}\tilde{h}}{\partial x^{2}}-\frac{3}{2}\partial_{\mu}\partial^{\mu}\phi+30k^{2}\phi-12k\phi\tilde{\delta}=0.$
WYGIWYS	\frac{3}{2}(\partial^{\mu}\partial^{\nu}h_{\mu\nu}-\partial_{\mu}\partial^{\mu}\tilde{h})-\frac{15}{2}\partial_{4}\sigma\partial_{4}\tilde{de}{h}-\frac{3}{2}\frac{\partial^{2}\tilde{h}}{\partial x}^{4})-\frac{3}{2}\partial_{\mu}\partial^{\mu}\phi+30k^{2}\phi-12k\phi\tilde{delta}=0.	$\frac{3}{2}\left(\partial^{\mu}\partial^{\nu}h_{\mu\nu}-\partial_{\mu}\partial^{\mu}\tilde{h}\right)-\frac{15}{2}\partial_{4}\sigma\partial_{4}\tilde{h}-\frac{3}{2}\frac{\partial^{2}\tilde{h}}{\partial x^{2}}-\frac{3}{2}\partial_{\mu}\partial^{\mu}\phi+30k^{2}\phi-12k\phi\tilde{\delta}=0.$
Our Method	\frac{3}{2}(\partial^{\mu}\partial^{\nu}h_{\mu\nu}-\partial_{\mu}\partial^{\mu}\tilde{h})-\frac{15}{2}\partial_{4}\sigma\partial_{4}\tilde{de}{h}-\frac{3}{2}\frac{\partial^{2}\tilde{h}}{\partial x}^{4^{2}})-\frac{3}{2}\partial_{\mu}\partial^{\mu}\phi+30k^{2}\phi-12k\phi\tilde{delta}=0.	$\frac{3}{2}\left(\partial^{\mu}\partial^{\nu}h_{\mu\nu}-\partial_{\mu}\partial^{\mu}\tilde{h}\right)-\frac{15}{2}\partial_{4}\sigma\partial_{4}\tilde{h}-\frac{3}{2}\frac{\partial^{2}\tilde{h}}{\partial x^{2}}-\frac{3}{2}\partial_{\mu}\partial^{\mu}\phi+30k^{2}\phi-12k\phi\tilde{\delta}=0.$

Fig. 5. Examples of recognition results

6   Conclusion

In this paper, we proposed a deep neural encoder-decoder model with Graph Neural Network (GNN) to convert formula images into LaTeX sequences. To make the model focus on the relevant symbol regions, we bring in extra visual information of symbols for encoding. In addition, GNN is exploited to enrich symbol representations with visual relationships in the structured spatial graph. Extensive experiments conducted on IM2LATEX-100K dataset validate the proposed model. The results demonstrate better and more robust performance of this model than state-of-the-art approaches. We would like to incorporate graph construction into model and build an end-to-end framework to support hand-written formula recognition in the future.

Acknowledgements. This work is supported by the projects of National Natural Science Foundation of China (No. 61876003) and National Key R&D Program of China (2019YFB1406303), which is also a research achievement of Key Laboratory of Science, Technology and Standard in Press Industry (Key Laboratory of Intelligent Press Media Technology).

References

1. Yuan, K.: Multi-dimensional formula feature modeling for mathematical information retrieval. In: Proceedings of the 40th International ACM SIGIR Conference on Research and Development in Information Retrieval, p. 1381 (2017)
2. Nishizawa, G., Liu, J., Diaz, Y., Dmello, A., Zhong, W., Zanibbi, R.: MathSeer: a math-aware search interface with intuitive formula editing, reuse, and lookup. In: Jose, J.M., et al. (eds.) ECIR 2020. LNCS, vol. 12036, pp. 470–475. Springer, Cham (2020). https://doi.org/10.1007/978-3-030-45442-5_60
3. Peng, S., Yuan. K., Gao, L., et al.: MathBERT: a pre-trained model for mathematical formula understanding. arXiv e-prints (2021). arXiv: 2105.00377
4. Yuan, K., Gao, L., Wang, Y., et al.: A mathematical information retrieval system based on RankBoost. In: Proceedings of the 16th ACM/IEEE-CS on Joint Conference on Digital Libraries, pp. 259–260 (2016)
5. Anderson, R.H.: Syntax-directed recognition of hand-printed two-dimensional mathematics. In: Symposium on Interactive Systems for Experimental Applied Mathematics: Proceedings of the Association for Computing Machinery Inc., Symposium, pp. 436–459. ACM (1967)
6. Suzuki, M., Tamari, F., Fukuda, R., Uchida, S., Kanahori, T.: Infty: an integrated OCR system for mathematical documents. In: Proceedings of the 2003 ACM Symposium on Document Engineering, pp. 95–104. ACM (2003)
7. Girshick, R.: Fast R-CNN. In: Proceedings of the IEEE International Conference on Computer Vision, pp. 1440–1448 (2015)
8. Sharif Razavian, A., Azizpour, H., Sullivan, J., et al.: CNN features off-the-shelf: an astounding baseline for recognition. In: Proceedings of the IEEE Conference on Computer Vision and Pattern Recognition Workshops, pp. 806–813 (2014)
9. Yuan, K., He, D., Yang, X., et al.: Follow the curve: arbitrarily oriented scene text detection using key points spotting and curve prediction. In: 2020 IEEE International Conference on Multimedia and Expo (ICME), pp. 1–6. IEEE (2020)
10. Cho, K., Van Merriënboer, B., Gulcehre, C., et al.: Learning phrase representations using RNN encoder-decoder for statistical machine translation. arXiv preprint arXiv:1406.1078 (2014)
11. Yuan, K., He, D., Jiang, Z., et al.: Automatic generation of headlines for online math questions. In: Proceedings of the AAAI Conference on Artificial Intelligence, vol. 34, no. (05), pp. 9490–9497 (2020)
12. Hochreiter, S., Schmidhuber, J.: Long short-term memory. Neural Comput. **9**(8), 1735–1780 (1997)
13. Zanibbi, R., Blostein, D.: Recognition and retrieval of mathematical expressions. Int. J. Doc. Anal. Recogn. (IJDAR) **15**(4), 331–357 (2012)
14. Yan, Z., Zhang, X., Gao, L., et al.: ConvMath: a convolutional sequence network for mathematical expression recognition. In: 2020 25th International Conference on Pattern Recognition (ICPR), pp. 4566–4572. IEEE (2021)

15. Zhang, X., Gao, L., Yuan, K., et al.: A symbol dominance based formulae recognition approach for pdf documents. 2In: 017 14th IAPR International Conference on Document Analysis and Recognition (ICDAR), vol. 1, pp. 1144–1149. IEEE (2017)
16. Twaakyondo, H.M., Okamoto, M.: Structure analysis and recognition of mathematical expressions. In: Proceedings of 3rd International Conference on Document Analysis and Recognition, vol. 1, pp. 430–437. IEEE (1995)
17. Zanibbi, R., Blostein, D., Cordy, J.R.: Recognizing mathematical expressions using tree transformation. IEEE Trans. Pattern Anal. Mach. Intell. 24(11), 1455–1467 (2002)
18. Yamamoto, R., Sako, S., Nishimoto, T., Sagayama, S.: On-line recognition of handwritten mathematical expressions based on stroke-based stochastic context-free grammar. In: tenth International Workshop on Frontiers in Handwriting (2006)
19. Prusa, D., Hlaváč, V.: Mathematical formulae recognition using 2D grammars. In: Ninth International Conference on Document Analysis and Recognition, ICDAR 2007, vol. 2, pp. 849–853. IEEE (2007)
20. Alvaro, F., Benedi, J.M., et al.: Recognition of printed mathematical expressions using two-dimensional stochastic context-free grammars. In: 2011 International Conference on Document Analysis and Recognition (ICDAR), pp. 1225–1229. IEEE (2011)
21. Ciresan, D.C., et al.: Deep, big, simple neural nets for handwritten digit recognition. Neural Comput. 22(12), 3207–3220 (2010)
22. Jaderberg, M., Simonyan, K., Vedaldi, A., Zisserman, A.: Deep structured output learning for unconstrained text recognition. In: ICLR (2015)
23. Karpathy, A., Fei-Fei, L.: Deep visual-semantic alignments for generating image descriptions. In: Proceedings of the IEEE Conference on Computer Vision and Pattern Recognition, pp. 3128–3137 (2015)
24. Zhang, J., Du, J., Dai, L.: A GRU-based encoder-decoder approach with attention for online handwritten mathematical expression recognition. In: 2017 14th IAPR International Conference on Document Analysis and Recognition, vol. 1, pp. 902–907. IEEE (2017)
25. Deng, Y., Kanervisto, A., Rush, A.M.: What you get is what you see: a visual markup decompiler. arXiv preprint arXiv:1609.04938, vol. 10, pp. 32–37 (2016)
26. Wang, J., Sun, Y., Wang, S.: Image to Latex with DenseNet encoder and joint attention. Procedia Comput. Sci. 147, 374–380 (2019)
27. Huang, G., Liu, Z., Van Der Maaten, L., Weinberger, K.Q.: Densely connected convolutional networks. In: Proceedings of the IEEE Conference on Computer Vision and Pattern Recognition, pp. 4700–4708 (2017)
28. Chen, L., et al.: SCA-CNN: spatial and channel-wise attention in convolutional networks for image captioning. In: Proceedings of the IEEE Conference on Computer Vision and Pattern Recognition, pp. 5659–5667 (2017)
29. Mahdavi, M., Condon, M., Davila, K., Zanibbi, R.: LPGA: line-of-sight parsing with graph-based attention for math formula recognition. In: Proceedings of the International Conference on Document Analysis and Recognition (2019)
30. Deng, J., Guo, J., Xue, N., Zafeiriou, S.: ArcFace: additive angular margin loss for deep face recognition. arXiv preprint arXiv:1801.07698 (2018)
31. Li, Y., Tarlow, D., Brockschmidt, M., Zemel, R.: Gated graph sequence neural networks. arXiv preprint arXiv:1511.05493 (2015)
32. Levy, O., Goldberg, Y.: Neural word embedding as implicit matrix factorization. In: Advances in Neural Information Processing Systems, pp. 2177–2185 (2014)
33. Luong, M., Pham, H., Manning, C.D.: Effective approaches to attention-based neural machine translation. arXiv preprint arXiv: 1508.04025 (2015)

34. Graves, A.: Sequence transduction with recurrent neural networks. arXiv preprint arXiv:1211.3711 (2012)
35. Papineni, K., Roukos, S., Ward, T., Zhu, W.J.: BLEU: a method for automatic evaluation of machine translation. In: Proceedings of the 40th Annual Meeting on Association for Computational Linguistics. Association for Computational Linguistics, pp. 311–318 (2002)
36. Zhang, W., Bai, Z., Zhu, Y.: An improved approach based on CNN-RNNs for mathematical expression recognition. In: Proceedings of the 2019 4th International Conference on Multimedia Systems and Signal Processing, pp. 57–61. ACM (2019)
37. Zhang, J., et al.: Watch, attend and parse: an end-to-end neural network based approach to handwritten mathematical expression recognition. Pattern Recogn. **71**, 196–206 (2017)
38. Zhang, J., Jun, D., Dai, L.: Track, attend, and parse (tap): an end-to-end framework for online handwritten mathematical expression recognition. IEEE Trans. Multimedia **21**(1), 221–233 (2018)
39. Zhang, J., et al.: A tree-structured decoder for image-to-markup generation. In: International Conference on Machine Learning. PMLR (2020)
40. Yan, Z., Ma, T., Gao, L., et al.: Persistence homology for link prediction: an interactive view. arXiv preprint arXiv:2102.10255 (2021)
41. Wu, J., Yin, F., et al.: Graph-to-graph: towards accurate and interpretable online handwritten mathematical expression recognition. In: AAAI (2021)

NLP for Document Understanding

A Novel Method for Automated Suggestion of Similar Software Incidents Using 2-Stage Filtering: Findings on Primary Data

Badal Agrawal[1]($\boxtimes$), Mohit Mishra[2]($\boxtimes$), and Varun Parashar[2]($\boxtimes$)

[1] Microsoft India, Hyderabad, India
[2] Indian Institute of Information Technology Guwahati, Guwahati, India
mohitmishra@ieee.org

Abstract. Advancements in software technology have resulted in a sharp increase in the complexity of software with an equally increasing number of bugs reported every day. Some of these bugs have a high severity and can often lead to significant business impacts. Thus, they need to be resolved by the developers at the earliest. Many of these reports are similar to the bug reports that were reported and resolved in the past. By suggesting similar incidents, the developers can refer to the troubleshooting information, thus effectively reducing the TTM (Time to Mitigate) of the software bugs. The developers also spend a significant amount of time and effort in triaging the bugs into their respective areas. Previous studies have mainly relied on unsupervised learning techniques for the detection of duplicate reports and ignored some key aspects of the bug reports. We conducted comprehensive research on real bugs reported for Microsoft Dynamics 365 Application Software. Our research presents a novel two-phase approach for suggesting similar incidents. The first phase called Binning involves the creation of a labelled dataset for employing a supervised learning algorithm for triaging the software incidents into multiple categories. Thus, the first phase also presents a solution for automating the process of triaging the incidents in addition to the first stage of filtering. The second phase introduces the use of error execution information and acknowledgment information for the calculation of similarity scores which has largely been ignored in the previous studies. The evaluation results show that the precision rate of our proposed approach reaches up to 95.8% while the model achieves recall rates of 67%–93.5%.

Keywords: Duplicate bug report · TTM · Word2Vec · Bug classification · Supervised learning

1 Introduction

The complexity of software design has resulted in an increasing number of software bugs being encountered by the developers in various organizations. The end-users of the software report these bugs in the form of support tickets which

J. Lladós et al. (Eds.): ICDAR 2021, LNCS 12822, pp. 667–682, 2021.
https://doi.org/10.1007/978-3-030-86331-9_43

we will henceforth refer to as 'Software Incidents'. Some of the features or functionalities may not work as expected because of these bugs which can in turn lead to significant business impact incurring huge loss. Thus, the developers need to resolve these incidents at the earliest. Some of the software incidents that are reported may have similar incidents that have already been resolved in the past. We call these incidents as 'Duplicate Incidents'. TTM (Time to Mitigate) is defined as the time taken by the developers to resolve a software incident. The developers can refer to the troubleshooting information of the suggested similar incidents thus effectively reducing the TTM. Our definition of duplicate is guided by the sole motive to reduce the TTM and might differ slightly from previous researches which is explained in this section. The software incidents have a huge amount of data in the form of Title, Summary, Discussions and Execution Information. Jalbert and Weimer [1] found that title similarity is the most important element for the detection of duplicate bug reports in their study. While our research supports the result that title similarity is an important measure, we found that the 'Error Execution Information' is an equally important measure that must not be ignored. To provide a better understanding of our analysis we present an example:

Table 1. Incidents table

Incident id	Incident title	Error message
13456	Qualify lead is not working	The account is duplicated after qualifying the lead
14563	User is not able to qualify lead	There is no active transaction

Consider the incidents in Table 1. Calculating the similarity of the incidents on the basis of title only will classify these two incidents as duplicates with a high degree of confidence as they both represent the same issue in which the user is not able to qualify a lead through the Dynamics 365 application software. However, the error execution message of the incidents suggest that these are in fact, not similar.

Wang et al. [2] proposed the use of execution information in addition to natural language information for comparing the similarity of the bug reports. We conducted an independent research that validates the use of execution information as an important factor for comparing bug report similarity. Execution information reflects the internal error in the underlying codebase which provides a better understanding to the developers about the issue. In addition, it is not affected by the variety of natural languages that arise due to the difference in the logging traits of the end-users or developers while raising a support ticket. However, it is possible for two incidents to have similar error execution information such as NULL reference but they should still not be categorized as duplicates as they might belong to two completely different domains.

Some of the previous studies have also used the summary and discussions of the incidents for topic modelling and learning the textual and semantic features of the data. While we agree that this does help in improving the performance of the model to a certain extent, we found that the content of summary and discussions has a lot of irrelevant data that might introduce noise and thus negatively affect the performance. We also found that the first two comments of the discussions add a lot more context to the issue than the title alone. When a new incident arrives, the developers acknowledge the incidents by logging comments in the discussions that often add more context to the information present in the title of the incidents. Our analysis of a considerable number of incidents found that either the first or the second comment added more context to the issue in most cases. We will henceforth refer the first two comments as Ack information (Acknowledgment information). After careful consideration of all the above factors, we propose a novel two phase method (Fig. 1) for detection of duplicate incidents in which the first phase of filtering occurs on the basis of title (short description) and the second phase of filtering occurs on the basis of error execution information and Ack information. The first phase involves Binning of the software incidents using an optimized Vector Space Model (VSM) and ensemble learning to classify the incidents into multiple fields. This helps in learning the categorical features of the software incidents. The second phase involves calculating its similarity with all the past incidents that have been classified in the same category as the new incident based on error execution information and Ack information. The final suggestions are made by combining the similarity

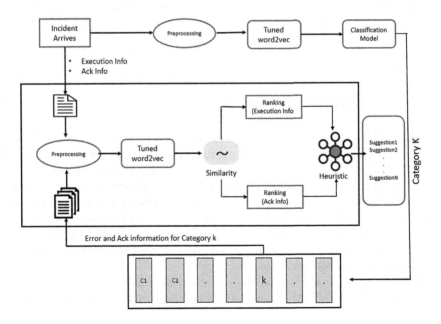

Fig. 1. Schematic flow for the two phase filtering method.

scores obtained on the basis of error execution information and Ack information by defining an appropriate heuristic described in Sect. 3.7. Our model makes a highly efficient use of the noisy data present in the form of summary and discussions which we will refer to as 'Bad Data' for developing our own domain specific text corpus to learn the semantic correlations, word order and contextual connections. We generated the word embeddings using word2vec model, fine-tuned to our vocabulary by training the model on the said corpus. The text corpus is periodically enriched thus improving the traditional VSM by generating appropriate dynamic vector representations of the title, error execution information and Ack information used in the above two phases. The major contributions of this paper are as follows:

- After careful analysis of the data at hand, we found that title and execution information are equally important for duplicate detection and used them to propose a novel two phase filtering method. We also introduced the use of Ack information by defining an appropriate heuristic. We made a highly efficient use of the Bad Data present in the form of summary and discussions for developing a domain specific text corpus to learn the textual, semantic and contextual regularities.
- We improved the traditional VSM by using state-of-the-art methods such as word2vec and fine-tuned it according to our own domain specific text corpus using Transfer Learning. This helped in generating better word representations as compared to the representations generated using pre-trained weights.
- We took the categorical features into consideration by using supervised ensemble learning instead of using unsupervised techniques such as topic modelling used in the previous researches.

The proposed method has been adopted by multiple squads of the Dynamics 365 Core Sales Team.

2   Related Work

The increasing complexity of software design and technology has attracted many researchers in the field of duplicate detection of software bug reports. The early works in this field have employed traditional Vector Space Model (VSM) and Statistical Information Retrieval techniques to detect duplicate bug reports. Anviket et al. [3] in 2005, were able to detect 28% of all duplicate bug reports by using a statistical model based on traditional VSM. One major issue with VSM is the size of the corpus occasionally exceeding more than 100,000 documents. The first phase of filtering in our proposed method reduces the time complexity as a function of the number of classes.

Wang et al. [2] made use of the execution information in addition to the natural language information and defined several heuristics for combining the similarities based on the above two factors. They were able to detect 67%–93% of duplicate bug reports in the Firefox project, compared to 43%–72% using only natural language information. Our research validates their study involving the use of execution information, but differs with their study with respect to the use of error execution information in the second phase of filtering only, after the incidents have been classified into their respective categories.

Jalbert and Weimar [1] proposed a system that uses surface features, textual semantics and graph clustering to predict the duplicate status of the bug reports. They concluded that inverse document frequency is not useful for duplicate detection and were able to reduce the development cost by filtering out 8% of the duplicate bug reports. Sureka and Jalote [4] used character N-gram based model that uses character N-Grams as low level features to represent title and detailed description of the incidents. The evaluation results show that the recall rate for the top 50 results is 33.92% for 1100 randomly selected samples and 61.94% for 2270 randomly selected samples with title to title similarity of more than 50(threshold). Zhang et al. [5] built a discriminatory SVM based model and used smoothed UM and KL convergence for detecting textual similarity between bug reports and concluded the superiority of their method over previously employed methods.

Nguyenet et al. [6] combined topic modelling and information retrieval to propose the DBTM model. Their evaluation concluded that the proposed method can improve the accuracy of the then state-of-the-art methods by 20%. A similar approach was proposed by [7] called the LNG model in which they used a combination of LDA topic modelling and N-Gram similarity algorithm. The results on more than 213,000 shows that LNG outperforms the LDA and N-Gram models in terms of precision and recall rate. As compared to DBTM, LNG improves the recall rate by up to 10.52%. While these methods helped in improving the accuracy of the duplicate detection of bug reports to a certain degree, they did not take some important factors into consideration such as error execution information, ack information and summary. Moreover, topic models such as LDA suffers from some major drawbacks because the number of topics is fixed and should be known ahead of time. In addition, there is no evolution of topics over time. The previously used text vectorization techniques such as term frequency and N-gram extraction fail to learn the semantic, contextual and categorical regularities of the data.

Our proposed method performs a concept matching of the data in the incidents to their best possible use and thus takes all the factors into consideration. The model learns the categorical features using supervised ensemble learning instead of using unsupervised techniques such as topic modelling. We took the semantic and contextual features into consideration by training an optimized dynamic VSM on our own domain specific text corpus.

3   The Proposed Method

PHASE 1: The first phase of the proposed method uses a supervised machine learning algorithm for classifying the incidents into their respective categories. Moreover, it automates the triaging process performed manually until now.

3.1   Class Labelling and Dataset Creation

We made a thorough analysis of the 3984 incidents and identified 10 broad categories for the classification of the incidents. The categories were chosen according to the categorical features of the incidents by experienced developers. We labelled the incidents according to their respective classes to create a dataset for training a supervised learning algorithm.

3.2   Data Preprocessing

The data present in the various fields of the incidents has a lot of noise in the form of numbers, URLs, stack traces, hyperlinks etc. The data is textual in nature and is logged by different developers. The data thus contains spelling errors, grammatical errors and has a lot of stop words that add no real meaning to the context. We cleaned the data using advanced preprocessing techniques such as Lemmatization, Stemming, Tokenization, Case Normalization, Removal of stop words, Removal of numbers (Replaced with $ sign in order to catch the semantic information of the syntax), punctuations and special characters.

3.3   Text Vectorization

The data in the fields of title and error execution information is logged by different developers and is thus affected by the variety of natural language. Previous studies have used the traditional VSM that uses one-hot frequency, term frequency or N-gram extraction for generating the vector representations of the textual data. But text data often has semantic and contextual regularities which are not taken into consideration by these methods. We developed a highly optimized and dynamic version of the VSM for generating appropriate vector representations of the data that considers all the factors including textual, syntactic, semantic and contextual regularities of the data using word2vec.

Word2Vec
Word2Vec is used to generate word embeddings using a shallow neural network. The embeddings can be obtained using two methods: CBOW (Common Bag of words) and skip-gram. We have used the skip-gram model as it is known to work well with small amount of data and generates appropriate representations of even the words whose frequency of occurrence is very less, as suggested by Mikolov et al. [8].

The algorithm uses a shallow neural network in order to learn the vector representations of the words in the vocabulary. The target size of the word vectors is chosen as 300. Also, there are more than 1 million words in the vocabulary. Thus, both the hidden layer and the output layer would have a weight matrix having 3 billion weights. The computational cost for training such a large number of weights is very high. Thus, we have used the skip-gram model with negative sampling and subsampling of frequent words (Words that are too frequent might not be able to differentiate the context) to reduce the computational cost.

The model was initialized with the pre-trained weights obtained by training it on the Google News text corpus (3bn running words). The model was then fine-tuned by training it on our own text corpus (1 million running words). This is an important step for extending the proposed model to other domain specific scenarios because every organization has a different vocabulary. The model was trained over a series of 15 epochs over the text corpus. A major drawback of word2vec is that it does not generate embeddings for out-of-vocabulary words which was resolved by updating the vocabulary with the google news text corpus. The text corpus was generated by extracting every usable piece of information present in the incidents. The data in the summary and discussions is very noisy and has a lot of irrelevant information and is thus not significantly useful for the classification or calculation of similarity scores. The enormous amount of data present in these fields is therefore used for generating our own domain specific vocabulary. The corpus is enriched with more data periodically to further improve the performance which makes the model truly dynamic.

The word vectors were converted to sentence vectors using tf-idf weighting as the combination of the two can outperform either individually [9].

3.4   Removal of Imbalance

One of the major challenges involved in multi-class classification problem is to handle class imbalance. The labelled dataset used for training the classification model in the first phase has a highly imbalanced class distribution which means that the occurrence of some the classes is significantly greater as compared to the occurrence of some other classes. This happens because of the general trend of the frequency of software incidents reported in each of the incoming areas. Hence, the predictive models developed using conventional machine learning algorithms can be biased or inaccurate in favor of the majority class. We have used a combination of Synthetic Minority Oversampling Technique (SMOTE) to oversample the minority classes and random under-sampling. A combination of SMOTE and under-sampling performs better as compared to plain under-sampling (reduces the dataset) or oversampling [10].

A general downside of using SMOTE for multi-class imbalance is that synthetic samples are generated without taking the majority class into consideration which could lead to ambiguous results in case of overlapping between the classes. We have specified the percentage of examples to be oversampled in each of the minority class by identifying a local majority class for each of the minority classes. Thus, we successfully extended the SMOTE library for binary classification to work for multi-class imbalanced classification without causing any ambiguity. In addition to removing the class imbalance, SMOTE also performs the much needed data augmentation. The number of instances were increased from 3984 to 5048, after reducing the class imbalance. The class distribution before and after removing the imbalance is shown in Fig. 2 and Fig. 3 respectively.

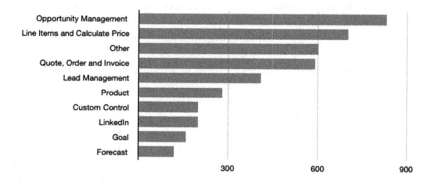

Fig. 2. Class Distribution before removing imbalance.

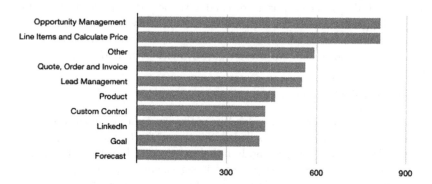

Fig. 3. Class Distribution after removing imbalance.

3.5   Classification Model

Ensemble Learning is a technique that combines multiple base models into one predictive model in order to decrease the bias, variance or improve predictions. Previous studies have shown that the performance of a single classifier for a multi-class classification problem decreases with the existence of imbalanced data [11]. Thus, using an ensemble of classifiers has increasingly been adopted to increase the overall predictive power of the model which might outperform any single sophisticated classifier. The general trend in the frequency of incidents reported in each of the categories introduces a perpetual class imbalance in our dataset that can be best resolved using Ensemble Learning. *(Note: The class imbalance will not increase significantly even in the future as past trends have shown that the class distribution does not vary significantly as more incidents are reported).*

We have used the Boosted Decision Tree Algorithm [12] as the underlying binary classification algorithm for ensemble learning. Boosting is a sequential process in which each subsequent model tries to correct the errors of the previous model. If an observation was classified incorrectly, it tries to increase the weight of this observation and vice versa.

Each tree is created iteratively. The tree's output (h(x)) is given a weight (w) relative to its accuracy. The ensemble output is the weighted sum:

$$\hat{y}(x) = \sum_t w_t h_t(x) \tag{1}$$

Each data sample is assigned a weight based on its misclassification after each iteration. The degree of importance of a data sample increases with the rate of misclassification. We define the objective function O(x) as:

$$O(x) = \sum_i l(\hat{y}_i, y_i) + \sum_t \Omega(f_t) \tag{2}$$

We use binary logistic as the loss function and XGBoost as the boosting algorithm. The goal is to minimize the objective function. The hyperparameters of the model include maximum number of nodes per tree, minimum number of samples per leaf node, learning rate and number of trees constructed. We use the following representation of the values for the above hyperparameters: $[\alpha, \beta, \gamma, \eta]$. The model hyperparameters were initialized to the following values: [10, 15, 0.1, 100].

3.6 Classification Performance Analysis

We summarize the performance metrics of the classification model in the following section. The split ratio was set to 7:3 for the training and testing sets, respectively.

The model was trained using the One-Vs-All multi-class classifier with 2-class boosted decision tree as the underlying algorithm. The hyperparameters of the algorithm were initialized to [10, 15, 0.1, 100] and constantly adjusted over a series of 40 epochs which helped in achieving a percentage accuracy of 76.94 for the following set of hyperparameter values: [25, 25, 0.167, 125]. The proposed model is validated by providing the following evaluation metrics.

– Precision = $\frac{TP}{TP+FP}$ = 76.94   – Recall = $\frac{TP}{TP+FN}$ = 77.044

The evaluation metrics stand true for error classification into 10 classes which can be validated through the corresponding confusion matrix shown in Fig. 4.

Actual / Predicted	Custom Controls	Forecast	Goal	Lead Management	Line Items and Calculate Price	LinkedIn	Opportunity Management	Other	Product	Quote, Order and Invoice
Custom Controls	95.8	0	0	0	1.4	0	0.7	0.7	0.7	0.7
Forecast	0	91.1	0	0	0	2.4	1.6	4.1	0	0.8
Goal	1.8	1.8	91.9	0.9	0	0.9	0.9	0	0.9	0.9
Lead Management	1.6	1.6	0.8	84.5	2.5	0.8	4.1	3.3	0.8	0
Line Items and Calculate Price	4.7	0	2.8	0.9	68	0	5.2	3.8	8.0	6.6
LinkedIn	1.6	1.6	0	0	0.5	91.7	0.5	3.1	0.5	0.5
Opportunity Management	2.3	2.8	0.9	5.5	9.7	0.9	67.4	5.5	1.8	3.2
Other	5.0	5.0	1.7	4.4	6.6	5.0	10.5	56.2	1.7	3.9
Product	4.3	1.4	0	0	14.3	1.4	8.6	5.7	61.4	2.9
Quote, Order and Invoice	4.2	2.8	3.5	0	12.7	1.4	7.0	4.9	2.1	61.4

Fig. 4. Confusion Matrix

Note: The accuracy of the classification model is independent to the accuracy of the overall model. The above evaluation is valid for an independent automated system for the triaging of the incidents thus reducing the manual effort of the triagers.

PHASE 2: Consider the event of occurrence of a new incident having a category say k. The second phase of the proposed method involves the calculation of similarity of all the incidents belonging to category k, with the incident under investigation based on error execution information and Ack information.

The error execution information and Ack information of the new incident and all the incidents in that category (k) goes through the same preprocessing steps and their word embeddings are generated by using the fine-tuned word2vec model trained in Sect. 3.3.1. The word vectors are then converted into sentence vectors using tf-idf weighting. The similarity scores are calculated using cosine similarity. Consider two incidents and let $\hat{v}1$ and $\hat{v}2$ denote the vector representations of the two incidents. Let θ denote the angle between $\hat{v}1$ and $\hat{v}2$. Then the similarity (S) is defined as:

$$S = \cos\theta = \hat{v}1.\hat{v}2/|\hat{v}1||\hat{v}2| \tag{3}$$

3.7   Heuristic Definition

We denote the similarity scores obtained on the basis of execution information as S_e while the similarity scores obtained on the basis of Ack information are denoted as S_a. Wang et al. [2] introduced a classification based heuristic for ranking the documents on the basis of execution information or natural language information by distinguishing the dominant source of information. We extended the theory for error execution information and Ack information for ranking the incidents by making appropriate changes based on the dynamics of the above two factors of similarity calculation. We assigned a higher priority to the error execution information as compared to Ack information as our heuristic is guided by the motive to reduce the TTM. We define two thresholds T_e and T_a on the similarity values to denote the credibility thresholds for S_e and S_a respectively. For an existing incident, the rules for ranking the incidents are defined as follows:

- If $S_e > T_e$ incidents are ranked according to S_e and belong to Class 1.
- If $S_e < T_e$ and $S_a > T_a$, incidents are ranked according to S_a and belong to Class 2.
- If $S_a < T_a$ and $S_e < T_e$, incidents belong to Class 3.

The ranking order for the classes are as follows.

$$Class1 > Class2 > Class3$$

A high similarity score between the incidents binned in the same category after the first stage of filtering on the basis of error execution information indicates that the incidents are duplicate with a high degree of confidence. However, if the similarity score is less that the specified threshold T_e, we calculate the similarity scores based on the Ack information. The error execution information is given a higher priority as compared to Ack information because execution information directly represents the underlying error such as runtime errors, patch reports and feature errors. In addition, Ack information might contain irrelevant data in some cases. If the similarity score based on the execution information and Ack information are both less than their respective thresholds, the incidents are declared as NOT similar and belong to Class 3. The optimum values for T_e and T_a are calculated by using an experimental setup.

3.8   Calculating Thresholds

The calculation of the credibility thresholds T_e and T_a is based on experimental results. Our technique is based on the analysis of pre-existing incidents. We create a test sample space H using random sampling of the entire sample space. The test sample consists of d pair of duplicates. We calculate the similarity scores of all the pairs of duplicates on the basis of error execution information and take their average.

$$S_e avg = \frac{\sum_{i=1}^{d}(S_e i)}{d} \tag{4}$$

In order to increase the degree of generalization, we take the mean of all the average similarity scores obtained on the basis of error execution information to set the required threshold T_e, for k test samples.

$$T_e = \frac{\sum_{j=1}^{k}((S_e avg)_j)}{k} \tag{5}$$

We repeat the above process for calculating the value of T_a as well.

4   Experimental Setup and Evaluation

In our experiment, we use the bug repository of the Dynamics 365 application software which has millions of users worldwide. We discarded the invalid bug reports for increasing the validity of our experiment. We did not consider the most recent incidents for the purpose of our experiment as they may have been erroneously marked as duplicates or misclassified and the mistakes might not have been corrected. For the purpose of our experiment we need the title, error execution information and Ack information of the incidents. The incident reports do not have the error execution information. Thus, we created the error execution information for each of the reports used in our study.

We collected about 3984 valid bug reports reported from 2018 to 2020 for the purpose of evaluation. Out of the 3984 bug reports, 400 were marked as duplicates which is approximately 10% of the total number of valid bug reports. We construct a test sample space for validating the results of our experiment. The ratio of duplicates to non-duplicates in our test space must be same as that of the entire sample space. We thus select 30 bug reports as duplicates randomly which is 10% of the test sample space (having 270 non-duplicates). To ensure that there are enough duplicates in our sample space, we add 30 more duplicate reports using random selection. We call these incidents as positive samples. Thus, the test sample space consists of 330 bug reports. The remaining 270 incidents are also selected randomly from the set of non-duplicate incidents. We call these incidents as negative samples.

The values of the thresholds T_e and T_a are calculated as described in Sect. 3.8. The value of T_e was found out to be 0.9 while that of T_a was found out to be 0.61. The experimental setup for the calculation of T_e consisted of 300 bug reports with 42 pairs of duplicate bug reports while that of T_a consisted of 280 bug reports with 38 pairs of duplicates. The value of k was chosen as 20. The experimental results were obtained for the sample test space with the above calculated values of T_e and T_a and a suggested list size of 4. The effectiveness of the proposed model is evaluated by using accuracy, precision and recall as the evaluation metrics. The values for the precision, accuracy and recall obtained from the experimental results are 93.1%, 96.9% and 90% respectively which is a significant improvement over the previously employed methods.

To further evaluate the effectiveness of the proposed model, we repeated the above experiments for different values of suggested list size. Figure 5 shows the variation of the precision and recall rates with the suggested list size. From Fig. 5, we make the following observations. There is a sharp decrease in the precision rate when the size of the suggested list increases beyond 5. This is probably because of an increase in the number of false positives. The recall rate increases with the increase in suggested list size and then becomes almost constant. The precision rate and the recall rate reached up to a maximum value of 95% and 93.5% respectively with the variation in the size of the suggested list. This is an improvement over the recall rates of 67%–93% achieved by the model proposed by Wang et al. [2] and the recall rate of 93.1% achieved by the CPwT approach.

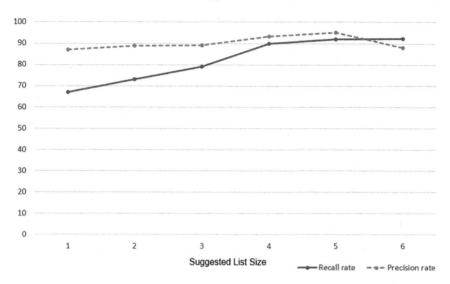

Fig. 5. Variation of precision and recall rates with suggested list size.

By calculating the values of T_e and T_a, based on Eq. (5), we provide a fairly accurate approximation for the thresholds thus decreasing the error margin for the two thresholds. We therefore further evaluate the performance of the model, by varying T_e and T_a within a small error margin of $\pm(0.02)p$ where $p = 1, 2, 3$ and monitor the corresponding change in the precision and recall rates. The graphs in Fig. 6 and Fig. 7 show the variation of the precision and the recall rates by changing the values of T_e (T_a constant) and T_a (T_e constant) individually, within the above defined error margin.

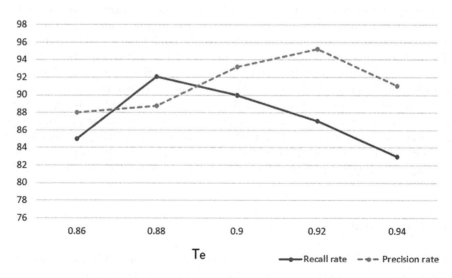

Fig. 6. Variation of precision and recall rates with T_e.

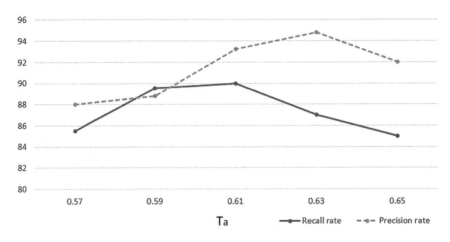

Fig. 7. Variation of precision and recall rates with T_a.

We make the following observations from Fig. 6 and Fig. 7. The precision reaches up to a maximum of 95.8% while the recall reaches up to 92.2% with the variation in T_e. However, with the variation in T_a, the maximum value of recall rate does not increase beyond 90% while the precision rate reaches up to 94.8%. Thus, the precision rate of the proposed model reaches higher than the precision rate of 89.2% achieved by the CPwT approach [5], and reached equal to the LNG model, if not outperform it.

4.1   TTM Reduction

The proposed model was deployed and adopted by the various squads of the Dynamics 365 Sales Team. The monitoring of the proposed model by the developers suggested that the model was effective in reducing the TTM significantly. The performance of the model with respect to TTM was tracked from 2019 to 2020 among multiple squads. The results show that the model was effective in reducing the average TTM by 10–30%. The improvement is significant especially in case of incidents having a high severity. A Sev2 incident is expected to be resolved within 24 h of raising a support ticket and thus demands immediate attention of the developers. By referring to the troubleshooting information of the suggested similar incidents, the developers can resolve the incidents within the expected timeframe. Figure 8 compares the average TTM for the year 2019 and 2020.

5   Discussions

The proposed model has been validated on the Dynamics 365 bug repository. The first stage of filtering requires information about the various categories to which the incidents belong. The lack of availability of this information for open

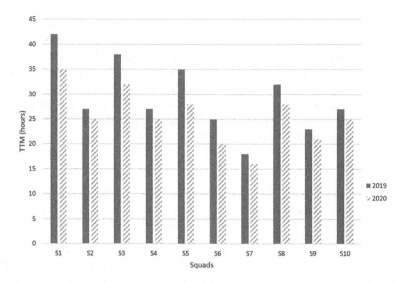

Fig. 8. Comparison chart of TTM for multiple squads.

bug repositories such as Mozilla, Firefox and Eclipse limits us to validate our findings on these open source repositories. However, we encourage the researchers to adopt our proposed method by identifying the right categories for popular bug repositories and we are certain that the results can be overwhelming.

Moreover, we would like to reiterate that the precision and the recall obtained for the classification of incidents in Sect. 3.6 does not affect the overall precision and recall for the duplicate detection. This is because the labels for the misclassified incidents are manually corrected by the developers. The experimental setup and evaluation in Sect. 4 is fail proof as we have discarded the most recent incidents for the purpose of the experiment as they might have been misclassified and the mistakes might not have been corrected.

Creation of a domain specific corpus for the generation of the word-embeddings does NOT restrict the generalization of the proposed method, but rather improves the quality of the word-embeddings by eliminating words that are out of scope with respect to the domain in which the proposed method is being utilized.

6   Conclusion

In this paper, we presented a novel two phase approach for the suggestion of similar incidents to the developers which helped in reducing the TTM by 10%–30%. Moreover, the proposed method also relieves the triager from the task of the classification of incidents into their respective categories by automating the triaging process. The proposed model learned the categorical features by using supervised ensemble learning to classify the incidents into their respective categories in the first phase. In the second phase, we introduced the use of execution information and Ack information for the calculation of similarity scores by defining an appropriate heuristic. Moreover, we improved the traditional VSM by generating fine-tuned word embeddings and dynamically enriching the text corpus. We provided an extensive evaluation of the proposed method and found that the precision rate of the model reached up to 95.8% and the model achieved recall rates of 67%–93.5% which is a significant improvement over the state-of-the-art approaches such as DBMT, CPwT and LNG.

References

1. Jalbert, N., Weimer, W.: Automated duplicate detection for bug tracking systems, In: IEEE International Conference on Dependable Systems and Networks With FTCS and DCC (DSN), Anchorage, AK, pp. 52–61 (2008). https://doi.org/10.1109/DSN.2008.4630070
2. Wang, X., Zhang, L., Xie, T., Anvik, J., Sun, J.: An approach to detecting duplicate bug reports using natural language and execution information. In: ACM/IEEE 30th International Conference on Software Engineering, Leipzig, pp. 461–470 (2008). https://doi.org/10.1145/1368088.1368151
3. Anvik, J., Hiew, L., Murphy, G.C.: Coping with an open bug repository. In: Proceedings of the OOPSLA workshop on Eclipse technology eXchange, New York, NY, USA, pp. 35–39. Association for Computing Machinery (2005). https://doi.org/10.1145/1117696.1117704
4. Sureka, A., Jalote, P.: Detecting duplicate bug report using character N-gram-based features. In: Asia Pacific Software Engineering Conference, Sydney, NSW, pp. 366–374 (2010). https://doi.org/10.1109/APSEC.2010.49
5. Zhang, T., Leo, B.: A novel technique for duplicate detection and classification of bug reports. IEICE Trans. Inf. Syst. **E97-D**, 1756–1768 (2014). https://doi.org/10.1587/transinf.E97.D.1756
6. Nguyen, A.T., Nguyen, T.T., Nguyen, T.N., Lo, D., Sun, C.: Duplicate bug report detection with a combination of information retrieval and topic modeling. In: Proceedings of the 27th IEEE/ACM International Conference on Automated Software Engineering, Essen, pp. 70-79 (2012). https://doi.org/10.1145/2351676.2351687
7. Zou, J., Xu, L., Yang, M., Zhang, X., Zeng, J., Hirokawa, S.: Automated duplicate bug report detection using multi-factor analysis. IEICE Trans. Inf. Syst. **E99.D**, 1762–1775 (2065). https://doi.org/10.1587/transinf.2016EDP7052
8. Mikolov, T., Chen, K., Corrado, G.S., Dean, J.: Efficient estimation of word representations in vector space. In: Proceedings of Workshop at ICLR (2013)
9. Zhao, J., Lan, M., Tian, J.F.: ECNU: using traditional similarity measurements and word embedding for semantic textual similarity estimation. In: Proceedings of the 9th International Workshop on Semantic Evaluation (SemEval 2015), p. 117 (2015)
10. Chawla, N.B., Hall, K., Kegelmeyer, L.: SMOTE: synthetic minority over-sampling technique. J. Artif. Intell. Res. (JAIR) **16**, 321–357 (2002). https://doi.org/10.1613/jair.953
11. Sainin, M.S., Alfred, R.: A direct ensemble classifier for imbalanced multiclass learning. In: 4th Conference on Data Mining and Optimization (DMO), Langkawi, pp. 59–66 (2012). https://doi.org/10.1109/DMO.2012.6329799
12. Drucker, H., Cortes, C.: Boosting decision trees. In: Advances in Neural Information Processing Systems, vol. 8, pp. 479–485 (1995)

Research on Pseudo-label Technology for Multi-label News Classification

Lianxi Wang[1,2], Xiaotian Lin[1], and Nankai Lin[1(✉)]

[1] School of Information Science and Technology, Guangdong University of Foreign Studies, Guangzhou, China
[2] Guangzhou Key Laboratory of Multilingual Intelligent Processing, Guangdong University of Foreign Studies, Guangzhou, China

Abstract. Multi-label news classification exerts a significant importance with the growing size of news containing multiple semantics. However, most of the existing multi-label classification methods rely on large-scale labeled corpus while publicly available resources for multi-label classification are limited. Although many researches have proposed the application of pseudo-label technology to expand the corpus, few studies explored it for multi-label classification since the number of labels is not prone to determine. To address these problems, we construct a multi-label news classification corpus for Indonesian language and propose a new multi-label news classification framework through using pseudo-label technology in this paper. The framework employs the BERT model as a pre-trained language model to obtain the sentence representation of the texts. Furthermore, the cosine similarity algorithm is utilized to match the text labels. On the basis of matching text labels with similarity algorithms, a pseudo-label technology is used to pick up the classes for unlabeled data. Then, we screen high-confidence pseudo-label corpus to train the model together with original training data, and we also introduce loss weights including class weight adjustment method and pseudo-label loss function balance coefficient to solve the problem of data with class imbalance, as well as reduce the impact of the quantity difference between labeled texts and pseudo-label texts on model training. Experiment results demonstrate that the framework proposed in this paper has achieved significant performance in Indonesian multi-label news classification, and each strategy can perform a certain improvement.

Keywords: Multi-label news classification · Pseudo-label · BERT · Indonesian

1 Introduction

With the growing size of news containing multiple semantics, how to enable the users to accurately classify news has become a topic of increasing concern [1], while most previous studies on news classification regard it as a single-label supervised learning task. In fact, many things in real-world might be complicated

© Springer Nature Switzerland AG 2021
J. Lladós et al. (Eds.): ICDAR 2021, LNCS 12822, pp. 683–698, 2021.
https://doi.org/10.1007/978-3-030-86331-9_44

and have multiple semantic meanings simultaneously. For example, a news article often covers multiple categories such as politics and economy at the same time.

Indonesian, which about 30 million people use as their mother tongue, is one of the official languages in Indonesia. Although most of the existing researches on multi-label news classification focus on the general language, there is a lack of targeted research on non-universal languages such as Indonesian. Hence, multi-label news classification for Indonesian language seems particularly crucial.

So far, the existing research on multi-label classification tasks can be divided into two categories: problem transformation-based methods and algorithm adaptation-based methods [2]. In light of problem transformation-based methods, each label has its own classifier. According to whether a label classifier would be influenced by other label classifiers or not, two approaches are developed. The former takes the output of a classifier as the input of another label classifier, such as the classifier chain algorithm [3]. The latter, however, ignores such influence and hence the label classifiers work separately. As regards the adaptation-based methods, some classic algorithms are adjusted for the multi-label classification task. In the field of traditional machine learning, multi-label classification models include MLKNN [4], calibrated label ranking, etc. As for deep learning, the output layer of a neural network is often modified to allow multi-label classification. For example, Zhang et al. [5] proposed a hierarchical multi-input and output model based on the bi-directional recursive neural network.

Recently, many researchers also try to apply pseudo-label technology to multi-label classification tasks. Ahmadi et al. [6] proposed a fast linear label space dimension reduction method that transforms the labels into a reduced encoded space and train models on the obtained pseudo labels. Fan et al. [7] and Fu et al. [8] adopted clustering to estimate pseudo labels to train the CNN on the ReID dataset. Nevertheless, we notice that most of the researches are single-label classification task and concentrated in the image field. To our best knowledge, few studies explore the pseudo-label technology with multiple labels since the number of labels is difficult to determine. For instance, when we utilize the model to identify the label probability of a piece of news and the probabilities of labeling as politics and labeling as economy are extremely close, we cannot determine whether the news contains only one label or both two labels.

To tackle these gaps, we construct a multi-label news classification corpus for Indonesian language and propose a new multi-label news classification framework by using pseudo-label technology. The framework employs the BERT model as a pre-trained language model to obtain the sentence representation of the texts. Through matching text labels with similarity algorithms, a pseudo-label technology is used to pick up the classes for unlabeled data, as a consequence of which we can screen high-confidence pseudo-label corpus to train the model together with original training data. At the same time, in order to address the problem of data with class imbalance as well as reduce the impact of the difference between labeled texts and pseudo-label texts on model training, we further introduce loss weights including class weight adjustment methods and pseudo-label loss function balance coefficients.

The main contributions of this paper are as follows:

(1) We construct an Indonesian multi-label news classification dataset, which could make up for the lack of resources in the Indonesian multi-label classification field.

(2) We propose a pseudo-label generation strategy based on similarity calculation for multi-label news classification.

(3) We freeze the class weights and introduce the pseudo-label loss function balance coefficient to solve the problem for imbalanced data and reduce the impact of the quantity difference between labeled texts and pseudo-label texts on model training.

(4) The experimental results demonstrate that the model proposed in this paper has achieved significant performance on Indonesian multi-label news classification with the Macro-F1 of 70.82%, which is 1.5% higher than the original Indonesian-BERT model.

2   Related Work

Multi-label Text Classification. In recent years, text classification tasks have been extensively studied, while there are relatively few studies aiming to multi-label text classification. Read et al. [9] proposed the classifier chains (CC) algorithm which effectively extracts the correlation of labels. However, the CC algorithm requires the nature of Markov chains among labels and is more sensitive to the order of the labels. Aiming to solve this problem, Hu et al. [1] further developed a bi-directional classifier chain algorithm based on deep neural networks. This method utilized a forward classifier to obtain the correlation between each label and all previous labels, besides, a backward classifier chain is involved, starting from the output of the last base classifier in the forward classifier chain. Shehab et al. [10] conducted experiments on three multi-label classifiers (DT, RF, KNN) on the Arabic dataset, and the results showed that the DT model outperformed the other two models. Nowadays, deep learning-based methods are also applied in this field. For example, You et al. [11] developed Attention XML model which used a recurrent neural network (RNN) and the multi-label attention to address the problem of tagging each text with the most relevant labels from an extreme-scale label set. Nevertheless, this model can hardly deal with extreme-scale (hundreds of thousands labels) problem. Hence, based on this research, they [12] further proposed HAXMLNet which used an efficient and effective hierarchical structure with the multi-label attention to tackle this issue.

Indonesian Multi-label News Classification. Isnaini et al. [13] proposed to use the K-Nearest Neighbor classification algorithm for multi-label classification of Indonesian news. Based on it, Pambudi et al. [14] further employed the Pseudo

Nearest Neighbor Rule (PNNR) algorithm, which is a variant of the K-Nearest Neighbor algorithm. Rahmawati et al. [15] combined the four factors to explore the best combination of multi-label classification of Indonesian news articles, i.e., feature weighting method, feature selection method, multi-label classification approach, and single label classification algorithm. What's more, the experiments show that the best performer for multi-label classification of Indonesian news articles is the combination of TF-IDF feature weighting method, symmetrical uncertainty feature selection method, and Calibrated Label Ranking. Irsan et al. [16] employed some potential methods to improve performance of hierarchical multi-label classifier for Indonesian news article. They developed the calibrated label ranking as a multi-label classification method to build the model and utilized the Naive Bayes algorithm for training. Besides, multiplication of word term frequency and the average of word vectors were also used to build these classifiers.

Pseudo-label Technology. Lee [17] first proposed to generate pseudo-labeled data using a neural network model trained on labeled data. Shi et al. [18] extended the idea of pseudo-label, and scored the confidence of unlabeled data based on the density of the local domain. Xie et al. [19] utilized the Teacher network of the Teacher-Student mechanism to generate pseudo-label data, mixing them together with the original data to train the Student Network, which achieved good results on the ImageNet dataset. Haase-Schtz et al. [20] divided the unlabeled dataset and trained the model in each sub-dataset. They used previously trained networks to filter the labels used for training newer networks.

3   Method

As shown in Fig. 1, the model proposed in this paper mainly consists of three modules: 1) Text representation module. 2) Text label matching module. 3) Pseudo-label generation module.

In the text representation module, we employ the BERT model to obtain the sentence representation of the texts. As regards the text label matching module, the labels of the news text are obtained through calculating the cosine similarity between its sentence vector and sentence vectors of all training texts. During the pseudo-label generation module, based on the text label matching module, a matching threshold is set to score the confidence of unlabeled data to filter high-confidence pseudo-label corpus assisting training. In addition, in order to solve the problem of data with class imbalance as well as reduce the impact of the difference between labeled texts and pseudo-label texts on model training, we further introduce loss weights including class weight adjustment method and pseudo-label loss function balance coefficient.

3.1   Text Representation

BERT [21] is a language model based on bidirectional encoder characterization. Through abandoning the traditional CNN and RNN models, its entire network structure is completely composed of attention mechanism, which makes the model well address the long-distance dependence and parallelism task. For classification problems, the first position in the input sentence will be given a special token [CLS], and the final hidden state corresponding to the token is usually represented as the aggregation sequence of the classification task. While each token in a given sentence, its input representation is constructed by summing the corresponding token, segment and position embeddings.

Therefore, this paper employs the BERT model as the encoder of language features. Besides, the final hidden vector corresponding to the first input token is used to obtain the sentence vector representation of the text. Specially, for the i-th sentence, the sentence vector is expressed as follows.

$$S_i = BERT(a_i, b_i, c_i) \tag{1}$$

where a_i, b_i, c_i are the token embeddings, the segmentation embeddings and the position embeddings respectively.

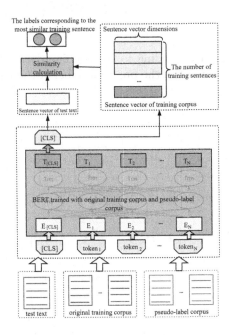

Fig. 1. The structure of our model.

3.2   Text Label Matching

As mentioned above, for applying deep learning methods to multi-label classification tasks, the main idea is to set a classification threshold for the neural network with activation function layer. However, the classification threshold is not prone to determine. Therefore, we propose a text label matching method with cosine similarity algorithm shown in Fig. 2 to solve such problems.

Suppose the sentence vectors of all training data outputted by the BERT pre-trained language model are expressed as $H = (H_1, H_2, H_3, \cdots, H_l)$, which $H_j \in R^m$ represents the sentence vector of the $j-th$ training data, where m is the sentence vectors dimension and l is the number of texts in the training dataset. Specially, given a sequence whose token embedding, segmentation embedding and position embedding are respectively expressed as $\alpha = (\alpha_1, \alpha_2, \alpha_3, \cdots, \alpha_n)$, $\beta = (\beta_1, \beta_2, \beta_3, \cdots, \beta_n)$, $\gamma = (\gamma_1, \gamma_2, \gamma_3, \cdots, \gamma_n)$, its sentence vector is calculated as follows

$$S = BERT(\alpha, \beta, \gamma) \tag{2}$$

$$H_j = BERT(a_j, b_j, c_j) \tag{3}$$

Later, in order to show the similarity among texts, the cosine similarity will be calculated between the sentence vector S and sentence vectors of all training data H.

$$cos(S, H_j) = c_j = |SH_j^T|/(\sqrt{SS^T}\sqrt{H_j H_j^T}) \tag{4}$$

$$C = (c_1, c_2, c_3, \cdots, c_l) \tag{5}$$

Ultimately, the label corresponding to the training data with the largest similarity value will be regarded as the predicted label of the text sequence.

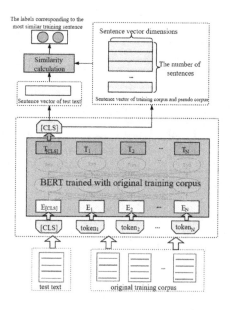

Fig. 2. Text label matching model.

3.3   Pseudo-label Generation

Most existing multi-label news classification algorithms rely on large-scale annotated corpus, while publicly available resources for multi-label classification are
limited. Besides, unlabeled data is not only prone to obtain, but also can enhance
the robustness of the model through more precise decision boundaries. In the
field of classification, some researchers proposed to solve this problem by using
pseudo label technology [17,19]. As for the field of multi-label news classification,
there is still a lack of targeted research. Therefore, we leverage a pseudo-label
generation technology for the multi-label news classification task.

Multi-BERT with Matching Thresholds. On the basis of using BERT to
encode sentences and Softmax activation function to normalize label probabilities, matching thresholds α and β are further introduced. The specific screening
process is shown in Fig. 3.

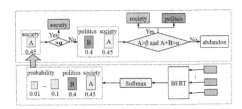

Fig. 3. The screening process of the labels.

Initially, aiming to predict the probability of each pseudo-label in unlabeled
texts, we train a BERT model with Softmax activation function layer using
the original training corpus. Then, the maximum value A among all the label
probabilities is compared with the classification threshold α. Besides, the label
corresponding to this probability will be regarded as the pseudo-label of the
unlabeled texts and added to the pseudo-label candidate corpus if A is greater
than α. On the contrary, if the probability is less than the classification threshold
α, the maximum probability value A and the sub-maximum one B are further
compared with the classification threshold α and β. Eventually, if $A + B$ greater
than β and A greater than α, the text will also be added to the pseudo-label
candidate corpus.

Text Label Matching with Matching Threshold. When the maximum
label probability and the second largest label probability are very close, it is
difficult to ascertain the number of the labels and the size of the classification
threshold. Hence, we further introduce the matching threshold μ on the basis
of the model proposed in Sect. 3.2. The specific screening process is shown in
Fig. 4.

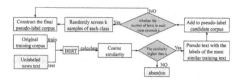

Fig. 4. The screening process of the pseudo-label data.

Based on the text label matching method proposed in Sect. 3.2, we train a BERT model with the original training corpus to obtain the sentence vectors representation of all training corpus, using it to predict a large number of pseudo-label news texts aiming to obtain the sentence vectors of pseudo-label news texts. For each unlabeled news text, the cosine similarity is calculated between its sentence vector and the sentence vectors of all training corpus. Then, the maximum value is compared with the classification threshold μ, the labels corresponding to this value will be regarded as the label of that pseudo-label news text and the text with the labels will be added to the pseudo-label candidate corpus if the value is greater than μ.

3.4   Weighted Loss

In order to address the problem of data with class imbalance as well as reduce the impact of the quantity difference between original training data and pseudo-label data on model training, the loss function of the pseudo-label data is added for training on the loss function $L(x_i, y_i)$ in this paper. Additionally, we further introduce loss weights, proposing category weight adjustment method and pseudo-label loss function balance coefficient λ.

$$L(x_i, y_i) = y_i \log(e^{x_i}/(1 + e^{x_i})) + (1 - y_i) \log(1/(1 + e^{x_i})) \tag{6}$$

$$L = \sum_{m=1}^{n} (\sum_{i=1}^{C} (L(x_i^m, y_i^m))) + \sum_{m=1}^{n} (\sum_{i=1}^{C} (L(x_i^{'m}, y_i^{'m}))) \tag{7}$$

where n, C are the number of the original data and labels respectively, n' for pseudo-label data, y_i^m refers to the output units of m's sample in labeled data, x_i^m is the label of that, $y_i^{'m}$ for pseudo-label data, $x_i^{'m}$ is the pseudo-label of that pseudo-label data.

Class Weight. To our best knowledge, class weights are mainly used to solve the problem of high cost of misclassification as well as high imbalance of samples. In this paper, we propose a class weight adjustment method for freezing the class weight to tackle the problem of data with class imbalance. Assuming that the given labels $c = (c_1, c_2, \cdots, c_k)$, the class weight w_i of the i-th label is calculated as follows.

$$w_i = \log_z(\sum_{i=1}^{n} c_i/c_i) \tag{8}$$

where the value of z is e and c_i is the number of samples of the i-th label. Therefore, the new loss function with class weight is expressed as

$$L_n(x_i, y_i) = y_i \log(e^{g(x_i)}/(1 + e^{g(x_i)})) + (1 - y_i) \log(1/(1 + e^{g(x_i)})) \qquad (9)$$

$$L_n = \sum_{m=1}^{n} (\sum_{i=1}^{C} (L_n(x_i^m, y_i^m))) + \sum_{m=1}^{n} (\sum_{i=1}^{C} (L_n(x_i^{'m}, y_i^{'m}))) \qquad (10)$$

$$g(x_i) = w_i x_i \qquad (11)$$

Balance Coefficient. Since the total number of original training data and pseudo-label data is quite different, and the training balance between them is extraordinary important for network performance. Hence, Lee [17] introduced a loss function balance coefficient between original training data and pseudo-label data to alleviate this problem.

$$L_n' = \sum_{m=1}^{n} (\sum_{i=1}^{C} (L_n(x_i^m, y_i^m))) + (1/\lambda) \sum_{m=1}^{n} (\sum_{i=1}^{C} (L_n(x_i^{'m}, y_i^{'m}))) \qquad (12)$$

Table 1. The data distribution of dataset.

Class	Train	Valid	Test	Pseudo-label data1	Pseudo-label data2
Others	716	180	387	200	200
Society	1456	366	783	200	200
Politics	519	130	279	200	200
Economy	256	66	140	200	200
Technology	48	12	27	8	30
Military	15	4	9	2	5
Environment	15	4	9	3	3
Culture	12	4	9	1	1
Politics, Society	5	30	63	49	121
Economy, Society	91	24	51	33	85
Environment, Society	43	12	24	/	/
Culture, Society	47	12	27	28	45
Politics, Economy	6	2	6	/	/
Politics, Military	5	2	4	/	/
Technology, Society	5	2	5	1	1
Technology, Economy	2	2	3	/	/
Military, Society	4	2	3	2	5
Society, Others	15	4	9	1	6
Culture, Others	3	2	3	/	/
Total	3373	860	1841	933	1119

4   Experiments

4.1   Datasets

As the foundation of natural language processing research, the construction of corpus exerts a significant importance. However, most of the existing corpus concerning multi-label news classification are constructed from common languages such as English, which causes the existing Indonesian corpus cannot meet the development of deep learning technology in terms of scale and quality. Therefore, we build a corpus containing 6074 samples based on the Indonesian news texts crawled from Antara News[1] and Kompas News[2]. In the annotating process, we stipulate that each sample labeled by two persons can have up to 5 labels among 8 labels (i.e. Society, Politics, Economy, Technology, Military, Environment, Culture, and Others). Then samples with the same labeling results are added to the corpus. Instead, samples with different annotation results will be annotated again by a third person. If the annotation results are the same as one of the first two persons, they will also be added to the corpus. Furthermore, we generated two pseudo-label corpus by predicting 50,000 and 100,000 samples respectively. After that, in order to make the pseudo-label corpus pay more attention to the news texts of fewer categories, we further randomly select 200 pseudo-label news texts of this class if the number of pseudo-label texts for a certain category exceeds 200 while keep all samples if not exceed. Therefore, the second pseudo-label corpus has more samples of less-labeled categories. As shown in Table 1, the news articles in the first eight rows are tagged with one "class-label", and the news articles in the last eleven rows are tagged with two or more "class-labels".

4.2   Experimental Setting

The framework which uses pytorch[3], pytorch-lightning[4] and transformers[5] to implement is compared with SGM [22], SU4MLC [23], CNN [24], Bi-LSTM [25, 26], Bi-LSTM-Attention [27,28] where SGM and SU4MLC adopt the optimal parameters given in the paper that proposed these two models and the specific parameters of other models are shown in Table 2. At the same time, Macro-F1, Micro-F1 and Hamming Loss (HL) are utilized as the evaluation indicators. Due to the imbalance in the number of samples among different categories in the multi-label classification task, we use Macro-F1 as the first evaluation indicator.

---

[1] https://www.antaranews.com/.
[2] https://www.kompas.com/.
[3] https://github.com/pytorch/pytorch.
[4] https://github.com/PyTorchLightning/pytorch-lightning.
[5] https://github.com/huggingface/transformers.

Table 2. The hyperparameters used in our experiments.

	Bi-LSTM/Bi-LSTM-attention	CNN	MBERT/BERT
The number of filters	/	32	/
The size of filter	/	[3,4,5]	/
Learning rate	0.001		5.00E-06
The number of neurons	128		/
The number of epochs	100		/
The number of batches	64		/
Embedding size	100		/
Learning rate decay	/		0.001
The rate of dropout	0.5		
Max length	64		

5   Result and Analysis

5.1   Model Comparison

Based on the dataset constructed in this paper, we compare the effects of the pre-trained language models: MBERT[6] and Indonesian-BERT[7], where the classification threshold γ of the activation function is set from 0.25 to 0.45 in increments of 0.05. The results shown in Table 3 demonstrate that when the classification threshold is 0.25, Indonesian-BERT achieved the best performance.

Table 3. The results of the MBERT and Indonesian-BERT.

Model	γ	Macro-F1	Micro-F1	Hamming Loss
MBERT	0.25	66.56%	81.85%	0.0486
	0.3	66.54%	81.80%	0.0485
	0.35	66.24%	81.66%	0.0487
	0.4	64.96%	81.38%	0.0492
	0.45	64.88%	81.25%	0.0495
Indonesian-BERT	0.25	**67.57%**	**84.53%**	**0.0414**
	0.3	67.49%	84.34%	0.0418
	0.35	67.07%	84.09%	0.0423
	0.4	66.19%	83.79%	0.0429
	0.45	65.93%	83.81%	0.0428

---

[6] https://huggingface.co/bert-base-multilingual-cased.
[7] https://huggingface.co/cahya/bert-base-indonesian-1.5G.

Table 4. The results of baselines.

Model	γ	Macro-F1	Micro-F1	Hamming Loss
SGM	–	44.08%	74.24%	0.0706
SU4MLC	–	40.38%	75.66%	0.0658
Bi-LSTM	0.25	43.34%	75.33%	0.0685
CNN	0.25	40.44%	77.99%	0.0616
Bi-LSTM-Attention	0.25	46.48%	74.87%	0.0698
Indonesian-BERT	0.25	**67.57%**	**84.53%**	**0.0414**

Table 5. The results of the matching thresholds.

Model	α/μ	β	Num. of identified samples	Num. of correctly identified samples	Accuracy
Indonesian-BERT+ loss weight	0.98	0.3	1552	1311	84.47%
	0.98	0.35	1541	1309	84.94%
	0.98	0.4	1529	1307	85.48%
	0.99	0.3	1476	1267	85.84%
	0.99	0.35	1467	1265	86.23%
	0.99	0.4	1456	1264	86.81%
Indonesian-BERT+ Sim + loss weight	0.96	–	1422	1195	84.04%
	0.97	–	1250	1099	87.92%
	0.98	–	942	878	93.21%
	0.99	–	464	458	**98.71%**

We further compare the Indonesian-BERT pre-trained language model with the most competitive models (SGM and SU4MLC) in current multi-label classification tasks and other commonly used deep learning classification models (Bi-LSTM, CNN, and Bi-LSTM-Attention). The results shown in Table 4 demonstrate that the pre-trained language model Indonesian-BERT outperforms than other classification models, which is also obvious to see that the Indonesian-BERT model can better learn the inner connection of continuous text and the expressive ability of language structure through adopting the attention mechanism network structure and pre-trained with a large amount of data when the training data is limited.

5.2   Pseudo-label Matching Threshold Screening

We verify the matching thresholds of the two pseudo-label strategies proposed in this paper. The results are shown in Table 5. It is obvious to see that when the matching threshold μ is 0.98 and use the pseudo-label generation strategy with text label matching method, the amount of news texts identified and accuracy have achieved relatively good performance. Although the accuracy with the matching threshold μ of 0.99 outperforms one with 0.98, the number of identified samples is too small.

5.3   Pseudo-label Generation Strategy Verification

Simultaneously, in order to verify our pseudo-label generation strategy and text label matching strategy, we train the model with mixed training data, which composes pseudo-label data generated by the text label matching method and the original training data. Then we compare it with the model trained only with the original training data. The experimental results in Table 6 demonstrate that the model performs the best when using pseudo-label generation strategy with text label matching method to select pseudo-label corpus for joint training. Among them, Macro-F1 reaches 70.82%, which is 1.5% higher than the pre-trained language model Indonesian-BERT.

Table 6. The result of the matching threshold.

Model	Pseudo data	γ	Macro-F1	Micro-F1	Hamming Loss
Indonesian-BERT	–	0.25	69.32%	84.40%	0.0420
Indonesian-BERT+Sim	–	–	69.61%	84.79%	0.0424
Indonesian-BERT	1	0.25	70.00%	**85.31%**	**0.0399**
Indonesian-BERT+Sim	1	–	70.31%	84.88%	0.0420
Indonesian-BERT	2	0.25	70.55%	85.17%	0.0400
Indonesian-BERT+Sim	2	–	**70.82%**	84.66%	0.0428

5.4   Pseudo-label Loss Weight Screening

The size of the pseudo-label loss weight is also verified. The experimental results shown in Table 7 demonstrate that when the value of the loss weight is 10^{-2}, the model performs best.

Table 7. The results of each strategy.

Model	Pseudo data	threshold	Macro-F1	Micro-F1	Hamming Loss
Indonesian-BERT	1	10^{-1}	70.65%	**85.31%**	**0.0399**
		10^{-2}	69.43%	84.86%	0.0407
		10^{-3}	69.42%	84.87%	0.0408
Indonesian-BERT	2	10^{-1}	70.55%	85.17%	0.0400
		10^{-2}	68.05%	84.37%	0.0420
		10^{-3}	69.92%	85.08%	0.0403
Indonesian-BERT+Sim	1	10^{-1}	68.74%	84.81%	0.0425
		10^{-2}	70.31%	84.88%	0.0420
		10^{-3}	70.00%	84.69%	0.0425
Indonesian-BERT+Sim	2	10^{-1}	70.65%	85.00%	0.0420
		10^{-2}	**70.82%**	84.66%	0.0428
		10^{-3}	70.13%	84.66%	0.0429

6 Conclusion

In this paper, we construct a multi-label news classification dataset for Indonesian language and propose a new multi-label news classification framework through using pseudo-label technology. Experiments show that the proposed framework has achieved significant results in Indonesian multi-label news classification, and each classification strategy has a certain improvement. In the future, we will expand the scale and quality of the corpus to improve the performance of the framework, and apply it to multiple languages.

Acknowledgement. This work was supported by the National Social Science Foundation Project (17CTQ045). The authors would like to thank the anonymous reviewers for their valuable comments and suggestions.

References

1. Hu, T., Wang, H., Yin, W.: Multi-label news classification algorithm based on deep bi-directional classifier chains. J. ZheJiang Univ. (Eng. Sci.) **53**, 2110–2117 (2019)
2. Zhang, M., Zhou, Z.: A review on multi-label learning algorithms. IEEE Trans. Knowl. Data Eng. **26**, 1819–1837 (2014). https://doi.org/10.1109/TKDE.2013.39
3. Madjarov, G., Kocev, D., Gjorgjevikj, D., et al.: An extensive experimental comparison of methods for multi-label learning. Pattern Recogn. **45**, 3084–3104 (2012). https://doi.org/10.1016/j.patcog.2012.03.004
4. Zhang, M.-L., Zhou, Z.-H.: ML-KNN: a lazy learning approach to multi-label learning. Pattern Recogn. **40**, 2038–2048 (2007). https://doi.org/10.1016/j.patcog.2006.12.019
5. Zhang, L., Zhou, Y., Duan, X., et al.: A hierarchical multi-input and output Bi-GRU model for sentiment analysis on customer reviews. IOP Conf. Ser. Mater. Sci. Eng. **322**, 62007 (2018). https://doi.org/10.1088/1757-899x/322/6/062007
6. Ahmadi, Z., Kramer, S.: A label compression method for online multi-label classification. Pattern Recogn. Lett. **111**, 64–71 (2018). https://doi.org/10.1016/j.patrec.2018.04.015

7. Fan, H., Zheng, L., Yan, C., et al.: Unsupervised person re-identification: clustering and fine-tuning. ACM Trans. Multimed. Comput. Commun. Appl. **14**, 1–18 (2018). https://doi.org/10.1145/3243316

8. Fu, Y., Wei, Y., Wang, G., et al.: Self-similarity grouping: a simple unsupervised cross do-main adaptation approach for person re-identification. In: Proceedings of the IEEE/CVF International Conference on Computer Vision, pp. 6112–6121 (2019)

9. Read, J., Pfahringer, B., Holmes, G., Frank, E.: Classifier chains for multi-label classification. In: Buntine, W., Grobelnik, M., Mladenić, D., Shawe-Taylor, J. (eds.) ECML PKDD 2009. LNCS (LNAI), vol. 5782, pp. 254–269. Springer, Heidelberg (2009). https://doi.org/10.1007/978-3-642-04174-7_17

10. Shehab, M.A., Badarneh, O., Al-Ayyoub, M., et al.: A supervised approach for multi-label classification of Arabic news articles. In: 2016 7th International Conference on Computer Science and Information Technology (CSIT), pp. 1–6 (2016). https://doi.org/10.1109/CSIT.2016.7549465

11. You, R., Zhang, Z., Wang, Z., et al.: AttentionXML: label tree-based attention-aware deep model for high-performance extreme multi-label text classification. arXiv preprint. arXiv:1811.01727 (2018)

12. You, R., Zhang, Z., Dai, S., et al.: HAXMLNet: hierarchical attention network for extreme multi-label text classification. arXiv preprint. arXiv:1904.12578 (2019)

13. Isnaini, N., Adiwijaya, Mubarok, M.S., Bakar, M.Y.A.: A multi-label classification on topics of Indonesian news using K-Nearest Neighbor. J. Phys. Conf. Ser. **1192**, 12027 (2019). https://doi.org/10.1088/1742-6596/1192/1/012027

14. Pambudi, R.A., Adiwijaya, Mubarok, M.S.: Multi-label classification of Indonesian news topics using Pseudo Nearest Neighbor Rule. J. Phys. Conf. Ser. **1192**, 12031 (2019). https://doi.org/10.1088/1742-6596/1192/1/012031

15. Rahmawati, D., Khodra, M.L.: Automatic multilabel classification for Indonesian news articles. In: 2015 2nd International Conference on Advanced Informatics: Concepts, Theory and Applications (ICAICTA), pp. 1–6 (2015). https://doi.org/10.1109/ICAICTA.2015.7335382

16. Irsan, I.C., Khodra, M.L.: Hierarchical multi-label news article classification with distributed semantic model based features. Int. J. Adv. Intell. Inform. **5**, 40–47 (2019)

17. Lee, D.: Pseudo-label: the simple and efficient semi-supervised learning method for deep neural networks. In: ICML 2013 Workshop: Challenges in Representation Learning (WREPL), pp. 1–6 (2013)

18. Shi, W., Gong, Y., Ding, C., et al.: Transductive semi-supervised deep learning using min-max features. In: Proceedings of the European Conference on Computer Vision (ECCV), pp. 299–315 (2018)

19. Xie, Q., Luong, M., Hovy, E., et al.: Self-training with noisy student improves ImageNet classification. In: 2020 IEEE/CVF Conference on Computer Vision and Pattern Recognition (CVPR), pp. 10684–10695 (2020). https://doi.org/10.1109/CVPR42600.2020.01070

20. Haase-Schtz, C., Stal, R., Hertlein, H., Sick, B.: Iterative label improvement: robust training by confidence based filtering and dataset partitioning. arXiv e-prints. arXiv-2002 (2020)

21. Devlin, J., Chang, M. W., Lee, K., et al.: BERT: pre-training of deep bidirectional transformers for language understanding. arXiv preprint. arXiv:1810.04805 (2018)

22. Yang, P., Sun, X., Li, W., et al.: SGM: sequence generation model for multi-label classification. arXiv e-prints. arXiv:1806.04822 (2018)

23. Lin, J., Su, Q., Yang, P., et al.: Semantic-unit-based dilated convolution for multi-label text classification. CoRR. abs/1808.08561 (2018)
24. Kim, Y.: Convolutional neural networks for sentence classification. In: Proceedings of the 2014 Conference on Empirical Methods in Natural Language Processing, pp. 174–1751 (2014)
25. Schuster, M., Paliwal, K.K.: Bidirectional recurrent neural networks. IEEE Trans. Signal Process. **45**, 2673–2681 (1997). https://doi.org/10.1109/78.650093
26. Yao, Y., Huang, Z.: Bi-directional LSTM recurrent neural network for Chinese word segmentation. In: International Conference on Neural Information Processing, pp. 345–353 (2016)
27. Wang, Y., Huang, M., Zhao, L.: Attention-based LSTM for aspect-level sentiment classification. In: Proceedings of the 2016 Conference on Empirical Methods in Natural Language Processing, pp. 606–615 (2016)
28. Zhou, P., Shi, W., Tian, J., et al.: Attention-based bidirectional long short-term memory networks for relation classification. In: Proceedings of the 54th Annual Meeting of the Association for Computational Linguistics, vol. 2, pp. 207–212 (2016)

Information Extraction from Invoices

Ahmed Hamdi[1]([✉])(iD), Elodie Carel[2](iD), Aurélie Joseph[2](iD), Mickael Coustaty[1](iD),
and Antoine Doucet[1](iD)

[1] Université de La Rochelle, L3i, Avenue Michel Crépeau, 17042 La Rochelle, France
{ahmed.hamdi,mickael.coustaty,antoine.doucet}@univ-lr.fr
[2] Yooz, 1 Rue Fleming, 17000 La Rochelle, France
{elodie.carel,aurelie.joseph}@getyooz.com

Abstract. The present paper is focused on information extraction from
key fields of invoices using two different methods based on sequence
labeling. Invoices are semi-structured documents in which data can be
located based on the context. Common information extraction systems
are model-driven, using heuristics and lists of trigger words curated by
domain experts. Their performances are generally high on documents
they have been trained for but processing new templates often requires
new manual annotations, which is tedious and time-consuming to produce. Recent works on deep learning applied to business documents
claimed a gain in terms of time and performance. While these systems
do not need manual curation, they nevertheless require a large amount
of data to achieve good results. In this paper, we present a series of
experiments using neural networks approaches to study the trade-off
between data requirements and performance in the extraction of information from key fields of invoices (such as dates, document numbers,
types, amounts...). The main contribution of this paper is a system that
achieves competitive results using a small amount of data compared to
the state-of-the-art systems that need to be trained on large datasets,
that are costly and impractical to produce in real-world applications.

Keywords: Invoices · Data extraction · Features · Neural networks

1 Introduction

Administrative documents (such as invoices, forms, payslips...) are very common
in our daily life, and are required for many administrative procedures. Among
those documents, invoices claim specific attention as they are related to the financial part of our activities, many of them are received every day, they generally
need to be paid shortly, and any error is not acceptable. Most of those procedures are nowadays dematerialized and performed automatically. In this aim,
information extraction systems extract key fields from documents such as identifiers and types, amounts, dates and so on. This automatic process of invoices
has been formalized by Poulain d'Andecy et al. [18] and requires some specific
features:

© Springer Nature Switzerland AG 2021
J. Lladós et al. (Eds.): ICDAR 2021, LNCS 12822, pp. 699–714, 2021.
https://doi.org/10.1007/978-3-030-86331-9_45

- handling the variability of layouts,
- minimizing the end-user effort,
- training and quickly adapt to new languages and new contexts.

Even if a formal definition has been proposed in the literature, current approaches rely on heuristics which describe spatial relationships between the data to be extracted and a list of trigger words through the use of dictionaries. Designing heuristics-based models is time-consuming since it requires human expertise. Furthermore, such models are dependent on the language and the templates they have been trained for, which requires annotating a large number of documents and labeling every piece of data from which to extract information. Thus, even once a system has reached a good level of performance, it is very tedious to integrate new languages and new templates. We can for instance cite recent data analysis approaches (CLOUDSCAN [22], SYPHT [9]) which consider information extraction as a classification problem. Based on its context, each word is either assigned to a specific class. Their results are very competitive but these systems require huge volumes of data to get good results.

The problem of extracting some specific entities from textual documents has been studied by the natural language processing field and is known as Named Entity Recognition (NER). NER is a subtask of information extraction that aims to find and mark up real word entities from unstructured text and then to categorize them into a set of predefined classes. Most NER tagsets define three classes to categorize named entities: persons, locations and organizations [16]. Taking advantage of the development of neural-based methods, the performance of NER systems has kept increasing since 2011 [3].

In this paper, we propose to adapt and compare two deep learning approaches to extract key fields from invoices which comprise regular invoices, receipts, purchase orders, delivery forms, accounts' statements and payslips. Both methods are language-independent and process invoices regardless of their templates. In the first approach, we formulate the hypothesis that key fields related to document analysis can extend name entity categories and be extracted in the same way. It is based on annotated texts where all the target fields are extracted and labeled into a set of predefined classes. The system is based on the context in which the word appears in the document and additional features that encode the spatial position of the word in the documents.

The second approach converts each word of the invoice into a vector of features which will be semantically categorized in a second stage. To reduce the annotation effort, the corpus comes from a real-life industrial workflow and the annotation is semi-supervised. The corpus has been tagged with an existing information extraction system and manually checked by experts. Compared to the state of the art, our experiments show that by adequately selecting part of the data, we can train competitive systems by using a significantly smaller amount of training data compared to the ones used in state-of-the-art approaches.

This paper first introduces prior works on information extraction from invoices (Sect. 2). We then describe the data used to assess our work (Sect. 3). Our two approaches based on named entity recognition and document analysis are detailed in Sect. 4. The experiments are described in Sect. 5 to compare our models to state-of-the-art methods, before discussions and future work (Sect. 6).

2 Related Work

Data extraction from administrative documents can be seen as a sequence labeling task. The text is extracted from documents using an OCR engine. Then, each extracted token is to be assigned a label. Early sequence labeling methods are rule-based [5,8]. These rules were defined by humans and based on trigger words, regular expressions and linguistic descriptors. For instance, amounts can be found in the neighborhood of words such as "total", "VAT", etc.

More recent sequence labeling methods are based on neural networks and have been shown to outperform rule-based algorithms [2,6,11,13]. Devlin *et al.* [7] introduced the bidirectional encoder representations from transformers (BERT), which have become popular among researchers. BERT is pre-trained on large data and known for its adaptability to new domains and tasks through fine-tuning. Aspects such as the easiness to fine-tune and high performance, have made the use of BERT widespread in many sequence labeling tasks.

However, administrative documents such as invoices and receipts contain only a few word sequences in natural language. Their content is structured and models can be designed to locate data. For this reason, many other systems combined the textual context with the structural one. Candidates for each field are generally selected and ranked based on a confidence value, and the system returns the best one. Rusiñol *et al.* [21] and their extended work, Poulain d'Andecy *et al.* [18], capture the context around data to be extracted in a star graph. The graph encodes pairwise relationships between the target and the rest of the words of the document which are weighted based on an adaptation of the TF-IDF metric. The system is updated incrementally. The more the system receives documents from a given supplier, the more the graph will be efficient. Once they have been designed for a domain, model-driven systems are often efficient in terms of performance. But processing a real-world document flow is challenging because its input is constantly evolving over time. Hence, extraction systems should cope with unseen templates. Eventually, they should also be able to cope with multiple languages. Designing heuristics requires an expert in the domain. In the production phase, this step has to be done for each new model which cannot be processed properly by the extraction system. It is time-consuming and error-prone. Updating the extraction system implies very tedious work to check the regressions in order to keep high performance. Some systems, such as that of Poulain d'Andecy *et al.* [18], try to limit user intervention by labeling only the target data. However, this process requires a pre-classification step to be able to recognize the supplier.

Recent works proposed deep learning based approaches to solve the extraction problem. Palm *et al.* [17] presented CLOUDSCAN, a commercial solution by Tradeshift. They train a recurrent neural network (RNN) model over 300k invoices to recognize eight key fields. This system requires no configuration and does not rely on models. Instead, it considers tokens and encodes a set of contextual features of different kinds: textual, spatial, related to the format, etc. They decided to apply a left-to-right order. However, invoices are often written in both vertical and horizontal directions. Other works have been inspired by CLOUDSCAN. In their work, they compare their system to alternative extraction systems and they claim an absolute accuracy gain of 20% across compared fields. Lohani *et al.* [12] built a system based on graph convolutional networks to extract 27 entities of interest from invoices. Their system learns structural and semantic features for each entity and then uses surrounding information to extract them. The evaluation of the system showed good performances and achieves an overall F1-score of 0.93. In the same context, Zhao *et al.* [26] proposed the CUTIE (Convolutional Universal Text Information Extractor) model. It is based on spatial information to extract key text fields. CUTIE converts the documents to gridded texts using positional mappings and uses a convolutional neural network (CNN). The proposed model concatenates the CNN network with a word embedding layer in order to simultaneously look into spatial and contextual information. This method allowed them to reach state-of-the-art results on key information extraction.

In this work, similarly to [11,13], we combine sequence labeling methods. However, we add neural network layers to our systems so as to encode engineered textual and spatial features. One of these methods is based on named entity recognition using the Transformer architecture [24] and BERT [7] that, to our knowledge, have not been reported in previous research on the specific tasks of processing administrative documents. The second method is based on word classification that, unlike previous works, do not require neither pre- nor post-processing to achieve satisfactory results.

3   Dataset

Datasets from business documents are usually not publicly available due to privacy issues. Previous systems such as CLOUDSCAN [22] and SYPHT [9] use their own proprietary datasets. In the same line, this work is based on a private industrial dataset composed of French and English invoices coming from a real document workflow provided by customers. The dataset covers 10 types of invoices (orders, quotations, invoice notes, statements...) with diverse templates.

The dataset has been annotated in a semi-automatic way. A first list of fields was extracted from the system currently in use, and finally checked and completed by an expert. The main advantage of this process is its ability to get a large volume of documents. However, even if most of the returned values have been checked, we should take into account a part of noise which is expert-dependent. In other words, some fields may be missed by the system or by the expert, as for instance redundant information in a document are typically annotated only once.

The dataset includes two databases which respectively comprise 19,775 and 134,272 invoices. We refer to them by database-20k and database-100k. 8 key fields have been extracted and annotated from these invoices (cf. Fig. 1).

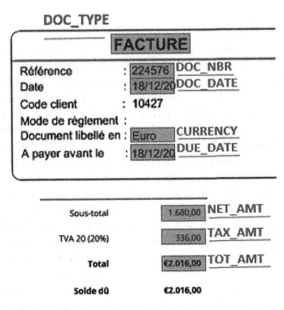

Fig. 1. Part of an invoice from our dataset. The blue text is the label of the field (blue boxes) (Color figure online)

Table 1 provides statistics on the databases. Each one was split into 70% for training and 30% for validation.

Table 1. Statistical description of the in-house dataset

Fields	Database-20k	Database-100k
DOC_TYPE	14,301	105,753
DOC_NBR	18,131	131,133
DOC_DATE	17,739	130,418
DUE_DATE	9,554	68,718
NET_AMT	7,323	63,064
TAX_AMT	12,509	91,216
TOT_AMT	15,505	113,372
Currency	12,111	87,965

As mentioned above, this dataset can be noisy, and we therefore decided to evaluate our methods on a manually checked subset. Thus, all the following experiments have been evaluated on a clean set of 4k French and English documents, that are not part of the training or validation datasets but come from the same workflow. They are similar in terms of content (i.e. invoices, multi-templates) and have the same ratio of key fields.

4   Methodology

As we mentioned in the introduction, we define and compare two different methods on information extraction that are generic and language-independent: a NER-based method and a word classification-based one (henceforth, we respectively denote them NER-based and class-based). To the best of our knowledge, no research study has adapted NER systems for invoices so far. The NER-based evaluates the ability of NLP mainstream approaches to extract fields from invoices by fine-tuning BERT to this specific task. BERT can capture the context from a large neighborhood of words. The class-based method adapted the features of CloudScan [17] in order to fit our constraints and proposed some extra features to automatically extracts the features, with no preprocessing step nor dictionary lookup. Thus, our methods can easily be adapted to any type of administrative documents. These features significantly reduced the processing of the class-based method on the one hand and allow BERT to deal with challenges related to semi-structured documents on the other hand. Both systems assign a class to a sequence of words. The classes are the key fields to be extracted. We assign the class "undefined" to each word which does not correspond to a key field. Both can achieve good performance with a small volume of training data compared to the state of the art. Each word of the sequence to be labeled is enriched with a set of features that encode contextual and spatial information of the words. We therefore extract such features prior to data labeling. The same features are used for both methods.

4.1   Feature Extraction

Previous research showed that spatial information is relevant to extract key fields from invoices [26]. We therefore attempt to combine classic semantic features to spatial ones, defined as follows.

- Textual features: refer to all the words that can define the context of the word to be labeled w_i in semi-structured documents. These words include the framing words of w_i such as the left and right words as well as the bottom and the top words. These features also include the closest words in the upper and lower lines.
- Layout features: they encode the position of the word in the document, block and line as well as its coordinates (left, top, right, bottom). These features additionally encode page descriptors such as the page's width, height and margin.
- Pattern features: each input word is represented by a normalised pattern built by substituting each character with a normalized one such as C for upper-cased letters, c for lower-cased ones and d for digits. Items such as emails and URLs are also normalised as <EMAIL> and <URL>.
- Logical features: these features have Boolean values indicating whether the word is a title (i.e. if the word is written in a larger font than other words) or a term/result of a mathematical operation (sum, product, factor) using trigrams of numbers vertically or horizontally aligned.

4.2   Data Extraction Using the NER-Based Method

The first contribution of this paper relies on the fine-tuning of BERT [7] to extract relevant information from invoices. The reason for using the BERT model is not only because it is easy to fine-tune, but it has also proved to be one of the most performing technologies in multiple tasks [4,19]. Nonetheless, despite the major impact of BERT, we aim in this paper to evaluate the ability of this model to deal with structured texts as with administrative documents.

BERT consists of stacked encoders/decoders. Each encoder takes as input the output of the previous encoder (except the first which takes as input the embeddings of the input words). According to the task, the output of BERT is a probability distribution that allows predicting the most probable output element. In order to obtain the best possible performance, we adapted this architecture to use both BERT word embeddings and our proposed features. At the input layer of the first encoder, we concatenate the word embedding vector with a fixed-size vector for features in order to combine word-level information with contextual information. The size of the word embedding vector is 572 (numerical values) for which we first concatenate another embedding vector that corresponds to the average embedding of the contextual features. The obtained vector of size 1,144 is then concatenated with a vector containing the logical features (Boolean) and the spatial features (numerical). The final vector size is 1,155.

As an embedding model, we rely on the large version of the pre-trained CamemBERT [14] model. For tokenization, we use CamemBERT's built-in tokenizer, which splits text based on white-spaces before applying a Byte Pair Encoding (BPE), based on WordPiece tokenization [25]. BPE can split words into character n-grams to represent recognized sub-words units. BPE allows managing words with OCR errors, for instance, 'in4voicem' becomes 'in', '##4', '##voi', '##ce', '##m'. This word fragment can usually handle out of vocabulary words and those with OCR errors, and still generates one vector per word.

At this point, the feature-level representation vector is concatenated with the word embedding vector to feed the BERT model. The output vectors of BERT are then used as inputs to the CRF top layer to jointly decode the best label sequence. To alleviate OCR errors, we add a stack of two transformer blocks (cf. Fig. 2) as recommended in [1], which should contribute to a more enriched representation of words and sub-words from long-range contexts.

The system converts the input sequences of words into sequences of fixed-size vectors $(x_1, x_2, ..., x_n)$, *i.e.* the word-embeddings part is concatenated to the feature embedding, and returns another sequence of vectors $(h_1, h_2, ..., h_n)$ that represents named entity labels at every step of the input. In our context, we aim to assign a sequence of labels for a given sequence of words. Each word gets a pre-defined label (e.g. DOCNBR for document number, DATE for dates, AMT for amounts ...) or O for words that are not to be extracted. According to this system, the example sentence[1] *"Invoice No. 12345 from 2014/10/31 for an amount of 525 euros."* should be labeled as follows: *"DOCTYPE O DOCNBR O DATE O O O O AMT CURRENCY"*.

---

[1] All the examples/images used in this paper are fake for confidentiality reasons.

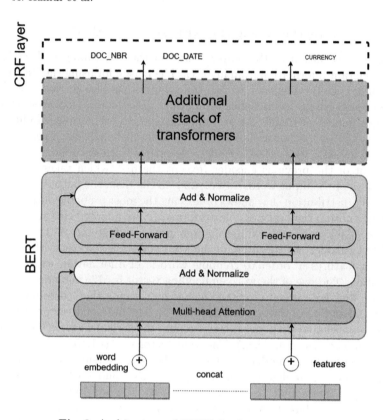

Fig. 2. Architecture of BERT for data extraction

We ran this tool over texts extracted from invoices using an OCR engine. The OCR-generated XML files contain lines and words grouped in blocks, with extracted text aligned in the same way as in regular documents (from top to bottom and from left to right). As OCR segmentation can lead to many errors with the presence of tables or difficult structures, we only kept words from OCR and rebuilt the lines based on word centroid coordinates. The left context therefore allows defining key fields (cf. Fig. 5). However, invoices, as all administrative documents, may have or contain particular structures which should rather be aligned vertically. In tables, for example, the context defining target fields can appear only in the headers. For this reason, we define sequences including the whole text of the document and ensure the presence of the context and the field to be extracted in the same sequence (Fig. 3).

Invoice No.	XX00000	Currency	EUR
Invoice Date	17/09/2019		
Delivery No.	YY00000		
Ref Sales Order No.	WW0000		
Customer No	ZZ00000		
Customer VAT Reg	FR----------		
To receive invoices by email, send your email to: ------@------			
Please reference your customer number			

Fig. 3. Sample of an invoice

4.3 Data Extraction Using the Class-Based Method

In parallel to NER experiments, and as our end goal is classification rather than sequence labeling, we decided to compare our work to more classical classification approaches from the document analysis community. Our aim is to predict the class of each word within its own context based on our proposed feature vector (textual, spatial, structural, logical). The output class is the category of the item, which can either one of the key fields or the undefined tag. Our work is similar to the CloudScan approach [17], which is currently the state of the art approach for invoice field extraction, our classification is mainly based on features. However, unlike them, our system is language-independent and does not require neither resources nor human actions as pre- and post-processing. Indeed, they build resources such as a lexicon of towns to define a pre-tagging. The latter is represented by Boolean features that check whether the word in processing corresponds to a town or a zip code. In the same way, they extract dates and amounts. In this work, our system is resourceless and avoids any pre-tagging. We define a pattern feature to learn and predict dates and amounts. In this way, we do not need to update lists nor detect language. We also added new Boolean features (cf. Sect. 4.1) to define titles and mathematical assertions (i.e.: isTitle, isTerm, isProduct).

In order to accelerate the process, we proposed a strategy to reduce the volume of data injected to train our models. To this end, we kept N-grams which are associated with one of the ground-truth fields and reduced the volume of undefined elements. In other terms, for each annotated field, we randomly select five words with "undefined" classes as counter-examples. This strategy allowed us to be close to the distribution of labeled terms in natural language processing tasks. For instance, in named entity recognition, it is estimated to 16% the ratio of labeled named entities compared to all the words in a text [20,23]. Furthermore, keeping only 40 n-grams for each document with 8 target fields to be extracted, showed better performance than the classification using all the words of every document.

Finally, our experiments were conducted on the Ludwig[2] open-source toolbox designed by Uber as an interface on top of TensorFlow [15]. This toolbox proposes

---

[2] https://uber.github.io/ludwig/.

a flexible, easy and open platform that facilitates the reproducibility of our work. From a practical point of view, a list of items combined with its feature vector is provided to Ludwig as input features. We worked at an n-gram level and the textual features were encoded with a convolutional neural network when they were related to the word itself. When they were spatially ordered (e.g. all words to the top, left, right, bottom) we used a bidirectional lstm-based encoder. A concat combiner provided the combined representation of all features to the output decoders. The model was trained with the Adam optimizer [10] using mini-batches of size 96 until the validation performance had not improved on the validation set for 5 epochs. The combiner was composed of 2 fully connected layers with 600 rectified linear units each. We applied a dropout fraction of 0.2 in order to avoid over-fitting.

5   Results

In order to evaluate our methods, we used two traditional metrics from the information retrieval field: precision and recall. While precision is the rate of predicted key fields correctly extracted and classified by the system, recall is the rate of fields present in the reference that are found and correctly classified by the system. The industrial context involves particular attention to the precision measure because false positives are more problematic to customers than missing answers. We therefore aim to reach a very high precision with a reasonable recall.

We report our results on 8 fields: the **docType** and **docNbr** respectively define the type (i.e. regular invoices, credit notes, orders, account statements, delivery notes, quotes, etc.) and the number of the invoice. The **docDate** and **dueDate** are respectively the date on which the invoice is issued and the due date by which the invoice amount should be paid. We additionally extract the net amount **netAmt**, the tax amount **taxAmt** and the total amount **totAmt** as well as **the currency**. Table 2 shows results of the first experiment which has been conducted using the NER-based model and the class-based system.

These first results show that the class-based system outperforms NER-based on most fields. Except for amounts, NER-based has a good precision for all while the class-based system rightly manages to find all the fields with high precision and recall. Despite the high performance of NER-based in the NER task, the system showed some limits over invoices which we explain by the ratio between undefined words and named entities which is much bigger in the case of invoices. Having many undefined tokens tends to disturb the system especially when the fields are redundant in the documents (i.e. amounts) unlike fields that appear once in the document, for which results are quite good. One particularity of the amount fields is that they often appear in a summary table which is a non-linear block that contains many irrelevant words.

Table 2. First results using the NER-based model and the class-based system over the database-20k. "Support" stands for the number of occurrences of each field class in the test set. Best results are in **bold**.

Fields	Support	Recall		Precision	
		NER-based	class-based	NER-based	class-based
docType	3856	0.79	**0.84**	**0.97**	**0.97**
docNbr	3841	0.66	**0.87**	0.74	**0.86**
docDate	3897	**0.78**	0.73	0.94	**0.95**
dueDate	2603	0.73	**0.78**	**0.94**	0.92
netAmt	3441	0.47	**0.72**	0.58	**0.78**
taxAmt	3259	0.47	**0.70**	0.65	**0.86**
totAmt	3889	0.45	**0.85**	0.59	**0.89**
Currency	2177	0.79	**0.97**	**0.83**	**0.83**

In order to improve the results, we firstly visualized the weights of the features in the attention layer at the last encoder of the NER-based neural network (cf. Fig. 4). These weights indicate relevant features for each target field.

Figure 4 indicates the weights of the best performing epoch in the NER-based model. We can notice from the figure that many features have weights close to zero (with ocean blue) for all the fields. Features such as the position of the word in the document, block and line are unused by the final model and considered as irrelevant. Furthermore, it is clear that the relative position of the word in the document page (rightMargin, bottomMargin) are high-weighted in the predictions of all the fields. For the amount fields, the logical features as well as the relative margin position of the word on the left and on the top are also relevant. It is shown using white or light blue colors. We therefore conducted a second experiment, keeping only the most relevant features. We trained new models without considering the right and the bottom relative positions of the word and its positions in the document, line and block.

Table 3 shows practically better results for all the target fields. Except for the recall of docDate using the class-based system which is considerably degraded, all the other results are either improved or kept good performance. Even if the results are improved using relevant features, the NER-based system nevertheless showed some limits to predict amounts. This is not totally unexpected given that NER systems are mainly adapted to extract information from unstructured texts while amounts are usually indicated at the end of tables with different reading directions as shown in Fig. 5. We assume that an additional feature defining fields in tables can particularly improve the amounts' fields.

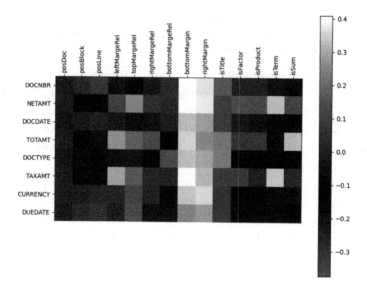

Fig. 4. Weights of features used by the NER based method. Features: position of the word in the document, line and block (table, paragraph or frame) **posDoc, posLine, posBlock**; relative position of the word compared to its neighbours **leftMargeRel, topMargeRel, rightMargeRel, bottomMargeRel**; relative position of the word in the document page **rightMargin, bottomMargin**; Boolean features for titles and mathematical operations **isTitle, isFactor, isProduct, isSum, isTerm**.

Table 3. Results of the NER-based model and the class-based system over the database-20k using relevant features. Best results are given in **bold**. * denotes better results compared to Table 2 (*i.e.* without feature selection)

Fields	Support	Recall		Precision	
		NER-based	class-based	NER-based	class-based
docType	3856	0.81*	**0.85***	**0.98***	0.97
docNbr	3841	0.67*	**0.86**	0.74	**0.86**
docDate	3897	**0.78**	0.33	**0.95***	0.92
dueDate	2603	**0.74***	0.70	**0.93**	0.91
netAmt	3441	0.47	**0.78***	0.58	**0.82***
taxAmt	3259	0.49*	**0.78***	0.66*	**0.87***
totAmt	3889	0.49*	**0.87***	0.61*	**0.89**
Currency	2177	0.82*	**0.96**	0.83	**0.83**

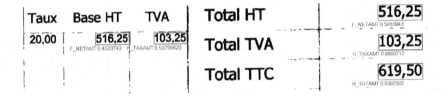

Fig. 5. Different reading directions on the same invoice

Finally, we evaluated the impact of the volume of documents used to train the systems. New models have been generated on the 100k-database documents. Table 4 summarises the performance measures. By increasing the volume, we also increase the number of different templates available in the dataset. We also report the results of the CLOUDSCAN tool [17] for comparative reasons. Even though we are not using the same dataset compared to CLOUDSCAN, we believe this can give a global idea of the performances achieved in the state of the art using an accurate system focusing on invoices and being evaluated in the same fields. Table 4 indicates the best CLOUDSCAN's results reached using a model trained on more than 300k invoices with the same templates as those in the test set.

Table 4. Results of the NER-based model and the class-based system over the database-100k using relevant features. Best results are given in **bold**.

Fields	Recall			Precision		
	CLOUDSCAN [17]	NER-based	class-based	CLOUDSCAN [17]	NER-based	class-based
docType	–	0.79	**0.90**	–	**0.99**	0.97
docNbr	0.84	0.69	**0.89**	0.88	0.85	**0.89**
docDate	0.77	0.78	**0.94**	0.89	**0.96**	**0.96**
dueDate	–	0.74	**0.90**	–	**0.96**	0.93
netAmt	**0.93**	0.47	0.81	**0.95**	0.62	0.86
taxAmt	**0.94**	0.49	0.79	**0.94**	0.70	0.90
totAmt	**0.92**	0.44	0.87	0.94	0.63	**0.95**
Currency	0.78	0.76	**0.98**	**0.99**	0.90	0.84

The results in Table 4 are quite promising regarding the small volume of data we used in these experiments. For some fields (e.g. document number), they can even be compared to our baseline. Unsurprisingly, the results are clearly improved for both the recall measure and the precision measure.

All in all, we can appreciate that, 100k sample is much less-sized than the corpus used to demonstrate the CLOUDSCAN system (more than 300k) and moreover, with only 20k training samples, the performance is yet very honorable.

6   Conclusion

This paper is dedicated to the extraction of key fields from semi-structured documents. We have conducted an evaluation on both NER-based approaches and word classification-based works, with a constraint of reduced training sets. Our main contribution considerably reduces the amount of required data by only selecting reliable data for the training and development. Our solution can easily cope with unseen templates and multiple languages. It neither requires parameter settings nor any other prior knowledge. We got comparable performances to the CLOUDSCAN system with a much smaller volume of training data. Reducing the amount of training data required is a significant result since constructing training data is an important cost in business document processing, to be replicated for every language and template.

We have implemented a network that uses neither heuristics nor post-processing related to the specific domain. Therefore, this system could easily be adapted for other documents such as payslips. Processing invoices raise issues because of the specific organization of the information in the documents, which rarely contain full sentences or even natural language. Key fields to be extracted are very often organized as label-value pairs in both the horizontal and the vertical direction. That makes the information extraction step particularly challenging for capturing contextual data. NLP approaches are seriously challenged in such a context. We also showed, in this paper, that it was possible to highly decrease the processing time to train a model which were still efficient and which fit our industrial context better. With only a small volume of data, we obtain promising results by randomly choosing some undefined items for each annotated field. As we consider the string format of the words, this kind of filtering is not dedicated to invoices. The experiment which has been conducted on a bigger volume shows an improvement of the recall measure. The precision value is also a bit better, although not significantly. Thus, it seems that a bigger volume of documents can make the system more generic because there are more different templates available. To make the precision value increase, we assumed that the more efficient way would be to work on the quality of the training data. Indeed, this neural network has been trained on an imperfect ground-truth, generated by the current system. In addition, we had to assume that the end-user had checked the output of the information extraction system, but this is not always true since in practice only the information required by his company is kept.

As future work, we are considering an interactive incremental approach in order to improve the performance with a minimal set of information. The main idea would be to use the current output to initiate the training process and then regularly train the network again. The more the end-user will check the output, the more the extraction will improve. Moreover, the error analysis showed that 2D information could improve performance and that 2D transformers are a promising prospect for a future work.

Acknowledgements. This work is supported by the Region Nouvelle Aquitaine under the grant number 2019-1R50120 (CRASD project) and AAPR2020-2019-8496610 (CRASD2 project), the European Union's Horizon 2020 research and innovation program under grant 770299 (NewsEye) and by the LabCom IDEAS under the grant number ANR-18-LCV3-0008.

References

1. Boroş, E., et al.: Alleviating digitization errors in named entity recognition for historical documents. In: Proceedings of the 24th Conference on Computational Natural Language Learning, pp. 431–441 (2020)
2. Chiu, J.P., Nichols, E.: Named entity recognition with bidirectional LSTM-CNNs. arXiv preprint arXiv:1511.08308 (2015)
3. Collobert, R., Weston, J., Bottou, L., Karlen, M., Kavukcuoglu, K., Kuksa, P.: Natural language processing (almost) from scratch. J. Mach. Learn. Res. **12**, 2493–2537 (2011)
4. Conneau, A., Lample, G.: Cross-lingual language model pretraining. In: Wallach, H., Larochelle, H., Beygelzimer, A., d' Alché-Buc, F., Fox, E., Garnett, R. (eds.) Advances in Neural Information Processing Systems, vol. 32, pp. 7059–7069. Curran Associates, Inc. (2019). http://papers.nips.cc/paper/8928-cross-lingual-language-model-pretraining.pdf
5. Dengel, A.R., Klein, B.: smartFIX: a requirements-driven system for document analysis and understanding. In: Lopresti, D., Hu, J., Kashi, R. (eds.) International Workshop on Document Analysis Systems, DAS 2002. LNCS, vol. 2423, pp. 433–444. Springer, Heidelberg (2002). https://doi.org/10.1007/3-540-45869-7_47
6. Dernoncourt, F., Lee, J.Y., Szolovits, P.: Neuroner: an easy-to-use program for named-entity recognition based on neural networks. arXiv preprint arXiv:1705.05487 (2017)
7. Devlin, J., Chang, M.W., Lee, K., Toutanova, K.: BERT: pre-training of deep bidirectional transformers for language understanding. arXiv preprint arXiv:1810.04805 (2018)
8. Grishman, R., Sundheim, B.M.: Message understanding conference-6: a brief history. In: COLING 1996 Volume 1: The 16th International Conference on Computational Linguistics (1996)
9. Holt, X., Chisholm, A.: Extracting structured data from invoices. In: Proceedings of the Australasian Language Technology Association Workshop 2018, pp. 53–59. Dunedin, New Zealand, December 2018
10. Kingma, D.P., Ba, J.: Adam: A method for stochastic optimization. arXiv preprint arXiv:1412.6980 (2014)
11. Lample, G., Ballesteros, M., Subramanian, S., Kawakami, K., Dyer, C.: Neural architectures for named entity recognition. arXiv preprint arXiv:1603.01360 (2016)
12. Lohani, D., Belaïd, A., Belaïd, Y.: An invoice reading system using a graph convolutional network. In: Carneiro, G., You, S. (eds.) ACCV 2018. LNCS, vol. 11367, pp. 144–158. Springer, Cham (2019). https://doi.org/10.1007/978-3-030-21074-8_12
13. Ma, X., Hovy, E.: End-to-end sequence labeling via bi-directional LSTM-CNNs-CRF. arXiv preprint arXiv:1603.01354 (2016)
14. Martin, L., et al.: Camembert: a tasty French language model. arXiv preprint arXiv:1911.03894 (2019)
15. Molino, P., Dudin, Y., Miryala, S.S.: Ludwig: a type-based declarative deep learning toolbox (2019)

16. Nadeau, D., Sekine, S.: A survey of named entity recognition and classification. Lingvisticae Investigationes **30**(1), 3–26 (2007)
17. Palm, R.B., Winther, O., Laws, F.: CloudScan - a configuration-free invoice analysis system using recurrent neural networks. CoRR abs/1708.07403 (2017), http://arxiv.org/abs/1708.07403
18. Poulain d'Andecy, V., Hartmann, E., Rusinol, M.: Field extraction by hybrid incremental and a-priori structural templates. In: 13th IAPR International Workshop on Document Analysis Systems, DAS 2018, Vienna, Austria, 24–27 April 2018, pp. 251–256, April 2018. https://doi.org/10.1109/DAS.2018.29
19. Radford, A., Narasimhan, K., Salimans, T., Sutskever, I.: Improving language understanding by generative pre-training (2018)
20. Reimers, N., Eckle-Kohler, J., Schnober, C., Kim, J., Gurevych, I.: GermEVAL-2014: nested named entity recognition with neural networks (2014)
21. Rusiñol, M., Benkhelfallah, T., D'Andecy, V.P.: Field extraction from administrative documents by incremental structural templates. In: ICDAR, pp. 1100–1104. IEEE Computer Society (2013). http://dblp.uni-trier.de/db/conf/icdar/icdar2013.html#RusinolBD13
22. Sage, C., Aussem, A., Elghazel, H., Eglin, V., Espinas, J.: Recurrent neural network approach for table field extraction in business documents. In: 2019 International Conference on Document Analysis and Recognition, ICDAR 2019, Sydney, Australia, 20–25 September 2019, pp. 1308–1313, September 2019. https://doi.org/10.1109/ICDAR.2019.00211
23. Sang, E.F., De Meulder, F.: Introduction to the CoNLL-2003 shared task: language-independent named entity recognition. arXiv preprint cs/0306050 (2003)
24. Vaswani, A., et al.: Attention is all you need. arXiv preprint arXiv:1706.03762 (2017)
25. Wu, Y., et al.: Google's neural machine translation system: bridging the gap between human and machine translation. arXiv preprint arXiv:1609.08144 (2016)
26. Zhao, X., Niu, E., Wu, Z., Wang, X.: Cutie: learning to understand documents with convolutional universal text information extractor. arXiv preprint arXiv:1903.12363 (2019)

Are You Really Complaining?
A Multi-task Framework for Complaint Identification, Emotion, and Sentiment Classification

Apoorva Singh[(✉)] and Sriparna Saha

Department of Computer Science and Engineering, Indian Institute of Technology
Patna, Bihta, India
{apoorva_1921cs19,sriparna}@iitp.ac.in

Abstract. In recent times, given the competitive nature of corporates, customer support has become the core of organizations that can strengthen their brand image. Timely and effective settlement of customer's complaints is vital in improving customer satisfaction in different business organizations. Companies experience difficulties in automatically identifying complaints buried deep in enormous online content. Emotion detection and sentiment analysis, two closely related tasks, play very critical roles in complaint identification. We hypothesize that the association between emotion and sentiment will provide an enhanced understanding of the state of mind of the tweeter. In this paper, we propose a Bidirectional Encoder Representations from Transformers (BERT) based shared-private multi-task framework that aims to learn three closely related tasks, *viz.* complaint identification (primary task), emotion detection, and sentiment classification (auxiliary tasks) concurrently. Experimental results show that our proposed model obtains the highest macro-F1 score of 87.38%, outperforming the multi-task baselines as well as the state-of-the-art model by indicative margins, denoting that emotion awareness and sentiment analysis facilitate the complaint identification task when learned simultaneously.

Keywords: Complaint mining · Multi-task learning · Deep learning

1 Introduction

In linguistics research [14], the word complaining denotes contravention of expectations, i.e., the consumer expressing the complaint is irked or disappointed by the organization or party. Due to the enormous number of consumer complaints related to different products and services present online, a vast amount of complaint texts is generated. Thus, quick and effective identification of complaint texts in natural language is of extreme importance for:

- linguists to acquire an improved understanding of the specific context, intent [23]

© Springer Nature Switzerland AG 2021
J. Lladós et al. (Eds.): ICDAR 2021, LNCS 12822, pp. 715–731, 2021.
https://doi.org/10.1007/978-3-030-86331-9_46

- developers of natural language processing (NLP) downstream applications, such as dialogue systems [26]
- businesses and counsels to aid their customer services by analyzing and tending to customer complaints, thus promoting commercial value and brand image [25].

Sentiment analysis is a domain that evaluates an individual's sentiments from unstructured text [12]. On the contrary, emotion recognition analyzes the varied diversions in the state of mind of an individual. Hence, a positive sentiment label could be associated with multiple emotion classes like happiness, joy, pride, gratitude, and so forth. Previously, complaints were termed as minor deflections from sentiment analysis and often quoted as "sentence with negative undertone with additional information" [23]. Consider the following sentences about the *Samsung Galaxy S20 Ultra:*

1. *The Samsung Galaxy S20 Ultra is way too expensive and beyond my liking!*
2. *Samsung Galaxy S20 Ultra is deplorable!*

In the conventional sense, the two sentences express negative sentiment and anger as the emotion. However, sentence (1) additionally reveals a complaint about the price of a product. Sentence (2) does not have any reasonable sign to indicate any breach of expectation.

The two main challenges in classifying complaint texts given the existing text classification methods are:

1. Complaint texts generally have palpable emotions and sentiments attached to them, as few customers remain happy when expressing complaints. Efficient usage of these expressed emotion and sentiment features can boost the classification.
2. Feasibility of labeled/annotated complaint data is scant, which further requires human intervention, and is a time-consuming process.

In this work, we intend to analyze the role of emotion and sentiment in boosting the identification of complaints in tweets. Multi-task learning has frequently proven to be decisive when working with closely related tasks. It strengthens the generalization process by using the domain-specific information within the related tasks [8]. To explore this further, we learn the tasks of complaint identification, emotion recognition, and sentiment classification in a multi-task environment. This methodology benefits the prevailing single-task frameworks especially for the tasks where the availability of the data is limited. We train a Bidirectional Encoder Representations from Transformers (BERT) [11] based multi-task model and our best model achieves a macro-F1 score of 87.38% on the complaint identification task. We evaluate our best performing multi-task model by collating its per-task performance with the single-task models as well as multi-task baseline models. The experiments are performed on the *Complaints* dataset[1] [17]. The main contributions of our proposed work are summarized below:

---

[1] https://github.com/danielpreotiuc/complaints-social-media.

- We propose a BERT-based shared-private attentive multi-task framework that achieves overall superior performance for the complaint classification task on the complaint dataset.
- Our experimentation illustrates that the performance of the primary task of complaint classification gets boosted by the presence of two auxiliary tasks, namely emotion and sentiment classification.
- We perform a qualitative analysis of the proposed framework on the complaint identification task.

2   Related Work

Social media has become a mainstream platform for expressing opinions, where customers can explicitly address organizations in regards to issues with products, events, or services. Due to the constraints on text size, usage of sarcasm, accusations, condemnation, and even threats, identifying complaints in social media is a difficult task [14]. Complaints can be of two types either directly focused on the organization responsible for contravention of expectations (direct grievances) or indirectly imply the organization (indirect grievances). The term 'face' is used to define the notion of being embarrassed, deceived, or losing face [6]. Generally, people interact in personal and public relations and solicit cooperation in maintaining face. Complaints are usually considered as the face-threatening act [6], acts that may jeopardize the face of the company or individual it is directed at. Pryzant et al. [18] investigated financial complaints submitted to the Consumer Financial Protection Bureau based on the criteria of whether a complaint would receive a swift response or not. Complaints have been categorized based on possible indicators for recall of products [4]. Quite recently, customer service discourses were classified based on whether the complaints would be raised with the administration or made available on public platforms [26]. Preotiuc-Pietro et al. [17] introduced a new annotated dataset on Twitter data and proposed using several features in supervised machine learning models. Results are exhibited across nine domains attaining a predictive performance of up to 78% F1 using distant supervision. This approach is established on a broad range of syntactic and semantic linguistic features.

Simultaneous learning of tasks having logical similarities often reaps benefits from concurrent learning [5]. Recently, multi-task learning configurations have become robust in solving various NLP tasks [5]. Some of the reasons for this accomplishment are:

- knowledge transfer across tasks that produces better representations
- acts as a regularization process that prevents overfitting by preserving the performance across individual tasks [7]

Moreover, the emotional state and the closely associated sentiment perception of a tweeter has a direct impact on the content expressed online [20]. Akhtar et al. [2] proposed a multi-task ensemble framework that concurrently learns multiple problems. Their multi-task framework works on four problems (classification and regression) related to emotion and sentiment analysis. The authors of [21] put forward a twitter data-based framework that identifies and examines sentiment and emotion conveyed by people in their twitter posts and further utilizes them for developing generalized and personalized recommendation systems based on respective twitter interactions. The authors of [1] proposed a stacked ensemble model which merges the outputs from deep learning and feature-engineering-based models utilizing a multi-layer perceptron network for emotion and sentiment intensity prediction. These studies prompted us in exploring the role of emotion and sentiment in aiding the identification of complaints in tweets.

Yi et al. [27] proposed the Bidirectional Encoder Representations from Transformer with Bidirectional Gated Recurrent Unit (BERT-GRU) model to acquire the underlying linguistic information for relation extraction without using any extensive linguistic features extracted with the help of NLP tools. Previous studies on complaint identification have used supervised machine learning models with an extensive assortment of linguistic features drawn out from the text. Customizing a state-of-the-art pre-trained neural language model based on transformer networks [24] such as BERT with the multi-task framework is yet to be analyzed. Considering the unique characteristics of complaint texts and the scarcity of prevailing methods in identifying complaints, we propose a deep multi-task system that is strengthened by the usage of a pre-trained language model for enhancing the learning process. On comprehensive evaluation, we observe that our proposed system has outperformed the state-of-the-art system by an improvement of 9.38% in the macro-F1 score.

3   Proposed Methodology

In this section, we formulate our problem and discuss the details of the BERT-based multi-task framework for complaint identification. The different segments of the architecture are discussed in the accompanying sections.

3.1   Problem Definition

In the current work, we develop an end-to-end multi-task deep learning system to identify complaints and classify emotion and sentiment from the tweets, learning the three tasks jointly. To be more specific, let $(p_x, s_x, e_x, c_x)_{x=1}^{P}$ be a set of P tweets where s_x, e_x, and c_x represent the matching sentiment, emotion and complaint labels for p_x^{th} tweet, respectively. Here, $p_x \epsilon P, s_x \epsilon S$ (sentiment classes), $e_x \epsilon E$ (emotion classes) and $c_x \epsilon C$ (complaint classes). The objective of our multi-task learning system is to fit a function that draws an unknown tweet p_x (from relevant domain) to its fitting sentiment label s_x, emotion label e_x and complaint label c_x simultaneously.

3.2 BERT-Based Shared Private Multi-task Framework (BERT-SPMF)

We explore the idea of a BERT-based shared-private attentive multi-task archi-tecture comprising of a mix of shared (S) and private (P) layers. The proposed architecture is illustrated in Fig. 1. The partitioning of the shared and private layers is denoted by the color coded orange dashed line.

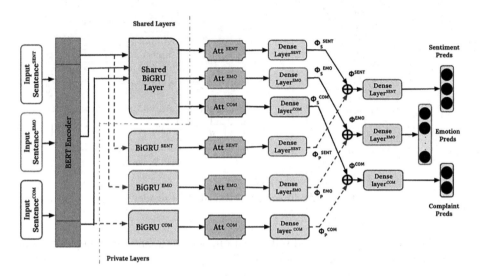

Fig. 1. Multi-task Framework for Complain, Emotion and Sentiment Detection (BERT-SPMF). The layers to the left of the dashed line are the shared layers whereas to its right are the private layers.

Fine-Tuning with Bidirectional Encoder Representations from Trans-formers (BERT). BERT model architecture is a multi-layer bidirectional Transformer encoder [11] and is established on the original work illustrated in [24]. BERT-BASE [11] encoder is a stack of identical encoders (N = 12) and each encoder consists of two sub-layers. The multi-head self-attention system is the first sub-layer and a position-wise fully connected feed-forward network operates as the second sub-layer, as shown on the left side of the dotted line in Fig. 2. The multi-head self-attention is defined as:

$$MultiHead(Q, K, V) = Concat(head_1, ..., head_h)W^O, \tag{1}$$

where Q is the matrix of queries, K is the matrix of keys, V is the matrix of values, W^O is the weight matrix that is trained jointly with the model and

$$head_i = Attention(Qw_i^Q, Kw_i^K, Vw_i^V). \tag{2}$$

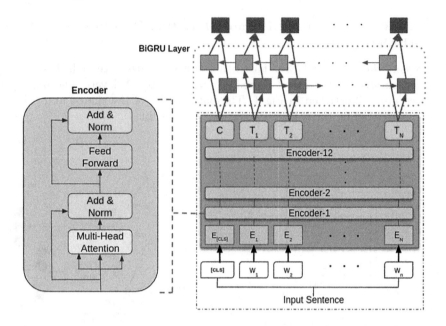

Fig. 2. Architecture of the BERT Encoder.

With respect to multi-head self-attention, we also define the scaled dot product attention. It is realized as follows:

$$Attention(Q, K, V) = softmax(\frac{QK^T}{\sqrt{d_k}})V. \qquad (3)$$

where d_k is the dimension of the Q and K matrices.

We employ the BERT-BASE pre-trained model (Uncased: 12-layer, 768-hidden, 12-heads, 110 M parameters) to produce a sequence of tokens. The sequence of tokens is defined as follows:

$$[CLS], w_1, w_2, w_3, ..., w_N, [SEP]$$

where $[CLS]$ is always the first token of the sequence that indicates the beginning of the sequence, the token $[SEP]$ is used to separate a pair of sentences from each other and also indicates the end of a sentence and each w_i represents a token of the sequence. Every token passes through the token Embedding layer, sentence Embedding layer, and the position Embedding layer of the BERT. Subsequently, the vectors generated from these three embedding layers are concatenated and passed as input to the BERT encoder. We denote the concatenated embeddings as E_i and we depict the final hidden vector for the i-th input token as T_i (BERT output of the input sequence). The architecture of the BERT Encoder is shown in Fig. 2.

Shared-Private Bidirectional GRU (BiGRU) Layers. The BERT Encoder output corresponding to each task-specific input sentence is fed to a shared BiGRU [9] layer (256 neurons) and also through three task-specific BiGRU layers. The shared sub-system consists of the BERT Encoder and a BiGRU layer followed by three task-specific attention layers and fully connected layers. Parallel to the shared BiGRU layer, we employ three task-specific (private) BiGRU layers that work on the task-specific word vectors from the embedding layer. BiGRU layer holds on to contextual information from both directions, forward ($\overrightarrow{GRU}$) and backward ($\overleftarrow{GRU}$) time steps, and generates a hidden representation of each word in the sentence.

During training, the weights of the Shared BiGRU (SBiGRU) layer are updated in accordance with all the three tasks which result in the generation of mutually dependent task-specific representations (common to all the tasks) from this BiGRU layer and the corresponding attention layers emphasize on this generic representation. On the other hand, the private BiGRU (PBiGRU) layers generate mutually independent task-specific representations of the input sentences, and correspondingly the following attention layers emphasize these representations. Further on, the task-specific outputs from the shared subsystem and private subsystem are concatenated before being fed to three task-specific fully connected (FC) and dropout layers, the output of which is passed to the softmax layer. Each of the three output layers predicts the output for the three tasks namely complaint, emotion, and sentiment.

Attention Layer. In order to focus on the words that contribute the most in the sentence meaning, we incorporate the attention technique [3] (Att) following the BiGRU layers in the model. We employ three independent task-specific attention layers after the Shared BiGRU layer and one attention layer each after the task-specific BiGRU layers. Each attention layer is followed by a fully connected layer of 100 neurons.

For each task, we linearly concatenate the output from their shared (ϕ_S^j)and private (ϕ_P^j) fully-connected layers and pass the concatenated vector (ϕ^j) through the final task-specific fully-connected layer and output layer. The overall flow of information through the network can be realized by the following equations:

$$\Phi_S^{\text{SENT}} = FC_S^{\text{SENT}}(Att_S^{\text{SENT}}(SBiGRU(C, T_1, T_2, ..., T_N))) \qquad (4)$$

$$\Phi_S^{\text{EMO}} = FC_S^{\text{EMO}}(Att_S^{\text{EMO}}(SBiGRU(C, T_1, T_2, ..., T_N))) \qquad (5)$$

$$\Phi_S^{\text{COM}} = FC_S^{\text{COM}}(Att_S^{\text{COM}}(SBiGRU(C, T_1, T_2, ..., T_N))) \qquad (6)$$

$$\Phi_P^{\text{SENT}} = FC_P^{\text{SENT}}(Att_P^{\text{SENT}}(PBiGRU^{\text{SENT}}(C, T_1, T_2, ..., T_N))) \qquad (7)$$

$$\Phi_P^{\text{EMO}} = FC_P^{\text{EMO}}(Att_P^{\text{EMO}}(PBiGRU^{\text{EMO}}(C, T_1, T_2, ..., T_N))) \qquad (8)$$

$$\Phi_P^{\text{COM}} = FC_P^{\text{COM}}(Att_P^{\text{COM}}(PBiGRU^{\text{COM}}(C, T_1, T_2, ..., T_N))) \qquad (9)$$

$$\Phi^{\text{SENT}} = [\Phi_S^{\text{SENT}}; \Phi_P^{\text{SENT}}] \qquad (10)$$

$$\Phi^{EMO} = [\Phi_S^{EMO}; \Phi_P^{EMO}] \tag{11}$$

$$\Phi^{COM} = [\Phi_S^{COM}; \Phi_P^{COM}] \tag{12}$$

$$s_i = argmaxsoftmax(FC^{SENT}(\Phi^{SENT})) \tag{13}$$

$$e_i = argmaxsoftmax(FC^{EMO}(\Phi^{EMO})) \tag{14}$$

$$c_i = argmaxsoftmax(FC^{COM}(\Phi^{COM})) \tag{15}$$

';' indicates linear concatenation.

Calculation of Loss: We calculate the categorical-cross entropy losses for the complaint, emotion, and sentiment tasks. The combined loss function of our proposed BERT-SPMF system is realized as follows:

$$L = a * L_{CE}^{C} + b * L_{CE}^{E} + c * L_{CE}^{S} \tag{16}$$

We compute the weighted sum of the different losses from the three tasks to produce the overall loss. Here, a, b, and c represent the constants which are in the range 0 to 1 that set the loss weights depicting the per task loss-share to the overall loss.

4   Experiments, Results, and Analysis

4.1   Dataset

Complaint Dataset: We have considered the recently introduced *Complaints-Data* dataset for our experiments. It is an assortment of 2217 tweets of non-complaint category and 1232 tweets of complaint category. Besides the dataset is also grouped in various high-level domains[2] based on various industries and areas of activity, to validate analyzing complaints by disciplines.

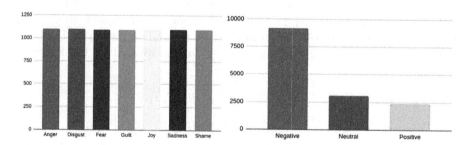

Fig. 3. a) Class wise distribution of Emotion dataset. b) Class wise distribution of Sentiment dataset. Here x-axis represents the classes for the respective auxiliary tasks. y-axis represents the number of instances for each class.

---

[2] *Food & beverage, apparel, software, electronics, services, retail, transport, cars, other.*

Emotion Dataset: The International Services on Emotion Antecedents and Reactions (ISEAR)[3] is a well established English emotion dataset. The ISEAR project was contributed by a congregation of psychologists and non-psychologist student interviewees. Both were directed to report different emotional instances in which they experienced the 7 major emotions (anger, disgust, fear, shame, guilt, sadness, and joy). The dataset comprises of 7666 records on seven emotions each having close to 1090 instances. The distributions of emotion classes across the test set are shown in Fig. 3.

Sentiment Dataset: The Twitter US Airline Sentiment dataset[4] is based on how travelers expressed their feelings about different US airlines in February 2015. It comprises of 14640 tweets of which 9178 tweets are of negative polarity, 3099 are of neutral polarity and 2363 tweets are of positive polarity. The distributions of sentiment classes across the test set are shown in Fig. 3.

We have down-sampled the Emotion and Sentiment datasets to match the size of the Complaint dataset for our experiments[5]. To maintain a balanced dataset for the experiments, an equal number of instances for all the classes (493 instances for each of the 7 emotion classes and 1150 instances for each of the 3 sentiment classes) were considered.

Data Pre-processing and Annotation: We have split the Complaint dataset in training (90%) and test sets (10%) and then annotated the test data for Emotion and Sentiment labels which we consider as the gold standard to evaluate our model's performances on all the tasks. Essential cleaning and pre-processing (excess blank spaces removal, discarding of punctuation marks, lowercase conversion, compaction handling ('I'm' for 'I am', 'shan't' for 'shall not', etc.), etc.) were carried out and the consequent sentences were given for annotation. All the sentences in the test set were annotated by 3 different annotators, having post-graduate academic degrees and well-acquainted in performing similar annotations. They were asked to assign, for each sentence, with at most one emotion label from the 7-emotion tag set as in the ISEAR dataset and at most one sentiment label from the 3-sentiment tag set as in the Twitter US Airline Sentiment dataset. Final emotion and sentiment labels for each sentence were decided by majority voting technique among the 3 annotators. Sentences marked with no common emotion or sentiment label (by the 3 annotators) were not considered in the final annotated dataset. We have calculated Cohen-Kappa scores on the annotated dataset to measure the inter-rater agreement among the 3 annotators.

---

[3] https://www.unige.ch/cisa/research/materials-and-online-research/research-material/.

[4] https://www.kaggle.com/crowdflower/twitter-airline-sentiment.

[5] We used the random module's inbuilt function sample() in Python which returns a particular length list of items chosen from the sequence. We additionally performed the experiments with an up-sampled Complaint dataset but due to the redundant instances, the results were erroneous.

We attain agreement scores of 0.72 and 0.83 on the Emotion and Sentiment task, respectively, which indicate that the annotations are of good quality[6].

4.2 Deep Learning Baselines

- **Single-task systems:** We develop BERT-based single-task deep learning models for each of the complaint (BERT-ST$_{COM}$), emotion (BERT-ST$_{EMO}$), and sentiment tasks (BERT-ST$_{SENT}$). The BiGRU output in each of the models passes through the attention, dense and outer layers (task-specific).
- **Basic Multi-task System (MT$_{Glove}$)** [19]: This baseline is a basic multi-task framework for complaint, emotion, and sentiment classification tasks. The embeddings are obtained from the pre-trained GloVe[7] [16] word embedding. The output of the embedding layer is passed to the word sequence encoder to capture the contextual knowledge from the sentence. The system consists of one fully-shared BiGRU [9] layer (256 units) followed by a shared attention layer. The attention layer's output is fed to the three task-specific dense layers followed by their output layers.
- **BERT-based Basic Multi-task System (BERT-MT)** [27]: This model is a BERT-based multi-task framework for complaint, emotion, and sentiment tasks which consists of one fully-shared BiGRU layer (256 units) followed by a shared attention layer. The output of the attention layer is fed to the three task-specific dense layers followed by their output layers.
- **BERT-SPMF$_{CE}$:** We build a BERT-based dual-task framework for complaint and emotion tasks. The framework is analogous to the BERT-SPMF system except for the fact that this model is built for two tasks (complaint and emotion) only. This baseline helps in understanding the role of the emotion task in aiding the primary task in our proposed framework.
- **BERT-SPMF$_{CS}$:** Similarly, to understand the impact of sentiment prediction on complaint classification, we build a dual-task framework for the two tasks. The architecture is similar to the BERT-SPMF system.

4.3 Experimental Setup

We implement our proposed framework and all the baselines on Python-based libraries, Keras[8] [10] and Scikit-learn[9] [15]. We perform all the experiments on a single GeForce GTX 1080 Ti GPU. We report the precision, recall, macro-F1 score, and accuracy for the complaint identification task. We also assess the performance of emotion detection and sentiment classification tasks in terms of

---

[6] In the case where emotion expressed in the tweet is not in the seven categories (anger, disgust, fear, shame, guilt, sadness, and joy) the annotators label it to the next closest emotion associated with the tweet.

[7] GloVe: http://nlp.stanford.edu/data/wordvecs/glove.840B.300d.zip.

[8] https://keras.io/.

[9] https://scikit-learn.org/stable/.

accuracy metrics. We employ the python-based *keras-bert*[10] library to implement the BERT-BASE model. We fine-tune the BERT-based models with the following set of hyperparameters: $\beta 1 = 0.9$, $\beta 2 = 0.999$, $\epsilon = 1e-7$. We apply Grid Search[11] technique to set the epoch size = 4 and learning rate = 1e−3 and train our network using Rectified-Adam [13]. For the multi-task experiments (BERT-SPMF, BERT-MT), we weigh the losses[12] obtained from the complaints, emotion and sentiment tasks by order of 1, 0.5 and 0.5, respectively. For the GloVe-based multi-task experiment (MT_{GloVe}), we set the loss weights for the complaints, emotion, and sentiment tasks by order of 1, 0.2, and 0.2, respectively. Furthermore, for the ablation experiment (BERT-SPMF$_{CE}$), the loss weights for the complaints, emotion tasks are set as 1 and 0.7, respectively. For the BERT-SPMF$_{CS}$ model the loss weights for the complaints, sentiment tasks are set as 1 and 0.5, respectively. We use the Grid Search method from the Scikit-learn library to tune the loss weights for all the experiments. We employ *ReLU* activation in the dense layers and the output layers for complaint, emotion, and sentiment classification tasks utilize *Softmax* activation with 2, 7, and 3 neurons, respectively. In the GRU units, we use *Tanh* activation and also apply a *dropout* [22] of 25% each. We employ a *dropout* of 25% following the attention layer as well as after every linear layer, to decrease the risk of overfitting. We clip and pad input sentences to set the maximum sequence length as 128. For deep learning baseline (MT_{GloVe}) we also used pre-trained GloVe[13] [16] word embedding which is trained on *Common Crawl* (840 billion tokens) corpus to get the word embedding representations.

4.4   Results and Discussion

The classification results from the different experiments are shown in Table 1. Our proposed framework (BERT-SPMF) outperforms the single-task classifier for the complaint task with substantial improvements of 3.9% and 4.07% in terms of accuracy and macro-F1 metrics, respectively. Additionally, the BERT-SPMF model surpasses the single task classifier's accuracy for the auxiliary tasks with improvements of 4.71% and 2.03% for emotion and sentiment tasks, respectively. We also perform a couple of ablation studies (BERT-SPMF$_{CE}$, BERT-SPMF$_{CS}$) to investigate the role of the auxiliary tasks (emotion, sentiment) in aiding the primary task (complaint task) in our proposed framework. Our proposed BERT-SPMF model performs better in comparison to the BERT-SPMF$_{CE}$ and BERT-SPMF$_{CS}$ models for emotion and sentiment tasks by attaining improvements of 1.62% and 0.5% in terms of accuracy, respectively. Moreover, the BERT-SPMF system accomplishes better results as compared to the MT_{GloVe} with the best-achieved accuracies of 88.53%, 53.21% and 69.27% for complaint, emotion, and sentiment classification tasks, respectively. This signifies that the incorporation of a pre-trained BERT encoder aids the multi-task model's efficacy.

---

[10] https://github.com/CyberZHG/keras-bert.

[11] We experimented with epochs = [3, 4, 5] and learning rates = [1e−3, 2e−3, 3e−5].

[12] Using *loss_weights* parameter of Keras *compile* function.

[13] http://nlp.stanford.edu/data/wordvecs/glove.840B.300d.zip.

It can be observed from the results that awareness about both emotion and sentiment enhances the performance of the complaint identification task. We also observe that leveraging the shared-private module on-top of the BERT-MT model improves the overall performance of the system.

Due to the down-sampling of emotion and sentiment datasets, the number of instances required for proper training of different single-task variants, multi-task baselines, and the proposed BERT-SPMF model are relatively less, owing to which the accuracies of our auxiliary tasks, emotion and sentiment classification are considerably low. The proposed BERT-SPMF framework is found to be statistically significant[14] over the next best performing system, BERT-SPMF$_{CE}$ when tested against the null hypothesis with p-esteem 0.04.

Table 1. Overall classification results including single-task variants, multi-task baselines (MT$_{GloVe}$, BERT-MT), ablations (BERT-SPMF$_{CE}$, BERT-SPMF$_{CS}$), and our proposed model (BERT-SPMF). P, R, F1, A ($\pm$ standard deviation) metrics are given in % and respectively, correspond to Precision, Recall, macro-F1 score and Accuracy. Bold faced values indicate the maximum scores attained. The * mark indicates these results are statistically significant.

Model	Complaint				Emotion	Sentiment
	P	R	F1	A	A	A
SOTA	–	–	78	80.5	–	–
Deep learning Baselines						
BERT-ST$_{COM}$	84	82.64	83.31	84.63 $\pm$ 1.04	–	–
BERT-ST$_{EMO}$	–	–	–	–	48.50 $\pm$ 1.26	–
BERT-ST$_{SENT}$	–	–	–	–	–	67.24 $\pm$ 1.31
MT$_{GloVe}$	83.58	82.33	82.95	84.31 $\pm$ 1.63	41.45 $\pm$ 1.31	67.52 $\pm$ 1.5
BERT-MT	84.79	83.86	84.32	86.37 $\pm$ 1.22	49.71 $\pm$ 1.07	68.90 $\pm$ 1.09
BERT-SPMF$_{CE}$	86.41	85.98	86.19	86.95 $\pm$ 1.45	51.59 $\pm$ 1.33	–
BERT-SPMF$_{CS}$	87	84.12	85.53	86.66 $\pm$ 1.6	–	68.77 $\pm$ 1.08
Proposed approach						
BERT-SPMF	**87.89***	**86.88***	**87.38***	**88.53*** $\pm$ 1.02	**53.21*** $\pm$ 1.32	**69.27*** $\pm$ 1.14

Comparison with State-of-the-Art Technique (SOTA): We have compared our proposed model with the state-of-the-art technique [17] (Table 1). SOTA depicts a thorough analysis of identifying complaints in social media with a wide range of feature engineering. Their supervised learning model attains the best predictive performance of 80.5% accuracy and 78% macro-F1 using a logistic regression classifier with a variety of generic linguistic features and complaint-specific features. The reported results in Table 1 signify that the BERT-SPMF framework surpasses the studied state-of-the-art baseline remarkably for the complaint classification task (primary task).

---

[14] We perform *Student's t-test* for assessing the statistical significance.

4.5 Error Analysis

Detailed analysis of the results generated by our proposed model brought to light the possible contexts where the model failed to perform, some of which are discussed below:

Imbalanced Dataset: The imbalanced distribution of the complaint dataset affects the predictions of the proposed model. For example, *'@BestBuysSupport it says order out of stock when I pre-ordered a month in advance'*. Predicted class: non-complaint. For the given example the correct class is complaint but due to the over-representation of non-complaint instances in the training dataset, the model misclassifies it as non-complaint.

Table 2. Comparative study of the performance of the BERT-SPMF system with its single-task variants (BERT-ST$_{COM}$, BERT-ST$_{EMO}$, BERT-ST$_{SENT}$) and other multi-task baselines through qualitative analysis. These examples show that emotion and sentiment as auxiliary tasks enhance the performance of BERT-SPMF framework. The values in bold indicate actual labels. Non-Com: Non-Complaint, Com: Complaint

Sentence	Model	Complaint	Emotion	Sentiment
@Uber_Support my driver on saturday was doing 52 mph in a 20 mph zone. i was terrified. How can you allow this??	Single task	Non-Com	Anger	**Negative**
	BERT-SPMF$_{CE}$	**Com**	**Fear**	–
	BERT-SPMF$_{CS}$	**Com**	–	**Negative**
	MT$_{GloVe}$	Non-Com	Anger	Neutral
	BERT-MT	**Com**	Anger	Neutral
	BERT-SPMF	**Com**	**Fear**	**Negative**
@WhirlpoolCare 20 odd mins hold music, somebody trying to sell me a service plan for my new machine... can I book an engineer?	Single Task	Non-Com	Sadness	Neutral
	BERT-SPMF$_{CE}$	**Com**	**Anger**	–
	BERT-SPMF$_{CE}$	Non-Com	–	**Negative**
	MT$_{GloVe}$	**Com**	Disgust	**Negative**
	BERT-MT	**Com**	**Anger**	**Negative**
	BERT-SPMF	**Com**	**Anger**	**Negative**
@NBASTORE_Support i accidentally ordered a jersey and shipped to the wrong address. can i change it please?	Single task	**Non-Com**	Anger	**Neutral**
	BERT-SPMF$_{CE}$	Com	**Sadness**	–
	BERT-SPMF$_{CS}$	**Non-Com**	–	Positive
	MT$_{GloVe}$	**Non-Com**	Guilt	**Neutral**
	BERT-MT	Com	**Sadness**	**Neutral**
	BERT-SPMF	**Non-Com**	**Sadness**	**Neutral**
thanks @VW for your fantastic car seizing up while driving causing me to spin out and almost die	Single task	Non-Com	Fear	Positive
	BERT-SPMF$_{CE}$	**Com**	**Anger**	–
	BERT-SPMF$_{CS}$	**Com**	–	Neutral
	MT$_{GloVe}$	Non-Com	Sadness	Neutral
	BERT-MT	**Com**	Sadness	**Negative**
	BERT-SPMF	**Com**	**Anger**	**Negative**

Mis-classification of Emotion Labels: Mis-classification of the BERT-SPMF framework can be mostly accredited to the misclassification of the emotion class for a particular sentence. The down-sampled emotion dataset used is not sufficient to train the model properly on the emotion task. For example, *'Fuck me live chat with BT and Vodafone is worse than the VWR!'*. Predicted emotion label: sadness. Predicted class: non-complaint. For the given example the correct emotion label is *anger* which is misclassified by the BERT-SPMF model for emotion task as *sadness* leading to misclassification of the primary task. Our model is slightly biased towards emotions such as sadness and anger. The primary reason could be the explicit nature (explicit words such as rage, annoyance) and intensifiers (very, too, etc.) of most of the instances in these classes. On the other hand, fear and shame are less explicit and these emotions are submissive in nature in comparison to anger and sadness classes; this leads to the incorrect classification.

Prejudiced Towards Negative Sentiment: Instances posing open-ended questions or consisting of sarcastic comments where the underlying tone may be negative but the sentence is of non-complaint type, the proposed BERT-SPMF system incorrectly predicts such instances as complaint. For example, *'@ChryslerCares is it fair to ask for a new case manager if I feel mine is inadequate?'*. Predicted class: complaint. For the given example, the actual class is non-complaint but the negative tone and underlying sarcasm are interpreted as a complaint by the BERT-SPMF model.

Compound Sentences: Long sentences with multiple phrases or sentences often pose a problem for the BERT-SPMF model to predict the correct class as complaint or non-complaint. For example, *'@JetBlue please help. we love JetBlue! just got back from LA-NYC (flight424) and my wife left her purse on the plane! what do we do? #trueblue #jetblue'*; predicted class: complaint. For the given example the correct class is non-complaint but owing to the composite nature and contrasting context of the sentence the BERT-SPMF model misclassifies it as a complaint.

Table 2 shows some sample predictions from all the developed models. The 1st, 3rd and 4th instances in Table 2 are some of the instances where the proposed BERT-SPMF system surpasses the results of different single-task and multi-task baselines. Likewise, for the 2nd instance, BERT-SPMF model correctly predicts the instance as *complaint*, the emotion being displayed as *anger* and also understands the sentiment as *negative* polarity. Moreover, the BERT-MT classifier is also able to classify the primary and the auxiliary tasks correctly. Lastly, for the 4th sentence, even though this sentence contains ironical comment, the BERT-SPMF model is able to grasp the *complaint* context, the stronger emotion being *anger* and the *negative* polarity of the sentence.

5   Conclusion and Future Work

In this paper, we propose a BERT-based shared-private multi-task (*BERT-SPMF*) architecture for complaint identification, emotion analysis, and sentiment classification. We develop five multi-task frameworks starting from four multi-task baseline models and moving on to a more sophisticated BERT-based shared-private multi-task framework for superior performance on the primary task of complaint identification. Extensive experiments show that the proposed BERT-SPMF model exhibits superlative performance in comparison to other techniques with the best-attained results of 88.53% accuracy and 87.38% macro-F1 score for the complaint identification task and accuracies of 53.21% and 69.27% for the emotion and sentiment classification task, respectively. We have also performed a detailed error analysis in Sect. 4.5. Finally, the results of our proposed framework firmly depict the correlations between emotion recognition, sentiment detection, and complaint classification tasks thus profiting from each other when learned concurrently. For future work, we want to explore other robust architectures to identify complaint texts and also intend to include information from other modalities into the architecture.

Acknowledgement. This publication is an outcome of the R&D work undertaken in the project under the Visvesvaraya Ph.D. Scheme of Ministry of Electronics & Information Technology, Government of India, being implemented by Digital India Corporation (Formerly Media Lab Asia).

References

1. Akhtar, M.S., Ekbal, A., Cambria, E.: How intense are you? Predicting intensities of emotions and sentiments using stacked ensemble [application notes]. IEEE Comput. Intell. Mag. **15**(1), 64–75 (2020)
2. Akhtar, S., Ghosal, D., Ekbal, A., Bhattacharyya, P., Kurohashi, S.: All-in-one: emotion, sentiment and intensity prediction using a multi-task ensemble framework. IEEE Trans. Affect. Comput. (2019)
3. Bahdanau, D., Cho, K., Bengio, Y.: Neural machine translation by jointly learning to align and translate. In: Bengio, Y., LeCun, Y. (eds.) 3rd International Conference on Learning Representations, ICLR 2015, San Diego, CA, USA, 7–9 May 2015, Conference Track Proceedings (2015). http://arxiv.org/abs/1409.0473
4. Bhat, S., Culotta, A.: Identifying leading indicators of product recalls from online reviews using positive unlabeled learning and domain adaptation. In: Proceedings of the International AAAI Conference on Web and Social Media, vol. 11 (2017)
5. Bingel, J., Søgaard, A.: Identifying beneficial task relations for multi-task learning in deep neural networks. In: Lapata, M., Blunsom, P., Koller, A. (eds.) Proceedings of the 15th Conference of the European Chapter of the Association for Computational Linguistics, EACL 2017, Valencia, Spain, 3–7 April 2017, Volume 2: Short Papers, pp. 164–169. Association for Computational Linguistics (2017). https://doi.org/10.18653/v1/e17-2026
6. Brown, P., Levinson, S.C., Levinson, S.C.: Politeness: Some Universals in Language Usage, vol. 4. Cambridge University Press, Cambridge (1987)

7. Caruana, R.: Multitask learning. Mach. Learn. **28**(1), 41–75 (1997)
8. Caruana, R., De Sa, V.R.: Promoting poor features to supervisors: some inputs work better as outputs. In: Advances in Neural Information Processing Systems, pp. 389–395 (1997)
9. Cho, K., van Merrienboer, B., Bahdanau, D., Bengio, Y.: On the properties of neural machine translation: encoder-decoder approaches. In: Wu, D., Carpuat, M., Carreras, X., Vecchi, E.M. (eds.) Proceedings of SSST@EMNLP 2014, Eighth Workshop on Syntax, Semantics and Structure in Statistical Translation, Doha, Qatar, 25 October 2014, pp. 103–111. Association for Computational Linguistics (2014). https://doi.org/10.3115/v1/W14-4012. https://www.aclweb.org/anthology/W14-4012/
10. Chollet, F., et al.: keras (2015)
11. Devlin, J., Chang, M., Lee, K., Toutanova, K.: BERT: pre-training of deep bidirectional transformers for language understanding. In: Burstein, J., Doran, C., Solorio, T. (eds.) Proceedings of the 2019 Conference of the North American Chapter of the Association for Computational Linguistics: Human Language Technologies, NAACL-HLT 2019, Minneapolis, MN, USA, 2–7 June 2019, Volume 1 (Long and Short Papers), pp. 4171–4186. Association for Computational Linguistics (2019). https://doi.org/10.18653/v1/n19-1423. https://doi.org/10.18653/v1/n19-1423
12. Liu, B., Zhang, L.: A survey of opinion mining and sentiment analysis. In: Aggarwal, C., Zhai, C. (eds.) Mining text data, pp. 415–463. Springer, Boston (2012). https://doi.org/10.1007/978-1-4614-3223-4_13
13. Liu, L., et al.: On the variance of the adaptive learning rate and beyond. In: 8th International Conference on Learning Representations, ICLR 2020, Addis Ababa, Ethiopia, 26–30 April 2020. OpenReview.net (2020). https://openreview.net/forum?id=rkgz2aEKDr
14. Olshtain, E., Weinbach, L.: 10. complaints: a study of speech act behavior among native and non-native speakers of Hebrew. In: The pragmatic perspective, p. 195. John Benjamins (1987)
15. Pedregosa, F., et al.: Scikit-learn: machine learning in python. J. Mach. Learn. Res. **12**, 2825–2830 (2011)
16. Pennington, J., Socher, R., Manning, C.: Glove: global vectors for word representation. In: Proceedings of the 2014 Conference on Empirical Methods in Natural Language Processing (EMNLP), pp. 1532–1543 (2014)
17. Preotiuc-Pietro, D., Gaman, M., Aletras, N.: Automatically identifying complaints in social media. In: Korhonen, A., Traum, D.R., Màrquez, L. (eds.) Proceedings of the 57th Conference of the Association for Computational Linguistics, ACL 2019, Florence, Italy, 28 July–2 August 2019, Volume 1: Long Papers, pp. 5008–5019. Association for Computational Linguistics (2019). https://doi.org/10.18653/v1/p19-1495. https://doi.org/10.18653/v1/p19-1495
18. Pryzant, R., Shen, K., Jurafsky, D., Wagner, S.: Deconfounded lexicon induction for interpretable social science. In: Proceedings of the 2018 Conference of the North American Chapter of the Association for Computational Linguistics: Human Language Technologies, Volume 1 (Long Papers), pp. 1615–1625 (2018)
19. Qureshi, S.A., Dias, G., Hasanuzzaman, M., Saha, S.: Improving depression level estimation by concurrently learning emotion intensity. IEEE Comput. Intell. Mag. **15**(3), 47–59 (2020)
20. Saha, T., Patra, A., Saha, S., Bhattacharyya, P.: Towards emotion-aided multimodal dialogue act classification. In: Proceedings of the 58th Annual Meeting of the Association for Computational Linguistics, pp. 4361–4372 (2020)

21. Sailunaz, K., Alhajj, R.: Emotion and sentiment analysis from twitter text. J. Comput. Sci. **36**, 101003 (2019)
22. Srivastava, N., Hinton, G., Krizhevsky, A., Sutskever, I., Salakhutdinov, R.: Dropout: a simple way to prevent neural networks from overfitting. J. Mach. Learn. Res. **15**(1), 1929–1958 (2014)
23. Vásquez, C.: Complaints online: the case of tripadvisor. J. Pragmat. **43**(6), 1707–1717 (2011)
24. Vaswani, A., et al.: Attention is all you need. arXiv preprint arXiv:1706.03762 (2017)
25. Wang, S., Wu, B., Wang, B., Tong, X.: Complaint classification using hybrid-attention GRU neural network. In: Yang, Q., Zhou, Z.-H., Gong, Z., Zhang, M.-L., Huang, S.-J. (eds.) PAKDD 2019. LNCS (LNAI), vol. 11439, pp. 251–262. Springer, Cham (2019). https://doi.org/10.1007/978-3-030-16148-4_20
26. Yang, W., et al.: Detecting customer complaint escalation with recurrent neural networks and manually-engineered features. In: Proceedings of the 2019 Conference of the North American Chapter of the Association for Computational Linguistics: Human Language Technologies, Volume 2 (Industry Papers), pp. 56–63 (2019)
27. Yi, R., Hu, W.: Pre-trained BERT-GRU model for relation extraction. In: Proceedings of the 2019 8th International Conference on Computing and Pattern Recognition, pp. 453–457 (2019)

Going Full-TILT Boogie on Document Understanding with Text-Image-Layout Transformer

Rafał Powalski[1], Łukasz Borchmann[1,2(✉)], Dawid Jurkiewicz[1,3],
Tomasz Dwojak[1,3], Michał Pietruszka[1,4], and Gabriela Pałka[1,3]

[1] Applica.ai, Warsaw, Poland
[2] Poznan University of Technology, Poznań, Poland
{rafal.powalski,lukasz.borchmann,dawid.jurkiewicz,tomasz.dwojak,
michal.pietruszka,gabriela.palka}@applica.ai
[3] Adam Mickiewicz University, Poznań, Poland
[4] Jagiellonian University, Cracow, Poland

Abstract. We address the challenging problem of Natural Language Comprehension beyond plain-text documents by introducing the TILT neural network architecture which simultaneously learns layout information, visual features, and textual semantics. Contrary to previous approaches, we rely on a decoder capable of unifying a variety of problems involving natural language. The layout is represented as an attention bias and complemented with contextualized visual information, while the core of our model is a pretrained encoder-decoder Transformer. Our novel approach achieves state-of-the-art results in extracting information from documents and answering questions which demand layout understanding (DocVQA, CORD, SROIE). At the same time, we simplify the process by employing an end-to-end model.

Keywords: Natural Language Processing · Transfer learning · Document understanding · Layout analysis · Deep learning · Transformer

1 Introduction

Most tasks in Natural Language Processing (NLP) can be unified under one framework by casting them as triplets of the question, context, and answer [26, 29,39]. We consider such unification of Document Classification, Key Information Extraction, and Question Answering in a demanding scenario where context extends beyond the text layer. This challenge is prevalent in business cases since contracts, forms, applications, and invoices cover a wide selection of document types and complex spatial layouts.

R. Powalski, L. Borchmann, D. Jurkiewicz, T. Dwojak and M. Pietruszk—Contributed equally.

J. Lladós et al. (Eds.): ICDAR 2021, LNCS 12822, pp. 732–747, 2021.
https://doi.org/10.1007/978-3-030-86331-9_47

Importance of Spatio-visual Relations. The most remarkable successes achieved in NLP involved models that map raw textual input into raw textual output, which usually were provided in a digital form. An important aspect of real-world oriented problems is the presence of scanned paper records and other analog materials that became digital.

Consequently, there is no easily accessible information regarding the document layout or reading order, and these are to be determined as part of the process. Furthermore, interpretation of shapes and charts beyond the layout may help answer the stated questions. A system cannot rely solely on text but requires incorporating information from the structure and image.

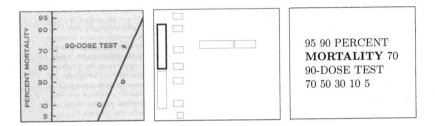

Fig. 1. The same document perceived differently depending on modalities. Respectively: its visual aspect, spatial relationships between the bounding boxes of detected words, and unstructured text returned by OCR under the detected reading order.

Thus, it takes three to solve this fundamental challenge—the extraction of key information from richly formatted documents lies precisely at the intersection of NLP, Computer Vision, and Layout Analysis (Fig. 1). These challenges impose extra conditions beyond NLP that we sidestep by formulating layout-aware models within an encoder-decoder framework.

Limitations of Sequence Labeling. Sequence labeling models can be trained in all cases where the token-level annotation is available or can be easily obtained. Limitations of this approach are strikingly visible on tasks framed in either key information extraction or property extraction paradigms [9,19]. Here, no annotated spans are available, and only property-value pairs are assigned to the document. Occasionally, it is expected from the model to mark some particular subsequence of the document. However, problems where the expected value is not a substring of the considered text are unsolvable assuming sequence labeling methods.[1] As a result, authors applying state-of-the-art entity recognition models were forced to rely on human-made heuristics and time-consuming rule engineering.

Take, for example, the total amount assigned to a receipt in the SROIE dataset [19]. Suppose there is no exact match for the expected value in the

---

[1] Expected values have always an exact match in CoNLL, but not elsewhere, e.g., it is the case for 20% WikiReading, 27% Kleister, and 93% of SROIE values.

document, e.g., due to an OCR error, incorrect reading order or the use of a different decimal separator. Unfortunately, a sequence labeling model cannot be applied off-the-shelf. Authors dealing with property extraction rely on either manual annotation or the heuristic-based tagging procedure that impacts the overall end-to-end results [11,18,36,52,55,56]. Moreover, when receipts with one item listed are considered, the total amount is equal to a single item price, which is the source of yet another problem. Precisely, if there are multiple matches for the value in the document, it is ambiguous whether to tag all of them, part or none.

Another problem one has to solve is which and how many of the detected entities to return, and whether to normalize the output somehow. Consequently, the authors of Kleister proposed a set of handcrafted rules for the final selection of the entity values [52]. These and similar rules are either labor-intensive or prone to errors [40].

Finally, the property extraction paradigm does not assume the requested value appeared in the article in any form since it is sufficient for it to be inferable from the content, as in document classification or non-extractive question answering [9].

Resorting to Encoder-Decoder Models. Since sequence labeling-based extraction is disconnected from the final purpose the detected information is used for, a typical real-world scenario demands the setting of Key Information Extraction.

To address this issue, we focus on the applicability of the encoder-decoder architecture since it can generate values not included in the input text explicitly [16] and performs reasonably well on all text-based problems involving natural language [44]. Additionally, it eliminates the limitation prevalent in sequence labeling, where the model output is restricted by the detected word order, previously addressed by complex architectural changes (Sect. 2).

Furthermore, this approach potentially solves all identified problems of sequence labeling architectures and ties various tasks, such as Question Answering or Text Classification, into the same framework. For example, the model may deduce to answer *yes* or *no* depending on the question form only. Its end-to-end elegance and ease of use allows one to not rely on human-made heuristics and to get rid of time-consuming rule engineering required in the sequence labeling paradigm.

Obviously, employing a decoder instead of a classification head comes with some known drawbacks related to the autoregressive nature of answer generation. This is currently investigated, e.g., in the Neural Machine Translation context, and can be alleviated by methods such as lowering the depth of the decoder [24,47]. However, the datasets we consider have target sequences of low length; thus, the mentioned decoding overhead is mitigated.

The specific contribution of this work can be better understood in the context of related works (Fig. 2).

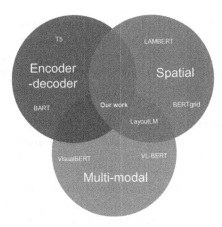

Fig. 2. Our work in relation to encoder-decoder models, multi-modal transformers, and models for text that are able to comprehend spatial relationships between words.

2   Related Works

We aim to bridge several fields, with each of them having long-lasting research programs; thus, there is a large and varied body of related works. We restrict ourselves to approaches rooted in the architecture of Transformer [54] and focus on the inclusion of spatial information or different modalities in text-processing systems, as well as on the applicability of encoder-decoder models to Information Extraction and Question Answering.

Spatial-Aware Transformers. Several authors have shown that, when tasks involving 2D documents are considered, sequential models can be outperformed by considering layout information either directly as positional embeddings [11,17,56] or indirectly by allowing them to be contextualized on their spatial neighborhood [6,15,57]. Further improvements focused on the training and inference aspects by the inclusion of the area masking loss function or achieving independence from sequential order in decoding respectively [18,20]. In contrast to the mentioned methods, we rely on a bias added to self-attention instead of positional embeddings and propose its generalization to distances on the 2D plane. Additionally, we introduce a novel word-centric masking method concerning both images and text. Moreover, by resorting to an encoder-decoder, the independence from sequential order in decoding is granted without dedicated architectural changes.

Encoder-Decoder for IE and QA. Most NLP tasks can be unified under one framework by casting them as Language Modeling, Sequence Labeling or Question Answering [25,43]. The QA program of unifying NLP frames all the problems as triplets of question, context and answer [26,29,39] or item, property name and answer [16]. Although this does not necessarily lead to the use of encoder-decoder models, several successful solutions relied on variants of Transformer architecture [9,34,44,54]. The T5 is a prominent example of large-scale Transformers achieving state-of-the-art results on varied NLP benchmarks [44]. We extend this approach beyond the text-to-text scenario by making it possible to consume a multimodal input.

Multimodal Transformers. The relationships between text and other media have been previously studied in Visual Commonsense Reasoning, Video-Grounded Dialogue, Speech, and Visual Question Answering [3,13,32]. In the context of images, this niche was previously approached with an image-to-text cross-attention mechanism, alternatively, by adding visual features to word embeddings or concatenating them [33,35,37,53,56]. We differ from the mentioned approaches, as in our model, visual features added to word embeddings are already contextualized on an image's multiple resolution levels (see Sect. 3).

3   Model Architecture

Our starting point is the architecture of the Transformer, initially proposed for Neural Machine Translation, which has proven to be a solid baseline for all generative tasks involving natural language [54].

Let us begin from the general view on attention in the first layer of the Transformer. If n denotes the number of input tokens, resulting in a matrix of embeddings X, then self-attention can be seen as:

$$\text{softmax}\left(\frac{Q_X K_X^\top}{\sqrt{n}} + B\right) V_X \tag{1}$$

where Q_X, K_X and V_X are projections of X onto query, keys, and value spaces, whereas B stands for an optional attention bias. There is no B term in the original Transformer, and information about the order of tokens is provided explicitly to the model, that is:

$$X = S + P \qquad B = 0_{n \times d}$$

where S and P are respectively the semantic embeddings of tokens and positional embedding resulting from their positions [54]. $0_{n \times d}$ denote a zero matrix.

In contrast to the original formulation, we rely on relative attention biases instead of positional embeddings. These are further extended to take into account spatial relationships between tokens (Fig. 3).

Spatial Bias. Authors of the T5 architecture disregarded positional embeddings [44], by setting $X = S$. They used relative bias by extending self-attention's equation with the sequential bias term $B = B^{1D}$, a simplified form of positional signal inclusion. Here, each logit used for computing the attention head weights has some learned scalar added, resulting from corresponding token-to-token offsets.

We extended this approach to spatial dimensions. In our approach, biases for relative horizontal and vertical distances between each pair of tokens are calculated and added to the original sequential bias, i.e.:

$$B = B^{1D} + B^H + B^V$$

Such bias falls into one of 32 buckets, which group similarly-distanced token-pairs. The size of the buckets grows logarithmically so that greater token pair distances are grouped into larger buckets (Fig. 4).

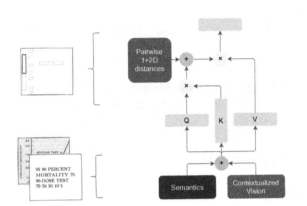

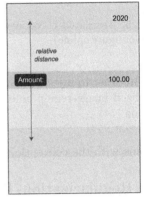

Fig. 3. T5 introduces sequential bias, separating semantics from sequential distances. We maintain this clear distinction, extending biases with spatial relationships and providing additional *image semantics* at the input.

Fig. 4. Document excerpt with distinguished vertical buckets for the *Amount* token.

Contextualized Image Embeddings. Contextualized *Word* Embeddings are expected to capture context-dependent semantics and return a sequence of vectors associated with an entire input sequence [10]. We designed Contextualized *Image* Embeddings with the same objective, i.e., they cover the image region semantics in the context of its entire visual neighborhood.

To produce image embeddings, we use a convolutional network that consumes the whole page image of size 512×384 and produces a feature map of $64 \times 48 \times 128$. We rely on U-Net as a backbone visual encoder network [48] since this architecture provides access to not only the information in the near neighborhood of the token, such as font and style but also to more distant regions of the page, which is useful in cases where the text is related to other structures, i.e., is the description of a picture. This multi-scale property emerges from the skip connections within chosen architecture (Fig. 5). Then, each token's bounding box is used to extract features from U-Net's feature map with ROI pooling [5]. The obtained vector is then fed into a linear layer which projects it to the model embedding dimension.

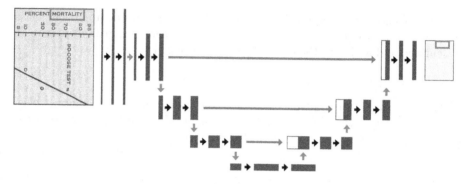

Fig. 5. Truncated U-Net network. ■ conv ■ max-pool ■ up-conv ■ residual (Color figure online)

In order to inject visual information to the Transformer, a matrix of contextualized image-region embeddings U is added to semantic embeddings, i.e. we define

$$X = S + U$$

in line with the convention from Sect. 3 (see Fig. 3).

4 Regularization Techniques

In the sequence labeling scenario, each document leads to multiple training instances (token classification), whereas in Transformer sequence-to-sequence models, the same document results in one training instance with feature space of higher dimension (decoding from multiple tokens).

Since most of the tokens are irrelevant in the case of Key Information Extraction and contextualized word embeddings are correlated by design, one can suspect our approach to overfit easier than its sequence labeling counterparts. To improve the model's robustness, we introduced a regularization technique for each modality.

Case Augmentation. Subword tokenization [28,49] was proposed to solve the word sparsity problem and keep the vocabulary at a reasonable size. Although the algorithm proved its efficiency in many NLP fields, the recent work showed that it performs poorly in the case of an unusual casing of text [42], for instance, when all words are uppercased. The problem occurs more frequently in formated documents (FUNSD, CORD, DocVQA), where the casing is an important visual aspect. We overcome both problems with a straightforward regularization strategy, i.e., produce augmented copies of data instances by lower-casing or upper-casing both the document and target text simultaneously.

Spatial Bias Augmentation. Analogously to Computer Vision practices of randomly transforming training images, we augment spatial biases by multiplying the horizontal and vertical distances between tokens by a random factor.

Such transformation resembles stretching or squeezing document pages in horizontal and vertical dimensions. Factors used for scaling each dimension were sampled uniformly from range $[0.8, 1.25]$.

Affine Vision Augmentation. To account for visual deformations of real-world documents, we augment images with affine transformation, preserving parallel lines within an image but modifying its position, angle, size, and shear. When we perform such modification to the image, the bounding box of every token is updated accordingly. The exact hyperparameters were subject to an optimization. We use 0.9 probability of augmenting and report the following boundaries for uniform sampling work best: $[-5, 5]$ degrees for rotation angle, $[-5\%, 5\%]$ for translation amplitude, $[0.9, 1.1]$ for scaling multiplier, $[-5, 5]$ degrees for the shearing angle.

5   Experiments

Our model was validated on series of experiments involving Key Information Extraction, Visual Question Answering, classification of rich documents, and Question Answering from layout-rich texts. The following datasets represented the broad spectrum of tasks and were selected for the evaluation process (see Table 1 for additional statistics).

The CORD dataset [41] includes images of Indonesian receipts collected from shops and restaurants. The dataset is prepared for the information extraction task and consists of four categories, which fall into thirty subclasses. The main goal of the SROIE dataset [19] is to extract values for four categories (company, date, address, total) from scanned receipts. The DocVQA dataset [38] is focused on the visual question answering task. The RVL-CDIP dataset [14] contains gray-scale images and assumes classification into 16 categories such as letter, form, invoice, news article, and scientific publication. For DocVQA, we relied on Amazon Textract OCR; for RVL-CDIP, we used Microsoft Azure OCR, for SROIE and CORD, we depended on the original OCR.

5.1   Training Procedure

The training procedure consists of three steps. First, the model is initialized with vanilla T5 model weights and is pretrained on numerous documents in an unsupervised manner. It is followed by training on a set of selected supervised tasks. Finally, the model is finetuned solely on the dataset of interest. We trained two size variants of TILT models, starting from T5-Base and T5-Large models. Our models grew to 230M and 780M parameters due to the addition of Visual Encoder weights.

Unsupervised Pretraining. We constructed a corpus of documents with rich structure, based on RVL-CDIP (275k docs), UCSF Industry Documents Library

Table 1. Comparison of datasets considered for supervised pretraining and evaluation process. Statistics given in thousands of documents or questions.

Dataset	Data type	Image	Docs (k)	Questions (k)
CORD [41]	Receipts	+	1.0	—
SROIE [19]	Receipts	+	0.9	—
DocVQA [38]	Industry documents	+	12.7	50.0
RVL-CDIP [14]	Industry documents	+	400.0	—
DROP [8]	Wikipedia pages	−	6.7	96.5
QuAC [2]	Wikipedia pages	−	13.6	98.4
SQuAD 1.1 [45]	Wikipedia pages	−	23.2	107.8
TyDi QA [4]	Wikipedia pages	−	204.3	204.3
Natural Questions [30]	Wikipedia pages	−	91.2	111.2
WikiOps [1]	Wikipedia tables	−	24.2	80.7
CoQA [46]	Various sources	−	8.4	127.0
RACE [31]	English exams	−	27.9	97.7
QASC [27]	School-level science	−	—	10.0
FUNSD [21]	RVL-CDIP forms	+	0.1	—
Infographics VQA	infographics	+	4.4	23.9
TextCaps [50]	Open Images	+	28.4	—
DVQA [22]	Synthetic bar charts	+	300.0	3487.2
FigureQA [23]	Synthetic, scientific	+	140.0	1800.0
TextVQA [51]	Open Images	+	28.4	45.3

(480k),[2] and PDF files from Common Crawl (350k). The latter were filtered according to the score obtained from a simple SVM business document classifier.

Then, a T5-like masked language model pretraining objective is used, but in a salient span masking scheme, i.e., named entities are preferred rather than random tokens [12,44]. Additionally, regions in the image corresponding to the randomly selected text tokens are masked with the probability of 80%. Models are trained for $100,000$ steps with batch size of 64, AdamW optimizer and linear scheduler with an initial learning rate of $2e-4$.

Supervised Training. To obtain a general-purpose model which can reason about documents with rich layout features, we constructed a dataset relying on a large group of tasks, representing diverse types of information conveyed by a document (see Table 1 for datasets comparison). Datasets, which initially had been plain-text, had their layout produced, assuming some arbitrary font size and document dimensions. Some datasets, such as *WikiTable Questions*, come

---

[2] http://www.industrydocuments.ucsf.edu/.

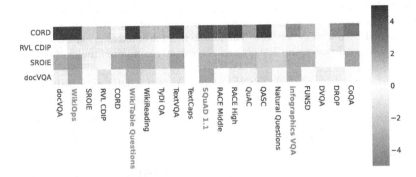

Fig. 6. Scores on CORD, DocVQA, SROIE and RVL-CDIP compared to the baseline without supervised pretraining. The numbers represent the differences in the metrics, orange text denote datasets chosen for the final supervised pretraining run.

with original HTML code – for the others, we render text alike. Finally, an image and computed bounding boxes of all words are used.

At this stage, the model is trained on each dataset for 10,000 steps or 5 epochs, depending on the dataset size: the goal of the latter condition was to avoid a quick overfitting.

We estimated each dataset's value concerning a downstream task, assuming a fixed number of pretraining steps followed by finetuning. The results of this investigation are demonstrated in Fig. 6, where the group of WikiTable, WikiOps, SQuAD, and infographicsVQA performed robustly, convincing us to rely on them as a solid foundation for further experiments.

Model pretrained in unsupervised, and then supervised manner, is at the end finetuned for two epochs on a downstream task with AdamW optimizer and hyperparameters presented in Table 2.

Table 2. Parameters used during the finetuning on a downstream task. Batch size, learning rate and scheduler were subject of hyperparameter search with considered values of respectively $\{8, 16, ..., 2048\}$, $\{5e - 5, 2e - 5, 1e - 5, 5e - 4, ..., 1e - 3\}$, $\{constant, linear\}$. We have noticed that the classification task of RVL-CDIP requires a significantly larger bath size. The model with the highest validation score within the specified steps number limit was used.

Dataset	Batch size	Steps	Learning rate	Scheduler
SROIE	8	6,200	1e−4	Constant
DocVQA	64	100,000	2e−4	Linear
CORD	8	36,000	2e−4	Linear
RVL-CDIP	1,024	12,000	1e−3	Linear

Table 3. Results of selected methods in relation to our base and large models. Bold indicates the best score in each category. All results on the test set, using the metrics proposed by dataset's authors. The number of parameters given for completeness thought encoder-decoder and LMs cannot be directly compared under this criterion.

Model	CORD F1	SROIE F1	DocVQA ANLS	RVL-CDIP Accuracy	Size variant (Parameters)
LayoutLM [56]	94.72	94.38	69.79	94.42	Base (113–160M)
	94.93	95.24	72.59	94.43	Large (343M)
LayoutLMv2 [55]	94.95	96.25	78.08	95.25	Base (200M)
	96.01	97.81	86.72	**95.64**	Large (426M)
LAMBERT [11]	96.06	**98.17**	—	—	Base (125M)
TILT (our)	95.11	97.65	83.92	95.25	Base (230M)
	96.33	**98.10**	**87.05**	95.52	Large (780M)

5.2   Results

The TILT model achieved state-of-the-art results on three out of four considered tasks (Table 3). We have confirmed that unsupervised layout- and vision-aware pretraining leads to good performance on downstream tasks that require comprehension of tables and other structures within the documents. Additionally, we successfully leveraged supervised training from both plain-text datasets and these involving layout information.

DocVQA. We improved SOTA results on this dataset by 0.33 points. Moreover, detailed results show that model gained the most in table-like categories, i.e., forms (89.5 → 94.6) and tables (87.7 → 89.8), which proved its ability to understand the spatial structure of the document. Besides, we see a vast improvement in the yes/no category (55.2 → 69.0).[3] In such a case, our architecture generates simply *yes* or *no* answer, while sequence labeling based models require additional components such as an extra classification head. We noticed that model achieved lower results in the image/photo category, which can be explained by the low presence of image-rich documents in our datasets.

RVL-CDIP. Part of the documents to classify does not contain any readable text. Because of this shortcoming, we decided to guarantee there are at least 16 image tokens that would carry general image information. Precisely, we act as there were tokens with bounding boxes covering 16 adjacent parts of the document. These have representations from U-Net, exactly as they were regular text tokens. Our model places second, 0.12 below the best model, achieving the similar accuracy of 95.52.

---

[3] Per-category test set scores are available after submission on the competition web page: https://rrc.cvc.uab.es/?ch=17&com=evaluation&task=1.

Table 4. Results of ablation study. The minus sign indicates removal of the mentioned part from the base model.

Model	Score	Relative change
TILT-Base	82.9 ± 0.3	—
– Spatial Bias	81.1 ± 0.2	-1.8
– Visual Embeddings	81.2 ± 0.3	-1.7
– Case Augmentation	82.2 ± 0.3	-0.7
– Spatial Augmentation	82.6 ± 0.4	-0.3
– Vision Augmentation	82.8 ± 0.2	-0.1
– Supervised Pretraining	81.2 ± 0.1	-1.7

CORD. Since the complete inventory of entities is not present in all examples, we force the model to generate a *None* output for missing entities. Our model achieved SOTA results on this challenge and improved the previous best score by 0.3 points. Moreover, after the manual review of the model errors, we noticed that model's score could be higher since the model output and the reference differ insignificantly e.g. "2.00 ITEMS" and "2.00".

SROIE. We excluded OCR mismatches and fixed total entity annotations discrepancies following the same evaluation procedure as Garncarek et al. [11].[4] We achieved results indistinguishable from the SOTA (98.10 vs. 98.17). Significantly better results are impossible due to OCR mismatches in the test-set.

Though we report the number of parameters near the name of the model size variant, note it is impossible to compare the TILT encoder-decoder model to language models such as LayoutLMs and LAMBERT under this criterion. In particular, it does not reflect computational cost, which may be similar for encoder-decoders twice as big as some language model [44, Section 3.2.2]. Nevertheless, it is worth noting that our Base model outperformed models with comparable parameter count.

6   Ablation Study

In the following section, we analyze the design choices in our architecture, considering the base model pretrained in an unsupervised manner and the same hyperparameters for each run. The DocVQA was used as the most representative and challenging for Document Intelligence since its leaderboard reveals a large gap to human performance. We report average results over two runs of each model varying only in the initial random seed to account for the impact of different initialization and data order [7].

---

[4] Corrections can be obtained by comparing their two public submissions.

Significance of Modalities. We start with the removal of the 2D layout positional bias. Table 4 demonstrates that information that allows models to recognize spatial relations between tokens is a crucial part of our architecture. It is consistent with the previous works on layout understanding [11,55]. Removal of the UNet-based convolutional feature extractor results in a less significant ANLS decrease than the 2D bias. This permits the conclusion that contextualized image embeddings are beneficial to the encoder-decoder.

Justifying Regularization. Aside from removing modalities from the network, we can also exclude regularization techniques. To our surprise, the results suggest that the removal of case augmentation decreases performance most severely. Our baseline is almost one point better than the equivalent non-augmented model. Simultaneously, model performance tends to be reasonably insensitive to the bounding boxes' and image alterations. It was confirmed that other modalities are essential for the model's success on real-world data, whereas regularization techniques we propose slightly improve the results, as they prevent overfitting.

Impact of Pretraining. As we exploited supervised pretraining similarly to previous authors, it is worth considering its impact on the overall score. In our ablation study, the model pretreated in an unsupervised manner achieved significantly lower scores. The impact of this change is comparable to the removal of spatial bias or visual embeddings. Since authors of the T5 argued that pretraining on a mixture of unsupervised and supervised tasks perform equally good with higher parameter count, this gap may vanish with larger variants of TILT we did not consider in the present paper [44].

7   Summary

In the present paper, we introduced a novel encoder-decoder framework for layout-aware models. Compared to the sequence labeling approach, the proposed method achieves better results while operating in an end-to-end manner. It can handle various tasks such as Key Information Extraction, Question Answering or Document Classification, while the need for complicated preprocessing and postprocessing steps is eliminated.

Although encoder-decoder models are commonly applied to generative tasks, both DocVQA, SROIE, and CORD we considered are extractive. We argue that better results were achieved partially due to the independence from the detected word order and resistance to OCR errors that the proposed architecture possesses. Consequently, we were able to achieve state-of-the-art results on two datasets (DocVQA, CORD) and performed on par with the previous best scores on SROIE and RVL-CDIP, albeit having a much simpler workflow.

Spatial and image enrichment of the Transformer model allowed the TILT to combine information from text, layout, and image modalities. We showed that the proposed regularization methods significantly improve the results.

Acknowledgments. The authors would like to thank Filip Graliński, Tomasz Stanisławek, and Łukasz Garncarek for fruitful discussions regarding the paper and our managing directors at Applica.ai. Moreover, Dawid Jurkiewicz pays due thanks to his son for minding the deadline and generously coming into the world a day after.

The Smart Growth Operational Programme supported this research under project no. POIR.01.01.01-00-0877/19-00 (*A universal platform for robotic automation of processes requiring text comprehension, with a unique level of implementation and service automation*).

References

1. Cho, M., Amplayo, R., Hwang, S.W., Park, J.: Adversarial TableQA: attention supervision for question answering on tables. In: PMLR (2018)
2. Choi, E., et al.: QuAC: question answering in context. In: EMNLP (2018)
3. Chuang, Y., Liu, C., Lee, H., Lee, L.: SpeechBERT: an audio-and-text jointly learned language model for end-to-end spoken question answering. In: ISCA (2020)
4. Clark, J.H., et al.: TyDi QA: a benchmark for information-seeking question answering in typologically diverse languages. TACL **8**, 454–470 (2020)
5. Dai, J., Li, Y., He, K., Sun, J.: R-FCN: object detection via region-based fully convolutional networks. In: NeurIPS (2016)
6. Denk, T.I., Reisswig, C.: BERTgrid: contextualized embedding for 2d document representation and understanding. arXiv preprint (2019)
7. Dodge, J., Ilharco, G., Schwartz, R., Farhadi, A., Hajishirzi, H., Smith, N.A.: Fine-tuning pretrained language models: weight initializations, data orders, and early stopping. arXiv preprint (2020)
8. Dua, D., Wang, Y., Dasigi, P., Stanovsky, G., Singh, S., Gardner, M.: DROP: a reading comprehension benchmark requiring discrete reasoning over paragraphs. In: NAACL-HLT (2019)
9. Dwojak, T., Pietruszka, M., Borchmann, Ł., Chłedowski, J., Graliński, F.: From dataset recycling to multi-property extraction and beyond. In: CoNLL (2020)
10. Ethayarajh, K.: How contextual are contextualized word representations? comparing the geometry of BERT, ELMo, and GPT-2 embeddings. In: EMNLP-IJCNLP (2019)
11. Garncarek, Ł., et al.: LAMBERT: layout-aware (language) modeling using BERT for information extraction. In: Llads, J. et al. (eds.) ICDAR 2021. LNCS, vol. 12822, pp. 532–547 (2021). Accepted to ICDAR 2021
12. Guu, K., Lee, K., Tung, Z., Pasupat, P., Chang, M.: Retrieval augmented language model pre-training. In: ICML (2020)
13. Han, K., et al.: A survey on visual transformer. arXiv preprint (2021)
14. Harley, A.W., Ufkes, A., Derpanis, K.G.: Evaluation of deep convolutional nets for document image classification and retrieval. In: ICDAR (2015)
15. Herzig, J., Nowak, P.K., Müller, T., Piccinno, F., Eisenschlos, J.: TaPas: weakly supervised table parsing via pre-training. In: ACL (2020)
16. Hewlett, D., et al.: WikiReading: a novel large-scale language understanding task over Wikipedia. In: ACL (2016)
17. Ho, J., Kalchbrenner, N., Weissenborn, D., Salimans, T.: Axial attention in multi-dimensional transformers. arXiv preprint (2019)
18. Hong, T., Kim, D., Ji, M., Hwang, W., Nam, D., Park, S.: BROS: a pre-trained language model for understanding texts in document. openreview.net preprint (2021)

19. Huang, Z., et al.: ICDAR2019 competition on scanned receipt OCR and information extraction. In: ICDAR (2019)
20. Hwang, W., Yim, J., Park, S., Yang, S., Seo, M.: Spatial dependency parsing for semi-structured document information extraction. arXiv preprint (2020)
21. Jaume, G., Ekenel, H.K., Thiran, J.P.: FUNSD: a dataset for form understanding in noisy scanned documents. In: ICDAR-OST (2019)
22. Kafle, K., Price, B.L., Cohen, S., Kanan, C.: DVQA: understanding data visualizations via question answering. In: CVPR (2018)
23. Kahou, S.E., Michalski, V., Atkinson, A., Kádár, Á., Trischler, A., Bengio, Y.: FigureQA: an annotated figure dataset for visual reasoning. In: ICLR (2018)
24. Kasai, J., Pappas, N., Peng, H., Cross, J., Smith, N.A.: Deep encoder, shallow decoder: reevaluating the speed-quality tradeoff in machine translation. arXiv preprint (2020)
25. Keskar, N., McCann, B., Xiong, C., Socher, R.: Unifying question answering and text classification via span extraction. arXiv preprint (2019)
26. Khashabi, D., et al.: UnifiedQA: crossing format boundaries with a single QA system. In: EMNLP-Findings (2020)
27. Khot, T., Clark, P., Guerquin, M., Jansen, P., Sabharwal, A.: QASC: a dataset for question answering via sentence composition. In: AAAI (2020)
28. Kudo, T.: Subword regularization: improving neural network translation models with multiple subword candidates. In: ACL (2018)
29. Kumar, A., et al.: Ask me anything: dynamic memory networks for natural language processing. In: ICML (2016)
30. Kwiatkowski, T., et al.: Natural questions: a benchmark for question answering research. TACL **7**, 453–466 (2019)
31. Lai, G., Xie, Q., Liu, H., Yang, Y., Hovy, E.: RACE: large-scale ReAding comprehension dataset from examinations. In: EMNLP (2017)
32. Le, H., Sahoo, D., Chen, N., Hoi, S.: Multimodal transformer networks for end-to-end video-grounded dialogue systems. In: ACL (2019)
33. Lee, K.-H., Chen, X., Hua, G., Hu, H., He, X.: Stacked cross attention for image-text matching. In: Ferrari, V., Hebert, M., Sminchisescu, C., Weiss, Y. (eds.) ECCV 2018. LNCS, vol. 11208, pp. 212–228. Springer, Cham (2018). https://doi.org/10.1007/978-3-030-01225-0_13
34. Lewis, M., et al.: BART: denoising sequence-to-sequence pre-training for natural language generation, translation, and comprehension. In: ACL (2020)
35. Li, L.H., Yatskar, M., Yin, D., Hsieh, C.J., Chang, K.W.: VisualBERT: a simple and performant baseline for vision and language. arXiv preprint (2019)
36. Liu, X., Gao, F., Zhang, Q., Zhao, H.: Graph convolution for multimodal information extraction from visually rich documents. In: NAACL-HLT (2019)
37. Ma, J., Qin, S., Su, L., Li, X., Xiao, L.: Fusion of image-text attention for transformer-based multimodal machine translation. In: IALP (2019)
38. Mathew, M., Karatzas, D., Jawahar, C.: DocVQA: a dataset for VQA on document images. In: WACV, pp. 2200–2209 (2021)
39. McCann, B., Keskar, N.S., Xiong, C., Socher, R.: The natural language decathlon: multitask learning as question answering. arXiv preprint (2018)
40. Palm, R.B., Winther, O., Laws, F.: CloudScan - a configuration-free invoice analysis system using recurrent neural networks. In: ICDAR (2017)
41. Park, S., et al.: CORD: a consolidated receipt dataset for post-OCR parsing. In: Document Intelligence Workshop at NeurIPS (2019)
42. Powalski, R., Stanislawek, T.: UniCase - rethinking casing in language models. arXiv prepint (2020)

43. Radford, A., Wu, J., Child, R., Luan, D., Amodei, D., Sutskever, I.: Language models are unsupervised multitask learners. Technical report (2019)
44. Raffel, C., et al.: Exploring the limits of transfer learning with a unified text-to-text transformer. JMRL (2020)
45. Rajpurkar, P., Zhang, J., Lopyrev, K., Liang, P.: SQuAD: 100,000+ questions for machine comprehension of text. In: EMNLP (2016)
46. Reddy, S., Chen, D., Manning, C.D.: CoQA: a conversational question answering challenge. TACL **7**, 249–266 (2019)
47. Ren, Y., Liu, J., Tan, X., Zhao, Z., Zhao, S., Liu, T.Y.: A study of non-autoregressive model for sequence generation. In: ACL (2020)
48. Ronneberger, O., Fischer, P., Brox, T.: U-Net: convolutional networks for biomedical image segmentation. In: Navab, N., Hornegger, J., Wells, W.M., Frangi, A.F. (eds.) MICCAI 2015. LNCS, vol. 9351, pp. 234–241. Springer, Cham (2015). https://doi.org/10.1007/978-3-319-24574-4_28
49. Sennrich, R., Haddow, B., Birch, A.: Neural machine translation of rare words with subword units. In: ACL (2016)
50. Sidorov, O., Hu, R., Rohrbach, M., Singh, A.: TextCaps: a dataset for image captioning with reading comprehension. In: Vedaldi, A., Bischof, H., Brox, T., Frahm, J.-M. (eds.) ECCV 2020. LNCS, vol. 12347, pp. 742–758. Springer, Cham (2020). https://doi.org/10.1007/978-3-030-58536-5_44
51. Singh, A., et al.: Towards VQA models that can read. In: CVPR (2019)
52. Stanisławek, T., et al.: Kleister: key information extraction datasets involving long documents with complex layouts. In: Llads, J. et al. (eds.) ICDAR 2021. LNCS, vol. 12822, pp. 564–579 (2021). Accepted to ICDAR 2021
53. Su, W., Zhu, X., Cao, Y., Li, B., Lu, L., Wei, F., Dai, J.: VL-BERT: pre-training of generic visual-linguistic representations. In: ICLR (2020)
54. Vaswani, A., et al.: Attention is all you need. In: NeurIPS (2017)
55. Xu, Y., et al.: LayoutLMv2: multi-modal pre-training for visually-rich document understanding. arXiv preprint (2020)
56. Xu, Y., Li, M., Cui, L., Huang, S., Wei, F., Zhou, M.: LayoutLM: pre-training of text and layout for document image understanding. In: KDD (2020)
57. Yin, P., Neubig, G., Yih, W.t., Riedel, S.: TaBERT: pretraining for joint understanding of textual and tabular data. In: ACL (2020)

Data Centric Domain Adaptation
for Historical Text with OCR Errors

Luisa März[1,2,4(✉)] ⓘ, Stefan Schweter[3] ⓘ, Nina Poerner[1] ⓘ, Benjamin Roth[2] ⓘ,
and Hinrich Schütze[1] ⓘ

[1] Center for Information and Language Processing, Ludwig Maximilian University,
Munich, Germany
`maerz@cis.lmu.de, inquiries@cislmu.org`
[2] Digital Philology, Research Group Data Mining and Machine Learning, University
of Vienna, Vienna, Austria
[3] Bayerische Staatsbibliothek München, Digital Library/Munich Digitization Center,
Munich, Germany
[4] NLP Expert Center, Data:Lab, Volkswagen AG, Munich, Germany
`https://www.cis.uni-muenchen.de/`

Abstract. We propose new methods for in-domain and cross-domain
Named Entity Recognition (NER) on historical data for Dutch and
French. For the cross-domain case, we address domain shift by integrating
unsupervised in-domain data via contextualized string embeddings; and
OCR errors by injecting synthetic OCR errors into the source domain
and address data centric domain adaptation. We propose a general approach to imitate OCR errors in arbitrary input data. Our cross-domain
as well as our in-domain results outperform several strong baselines and
establish state-of-the-art results. We publish preprocessed versions of the
French and Dutch Europeana NER corpora.

Keywords: Named Entity Recognition · Historical data · FLAIR

1 Introduction

Neural networks achieve good NER accuracy on high-resource domains such as
modern news text or Twitter [2,4]. But on historical text, NER often performs
poorly. This is due to several challenges: i) Domain shift: Entities in historical
texts can be different from contemporary entities, this makes it difficult for
modern taggers to work with historical data. ii) OCR errors: historical texts –
usually digitized by OCR – contain systematic errors not found in non-OCR text
[14]. In addition these errors can change the surface form of entities. iii) Lack of
annotation: Some historical text is now available in digitized form, but without
labels, and methods are required for beneficial use of such data [16].

In this paper, we address *data centric domain adaptation* for NER tagging
on historical French and Dutch data. Following Ramponi and Plank [20], data
centric approaches do not adapt the model but the training data in order to

© Springer Nature Switzerland AG 2021
J. Lladós et al. (Eds.): ICDAR 2021, LNCS 12822, pp. 748–761, 2021.
https://doi.org/10.1007/978-3-030-86331-9_48

improve generalization across domains. We address both in-domain and cross-domain NER. In the cross-domain setup, we use supervised contemporary data and integrate unsupervised historical data via contextualized embeddings. We introduce artificial OCR errors into supervised modern data and find a way to perturb corpora in a general and robust way – independent of language or linguistic properties.

In the cross-domain setup as well as in-domain, our system outperforms neural and statistical state-of-the-art methods, achieving 69.3% F_1 for French and 63.4% for Dutch. With the in-domain setup, we achieve 77.9% for French and 84.2% for Dutch. If we only consider named entities that contain OCR errors, our domain-adapted cross-domain tagger even performs better (83.5% French/ 46.2% Dutch) than in-domain training (77.1% French/ 43.8% Dutch). Our main contributions are:

- Release of the preprocessed French and Dutch NER corpora[1];
- Developing synOCR to mimic historical data while exploiting the annotation of modern data;
- Training historical embeddings on a large amount of unlabeled historical data;
- Ensembling a NER system that establishes SOTA results for both languages and scenarios.

2 Methods

2.1 Architecture

We use the FLAIR NLP framework [1]. FLAIR taggers achieve SOTA results on various benchmarks and are well suited for NER. Secondly, there are powerful FLAIR embeddings. They are trained without explicit notion of words and model words as character sequences depending on their context. These two properties contribute to making atypical entities - even those with distorted surface - easier to recognize. In all our experiments, word embeddings are generated by a character-level RNN and passed to a word-level bidirectional LSTM with a CRF as the final layer. Depending on the experiment, we concatenate ($\odot$) additional embeddings and refer to that as *ensembling process*.

2.2 Noise Methods

Since digitizing by OCR introduces a lot of noise into the data, we recreate some of those phenomena in the modern corpora that we use for training. Our goal is to increase the similarity of historical (OCR'd) and modern (clean) data. An example drawn from the dutch training corpora can be found in Fig. 1. Words that are different from the original text are indicated in bold font.

Generation of Synthetic OCR (synOCR) Errors. This method processes every sentence by assigning a randomly selected font and a font size between 6

---

[1] https://github.com/stefan-it/historic-domain-adaptation-icdar

and 11 pt. Batches of 150 sentences are printed to PDF documents and then converted to PNG images. The images are perturbed using imgaug[2] with the following steps: (i) rotation, (ii) Gaussian blur and (iii) white or black pixel dropout. The resulting image is recognized using tesseract version 0.2.6.[3]. We re-align the recognized sentences with the clean annotated corpus to transfer the NER tags. For the alignment between original and degraded text we select a window of the bitext and calculate a character-based alignment cost. We then use the Wagner-Fisher algorithm [27] to obtain the best alignment path through the window and the lowest possible cost. If the cost is below a threshold, we shift the window to the mid-point of the discovered path. Otherwise, we iteratively increase the window size and re-align, until the threshold criterion is met. This procedure allows us to find an alignment with reasonable time and space resources, without risking to lose the optimal path in low-quality areas. Finally this results in an OCR-error enhanced annotated corpus with a range of recognition quality, from perfectly recognized to fully illegible. We refer to OCR-corrupted data as *synOCR'd data*.

Generation of Synthetic Corruptions. This method is applied to our modern corpora, again to introduce noise as we find it in historical data. Similar to [21], we randomly corrupt 20% of all words by (i) inserting a character or (ii) removing a character or (iii) transposing two characters. Therefore, we use the standard alphabet of French/Dutch. We re-align the corrupted tokens with the clean annotated tokens while maintaining the sentence boundaries to transfer the NER tags. Since the corruption method does not break the word boundaries we can simply map each corrupted word to the original one and retrieve the corresponding NER tag. We refer to synthetically corrupted data as *corrupted data*.

De tekst van het arrest is nog niet schriftelijk beschikbaar maar het bericht werd alvast   *Original*
bekendgemaakt door een communicatiebureau dat Floralux inhuurde .

De :eks: van het attest is nog **hie: schiiizeujk beschikbaaz** maar **bet helicht weéd alvas:**   *synOCR*
bekendqemaakt door **sen' communicacisbuxeau dac** Floralux inhuurde , ' *———

De tekst van het arrest is nog niet schriftelijk beschikbaar maar **ht bCericht** werd alvast   *corrupted*
bekendgemaakt door **eeRn** communicatiebureau dat Floralux inhuurde .

Fig. 1. Example from the dutch train set. Text in its original, the synOCR'd and the corrupted form.

---

[2] https://github.com/aleju/imgaug
[3] https://github.com/tesseract-ocr/

2.3   Embeddings

We experiment with various common embeddings and integrate them in our neural system. Some of them are available in the community and some others we did train on data described in Sect. 3.1.

Flair Embeddings. [3] present contextual string embeddings which can be extracted from a neural language model. FLAIR embeddings use the internal states of a trained character language model at token boundaries. They are contextualized because a word can have different embeddings depending on its context. These embeddings are also less sensitive to misspellings and rare words and can be learned on unlabeled corpora. We also use *multilingual* FLAIR *embeddings*. They were trained on a mix of corpora from different domains (Web, Wikipedia, Subtitles, News) and languages.

Historical Embeddings. We train FLAIR embeddings on large unlabeled historical corpora from a comparable time period (see Sect. 3.1) and refer to them as *historical embeddings*.

BERT Embeddings. Since *BERT embeddings* [9] produce state-of-the-art results for a wide range of NLP tasks, we also experiment with multilingual BERT embeddings[4]. BERT embeddings are subword embeddings based on a bidirectional transformer architecture and can model the context of a word. For NER on CoNLL-03 [25], BERT embeddings do not perform as well as on other tasks [9] and we want to examine if this observation holds for a cross-domain scenario with different data.

FastText Embeddings. We do also use *FastText embeddings* [6] which are widely used in NLP. They can be efficiently trained and address character-level phenomena. Subwords are used to represent the target word (as a sum of all its subword embeddings). We use pre-trained FastText embeddings for French/Dutch[5].

Character-Level Embeddings. Due to the OCR errors out-of-vocabulary problems occur. Lample et al. [15] create *character embeddings*, passing all characters in a sentence to a bidirectional LSTM. To obtain word representations, the forward and backward representations of all the characters of the word from this LSTM are concatenated. Having the character embedding, every single words vector can be formed even if it is out-of-vocabulary. Therefore, we do also compute these embeddings for our experiments.

3   Experiments

In the cross-domain setup, we train on modern data (clean or synOCR'd) and test on historical data (OCR'd). In the in-domain setup, we train and test on a set of historical data (OCR'd). We do use different combinations of embeddings and also use our noise methods in the experiments.

---

[4] We use the cased variant from https://huggingface.co/bert-base-multilingual-cased
[5] https://fasttext.cc/docs/en/crawl-vectors.html

3.1   Data

We use different data sources for our experiments from which some are openly available and some historical data come from an in-house project. For an overview of different properties (domain, labeling, size, language) see Table 1.

Annotated Historical Data. Our annotated historical data comes from the Europeana Newspapers collection[6], which contains historical news articles in 12 languages published between 1618 and 1990. Parts of the German, Dutch and French data were manually annotated with NER tags in IO/IOB format for PER (person), LOC (location), ORG (organization) by Neudecker [17]. Each NER corpus contains 100 scanned pages (with OCR accuracy over 80%), amounting to 207K tokens for French and 182K tokens for Dutch.

   We preprocess the data as follows. We perform sentence splitting, filter out metadata, re-tokenize punctuation and convert all annotations to IOB1 format. We split the data 80/10/10 into train/dev/test. We will make this preprocessed version available in CoNLL format.

Annotated Modern Data. For the French cross-domain experiments, we use the French WikiNER corpus [18]. WikiNER is tagged in IOB format with an additional MISC (miscellaneous) category; we convert the tags to our Europeana format. For better comparability we downsample (sentence-wise) the corpus from 3.5M to 525K tokens. Therefore, entire sentences were sampled uniformly at random without replacement. For Dutch, we use the CoNLL-02 corpus [24], which consists of four editions of the Belgian Dutch newspaper "De Morgen" from the year 2000. The data comprises 309K tokens and is annotated for PER, ORG, LOC and MISC. We convert the tags to our Europeana format.

Unlabeled Historical Data. For historical French, we use "Le Temps", a journal published between 1861 and 1942 (initially under a different name), a similar time period as the Europeana Newspapers. The corpus contains 977M tokens and is available from the National Library of France.[7] For historical Dutch, we use data from an in-house OCR project. The data is from the 19th century and it consists of 444M tokens. We use the unlabeled historical data to pre-train historical embeddings (see Sect. 2.3).

3.2   Baselines

We experiment with three baselines. (i) The Java implementation[8] of the Stanford NER tagger [12]. (ii) A version of Stanford NER published by Neudecker [17][9] that was trained on Europeana. In contrast to our system they trained

---

[6] http://www.europeana-newspapers.eu/
[7] https://www.bnf.fr/fr
[8] https://nlp.stanford.edu/software/CRF-NER.html
[9] https://lab.kb.nl/dataset/europeana-newspapers-ner

Table 1. Number of tokens per dataset in our experiments.

	Domain	Data	Labeled	Size
French	Historical	Europeana NER	+	207K
	Modern	WikiNER	+	525K
	Historical	"Le Temps"	−	977M
Dutch	Historical	Europeana NER	+	182K
	Modern	CoNLL-02	+	309K
	historical	in-house OCR	−	444M

theirs on the entire amount of the labeled Europeana corpora with 4-fold cross validation. (iii) NN base. The neural network (see Sect. 2.1) with FastText, character and multilingual FLAIR embeddings, as recommended in Akbik et al. [1]. For French, we also list the result reported by Çavdar [8]. Since we do not have access to their implementation and could not confirm that their data splits conform to ours, we could not compute the combined F_1 score or test for significance.

Table 2. Results (F_1 scores on French/Dutch Europeana test set) of training on Europeana French/Dutch training set. *Hist. Embs.* are historical embeddings. Scores marked with * are significantly lower than *NN base $\odot$ hist. Es.*

French models	Overall	PER	ORG	LOC
Çavdar		0.68	0.37	0.68
Stanford NER tagger	0.662*	0.569*	0.335*	0.753*
Stanford Neudecker	0.750*	0.750*	**0.505**	0.826*
NN base	0.741*	0.703*	0.320*	0.813*
NN base $\odot$ hist. Embs	**0.779**	**0.759**	0.498	**0.832**
Dutch models	Overall	PER	ORG	LOC
Stanford NER tagger	0.696*	0.640*	0.333*	0.794*
Stanford Neudecker	0.623*	0.700*	0.253*	0.702*
NN base	0.818*	0.809*	0.442*	0.871*
NN base $\odot$ hist. Embs	**0.842**	**0.833**	**0.480**	**0.891**

4   Results and Discussion

We evaluate our systems using the CoNLL-2000 evaluation script[10], with F_1 score. To check statistical significance we use randomized testing [28] and results are considered significant if $p < 0.05$.

---

[10] https://www.clips.uantwerpen.be/conll2000/chunking/conlleval.txt

4.1   In-domain Setup

For both languages we achieve the best results with NN base ⊙ historical embeddings. With this setup we can produce F_1 scores of around 80% for both languages, which outperforms all three baselines in the overall performance significantly. The results are presented in Table 2. For French, the overall F_1 score as well as the F_1 for LOC and ORG is best with NN base ⊙ historical embeddings. For ORG the pre-trained tagger of Neudecker [17] works best, which could be due to the gazetteer information they included and of course due to the fact that they train with the entire Europeana data. We hypothesize that the category with the most structural changes over time is ORG. In the military or ecclesiastical context in particular, there are a number of names that no longer exist (in this form). For Dutch we observe the best overall performance with NN base ⊙ historical embeddings except for all entity types.

4.2   Cross-Domain Setup

As shown in Table 3, NN base performs better than the statistical Stanford NER baseline, which is in line with the observations for the in-domain training. We experimented with concatenating BERT embeddings to NN base. For both languages this increases the performance (Table 3, NN base ⊙ BERT). The usage of the historical embeddings is also very beneficial for both languages. We can achieve our best results by using BERT for Dutch and by using historical embeddings for French. We conclude that the usage of modern pre-trained language models is crucial for the performance of NER taggers.

We generated synthetic corruptions for the WikiNER/CoNLL corpus. This could not outperform NN base for both languages. The training on synOCR'd WikiNER/CoNLL gives slightly worse results than NN base too. The corruption of the training data without the usage of any embeddings seems to harm performance drastically, what is in line with the observation of Hamdi et al. [13]. It is striking that the training on corrupted/synOCR'd Dutch gives especially bad results for PER compared to French. A look at the Dutch test set shows that many entities are abbreviated first names (e.g. in *A J van Roozendal*) and are often misrecognized what leads to a performance decrease. For French the combination of NN base and historical embeddings, trained on corrupted data or on synOCR'd (*NN ensemble corrupted/ NN ensemble synOCR'd*) gives the best results and outperforms all other systems. For Dutch *NN ensemble corrupted* and *NN ensemble synOCR* give slightly worse results than NN base ⊙ BERT and NN base ⊙ historical embeddings, but performs better than the tagger trained on synOCR'd or corrupted data only (Table 3, NN ensemble).

Ablation Study. We analyze our results and examine the composition of NN ensemble synOCR more closely (since the results for NN ensemble corrupted are very similar we perform the analysis for NN ensemble synOCR as a representative for both NN ensemble).

The ablation study (see Table 4) shows that NN ensemble benefits from different information in combination. For French NN ensemble gives the best results only for PER. The overall performance increases if we do not use character level embeddings. There is a big performance loss if we omit the historical embeddings. If we do not train on synOCR'd data the performance decreases. For Dutch we can observe these facts even more clearly. If we do not train on synOCR'd data the F_1 score even increases. If omitting the historical embeddings we loose performance as well.

Table 3. Results of training on WikiNER/CoNLL corpus. Scores marked with * are significantly lower than *NN ensemble*.

French models	Overall	PER	ORG	LOC
Çavdar		0.48	0.11	0.56
Stanford NER tagger	0.536*	0.451*	0.059*	0.618*
NN base	0.646*	0.636*	0.096*	0.721*
NN base ⊙ BERT	0.660	0.639*	**0.163**	0.725*
NN base ⊙ hist. Embs.	0.672	0.661*	0.015*	0.748
Corrupted WikiNER	0.627*	0.635*	0.085*	0.710*
synOCR'd WikiNER	0.619*	0.590*	0.078	0.710
NN ensemble corrupted	**0.693**	0.624	0.063	**0.783**
NN ensemble synOCR	0.684	**0.710**	0.111	0.744
Dutch models	Overall	PER	ORG	LOC
Stanford NER tagger	0.371*	0.217*	0.083*	0.564*
NN base	0.567*	0.493*	0.085*	0.700*
NN base ⊙ BERT	**0.634**	**0.572**	**0.250**	0.771
NN base ⊙ hist. Embs.	0.632	0.568	0.084	0.738*
Corrupted CoNLL	0.535*	0.376*	0.155*	0.717*
synOCR'd CoNLL	0.521*	0.327*	0.061*	0.721*
NN ensemble corrupted	0.606*	0.439*	0.158	**0.799**
NN ensemble synOCR	0.614	0.481	0.157	0.775

To find out why our implementation of the assumption – synOCR increases the similarity of the data and improves results – does not have the expected effect, we analyze the test sets. It shows, that only 10% of the French and 6% of the Dutch entities contain OCR errors. Therefore the wrong predictions are mostly not due to the OCR errors, but due to the inherent difficulty of recognizing entities cross-domain. This also explains why synthetic noisyfication does not consistently improve the system. In addition there are some illegible lines in the synOCR'd corpora consisting of dashes and metasymbols, what is not similar to real OCR errors.

Table 4. Ablation study. Results of training on the clean and the synOCR'd WikiNER/CoNLL corpus.

French models	Overall	PER	ORG	LOC
NN ensemble synOCR	0.684	**0.710**	0.011	0.744
- char	**0.693**	0.681	0.078	**0.758**
- word	0.686	0.664	0.080	0.756
- hist. Embs.	0.619	0.590	0.078	0.710
- synOCR'd data	0.672	0.661	**0.015**	0.748
Dutch models	Overall	PER	ORG	LOC
NN ensemble synOCR	0.614	0.382	**0.157**	0.775
- char	0.600	0.404	0.119	**0.780**
- word	0.584	0.430	0.102	0.745
- hist. Embs.	0.521	0.327	0.061	0.721
- synOCR'd data	**0.632**	**0.568**	0.084	0.738

D' **Arras** <LOC> d'où nous avons débouché , nous sommes arrivés aux   *NN_ensemble*
premières maisons de **Siiint-lûrenl-Blangy** <LOC>.

D' **Arras** <LOC> d'où nous avons débouché , nous sommes arrivés aux   *Stanford NER tagger*
premières maisons de **Siiint-lûrenl-Blangy** <O>.

Fig. 2. Example sentence from the French test set.

To verify our assumption we also compare the different systems only on the entities with OCR errors. Here NN ensemble outperforms both of the cross-domain baselines (Table 5, Stanford NER tagger, NN base cross-domain). Compared to the French results Dutch is a lot worse. A look at the entities shows that in the Dutch test set there are many hyphenated words where both word parts are labeled. However, if looking at the parts of the word individually, a clear assignment to an entity type cannot be made, which leads to difficulties with tagging. Though it is plausible that NN ensemble can capture specific phenomena in the historical data better, since the difference between the domains is reduced by the synthetic noisyfication and the historical embeddings. The example in Fig. 2 drawn from the test set shows, that NN ensemble can handle noisy entities well in contrast to e.g. the Stanford NER tagger. Thus in a scenario with many OCR errors the NN ensemble performs well.

5   Related Work

There is some research on using natural language processing for improving OCR for historical documents [5, 26] and also on NER for historical documents [11]. In the latter - a shared task for Named Entity Processing in historical documents - Ehrmann et al. find that OCR noise drastically harms systems performance. Like

Table 5. Results on entities with OCR errors in the French/Dutch test set. Scores marked with * are significantly lower than *NN ensemble*.

Models	French	Dutch
Stanford NER tagger	0.661*	0.207*
NN base in-domain	0.771	0.438
NN base cross-domain	0.783	0.200*
NN ensemble synOCR	**0.835**	**0.462**

us several participants (e.g. [7,23]) also use language models that were trained on historical data to boost the performance of NER taggers. Schweter and Baiter [22] explore NER for historical German data in a cross-domain setting. Like us, they train a language model on unannotated in-domain data and integrate it into a NER tagger. In addition to the above mentioned work, we employ "OCR noisyfication" (Sect. 2.2) and examine the influence of different pretrained embeddings systematically. Çavdar [8] addresses NER and relation extraction on the French Europeana Newspaper corpus. Ehrmann et al. [10] investigate the performance of NER systems on Swiss historical Newspapers and show that historical texts are a great challenge compared to contemporary texts. They find that the LOC class entities causes the most difficulties in the recognition of named entities. The recent work of Hamdi et al. [13] investigates the impact of OCR errors on NER. To do so, they also perturb modern corpora synthetically with different degrees of error rates. They experiment with Spanish, Dutch and English. Like us they perturb the Dutch CoNLL corpus and train NER taggers on that data. Unlike us they do also train on a subset of the perturbed corpus. We test on a subset of the Dutch Europeana corpus. Hamdi et al. [13] show that neural taggers perform better compared to other taggers like the Stanford NER tagger and they also prove that performance decreases drastically if the OCR error rate increases. Piktus et al. [19] learn misspelling-oblivious FastText embeddings from synthetic misspellings generated by an error model for part-of-speech tagging. We use a similar corruption method, but we also use synOCR and historical embeddings for NER.

6   Conclusion

We proposed new methods for in-domain and cross-domain Named Entity Recognition (NER) on historical data and addressed data centric domain adaptation. For the cross-domain case, we handle domain shift by integrating non-annotated historical data via contextualized string embeddings; and OCR errors by injecting synthetic OCR errors into the modern data. This allowed us to get good results when labeled historical data is not available and the historical data is noisy. For training on contemporary corpora and testing on historical corpora we achieve new state-of-the-art results of 69.3% on French and 63.4% on Dutch. For the in-domain case we obtain state-of-the-art results of 77.9% for French and

84.2% for Dutch. There is an increasing demand for advancing the digitization of the world's cultural heritage. High quality digitized historical data, with reliable meta information, will facilitate convenient access and search capabilities, and allow for extensive analysis, for example of historical linguistic or social phenomena. Since named entity recognition is one of the most fundamental labeling tasks, it would be desirable that advances in this area translate to other labeling tasks in processing of historical data as well.

Acknowledgement. This work was funded by the European Research Council (ERC #740516).

A   Appendix

Detailed Information About Experiments and Data

The computing infrastructure we use for all our experiments is one GeForce GTX 1080Ti GPU with an average runtime of 12 h per experiment. For the French and Dutch baseline model *NN base* we count 15,895,683 parameters each. For the French *NN ensemble* model there are 88,264,777 parameters and 96,895,161 parameters for the Dutch *NN ensemble*.

The Europeana Newspaper Corpus is split 80/10/10 into train/dev/test (Table 6). The downsampled French WikiNER corpus is split 70/15/15 into train/dev/test and the Dutch CoNLL-02 corpus is already split in its original version. The downloadable version of the data can be found here: https://github.com/stefan-it/historic-domain-adaptation-icdar.

Table 6. Number of tokens for each datasplit.

Dataset	Train	dev	Test
French Europeana	167,723	18,841	20,346
Dutch Europeana	147,822	16,391	18,218
French WikiNER	411,687	88,410	88,509
Dutch ConNLL-02	202,930	68,994	37,761

References

1. Akbik, A., Bergmann, T., Blythe, D., Rasul, K., Schweter, S., Vollgraf, R.: FLAIR: an easy-to-use framework for state-of-the-art NLP. In: Proceedings of the 2019 Conference of the North American Chapter of the Association for Computational Linguistics (Demonstrations), pp. 54–59. Association for Computational Linguistics, Minneapolis (June 2019). https://www.aclweb.org/anthology/N19-4010

2. Akbik, A., Bergmann, T., Vollgraf, R.: Pooled contextualized embeddings for named entity recognition. In: Proceedings of the 2019 Conference of the North American Chapter of the Association for Computational Linguistics: Human Language Technologies (Long and Short Papers), vol. 1, pp. 724–728. Association for Computational Linguistics, Minneapolis (June 2019)

3. Akbik, A., Blythe, D., Vollgraf, R.: Contextual string embeddings for sequence labeling. In: 27th International Conference on Computational Linguistics, COLING 2018, pp. 1638–1649 (2018)

4. Baevski, A., Edunov, S., Liu, Y., Zettlemoyer, L., Auli, M.: Cloze-driven pretraining of self-attention networks. In: Proceedings of the 2019 Conference on Empirical Methods in Natural Language Processing and the 9th International Joint Conference on Natural Language Processing (EMNLP-IJCNLP), pp. 5359–5368. Association for Computational Linguistics, Hong Kong (November 2019). https://www.aclweb.org/anthology/D19-1539

5. Berg-Kirkpatrick, T., Durrett, G., Klein, D.: Unsupervised transcription of historical documents. In: Proceedings of the 51st Annual Meeting of the Association for Computational Linguistics (Volume 1: Long Papers), pp. 207–217. Association for Computational Linguistics, Sofia (August 2013). https://www.aclweb.org/anthology/P13-1021

6. Bojanowski, P., Grave, E., Joulin, A., Mikolov, T.: Enriching word vectors with subword information. Trans. Assoc. Comput. Linguist. **5**, 135–146 (2017). https://www.aclweb.org/anthology/Q17-1010

7. Boros, E., et al.: Robust named entity recognition and linking on historical multilingual documents. In: Conference and Labs of the Evaluation Forum (CLEF 2020). Working Notes of CLEF 2020 - Conference and Labs of the Evaluation Forum, vol. 2696, pp. 1–17. CEUR-WS Working Notes, Thessaloniki (September 2020). https://hal.archives-ouvertes.fr/hal-03026969

8. Çavdar, M.: Distant supervision for French relation extraction (2017)

9. Devlin, J., Chang, M.W., Lee, K., Toutanova, K.: BERT: pre-training of deep bidirectional transformers for language understanding. In: Proceedings of the 2019 Conference of the North American Chapter of the Association for Computational Linguistics: Human Language Technologies (Long and Short Papers), vol. 1, pp. 4171–4186. Association for Computational Linguistics, Minneapolis (June 2019). https://www.aclweb.org/anthology/N19-1423

10. Ehrmann, M., Colavizza, G., Rochat, Y., Kaplan, F.: Diachronic evaluation of NER systems on old newspapers. In: KONVENS (2016)

11. Ehrmann, M., Romanello, M., Flückiger, A., Clematide, S.: Extended overview of CLEF HIPE 2020: named entity processing on historical newspapers. Zenodo (October 2020)

12. Finkel, J.R., Grenager, T., Manning, C.: Incorporating non-local information into information extraction systems by Gibbs sampling. In: Proceedings of the 43rd Annual Meeting of the Association for Computational Linguistics (ACL 2005), pp. 363–370. Association for Computational Linguistics, Ann Arbor (June 2005). https://www.aclweb.org/anthology/P05-1045

13. Hamdi, A., Jean-Caurant, A., Sidère, N., Coustaty, M., Doucet, A.: Assessing and minimizing the impact of OCR quality on named entity recognition. In: Hall, M., Merčun, T., Risse, T., Duchateau, F. (eds.) TPDL 2020. LNCS, vol. 12246, pp. 87–101. Springer, Cham (2020). https://doi.org/10.1007/978-3-030-54956-5_7

14. Jean-Caurant, A., Tamani, N., Courboulay, V., Burie, J.: Lexicographical-based order for post-OCR correction of named entities. In: 14th IAPR International

Conference on Document Analysis and Recognition, ICDAR 2017, Kyoto, Japan, November 9–15, 2017, pp. 1192–1197. IEEE (2017). https://doi.org/10.1109/ICDAR.2017.197

15. Lample, G., Ballesteros, M., Subramanian, S., Kawakami, K., Dyer, C.: Neural architectures for named entity recognition. In: Proceedings of the 2016 Conference of the North American Chapter of the Association for Computational Linguistics: Human Language Technologies, pp. 260–270. Association for Computational Linguistics, San Diego (June 2016). https://www.aclweb.org/anthology/N16-1030

16. Martinek, J., Lenc, L., Král, P., Nicolaou, A., Christlein, V.: Hybrid training data for historical text OCR, pp. 565–570 (September 2019). https://doi.org/10.1109/ICDAR.2019.00096

17. Neudecker, C.: An open corpus for named entity recognition in historic newspapers. In: Proceedings of the Tenth International Conference on Language Resources and Evaluation (LREC 2016), pp. 4348–4352. European Language Resources Association (ELRA), Portorož (May 2016). https://www.aclweb.org/anthology/L16-1689

18. Nothman, J., Ringland, N., Radford, W., Murphy, T., Curran, J.R.: Learning multilingual named entity recognition from Wikipedia. Artif. Intell. **194**, 151–175 (2013)

19. Piktus, A., Edizel, N.B., Bojanowski, P., Grave, E., Ferreira, R., Silvestri, F.: Misspelling oblivious word embeddings. In: Proceedings of the 2019 Conference of the North American Chapter of the Association for Computational Linguistics: Human Language Technologies (Long and Short Papers), vol. 1, pp. 3226–3234. Association for Computational Linguistics, Minneapolis (June 2019). https://www.aclweb.org/anthology/N19-1326

20. Ramponi, A., Plank, B.: Neural unsupervised domain adaptation in NLP–a survey (2020)

21. Schick, T., Schütze, H.: Rare words: a major problem for contextualized embeddings and how to fix it by attentive mimicking. CoRR abs/1904.06707 (2019). http://arxiv.org/abs/1904.06707

22. Schweter, S., Baiter, J.: Towards robust named entity recognition for historic German. In: Proceedings of the 4th Workshop on Representation Learning for NLP (RepL4NLP-2019), pp. 96–103. Association for Computational Linguistics, Florence (August 2019). https://www.aclweb.org/anthology/W19-4312

23. Schweter, S., März, L.: Triple E - effective ensembling of embeddings and language models for NER of historical German. In: Cappellato, L., Eickhoff, C., Ferro, N., Névéol, A. (eds.) Working Notes of CLEF 2020 - Conference and Labs of the Evaluation Forum, Thessaloniki, Greece, September 22–25, 2020. CEUR Workshop Proceedings, vol. 2696. CEUR-WS.org (2020). http://ceur-ws.org/Vol-2696/paper_173.pdf

24. Tjong Kim Sang, E.F.: Introduction to the CoNLL-2002 shared task: language-independent named entity recognition. In: COLING-02: The 6th Conference on Natural Language Learning 2002 (CoNLL-2002) (2002). https://www.aclweb.org/anthology/W02-2024

25. Tjong Kim Sang, E.F., De Meulder, F.: Introduction to the CoNLL-2003 shared task: language-independent named entity recognition. In: Proceedings of the Seventh Conference on Natural Language Learning at HLT-NAACL 2003, pp. 142–147 (2003). https://www.aclweb.org/anthology/W03-0419

26. Vobl, T., Gotscharek, A., Reffle, U., Ringlstetter, C., Schulz, K.U.: Pocoto - an open source system for efficient interactive postcorrection of ocred historical texts. In: Proceedings of the First International Conference on Digital Access to Textual

Cultural Heritage, DATeCH 2014, pp. 57–61. ACM, New York (2014). http://doi.acm.org/10.1145/2595188.2595197

27. Wagner, R.A., Fischer, M.J.: The string-to-string correction problem. J. ACM **21**(1), 168–173 (1974). https://doi.org/10.1145/321796.321811

28. Yeh, A.: More accurate tests for the statistical significance of result differences. In: The 18th International Conference on Computational Linguistics, COLING 2000, vol. 2 (2000). https://www.aclweb.org/anthology/C00-2137

Temporal Ordering of Events via Deep Neural Networks

Nafaa Haffar[1,2]([⊠]), Rami Ayadi[3,4], Emna Hkiri[2], and Mounir Zrigui[2]

[1] ISITCom, University of Sousse, 4011 Hammam Sousse, Tunisia
nafaa.haffar.5@gmail.com
[2] Research Laboratory in Algebra, Numbers Theory and Intelligent Systems,
University of Monastir, Monastir, Tunisia
mounir.zrigui@fsm.rnu.tn
[3] Computer Science Department, College of Science and Arts in Al Qurayyat,
Jouf University, Sakakah, Kingdom of Saudi Arabia
rayadi@ju.edu.sa
[4] Higher Institute of Computer of Medenine, University of Gabes, Gabès, Tunisia

Abstract. Ordering events with temporal relations in texts remains a challenge in natural language processing. In this paper, we introduce a new combined neural network architecture that is capable of classifying temporal relations between events in an Arabic sentence. Our model consists of two branches: the first one extracts the syntactic information and identifies the orientation of the relation between the two given events based on a Shortest Dependency Path (SDP) layer with Long and Short Memory (LSTM), and the second one encourages the model to focus on the important local information when learning sentence representations based on a Bidirectional-LSTM (BiLSTM) attention layer. The experiments suggest that our proposed model outperforms several previous state-of-the-art methods, with an F1-score equal to 86.40%.

Keywords: Ordering events · Temporal relation · TimeML · Attention mechanism · Semantic features

1 Introduction

In many Natural Language Processing (NLP) tasks such as military Field [17], the chronological order between events in a text has become a crucial element. This step is mainly responsible for identifying entities from text and extracting semantic relations between them.

Over the past few years, the classification of temporal relations in a text has been the objective of existing approaches. Feature-based methods [21,22], such as the Support Vector Machine (SVM), the naive Bayes classifier and other machine learning algorithms, have shown good performance in this area. Most of these methods require the development of a set of lexical, syntactic or semantic characteristics of words. Other researchers [5,31] have used methods based on the dependency between entities with the SDP in the dependency tree. This

© Springer Nature Switzerland AG 2021
J. Lladós et al. (Eds.): ICDAR 2021, LNCS 12822, pp. 762–777, 2021.
https://doi.org/10.1007/978-3-030-86331-9_49

phenomenon has reached high levels of performance but has shown disadvantages in the phase of extracting the relevant information for the task of classifying relations. Although feature engineering is a crucial and efficient element for this task, it is still labor intensive and the extracted features cannot effectively encode the semantic and syntactic information of the words.

Recently, deep Neural Networks (NNs) [28,34,39] have shown good capabilities in the task of classifying temporal relations based on their ability to minimize the effort of feature engineering in NLP tasks. These researchs have recently used recursive NNs (RNN) models based on LSTM along the SDP to learn compositional vector representations for sentences of various syntactic types and variable length. It is noticeable that all these neural models consider all words in a sentence, which are equally important, and also contribute in the same way to the representation of the semantic meaning of the sentence. However, in several areas such as temporal relations, words in the sentence does not have the same impact on the type of relations and on the way to determine the primary semantic information. In this context, the attention mechanism [2] was used to enhance several tasks in the NLP domain by focusing on the most important semantic information for a specific task and to calculate the correlation between input and output [4]. Nevertheless, despite the potential of these systems in the task of classifying relations, the use of a variety of human knowledge is still essential and beneficial. In spite of this explosion of models, this task in Arabic language has received low attention compared to English and other Western languages.

In this paper, our objective is to present a model that classifies the temporal relations between two events in the same sentence for Arabic texts. Our idea is based on: (1) Combining of the sentence sequence representation and the SDP branch between two events in the same sentence with the use of the POS features and the dependency relation features. This idea is designed by, (2) Modeling of an LSTM branch to capture information from the SDP representation, and (3) Using an attention model with a BiLSTM mechanism in order to capture the most important information required to classify temporal relations existing between two events in the sentence sequence.

2   State of the Art

Our objective in this paper is to treat task C, which was proposed as part of the TempEval-3 shared task in UzZaman et al. [36]. This choice has been made due to the fact that TempEval-3 recognizing the complete set of temporal relations[1] that exist in the TimeML standard, instead of a reduced set, which increases the complexity of the task [12].

Several studies for the temporal relation classification task have been developed. Early methods used handcrafted features to classify the temporal relations between pairs of entities through a series of NLP tools. Most of these studies

---

[1] Temporal relations: After, Before, I-before, I-after, Begins, begun-by, Ends, Ends-by, During, During-INV, Includes, IS-Included, Simultaneous, Overlaps, overlapped by, Identity.

[8, 27] were based on characteristic-based classifiers but with a limited set of temporal relations, obtained by grouping the types of relations inverted into only one type. Other work in this context has shown good results in the TempEval-3 challenge, such as Bethard [3], Laokulrat et al. [22], Chambers [6] and Kolya et al. [21] who focused on the development of classifiers based on lexical and syntactic characteristics to classify temporal relations between small subsets of event pairs. Although these systems were performed well in the area of temporal relations, they also showed inefficiencies in several cases, such as the problem of symmetry and transitivity for temporalities between entities.

In this context, D'Souza and Do [11] implemented a set of 437 hand-coded rules combined with a classifier trained on deep lexical features. After a year, Chambers et al. [7] proposed a novel architecture based on sieve idea and incorporated the problem of transitivity when every classifier marked a pair of events. Similarly, Mirza and Tonnelli [29] put forward an architecture based on an SVM classifier but with a set of lexico-semantic features such as POS tags and dependency paths to extract semantic information that would respond to temporal relations. This approach demonstrated that the use of semantic and syntactic features such as dependency relationships as a sequence instead of using a dependency order could give good results.

Although fundamental machine learning models have been gradually improving, regardless of their function, they still need guidance. If an algorithm returns an inaccurate prediction, an engineer must intervene and make adjustments. For this purpose, a new field in supervised learning, called deep learning has been proposed. It can determine on its own, through a deep learning algorithm, whether a prediction is accurate or not through its own NN. This innovation has taken place in recent years with a vast explosion of work based on deep learning.

In 2016, Do and Jeong [10] presented a Convolutional NN (CNN) model combined with a feature extraction step at the lexical and sentence level. The evaluation of this architecture showed poorly results for the task of temporal relations. These problems were due to the variable length of the sentence, the representation of the inputs and the elimination of problems concerning the syntactic structure of the sentence. To solve some of these problems, Dligach et al. [9] suggested the idea of combining a CNN model with a mechanism that used LSTM. This combination utilized an XML tag representation input to mark the positions of medical events. These improvements provided more satisfactory results than using the CNN only and token position embedding.

In the same vein, several studies such as Meng et al. [28] and Tourille et al. [34] propounded new methods based on syntactic and semantic features with the use of LSTM or BiLSTM models. These methods used syntactic dependency and distance characteristics between words in the sentence based on the utilization of the SDP between entities to detect several types of temporal relations in the text. These methods gave good results and showed the potential to use the SDP in the field of temporal relation classification, but they removed some information from the sentence that was relevant to this task in several cases.

In 2018, an approach was put forward by Lim and Choi [24], which used LSTM and different activation functions such as sigmoid and ReLU to extract the temporal context in Korean sentence. A year later, Pandit et al. [30] suggested a method to automatically extract contextual and morphosyntactic characteristics using recurrent NNs between pairs of events. Similarly, Haffar et al. [16] proposed an approach based on two branches. The first branch used the SDP with a BiLSTM layer to capture the two target entities and the dependency relations between them. The second branch used the BiLSTM mechanism on the sentence sequence to capture the syntactic information between the words of the sentence. The problem with this work was that the combination method led to several problems in the orientation of the relations between the two entities.

All the previous methods showed good results for our task. We observe that the different words in a sentence are informative in a differentiated way. Furthermore, the importance of words depends strongly on the context; i.e., the same word can have different importance in the sentence depending on the context. Accordingly, the attention method has become an increasingly popular method in the fields of NLP and more particularly for the classification of relations. For example, Zhou et al. [41] introduced the BiLSTM method based on an attention mechanism to automatically select the characteristics that had a decisive effect on the classification of relations. Xiao and Liu [38] divided the sentence into three contextual sub-sequences according to two annotated entities and used two neural attention based BiLSTMs hierarchically. Wang et al. [37] proposed two levels of CNN attention in order to better discern patterns in heterogeneous context for the classification of relations. Furthermore, Shen and Huang [32] proposed an attention-based CNN using a word-level attention mechanism that could focus on more influential parts of the sentence. Zhang et al. [40] presented an architecture that combined RNN and CNN models after an attention layer. Some recent work, such as Tran et al. [35], used also the attention mechanism with the CNN combined with a BiLSTM model and the SDP to gain long-distance features.

All these methods demonstrated comparable results in the task of classifying relations between entities. Most of them are for the English language and several other languages, but with a low frequency for Arabic language in the field of temporal relations between events. Our task, therefore, is to classify temporal relations between (E−E) in Arabic intra-sentences.

3   Proposed Model

The overall architecture of our temporal classification model is shown in Fig. 1. We formulate this task as a multi-class classification problem which is based on two branches. The first branch is composed for an SDP word sequence representation and a word LSTM layer. The second branch consists of several parts: a word sequence representation, a word BiLSTM layer and an attention sentence layer.

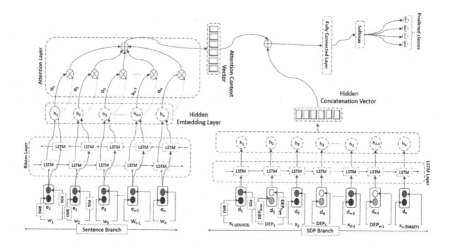

Fig. 1. Architecture of our proposed model.

3.1 Shortest Dependency Path Branch

Our goal in this step is to extract the information that creates a link from a source event to a target event (its dependents). Our idea is based on the fundamental notion that we can extract the dependency structure between the two events by the use of the SDP method which offers several syntactic and semantic advantages for the task of temporal relation classification between two events.

SDP Word Sequence Representation: An SDP is composed of a sequence of words and a dependency relation between each two consecutive words. This composed sequence begins with an event source and ends with an event target. Hence, an SDP sequence with length N is represented as follows:

$$R_{SDP} = (X_1, DEP_1, X_2, DEP_2, X_3, \ldots, X_{n-1}, DEP_{n-1}, X_n) \qquad (1)$$

Each word in the SDP is modeled by its word embedding (E_{Xi}), which is represented by the skip-gram model of the pre-trained distributed word embedding, called "AraVec" [33]. With this SDP representation, we consider also the POS representation tag $E_{POS_{Xi}}$ of each word to enrich the syntactic representation of the word and to improve the robustness. For that, the final representation of each word in the SDP is represented as follows:

$$R_{Xi} = [E_{Xi} \oplus E_{POS_{Xi}}] \qquad (2)$$

The representation vector of the dependency relation (E_{DEPi}) between each two words in the SDP is modeled by the concatenation of its type (DEP_{type}) and its direction (DEP_{dirc}), which is presented as follows:

$$E_{DEPi} = R_{type} \oplus R_{dirc} \qquad (3)$$

In our work, the POS tag, the SDP and each component on the dependency relation are parsed by udpipe[2] tool and its embedded representations are modeled by a look-up table of a randomly fixed size according to the needs of our model. The representation of the SDP sequence with **n** words is described as follows:

$$R_{f_{SDP}} = (R_{X_1}, E_{DEP_1}, R_{X_2}, E_{DEP_2}, R_{X_3}, ..., R_{X_{n-1}}, E_{DEP_{n-1}}, E_{X_n}) \quad (4)$$

Finally, the embedding representation of the SDP, where n is the number of tokens in $R_{f_{SDP}}$, is presented as follows: $E_{f_{SDP}} = \{(d_i)\}_{i=1}^n$.

LSTM Layer: Now, to pick up the information along the SDP sequence, each token in $E_{f_{SDP}}$ is passed through an LSTM unit, which is defined by the following equations:

$$Input\ Gates: \overrightarrow{i_t} = \sigma(\overrightarrow{w_i} \cdot [\overrightarrow{h_{t-1}}, x_t] + \overrightarrow{b_i}) \quad (5)$$

$$Forgot\ Gates: \overrightarrow{f_t} = \sigma(\overrightarrow{w_f} \cdot [\overrightarrow{h_{t-1}}, x_t] + \overrightarrow{b_f}) \quad (6)$$

$$Output\ Gates: \overrightarrow{o_t} = \sigma(\overrightarrow{w_o} \cdot [\overrightarrow{h_{t-1}}, x_t] + \overrightarrow{b_o}) \quad (7)$$

$$Transformation: \overrightarrow{\tilde{C}} = ReLu(\overrightarrow{w_c} \cdot [\overrightarrow{h_{t-1}}, x_t] + \overrightarrow{b_c}) \quad (8)$$

$$Cell\ states: \overrightarrow{C_t} = \overrightarrow{f_t} \odot \overrightarrow{C_{t-1}} + \overrightarrow{i_t} \odot \overrightarrow{\tilde{C}}) \quad (9)$$

$$Cell\ output: \overrightarrow{h_t} = \overrightarrow{o_t} \odot ReLu(\overrightarrow{C_t}) \quad (10)$$

where the weighted matrices parameters w_i, w_f, w_o and w_c ($w \in \mathbb{R}^{d*L}$, L is the size of the LSTM), and the bias parameters b_i, b_f, b_o and b_c ($b \in \mathbb{R}^d$) are updated throughout the training process, x_t is the inputs of the LSTM cell unit at time t, σ is the sigmoid function, and $\odot$ stands for element-wise multiplication.

This step generates a set of n hidden states $H^p = \{h_i^{p_i}\}_{i=1}^N$). This set of hidden states can be seen as the structural and semantic context representation of tokens in the SDP input. Finally, all the hidden state vectors are concatenated to obtain ne final vector, presented as follows:

$$V_{SDP} = Concat(h_1^{P_1}, h_2^{P_2}, h_3^{P_3}, ..., h_n^{P_n}) \quad (11)$$

3.2   Sentence Sequence Branch

In this phase, we aim to extract the relevant information that can affect the class of temporal relations

Word Sequence Representation: To extract the semantic relation between two events, the SDP is the most used method. This motivation is based on their simple structure, which usually contains the important information to identify the relations between them. On the counterpart, this simple SDP structure causes in some cases the loss of some important information, which can be leveraged to determine more precisely the type of temporal relations. The following example can briefly explain this problem:

---

[2] udpipe tool: http://lindat.mff.cuni.cz/services/udpipe/.

Thousands of people $[died]_{E1}$ after the $[explosion]_{E2}$ of the bomb.

القنبلة $[$إنفجار$]_{E2}$ بعد الأشخاص آلاف $[$مات$]_{E1}$

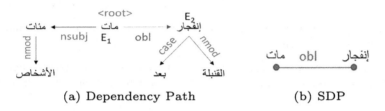

(a) Dependency Path   (b) SDP

Fig. 2. Example of dependency path for the given sentence, and SDP between E1 and E2.

Figure 2 shows the dependency tree of the given example and its SDP. we can see that the relation between E1 and E2 is of type **After**. This relation is usually classified by the particle (بعد, after) which is deleted in the SDP representation. Based on this problem, we think to extract the necessary information that strongly affects the temporal relation between the two given events. This process is done by representing the whole sequence of the sentence; i.e., we will use in this branch the whole given sentence (S) of length N which is represented as follows: $R_S = \{W_i\}_{i=1}^{N}$. Like the SDP branch, each word in the sentence is presented by its word embedding (E_{Wi}) and enriched by its POS representation tag $E_{POS_{Wi}}$. Therefore, the final representation of each word in the sentence branch is represented as follows:

$$R_{Wi} = [E_{Wi} \oplus E_{POS_{Wi}}] \tag{12}$$

Finally, the representation of the sentence, with N tokens, is presented as follows: $E_{f_S} = (e_1, e_2, e_3, ..., e_{N-1}, e_N) = \{e_i\}_{i=1}^{N}$.

BiLSTM Layer: In this layer, each embedding word in the sentence is passed through two LSTM cells. The first one is an LSTM forward which processes from the beginning to the end of the sentence $\overrightarrow{h_t} = LSTM(e_t, \overrightarrow{h_{t-1}})$. The second is an LSTM backward which processes from the end to the beginning $\overleftarrow{h_t} = LSTM(e_t, \overleftarrow{h_{t+1}})$. This combination solves a single LSTM cell problem that captures the previous context and neglects the future. With this architecture, the output layer is able to use information related to past and future context. For that, the concatenated output vector of the BiLSTM unit for each word vector is simply expressed as follows:

$$\overleftrightarrow{h_t} = BiLSTM(e_t, \overrightarrow{h_{t-1}}, \overleftarrow{h_{t+1}}) \tag{13}$$

$$h_t = [\overrightarrow{h_t} \oplus \overleftarrow{h_t}], h_t \in \mathbb{R}^{2L} \tag{14}$$

lastly, a set of hidden states is represented as a result of the BiLSTM layer for the given sentence: $H^{(E_{fs})} = (h_1^{e_1}, h_2^{e_2}, h_3^{e_3}, ..., h_N^{e_N}) = \{h_i^{e_i}\}_{i=1}^{N}$.

Attention Model Layer: Now, and to overcome the SDP problem, the attention mechanism is used. It pays attention only to certain input words in the sentence and it allows modelling the dependencies without considering account their distance in the sentence sequence. So, to identify the tokens in the sequence that are clues to classify correctly a temporal relation between two events, we introduce a sentence level context vector, and then we use this vector to measure the importance of each term in the sentence. In this context, each word feature hidden state $\{h_i^{e_i}\}_{i=1}^{N}$ generated previously from the BiLSTM layer is used to calculate a weight W_i through a softmax function. This weight is afterwards used to calculate a hidden sentence feature vector V_C by a weighted sum function, which can be seen as a high-level representation to distinguish the importance of different words on word sequences. Concretely, based on the previous explanations, the attention mechanism is computed as follows:

$$u_i = ReLU(W_h h_i + b_h), u_i \in [0,1] \tag{15}$$

$$W_i = \frac{exp(u_i)}{\sum_{t=1}^{M} exp(u_t)}, \sum_{i=1}^{N} W_i = 1 \tag{16}$$

$$V_C = \sum_{i=1}^{N} W_i h_i \tag{17}$$

where W_h and b_h are the weight and bias from the attention layer, and M is the number of temporal label types.

3.3   Classification Module

As previously mentioned, the objective of our architecture is to classify the temporal relations between two given events. This idea is founded on the combination of an SDP and the relevant information extracted from the entire sentence sequence. This fundamental idea is finally represented by a final vector V_F which is based on the combination of the two output vectors of the two branches:

$$V_F = [V_C \oplus V_{SDP}] \tag{18}$$

The representation of the final vector V_F will be used as an input in a fully connected layer to generate a context vector.

$$h_F = ReLU(V_F) \tag{19}$$

Finally, this context vector will be used in an output layer with the Softmax function to give the distribution $\hat{p}$ of the temporal relations between the two events (E−E) and to predict the correct class of the temporal relation:

$$\hat{p} = softmax(h_F \cdot W_{V_F}, b_{V_F}) \qquad (20)$$

where W_{VF} and b_{VF} are the weight parameters.

3.4   Training Model

We model the temporal relation as a multi-class classification task. We train our model in a supervised manner by minimizing the distance between the probability distributions predicted by the model and the true distribution of the gold-standard label. For that, the loss function is described as follows:

$$L = -\sum_i p_i log(\hat{p}_i) \qquad (21)$$

where p_i is the gold-standard temporal labels, and $\hat{p}_i$ is the predicted labels from the model.

4   Experiments

In this section, we present our dataset and describe our network implementation and training details.

4.1   Dataset

Arabic is a rich and complex language [18]. It is one of the poorly endowed languages [19]. Indeed, several attempts have been made to annotate an Arabic corpus of temporal information based on the specifications of TimeML. We find in the literature an Arabic TimeML corpus which was created by Haffar et al. [15]. It is an enrichment version of a previous TimeML Arabic corpus [14] and is based on the Arabic TimeML guideline [13], but with a small size for the learning step of our model. For this reason, we think of enriching this corpus with new temporal relations. Our objective here is to prepare a corpus that will be used to train and test our system. To do this, we set up a data augmentation phase composed of two steps: new sentence prediction and new relation extraction.

New Sentence Prediction: This step consists in predicting new sentences similar to the annotated original. Figure 3 shows the main steps for enriching the corpus with new annotated sentences.

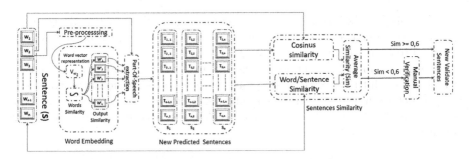

Fig. 3. Architecture of new sentence prediction step

Our proposed method in this step consists in analyzing and extracting each event in each previously annotated sentence and predicting new similar sentences as follows: Extract the events annotated in each sentence and transfer them to raw words; i.e., each tagged event is passed through a pre-processing step that removes the $<event><\backslash event>$ tags and these attributes. After that, each extracted event is used as an input in the word embedding model to transfer it into an associated vector and then extract the most similar words whose similarities are superior to a given threshold α. Next, we consolidate the representation of words using a skip-gram model, which is advantageous in predicting the context of words and training systems on a large corpus.

This step gives a list of words similar to the input word but with different types; i.e., if the given event is of type verb, we can find words with a similarity superior to a given threshold but with type verb, noun or other than the original type. In our work, we consider only words that have the same POS as the original word. The fundamental reason behind this idea is that it is more relevant to keep the syntactic structure of the original sentence. For that, each predicted word is passed through a POS step to select only those words that have the same POS as the original word. Each predicted word is placed in the same position as the original word to create new sentences. Thereafter, each new sentence is passed through a semantic similarity step. This method is based on the idea of preserving the semantics of the main sentence. An average similarity of two similarity models is calculated. The first model is the similarity cosine which calculates the similarity by measuring the angle cosine between two vectors. This model is frequently used in NLP as a similarity measure [25,26] and is presented as follows:

$$Cos(S_{new}, S_{old}) = \frac{V_{new}, V_{old}}{||V_{new}||, ||V_{old}||} = \frac{\sum_{i=1}^{n} V_{new-i} V_{old-i}}{\sqrt{\sum_{i=1}^{n}(V_{new-i})^2}\sqrt{\sum_{i=1}^{n}(V_{old-i})^2}} \quad (22)$$

The value of the Cosine(S_{new}, S_{old}) is in range [0,1], with 0 indicates non-similar (orthogonal) vectors, and 1 indicates similar vectors. Intermediate values are used to evaluate the similarity degree.

The second model is the max match similarity between each word in the new sentence and all the words in the original sentence. This method is based on the idea of determining the degree of resemblance between each word in the new sentence and the original sentence. It is calculated according to the following equation:

$$Sim(S_{new}, S_{old}) = \frac{\sum_{i=1}^{n} cos(W_i, S_{old})}{n} = \frac{\sum_{i=1}^{n} \sum_{j=1}^{n} cos(W_i, W_{S_{old j}})}{n} \quad (23)$$

The output of the two models is used to calculate the average similarity values of the predicted and original sentences, as follows:

$$Average_{Sim}(S_{new}, S_{old}) = \frac{Cosine(S_{new}, S_{old}) + similarity(S_{new}, S_{old})}{2} \quad (24)$$

After that, a verification similarity step is used by comparing between the value obtained in Eq. (24) and a given threshold β. If $Average_{Sim} \geq \beta$, then the predicted sentence is validated; otherwise, the new sentence is passed through a manual verification step. The latter is intended to judge, by two linguistic experts, whether the predicted sentence can be used as a similar sentence or not.

New Relations Extraction: In this step, the main goal is to extract new temporal relations from the existing relations in the corpus. We use for that several constraint propagation rules which take existing temporal relations in the corpus and derive new implicit relations, i.e. make explicit what was implicit. These rules are based on Allen's interval algebra [1] and use some rules of symmetry (see Table 1) and transitivity (see Table 2) to model the temporal relations between all pair events in the same sentence.

At the end of these steps, in addition to the previous annotated relations in the Arabic TimeML corpus, we obtain 82,374 temporal relations. Table 3 presents the distribution of our annotated temporal relation in the Arabic corpus.

Table 1. Symmetry rules.

(e1,e2)	⇔	(e2,e1)
before	⇔	After
Ibefore	⇔	Iafter
During	⇔	During_inv
begins	⇔	begin_by
ends	⇔	ends_by
Overlaps	⇔	overlapped_by
Includes	⇔	IS_Included
equals	⇔	equals

Table 2. Some transitivity rules.

If	And		Then
during(e1,e2)	OR	After(e2,e3)	After(e1,e3)
		Iafter(e2,e3)	
during(e1,e2)	OR	before(e2,e3)	before(e1,e3)
		Ibefore(e2,e3)	
Ends(e1,e2)	OR	After(e2,e3)	After(e1,e3)
		Iafter(e2,e3)	
Ends(e1,e2)	before(e2,e3)		before(e1,e3)
begins(e1,e2)	OR	before(e2,e3)	After(e1,e3)
		Ibefore(e2,e3)	
begins(e1,e2)	After(e2,e3)		After(e1,e3)
overlaps(e1,e2)	begins(e2,e3)		overlaps(e1,e3)
before(e1,e2)	OR	begins(e2,e3)	before(e1,e3)
		begin-by(e2,e3)	
		ends-by(e2,e3)	

Table 3. Statistics of our enriched corpus.

E−E	Training set	Validation set	Test set
82,374	71,000	8,000	3,374

To better explain this phase, the following example is proposed:

– **Original Sentence:**

مات الولد بعد إنفجار البركان

The boy was died after the explosion of the volcano.

- **Events:** (مات, died, e_1), (إنفجار, explosion, e_2).
- **Relations:** After(e_1, e_2), Before(e_2, e_1)

This example contains two events (مات, died, Event e_1) and (إنفجار, explosion, Event e_2) with the temporal relation After(e_1, e_2). With the application of the proposed method of data augmentation, we obtain 12 new similar sentences with 12 new similar events[3] with the same POS type of the original word and 25 new temporal relations in which we just operate with the event word (مات, Event e_1).

---

[3] Similar events list:(هلك, Perished), (فني, Perished), (قتل, killed), (إحترق, burned), (إستشهد, martyred), (توفي, died), (وبق, Perished), (زهق, died), (ردي, died), (ثوى, died), (ذأف, died), (فاض, Perished).

4.2 Model Implementation and Training

To evaluate our model, we use the enriched corpus[4] (see Sect. 4.1). To train our model perfectly, we use the k-fold cross validation, where k is equal to 10. During the k iterations, we combine the training and validation sets and run the algorithm once again. For all the following experiments, we use the activation function "ReLu" for each LSTM unit and for the fully connected layer in order to overcome the problem of the disappearance of gradients and to speed up the training [20]. Regarding the other parameters, we take those hyper-parameters that achieve the best performance of the development set. We use the Adam optimizer with a learning rate equal to 0.001 and a mini-batch size equal to 64. To evaluate the over-fitting of the model, we use a dropout equal to 0.4. The size of each LSTM is 128. The fully connected layer is set as 128 and the number of epochs is 50. The size of the embedding words is fixed as 300. The α and β thresholds are both equal to 0.6. All these parameters give the best results and are chosen after several attempts.

4.3 Results and Model Analysis

In this work, our method is compared to other methods performed under the same dataset. For this, different experiments are implemented and tested. For each evaluation, we use the Precision (P), the Recall (R) and the F1 score (F1).

Main Results: To demonstrate the performance of each component in a proposed model, we put forward several configurations described in Table 4.

Table 4. Evaluation results of our model.

	P	R	F1
$SDP\_only_{/BiLSTM}$	0.78	0.74	0.76
$SDP\_only_{/LSTM}$	0.839	0.764	0.81

	P	R	F1
Sentence+$SDP_{/LSTM}$	0.86	0.79	0.823
+DEP	0.88	0.80	0.84
+POS	0.86	0.793	0.825
+POS+DEP	0.89	0.82	0.85
+POS+DEP +Attention	**0.92**	**0.815**	**0.864**

From these configurations, we can see that configuration $SDP_{only/BiLSTM}$ has a good effect on the temporal relation classification task, compared to the use of BiLSTM networks. From this comparison, it is possible that the BiLSTM networks can obtain more semantic information, but our work is based mostly on the direction of the relation dependency between the two given events. For example, if the relation from E_1 to E_2 is of type "AFTER", then the inverted relation from E_2 to E_1 is of type "BEFORE". Consequently, configuration ($SDP_{only/LSTM}$) is more helpful for our work. We can see also that

---

[4] The enriched corpus can be found at: https://github.com/nafaa5/Enriched-Arabic-TimeML-Corpus.

the combination of the SDP branch and the sentence sequence branch without using external features improves the results. We can therefore conclude that this combination shows the robustness of our proposed method and its potential to classify temporal relations between intra-sentence events.

Compared with the first configuration, The results show that the use of POS and dependency relations as well as the attention mechanism, play an important and crucial role in improving the performance of the first configuration $SDP_{/LSTM}$ by almost 4%. Based on all these configurations and results, we conclude that the proposed model is effective for our task.

Comparisons with State of the Art: We compare our model to the work of Zhou et al. [41] and Zhang et al. [40] for the relations classification task (see Table 5). For that, all these methods are re-implemented and executed on our dataset for Arabic, but with the same parameters indicated in Sect. 4.2 and other parameters from their work.

Table 5. Comparison of different relations classification models. WV and TP stand for Word Vector and Tag Position.

Models	Features	F1
BiLSTM+Attention [41]	WV, TP	0.842
CNN+BiLSTM [40]	WV, TP	0.813
Sentence$_{BiLSTM}$+ATTENTION +SDP$_{LSTM}$ (Our Model)	WV, DEP POS	0.864

5   Conclusion and Future Work

In this paper, we have put forward a novel attention NN model for the temporal relation classification task in Arabic texts. The strength of this model resides in the combination of locally important characteristics in the sentence for the type of temporal relations and the orientation of the relations between the two events given in the SDP. We have also used the POS features and dependency relation features. The effectiveness of our proposed model has been demonstrated by its evaluation on the created Arabic TimeBank corpus.

We expect that this type of architecture will also be of interest beyond the specific task of classifying temporal relations, which we intend to explore in future work. We would also like to use this architecture in solving many other NLP problems such as automatic question-answering and identification of causal relations between entities. We plan to use the Seq2Seq model [23] and transformer-based methods in future works.

References

1. Allen, J.F.: Maintaining knowledge about temporal intervals. Commun. ACM **26**, 832–843 (1983)

2. Bahdanau, D., Cho, K., Bengio, Y.: Neural machine translation by jointly learning to align and translate. In: ICLR (2015)
3. Bethard, S.: Cleartk-timeml: a minimalist approach to tempeval 2013. In: SemEval@NAACL-HLT, vol. 2, pp. 10–14 (2013)
4. Bsir, B., Zrigui, M.: Document model with attention bidirectional recurrent network for gender identification. In: Rojas, I., Joya, G., Catala, A. (eds.) IWANN 2019. LNCS, vol. 11506, pp. 621–631. Springer, Cham (2019). https://doi.org/10.1007/978-3-030-20521-8_51
5. Bunescu, R.C., Mooney, R.J.: A shortest path dependency kernel for relation extraction. In: HLT/EMNLP, pp. 724–731 (2005)
6. Chambers, N.: Navytime: event and time ordering from raw text. In: SemEval@NAACL-HLT, vol. 2, pp. 73–77 (2013)
7. Chambers, N., Cassidy, T., McDowell, B., Bethard, S.: Dense event ordering with a multi-pass architecture. Trans. ACL **2**, 273–284 (2014)
8. Chambers, N., Wang, S., Jurafsky, D.: Classifying temporal relations between events. In: ACL (2007)
9. Dligach, D., Miller, T., Lin, C., Bethard, S., Savova, G.: Neural temporal relation extraction. In: EACL, vol. 2 (2017)
10. Do, H.W., Jeong, Y.S.: Temporal relation classification with deep neural network. In: BigComp, pp. 454–457 (2016)
11. D'Souza, J., Ng, V.: Classifying temporal relations with rich linguistic knowledge. In: HLT-NAACL (2013)
12. Glavas, G., Šnajder, J.: Construction and evaluation of event graphs. Nat. Lang. Eng. **21**(4), 607–652 (2015)
13. Haffar, N., Hkiri, E., Zrigui, M.: Arabic linguistic resource and specifications for event annotation. In: IBIMA, pp. 4316–4327 (2019)
14. Haffar, N., Hkiri, E., Zrigui, M.: TimeML annotation of events and temporal expressions in Arabic texts. In: Nguyen, N.T., Chbeir, R., Exposito, E., Aniorté, P., Trawiński, B. (eds.) ICCCI 2019. LNCS (LNAI), vol. 11683, pp. 207–218. Springer, Cham (2019). https://doi.org/10.1007/978-3-030-28377-3_17
15. Haffar, N., Hkiri, E., Zrigui, M.: Enrichment of Arabic TimeML corpus. In: Nguyen, N.T., Hoang, B.H., Huynh, C.P., Hwang, D., Trawiński, B., Vossen, G. (eds.) ICCCI 2020. LNCS (LNAI), vol. 12496, pp. 655–667. Springer, Cham (2020). https://doi.org/10.1007/978-3-030-63007-2_51
16. Haffar, N., Hkiri, E., Zrigui, M.: Using bidirectional LSTM and shortest dependency path for classifying Arabic temporal relations. KES. Procedia Comput. Sci. **176**, 370–379 (2020)
17. Hkiri, E., Mallat, S., Zrigui, M.: Events automatic extraction from Arabic texts. Int. J. Inf. Retr. Res. **6**(1), 36–51 (2016)
18. Hkiri, E., Mallat, S., Zrigui, M.: Integrating bilingual named entities lexicon with conditional random fields model for Arabic named entities recognition. In: ICDAR, pp. 609–614. IEEE (2017)
19. Hkiri, E., Mallat, S., Zrigui, M., Mars, M.: Constructing a lexicon of Arabic-English named entity using SMT and semantic linked data. Int. Arab J. Inf. Technol. **14**(6), 820–825 (2017)
20. Kang, Y., Wei, H., Zhang, H., Gao, G.: Woodblock-printing Mongolian words recognition by bi-LSTM with attention mechanism. In: The International Conference on Document Analysis and Recognition (ICDAR), pp. 910–915 (2019)
21. Kolya, A.K., Kundu, A., Gupta, R., Ekbal, A., Bandyopadhyay, S.: JU_CSE: a CRF based approach to annotation of temporal expression, event and temporal relations. In: SemEval@NAACL-HLT, vol. 2, pp. 64–72 (2013)

22. Laokulrat, N., Miwa, M., Tsuruoka, Y., Chikayama, T.: Uttime: temporal relation classification using deep syntactic features. In: SemEval@NAACL-HLT, vol. 2, pp. 88–92 (2013)
23. Li, Z., Cai, J., He, S., Zhao, H.: Seq2seq dependency parsing. In: The 27th International Conference on Computational Linguistics, pp. 3203–3214 (2018)
24. Lim, C.G., Choi, H.J.: LSTM-based model for extracting temporal relations from Korean text. In: BigComp, pp. 666–668 (2018)
25. Mahmoud, A., Zrigui, M.: Sentence embedding and convolutional neural network for semantic textual similarity detection in Arabic language. Arab. J. Sci. Eng. 44(11), 9263–9274 (2019)
26. Mahmoud, A., Zrigui, M.: BLSTM-API: Bi-LSTM recurrent neural network-based approach for Arabic paraphrase identification. Arab. J. Sci. Eng. 46(4), 4163–4174 (2021)
27. Mani, I., Verhagen, M., Wellner, B., Lee, C.M., Pustejovsky, J.: Machine learning of temporal relations. In: ACL (2006)
28. Meng, Y., Rumshisky, A., Romanov, A.: Temporal information extraction for question answering using syntactic dependencies in an LSTM-based architecture. In: EMNLP (2017)
29. Mirza, P., Tonelli, S.: Classifying temporal relations with simple features. In: EACL, pp. 308–317 (2014)
30. Pandit, O.A., Denis, P., Ralaivola, L.: Learning rich event representations and interactions for temporal relation classification. In: ESANN (2019)
31. Plank, B., Moschitti, A.: Embedding semantic similarity in tree kernels for domain adaptation of relation extraction. In: ACL, vol. 1, pp. 1498–1507 (2013)
32. Shen, Y., Huang, X.: Attention-based convolutional neural network for semantic relation extraction. In: COLING, pp. 2526–2536 (2016)
33. Soliman, A.B., Eissa, K., El-Beltagy, S.R.: Aravec: a set of Arabic word embedding models for use in Arabic NLP. In: ACLING, vol. 117, pp. 256–265 (2017)
34. Tourille, J., Ferret, O., Névéol, A., Tannier, X.: Neural architecture for temporal relation extraction: a Bi-LSTM approach for detecting narrative containers. In: ACL, vol. 2, pp. 224–230 (2017)
35. Tran, V.H., Phi, V.T., Shindo, H., Matsumoto, Y.: Relation classification using segment-level attention-based CNN and dependency-based RNN. In: NAACL-HLT, vol. 1, pp. 2793–2798 (2019)
36. UzZaman, N., Llorens, H., Derczynski, L., Allen, J., Verhagen, M., Pustejovsky, J.: SemEval-2013 task 1: TempEval-3: evaluating time expressions, events, and temporal relations. In: (SemEval-2013), vol. 1 (2013)
37. Wang, L., Cao, Z., de Melo, G., Liu, Z.: Relation classification via multi-level attention CNNs. In: ACL, vol. 1, pp. 1298–1307 (2016)
38. Xiao, M., Liu, C.: Semantic relation classification via hierarchical recurrent neural network with attention. In: COLING, pp. 1254–1263 (2016)
39. Xu, Y., Mou, L., Li, G., Chen, Y., Peng, H., Jin, Z.: Classifying relations via long short term memory networks along shortest dependency paths. In: EMNLP, pp. 1785–1794 (2015)
40. Zhang, X., Chen, F., Huang, R.: A combination of RNN and CNN for attention-based relation classification. Procedia Comput. Sci. 131, 911–917 (2018)
41. Zhou, P., et al.: Attention-based bidirectional long short-term memory networks for relation classification. In: ACL, vol. 2, pp. 207–212 (2016)

Document Collection Visual Question Answering

Rubèn Tito[(✉)], Dimosthenis Karatzas, and Ernest Valveny

Computer Vision Center, UAB, Barcelona, Spain
{rperez,dimos,ernest}@cvc.uab.es

Abstract. Current tasks and methods in Document Understanding aims to process documents as single elements. However, documents are usually organized in collections (historical records, purchase invoices), that provide context useful for their interpretation. To address this problem, we introduce Document Collection Visual Question Answering (DocCVQA) a new dataset and related task, where questions are posed over a whole collection of document images and the goal is not only to provide the answer to the given question, but also to retrieve the set of documents that contain the information needed to infer the answer. Along with the dataset we propose a new evaluation metric and baselines which provide further insights to the new dataset and task.

Keywords: Document collection · Visual Question Answering

1 Introduction

Documents are essential for humans since they have been used to store knowledge and information over the history. For this reason there has been a strong research effort on improving the machine understanding of documents. The research field of Document Analysis and Recognition (DAR) aims at the automatic extraction of information presented on paper, initially addressed to human comprehension. Some of the most widely known applications of DAR involve processing office documents by recognizing text [14], tables and forms layout [8], mathematical expressions [25] and visual information like figures and graphics [30]. However, even though all these research fields have progressed immensely during the last decades, they have been agnostic to the end purpose they can be used for. Moreover, despite the fact that document collections are as ancient as documents themselves, the research in this scope has been limited to document retrieval by lexical content in word spotting [22,27], blind to the semantics and ignoring the task of extracting higher level information from those collections.

On the other hand, over the past few years Visual Question Answering (VQA) has been one of the major relevant tasks as a link between vision and language. Even though the works of [6] and [31] start considering text in VQA by requiring the methods to read the text in the images to answer the questions, they constrained the problem to natural scenes. It was [24] who first introduced VQA on

© Springer Nature Switzerland AG 2021
J. Lladós et al. (Eds.): ICDAR 2021, LNCS 12822, pp. 778–792, 2021.
https://doi.org/10.1007/978-3-030-86331-9_50

documents. However, none of those previous works consider the image collection perspective, neither from real scenes nor documents.

In this regard, we present Document Collection Visual Question Answering (DocCVQA) as a step towards better understanding document collections and going beyond word spotting. The objective of DocCVQA is to extract information from a document image collection by asking questions and expecting the methods to provide the answers. Nevertheless, to ensure that those answers have been inferred using the documents that contain the necessary information, the methods must also provide the IDs of the documents used to obtain the answer in the form of a confidence list as answer evidence. Hence, we design this task as a retrieval-answering task, for which the methods should be trained initially on other datasets and consequently, we pose only a set of 20 questions over this document collection. In addition, most of the answers in this task are actually a set of words extracted from different documents for which the order is not relevant, as we can observe in the question example in Fig. 1. Therefore, we define a new evaluation metric based on the Average Normalized Levenshtein Similarity (ANLS) to evaluate the answering performance of the methods in this task. Finally, we propose two baseline methods from very different perspectives which provide some insights on this task and dataset.

The dataset, the baselines code and the performance evaluation scripts with an online evaluation service are available in https://docvqa.org.

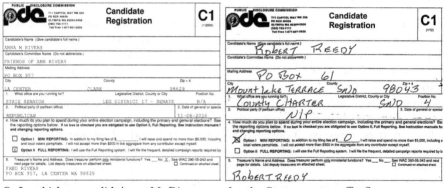

Q: In which years did Anna M. Rivers run for the State senator office?
A: [2016, 2020]
E: [454, 10901]

Fig. 1. Top: Partial visualization of sample documents in DocCVQA. The left document corresponds to the document with ID 454, which is one of the relevant documents to answer the question below. Bottom: Example question from the sample set, its answer and their evidences. In DocCVQA the evidences are the documents where the answer can be inferred from. In this example, the correct answer are the years 2016 and 2020, and the evidences are the document images with ids 454 and 10901 which corresponds to the forms where Anna M. Rivers presented as a candidate for the State senator office.

2   Related Work

2.1   Document Understanding

Document understanding has been largely investigated within the document analysis community with the final goal of automatically extracting relevant information from documents. Most works have focused on structured or semi-structured documents such as forms, invoices, receipts, passports or ID cards, e-mails, contracts, etc. Earlier works [9,29] were based on a predefined set of rules that required the definition of specific templates for each new type of document. Later on, learning-based methods [8,26] allowed to automatically classify the type of document and identify relevant fields of information without predefined templates. Recent advances on deep learning [20,36,37] leverage natural language processing, visual feature extraction and graph-based representations in order to have a more global view of the document that take into account word semantics and visual layout in the process of information extraction.

All these methods mainly focus on extracting key-value pairs, following a bottom-up approach, from the document features to the relevant semantic information. The task proposed in this work takes a different top-down approach, using the visual question answering paradigm, where the goal drives the search of information in the document.

2.2   Document Retrieval

Providing tools for searching relevant information in large collections of documents has been the focus of document retrieval. Most works have addressed this task from the perspective of word spotting [27], i.e., searching for specific query words in the document collection without relying on explicit noisy OCR. Current state-of-the-art on word spotting is based on similarity search in an common embedding space [1] where both the query string and word images can be projected using deep networks [17,32] In order to search for the whole collection, these representations are combined with standard deep learning architectures for object detection in order to find all instances of a given word in the document [17,34].

Word spotting only allows to search for the specific instances of a given word in the collection without taking into account the semantic context where that word appears. On the contrary, the task proposed in this work does not aim to find specific isolated words, but to make a semantic retrieval of documents based on the query question.

2.3   Visual Question Answering

Visual Question Answering (VQA) is the task where given an image and a natural language question about that image, the objective is to provide an accurate natural language answer. It was initially introduced in [21,28] and [3] proposed the first large scale dataset for this task. All the images from those works are real scenes and the questions mainly refer to objects present in the images. Nonetheless, the field became very popular and several new datasets were released exploring new challenges like ST-VQA [6] and TextVQA [31], which were the first datasets that considered the text in the scene. In the former dataset, the answers are always contained within the text found in the image while the latter requires to read the text, but the answer might not be a direct transcription of the recognized text. The incorporation of text in VQA posed two main challenges. First, the number of classes as possible answers grew exponentially and second, the methods had to deal with a lot of out of vocabulary (OOV) words both as answers or as input recognized text. To address the problem of OOV words, embeddings such as Fasttext [7] and PHOC [1] became more popular, while in order to predict an answer, along with the standard fixed vocabulary with the most common answers a copy mechanism was introduced by [31] which allowed to propose an OCR token as an answer. Later [13] changed the classification output to a decoder that outputs a word from the fixed vocabulary or from the recognized text at each timestep, and provided more flexibility in complex and longer answers.

Concerning documents, FigureQA [16] and DVQA [15] focused on complex figures and data representation like different kinds of charts and plots by proposing synthetic datasets and corresponding questions and answers over those figures. More recently, [23] proposed DocVQA, the first VQA dataset over document images, where the questions also refer to figures, forms or tables but also text in complex layouts. Along with the dataset they proposed some baselines based on NLP and scene text VQA models. In this sense, we go a step further extending this work for document collections.

Finally, one of the most relevant works for this paper is ISVQA [4] where the questions are asked over a small set of images which consist of different perspectives of the same scene. Notice that even though the set up might seem similar, the methods to tackle this dataset and the one we propose are very different. For ISVQA all the images share the same context, which implies that finding some information in one of the images can be useful for the other images in the set. In addition, the image sets are always small sets of 6 images, in contrast to the whole collection of DocCVQA and finally, the images are about real scenes which don't even consider the text. As an example, the baselines they propose are based on the HME-VideoQA [11] and standard VQA methods stitching all the images, or the images features. Which are not suitable to our problem.

3   DocCVQA Dataset

In this section we describe the process for collecting images, questions and answers, an analysis of the collected data and finally, we describe the metric used for the evaluation of this task.

3.1   Data Collection

Images: The DocCVQA dataset comprises 14, 362 document images sourced from the Open Data portal of the Public Disclosure Commission (PDC), an agency that aims to provide public access to information about the financing of political campaigns, lobbyist expenditures, and the financial affairs of public officials and candidates. We got the documents from this source for various reasons. First, it's a live repository that is updated periodically and therefore, the dataset can be increased in size in the future if it's considered necessary or beneficial for the research. In addition, it contains a type of documents in terms of layout and content that makes sense and can be interesting to reason about an entire collection. Moreover, along with the documents, they provide their transcriptions in the form of CSV files which allows us to pose a set of questions and get their answers without the costly process of annotation. From the original collection of document images, we discarded all the multi-page documents and documents for which the transcriptions were partially missing or ambiguous. Thus, all documents that were finally included in the dataset were sourced from the same document template, the US Candidate Registration form, with slight design differences due to changes over the time. However, these documents still pose some challenges since the proposed methods will need to understand its complex layout, as well as handwritten and typewritten text at the same time. We provide some document examples in Fig. 1.

Questions and Answers: Considering that DocCVQA dataset is set up as a retrieval-answering task and documents are relatively similar we pose only a set of 20 natural language questions over this collection. To gather the questions and answers, we first analyzed which are the most important fields in the document form in terms of complexity (numbers, dates, candidate's names, checkboxes and different form field layouts) and variability (see Sect. 3.2). We also defined different types of constraints for the types of questions since limiting the questions to find a specific value would place this in a standard word spotting scenario. Thus, we defined different constraints depending on the type of field related to the question: for dates, the questions will refer to the document before, after and between specific dates, or specific years. For other textual fields the questions refer to documents with specific values (candidates from party P), to documents that do not contain specific values (candidates which do not represent the party P), or that contains a value from a set of possibilities (candidates from parties P, Q or R). For checkboxes we defined constraints regarding if a value is checked or not. Finally, according to the fields and constraints defined we posed the questions in natural language referring to the whole collection, and asking for particular values instead of the document itself. We provide the full list of questions in the test set in Table 1.

Once we had the question, we got the answer from the annotations downloaded along with the images. Then, we manually checked that those answers were correct and unambiguous, since some of the original annotations were wrong. Finally, we divided the questions into two different splits; the sample

Table 1. Questions in test set.

ID	Question
8	Which candidates in 2008 were from the Republican party?
9	Which candidates ran for the State Representative office between 06/01/2012 and 12/31/2012?
10	In which legislative counties did Gary L. Schoessler run for County Commissioner?
11	For which candidates was Danielle Westbrook the treasurer?
12	Which candidates ran for election in North Bonneville who were from neither the Republican nor Democrat parties?
13	Did Valerie I. Quill select the full reporting option when she ran for the 11/03/2015 elections?
14	Which candidates from the Libertarian, Independent, or Green parties ran for election in Seattle?
15	Did Suzanne G. Skaar ever run for City Council member?
16	In which election year did Stanley J Rumbaugh run for Superior Court Judge?
17	In which years did Dean A. Takko run for the State Representative office?
18	Which candidates running after 06/15/2017 were from the Libertarian party?
19	Which reporting option did Douglas J. Fair select when he ran for district court judge in Edmonds? Mini or full?

set with 8 questions and the test set with the remaining 12. Given the low variability of the documents layout, we ensured that in the test set there were questions which refer to document form fields or that had some constraints that were not seen in the sample set. In addition, as depicted in Fig. 2 the number of relevant documents is quite variable among the questions, which poses another challenge that methods will have to deal with.

3.2 Statistics and Analysis

We provide in Table 2 a brief description of the document forms fields used to perform the questions and expected answers with a brief analysis of their variability showing the number of values and unique values in their annotations.

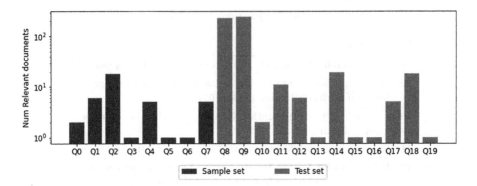

Fig. 2. Number of relevant documents in ground truth for each question in the sample set (blue) and the test set (red). (Color figure online)

3.3 Evaluation Metrics

The ultimate goal of this task is the extraction of information from a collection of documents. However, as previously demonstrated, and especially in unbalanced datasets, models can learn that specific answers are more common to specific questions. One of the clearest cases is the answer *Yes*, to questions that are answered with *Yes* or *No*. To prevent this, we not only evaluate the answer to the question, but also if the answer has been reasoned from the document that contains the information to answer the question, which we consider as evidence. Therefore, we have two different evaluations, one for the evidence which is based on retrieval performance, and the other for the answer, based on text VQA performance.

Table 2. Description of the document forms fields with a brief analysis of their variability showing the number of values and unique values in their annotations.

Field	Type	# Values	# Unique values
Candidate name	Text	14362	9309
Party	Text	14161	10
Office	Text	14362	43
Candidate city	Text	14361	476
Candidate county	Text	14343	39
Election date	Date	14362	27
Reporting option	Checkbox	14357	2
Treasurer name	Text	14362	10197

Evidences: Following standard retrieval tasks [19] we use the Mean Average Precision (MAP) to assess the correctness of the positive evidences provided by the methods. We consider as positive evidences the documents in which the answer to the question can be found.

Answers: Following other text based VQA tasks [5,6] we use the Average Normalized Levenshtein Similarity (ANLS) which captures the model's reasoning capability while smoothly penalizing OCR recognition errors. However, in our case the answers are a set of items for which the order is not relevant, in contrast to common VQA tasks where the answer is a string. Thus, we need to readapt this metric to make it suitable to our problem. We name this adaptation as Average Normalized Levenshtein Similarity for Lists (ANLSL), formally described in Eq. 1. Given a question Q, the ground truth list of answers $G = \{g_1, g_2 \ldots g_M\}$ and a model's list predicted answers $P = \{p_1, p_2 \ldots p_N\}$, the ANLSL performs the Hungarian matching algorithm to obtain a k number of pairs $U = \{u_1, u_2 \ldots u_K\}$ where K is the minimum between the ground truth and the predicted answer lists lengths. The Hungarian matching (Ψ) is performed according to the Normalized Levenshtein Similarity (NLS) between each ground truth element $g_j \in G$ and each prediction $p_i \in P$. Once the matching is performed, all the NLS scores of the $u_z \in U$ pairs are summed and divided for the maximum length of both ground truth and predicted answer lists. Therefore, if there are more or less ground truth answers than the ones predicted, the method is penalized.

$$U = \Psi(NLS(G, P))$$

$$ANLSL = \frac{1}{\max(M, N)} \sum_{z=1}^{K} NLS(u_z) \tag{1}$$

4 Baselines

This section describes the two baselines that are employed in the experiments. Both baselines breakdown the task into two different stages. First, they rank the documents according to the confidence of containing the information to answer a given a question and then, they get the answers from the documents with the highest confidence. The first baseline combines methods from the word spotting and NLP Question Answering fields to retrieve the relevant documents and answer the questions. We name this baseline as *Text spotting + QA*. In contrast, the second baseline is an ad-hoc method specially designed for this task and data, which consist on extracting the information from the documents and map it in the format of key-value relations. In this sense it represents the collection similar as databases do, for which we name this baseline as *Database*. These baselines allows to appreciate the performance of two very different approaches.

4.1   Text Spotting + QA

The objective of this baseline is to set a starting performance result from the combination of two simple but generic methods that will allow to assess the improvement of future proposed methods.

Evidence Retrieval: To retrieve the relevant documents we apply a text spotting approach, which consist on ranking the documents according to a confidence given a query, which in our case is the question. To obtain this confidence, we first run a Part Of Speech (POS) tagger over the question to identify the most relevant words in it by keeping only nouns and digits, and ignore the rest of the words. Then, as described in Eq. 2, given a question Q, for each relevant word in the question $qw_i \in Q$ we get the minimum Normalized Levenshtein Distance (NLD) [18], between all recognized words rw_j extracted through an OCR in the document and the question word. Then, we average over all the distances and use the result as the confidence c for which the document d is relevant to the question.

$$c_d = \frac{1}{|Q|} \sum_{i=1}^{|Q|} \min_{j=1}^{|OCR|} \{NLD(qw_i, rw_j)\} \tag{2}$$

Notice that removing only stopwords is not enough, like in the question depicted in Fig. 1, where the verb *run* is not considered as stopword, but can't be found in the document and consequently would be counterproductive.

Answering: Once the documents are ranked, to answer the given questions we make use of BERT [10] question answering model. BERT is a task agnostic language representation based on transformers [33] that can be afterwards used in other downstream tasks. In our case, we use extractive question answering BERT models which consist on predicting the answer as a text span from a context, usually a passage or paragraph by predicting the start and end indices on that context. Nonetheless, there is no such context in the DocCVQA documents that encompasses all the textual information. Therefore, we follow the approach of [23] to build this context by serializing the recognized OCR tokens on the document images to a single string separated by spaces following a top-left to bottom-right order. Then, following the original implementation of [10] we introduce a start vector $S \in \mathbb{R}^H$ and end vector $E \in \mathbb{R}^H$. The probability of a word i being the start of the answer span is obtained as the dot product between the BERT word embedding hidden vector T_i and S followed by a softmax over all of the words in the paragraph. The same formula is applied to compute if the word i is the end token by replacing the start vector S with the end vector E. Finally, the score of a candidate span from position i to position j is defined as $S \cdot T_i + E \cdot T_j$, and the maximum scoring span where $j \geq i$ is used as a prediction.

4.2   Database Approach

The objective of proposing this baseline is to showcase which is the performance of an ad-hoc method using heuristics and commercial software to achieve the best possible performance. Since obtaining a human performance analysis is near impossible because it would mean that the people involved in the experiment should check more than $14k$ documents for each question, we see this baseline as a performance to beat in a medium-long term.

This approach also breakdown the task in the same retrieval and answering stages. However, in this case the ranking of the results is binary rather indicating if a document is relevant or not. For that, we first run a commercial OCR over the document collection, extracting not only the recognized text, but also the key-value relationship between the field names and their values, including checkboxes. This is followed by a process to correct possible OCR recognition errors for the fields with low variability (field names, parties and reporting options) and normalize all the dates by parsing them to the same format. Finally, we map the key-value pairs into a database like data structure. At the time of answering a question, the fields in the query are compared with those in the stored records. If all the constraints are met, that document is considered relevant and is given a confidence of 1, while otherwise it is given a confidence of 0. Finally, the requested value in the question is extracted from the records of the relevant documents.

It is very important to consider two relevant aspects on this baseline. First, it is a very rigid method that does not allow any modification in the data and therefore, is not generalizable at all. Moreover, it requires a preprocessing that is currently done manually to parse the query from Natural Language to a Structured Query Language (SQL).

5   Results

5.1   Evidences

To initially assess the retrieval performance of the methods, we first compare two different commercial OCR systems that we are going to use for text spotting, Google OCR [12] and Amazon Textract [2]. As reported in Table 3 the performance on text spotting with the latter OCR is better than Google OCR, and is the only one capable of extracting the key-value relations for the database approach. For this reason we use this as the standard OCR for the rest of the text spotting baselines.

Compared to text spotting, the database retrieval average performance is similar. However, as depicted in Fig. 3 we can appreciate that performs better for all the questions but the number 11 where it gets a MAP of 0. This is the result from the fact that the key-value pair extractor is not able to capture the relation between some of the forms fields, in this case the treasurer name, and consequently it catastrophically fails at retrieving documents with specific values on those fields, one of the main drawbacks of such rigid methods. On the other

Table 3. Performance of different retrieval methods.

Retrieval method	MAP
Text spotting (google)	71.62
Text spotting (textract)	72.84
Database	71.06

hand, the questions where the database approach shows a greater performance gap are those where in order to find the relevant documents the methods must search not only documents with a particular value, but understand more complex constraints such as the ones described in Sect. 3.1, which are finding documents between two dates (question 9), after a date (question 18), documents that do not contain a particular value (question 12), or where several values are considered as correct (question 14).

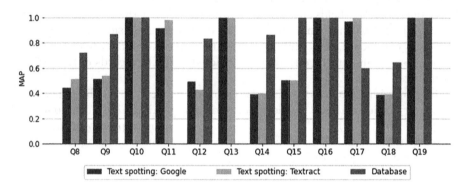

Fig. 3. Evidence retrieval performance of the different methods reported by each question in the test set.

5.2   Answers

For the BERT QA method we use the pretrained weights bert-large-uncased-whole-wordmasking-finetuned-squad from the Transformers library [35]. This is a pretrained model finetuned on SQuAD 1.1 question answering task consisting on more than 100,000 questions over 23,215 paragraphs. Then, we finetune it again on the DocVQA dataset for 2 epochs following [23] to teach the model reason about document concepts as well as adapting the new context style format. Finally, we perform a third finetunning phase on the DocCVQA sample set for 6 epochs. Notice that the sample set is specially small and during these 6 epochs the model only see around 80 samples. Nonetheless, this is sufficient to improve the answering performance without harming the previous knowledge.

Given the collection nature of DocCVQA, the answer to the question usually consists on a list of texts found in different documents considered as relevants. In our case, we consider a document as relevant when the confidence provided for the retrieval method on that document is greater than a threshold. For the text spotting methods we have fixed the threshold through an empirical study where we have found that the best threshold is 0.9. In the case of the database approach, given that the confidence provided is either 0 or 1, we consider relevant all positive documents.

In the experiments we use the BERT answering baseline to answer the questions over the ranked documents from the text spotting and the database retrieval methods. But we only use the database method to answer the ranked documents from the same retrieval approach. As reported in Table 4 the latter is the one that performs the best. The main reason for this is that the wrong retrieval of the documents prevents the answering methods to find the necessary information to provide the correct answers. Nevertheless, the fact of having the key-value relations allows the database method to directly output the value for the requested field as an answer while BERT needs to learn to extract it from a context that has partially lost the spatial information of the recognized text when at the time of being created, the value of a field might not be close to the field name, loosing the semantic connection between the key-value pair. To showcase the answering performance upper bounds of the answering methods we also provide their performance regardless of the retrieval system, where the documents are ranked according to the test ground truth.

Table 4. Baselines results comparison.

Retrieval method	Answering method	MAP	ANLSL
Text spotting	BERT	72.84	0.4513
Database	BERT	71.06	0.5411
Database	Database	71.06	0.7068
GT	BERT	100.00	0.5818
GT	Database	100.00	0.8473

As depicted in Fig. 4, BERT does not perform well when the answer are candidate's names (questions 8, 9, 11, 14 and 18). However, it has a better performance when asking about dates (questions 16 and 17) or legislative counties (question 10). On the other hand, the database approach is able to provide the required answer, usually depending solely on whether the text and the key-value relationships have been correctly recognized.

The most interesting question is the number 13, where none of the methods are able to answer the question regardless of a correct retrieval. This question asks if a candidate selected a specific checkbox value. The difference here is that the answer is *No*, in contrast to the sample question number 3. Then,

BERT can't answer because it lacks of a document collection point of view, and moreover, since it is an extractive QA method, it would require to have a *No* in the document surrounded with some context that could help to identify that word as an answer. On the other hand, the database method fails because of its logical structure. If there is a relevant document for that question, it will find the field for which the query is asking for, or will answer '*Yes*' if the question is a Yes/No type.

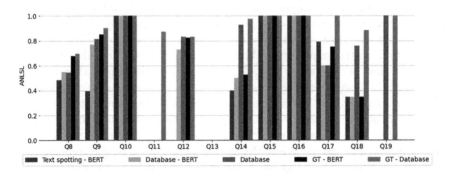

Fig. 4. Answering performance of the different methods reported by each question in the test set.

6   Conclusions and Future Work

This work introduces a new and challenging task to both the VQA and DAR research fields. We presented the DocCVQA that aims to provide a new perspective to Document understanding and highlight the importance and difficulty of contemplating a whole collection of documents. We have shown the performance of two different approaches. On one hand, a text spotting with an extractive QA baseline that, although has lower generic performance it is more generic and could achieve similar performance in other types of documents. And on the other hand, a baseline that represents the documents by their key-value relations that despite of achieving quite good performance, is still far from being perfect and because of its design is very limited and can't generalize at all when processing other types of documents. In this regard, we believe that the next steps are to propose a method that can reason about the whole collection in a single stage, being able to provide the answer and the positive evidences.

Acknowledgements. This work has been supported by the UAB PIF scholarship B18P0070 and the Consolidated Research Group 2017-SGR-1783 from the Research and University Department of the Catalan Government.

References

1. Almazán, J., Gordo, A., Fornés, A., Valveny, E.: Word spotting and recognition with embedded attributes. IEEE Trans. Pattern Anal. Mach. Intell. **36**(12), 2552–2566 (2014)
2. Amazon: Amazon textract (2021). https://aws.amazon.com/es/textract/. Accessed 11 Jan 2021
3. Antol, S., et al.: VQA: visual question answering. In: Proceedings of the IEEE International Conference on Computer Vision, pp. 2425–2433 (2015)
4. Bansal, A., Zhang, Y., Chellappa, R.: Visual question answering on image sets. In: Vedaldi, A., Bischof, H., Brox, T., Frahm, J.-M. (eds.) ECCV 2020, Part XXI. LNCS, vol. 12366, pp. 51–67. Springer, Cham (2020). https://doi.org/10.1007/978-3-030-58589-1_4
5. Biten, A.F., et al.: ICDAR 2019 competition on scene text visual question answering. In: 2019 International Conference on Document Analysis and Recognition (ICDAR), pp. 1563–1570. IEEE (2019)
6. Biten, A.F., et al.: Scene text visual question answering. In: Proceedings of the IEEE/CVF International Conference on Computer Vision, pp. 4291–4301 (2019)
7. Bojanowski, P., Grave, E., Joulin, A., Mikolov, T.: Enriching word vectors with subword information. Trans. Assoc. Comput. Linguist. **5**, 135–146 (2017)
8. Coüasnon, B., Lemaitre, A.: Recognition of tables and forms (2014)
9. Dengel, A.R., Klein, B.: *smartFIX*: a requirements-driven system for document analysis and understanding. In: Lopresti, D., Hu, J., Kashi, R. (eds.) DAS 2002. LNCS, vol. 2423, pp. 433–444. Springer, Heidelberg (2002). https://doi.org/10.1007/3-540-45869-7_47
10. Devlin, J., Chang, M.W., Lee, K., Toutanova, K.: BERT: pre-training of deep bidirectional transformers for language understanding. ACL (2019)
11. Fan, C., Zhang, X., Zhang, S., Wang, W., Zhang, C., Huang, H.: Heterogeneous memory enhanced multimodal attention model for video question answering. In: Proceedings of the IEEE/CVF Conference on CVPR, pp. 1999–2007 (2019)
12. Google: Google OCR (2020). https://cloud.google.com/solutions/document-ai. Accessed 10 Dec 2020
13. Hu, R., Singh, A., Darrell, T., Rohrbach, M.: Iterative answer prediction with pointer-augmented multimodal transformers for TextVQA. In: Proceedings of the IEEE/CVF Conference on Computer Vision and Pattern Recognition (2020)
14. Hull, J.J.: A database for handwritten text recognition research. IEEE Trans. Pattern Anal. Mach. Intell. **16**(5), 550–554 (1994)
15. Kafle, K., Price, B., Cohen, S., Kanan, C.: DVQA: understanding data visualizations via question answering. In: Proceedings of the IEEE Conference on Computer Vision and Pattern Recognition, pp. 5648–5656 (2018)
16. Kahou, S.E., Michalski, V., Atkinson, A., Kádár, Á., Trischler, A., Bengio, Y.: FigureQA: an annotated figure dataset for visual reasoning. arXiv preprint arXiv:1710.07300 (2017)
17. Krishnan, P., Dutta, K., Jawahar, C.V.: Word spotting and recognition using deep embedding. In: 2018 13th IAPR International Workshop on Document Analysis Systems (DAS), pp. 1–6 (2018)
18. Levenshtein, V.I.: Binary codes capable of correcting deletions, insertions, and reversals. In: Soviet Physics doklady, pp. 707–710. Soviet Union (1966)
19. Liu, T.Y.: Learning to rank for information retrieval. Found. Trends Inf. Retr. **3**(3), 225–331 (2009)

20. Liu, X., Gao, F., Zhang, Q., Zhao, H.: Graph convolution for multimodal information extraction from visually rich documents. In: Proceedings of the 2019 Conference of the North American Chapter on Computational Linguistics, pp. 32–39 (2019)
21. Malinowski, M., Fritz, M.: A multi-world approach to question answering about real-world scenes based on uncertain input. arXiv preprint arXiv:1410.0210 (2014)
22. Manmatha, R., Croft, W.: Word spotting: indexing handwritten archives. In: Intelligent Multimedia Information Retrieval Collection, pp. 43–64 (1997)
23. Mathew, M., Karatzas, D., Jawahar, C.: DocVQA: a dataset for VQA on document images. In: Proceedings of the IEEE/CVF WACV, pp. 2200–2209 (2021)
24. Mathew, M., Tito, R., Karatzas, D., Manmatha, R., Jawahar, C.: Document visual question answering challenge 2020. arXiv e-prints, pp. arXiv-2008 (2020)
25. Mouchère, H., Viard-Gaudin, C., Zanibbi, R., Garain, U.: ICFHR 2016 CROHME: competition on recognition of online handwritten mathematical expressions. In: 2016 15th International Conference on Frontiers in Handwriting Recognition (ICFHR) (2016)
26. Palm, R.B., Winther, O., Laws, F.: Cloudscan - a configuration-free invoice analysis system using recurrent neural networks. In: 2017 14th IAPR International Conference on Document Analysis and Recognition (ICDAR), pp. 406–413 (2017)
27. Rath, T.M., Manmatha, R.: Word image matching using dynamic time warping. In: 2003 IEEE Computer Society Conference on Computer Vision and Pattern Recognition, 2003. Proceedings, vol. 2, pp. II-II. IEEE (2003)
28. Ren, M., Kiros, R., Zemel, R.: Exploring models and data for image question answering. arXiv preprint arXiv:1505.02074 (2015)
29. Schuster, D., et al.: Intellix - end-user trained information extraction for document archiving. In: 2013 12th ICDAR (2013)
30. Siegel, N., Horvitz, Z., Levin, R., Divvala, S., Farhadi, A.: FigureSeer: parsing result-figures in research papers. In: Leibe, B., Matas, J., Sebe, N., Welling, M. (eds.) ECCV 2016, Part VII. LNCS, vol. 9911, pp. 664–680. Springer, Cham (2016). https://doi.org/10.1007/978-3-319-46478-7_41
31. Singh, A., et al.: Towards VQA models that can read. In: Proceedings of the IEEE/CVF Conference on Computer Vision and Pattern Recognition (2019)
32. Sudholt, S., Fink, G.A.: Evaluating word string embeddings and loss functions for CNN-based word spotting. In: 2017 14th IAPR International Conference on Document Analysis and Recognition (ICDAR), vol. 01, pp. 493–498 (2017)
33. Vaswani, A., et al.: Attention is all you need. In: NIPS (2017)
34. Wilkinson, T., Lindström, J., Brun, A.: Neural ctrl-f: segmentation-free query-by-string word spotting in handwritten manuscript collections. In: 2017 IEEE International Conference on Computer Vision (ICCV), pp. 4443–4452 (2017)
35. Wolf, T., et al.: Transformers: state-of-the-art natural language processing. In: Proceedings of the 2020 Conference on Empirical Methods in Natural Language Processing: System Demonstrations. Association for Computational Linguistics (2020)
36. Xu, Y., Li, M., Cui, L., Huang, S., Wei, F., Zhou, M.: LayoutLM: pre-training of text and layout for document image understanding. In: KDD 2020, pp. 1192–1200 (2020)
37. Zhang, P., et al.: TRIE: end-to-end text reading and information extraction for document understanding, pp. 1413–1422 (2020)

Dialogue Act Recognition Using Visual Information

Jiří Martínek[1,2]([✉]), Pavel Král[1,2], and Ladislav Lenc[1,2]

[1] Department of Computer Science and Engineering, University of West Bohemia, Plzeň, Czech Republic
[2] NTIS - New Technologies for the Information Society, University of West Bohemia, Plzeň, Czech Republic
{jimar,pkral,llenc}@kiv.zcu.cz

Abstract. Automatic dialogue management including dialogue act (DA) recognition is usually focused on dialogues in the audio signal. However, some dialogues are also available in a written form and their automatic analysis is also very important.

The main goal of this paper thus consists in the dialogue act recognition from printed documents. For visual DA recognition, we propose a novel deep model that combines two recurrent neural networks.

The approach is evaluated on a newly created dataset containing printed dialogues from the English VERBMOBIL corpus. We have shown that visual information does not have any positive impact on DA recognition using good quality images where the OCR result is excellent. We have also demonstrated that visual information can significantly improve the DA recognition score on low-quality images with erroneous OCR.

To the best of our knowledge, this is the first attempt focused on DA recognition from visual data.

Keywords: Dialogue act recognition · Multi-modal · OCR · RNN · Visual information

1 Introduction

Dialogue Act (DA) recognition is a task to segment a dialogue into sentences (or their parts) and to assign them appropriate labels depending on their function in the dialogue [1]. These labels are defined by several taxonomies [2] (e.g. questions, commands, backchannels, etc.).

The standard input is a speech signal which is usually converted into textual representation using an Automatic Speech Recognition (ASR) system [3]. The combination of the following information sources is often considered for recognition: lexical (words in the sentence), prosodic (sentence intonation), and the dialogue history (sequence of the DAs) [4].

However, dialogues are also available in a written form (books and comics), and their automatic analysis is also beneficial for further text analysis. Hence,

© Springer Nature Switzerland AG 2021
J. Lladós et al. (Eds.): ICDAR 2021, LNCS 12822, pp. 793–807, 2021.
https://doi.org/10.1007/978-3-030-86331-9_51

the main goal of this paper is the DA recognition from the documents in a printed form.

Similarly, as in the DA recognition from the audio signal, we first convert the images into a lexical representation using Optical Character Recognition (OCR) methods. We assume that the image form (as the speech signal) can contain some additional information.

Therefore, the main contribution of this paper lies in the usage of visual information for automatic DA recognition from printed dialogues. To the best of our knowledge, there is no prior work that focuses on the DA recognition from printed/handwritten documents.

For evaluation, we create a novel image-based DA recognition dataset from written dialogues. This corpus is based on the dialogues from the VERBMOBIL corpus [5] and the scripts for the creation of such a dataset are available online. These scripts represent another contribution of this work.

We further assume that with the decreasing quality of the printed documents, the importance of the visual text representation will play a more important role for DA recognition, since a recognized text contains greater amount of OCR errors. We will also evaluate this hypothesis using four different image quality in the corpus.

For visual DA recognition, we propose a deep neural network model that combines Convolutional Recurrent Neural Network (CRNN) and Recurrent Neural Network (RNN). We utilize the Bidirectional Long Short-term Memory (BiLSTM) as a recurrent layer in both architectures.

2   Related Work

This section first briefly outlines the DA recognition field and presents popular datasets. Then, we describe recent multi-modal methods that use text and image inputs to improve the performance of a particular task.

Usually, the research in the DA recognition field is evaluated on monolingual standard datasets such as Switchboard (SwDA) [6], Meeting Recorder Dialogue Act (MRDA) [7] or DIHANA [8]. Colombo et al. [9] proposed a *seq2seq* deep learning model with the attention and achieved excellent results that are comparable or even better than current state-of-the-art results.

Shang et al. [10] presented experiments with a deep (BiLSTM-CRF) architecture with an additional extra input representing speaker-change information. The evaluation was conducted on SwDA dataset.

The VERBMOBIL Dialogue Acts corpus [5,11] has been used in the past as a representative of the multi-lingual corpus (see e.g. Reithinger and Klesen [12], Samuel et al. [13] or Martínek et al. in [14]).

Recently, experiments on joining DA recognition and some other Natural Language Processing (NLP) tasks have begun to emerge. Cerisara et al. in [15] presented a multi-task hierarchical recurrent network on joint sentiment and dialogue act recognition. A multi-task recognition which is related to the DA recognition is presented by Li et al. [16]. They utilize the *DiaBERT* model for

DA recognition and sentiment classification and evaluate their approach on two benchmark datasets.

There are efforts to join the visual information with text and improve to some extent text-based NLP tasks. Zhang et al. [17] investigated Named Entity Recognition (NER) in tweets containing also images. They showed that visual information is valuable in the name entity recognition task because some entity word may refer directly to the image included.

Audebert et al. [18] present a combination of image and text features for document classification. They utilize Tesseract OCR together with *FastText* [19] to create character-based embeddings and, in the sequel, the whole document vector representation. For the extraction of image features, they use the *MobileNetv2* [20]. The final classification approach combines both features.

A very nice approach for multi-modal document image classification has been presented by Jain and Wigington in [21]. Their fusion of visual features and semantic information improved the classification of document images.

3   Model Architectures

We describe gradually three models we use for the DA recognition. First of all, we present the visual model that we use for DA recognition based only on image features. Next, we describe our text model and, finally, the joint model that combines both image and text inputs.

3.1   Visual Model

The key component in this model is the Convolutional Recurrent Neural Network (CRNN) that has been successfully utilized for OCR (e.g. [22,23]) and also for image classification [24].

For the visual DA recognition, the input is the image of an entire page of a dialogue where each text line represents an utterance. This page is processed by the *Utterance Segmentation* module that produces segmented images of text lines. These images are fed into the CRNN that maps each utterance to the predicted label. The scheme of this approach is depicted in Fig. 1.

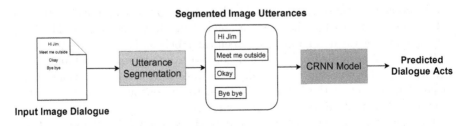

Fig. 1. Visual DA recognition model

Convolutional layers within CRNN create feature maps with relation to the specific receptive fields in the input image. Due to the pooling layers, dimensionality is reduced, and significant image features are extracted, which are further processed by recurrent layers. The recurrent layers are fed by feature sequences (the feature vectors in particular frames in the image).

The CRNN model is depicted in Fig. 2 in two forms: the original model proposed by Shi et al. [22] for OCR and our adapted version for DA recognition.

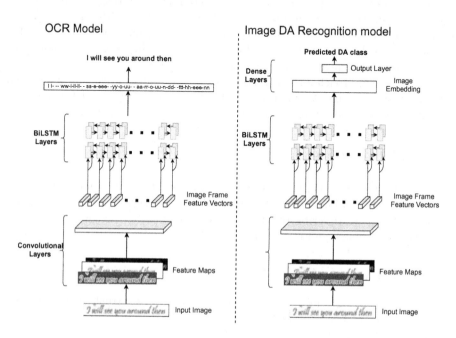

Fig. 2. CRNN models: OCR model proposed by Shi et al. [22] (left); our modified version used for visual DA recognition (right)

The inputs of both models are segmented images of utterances. The activation function of convolutional and recurrent layers is ReLU and we employed the *Adamax* optimizer.

The crucial part of the OCR model is the connectionist temporal classification (CTC) loss function which has been presented by Graves et al. [25]. The CTC is designed to create an alignment between the labels and specific image frames. It allows to use a simple form of annotation, for example, image and annotation text without the necessity of providing the precise character positions in the image. The output of the BiLSTM is given to the output which represents a probability distribution of characters per image frame.

The right part of Fig. 2 visualizes our modified version for the image-based DA recognition. It doesn't utilize the CTC loss function but we use the *categorical cross-entropy* since the output is a vector of probabilities indicating the membership in the particular DA class. The size of the output layer corresponds to the number of recognized DA categories.

3.2 Text Model

The centerpiece of this model is the Bidirectional Long Short-Term memory [26] (BiLSTM). The input utterance is aligned to 15 words, so the utterances with less than 15 words are padded with a special token while the longer ones are shortened. We chose *Word2Vec* [27] embeddings as a representation of the input text. The word vectors (with the dimension equal to 300) are fed into the BiLSTMlayer and the final states of both LSTMs are connected to a dense layer with size 400. Then a DA label is predicted through the softmaxed output layer. The model is depicted in Fig. 3.

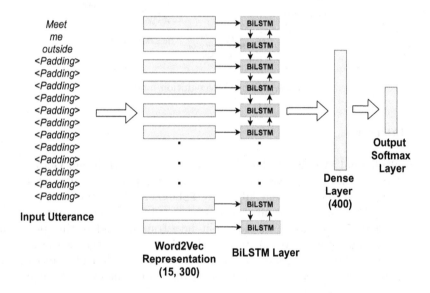

Fig. 3. Text DA recognition model

3.3 Joint Model

The second employment of the CRNN is in the combination with the text model presented in the previous section. The objective is to create a joint model that takes multi-modal input (segmented utterance image and simultaneously the text of an utterance). Figure 4 shows the Joint model with both inputs.

Since the input text doesn't have to be well-recognized, some words which are out of vocabulary might appear resulting in a worse performance of the text model. In such a case, the Image Embedding input should help to balance this loss of text information.

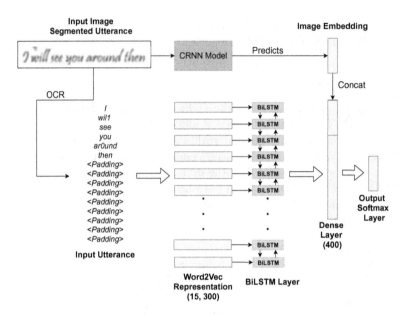

Fig. 4. Joint DA recognition model

4   Dataset

The multi-lingual VERBMOBIL dataset [5,11] contains English and German dialogues, but we limit ourselves only to the English part. The dataset is very
unbalanced. The most frequent labels are FEEDBACK (34%) and INFORM (24%)
while the eight least frequent labels occur in only 1% of utterances or less.

The VERBMOBIL data are already split into training and testing parts and
stored in CONLL format. We created a validation data by taking the last 468
dialogues from the training part. To summarize, we have 8921 utterances in the
training part, 667 utterances as validation data and, finally, 1420 utterances
serve as our test dataset.

4.1   Image Dataset Acquisition

For each dialogue, we have created four pages with image backgrounds of different noise level and programmatically rendered the utterances.

The first background (*noise_0*) contains no noise (perfectly scanned blank
piece of paper) while the fourth level (*noise_3*) contains significant amount of
noise[1].

---

[1] The noise is not artificial (i.e. we didn't perform any image transformation), but we
have created the noise by real usage of the scanner. We put a blank piece of paper
in the scanner and we changed the scanning quality by different scanning options
and the amount of light.

Each rendered utterance is considered as a paragraph. We must take into account, though, the utterances that are too long to fit the page width. In such a case, it continues on the next line and we would struggle with the situation where the beginning of the next utterance and continuing of the current utterance would be indistinguishable. Therefore, we increased the vertical space between paragraphs and we employed the intending of the first line of paragraphs. These two precautions together solve the above-mentioned potential problem and make the segmentation easier.

Another parameter that can be used to adjust the dataset difficulty is the font. We chose the *Pristina Font* which is a hybrid between printed and handwritten font.

Summing up, four steps of the acquisition of the image dataset are as follows.

1. Split original VERBMOBIL CONLL files to the individual dialogues;
2. Create the realistic scanned noisy background;
3. Choose a font;
4. Render the dialogues according to the above-mentioned scenario.

Figure 5 shows the examples of each dataset.

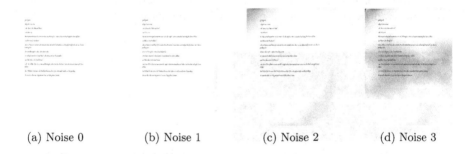

(a) Noise 0   (b) Noise 1   (c) Noise 2   (d) Noise 3

Fig. 5. Page examples from all four datasets

We have also created a second version of each of the four datasets. We have artificially applied random image transformations (rotation, blurring, and scaling). These transformations significantly increase the difficulty of our task because the segmentation and OCR will become harder to perform. We call this version the *transformed dataset* and in the following text, it will be labeled as follows: (*noise_0_trans, noise_1_trans, noise_2_trans, noise_3_trans*).

So in total, we have eight datasets of different noise levels and difficulties. Scripts for the dataset creation are available online[2].

---

[2] https://github.com/martinekj/image-da-recognition.

4.2   Utterance Segmentation

This section describes the algorithm we used for segmentation of the entire page into individual text line images – utterances. We utilized a simple segmentation algorithm based on the analysis of connected components.

We first employed the Sauvola thresholding [28] to binarize the input image that is a necessary step to perform the connected components analysis. Before getting to that, though, we carry out the morphological dilation to merge small neigbouring components that represent fragments of words or individual characters (see Fig. 6). The ideal case is if one text line is one connected component.

Fig. 6. Example of the morphological dilation with kernel $(2, 10)$

Thereafter, the analysis of the connected components is conducted. Figure 7 shows the output of this algorithm. The left part of the image shows the binarized image after morphological dilation while the right part depicts the bounding boxes detected by the analysis of connected components.

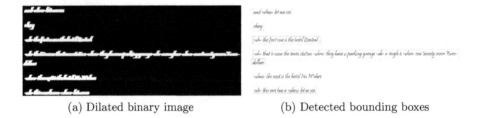

(a) Dilated binary image   (b) Detected bounding boxes

Fig. 7. Utterance segmentation

Once the bounding boxes are obtained, we crop these regions from the image and resize them to the common shape (1475×50). To maintain the image quality of narrow images, we perform the image expansion to the desired width by padding with a white background.

5   Experiments

Within this section, we first present the comparison with state-of-the-art (SoTA) results and then we quantify the difficulties of our datasets by measuring the OCR performance. The next experiment presents results with various sizes of Image Embedding within the visual model and their influences on the overall success rate.

We split the remaining experiments into three scopes to investigate the impact of the visual information in the DA recognition task. The first scope is "image-only" and its goal is to verify the performance of the visual model presented in Sect. 3.1. The second scope is called "text-only" and similarly as the first scope, the goal is to evaluate our text model (see Sect. 3.2). The purpose of the final scope is to find the best joint model which is robust enough to be able to respond to the deteriorating quality of text input. The joint model was presented in Sect. 3.3.

For all experiments, we employed the *Early Stopping* that checks the value of the validation loss to avoid over-fitting. We ran every experiment 5 times and we present average Accuracy, Macro F1-Score, and also Standard Deviation of each run evaluated on the testing part of each dataset.

5.1 Comparison with SoTA

Table 1 compares the results of our text model with state-of-the-art approaches on the testing part of the English VERBMOBIL dataset. This table shows that our results are comparable, but we need to take into account that some approaches in the table utilize the information about the label of the previous utterance. In this work, we did not use this information, since the utterance segmentation from the image is not perfect. Some utterances may be skipped or merged that results in jeopardizing the continuity of the dialogue.

Table 1. Comparison with the state of the art [accuracy in %].

Method	Accuracy
n-grams + complex features [12]	74.7
TBL + complex features [13]	71.2
LSTM + Word2vec features [4]	74.0
CNN + Word2vec features [14]	74.5
Bi-LSTM + Word2vec features [14]	**74.9**
Text model (proposed)	73.9

5.2 OCR Experiment

We use Tesseract as the OCR engine within this work. We measured the OCR performance by calculating the Word Error Rate (WER) and Character Error Rate (CER) against ground truth text in CONLL files.

Tesseract was employed on the testing part (1420 utterances) of each 8 datasets. The results are presented in Table 2 and depicted in Fig. 8. We can conclude that with the increasing difficulty, the WER and CER values are increasing as expected.

Table 2. OCR experiment – Word Error Rate (WER) and Character Error Rate (CER) over all datasets

	Dataset noise level							
	0	1	2	3	0_trans	1_trans	2_trans	3_trans
WER	0.132	0.149	0.132	0.143	0.319	0.322	0.306	0.325
CER	0.049	0.053	0.049	0.053	0.131	0.128	0.128	0.168

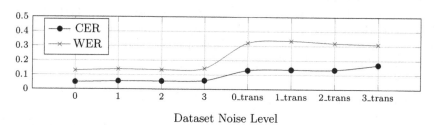

Dataset Noise Level

Fig. 8. Average Word Error Rate (WER) and Character Error Rate (CER) of all datasets

5.3 Image Embedding Dimension

The goal of this experiment is to find the optimal dimension of the Image Embedding (a.k.a the size of the penultimate dense layer in the Visual model).

For this purpose, we limited ourselves only on the dataset with the poorest quality (*noise_3_trans*). We started at dimension equal to 100 and this value was gradually increased by 100. Within each run, a new model with particular embedding size was trained and evaluated. Figure 9 shows the results. We present Accuracy as the evaluation metric. The number of epochs that are needed for training was in the range 8–16 depending on the Early Stopping.

We have here an interesting observation that the amount of information is not increasing with the higher dimension. The best results were obtained with values 400 and 500. So for the next set of experiments, we chose the value of the Image Embedding dimension equal to 500.

5.4 Visual Model Experiment

Table 3 shows the performance of the Visual model. This table illustrates that the results are relatively consistent for all given datasets.

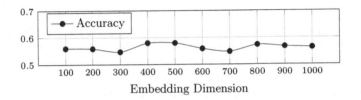

Fig. 9. Experiment to determine the optimal Image Embedding dimension. The standard deviation did not exceed the 0.006 for all runs.

Table 3. Visual model DA recognition results [in %]

| | Dataset noise level | | | | | | | |
	0	1	2	3	0_trans	1_trans	2_trans	3_trans
Macro F1	46.8	54.5	47.2	43.7	47.3	43.7	43.2	40.2
Accuracy	56.6	54.9	57.8	54.6	59.1	56.5	59.4	55.9
Std. Dev.	1.1	0.2	0.6	1.0	0.5	0.2	1.1	1.5

5.5 Text Model Experiment

Within a training phase, the model is fed with a text from the VERBMOBIL dataset while in the evaluation (prediction) phase the input utterances are provided by the OCR. Our intention is to create a real situation where only images with rendered text will be available and the only way to acquire the text itself is to use OCR methods. Table 4 shows Accuracies and Macro F1-scores for all datasets.

Table 4. Text Model DA recognition results [in %]

| | Dataset noise level | | | | | | | | GT |
	0	1	2	3	0_trans	1_trans	2_trans	3_trans	
Macro F1	59.2	59.9	54.8	59.6	50.9	54.2	56.6	52.2	61.6
Accuracy	71.9	70.4	71.8	71.2	56.2	56.4	58.1	56.0	73.9
Std. Dev.	0.5	0.4	0.9	1.1	0.2	1.3	1.3	0.6	0.5

The left part of the table presents the results on not transformed datasets (*Noise_0 – Noise_3*). For these datasets, the OCR results turned out well (see Sect. 5.2 – the average CER value around 0.05), which corresponds to Accuracy exceeding 0.7.

The results on transformed datasets (*Noise_0_trans – Noise_3_trans*) are presented in the right part of the table. The OCR performed significantly worse (average CER in range 0.12–0.16). Hence, the results are worse as well.

For completeness and comparison the rightmost column of Table 4 shows the results when the perfect ground truth text (from the CONNL VERBMOBIL files) is used instead of the recognized text. This Accuracy is used for the comparison with state of the art (Sect. 5.1).

Last but not least, for transformed datasets, in terms of Accuracy, the Image and Text model performed similarly. For datasets without transformation, the Text model was significantly better, primarily due to the less amount of recognition errors.

5.6   Joint Model Experiment

The fact that it is possible to successfully train a Visual DA recognition model based solely on images with reasonable results brought us to the idea to use learned image features in combination with text to create the joint model. We assume that it might have better adaption to recognized text with a significant amount of errors.

Similar to the text model, to simulate the real situation, the ground truth text from the VERBMOBIL dataset is used to train the model while a recognized text from the OCR is used to test the model to verify its generalization. As long as the very same text is used in both text and joint model, it is very easy to verify and measure the positive impact and the contribution of the visual information.

Our final experiment shows, among other things, the impact of the information which was embedded into a single image feature vector (Image Embedding) by the CRNN model. Based on the preliminary experiment, we chose the dimension of embedding equal to 500.

We have eight stored CRNN models that have been trained separately on particular datasets. We remind that the training of the joint model was carried out in the same way as the training of the text model. The only difference from the previous Text Model experiment is the usage of an auxiliary image input which is predicted by the CRNN model as depicted in Fig. 4. We present the results in Table 5.

Table 5. DA recognition results with Joint model [in %]

	Dataset noise level							
	0	1	2	3	0_trans	1_trans	2_trans	3_trans
Macro F1	49.6	56.8	50.3	51.9	48.3	46.8	54.2	49.3
Accuracy	60.2	60.6	61.1	63.3	61.2	59.9	66.4	60.1
Std. Dev.	0.5	0.2	0.4	0.2	0.4	0.6	0.6	3.1

As you can notice, the results no longer oscillate so much across all datasets. Another important observation is that some transformed dataset results outperformed results based on the not transformed datasets (compare *Noise_0* and

Noise_2 with their transformed versions). The help of auxiliary image input has a bigger impact on transformed datasets where the amount of noise is massive and vice versa.

Figure 10 shows the visual comparison of all models we used in our experiments. The blue curve shows Visual Model results, the red line represents Text Model results and green line depicts the performance of the Joint model (text and image input).

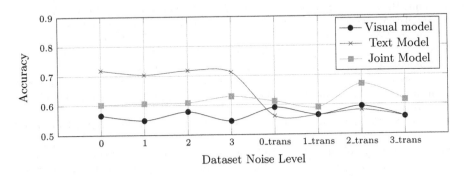

Fig. 10. Depicted results and comparison of all models

As expected, in the case of a better quality of recognized text (*Noise_0 – Noise_3*), the performance of the text model is the best. However, if the quality of the recognized text is low (*Noise_0_trans – Noise_3_trans*), Accuracy and Macro F1-score decrease.

6 Conclusions

The paper has dealt with the task of dialogue act recognition in the written form using a model with multi-modal inputs. The goal of this paper has been twofold.

First, we have successfully employed the CRNN model as the visual model for image-based DA recognition. We have shown that despite employing only visual features it is possible to obtain reasonable results in the task that is dominantly text-based.

Second, we have carried out a set of experiments where we have used the same CRNN model as an image feature extractor and we have combined it with BiLSTM text model for handling both text input (obtained by OCR) and image input. We have successfully extracted the hidden layer representation of the CRNN model (Image Embedding) and together with the text model we have created the joint model. For poor-quality datasets, where the OCR success rate is low, we have outperformed the text model that uses solely text input.

Hereby, we have shown that the visual information is beneficial and the loss of the text information is partially compensated. The impact of such image features results in improving Accuracy (4%–10%) depending on the noise level in the particular dataset.

Acknowledgements. This work has been partly supported from ERDF "Research and Development of Intelligent Components of Advanced Technologies for the Pilsen Metropolitan Area (InteCom)" (no.: CZ.02.1.01/0.0/0.0/17_048/0007267) and by Grant No. SGS-2019-018 Processing of heterogeneous data and its specialized applications. We would like to thank also Mr. Matěj Zeman for some implementation work.

References

1. Bunt, H.: Context and dialogue control. Think Quarterly **3**(1), 19–31 (1994)
2. Stolcke, A., et al.: Dialogue act modeling for automatic tagging and recognition of conversational speech. Comput. Linguist. **26**(3), 339–373 (2000)
3. Frankel, J., King, S.: ASR-articulatory speech recognition. In: Seventh European Conference on Speech Communication and Technology (2001)
4. Cerisara, C., Král, P., Lenc, L.: On the effects of using word2vec representations in neural networks for dialogue act recognition. Comput. Speech Lang. **47**, 175–193 (2018)
5. Jekat, S., Klein, A., Maier, E., Maleck, I., Mast, M., Quantz, J.J.: Dialogue acts in verbmobil (1995)
6. Godfrey, J.J., Holliman, E.C., McDaniel, J.: Switchboard: telephone speech corpus for research and development. In: Proceedings of the 1992 IEEE International Conference on Acoustics, Speech and Signal Processing - Volume 1, ser. ICASSP 1992, pp. 517–520. IEEE Computer Society, USA (1992)
7. Shriberg, E., Dhillon, R., Bhagat, S., Ang, J., Carvey, H.: The ICSI meeting recorder dialog act (MRDA) corpus. Technical report, International Computer Science Institute, Berkely (2004)
8. Benedı, J.-M., et al.: Design and acquisition of a telephone spontaneous speech dialogue corpus in Spanish: Dihana. In: Fifth International Conference on Language Resources and Evaluation (LREC), pp. 1636–1639 (2006)
9. Colombo, P., Chapuis, E., Manica, M., Vignon, E., Varni, G., Clavel, C.: Guiding attention in sequence-to-sequence models for dialogue act prediction. In: Proceedings of the AAAI Conference on Artificial Intelligence, vol. 34, no. 05, pp. 7594–7601 (2020)
10. Shang, G., Tixier, A.J.-P., Vazirgiannis, M., Lorré, J.-P.: Speaker-change aware CRF for dialogue act classification, arXiv preprint arXiv:2004.02913 (2020)
11. Alexandersson, J., et al.: Dialogue acts in Verbmobil 2. DFKI Saarbrücken (1998)
12. Reithinger, N., Klesen, M.: Dialogue act classification using language models. In: Fifth European Conference on Speech Communication and Technology (1997)
13. Samuel, K., Carberry, S., Vijay-Shanker, K.: Dialogue act tagging with transformation-based learning, arXiv preprint cmp-lg/9806006 (1998)
14. Martínek, J., Král, P., Lenc, L., Cerisara, C.: Multi-lingual dialogue act recognition with deep learning methods, arXiv preprint arXiv:1904.05606 (2019)
15. Cerisara, C., Jafaritazehjani, S., Oluokun, A., Le, H.: Multi-task dialog act and sentiment recognition on mastodon, arXiv preprint arXiv:1807.05013 (2018)
16. Li, J., Fei, H., Ji, D.: Modeling local contexts for joint dialogue act recognition and sentiment classification with bi-channel dynamic convolutions. In: Proceedings of the 28th International Conference on Computational Linguistics, pp. 616–626 (2020)
17. Zhang, Q., Fu, J., Liu, X., Huang, X.: Adaptive co-attention network for named entity recognition in tweets. In: Proceedings of the AAAI Conference on Artificial Intelligence, vol. 32, no. 1 (2018)

18. Audebert, N., Herold, C., Slimani, K., Vidal, C.: Multimodal deep networks for text and image-based document classification. In: Cellier, P., Driessens, K. (eds.) ECML PKDD 2019. CCIS, vol. 1167, pp. 427–443. Springer, Cham (2020). https:// doi.org/10.1007/978-3-030-43823-4_35
19. Joulin, A., Grave, E., Bojanowski, P., Douze, M., Jégou, H., Mikolov, T.: Fasttext.zip: compressing text classification models, arXiv preprint arXiv:1612.03651 (2016)
20. Sandler, M., Howard, A., Zhu, M., Zhmoginov, A., Chen, L.-C.: Mobilenetv2: inverted residuals and linear bottlenecks. In: Proceedings of the IEEE Conference on Computer Vision and Pattern Recognition, pp. 4510–4520 (2018)
21. Jain, R., Wigington, C.: Multimodal document image classification. In: 2019 International Conference on Document Analysis and Recognition (ICDAR), pp. 71–77. IEEE (2019)
22. Shi, B., Bai, X., Yao, C.: An end-to-end trainable neural network for image-based sequence recognition and its application to scene text recognition. IEEE Trans. Pattern Anal. Mach. Intell. **39**(11), 2298–2304 (2017)
23. Martínek, J., Lenc, L., Král, P., Nicolaou, A., Christlein, V.: Hybrid training data for historical text OCR. In: 2019 International Conference on Document Analysis and Recognition (ICDAR), pp. 565–570. IEEE (2019)
24. Han, L., Kamdar, M.R.: MRI to MGMT: predicting methylation status in glioblastoma patients using convolutional recurrent neural networks (2017)
25. Graves, A., Fernández, S., Gomez, F., Schmidhuber, J.: Connectionist temporal classification: labelling unsegmented sequence data with recurrent neural networks. In: Proceedings of the 23rd International Conference on Machine Learning, pp. 369–376. ACM (2006)
26. Hochreiter, S., Schmidhuber, J.: Long short-term memory. Neural Comput. **9**(8), 1735–1780 (1997)
27. Mikolov, T., Chen, K., Corrado, G., Dean, J.: Efficient estimation of word representations in vector space, arXiv preprint arXiv:1301.3781 (2013)
28. Sauvola, J., Pietikäinen, M.: Adaptive document image binarization. Pattern Recogn. **33**(2), 225–236 (2000)

Are End-to-End Systems Really Necessary for NER on Handwritten Document Images?

Oliver Tüselmann[✉][ID], Fabian Wolf[ID], and Gernot A. Fink[ID]

Department of Computer Science, TU Dortmund University,
44227 Dortmund, Germany
{oliver.tueselmann,fabian.wolf,gernot.fink}@cs.tu-dortmund.de

Abstract. Named entities (NEs) are fundamental in the extraction of information from text. The recognition and classification of these entities into predefined categories is called Named Entity Recognition (NER) and plays a major role in Natural Language Processing. However, only a few works consider this task with respect to the document image domain. The approaches are either based on a two-stage or end-to-end architecture. A two-stage approach transforms the document image into a textual representation and determines the NEs using a textual NER. The end-to-end approach, on the other hand, avoids the explicit recognition step at text level and determines the NEs directly on image level. Current approaches that try to tackle the task of NER on segmented word images use end-to-end architectures. This is motivated by the assumption that handwriting recognition is too erroneous to allow for an effective application of textual NLP methods. In this work, we present a two-stage approach and compare it against state-of-the-art end-to-end approaches. Due to the lack of datasets and evaluation protocols, such a comparison is currently difficult. Therefore, we manually annotated the known IAM and George Washington datasets with NE labels and publish them along with optimized splits and an evaluation protocol. Our experiments show, contrary to the common belief, that a two-stage model can achieve higher scores on all tested datasets.

Keywords: Named entity recognition · Document image analysis · Information retrieval · Handwritten documents

1 Introduction

Named entities (NEs) are objects in the real world, such as persons, places, organizations and products, that can be referred to by a proper name. They are known to play a fundamental role in the extraction of information from text. Thereby, NEs are not only used as a first step in question answering, search engines or topic modeling, but they are also one of the most important information for indexing documents in digital libraries [32]. The extraction of

© Springer Nature Switzerland AG 2021
J. Lladós et al. (Eds.): ICDAR 2021, LNCS 12822, pp. 808–822, 2021.
https://doi.org/10.1007/978-3-030-86331-9_52

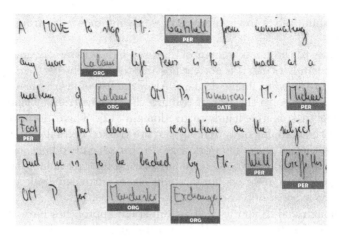

Fig. 1. An example for named entities on a handwritten document image from the IAM database.

this information from text is called Named Entity Recognition (NER) and is an important field of research in Natural Language Processing (NLP). Over the past few years, impressive progress has been made in this area [33]. The Document Image Analysis community, on the other hand, focuses on the recognition and retrieval of handwritten document images and less on the extraction of information from those. Therefore, there are only a few publications that try to tackle NER on document images. Figure 1 shows an example for NEs on a handwritten document image.

An intuitive approach for NER on document images is to combine the advances from the Document Image Analysis and NLP domain, using a two-stage model. The first stage is the transformation of the document into a textual representation. After that, the relevant NEs are extracted using an NLP model. Unfortunately, despite advances in machine learning, the recognition approaches are still not perfect and could produce high Character Error Rates (CERs) and Word Error Rates (WERs), especially on handwritten documents. An interesting question is whether and how errors from the recognition stage affect the performance of an NLP model. Especially in the last few years, there have been several publications that have studied the effect of Optical Character Recognition (OCR) errors on NLP tasks [6,13,14]. It is generally agreed that OCR errors have a negative impact on the performance of NLP models and that their performance degrades when CERs and WERs increase. To overcome the problem of error propagation, an end-to-end architecture is often used. Such an approach avoids the explicit recognition step and predicts the NEs directly from word images. Even though this approach can solve the error propagation problem, the architecture has the fundamental drawback of not using the most powerful advantages from the NLP domain. These are the use of pre-trained word embeddings on large datasets and transfer learning [33]. A word embedding is an encoded knowledge base containing semantic relations between words.

Transfer learning refers to the process of applying knowledge from one task to another.

Mainly due to the small amount of digitized handwritten documents and the variability of handwriting, it is not yet possible to apply the performance of word embeddings in the word image domain. Also, the use of synthetically generated documents cannot solve the lack of training data, since transfer learning is still a challenging task in the handwriting domain [12]. There are already works dealing with the mapping of word images to pre-trained textual word embeddings [16,30]. Nevertheless, they only work with static word embeddings, being limited. Therefore, end-to-end approaches can only learn from the available training data in the image domain and take into account either extremely limited or no prior semantic knowledge about words.

Even though end-to-end as well as two-stage approaches have their advantages and disadvantages for NER on word-segmented handwritten document images containing unstructured text, only end-to-end architectures are published so far. This is motivated by the assumption that handwriting recognition is too erroneous to allow for an effective application of textual NLP methods. However, Hamdi et al. successfully applied a two-stage approach in [14] and showed recently on a machine-printed Finnish historical newspaper with a CER of 6.96% and a WER of 16.67% that there is only a marginal decrease in the F1-score compared to the original text (89.77% to 87.40%). Since such error rates can also be achieved with a state-of-the-art recognizer on most handwritten datasets, we investigate in this work a two-stage approach and compare it to end-to-end methods from the literature. Such a comparison is not straightforward as there are almost no established datasets and evaluation protocols. Most approaches from the literature use their own, unpublished annotations and evaluation protocols, making a direct comparison impossible. In order to provide comparability, we also present and publish NE annotations for the well-known George Washington (GW) and IAM datasets as well as an evaluation protocol[1].

2 Related Work

Traditional NER methods are mainly implemented based on handcrafted rules, dictionaries, orthographic features or ontologies [33]. Further progress in this field has been achieved using statistical-based methods such as Hidden Markov Models and Conditional Random Fields (CRFs) [31]. In recent years, a large number of deep neural network approaches have emerged, which greatly improved the recognition accuracy. In this regard, combinations of recurrent neural networks and CRFs have been particularly successful [17]. The state-of-the-art methods show that word embeddings have a fundamental influence on the performance [21]. For a detailed overview of NER in the textual domain, see [33].

The extraction of information from document images has so far been rather a minor field of research in Document Image Analysis. However, the number of publications in recent years show that the interest in this topic has increased

---

[1] https://patrec.cs.tu-dortmund.de/cms/en/home/Resources/index.html.

considerably [1,2,7,9,10,26,29]. Publications can be grouped according to their focus on machine-printed or handwritten document images. On modern machine-printed document images, recognition errors are usually so small that the application of NLP tasks can be considered solved. The situation is different with respect to historical documents, such as old newspapers. In these cases, recognition quality influences the results considerably and, therefore, there is still an active field of research [13,14]. Due to the high variability of handwriting, recognition of handwritten document images can generally be considered a more difficult problem compared to machine-printed ones. This problem is also true for the task of information extraction. Therefore, the current trend in contrast to approaches on machine-printed documents is to use the presumably more robust end-to-end architectures. The NER approaches on handwritten document images can be divided into segmentation-free (i.e., entire document images are used without any segmentation) [7,9,10] and segmentation-based (i.e., a word or line level segmentation is assumed) [1,2,26,29]. One of the first approaches for NER on handwritten word images was provided by Adak et al. in [1]. In their approach, they use handcrafted features to decide whether a word image is a NE or not. Their approach only allows for detecting NEs and not for classifying them into predefined classes. At the International Conference on Document Analysis and Recognition in 2017, there was the Information Extraction in Historical Handwritten Records (IEHHR) competition [11]. It focuses on the automatic extraction of NEs on semi-structured historical handwritten documents. Approaches that initially performed recognition have won on both word and line segmentation. After the competition, Toledo et al. proposed two end-to-end architectures in [29] and evaluated them on the competition dataset. Their Bidirectional Long Short-Term Memory (BLSTM)-based approach is able to outperform the state-of-the-art results, presented in the competition. The end-to-end approach of Rowtula et al. [26] focuses on the extraction of NEs from handwritten documents containing unstructured text and has been trained and evaluated on automatically generated NE tags for the IAM database. They observed that NEs are related to the position and distribution of Part-of-Speech (PoS) tags in a sentence. Therefore, they first train their model on PoS tag prediction using the CoNLL2000 dataset [27] and synthetically generated word images. Then, they specialize the pre-trained model towards the real data, beginning with the prediction of PoS tags and lastly predicting the NE tags. Recently, Adak et al. proposed an approach for word images from Bengali manuscripts in [2]. They extract patches from a single word image using a sliding window approach. Then, they extract a feature representation for each patch using their self designed convolutional architecture. The features are further encoded using a BLSTM. They apply attention weights in order to concentrate on the relevant patches. For each patch, a distribution over the NE classes is predicted and finally averaged across all patches. The highest score in the averaged distribution is then predicted as the NE class for the word image.

3   Datasets

The semi-structured dataset, used in the IEHHR competition (Sect. 3.1), is so far the only established dataset in the field of NER on handwritten word images. Recently, Carbonell et al. evaluated two more datasets in [9], namely War Refugees and synthetic Groningen Meaning Bank (Sect. 3.2). However, due to privacy reasons, only the synthetically generated dataset is publicly available. In order to enable evaluation on unstructured handwritten text, we created NE annotations for the George Washington dataset (Sect. 3.3) and the IAM database (Sect. 3.4). The main difference compared to the datasets in [1] and [26] is that the annotations were generated entirely manually, which avoids the errors caused by automatic taggers. In this section, we also show that the commonly known partitionings into training, validation and test set on GW and IAM are unsuitable for NER and present optimized splits. It is important to note that beside for the dataset in the IEHHR competition, all tag sets have a default class called O. This class is assigned to every word image that is not part of the predefined categories. Usually, there exists a huge class imbalance in every NE dataset with around 90% towards the O class.

3.1   Esposalles

The ESPOSALLES database [25] is an excerpt of a larger collection of historical handwritten marriage license books at the archives of the Cathedral of Barcelona. The corpus is written in old Catalan by a single writer in the 17th century. For the database, both line and word segmentations are available. The marriage records generally have a fixed structure, although there are variations in some cases. Therefore, the dataset can be considered semi-structured. For the IEHHR competition [11], 125 pages of this database were annotated with semantic information. There is an official partitioning into training and test data, containing 968 training and 253 test records. Each word is labeled with a category *(name, surname, occupation, location, state, other)* and a person *(husband, wife, husbands_father, husbands_mother, wifes_father, wifes_mother, other_person, none)*.

3.2   Synthetic Groningen Meaning Bank

The synthetic Groningen Meaning Bank (sGMB) dataset [9] consists of synthetically generated handwritten document pages obtained from the corpus of the Groningen Meaning Bank [8]. It contains unstructured English text mainly from a newspaper, whereby the words have been labeled with the following categories: *Geographical Entity, Organization, Person, Geopolitical Entity and Time indicator*. There is an official split containing 38048 training, 5150 validation and 18183 test word images. A possible disadvantage for the identification of NEs is the absence of punctuation marks in this dataset.

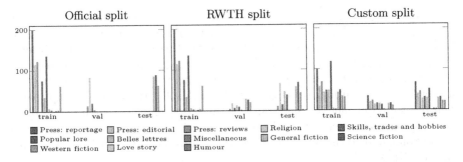

Fig. 2. The amount of document pages per genre in training, validation and test set. The histograms are shown for the official, RWTH and custom split of the IAM database.

3.3 George Washington

The George Washington (GW) dataset [23] has become the de-facto standard benchmark for word spotting. It consists of 20 pages of correspondences between George Washington and his associates dating from 1755. The documents were written by a single person in historical English. There are no publicly available semantic annotations for this dataset. Therefore, we created those manually and make them available for the community. The word images are labeled with the following categories: *Cardinal, Date, Location, Organization* and *Person*. Multiple splits have been published [5,24]. Unfortunately, they are unsuitable for our task, since NER is a sequence labeling problem and some sentences contain both training and validation words. Therefore, we present a more suitable split, which divides the documents into twelve training, two validation and six test pages. The partitioning was formulated as an optimization problem and solved using Answer Set Programming [18]. This involved dividing the pages from the dataset into training, validation and test data such that the NE categories are best split with respect to the ratio of 6:1:3. For this dataset, the transcription of a word image is only given in lowercase characters and does not contain punctuation. This could be a challenge for NER, since capitalization and punctuation are presumably important features for the identification of NEs.

3.4 IAM DB

The IAM Database [20] is a major benchmark for handwriting recognition and word spotting. The documents contain modern English sentences from the Lancaster - Oslo - Bergen Corpus [28] and were written by a total of 657 different people. The pages contain text from the genres listed in Fig. 2. The database consists of 1539 scanned text pages containing a total of 13353 text lines and 115320 words. The official partitioning splits the database in 6161 lines for training, 1840 for validation and 1861 for testing. These partitions are writer-independent, such that each writer contributed to only one partition (either training, validation or test). As in the GW dataset, the official word-level partitioning unfortunately has the disadvantage that some sentences contain both training and validation data, which is unsuitable for NER. The official line segmentation does not have that problem. However, Fig. 2 shows that the distribution of text categories strongly differs from split to split. Therefore, the training data is not representative for the test and validation data. There is also another split specifically designed for handwriting recognition, referred to as RWTH split. Here the lines are partitioned into writer-independent training, validation and test partitions of 6161, 966 and 2915 lines, respectively. Even if the distribution of genres is considerably better compared to the official partitioning, Fig. 3 shows that it is a suboptimal split for NER.

Since it is essential for the training data to be representative for the test, which is not the case with the available splits, we propose a novel split. The partitioning was formulated as an optimization problem such that the documents from each text category are split as best as possible in the ratio of 3:1:2 between training, validation and test, while remaining writer-independent. Figure 3 shows that optimizing this criteria also improves the 3:1:2 ratio within the NE categories considerably. For annotating the IAM database, the same tag set was used as in OntoNotes Release 5.0 [22]. The tag set contains 18 categories that are well-defined in their published annotation guideline[2]. The categories are: *Cardinal, Date, Event, FAC, GPE, Language, Law, Location, Money, NORP, Ordinal, Organization, Person, Percent, Product, Quantity, Time* and *Work_of_art*. As there is a relatively small training set compared to the datasets used in the NLP domain, only a few examples exist for most categories. To overcome that problem, we also developed a smaller tag set that summarizes categories as best as possible and removes severely underrepresented categories. This results in a tag set consisting of only six categories: *Location (FAC, GPE, Location), Time (Date, Time), Cardinal (Cardinal, Ordinal, Percent, Quantity, Money), NORP, Person* and *Organization*. Furthermore, we use the official sentence segmentation of the dataset for all splits.

4   Two-Staged Named Entity Recognition

For the comparison of end-to-end and two-stage approaches with respect to NER on segmented word images, we propose a two-stage approach. This approach is based on a state-of-the-art handwriting recognizer (HTR) and a NER model. The HTR was proposed by Kang et al. [15] and it is an attention-based sequence-to-sequence model for handwritten word recognition. It works on character-level and does not require any information about the language, except for an alphabet. The approach also has the advantage that it does not require any dataset-specific pre-processing steps. Therefore, the model produces satisfying results on most datasets and not only on a specific one. The NER model roughly follows the state-of-the-art architecture proposed by Lample et al. [17]. Here, the input words are first converted into a vector representation using a pre-trained word embedding model. For our datasets, the pre-trained RoBERTa base model of Huggingface

---

[2] Annotation guideline available at: https://bit.ly/3pyte8Q.

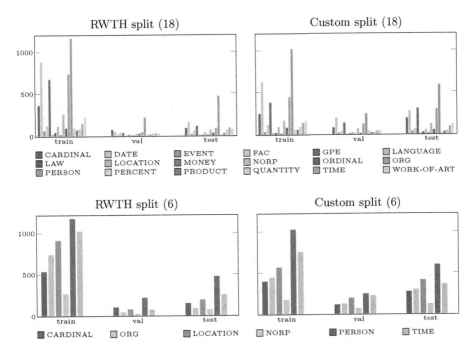

Fig. 3. The number of named entities in the training, validation and test partitions for the IAM database. Histograms are shown for the different combinations of splits (RWTH, Custom) and tag sets (six, eighteen).

[19] was shown to give the best results. Afterwards, these representations are encoded using a one layered BLSTM with a hidden layer size of 256. Finally, a CRF is used to predict NE tags for each word based on its encoding from the hidden layer. We implemented the NER model using the Flair framework [3].

The first step of our approach is to feed all word images into the recognizer in their order of occurrence on the document pages. If a sentence segmentation is available, the recognition results are divided accordingly. Otherwise, the entire page is defined as a single sentence. Finally, the sentences are processed sequentially by the NER model, which assigns a tag to each word image.

For the recognition model, we do not make any changes regarding the hyperparameters. We only customize the size of the input images, the maximum word length and the alphabet for each dataset. For the training of the NLP models in our approach, we use a mini batch size of 64. We update the model parameters using the SGD optimization procedure and the CRF loss. The learning rate is initially set to 0.1 and decreased by a factor of two whenever the F1-score on the validation data is not improving for five iterations. Since each word is to be represented by exactly one vector and the RoBERTa model uses subword tokenization [19], we require a pooling strategy. This is especially important for the HTR results, because the RoBERTa model divides words that are not part of its vocabulary, known as Out-of-Vocabulary (OOV) words, into several subwords. Therefore, we represent a word as the average of all its subword representations. We also use a technique called scalar mix [21] that computes a parameterized scalar mixture of Transformer model layers. This technique is very useful, because for some downstream tasks like NER or PoS tagging it can be unclear, which layers of a Transformer-based model perform well. Another design decision is to use a trainable linear mapping on top of the embedding layer. This mapping ensures that the input to the BLSTM is learnable and does not come directly from the word embedding model. Given that the text in the Esposalles dataset is written in Spanish, we adapt the embedding approach by using a Spanish pre-trained flair embedding [4].

5   Experiments

In this section, we evaluate our two-stage approach on the four datasets introduced in Sect. 3. We first describe the evaluation protocol in Sect. 5.1. We then show the results in Sect. 5.2 and compare our approach with two state-of-the-art end-to-end approaches for NER on segmented word images. For a fair comparison, we replicated the end-to-end models as best as possible and evaluate them with the same protocol and data. Finally, we discuss some potential methods for improving the robustness of our two-stage approach in Sect. 5.3.

5.1   Evaluation Protocol

The F1-score is a suitable measure for evaluating NER models. However, there are several definitions of this measure, with macro and micro F1 being the most popular ones. In our experiments, we use the macro F1-score, which first computes the metric independently for each class and we finally average these scores using the arithmetic mean. Therefore, all classes are considered equally, preventing the score from being dominated by a majority class. It is important to note that we exclude the O class in our evaluation. The F1-score can be interpreted as a weighted average of the precision (P) and the recall (R) and is formally defined as shown in Eq. 1. Precision is the number of correctly predicted labels for a class divided by the number of all predicted labels for that class and recall is the number of correctly predicted labels for a class divided by the number of relevant labels for that class. Precision, recall and F1-scores are calculated per class and are finally averaged. It may happen that there is no element in the test set for a class, but the class was predicted for an element. In this case, the recall is to be defined as 0. It is also possible that there are no predicted labels for a class. In this case, the precision score should be set to 0.

$$F1 = 2 * \frac{precision * recall}{precision + recall} \qquad (1)$$

Table 1. Handwriting recognition rates measured in Character Error Rate (CER) and Word Error Rate (WER) for the IAM, Esposalles, GW and sGMB datasets. For the IAM database, error rates are reported for both the RWTH and custom split.

Dataset	Dictionary free		Dictionary	
	CER	WER	CER	WER
IAM (RWTH)	6.80	17.67	6.20	10.70
IAM (Custom)	7.05	18.66	6.36	10.76
Esposalles	2.47	5.15	2.27	3.28
GW	5.24	14.52	4.10	6.05
sGMB	4.93	15.46	4.09	7.38

5.2   Results

In our comparison, we consider two state-of-the-art end-to-end models for NER on segmented word images. The first approach is the BLSTM-based model proposed by Toledo et al. [29]. We select this approach because it achieves state-of-the-art results on the IEHHR challenge dataset and it is able to outperform two-stage approaches. The other model was proposed by Rowtula et al. [26] and is currently the state-of-the-art approach on unstructured English word images.

In order to get an impression of the scores that can be achieved on the datasets and to obtain an indication of how the HTR errors affect the task, we also evaluate the NER model of our approach on perfect recognition results. For this purpose, the model receives the text annotations of the word images instead of the HTR results as input. In the following, we denote this approach as Annotation-NER.

As mentioned before, the HTR model used in our two-stage approach does not use any linguistic resources during recognition, such as language models or dictionaries. While this approach has the advantage of not penalizing OOV words, it also often helps with minor recognition errors. Since it is very unlikely to have no linguistic information about a given dataset in a real situation, we evaluate the other extreme case where there is a fixed vocabulary. For a dataset, this consists of its training, validation and test words. In the following, we denote our two-stage approach with HTR-NER if we do not use a dictionary and with HTR-D-NER otherwise.

The first step of our approach is to perform recognition for all word images in a given dataset. Table 1 shows the CERs and WERs of the test data for the four datasets. We report similar error rates as published in the literature [15]. Improvements are obviously possible with further optimizations and dataset-specific adaptations. However, this is not crucial for our comparison.

IAM. The results in Table 2 show that our custom split leads to a considerable performance increase in comparison to the RWTH split. This could be expected, as the training data is more representative regarding the validation and test

Table 2. Named Entity Recognition performances for the IAM database measured in precision (P), recall (R) and macro-F1 (F1) scores. Results are shown for the different combinations of splits (RWTH, custom) and tag sets (six, eighteen).

Method	RWTH (6)			Custom (6)			RWTH (18)			Custom (18)		
	P	R	F1	P	R	F1	P	R	F1	P	R	F1
Annotation-NER	83.8	77.5	80.1	87.3	87.6	87.5	63.6	57.6	59.8	68.5	61.0	63.5
HTR-D-NER	78.6	73.0	75.4	83.7	78.7	81.0	60.4	50.9	54.2	62.5	53.3	56.3
HTR-NER	77.3	65.9	70.7	83.3	71.0	76.4	55.8	50.1	52.0	64.8	47.5	53.6
Rowtula et al. [26]	58.8	41.3	47.4	65.5	47.6	54.6	33.8	30.9	32.3	36.9	28.0	30.3
Toledo et al. [29]	45.3	28.8	34.0	50.2	31.4	37.4	26.4	10.8	14.9	35.4	13.4	18.0

sets. Our experiments show that the prediction of 18 categories constitutes a hard task, resulting in low F1-scores. This is probably due to the small size of the training data, which leads to very few examples for some categories. Thus, the training set is not representative for the categories and the prediction of these is extremely difficult. Good results can be obtained with the tag set consisting of six categories and the Annotation-NER approach. Based on the results, it is also obvious that our two-stage approaches have large drops in comparison to the NER model working on the annotations. The additional use of a dictionary is able to reduce the errors in recognition and thus achieves better results for NER. The dictionary, thereby, greatly increases the recall scores, but shows similar performance in terms of precision. The two end-to-end approaches could only achieve comparatively low scores. We assume that the approach proposed by Toledo et al. performs rather poorly because it was developed for semi-structured data and cannot handle the strong imbalance on unstructured data. The model of Rowtula et al. is able to deliver considerably better scores on the IAM database. However, their method is optimized exactly for this dataset. The deviation between the scores in our analysis (see Table 2) and their published results can be explained by their training and evaluation on different data and their consideration of the O class during evaluation. However, the most crucial point for the high scores in their publication is probably due to the use of automatic generated NE tags with spaCy. This is an NLP framework that uses the same features to predict NE and PoS tags and, therefore, presumably creates a much stronger correlation between the two tag types compared to manually labeled ones. Since Rowtula et al. exploit exactly this correlation in their approach and also pre-train on a large NLP PoS dataset, the differences between the scores are reasonable. In addition, they only use pages that meet a predefined sentence segmentation in both training and evaluation.

Esposalles. The results for the dataset from the IEHHR challenge are shown in Table 3a. As the IEHHR challenge does not only consist of the correct prediction of the two tags *person* and *category*, but also of the correct recognition of the text, we use our own evaluation protocol instead of the one used in the competition. The results show that the end-to-end as well as two-stage approaches perform

Table 3. Named Entity Recognition performances for the (a) Esposalles, (b) GW and (c) sGMB datasets measured in precision (P), recall (R) and macro-F1 (F1) scores. For Esposalles, the results are presented for each of the tag sets (person, category).

(a) Esposalles

Method	Person			Category		
	P	R	F1	P	R	F1
Annotation-NER	99.3	99.3	99.3	98.8	98.8	98.8
HTR-D-NER	99.3	99.2	99.3	98.5	98.2	98.3
HTR-NER	99.1	99.3	99.2	98.0	98.1	98.1
Rowtula et al. [26]	97.0	96.2	96.6	97.1	97.0	97.0
Toledo et al. [29]	98.5	97.8	98.1	98.5	97.8	98.1

(b) GW

Method	P	R	F1
Annotation-NER	96.5	84.7	89.6
HTR-D-NER	86.1	80.1	82.1
HTR-NER	86.9	78.3	81.3
Rowtula et al. [26]	76.4	59.8	66.6
Toledo et al. [29]	72.5	33.5	45.3

(c) sGMB

Method	P	R	F1
Annotation-NER	81.9	79.2	80.2
HTR-D-NER	81.8	75.9	78.4
HTR-NER	80.1	72.7	75.8
Rowtula et al. [26]	62.7	58.1	60.1
Toledo et al. [29]	44.3	35.3	38.8

well on semi-structured data. Thereby, all methods from our analysis are able to achieve comparable results.

GW. Table 3b shows the results for the GW dataset. It is by far the smallest dataset and has few examples of each tag in the training set compared to the other datasets. However, the scope of context in the data is quite limited, making the training set highly representative for the validation and test data. This probably makes it possible to still obtain good results even under the limited amount of training material. Also, the end-to-end models achieve comparably good scores on this dataset.

sGMB. Table 3c presents the results for the sGMB dataset and shows that the difference in F1-score between Annotation-NER and our two-stage approaches is small. A possible explanation could be that the NER model is optimized for segmented sentences and no sentence-level segmentation is available. In addition, the missing punctuation marks could also have a negative impact on the performance of the Annotation-NER model. Even though the difference in F1-scores between the Rowtula et al. approach and ours is smallest on this dataset, there is still an obvious difference.

5.3   Discussion

In this section, we discuss some potential methods for improving the robustness of a two-stage approach with respect to the task of NER on document images. In our experiments, we train the recognizer as well as the NLP model on the same training data, which leads to a sort of over-adaptation. As a result, the inputs to the NLP model have a considerably lower CER and WER during training compared to validation and test. This also implies that the NLP model has not been optimized for robustness with respect to HTR errors. Therefore, it is not surprising that the F1-score degrades considerably with increasing recognition errors for most datasets. We assume that the performance of our approach could improve when the training and test data have comparable errors with respect to recognition. Another possible improvement would be to adjust the pre-trained word embedding such that words and their erroneous HTR variants are close to each other in the vector space. Furthermore, HTR post-processing can presumably further close the gap between two-stage systems working on annotation and recognition results. However, it is quite remarkable that the NLP models without being specialized for HTR errors still perform better than models developed specifically for this task. We observe that NLP models work well under reasonable and easy to achieve recognition error rates, making two-stage approaches an interesting option. We do not state that the two-stage approach is generally more suitable for NER on word images compared to an end-to-end approach. However, our experiments show that there are currently more advantages for the two-stage models and they still show promising improvements in terms of tackling the task. In order to close the gap between image and text level, we believe that methods are needed that can provide semantic information of a word in an image. Furthermore, it must be possible to handle strongly unbalanced data during training. If it was possible to adapt the advantages gained from the text level to the image domain, end-to-end approaches could be more promising again.

6   Conclusion

In this work, we propose and investigate a two-stage approach for Named Entity Recognition on word-segmented handwritten document images. In our experiments, we are able to outperform state-of-the-art end-to-end approaches on all four tested datasets, only using an unspecialized, standard handwriting recognizer from the literature and a textual Named Entity Recognition model. We demonstrate that due to the advantages from text level, two-stage architectures achieve considerably higher scores compared to end-to-end approaches for this task and have still potential for optimization. However, a final statement regarding the type of architecture for Named Entity Recognition on document images can not be derived based on our experiments. We also present and publish the first named entity tagged datasets on unstructured English text along with optimized splits as well as a suitable evaluation protocol.

References

1. Adak, C., Chaudhuri, B.B., Blumenstein, M.: Named entity recognition from unstructured handwritten document images. In: DAS, Santorini, Greece, pp. 375–380 (2016)
2. Adak, C., Chaudhuri, B.B., Lin, C., Blumenstein, M.: Detecting named entities in unstructured Bengali manuscript images. In: ICDAR, Sydney, Australia, pp. 196–201 (2019)
3. Akbik, A., Bergmann, T., Blythe, D., Rasul, K., Schweter, S., Vollgraf, R.: FLAIR: an easy-to-use framework for state-of-the-art NLP. In: NAACL, Minneapolis, MN, USA, pp. 54–59 (2019)
4. Akbik, A., Blythe, D., Vollgraf, R.: Contextual string embeddings for sequence labeling. In: COLING, Santa Fe, NM, USA, pp. 1638–1649 (2018)
5. Almazán, J., Gordo, A., Fornés, A., Valveny, E.: Word spotting and recognition with embedded attributes. IEEE Trans. Pattern Anal. Mach. Intell. **36**(12), 2552–2566 (2014)
6. Boros, E., et al.: Alleviating digitization errors in named entity recognition for historical documents. In: CoNNL, pp. 431–441 (2020)
7. Boros, E., et al.: A comparison of sequential and combined approaches for named entity recognition in a corpus of handwritten medieval charters. In: ICFHR, Dortmund, Germany, pp. 79–84 (2020)
8. Bos, J., Basile, V., Evang, K., Venhuizen, N., Bjerva, J.: The groningen meaning bank. In: Proceedings of Joint Symposium on Semantic Processing, pp. 463–496 (2017)
9. Carbonell, M., Fornés, A., Villegas, M., Lladós, J.: A neural model for text localization, transcription and named entity recognition in full pages. Pattern Recogn. Lett. **136**, 219–227 (2020)
10. Carbonell, M., Villegas, M., Fornés, A., Lladós, J.: Joint recognition of handwritten text and named entities with a neural end-to-end model. In: DAS, Vienna, Austria, pp. 399–404 (2018)
11. Fornés, A., et al.: ICDAR2017 competition on information extraction in historical handwritten records. In: ICDAR, Kyoto, Japan, pp. 1389–1394 (2017)
12. Gurjar, N., Sudholt, S., Fink, G.A.: Learning deep representations for word spotting under weak supervision. In: DAS, Vienna, Austria, pp. 7–12 (2018)
13. Hamdi, A., Jean-Caurant, A., Sidere, N., Coustaty, M., Doucet, A.: An analysis of the performance of named entity recognition over OCRed documents. In: Joint Conference on Digital Libraries, Champaign, IL, USA, pp. 333–334 (2019)
14. Hamdi, A., Jean-Caurant, A., Sidère, N., Coustaty, M., Doucet, A.: Assessing and minimizing the impact of OCR quality on named entity recognition. In: International Conference on Theory and Practice of Digital Libraries, Lyon, France, pp. 87–101 (2020)
15. Kang, L., Toledo, J.I., Riba, P., Villegas, M., Fornés, A., Rusiñol, M.: Convolve, attend and spell: an attention-based sequence-to-sequence model for handwritten word recognition. In: GCPR, Stuttgart, Germany, pp. 459–472 (2018)
16. Krishnan, P., Jawahar, C.V.: Bringing semantics into word image representation. Pattern Recogn. **108**, 107542 (2020)
17. Lample, G., Ballesteros, M., Subramanian, S., Kawakami, K., Dyer, C.: Neural architectures for named entity recognition. In: NAACL, San Diego, CA, USA, pp. 260–270 (2016)

18. Lifschitz, V.: What is answer set programming? In: Proceedings of AAAI Conference on Artificial Intelligence, Chicago, IL, USA, pp. 1594–1597 (2008)
19. Liu, Y., et al.: RoBERTa: a robustly optimized BERT pretraining approach. arXiv (2019)
20. Marti, U., Bunke, H.: The IAM-database: an English sentence database for offline handwriting recognition. IJDAR **5**(1), 39–46 (2002)
21. Peters, M.E., et al.: Deep contextualized word representations. In: NAACL, New Orleans, LA, USA, pp. 2227–2237 (2018)
22. Pradhan, S., et al.: Towards robust linguistic analysis using OntoNotes. In: CoNNL, Sofia, Bulgaria, pp. 143–152 (2013)
23. Rath, T.M., Manmatha, R.: Word spotting for historical documents. IJDAR **9**(2–4), 139–152 (2007)
24. Retsinas, G., Louloudis, G., Stamatopoulos, N., Gatos, B.: Efficient learning-free keyword spotting. IEEE Trans. Pattern Anal. Mach. Intell. **41**(7), 1587–1600 (2019)
25. Romero, V., et al.: The ESPOSALLES database: an ancient marriage license corpus for off-line handwriting recognition. Pattern Recogn. **46**, 1658–1669 (2013)
26. Rowtula, V., Krishnan, P., Jawahar, C.V.: PoS tagging and named entity recognition on handwritten documents. In: ICON, Patiala, India (2018)
27. Sang, E.F.T.K., Buchholz, S.: Introduction to the CoNLL-2000 shared task chunking. In: CoNLL, Lisbon, Portugal, pp. 127–132 (2000)
28. Stig, J., Leech, G., Goodluck, H.: Manual of information to accompany the Lancaster-Oslo-Bergen Corpus of British English, for use with digital computers (1978). http://korpus.uib.no/icame/manuals/LOB/INDEX.HTM
29. Toledo, J.I., Carbonell, M., Fornés, A., Lladós, J.: Information extraction from historical handwritten document images with a context-aware neural model. Pattern Recogn. **86**, 27–36 (2019)
30. Tüselmann, O., Wolf, F., Fink, G.A.: Identifying and tackling key challenges in semantic word spotting. In: ICFHR, Dortmund, Germany, pp. 55–60 (2020)
31. Wen, Y., Fan, C., Chen, G., Chen, X., Chen, M.: A survey on named entity recognition. In: Liang, Q., Wang, W., Liu, X., Na, Z., Jia, M., Zhang, B. (eds.) CSPS 2019. LNEE, vol. 571, pp. 1803–1810. Springer, Singapore (2020). https://doi.org/10.1007/978-981-13-9409-6_218
32. Wilde, M.D., Hengchen, S.: Semantic enrichment of a multilingual archive with linked open data. Digit. Humanit. Q. **11** (2017)
33. Yadav, V., Bethard, S.: A survey on recent advances in named entity recognition from deep learning models. In: COLING, Santa Fe, NM, USA, pp. 2145–2158 (2018)

Training Bi-Encoders for Word Sense Disambiguation

Harsh Kohli[✉] [iD]

Compass, Inc., Bangalore, India
harsh.kohli@compass.com

Abstract. Modern transformer-based neural architectures yield impressive results in nearly every NLP task and Word Sense Disambiguation, the problem of discerning the correct sense of a word in a given context, is no exception. State-of-the-art approaches in WSD today leverage lexical information along with pre-trained embeddings from these models to achieve results comparable to human inter-annotator agreement on standard evaluation benchmarks. In the same vein, we experiment with several strategies to optimize bi-encoders for this specific task and propose alternative methods of presenting lexical information to our model. Through our multi-stage pre-training and fine-tuning pipeline we further the state of the art in Word Sense Disambiguation.

Keywords: Word Sense Disambiguation · Embedding optimization · Transfer learning

1 Introduction

A long-standing problem in NLP, Word Sense Disambiguation or WSD, has seen several varied approaches over the years. The task is to determine the correct sense in which a word has been used from a predefined set of senses for a particular word. To put it another way, we try and derive the meaning of an ambiguous word from its surrounding context. Despite recent advances in neural language models, WSD continues to be a challenging task given the large number of fine-grained word senses and limited availability of annotated training data.

While the best performing systems today leverage BERT [4] or similar contextual embedding architectures, they are typically augmented with structural or relational knowledge from WordNet [19] as well as other information such as gloss definitions. Our approach builds upon these ideas, and we present our take on how to best incorporate this knowledge in our training.

We use Bi-Encoders to learn a unique representation for a sentence, with a set target word. This embedding can be used to disambiguate or classify the target word into one of its many synsets using metrics such as cosine similarity or euclidean distance. We experiment with different optimization objectives to tune our encoding architecture as well as methods of injecting useful prior knowledge into our training. Empirically, we arrive at the best combination of settings and

© Springer Nature Switzerland AG 2021
J. Lladós et al. (Eds.): ICDAR 2021, LNCS 12822, pp. 823–837, 2021.
https://doi.org/10.1007/978-3-030-86331-9_53

our model, at the time of writing, achieves SOTA results on popular evaluation benchmarks.

2   Related Work and Motivation

Unsupervised approaches to WSD such as the methods proposed by [13] and [21] successfully leveraged knowledge-graph information as well as synset definitions through semantic networks and similar techniques. These methods were initially popular owing to the fact that they required no annotated training corpus. Initial supervised approaches comprised of word-expert systems which involved training a dedicated classifier for each word [34]. Traditional supervised learning approaches were used atop features such as words within a context-window, part of speech (POS) tags etc. Given the large number of classifiers, one for each target lemma, these approaches were harder to scale. They were also difficult to adapt to lemmas not seen during training (the most frequent sense was usually picked in such cases).

The first neural WSD models including systems by [11] and [26] used attention-LSTM architectures coupled with additional hand-crafted features such as POS tags. More recently, GlossBERT [7] proposed an approach to imbue gloss information during training by creating a pairwise BERT [4] classifier using context sentences and synset definitions. The structured-logits mechanism used in EWISER [2] goes one step further to include relational information such as hypernymy and hyponymy. They also demonstrate the benefits of incorporating this information and their approach out-performs all other systems so far.

We build upon the general ideas presented in GlossBERT, and discuss methods to improve the model performance using relational information. Moreover, GlossBERT uses a pairwise classifier or Cross-encoder which, while performing a full self-attention over the input pairs, could potentially lead to prohibitive compute costs at inference time. This is because for a lemma with n distinct senses, the model has to infer over n context-gloss pairs which can be expensive especially when using large Transformer [30] based models.

Bi-Encoders, on the other hand, offer the flexibility to pre-index embeddings corresponding to a synset and utilize libraries such as FAISS [10] for fast vector similarity search during inference. We experiment with several optimization strategies to learn these synset embeddings and are able to improve upon the performance of GlossBERT, leading to gains both in terms of prediction time as well as model accuracy.

3   Datasets and Preprocessing

3.1   Source Datasets

Consistent with most recent work, we use SemCor 3.0 [20] as our primary training corpus. SemCor consists of 226k sense tags and forms the largest manually annotated training corpus available. In addition, [2] have shown that better results can be obtained when using both tagged and untagged WordNet examples as well as glosses. While the tagged glosses are central to approach, much like GlossBERT, we also add the tagged examples corpus to our final training dataset.

In our evaluation, we use the framework described by [25] for benchmarking on WSD. The SemEval-2007 corpus [24] is used as our dev set for tuning parameters and model checkpointing, whereas SemEval-2013 [23], SemEval-2015 [22], Senseval-2 [5], and Senseval-3 [29] comprise the remainder of our evaluation sets.

3.2   Data Preprocessing

Context Sentence	Synset Gloss Definition	Label
Styka blew his " nose " again.	nose : the organ of smell and entrance to the respiratory tract; the prominent part of the face of man or other mammals	1
Styka blew his " nose " again.	nose : search or inquire in a meddlesome way	0
Styka blew his " nose " again.	nose : defeat by a narrow margin	0

Context Sentence	Hypernym Gloss Definition	Label
Styka blew his " nose " again.	spout : an opening that allows the passage of liquids or grain	1
Styka blew his " nose " again.	search : search or seek	0
Styka blew his " nose " again.	get the better of : win a victory over	0

Context Sentence (Anchor)	Gloss Definition (Positive)	Gloss Definition (Negative)
Styka blew his " nose " again.	nose : the organ of smell and entrance to the respiratory tract; the prominent part of the face of man or other mammals	nose : search or inquire in a meddlesome way
Styka blew his " nose " again.	nose : the organ of smell and entrance to the respiratory tract; the prominent part of the face of man or other mammals	nose : defeat by a narrow margin

Fig. 1. Context-gloss pairs, context-hypernym pairs, and context-positive glossnegative gloss triplets with weak supervision

We adopt the same preprocessing approach described in GlossBERT. Context sentences from SemCor as well as WordNet examples contain a signal to help identify the target word - highlighted in the context sentence column in Fig. 1. In the gloss sentences, the lemma of the sense is prepended to the definition followed by a semicolon. This weak supervision helps better emphasize the target words in both the context as well as gloss sentences, and was also used in EWISER [2] as well as MTDNN+Gloss [12] in the data preprocessing step. We prepare three different datasets using the same weak supervision for our training experiments. Figure 1 contains example rows corresponding to each of the datasets described in the proceeding sections.

Context-Gloss Pairs. We follow GlossBERT [7] for preparing our context-gloss pairwise dataset. For each target word within a context sentence, we retrieve all senses and their corresponding gloss. Thus, for a target word with n different senses we have n pairwise examples for training. The pair corresponding to the correct sense (gold synsets) are positive examples (label 1) while all other pairs are negative examples (label 0) in our dataset.

Context-Hypernym Gloss Pairs. For each gold (correct) synset, we take the set of immediate hypernyms. The hypernym gloss is augmented with weak signals similar to the target synset gloss. These constitute the positive pairs in our context-hypernym gloss pairs dataset. Similarly, hypernyms of incorrect synsets for a target word in a context are negative samples in this data.

Context-Positive Gloss-Negative Gloss Triplets. We use another formulation of our datasets for optimizing on the triplet loss objective [28]. Here, the context sentence is used as an anchor, the gloss corresponding to the correct(gold) synset for the target word is the positive example, whereas gloss corresponding to each synset for the target lemma which is not the positive synset is used as a negative example. Thus, for a target word with n different senses we get $n - 1$ training triplets in the anchor-positive-negative format where the anchor and positive are the same across each triplet while the negative varies. We use all the incorrect senses for each anchor-positive pair while preparing this dataset.

3.3 Oversampling

Our final pairwise datasets are skewed towards the negative class as we take all the negative gloss for a target lemma, which form the negative examples in our dataset. Each context sentence, however, has only one positive example for a target word. [12] augmented the positive class by generating examples using simple as well as chained back-translation. On average, 3 synthetic examples were generated for each context sentence. While we do not augment our data similarly, we use the same oversampling ratio of 3. In other words, each positive example is repeated 3 times in our training set for the pairwise training sets - both context-gloss as well as context-hypernym. The triplet loss dataset, however, is not modified as it does not suffer the same class-imbalance problem. To our context-gloss datasets, the gloss definitions of different synsets with the same lemma are added to maximize distance between gloss definitions themselves.

4   Model Training

We experiment with different optimization objectives for training the encoder using the datasets described in the previous section. We augment the model using a multi-stage pre-training and fine-tuning pipeline.

4.1   Base Model

MTDNN+Gloss [12] has shown improved results over vanilla GlossBERT through a pre-training approach which used Glue data [32]. The MT-DNN [14] architecture is used to train on each individual Glue task in a multi-task training procedure. The trained weights are then used to initialize the encoder for the fine-tuning pipeline using context-gloss pairs. To benefit from a better initial sentence representation using Glue data, we use the nli-roberta-large model from sentence-transformers [27]. The sentence encoding model is a Siamese Network using the RoBERTa encoder [15] tuned on Natural Language Inference tasks such as SNLI [3] and MNLI [33] which are part of the Glue benchmark. The sentence-transformers library also contains functionality for the various training experiments described in this section. We also run one trial with nli-roberta-base to justify the usage of RoBERTa large (24 layers, 1024 dimension, 335 m parameters) over RoBERTa base (12 layers, 768 dimension, 110 m parameters).

4.2   Context-Gloss Training

We use the context-gloss pairwise dataset described in Sect. 3.2 to train the Bi-Encoder using various optimization strategies. Let us assume that u and v correspond to the pooled sentence embeddings for the context and gloss respectively, and $y \in (0,1)$ is the label for the particular training example.

Cosine Similarity Loss. Cosine similarity between u and v is computed and the loss is defined as the distance between the similarity and the true label y.

$$||y - cosine\_similarity(u,v)||$$

Here, $||.||$ corresponds to the distance metric. We use the MSE or squared error distance for our model.

Contrastive Loss. Next, we try the contrastive loss function [6] which selectively optimizes for negative examples when they are within a certain distance or margin of each other. Mathematically, it can be defined as follows:

$$\frac{1}{2}(1-y)(d(u,v))^2 + \frac{1}{2}y(max\{0, m - (d(u,v))^2\})$$

Here $d(u,v)$ is the cosine distance between the embeddings u and v.

$$d(u,v) = 1 - cosine\_similarity(u,v)$$

m is the margin for our contrastive loss which is set at 0.5 in our experiments.

Online Contrastive Loss. The Online Contrastive Loss defined in the sentence-transformers library is similar to the contrastive loss, except that it samples only the hard negative and positive examples from each batch. For a given batch, the maximum distance between any positive input pair (pos_max, say) and minimum distance between any negative pair (neg_min) is computed. Negative examples having distance less than pos_max and positive examples having distance greater than neg_min are isolated and these are treated as our hard negatives and hard positives respectively. Thereafter, contrastive loss is computed on the hard positives and negatives.

4.3   Context-Hypernym Gloss Training

Hypernyms to a synset correspond to a synset with a broader or more generalized meaning, of which the target synset is a subtype or hyponym. WordNet 3.0 consists of about 109k unique synsets and relational information between them such as their hypernyms, hyponyms, antonyms, entailment etc. In EWISER [2], experiments with both hypernyms and hyponyms are performed using the structured logits mechanism. Largest improvement over baselines are obtained by including hypernyms alone. The authors claim that hypernym information helps the model generalize better to synsets that are not present or under-represented in the train data (SemCor).

We only use hypernym information in our training. Immediate hypernyms for a synset are considered and the data is prepared as described in Sect. 3.2. Pairwise models for context-hypernym pairs are used primarily in our pre-training steps. After observing results for context-gloss training we train the siamese network using the contrastive loss objective.

4.4   Context-Gloss and Context-Hypernym Multi-Task Training

Recently approaches such as MT-DNN [14] have shown strong improvement over baselines by using multi-task learning. The multi-task training procedure uses shared encoders with output layers that are specific to task types - single sentence classification, pairwise classification, pairwise similarity, and pairwise ranking. Batches from all of glue data across task types are used to train this shared encoder. Later, the encoder is individually tuned on each individual dataset to yield best results.

While MT-DNN uses a Cross-Encoder, we try and adopt the same approach in our Bi-Encoder pre-training. Context-gloss and context-hypernym pairs are treated as separate tasks and the encoder is tuned on both of these simultaneously. Similar to the context-hypernym model, the multi-task model is used as a pre-training procedure atop which further tuning is conducted. These experiments are shown in Figs. 3 and 4. Like in the previous section, contrastive loss is used while training.

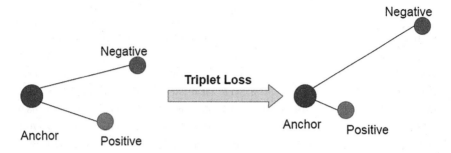

Fig. 2. The triplet loss objective aims to minimize distance between the anchor and positive while maximizing the distance between anchor and negative

4.5 Triplet Training

Finally, we train on the triplet dataset described in Sect. 3.2. The triplet loss [28] uses three reference embeddings often termed the anchor, positive and negative. The loss function, as depicted in Fig. 2, attempts to simultaneously increase the distance between the anchor and the negative, while minimizing the distance between anchor and positive. Let us consider E_a, E_p, and E_n to be the embeddings for the anchor, positive, and negative samples respectively. The triplet loss is then,

$$max\{||E_a - E_p||^2 - ||E_a - E_n||^2 + m, 0\}$$

Here $||.||$ denotes the euclidean distance between the embeddings. In our case, E_a, E_p, and E_n correspond to embeddings for the context, gloss corresponding to the gold synset, and gloss of other synsets having the same lemma respectively. m is the margin hyperparameter similar to the one in contrastive loss. We use $m = 5$ in our triplet loss training experiments.

5 Experiments

We do multiple iterations of training to empirically test the various optimization methods. We train both with and without the fine-tuning data using hypernym glosses (either directly or through multi-task pre-training). As alluded to earlier, we also train a smaller model (RoBERTa base) to observe the performance benefits of using a larger, more expressive model. At inference time, we compare the embedding of the weakly supervised context sentence with the target word against the gloss of all synsets corresponding to the target lemma. The synset with the best score (highest cosine similarity) is considered the predicted synset for the example.

Results from training the Bi-Encoder on the base model using contrastive loss, as well as larger models using cosine, contrastive, online contrastive as well as triplet loss are included in Table 1. We compare our results across evalution sets and POS types against recent neural approaches - Bi-LSTM [11], Bi-LSTM

Table 1. Final results

System	SE07	SE2	SE3	SE13	SE15	Noun	Verb	Adj	Adv	All
MFS Baseline	54.5	65.6	66.0	63.8	67.1	67.7	49.8	73.1	80.5	65.5
Lesk$_{ext+emb}$	56.7	63.0	63.7	66.2	64.6	70.0	51.1	51.7	80.6	64.2
Babelfly	51.6	67.0	63.5	66.4	70.3	68.9	50.7	73.2	79.8	66.4
IMS	61.3	70.9	69.3	65.3	69.5	70.5	55.8	75.6	82.9	68.9
IMS$_{+emb}$	62.6	72.2	70.4	65.9	71.5	71.9	56.6	75.9	84.7	70.1
Bi-LSTM	–	71.1	68.4	64.8	68.3	69.5	55.9	76.2	82.4	68.4
Bi-LSTM$_{+att.+LEX+POS}$	64.8	72.0	69.1	66.9	71.5	71.5	57.5	75.0	83.8	69.9
GAS$_{ext}$(Linear)	–	72.4	70.1	67.1	72.1	71.9	58.1	76.4	84.7	70.4
GAS$_{ext}$(Concatenation)	–	72.2	70.5	67.2	72.6	72.2	57.7	76.6	85.0	70.6
CAN	–	72.2	70.2	69.1	72.2	73.5	56.5	76.6	80.3	70.9
HCAN	–	72.8	70.3	68.5	72.8	72.7	58.2	77.4	84.1	71.1
SemCor,hyp	–	–	–	–	–	–	–	–	–	75.6
SemCor,hyp(ens)	69.5	77.5	77.4	76.0	78.3	79.6	65.9	79.5	85.5	76.7
SemCor+WNGC,hyp	–	–	–	–	–	–	–	–	–	77.1
SemCor+WNGC,hyp(ens)	73.4	79.7	77.8	78.7	82.6	81.4	68.7	**83.7**	85.5	79.0
BERT(Token-CLS)	61.1	69.7	69.4	65.8	69.5	70.5	57.1	71.6	83.5	68.6
GlossBERT(Sent-CLS)	69.2	76.5	73.4	75.1	79.5	78.3	64.8	77.6	83.8	75.8
GlossBERT(Token-CLS)	71.9	77.0	75.4	74.6	79.3	78.3	66.5	78.6	84.4	76.3
GlossBERT(Sent-CLS-WS)	72.5	77.7	75.2	76.1	80.4	79.3	66.9	78.2	86.4	77.0
MTDNN+Gloss	73.9	79.5	76.6	79.7	80.9	81.8	67.7	79.8	86.5	79.0
EWISER$_{hyper}$	75.2	80.8	**79.0**	80.7	81.8	82.9	69.4	83.6	86.7	80.1
EWISER$_{hyper+hypo}$	73.8	80.2	78.5	80.6	82.3	82.7	68.5	82.9	**87.6**	79.8
Bi-Enc$_{base}$ Contrastive	75.8	78.3	76.8	80.1	81.2	81.6	69.5	78.5	84.7	78.6
Bi-Enc Cosine	72.5	78.2	75.7	78.7	79.9	80.2	68.3	79.9	82.7	77.6
Bi-Enc Contrastive	76.3	80.4	77.0	80.7	81.7	82.5	70.3	80.3	85.0	79.5
Bi-Enc OnlineContrastive	75.4	79.4	78.3	80.3	82.2	81.8	71.3	81.1	85.1	79.4
Bi-Enc Multi-Task (MT)	72.3	78.9	75.6	78.4	81.3	80.5	68.6	80.0	83.8	77.9
Bi-Enc Triplet	75.4	80.4	77.6	80.7	**83.0**	82.1	70.9	82.7	85.8	79.8
Bi-Enc MT+Contrastive	76.3	80.1	77.8	80.6	81.7	82.3	70.4	81.3	86.1	79.6
Bi-Enc Hypernym+Triplet	**76.7**	**81.7**	77.5	**82.4**	82.4	**83.3**	**72.0**	80.7	87.0	**80.6**

+ att + lex +pos [26], CAN/HCAN [17], GAS [18], SemCor/SemCor+WNGC, hypernyms [31] and GlossBERT [7], as well as knowledge based - Lesk (ext+emb) [1] and Babelfly [21], and word expert models - IMS [34] and IMS+emb [9].

Of the large Bi-Encoder models without any transfer learning, we achieve best results using the triplet loss objective. While we do not employ any sampling heuristic, using all incorrect synset gloss for a target lemma in a context helps yield good negatives in our anchor-positive-negative formulation.

Of the pairwise approaches, we achieve best results with the contrastive loss objective. This further validates the efficacy of margin-based loss functions over traditional metrics like cosine distance for our problem. While the online contrastive loss objective often does better than simple contrastive loss, we see marginally better results with contrastive loss in this case. We reckon this is because of the inability to sample enough hard positives and negatives within a batch in a single iteration of training. While we believe that this might be mitigated and performance might improve by increasing the batch size, we are unable to test this hypothesis due to the large memory footprint of the model (335 m trainable parameters) and our inability to increase batch size due to resource constraints in the form of GPU memory. However, both simple and well as online contrastive loss both comfortably outperform the cosine distance based learning objective in our experiments. Under the same settings, the larger model (RoBERTa large with contrastive loss) does better than the base model. This is in contrast to GlossBERT [7], where the authors noted better results using BERT base model over BERT large.

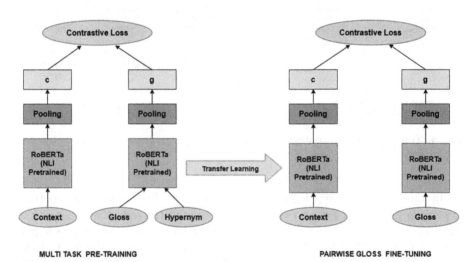

Fig. 3. Transfer learning: multi task pre-training followed by context-gloss fine-tuning

Next, we try to include relational information from WordNet in our pre-training step. First, we use multi-task learning as described in Sect. 4.3 and fine-tune using context-gloss pairs and contrastive loss. We also report results of running just the multi-task model on our evaluation sets. In another experiment, we train a pairwise model using the hypernym dataset from Sect. 3.2 using contrastive loss. We fine-tune this model using triplet data from Sect. 3.2. Architecture for the first transfer learning experiment using context-hypernym pairs and then triplet data is shown in Fig. 3. The second experiment using multi-task data and context-gloss pairs is depicted in Fig. 4.

Understandably, the vanilla multi-task model does not perform as well as the other pairwise Bi-Encoder models. Along with context-gloss pairs this model is also exposed to hypernym data making the training set less homogeneous and aligned with test sets. However, tuning the model with context-gloss pairs after pre-training in a multi-task setting we observe improvement in performance. This transfer learning setup, nevertheless, only gives a nominal improvement over a single-step training over context-gloss pairs.

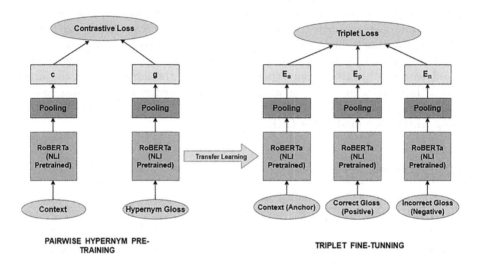

Fig. 4. Transfer learning: context-hypernym pre-training followed by triplet objective fine-tuning

In our other transfer learning experiment, we first train only context-hypernym pairs. Thus, we try and capture some of the inductive reasoning of hypernymy by trying to associate the target lemma in a context with its hypernym - a synset with a broader and less specific meaning. After the pre-training step we fine-tune on the triplet dataset. The pre-training step helps in this case and results in better model performance compared to the vanilla triplet training baseline. This arrangement also gives us the best results of all our experiments.

6   Performance on Unseen Synsets

In order to ascertain how the model performs on unseen synsets, we categorize each sample in our evaluation sets into one of two groups depending on whether the gold synset for the particular example is present in SemCor - our primary training corpus. In Table 2 we report results of our best models, both with and without transfer learning (TL).

There scores reveal that there is a considerable gap in performance between the seen split (S/TL-S) and unseen split (U/TL-U). Transfer learning does help model performance on the unseen set but not as much as it does in the case of the seen split. We hypothesize that this is because the hypernym data is constructed by associating hypernyms of gold synsets with the gloss in SemCor, thus leading to limited diversity in the train set even with hypernyms outside to SemCor examples. In Sect. 8, we discuss ideas on how to further enrich the training corpus with more relational information which might help model performance on unseen synsets.

Table 2. F1 scores on seen & unseen splits - U - triplet unseen, S - triplet seen, TL-U - triplet with hypernym pre-training unseen, TL-S - triplet with hypernym pre-training seen

Test set	SE07	SE2	SE3	SE13	SE15	ALL
U	47.2	76.0	65.8	75.7	71.0	72.3
S	79.1	81.2	78.9	82.0	85.6	81.2
TL-U	47.2	74.9	67.3	77.9	73.4	72.8
TL-S	80.6	83.1	78.8	83.5	84.4	82.1

7 Training Parameters

As mentioned previously, we use the sentence-transformers [27] library to conduct our experiments. For every run of training, including pre-training and fine-tuning steps wherever applicable, we train for 2 epochs and checkpoint at intervals of 10000 steps. We override the previous checkpoint only if results improve on our dev set.

A constant batch size of 32 is maintained across all our experiments. The AdamW [16] optimizer is used with a default learning rate of $2e^{-5}$. A linear warmup scheduler is applied to slowly increase the learning rate to a constant after 10% of training iterations (warm-up ratio = 0.1) to reduce volatility in the early iterations of training. A single Tesla V100 GPU (32 GB) is used for all training iterations.

8 Conclusion and Future Work

Through this paper, we present an approach to word sense disambiguation by training a siamese network on weakly supervised context-gloss pairs. We discuss strategies to optimize the network through various different learning objectives. We discuss methods to infuse relational information through optimizing on hypernym gloss in a distinct pre-training step. We also experiment with multi-task pre-training using hypernyms as well as gloss corresponding to the target lemma. This simple approach of pre-training with hypernyms helps us improve performance over vanilla models trained on context-gloss pairs or triplets and we achieve state of the art performance on our test sets.

While preparing our context-hypernym dataset, we only explore hypernym definitions that are one level up (one edge) from the target synset. However, it might also be worth exploring further levels to determine if pre-training with even more generalized gloss from farther synsets might help improve performance further. While hypernyms were specifically chosen as they have typically yielded the greatest improvement over baselines when included in training, we could also use additional structural features - including edges along the hyponyms, antonyms, entailment etc. In a multi-task pre-training setting, each of these datasets could be independent tasks or we could train each of them individually through a single-task, multi-stage pre-training pipeline. An ablation study might help us determine which of these features are useful. Of particular interest would be analyzing the impact of this data and additional relational information on model performance on unseen synsets.

Our best results, both with and without transfer learning, are achieved with the triplet loss objective. We employ a straight-forward strategy of creating triplets using all senses for a target lemma. However, we could also experiment with different hard or semi-hard sampling techniques [28] to determine their impact on performance. Negatives could be sampled from the synset pool at each iteration using one of these strategies (online sampling). Finally, Poly-encoders [8] have shown promising results on several tasks. Consisting of a transformer architecture with a novel attention mechanism, they outperform Bi-Encoders purely in terms of quality. In terms of speed, they are vastly more efficient than Cross-Encoders and comparable to Bi-Encoders when running an inference against up to 1000 samples. Since we do not have as many synsets to compare against for a target lemma, Poly-encoders might be a viable alternative to the Bi-Encoders described in this paper.

Some of the ideas discussed in this section - incorporating a richer relational information via pre-training along different edges and exploring farther beyond the immediate hypernym, better training and sampling, as well as different encoding architectures might aid in moving the needle further as we explore ways to improve performance on word sense disambiguation.

References

1. Basile, P., Caputo, A., Semeraro, G.: An enhanced Lesk word sense disambiguation algorithm through a distributional semantic model. In: Proceedings of COLING 2014, the 25th International Conference on Computational Linguistics: Technical Papers, Dublin, Ireland, pp. 1591–1600. Dublin City University and Association for Computational Linguistics, August 2014. https://www.aclweb.org/anthology/C14-1151
2. Bevilacqua, M., Navigli, R.: Breaking through the 80% glass ceiling: raising the state of the art in word sense disambiguation by incorporating knowledge graph information. In: Proceedings of the 58th Annual Meeting of the Association for Computational Linguistics, pp. 2854–2864. Association for Computational Linguistics, July 2020. https://doi.org/10.18653/v1/2020.acl-main.255. https://www.aclweb.org/anthology/2020.acl-main.255

3. Bowman, S.R., Angeli, G., Potts, C., Manning, C.D.: A large annotated corpus for learning natural language inference. In: Proceedings of the 2015 Conference on Empirical Methods in Natural Language Processing (EMNLP). Association for Computational Linguistics (2015)
4. Devlin, J., Chang, M.W., Lee, K., Toutanova, K.: BERT: pre-training of deep bidirectional transformers for language understanding. In: Proceedings of the 2019 Conference of the North American Chapter of the Association for Computational Linguistics: Human Language Technologies, Minneapolis, Minnesota, vol. 1 (Long and Short Papers), pp. 4171–4186. Association for Computational Linguistics, June 2019. https://doi.org/10.18653/v1/N19-1423. https://www.aclweb.org/anthology/N19-1423
5. Edmonds, P., Cotton, S.: SENSEVAL-2: overview. In: Proceedings of SENSEVAL-2 Second International Workshop on Evaluating Word Sense Disambiguation Systems, Toulouse, France, pp. 1–5. Association for Computational Linguistics, July 2001. https://www.aclweb.org/anthology/S01-1001
6. Hadsell, R., Chopra, S., LeCun, Y.: Dimensionality reduction by learning an invariant mapping. In: 2006 IEEE Computer Society Conference on Computer Vision and Pattern Recognition (CVPR 2006), vol. 2, pp. 1735–1742 (2006). https://doi.org/10.1109/CVPR.2006.100
7. Huang, L., Sun, C., Qiu, X., Huang, X.: Glossbert: bert for word sense disambiguation with gloss knowledge. arXiv abs/1908.07245 (2019)
8. Humeau, S., Shuster, K., Lachaux, M.A., Weston, J.: Poly-encoders: architectures and pre-training strategies for fast and accurate multi-sentence scoring. In: International Conference on Learning Representations (2020). https://openreview.net/forum?id=SkxgnnNFvH
9. Iacobacci, I., Pilehvar, M.T., Navigli, R.: Embeddings for word sense disambiguation: an evaluation study. In: Proceedings of the 54th Annual Meeting of the Association for Computational Linguistics (vol. 1: Long Papers), Berlin, Germany, pp. 897–907. Association for Computational Linguistics, August 2016. https://doi.org/10.18653/v1/P16-1085. https://www.aclweb.org/anthology/P16-1085
10. Johnson, J., Douze, M., Jégou, H.: Billion-scale similarity search with GPUs. arXiv preprint arXiv:1702.08734 (2017)
11. Kågebäck, M., Salomonsson, H.: Word sense disambiguation using a bidirectional LSTM. In: Proceedings of the 5th Workshop on Cognitive Aspects of the Lexicon (CogALex - V), Osaka, Japan, pp. 51–56. The COLING 2016 Organizing Committee, December 2016. https://www.aclweb.org/anthology/W16-5307
12. Kohli, H.: Transfer learning and augmentation for word sense disambiguation. In: Hiemstra, D., Moens, M.-F., Mothe, J., Perego, R., Potthast, M., Sebastiani, F. (eds.) ECIR 2021. LNCS, vol. 12657, pp. 303–311. Springer, Cham (2021). https://doi.org/10.1007/978-3-030-72240-1_29
13. Lesk, M.E.: Automatic sense disambiguation using machine readable dictionaries: how to tell a pine cone from an ice cream cone. In: SIGDOC 1986 (1986)
14. Liu, X., He, P., Chen, W., Gao, J.: Multi-task deep neural networks for natural language understanding. In: Proceedings of the 57th Annual Meeting of the Association for Computational Linguistics, Florence, Italy, pp. 4487–4496. Association for Computational Linguistics, July 2019. https://www.aclweb.org/anthology/P19-1441
15. Liu, Y., et al.: Roberta: a robustly optimized BERT pretraining approach. CoRR abs/1907.11692 (2019). http://arxiv.org/abs/1907.11692

16. Loshchilov, I., Hutter, F.: Decoupled weight decay regularization. In: International Conference on Learning Representations (2019). https://openreview.net/forum?id=Bkg6RiCqY7

17. Luo, F., Liu, T., He, Z., Xia, Q., Sui, Z., Chang, B.: Leveraging gloss knowledge in neural word sense disambiguation by hierarchical co-attention. In: Proceedings of the 2018 Conference on Empirical Methods in Natural Language Processing, Brussels, Belgium, pp. 1402–1411. Association for Computational Linguistics, Ocobert-November 2018. https://doi.org/10.18653/v1/D18-1170. https://www.aclweb.org/anthology/D18-1170

18. Luo, F., Liu, T., Xia, Q., Chang, B., Sui, Z.: Incorporating glosses into neural word sense disambiguation. In: Proceedings of the 56th Annual Meeting of the Association for Computational Linguistics (vol. 1: Long Papers), Melbourne, Australia, pp. 2473–2482. Association for Computational Linguistics, July 2018. https://doi.org/10.18653/v1/P18-1230. https://www.aclweb.org/anthology/P18-1230

19. Miller, G.A.: Wordnet: a lexical database for English. Commun. ACM **38**(11), 39–41 (1995)

20. Miller, G.A., Leacock, C., Tengi, R., Bunker, R.T.: A semantic concordance. In: Human Language Technology: Proceedings of a Workshop Held at Plainsboro, New Jersey, 21–24 March 1993 (1993). https://www.aclweb.org/anthology/H93-1061

21. Moro, A., Raganato, A., Navigli, R.: Entity linking meets word sense disambiguation: a unified approach. Trans. Assoc. Comput. Linguist. **2**, 231–244 (2014)

22. Moro, A., Navigli, R.: SemEval-2015 task 13: multilingual all-words sense disambiguation and entity linking. In: Proceedings of the 9th International Workshop on Semantic Evaluation (SemEval 2015), Denver, Colorado, pp. 288–297. Association for Computational Linguistics, June 2015. https://doi.org/10.18653/v1/S15-2049. https://www.aclweb.org/anthology/S15-2049

23. Navigli, R., Jurgens, D., Vannella, D.: SemEval-2013 task 12: multilingual word sense disambiguation. In: Second Joint Conference on Lexical and Computational Semantics (*SEM), vol. 2: Proceedings of the Seventh International Workshop on Semantic Evaluation (SemEval 2013), Atlanta, Georgia, USA, pp. 222–231. Association for Computational Linguistics, June 2013. https://www.aclweb.org/anthology/S13-2040

24. Pradhan, S., Loper, E., Dligach, D., Palmer, M.: SemEval-2007 task-17: English lexical sample, SRL and all words. In: Proceedings of the Fourth International Workshop on Semantic Evaluations (SemEval-2007), Prague, Czech Republic, pp. 87–92. Association for Computational Linguistics, June 2007. https://www.aclweb.org/anthology/S07-1016

25. Raganato, A., Camacho-Collados, J., Navigli, R.: Word sense disambiguation: a unified evaluation framework and empirical comparison. In: Proceedings of the 15th Conference of the European Chapter of the Association for Computational Linguistics: vol. 1, Long Papers, Valencia, Spain, pp. 99–110. Association for Computational Linguistics, April 2017. https://www.aclweb.org/anthology/E17-1010

26. Raganato, A., Delli Bovi, C., Navigli, R.: Neural sequence learning models for word sense disambiguation. In: Proceedings of the 2017 Conference on Empirical Methods in Natural Language Processing, Copenhagen, Denmark, pp. 1156–1167. Association for Computational Linguistics, September 2017. https://doi.org/10.18653/v1/D17-1120. https://www.aclweb.org/anthology/D17-1120

27. Reimers, N., Gurevych, I.: Sentence-bert: sentence embeddings using siamese bert-networks. In: Proceedings of the 2019 Conference on Empirical Methods in Natural Language Processing. Association for Computational Linguistics (2019). https://arxiv.org/abs/1908.10084

28. Schroff, F., Kalenichenko, D., Philbin, J.: Facenet: A unified embedding for face recognition and clustering. In: 2015 IEEE Conference on Computer Vision and Pattern Recognition (CVPR), pp. 815–823 (2015). https://doi.org/10.1109/CVPR.2015.7298682

29. Snyder, B., Palmer, M.: The English all-words task. In: Proceedings of SENSEVAL-3, the Third International Workshop on the Evaluation of Systems for the Semantic Analysis of Text, Barcelona, Spain, pp. 41–43. Association for Computational Linguistics, July 2004. https://www.aclweb.org/anthology/W04-0811

30. Vaswani, A., et al.: Attention is all you need (2017)

31. Vial, L., Lecouteux, B., Schwab, D.: Sense vocabulary compression through the semantic knowledge of wordnet for neural word sense disambiguation. CoRR abs/1905.05677 (2019). http://arxiv.org/abs/1905.05677

32. Wang, A., Singh, A., Michael, J., Hill, F., Levy, O., Bowman, S.: GLUE: a multi-task benchmark and analysis platform for natural language understanding. In: Proceedings of the 2018 EMNLP Workshop BlackboxNLP: Analyzing and Interpreting Neural Networks for NLP, Brussels, Belgium, pp. 353–355. Association for Computational Linguistics, November 2018. https://doi.org/10.18653/v1/W18-5446. https://www.aclweb.org/anthology/W18-5446

33. Williams, A., Nangia, N., Bowman, S.: A broad-coverage challenge corpus for sentence understanding through inference. In: Proceedings of the 2018 Conference of the North American Chapter of the Association for Computational Linguistics: Human Language Technologies, vol. 1 (Long Papers), pp. 1112–1122. Association for Computational Linguistics (2018). http://aclweb.org/anthology/N18-1101

34. Zhong, Z., Ng, H.T.: It makes sense: a wide-coverage word sense disambiguation system for free text. In: Proceedings of the ACL 2010 System Demonstrations, Uppsala, Sweden, pp. 78–83. Association for Computational Linguistics, July 2010. https://www.aclweb.org/anthology/P10-4014

DeepCPCFG: Deep Learning and Context Free Grammars for End-to-End Information Extraction

Freddy C. Chua$^{(\boxtimes)}$ and Nigel P. Duffy

Ernst and Young (EY) AI Lab, Palo Alto, CA, USA
{freddy.chua,nigel.p.duffy}@ey.com

Abstract. We address the challenge of extracting structured information from business documents without detailed annotations. We propose Deep Conditional Probabilistic Context Free Grammars (DeepCPCFG) to parse two-dimensional complex documents and use Recursive Neural Networks to create an end-to-end system for finding the most probable parse that represents the structured information to be extracted. This system is trained end-to-end with scanned documents as input and only relational-records as labels. The relational-records are extracted from existing databases avoiding the cost of annotating documents by hand. We apply this approach to extract information from scanned invoices achieving state-of-the-art results despite using no hand-annotations.

Keywords: 2D parsing · Document intelligence · Information extraction

1 Introduction

Extracting information from business documents remains a significant burden for all large enterprises. Existing extraction pipelines are complex, fragile, highly engineered, and yield poor accuracy. These systems are typically trained using annotated document images whereby human annotators have labeled the tokens or words that need to be extracted. The need for human annotation is the biggest barrier to the broad application of machine learning.

Business documents often have complex recursive structures whose understanding is critical to accurate information extraction, e.g. invoices contain a variable number of line-items each of which may have multiple (sometimes varying) fields to be extracted. This remains a largely unsolved problem.

Deep Learning has revolutionized sequence modeling resulting in systems that are trained end-to-end and have improved accuracy while reducing system complexity for a variety of applications [25]. Here we leverage these benefits in an end-to-end system for information extraction from complex business documents. This system does not use human annotations making it easily adaptable to new document types while respecting the recursive structure of many business documents.

© Springer Nature Switzerland AG 2021
J. Lladós et al. (Eds.): ICDAR 2021, LNCS 12822, pp. 838–853, 2021.
https://doi.org/10.1007/978-3-030-86331-9_54

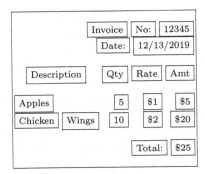

(a) Example of an invoice

Invoice Relational Table		
InvoiceID	*Date*	*TotalAmt*
12345	12/13/2019	$25
Line-items Relational Table					

InvoiceID	*Desc*	*Qty*	*Rate*	*Amt*
12345	Apples	5	$1	$5
12345	Chicken Wings	10	$2	$20

(b) Invoice record in relational tables

Fig. 1. Invoice image and invoice record

We apply this system to invoices which embody many of the challenges involved: the information they contain is highly structured, they are highly varied, and information is encoded in their 2D layout and not just in their language. A simplified example of an invoice is shown in Fig. 1a with the corresponding relational-record shown in Fig. 1b. The relational-record contains structured information including header fields that each appear once in the invoice and an arbitrary number of line-items each containing a description, quantity, rate and amount[1]. While human annotations of document images are rarely available, these relational-records are commonly stored in business systems such as Enterprise Resource Planning (ERP) systems.

Figure 1a illustrates bounding boxes surrounding each of the tokens in the invoice. Some of these bounding boxes, e.g. "Description" provide context by which other bounding boxes, e.g. "Apples" can be interpreted as corresponding to a description field. We assume that these bounding boxes are provided by an external Optical Character Recognition (OCR) process and given as input to our information extraction pipeline. Unlike alternative approaches which require annotated bounding boxes to train a bounding box classification model, our approach avoids the use of annotations for each box while only requiring relational-records.

A common approach to information extraction is to classify these bounding boxes individually using machine learning and then to use heuristics to group extracted fields to recover the recursive structure. This approach has several shortcomings: 1) It requires annotated bounding boxes, where each annotation is tagged with the coordinates of the box and a known label to train the classification model. 2) It requires heuristic post-processing to reconstruct the structured records and to resolve ambiguities. These post-processing pipelines tend to be brittle, hard to maintain, and highly tuned to specific document types.

---

[1] Other business documents are significantly more complicated often involving recursively defined structure.

Adapting these systems to new document types is an expensive process requiring new annotations and re-engineering of the post-processing pipeline. We address these issues using a structured prediction approach. We model the information to be extracted using a context free grammar (CFG) which allows us to capture even complex recursive document structures. The CFG is built by adapting the record schema for the information to be extracted. Information is extracted by parsing documents with this CFG in 2D. It is important to note that the CFG is not based on the layout of the document, as with previous work in this field, but on the structure of the information to be extracted. We do not have to fully describe the document, we only describe the information that needs to be extracted. This allows us to be robust to a wide variety of document layouts.

There may be many valid parses of a document. We resolve this ambiguity by extending Conditional Probabilistic Context Free Grammars (CPCFG) [23] in 3 directions. 1) We use deep neural networks to provide the conditional probabilities for each production rule. 2) We extend CPCFGs to 2D parsing. 3) We train them using structured prediction [4]. This results in a computationally tractable algorithm for handling even complex documents.

Our core contributions are: 1) A method for end-to-end training in complex information extraction problems requiring no post-processing and no annotated images or any annotations associated with bounding boxes. 2) A tractable extension of CFGs to parse 2D images where the grammar reflects the structure of the extracted information rather than the layout of the document. 3) We demonstrate state-of-the-art performance in an invoice reading system, especially the problem of extracting line-item fields and line-item grouping.

2   Related Work

We identify 3 areas of prior work related to ours: information extraction from invoices using machine learning, grammars for parsing document layout, and structured prediction using Deep Learning. The key innovation of our approach is that we use a grammar based on the information to be extracted rather than the document layout resulting in a system that requires no hand annotations and can be trained end-to-end.

There has been a lot of work on extracting information from invoices. In particular a number of papers investigate using Deep Learning to classify bounding boxes. [2] generated invoice data and used a graph convolutional neural network to predict the class of each bounding box. [26] uses BERT [6] that integrates 2D positional information to produce contextualized embeddings to classify bounding boxes and extract information from receipts[2]. [19] also used BERT and neighborhood encodings to classify bounding boxes. [5,15] use grid-like structures to encode information about the positions of bounding boxes and then to classify those bounding boxes. [2,19,26,27] predict the class of bounding boxes and then post-process to group them into records. [27] uses graph convolutional networks

---

[2] Receipts are a simplified form of invoice.

together with a graph learning module to provide input to a Bidirectional-LSTM and CRF decoder which jointly labels bounding boxes. This works well for flat (non-recursive) representations such as the SROIE dataset [12] but we are not aware of their application to hierarchical structures in more complex documents, especially on recursive structures such as line-items.

[13,14] provide a notable line of work which addresses the line-item problem. [13] reduce the 2D layout of the invoice to a sequence labeling task, then use Beginning-Inside-Outside (BIO) Tags to indicate the boundaries of line-items. [14] treats the line-item grouping problem as that of link prediction between bounding boxes. [14] jointly infers the class of each box, and the presence of a link between the pair of boxes in a line-item group. However, as with all of the above cited work in information extraction from invoices both [13,14] require hand annotation for each bounding box, and they require post-processing in order to group and order bounding boxes into records.

Documents are often laid out hierarchically and this recursive structure has been addressed using parsing. 2D parsing has been applied to images [20,29,30], where the image is represented as a 2D matrix. The regularity of the 2D matrix allows parsing to be extended directly from 1D. The 2D approaches to parsing text documents most related to ours are from [18,21,24]. [24] described a 2D-CFG for parsing document layout which parsed 2D regions aligned to a grid. [18] describe 2D-parsing of document layout based on context free grammars. Their Rectangle Hull Region approach is similar to our 2D-parsing algorithm but yields a $\mathcal{O}(n^5)$ complexity compared to our $\mathcal{O}(n^3)$. [21] extends [18] to use conditional probabilistic context free grammars based on Boosting and the Perceptron algorithm while ours is based on deep Learning with back-propagation through the parsing algorithm. Their work relies on hand-annotated documents. All of this work requires grammars describing the document layout and seeks to fully describe that layout. On the other hand, our approach provides end-to-end extraction from a variety of layouts simply by defining a grammar for the information to be extracted and without a full description for the document layout. To the best of our knowledge, no other work in the space of Document Intelligence takes this approach.

[17] provides the classical Probabilistic Context Free Grammar (PCFG) trained with the Inside-Outside algorithm, which is an instance of the Expectation-Maximization algorithm. This work has been extended in a number of directions that are related to our work. [23] extended PCFGs to the Conditional Random Field [16] setting. [7] use inside-outside applied to constituency parsing where deep neural networks are used to model the conditional probabilities. Like us they train using backpropagation, however, they use the inside-outside algorithm while we use structured prediction. Other work [1] considers more general applications of Deep Learning over tree structures. While [7,17,23] are over 1D sequences, here we use deep neural networks to model the conditional probabilities in a CPCFG over 2D structures.

Finally, we refer the reader to [22] for a recent survey on Deep Learning for document understanding.

3 Problem Description

We define an information extraction problem by a universe of documents D and a schema describing the structure of records to be extracted from each document $d \in D$. Each document is a single image corresponding to a page in a business document (extensions to multi-page documents are trivial). We assume that all documents d are processed (e.g. by OCR software) to produce a set of bounding boxes $b = (content, x_1, y_1, x_2, y_2)$ with top left coordinates (x_1, y_1) and bottom right coordinates (x_2, y_2).

The schema describes a tuple of named fields each of which contains a value. Values correspond to base types (e.g., an Integer), a list of values, or a recursively defined tuple of named fields. These schemas will typically be described by a JSON Schema, an XML Schema or via the schema of a document database. More generally, the schema is a context free grammar $G = (V, \Sigma, R, S)$ where Σ are the terminals in the grammar and correspond to the base types or tokens, V are the non-terminals and correspond to field names, R are the production rules and describe how fields are constructed either from base types or recursively, and S is the start symbol corresponding to a well-formed extracted record.

$$
\begin{aligned}
\textbf{Invoice} &:= \textbf{(InvoiceID Date LineItems TotalAmt)} \text{ !} \\
\textbf{InvoiceID} &:= \textbf{STRING} \mid \text{(N InvoiceID)} \text{ !} \mid \epsilon \\
\textbf{Date} &:= \textbf{STRING} \mid \textbf{Date Date} \mid \text{(N Date)} \text{ !} \mid \epsilon \\
\textbf{TotalAmt} &:= \textbf{MONEY} \mid \text{(N TotalAmt)} \text{ !} \mid \epsilon \\
\textbf{LineItems} &:= \textbf{LineItems LineItem} \mid \textbf{LineItem} \\
\textbf{LineItem} &:= \textbf{(Desc Qty Rate Amt)} \text{ !} \mid \text{(N LineItem)} \text{ !} \\
\textbf{Desc} &:= \textbf{STRING} \mid \textbf{Desc Desc} \mid \text{(N Desc)} \text{ !} \\
\textbf{Qty} &:= \textbf{NUMBER} \mid \text{(N Qty)} \text{ !} \mid \epsilon \\
\textbf{Rate} &:= \textbf{NUMBER} \mid \text{(N Rate)} \text{ !} \mid \epsilon \\
\textbf{Amt} &:= \textbf{MONEY} \mid \text{(N Amt)} \text{ !} \mid \epsilon \\
\text{N} &:= \text{N N} \mid \text{STRING}
\end{aligned}
$$

Fig. 2. Grammar

An example grammar is illustrated in Fig. 2. Reading only the content in **bold** gives the rules for G that represents the main information that we want to extract. Here $\Sigma = \{$STRING, NUMBER. MONEY$\}$, $V = \{$Invoice, InvoiceNumber, TotalAmount, LineItems, LineItem, Desc (Description), Qty (Quantity), Rate, Amt$\}$[3]. The goal of information extraction is to find the parse tree $t_d \in G$ corresponding to the record of interest.

4 Approach

We augment G to produce $G' = (V', \Sigma', R', S')$ a CPCFG whose parse trees t'_d can be traversed to produce a $t_d \in G$. Below we assume (without loss of

---

[3] Note that this is not in CNF but the conversion is straightforward [3].

generality) that all grammars are in Chomsky Normal Form (CNF) and hence all parse trees are binary.

The set of bounding boxes for a document d may be recursively partitioned to produce a binary tree. We only consider partitions that correspond to vertical or horizontal partitions of the document region so that each subtree corresponds to a rectangular region B of the document image[4]. Each such region contains a set of bounding boxes. We consider any two regions B_1 and B_2 equivalent if they contain exactly the same set of bounding boxes.

The leaves of a partition tree each contain single bounding boxes. We extend the tree by appending a final node to each leaf that is the bounding box itself. We refer to this extension as the partition tree for the remainder of the paper. The contents of these bounding boxes are mapped to the terminals $\Sigma' = \Sigma \cup \{\text{NOISE}, \epsilon\}$. The special NOISE token is used for any bounding box contents that do not map to a token in Σ. The special ϵ token is used to indicate that some of the Left-Hand-Side (LHS) non-terminals can be an empty string. Document fields indicated with ϵ are optional and can be missing from the document. We handle this special case within the parsing algorithm.

We augment the non-terminals $V' = V \cup \{\text{N}\}$ where non-terminal N captures regions of the image that contain no extractable information.

We augment the rules R by adding rules dealing with the N and NOISE symbols. Every rule $X \to YZ$ is augmented with a production $X \to NX$ and every rule $A \to \alpha$ is augmented with a rule $A \to NA$. In many cases the record information in a document may appear in arbitrary order. For example, the order of line-items in an invoice is irrelevant to the meaning of the data. We introduce the suffix "!" on a production to indicate that all permutations of the preceding list of non-terminals are valid. This is illustrated in Fig. 2 where the modifications are in not in **bold**.

Leaves of a partition tree are labeled with the terminals mapped to their bounding boxes. We label the internal nodes of a partition tree with non-terminals in V' bottom up. Node i corresponding to region B_i is labeled with any $X \in V'$ where $\exists (X \to YZ) \in R'$ such that the children of node i are labeled with Y and Z. We restrict our attention to document partition trees for which such a label exists and refer to the set of all labels of such trees as T'_d and a single tree as t'_d (with a minor abuse of notation). We recover a tree $t_d \in G$ from t'_d by removing all nodes labeled with N or NOISE and re-connecting in the obvious way. We say that such a t_d is compatible with t'_d.

By adding weights to each production, we convert G' to a CPCFG. Trees are assigned a score $s(t'_d)$ by summing the weights of the productions at each node. Here the weights are modeled by a deep neural network $m_{X \to YZ}(B_1, B_2)$ applied to a node labeled by X with children labeled by Y and Z and contained in regions B_1 and B_2 respectively.

---

[4] This approach is built on the assumption (by analogy with sentences) that documents are read contiguously. This assumption likely does not hold for general images where occlusion may require non-contiguous scanning to produce an interpretation.

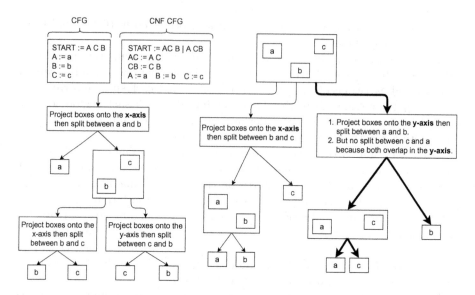

Fig. 3. An example of CYK 2D parsing on a simple document and grammar. The valid parse tree is the one on the left with **bolded arrows** →.

4.1   Parsing

We can now solve the information extraction problem by finding the tree t'_d with the highest score by running a chart parser.

A typical 1D chart parser creates a memo $c[\text{sentence span}][X]$ where $X \in V$ and "sentence span" corresponds to a contiguous span of the sentence being parsed and is usually represented by indexes i, j for the start and end of the span so that the memo is more typically described as $c[i][j][X]$. The memo contains the score of the highest scoring subtree that could produce the sentence span and be generated from the non-terminal X. The memo can be constructed (for a sentence of length n) top down starting from $c[0][n][S]$ with the recursion:

$$c[i][j][X] = \max_{(X \to YZ) \in R} \max_{i \le k < j} \left(w_{X \to YZ}(i, k, j) + c[i][k][Y] + c[k][j][Z] \right) \quad (1)$$

where $w_{X \to YZ}$ is the weight associated with the rule $X \to YZ$. It is easy to see that the worst-case time complexity of this algorithm is $O(n^3 |R'|)$.

We extend this algorithm to deal with 2D documents. In this case the memo $c[B][X]$ contains the score of the highest scoring sub-tree that could produce the region B of the document image from the non-terminal X. This results in a top down algorithm recursively defined as follows:

$$c[B][X] = \max_{(X \to YZ) \in R'} \max_{B_1, B_2 \in Part(B)} \left(m_{X \to YZ}(B_1, B_2) \right. \quad (2)$$
$$\left. + c[B_1][Y] + c[B_2][Z] \right)$$

Algorithm 1. The neural network model of the unary rules

---

function $m_{X \to x}(bb, lm)$ ▷ bb: the bounding box, lm: the language model
     $h_0 = \text{forward}(lm, bb)$ ▷ Get language model vector
     $h = \text{GELU}(W_0^{X \to x} h_0 + b_0^{X \to x})$ ▷ Hidden vector
     $s = W_1^{X \to x} h + b_1^{X \to x}$ ▷ Score for Non-terminal X
     return s, h
end function

Algorithm 2. The neural network model of the binary rules

---

function $m_{X \to YZ}(B, B_1, B_2, h_d, M)$ ▷ B: the region under consideration.
                     ▷ B_1, B_2: the sub-regions for one of the partitions of B.
                  ▷ h_d: the embedding representing the direction of the partition.
                       ▷ M: the memoization of the dynamic program.
     $h_1, h_2 = M[B_1][Y][h], M[B_2][Z][h]$
     $h = \text{GELU}(W_1^{X \to YZ} h_1 + W_2^{X \to YZ} h_2 + W_d^{X \to YZ} h_d + b_0^{X \to YZ})$ ▷ Hidden vector
     $s = W_3^{X \to YZ} h + b_3^{X \to YZ}$ ▷ Score for Non-terminal X
     return s, h
end function

---

where we consider $Part(B)$ defined as partitions of B obtained by splitting horizontally or vertically between adjacent pairs of bounding boxes in B, i.e. $B = B_1 \cup B_2, B_1 \cap B_2 = \emptyset$. There are $n - 1$ such horizontal splits and $n - 1$ such vertical splits. The worst-case time complexity of this algorithm is $O(n^3 |R'|)$[5].

Overloading notation, we say $s(d) = c[d][S']$ provides the score for the highest scoring tree for d. We can recover the tree itself by maintaining back-pointers as in a typical chart parser. We assign a score to $t_d \in G$ as the maximum score over all t_d' with which it is compatible.

One may refer to Fig. 3 for an illustration of how 2D parsing is done using the CYK algorithm.

4.2 Learning

Given training data consisting of pairs (d, t_d) with $d \in D$ a document and $t_d \in G$ our goal is to learn the parameters of the models m_r such that t_d is the highest scoring tree for d.

We achieve this following the structured prediction approach [4] and minimizing the structured prediction loss.

$$\sum_{d \in D} s(\hat{t}_d') - s(\bar{t}_d') \tag{3}$$

Where $\bar{t}_d'$ is the highest scoring tree compatible[6] with t_d (the correct tree) and $\hat{t}_d'$ is the highest scoring tree from the dynamic program. Intuitively we aim to increase the scores of correct trees and decrease scores of incorrect trees.

---

[5] The cost of calculating c can be further reduced by using beam search.
[6] We will address compatibility in Sect. 4.3.

We perform this minimization using gradient descent and back-propagation on Eq. 3. Each model m_r is a deep neural network and the score $s(t'_d)$ is computed recursively as a function of these models. We can back-propagate through this recursion to jointly train all of the m_r.

It remains to describe the models m_r. Each model outputs both a score for the production at a given tree node and an embedding meant to represent the sub-tree under that node. The models for terminal rules $X \rightarrow x$ where $x \in \Sigma$ take as input a target bounding box and any context that might be relevant to labeling that bounding box such as the coordinates of the bounding box. Intuitively these models predict which Non-terminal labels a given bounding box. For simplicity of presentation, we describe a relatively simple class of models.

Algorithm 1 shows the model $m_{X \rightarrow x}$ used in Eq. 1. The forward function in Algorithm 1 produces a vectorized representation of the given bounding box. One could either use a language model with pretrained weights or train the model end-to-end as part of the learning process. Many architectures are possible for such a model including ones based on language embeddings (e.g., BERT [6]) that embed only the contents of the bounding box, and ones which aim to take document image, layout, and format into account (e.g., LayoutLM [26]).

When we integrate our DeepCPCFG model and a language model, it results in an encoder-decoder architecture. The encoder is the language model (e.g. Layoutlm) which takes as input the bounding box coordinates and text, then produces a word embedding for each bounding box. The word embedding is given to the decoder, which is DeepCPCFG, that produces a parse tree reflecting the document hierarchy. Geometric information is captured explicitly by Layoutlm then implicitly again during 2D parsing.

The models for Non-terminal rules $X \rightarrow YZ$ take as inputs the sub-regions $B_1, B_2 \in Part(B)$ such that $B = B_1 \cup B_2, B_1 \cap B_2 = \emptyset$ where B, B_1, B_2 are labeled with X, Y, Z respectively and outputs a score for X and an embedding vector for B representing X.

Algorithm 2 shows the implementation of the function $m_{X \rightarrow YZ}$ (used in Eq. 2), as a derivation of a Tree Convolutional Block [10]. The matrices W_i^r with biases b_i^r are the learned parameters of each model m_r and $M[B][X][h]$ is the embedding for the best scoring sub-tree associated with B and generated by X. These embedding vectors are stored in the memo M of the chart parser.

4.3   Compatible Tree for Structured Prediction

We showed in Eq. 3 of Sect. 4.2 that a compatible tree $\bar{t}'_d$ with the annotation of d is required as part of learning the parameters for a Structured Prediction model. We define a tree $\bar{t}'_d$ to be most *compatible* if the tree gives the *smallest* tree edit-distance [28] as compared with the hierarchical structure derived from the relational-record of the document d. But ordering of columns (tree branches) within a relational database can interfere with how tree edit-distance is computed. Therefore, we propose to relax the ordering of fields within the document, and the fields within the recurrent line-items to derive a variant of tree edit-distance, which we call Hierarchical Edit-Distance (HED). In HED, we only require the ordering of line-items within a document and words within a field remain the same, while the ordering of fields within a line-item may be permuted without impacting the distance.

$$\text{HED}(x, y) = \sum_{f \in H} \text{SED}(x_f, y_f) + \text{LiSeqED}(x_{\text{li}}, y_{\text{li}}) \qquad (4)$$

H refers to the set of header fields: {InvoiceID, Date, TotalAmt}. SED stands for Levenshtein String Edit Distance. LiSeqED (Line-item sequence edit-distance) is defined by Eq. 5. x_{li} and y_{li} represent the line-items of x and y.

$$\text{LiSeqED}(x, y) = \begin{cases} \sum_{i=1}^{|x|} \text{LiED}(x_i, \emptyset) & \text{if } |y| = 0 \\ \sum_{i=1}^{|y|} \text{LiED}(\emptyset, y_i) & \text{if } |x| = 0 \\ \text{otherwise:} \\ \min \begin{cases} \text{LiED}(x_1, y_1) + \text{LiSeqED}(\text{tail}(x), \text{tail}(y)) \\ \text{LiED}(x_1, \emptyset) + \text{LiSeqED}(\text{tail}(x), y) \\ \text{LiED}(\emptyset, y_1) + \text{LiSeqED}(x, \text{tail}(y)) \end{cases} \end{cases} \qquad (5)$$

where LiED is defined by Eq. 6, tail is a function that returns the rest of the list except the first element in the list. $\emptyset$ represents an empty line-item.

$$\text{LiED}(x, y) = \sum_{f \in G} \text{SED}(x_f, y_f) \qquad (6)$$

where G represents the set of line-item fields {Desc, Qty, Rate, Amt}.

Using HED, we can obtain the compatible (smallest edit-distance) tree $\bar{t}'_d$ in the same way as we obtained the highest scoring parse tree $\hat{t}'_d$ by using HED as the scoring function then taking the minimum instead of maximum in Eqs. 1 and 2. We will re-use HED when evaluating the results of our experiments.

5   Experiments

Prior research on business documents, particularly invoices, has been limited by available public datasets. We are not aware of public datasets that provide structured extraction from business documents or invoices. In the FUNSD dataset [9] annotations reflect record linkages rather than hierarchical structures. The RVL-CDIP [11] dataset provides class labels for a document classification task rather than line-item annotations for information extraction. [8] provides a dataset of business documents for information extraction but its documents do not contain line-items. The receipts used in SROIE [12] have line-items but those are not annotated. [5,15,19,27] report results only on proprietary sets of invoices.

To evaluate our model on line-items we ran our experiments on three datasets summarized in Table 1. The first dataset from [13,14] consists of the CORD

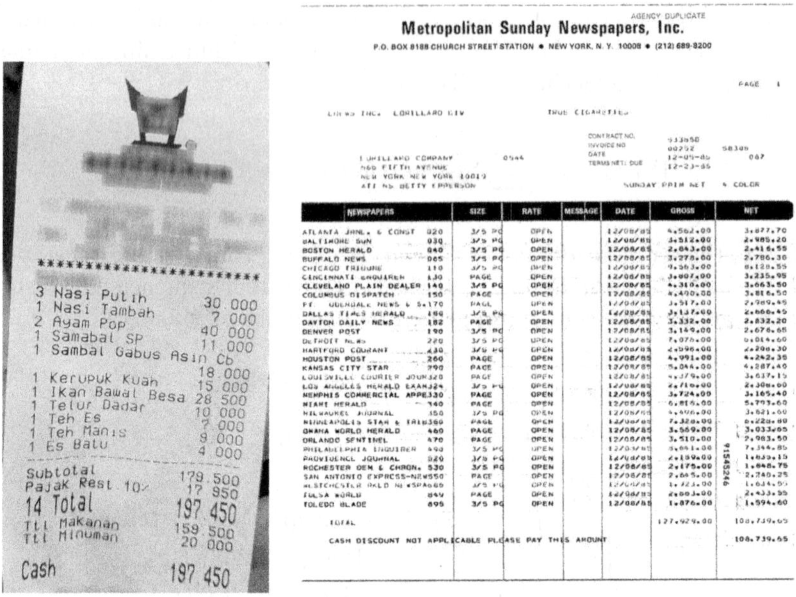

(a) CORD receipt   (b) RVL-CDIP invoice

Fig. 4. Examples of structured documents

receipt data. The second consists of proprietary invoices for which we have both hand annotations and relational-records. We created the third dataset by hand annotating invoices in the RVL-CDIP collection [11]. We use the RVL-CDIP invoices solely for testing and welcome other researchers to compare with our results[7].

Table 1. Dataset sizes

| | Receipts | Invoices | |
	CORD	Proprietary	RVL-CDIP
Training	800	17938	0
Validation	100	2085	0
Testing	100	2516	869

Our preliminary experiments demonstrated that the language model used in Algorithm 1 should be fine-tuned on relevant documents. In the experiments below we fine-tune Layoutlm [26] as follows: using the untrained CFG we derive compatible trees from the proprietary invoices' relational-records. For

---

[7] We release RVL-CDIP data and metric code at https://github.com/deepcpcfg.

each invoice d, we obtain the compatible tree $\bar{t}'_d$, which provides classes for each bounding box (token) from the leaves of the compatible trees. These leaves are used to fine-tune a LayoutLM model on the token classification task.

5.1   Results on CORD

Table 2 provides a comparison between [14] and DeepCPCFG. We compare against the best results reported in Table 9 of [14] using the SPADE metric described in the appendix of [14][8]. The SPADE (Spatial Dependency Parsing) metric can be seen as a special case of HED if SED in Eqs. 4 and 6 is implemented using exact string match as follows,

$$SED_{SPADE}(x, y) = \begin{cases} 0 & \text{if } x = y \\ 1 & \text{if } x \neq y \end{cases} \qquad (7)$$

Using the dataset provided by [14] we derive relational-records (see Fig. 1b) for our DeepCPCFG model. In Table 2, [14] is trained using hand annotations while DeepCPCFG makes no use of those annotations and is therefore at a significant disadvantage. Although DeepCPCFG can leverage hand annotations when they are present, we chose not to use them to emphasize the power of DeepCPCFG when trained end-to-end.

Overall DeepCPCFG achieves comparable results to [14] despite not being trained on hand annotations. Given that there are only 100 receipts in the hold-out test set, the numbers we report depend substantially on 1 or 2 receipts. When we inspect DeepCPCFG's errors we found rotated or distorted receipts (see Fig. 4a). DeepCPCFG's performance is compelling given that DeepCPCFG focuses on learning an end-to-end model for formally scanned business documents while [14] specializes in extraction from photos taken using handheld cameras.

5.2   Results on Invoices

Scanned invoices reflect the primary objective and motivation for our research, that is, information extraction from complex business documents. By comparison, the CORD receipt dataset is much smaller and simpler.

In Table 3 we report results of 3 models trained on our proprietary invoices. The models are evaluated on a holdout test set of proprietary invoices and on an unseen set of invoices from RVL-CDIP. In all cases we train only on relational-records and do not make use of hand annotations.

For invoices, we report results based on HED which allows for mismatches due to OCR errors. HED is implemented by tracking the number of unchanged characters (true positives), insertions (false negatives) and deletions (false positives)

---

[8] At the time of writing, [14] have not released their code. We re-implemented SPADE metric to the best of our understanding based on communication with the authors. We release the metric implementation and output files at https://github.com/deepcpcfg for anyone to compare or verify our results.

Table 2. F1 results on CORD using SPADE metric

Model	Overall	Desc	Qty	Rate	Amt	TotalAmt
Hwang et al. [14]	90.1	**91.6**	**92.1**	91.6	**93.4**	**96.9**
DeepCPCFG	**92.2**	88.7	90.6	**96.4**	91.7	95.0

Table 3. F1 results on proprietary invoices using HED metric

Model	Overall	Desc	Qty	Rate	Amt	InvoiceID	Date	TotalAmt
Proprietary invoices								
DeepCPCFG pre-trained LayoutLM (Epoch 3)	67.2	61.7	70.0	70.7	77.2	77.8	86.2	84.1
DeepCPCFG fine-tuned LayoutLM (Epoch 0)	73.5	68.7	81.2	83.4	81.8	92.8	86.3	80.9
DeepCPCFG fine-tuned LayoutLM (Epoch 1)	82.2	79.1	86.8	89.5	88.6	93.0	88.2	83.9
Compatible Trees	95.6	97.3	91.5	95.2	92.8	94.8	89.6	87.5
RVL-CDIP Invoices								
DeepCPCFG pre-trained LayoutLM (Epoch 3)	55.2	52.0	38.4	45.6	60.0	53.6	75.2	68.1
DeepCPCFG fine-tuned LayoutLM (Epoch 0)	63.1	60.0	48.7	57.2	66.8	75.5	83.7	68.5
DeepCPCFG fine-tuned LayoutLM (Epoch 1)	70.5	69.0	55.2	57.0	73.5	74.6	84.5	71.6
Compatible Trees	89.9	92.3	80.0	88.2	83.3	85.2	88.5	80.7

required to transform each prediction into its respective annotation. Replacements are treated as deletion/insertion pairs. These counts then allow us to derive precision, recall and f1 metrics.

First, we investigate DeepCPCFG with pre-trained Layoutlm as the language model. We compare this to DeepCPCFG using fine-tuned Layoutlm and note a significant improvement in performance on both datasets.[9]

We also examine the performance of untrained DeepCPCFG (Epoch 0) and observe that training DeepCPCFG provides a dramatic improvement in performance especially for the Desc field which is the most dependent on structure as it is composed of multiple bounding boxes.

The HED metric is sensitive to OCR or annotation errors. The rows "Compatible Trees" in Table 3 show the quality of the compatible trees derived using the relational-records annotations. These values reflect the best results possible on the holdout test set given OCR and annotation errors.

The RVL-CDIP dataset is rather old and its scanned images are typically noisy or of poor quality (see Fig. 4b) resulting in diminished OCR quality. This leads to deteriorated performance for the RVL-CDIP dataset in Table 3.

Relation to Bounding Box Classification. While our goal is to extract structured information from complex documents in an end-to-end fashion it is informative to compare against methods that classify bounding boxes based on hand annotations. In Table 4 we compare DeepCPCFG against a Layoutlm based bounding box classifier. Note that these results measure the number of bounding boxes where the model and the human annotator disagree on their label.

---

[9] In each case the optimal number of epochs was chosen using the validation set.

Table 4. F1 classification results on proprietary invoices evaluated on hand-annotations as ground truth

Model	Overall	Desc	Qty	Rate	Amt	InvoiceID	Date	TotalAmt
LayoutLM with token classification from Hand-annotations	93.3	92.0	96.4	95.3	96.5	96.6	97.2	89.3
Leaf-annotations	86.4	95.6	87.9	61.3	65.4	88.8	72.2	52.7
DeepCPCFG fine-tuned with LayoutLM using leaf-annotations (Epoch 1)	81.4	88.1	87.3	58.2	64.0	88.7	75.9	52.1

We first examine the performance of bounding box classifications produced by taking the leaves from the compatible tree ("Leaf-annotations"). This process uses the true relational-records on the test data and identifies the bounding boxes that best recover those records. This performs poorly particularly on fields like Rate, Amt, Date, and TotalAmt whose values may appear multiple times in an invoice, due to significant annotation errors. This illustrates the difference between "ground truth" when evaluating against relational-records rather than human annotations. Relational-records better reflect real-world objectives in most applications and this result suggests that human annotations are a rather poor proxy for evaluating these objectives.

Next, we examine the performance of DeepCPCFG which has been trained based on the compatible trees from the training data, as such it is penalized in this evaluation in the same way that the "Leaf-annotations" are. Notably Deep-CPCFG obtains results comparable to and sometimes better than the "Leaf-annotations" on the test data.

From these experiments we see that the problem of information extraction is substantially different from the problem of classifying bounding boxes. Despite this we see that DeepCPCFG is quite effective at classifying bounding boxes even when trained without any hand annotations.

6 Conclusion

We have described and demonstrated a method for end-to-end structured information extraction from business documents such as invoices and receipts. This work enhances existing capabilities by removing the need for brittle post-processing, and by reducing the need for annotated document images.

Our method "parses" a document image using a grammar that describes the information to be extracted. The grammar does not describe the layout of the document or significantly constrain that layout. This method yields compelling results. However, research on this important problem is limited by the lack of available benchmark data sets which has slowed development and stymied comparisons. In order to alleviate this, we released a new public evaluation set based on the RVL-CDIP data. Related "2D parsing" approaches have been previously explored for image analysis and we believe that the effectiveness of our approach suggests a broader re-examination of grammars in image understanding particularly in combination with Deep Learning.

We have gone beyond existing work in this space by [2,14,18,22,26,27]. The success of DeepCPCFG, together with earlier work in this space, shows the value of combining structured models with Deep Learning.

Our experimental results illustrate the significant gap between information extraction and recovering labels from hand annotations. We believe that evaluations based on recovering relational-records best reflect real-world use cases. In ongoing work, we are applying this technique to a wide variety of structured business documents including tax forms, proofs of delivery, and purchase orders.

Disclaimer. The views reflected in this article are the views of the authors and do not necessarily reflect the views of the global EY organization or its member firms.

Acknowledgements. The authors will like to thank the following colleagues: David Helmbold, Ashok Sundaresan, Larry Kite, Chirag Soni, Mehrdad Gangeh, Tigran Ishkhanov and Hamid Motahari.

References

1. Alvarez-Melis, D., Jaakkola, T.S.: Tree-structured decoding with doubly-recurrent neural networks. In: ICLR (2017)
2. Blanchard, J., Belaïd, Y., Belaïd, A.: Automatic generation of a custom corpora for invoice analysis and recognition. In: ICDARW (2019)
3. Cole, R.: Converting CFGs to CNF (Chomsky normal form) (2007)
4. Collins, M.: Discriminative training methods for hidden Markov models: theory and experiments with perceptron algorithms. In: EMNLP (2002)
5. Denk, T.I., Reisswig, C.: Bertgrid: contextualized embedding for 2D document representation and understanding. In: NeurIPS Document Intelligence Workshop (2019)
6. Devlin, J., Chang, M.W., Lee, K., Toutanova, K.: BERT: pre-training of deep bidirectional transformers for language understanding. In: ACL-HLT (2019)
7. Drozdov, A., Verga, P., Yadav, M., Iyyer, M., McCallum, A.: Unsupervised latent tree induction with deep inside-outside recursive auto-encoders. In: ACL-HLT (2019)
8. Gralinski, F., et al.: Kleister: a novel task for information extraction involving long documents with complex layout. arXiv abs/2003.02356 (2020)
9. Jaume, G., Ekenel, H.K., Thiran, J.P.: FUNSD: a dataset for form understanding in noisy scanned documents. In: ICDAR-OST (2019)
10. Harer, J., Reale, C., Chin, P.: Tree-transformer: a transformer-based method for correction of tree-structured data (2019)
11. Harley, A.W., Ufkes, A., Derpanis, K.G.: Evaluation of deep convolutional nets for document image classification and retrieval. In: ICDAR (2015)
12. Huang, Z., et al.: ICDAR 2019 competition on scanned receipt OCR and information extraction. In: ICDAR (2019)
13. Hwang, W., et al.: Post-OCR parsing: building simple and robust parser via bio tagging. arXiv (2019)

14. Hwang, W., Yim, J., Park, S., Yang, S., Seo, M.: Spatial dependency parsing for semi-structured document information extraction. In: ACL-IJCNLP (2021)

15. Katti, A.R., et al.: Chargrid: towards understanding 2D documents. In: EMNLP (2018)

16. Lafferty, J.D., McCallum, A., Pereira, F.C.N.: Conditional random fields: probabilistic models for segmenting and labeling sequence data. In: ICML (2001)

17. Lari, K., Young, S.: The estimation of stochastic context-free grammars using the inside-outside algorithm. Comput. Speech Lang. **4**(1), 35–56 (1990)

18. Liang, P., Narasimhan, M., Shilman, M., Viola, P.: Efficient geometric algorithms for parsing in two dimensions. In: ICDAR (2005)

19. Majumder, B.P., Potti, N., Tata, S., Wendt, J.B., Zhao, Q., Najork, M.: Representation learning for information extraction from form-like documents. In: ACL (2020)

20. Pedro, R.W.D., Nunes, F.L.S., Machado-Lima, A.: Using grammars for pattern recognition in images: a systematic review. ACM Comput. Surv. **46**(2), 1–34 (2013)

21. Shilman, M., Liang, P., Viola, P.: Learning non-generative grammatical models for document analysis. In: ICCV (2005)

22. Subramani, N., Matton, A., Greaves, M., Lam, A.: A survey of deep learning approaches for OCR and document understanding (2020)

23. Sutton, C., McCallum, A.: Conditional probabilistic context-free grammars. Ph.D. thesis, University of Massachusetts Amherst (2004)

24. Tomita, M.: Parsing 2-dimensional language. In: Tomita, M. (ed.) Current Issues in Parsing Technology, pp. 277–289. Springer, Boston (1991). https://doi.org/10.1007/978-1-4615-3986-5_18

25. Vaswani, A., et al.: Attention is all you need. In: NeurIPS (2017)

26. Xu, Y., Li, M., Cui, L., Huang, S., Wei, F., Zhou, M.: Layoutlm: pre-training of text and layout for document image understanding. In: SIGKDD (2020)

27. Yu, W., Lu, N., Qi, X., Gong, P., Xiao, R.: Pick: processing key information extraction from documents using improved graph learning-convolutional networks (2020)

28. Zhang, K., Shasha, D.: Simple fast algorithms for the editing distance between trees and related problems. SIAM J. Comput. **18**(6), 1245–1262 (1989)

29. Zhao, Y., Zhu, S.C.: Image parsing via stochastic scene grammar. In: NeurIPS (2011)

30. Zhu, L.L., Chen, Y., Lin, Y., Lin, C., Yuille, A.: Recursive segmentation and recognition templates for 2D parsing. In: NeurIPS (2008)

Consideration of the Word's Neighborhood in GATs for Information Extraction in Semi-structured Documents

Djedjiga Belhadj$^{(\boxtimes)}$ [ID], Yolande Belaïd [ID], and Abdel Belaïd [ID]

Université de Lorraine-LORIA, Campus Scientifique,
54500 Vandoeuvre-Lès-Nancy, France
{yolande.belaid,abdel.belaid}@loria.fr

Abstract. Most administrative documents take a semi-structured form (invoices, payslips, etc.). Extracting information from this type of document is still challenging because of the variability of its structure brought about by the change of layout style of the different administrations. In this work, we try to face this type of variation by using a multi-layer Graph Attention Network (GAT). We propose a general structure of a semi-structured document. Based on this latter, we adopt a star subgraph to exploit the surrounding context of words, allowing neighboring words to help locate the searched words and rank them. The GAT makes it possible to exploit this type of neighborhood and to highlight important neighboring words likely to be better identified. Each graph node contains at the same time textual and visual features. We experiment the multi-layer GAT on three different datasets: invoices and payslips (generated artificially), and receipts (issued from SROIE ICDAR competition). For the later dataset, we get an important F1 score of 0.892.

Keywords: Information extraction · Semi-structured document · Neighborhood · GAT

1 Introduction

Extracting information from administrative documents poses many research challenges due to their structure and to the variation in their content. In fact, regarding their structure, we can list them in three categories: structured, semi-structured and unstructured.

Structured documents, according to [11], are documents that have content presented with information placed in one fixed place. The various forms are examples of this type of document and the often used information extraction methods are of the type template matching [19]. In contrast, unstructured documents can contain all kinds of information that does not fit in a particular position, such as the content of emails, articles and so on. Information extraction methods require exploring the text and finding the right word associations, often using NLP techniques.

Supported by BPI DeepTech.

There remain the semi-structured documents which attract our attention in this work and which represent most of the documents used in administrations. There is no precise definition of semi-structured documents (SSD). However, based on our analysis of various examples of semi-structured documents (invoices, payslips, receipts, etc.), we find the following fact: A semi-structured document contains classes of information physically grouped together and identified by some indicative keywords in their vicinity (introductory words: date, invoice no., shipping and billing addresses, table headers, etc.). Some parts may be optional and appear in different places of the document [11,24]. They do not necessarily follow a predefined template [1], but the importance of the neighborhood is very characteristic (as for addresses, company headings and footnotes in invoices) [15].

Based on this neighborhood consideration and the importance of distinctive keywords, we propose as in [15] to use a graph to represent them. The nodes represent the document words (given by their textual and visual features) and the edges, the relationships between each word and its neighborhood. For each node, we build its neighborhood by a star sub-graph (see Fig. 5c). We limit this neighborhood to nearby words, at horizontal and vertical distances, to group together all the close words of the same information class and their keywords. A set of new annotated documents is exploited to make this calculation. The idea behind choosing this kind of neighborhood is to make sure that the word is connected to all the other words in the document that can help to classify it well. We propose a multi-layer Graph Attention Network (Multi-layer GAT) to assign importance to the most important neighboring words and classify these words into their appropriate classes. We test our approach on three databases: invoices, payslips and receipts to prove the performance of our method.

The remaining paper is structured as follows. Section 2 summarises studies of state of the art essentially focused on information extraction from documents, graph modeling and graph convolution. Section 3 exposes the proposed method giving details of the SSD structure, graph modeling, word modeling and the multi-layer GAT architecture. Section 4 shows the used datasets, the experiments and our results. Section 5 concludes the overall contribution of this research work.

2   Related Work

Many papers have addressed the task of understanding documents as the problem of sequence tagging and have used recurrent neural network models to extract entities. [12,14,16,23,25] all used a BLSTM followed by a CRF layer to respond to the Named Entity Recognition task. This type of neural network (BLSTM) has been shown to be efficient in sequence problems. While the CRF layer can use past and future tags efficiently to predict the current tag in a sentence. These works consider the sentence to be a sequence of words (features) and they tag sequence entities using the IOB format (short for inside, outside, beginning). What makes each approach different from the other is the choice of calculating the features of the text segments that represent the BLSTM's input. The authors in [16] use a CNN combination to calculate the character-level representation of words. The character representation vector is then concatenated with the word embedding before being fed into the BLSTM network. The authors in [14] concatenate graph embedding with token embedding and feed them into standard BLSTM-CRF. They use the Word2Vec vectors as token embedding and a convolutional graph architecture to output the graph embedding. This provides a visually rich context and adds contextual information to the input sequence. The authors in [25] used textual and visual information (color, layout, etc.) as well as position as input of the BLSTM. This approach integrates text recognition and information extraction into a single end-to-end framework. The authors in [23] combine text embedding and image embedding and add to them the result of a graph convolution and then feed the recurrent Neural network with this combination. The graph convolution tries to exploit the neighborhood of the text segments by learning a soft adjacent matrix. This type of architectures has already been shown to have a restricted capacity to learn the connection between distant words.

Some other approaches turn the task of understanding the document into a node classifying problem. They use a graph structure to represent the document and define a type of neighborhood to be considered in order to ensure a correct node classification of the text into some known classes. The authors in [15] model an invoice document as a graph. The nodes are the words of the document (represented by a 317 sized features vector). Each node can be connected to its 4 nearest neighbors in the four main directions (left, right, top and bottom). A GCN network get the constructed graph as input. The network has a Softmax layer as an output layer to classify the nodes into 28 possible classes. The authors in [8] model a document with a directed graph structure. Each node represents a text segment (text sequence) and is connected to all other text segments belonging to the same line and the previous line. They add a global node to connect all the local nodes together. Finally, they apply an attention-based graph neural network with a CRF layer to tag the text sequences.

Other works use the graph structure to model the document and suggest other approaches to extract information from the graph. The authors in [6] build a graph of color rectangles extracted from the document. The nodes of the graph are the rectangles described with 4 attributes (I, Q, W, H) including two color attributes and two geometric attributes. The connection between the nodes of the graph is determined by the visibility between the rectangles. Two nodes are visible if they share a sufficiently wide horizontal (or vertical) range and if they can be connected without crossing other rectangles. Finally, a subgraph isomorphism is used to locate a region of interest in a document with a subgraph as a query and a known graph as a target. Whereas the authors in [18] create a star graph for each field to extract from an invoice. The nodes of the graph are the words of the document and the edges are the spatial relations between the field to be extracted and the other words. Then, a voting scheme is applied to locate the position of the most plausible field to extract using the models (star graphs) learned during an initial learning step and stored in a database.

3   Proposed Method

In this section we give a detailed description of our described method. As shown in Fig. 1, we start by a pre-processing step. A sample of new labeled documents is generated from the original database. Those new labeled documents are used to calculate maximum distances (dv_max and dh_max) that are useful for the next steps. In the training step, we use the original labeled database with the distances calculated in the pre-processing step to build the graph and compute the word features. The output of the graph modeling is fed to an attention neural network in order to classify the words in their appropriate classes. In the analysis step, the document image is OCRed first, then the document graph is built and the words features are calculated from the OCR result. The trained multi-layer GAT model takes as input the document graph and outputs the classes of interest of the words. Finally, in a post processing step, we group the words of same classes to form entities. In the case of invoices, products attributes are grouped and enumerated according to the number of products in the table of products.

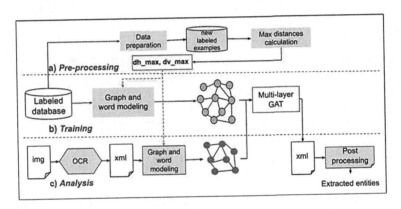

Fig. 1. Global architecture of the system.

The following subsections provide a detailed illustration of the SSD structure, the graph modeling approach, the word modeling and the graph attention network architecture.

3.1   SSD Representation

Based on the characteristics of a SSD, we propose an overall SSD structure as illustrated in Fig. 2. In this schema, we consider the SSD as a vertical sequence of blocks; each block is an horizontal sequence of sub-blocks that we call groups. A group, in turn, is a vertical sequence of elementary components (EC). The EC content can be adjusted horizontally (left, right or in the center of EC) or vertically (at the top, bottom or the center of EC). The EC content is a pair ("keywords", "information") or only "information" that could have several representations.

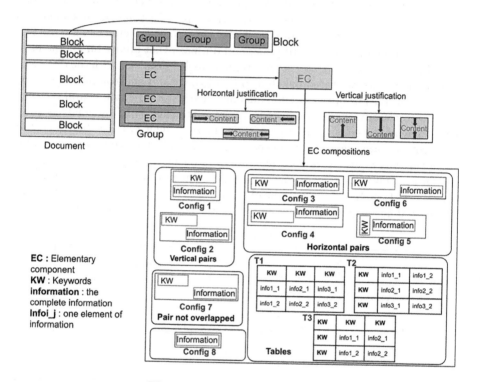

Fig. 2. Proposed SSD structure.

"keywords" can contain one or more words that introduce each distinct information in the SSD. They belong to a restricted vocabulary, for example to introduce billing address in an invoice, the keywords used are: "Bill to", "to", "Billing Address", "Billed to", etc.

"information" can be one word or a sequence of words found on one or more lines. It could also be separated into several elementary information giving a list of grouped words organized vertically (T1 in Fig. 2) or horizontally (T2 in Fig. 2), generally forming tables; In the case of T3, "infoi_j" belongs to the horizontal and the vertical pair at the same time.

Each distinct "information" can have a special format, for example: the totals have a floating value which can be followed by the currency; a French address is a multi-line information that includes the following details (some of them are optional): Recipient identification + apartment number, floor + building, residence + number and description of the street + locality + postcode and locality of destination + country. There is some information on the SSD that has a very distinctive format with which it can be easily recognized. The date is an example of this type of information; It has a limited set of formats such as: dd/mm/yyyy; dd-mm/yyyy, dd/mm/yyyy. This information can be found without any keyword and has a slightly variable position on the SSD (Config 8).

The words that we want to extract from a SSD are the "information" words. The characteristics of "information" and "keywords" mentioned below help classify each word into its correct class. As a result, for each word of "information": all words belonging to the same "information" and its corresponding "keywords" are called **indicative words**.

The possible locations of "information" in relation to "keywords" are:

- Next to the "keywords" forming horizontal pairs (Config 3, 4, 5 and 6);
- Below the "keywords" forming vertical pairs (Config 1 and 2).
- They can line up (Config 1 and 3), overlap (Config 2, 4, 5 and 6), or just be close (Config 7);

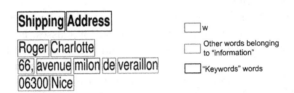

Fig. 3. An example of a word's indicative words in a shipping address

Based on the different configurations shown in the Fig. 2, we deduce that the common point between the indicative words of a word "w" (as can be seen in Fig. 3 for the word "milon") is the fact that they are at a limited vertical and horizontal distance from "w".

3.2   Graph Modeling

Data Preparation: To set the maximum horizontal and vertical distances in order to define the words neighborhood, we go through two steps:

- Firstly, we prepare a new labeled database from the original one. In the original database (see example (a) in Fig. 4), all the words belonging to the entities to extract are labeled with significant class names (INum, IDate, CNum, ONum), whereas the rest of the words in the SSD, including the keywords, get the class "undefined".

  In the new labeled database, the keywords of each information class get a new label by adding "K" (Keywords) to the name of the main class for example: "INum", "INumK" as shown in Fig. 4 (example (b)). If the information appears several times in the document (which is rare), we number the classes

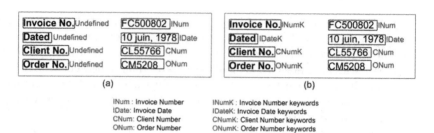

Fig. 4. a) An example from the original database b) An example from the new labeled database

to differentiate them, for example if the invoice number appears in the header and footer of a document, we assign the class "INum1", for the first INum and "INum2", for the second INum.

- Finally, we use the new labeling of the SSDs to calculate the horizontal and vertical distances (dh and dv) for the information classes. For each word of a class "C", we calculate the distance to its indicative words of the classes "C" and "CK" and then the maximum value (dv_max/dh_max) is registered. We take, at the end, the most occurring max distances recorded for all the words belonging to the different classes.

Graph Modeling Approach: We propose a document graph modeling that gathers all the words of the SSD, where a node neighborhood groups all the words present at a limited horizontal and vertical distance (dh_max and dv_max) thus forming star sub-graphs (each word and its neighbors form a star graph) as shown in Fig. 5c. We build an undirected graph G = (V, E) modeling each SSD. "V", the set of graph nodes, represent the features vectors of all the SSD words. The edges E correspond to the connections between the graph nodes.

Comparing to two other approaches that model a document as a graph: the approach in [18] (see Fig. 5a) takes into consideration all the document words that most of them do not give any important information for the word classification; While the proposition of [15] (see Fig. 5b) takes a maximum of 4 neighbors, which may exclude some important neighbors that could help classifying the document word. Our proposition tries to gather the maximum number of important neighbors for the node classification.

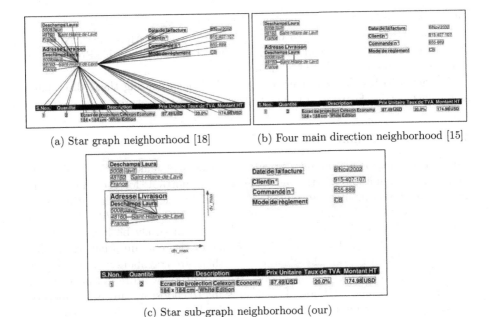

(a) Star graph neighborhood [18]   (b) Four main direction neighborhood [15]

(c) Star sub-graph neighborhood (our)

Fig. 5. Neighborhood configurations: a) The word is linked to all the other document words with a star graph [18]. b) The word is connected (at most) to its 4 nearest neighbors in the four main direction (here it is connected to 3 other words) [15]. c) Our proposition: the word is connected to the other words that are at a limited vertical and horizontal distance.

3.3 Word Modeling

To build the word representation, we combine text features, layout features and other features that represent the word's category (alphabetic, numeric, etc.).

Text Features: BPEmb [7] is a collection of pre-trained subword unit embeddings in 275 languages. In our case, we make use of the French and English BPEmb. It's based on Byte-Pair Encoding BPE. BPE is an unsupervised subword segmentation method that uses a sequence of symbols and the most common pair of symbols that iteratively merges to form a new symbol [20]. This turned out to be very useful in OCRed images as we could correctly deduce the meaning of a word even if there were OCR errors due to its subwords.

Layout Features: Layout features of the word consist on its normalized position in the document. They are two doubles calculated as follow: $x/width$ and $y/height$, where x and y are the word first coordinates in the document, whereas $width$ and $height$ are the document width and document height respectively.

Other Features: An 8-dimensional binary vector, proposed by [15], that indicates if the word belong to one of these categories: Alphabetic, Numeric, AlphaNumeric, NumberwithDecimal, RealNumber, Currency, hasRealandCurrency, mix or mixc (i.e. mix and currency word).

We get a 310 dimension features vector by concatenating these three parts as follows: Layout Features || Other features || Textual features.

3.4   Specific Graph: GAT

Using the attention mechanism in our approach will make the network learn which of the neighbors are the most and least important, for each node in the SSD graph. This will allow our model to focus on the relevant neighbouring nodes of each graph node, helping it make better node classification. With the type of neighborhood proposed: Indicative words/nodes will surely be included in the set of node neighbors. The set of neighbors could also contain other words that are not useful for the node classification. So, there are neighbors more important than others in the whole neighborhood. We cannot know a priori which are the most/least important neighbors. It is interesting to choose a convolution method which takes into account the proposed neighborhood set, and also which learns and then assigns different weights to these neighbors.

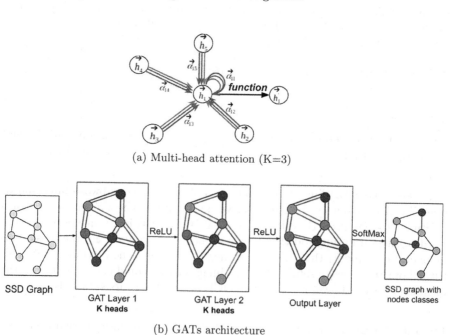

(a) Multi-head attention (K=3)

(b) GATs architecture

Fig. 6. Multi-layer GAT architecture with multi-head attention mechanism (function is concatenation in layer 1 and layer 2 and average in the output layer)

We propose a three GAT layers neural network architecture. The first two layers apply the multi-head attention with k heads (k varies according to the type of SSD and the number of entities to extract) with ReLU as a non linearity activation function. The output layer apply a Softmax function to classify the graph nodes. We build our multi-layer GAT architecture, as shown in Fig. 6, based on the basic GAT layer of [22]. The convolution in the graph is calculated using the multi-head attention mechanism. The input of our network is a set of features vectors (detailed in the sub-sequence Sect. 3.3) of all the graph nodes: $h = \{\vec{h_1}, \vec{h_2}, ...\vec{h_N}\}$ where $\vec{h_i} \in R^F$, N is the number of nodes, and F is the size of the features vector of each node. F corresponds to 310 in the first layer of our architecture. The layer outputs a new set of node features of size F', $h' = \{\vec{h'_1}, \vec{h'_2}, ...\vec{h'_N}\}$ where $\vec{h'_i} \in R^{F'}$ like its output.

The output features of GAT layer is calculated as follows: K attention mechanisms calculate K linear combinations of the features using normalized attention coefficients $\{\alpha_{ij}\}$, to which a non-linearity function σ is applied, then merge the final results, giving the following representation:

$$\vec{h'}_i = \|_{k=0}^K \sigma(\sum_{j \in N_i} \alpha_{ij}^k W^k \vec{h}_j) \tag{1}$$

where $\|$ represents the concatenation, α_{ij}^k are normalized attention coefficients calculated by the k-th attention mechanism (a^k), and W^k is the weight matrix of the corresponding input linear transformation. The final output returned, h', consist of KF' features (rather than F') for each node.

The coefficients calculated by the attention mechanism are as follows:

$$\alpha_{ij} = \frac{exp(LeakyReLU(\vec{a}^T[W\vec{h}_i\|W\vec{h}_j]))}{\sum_{k \in N_i} exp(LeakyReLU(\vec{a}^T[W\vec{h}_i\|W\vec{h}_j]))} \tag{2}$$

On the output (prediction) layer of our network, an average is used instead of concatenation to calculate the output features of the layer:

$$\vec{h'}_i = \sigma(\frac{1}{K}\sum_{k=1}^K \sum_{j \in N_i} \alpha_{ij}^k W^k \vec{h}_j) \tag{3}$$

4   Experiments

In this section, we present our results with different metrics: $Precision = \frac{TP}{TP+FP}$, $Recall = \frac{TP}{TP+FN}$, $Accuracy = \frac{TP+TN}{TP+FP+FN+TN}$ and $F1 = \frac{2*Precision*Recall}{Precision+Recall}$ where: TP: True Positive, TN: True Negative, FP: False Positive and FN: False Negative.

4.1 Datasets

Invoices: Generated artificially by a specific program developed in our team, this dataset contains 1480 images. It has 26 word classes to recognize, as reported in Table 1. This dataset mainly contains more than seven real cloned templates in English and French, making it a variable layout dataset. It is organized as follows: 70% of the images for training, 20% for validation and 10% for testing.

Payslips: Also artificially generated by our program and contains 1000 French payslips. Every payslip has 26 entities to extract as reported in Table 3. We randomly select 70 % of the data for training, 20% for validation and 10% for testing. This dataset and the previous one are available to researchers in order to make comparisons.

Receipts: It corresponds to the SROIE dataset [9]. It contain 626 receipt images for training and 347 receipt images for testing. Each receipt has four word classes to recognize: company, address, date, and total. The receipts are in English language. SROIE is also a variable layout dataset.

4.2 Implementation Details

The distances (dh_max, dv_max) calculated for the graph generation on invoices, payslips and receipts dataset are: (900, 300), (1050, 200) and (800, 30) respectively. We build our multi-layer graph attention network as shown in Fig. 6 based on the Keras tensorflow implementation of a GAT layer proposed by [5]. We apply a different number of attention heads for each type of SSD: 26 attention heads for the invoices and the payslips and 4 for the SROIE dataset. L2 regularization factor is fixed to 5.10^{-4}. The first GAT layer contains 16 channels and the second 32. The learning rate is set to 0.001. The model is trained using a mini-batch algorithm on a maximum of 2000 epochs and patience of 100 for early stopping.

4.3 Experimental Results

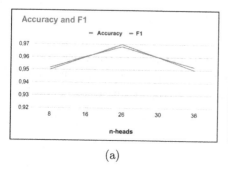

(a)

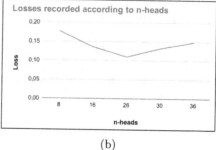

(b)

Fig. 7. a) Accuracy, F1 score and b) loss on variable n-heads in invoice dataset

GAT N-Heads: We conducted a study to analyze the effect of different numbers of n-heads on GAT's in entities extraction results. The results on the invoices dataset are shown in Fig. 7. As we can see in Fig. 7, the case of 26 n-heads gives the best results on the invoices dataset: Loss = **0.1097**, accuracy = **0.9682** and F1 = **0.97**. The results decrease by using a number of heads less than or greater than 26 which corresponds to the number of node classes. We assume that there must be a correlation between the n-heads and the number of node classes.

Detailed Results on the Three Datasets:

Invoices: Table 1 shows the results of our system compared to those of the invoice analysis system proposed by [15] on the same dataset of 1480 generated invoices. As can be seen in the table, our results on the invoices database are better than [15]'s results on most entities. For total tax amount and total without tax, the close spacing between them and the fact that they usually have the same values is what causes their confusion.

Table 1. F1 score on invoices dataset using GCN (4 neighbors) [15] system and ours

Entity	[15]	Ours	Entity	[15]	Ours
Invoice date	0.90	**0.99**	Invoice number	0.92	**0.98**
Order number	0.83	**0.91**	Payment mode	0.88	**1.00**
Company name	0.75	**0.87**	Company address	0.88	**0.94**
Company SIREN number	0.81	**0.93**	Company SIRET number	0.81	**0.92**
Company VAT number	0.91	**0.97**	Company APE code	0.85	**0.96**
Company Contact number	0.87	**0.96**	Company fax number	0.87	**0.97**
Client number	0.87	**0.90**	Client billing name	**0.90**	0.89
Client billing address	0.88	**0.90**	Client shipping name	**0.89**	0.88
Client shipping address	0.87	**0.90**	Product serial number	0.94	**0.99**
Product description	0.95	**0.98**	Product unit price	0.94	**1.00**
Product quantity	0.93	**0.97**	Product price without tax	**0.95**	0.91
Tax rate	0.99	**1.00**	Total without tax	**0.89**	0.77
Total tax amount	**0.93**	0.83	Net payable amount	**0.92**	0.91

We also tried to compare our method with other systems that extract entities from invoices in the Table 2. However, it is necessary to emphasize that each method extracts a different number of entities and gets the results on different invoice types. Our system succeeds in extracting 26 different entities from variable layout invoices with an accuracy of 0.963 an F1 score of 0.79 and surpasses the results of the other systems.

Table 2. Mean accuracy/F1 recorded on different information extraction systems: [2] applies patterns recognition; [18] proposes voting scheme on the bounding-boxes; [25] and [23] are based on Bidirectional-LSTM and [17] generate candidates based on neighboring words and knowledge of the type of fields to extract

Method	[2]	[18]	[25]	[23]	[17]	Ours
Entity Nb		3	9	6	7	26
Accuracy	0.92	0.938	0.932	–	–	**0.968**
F1		–	–	0.871	0.878	**0.97**

Payslips: The results shown in Table 3 prove the effectiveness of our system on the payslip database. Most classes score are higher than 0.91 (F1 score). Different companies utilize the same keywords to define different detailed and usually optional information. This could create confusion. Employee qualification, coefficient and earned leave are examples of those entities, that explains the scores obtained for them. Employee registration number is also confused with employee qualification, these two entities have the same format, could be found in similar positions and also the keywords that introduce them may look alike. The F1 score of a company's SIREN number is 0.57 because it is often confused with SIRET number and has similar properties. In the payslips, it is usually found without keywords, this increases more its confusion with the SIRET number.

Table 3. Recall (R), Precision (P) and F1 score (F1) on payslips dataset

Class	P	R	F1	Class	P	R	F1
Employee name	0.99	0.94	0.96	Employee name	0.99	0.94	0.96
Employee address	0.98	1.00	0.99	Employee code	0.92	0.96	0.94
Employee registration number	0.65	0.68	0.67	Employment	0.92	0.96	0.94
Employee classification	0.95	0.92	0.94	Social security number	1.00	1.00	1.00
Employee qualification	0.68	0.76	0.72	Coefficient	0.76	0.83	0.79
Seniority	0.94	0.87	0.91	Starting date	0.89	0.93	0.91
Payment period	0.99	0.99	0.99	Payment date	1.00	0.98	0.99
Earned leave	0.48	0.76	0.59	Company name	0.85	0.86	0.86
Company address	0.95	0.94	0.94	Company VAT number	0.93	0.95	0.94
Company SIRET number	0.73	0.93	0.82	Company APE code	0.94	0.85	0.89
Company SIREN number	0.91	0.41	0.57	Net salary	0.81	0.82	0.82
Gross salary	1.00	1.00	1.00	Net salary before taxes	0.80	0.83	0.81
Gross net pay	0.98	1.00	0.99	National collective agreement	0.98	0.95	0.97

Receipts (SROIE): We compare the average F1 score obtained to a BiL-STM+CRF method proposed by [10], an Attention-based LSTM decoder proposed by [13] and the top 4 methods of Task 3 of the ICDAR competition [9]. As can be seen in the Table 4, we got a high average F1 scores (0.892). Our method outperforms the BiLSTM+CRF [10] and the Attention-based LSTM [13] as well as the Top-4 (the "CLOVA OCR" method [3]) that proposes a sequence tagging approach and classify tokens in all text boxes using a BERT model. Our F1 score is very close to the 3 top methods recorded in the competition [9]: Top-1 is implemented by Ping An Property & Casualty Insurance Company. It applies different patterns of regular expression to extract the key information; Top-2 ("Entity detection" method) extracts the entities using a text classifier with RNN embedding; While Top-3 is the method of "H&H Lab" [16] which proposes a sequence labeling system based on BiLSTM-CNNs-CRF and they correct their results through restrictive rules. Some of these methods are specifically designed to solve the problem of extracting entities from receipts: Top-1 identifies the 4 entities with regular expressions; and Top-3 corrects the results by setting predetermined constraints. Although our results are those of a generic approach that can be applied to any type of SSD. In addition, we do not consider correcting errors in the post-processing step in these results.

Table 4. F1 score on SROIE dataset using: BiLSTM + CRF [10], Attention-based LSTM decoder [13], the top 4 methods for Task 3 in ICDAR competition [9] and our method

Method	[10]	[13]	Top-4	Top-3	Top-2	Top-1	Ours
F1	0.878	0.861	0.8905	0.8963	0.8970	**0.9049**	0.892

There are other systems that extract information from receipts, like [4] and [21] that test their methods on different datasets, which are unfortunately not public. [21]'s authors propose a receipt extraction ontology through data-frames. On 100 Japanese receipts, it reaches a F1-score of 0.85 over 13 recognized classes. While the approach of [4] suggests a U-Net convolutional architecture by taking into account both the semantic and geometrical aspects of the receipts content. Their method is tested on a dataset of 22670 receipt images. They only evaluated the total price field and the best-resulting accuracy was 0.878. As could be noticed, our results on SROIE still better than those of these two systems.

5   Conclusion

In this paper, we presented a node classification approach to extract entities from semi-structured documents. We have defined a general structure of a SSD that we have relied on to define a new type of node neighborhood in a document graph. We adopted a node neighborhood of star sub-graphs limited by maximum vertical and horizontal distances for each word in the document. The global document graph is fed to multi-layer graph attention network (GAT). This type of neural network has the ability to focus on the most important node neighbors using the attention mechanism. Our system extracts, with an interesting global F1 score of 0.96, 26 entities from both invoices and payslips and record a score of 0.892 in the SROIE dataset.

Acknowledgments. This work was carried out within the framework of the BPI DeepTech project, in partnership between the University of Lorraine (Ref. UL: GECO/2020/00331), the CNRS, the INRIA Lorraine and the company FAIR&SMART. The authors would like to thank all the partners for their fruitful collaboration.

References

1. Brown, J.: System and method for identification and extraction of data, US Patent 9,589,183, 7 March 2017
2. Dengel, A.R., Klein, B.: *smartFIX*: a requirements-driven system for document analysis and understanding. In: Lopresti, D., Hu, J., Kashi, R. (eds.) DAS 2002. LNCS, vol. 2423, pp. 433–444. Springer, Heidelberg (2002). https://doi.org/10.1007/3-540-45869-7_47
3. Devlin, J., Chang, M.W., Lee, K., Toutanova, K.: Bert: pre-training of deep bidirectional transformers for language understanding. arXiv preprint arXiv:1810.04805 (2018)
4. Gal, R., Morag, N., Shilkrot, R.: Visual-linguistic methods for receipt field recognition. In: Jawahar, C.V., Li, H., Mori, G., Schindler, K. (eds.) ACCV 2018. LNCS, vol. 11362, pp. 542–557. Springer, Cham (2019). https://doi.org/10.1007/978-3-030-20890-5_35
5. Grattarola, D., Alippi, C.: Graph neural networks in tensorflow and keras with spektral. arXiv preprint arXiv:2006.12138 (2020)
6. Hammami, M., Héroux, P., Adam, S., d'Andecy, V.P.: One-shot field spotting on colored forms using subgraph isomorphism. In: 2015 13th International Conference on Document Analysis and Recognition (ICDAR), pp. 586–590. IEEE (2015)
7. Heinzerling, B., Strube, M.: Bpemb: tokenization-free pre-trained subword embeddings in 275 languages. arXiv preprint arXiv:1710.02187 (2017)
8. Hua, Y., Huang, Z., Guo, J., Qiu, W.: Attention-based graph neural network with global context awareness for document understanding. In: Sun, M., Li, S., Zhang, Y., Liu, Y., He, S., Rao, G. (eds.) CCL 2020. LNCS (LNAI), vol. 12522, pp. 45–56. Springer, Cham (2020). https://doi.org/10.1007/978-3-030-63031-7_4
9. Huang, Z., et al.: ICDAR 2019 competition on scanned receipt OCR and information extraction. In: 2019 International Conference on Document Analysis and Recognition (ICDAR), pp. 1516–1520. IEEE (2019)
10. Huang, Z., Xu, W., Yu, K.: Bidirectional LSTM-CRF models for sequence tagging. arXiv preprint arXiv:1508.01991 (2015)
11. Kavas, I.: Analytic systems, methods, and computer-readable media for structured, semi-structured, and unstructured documents, US Patent 9,384,264, 5 July 2016
12. Lample, G., Ballesteros, M., Subramanian, S., Kawakami, K., Dyer, C.: Neural architectures for named entity recognition. arXiv preprint arXiv:1603.01360 (2016)
13. Le, A.D., Pham, D.V., Nguyen, T.A.: Deep learning approach for receipt recognition. In: Dang, T.K., Küng, J., Takizawa, M., Bui, S.H. (eds.) FDSE 2019. LNCS, vol. 11814, pp. 705–712. Springer, Cham (2019). https://doi.org/10.1007/978-3-030-35653-8_50

14. Liu, X., Gao, F., Zhang, Q., Zhao, H.: Graph convolution for multimodal information extraction from visually rich documents. arXiv preprint arXiv:1903.11279 (2019)
15. Lohani, D., Belaïd, A., Belaïd, Y.: An invoice reading system using a graph convolutional network. In: Carneiro, G., You, S. (eds.) ACCV 2018. LNCS, vol. 11367, pp. 144–158. Springer, Cham (2019). https://doi.org/10.1007/978-3-030-21074-8_12
16. Ma, X., Hovy, E.: End-to-end sequence labeling via bi-directional LSTM-CNNs-CRF. arXiv preprint arXiv:1603.01354 (2016)
17. Majumder, B.P., Potti, N., Tata, S., Wendt, J.B., Zhao, Q., Najork, M.: Representation learning for information extraction from form-like documents. In: Proceedings of the 58th Annual Meeting of the Association for Computational Linguistics, pp. 6495–6504 (2020)
18. Rusinol, M., Benkhelfallah, T., Poulain dAndecy, V.: Field extraction from administrative documents by incremental structural templates. In: 2013 12th International Conference on Document Analysis and Recognition, pp. 1100–1104. IEEE (2013)
19. Schuster, D., et al.: Intellix-end-user trained information extraction for document archiving. In: 2013 12th International Conference on Document Analysis and Recognition, pp. 101–105. IEEE (2013)
20. Sennrich, R., Haddow, B., Birch, A.: Neural machine translation of rare words with subword units. arXiv preprint arXiv:1508.07909 (2015)
21. Shen, Z., Tijerino, Y.: Ontology-based automatic receipt accounting system. In: 2012 IEEE/WIC/ACM International Conferences on Web Intelligence and Intelligent Agent Technology, vol. 3, pp. 236–239. IEEE (2012)
22. Veličković, P., Cucurull, G., Casanova, A., Romero, A., Liò, P., Bengio, Y.: Graph attention networks. In: International Conference on Learning Representations (2018)
23. Yu, W., Lu, N., Qi, X., Gong, P., Xiao, R.: Pick: processing key information extraction from documents using improved graph learning-convolutional networks. arXiv preprint arXiv:2004.07464 (2020)
24. Zhang, K., Li, J.Z., Hong, M.C., Yan, X.D., Song, Q.: A semantics enabled intelligent semi-structured document processor. In: Yuan, Y., Wu, X., Lu, Y. (eds.) ISCTCS 2013. CCIS, vol. 426, pp. 328–344. Springer, Heidelberg (2014). https://doi.org/10.1007/978-3-662-43908-1_41
25. Zhang, P., et al.: TRIE: end-to-end text reading and information extraction for document understanding. arXiv preprint arXiv:2005.13118 (2020)

Author Index